An Introduction to Space Robotics

Springer
London
Berlin
Heidelberg
New York
Barcelona
Hong Kong
Milan
Paris
Santa Clara
Singapore
Tokyo

Alex Ellery

An Introduction to Space Robotics

Springer

Published in association with
Praxis Publishing
Chichester

PRAXIS

Dr Alex Ellery
Space Division
Logica UK Ltd
London
UK

SPRINGER–PRAXIS BOOKS IN ASTRONOMY AND SPACE SCIENCES
SUBJECT *ADVISORY EDITOR*: John Mason, B.Sc., Ph.D.

ISBN 1-85233-164-X Springer-Verlag Berlin Heidelberg New York

British Library Cataloging in Publication Data
    Ellery, Alex
        An introduction to space robotics. - (Springer-Praxis books
        in astronomy and space sciences)
        1.Space robotics
        I.Title
        629.4'7

    ISBN 1-85233-164-X

Library of Congress Cataloging-in-Publication Data
    Ellery, Alex, 1963-
        An introduction to space robotics / Alex Ellery.
            p. cm. -- (Springer-Praxis books in astronomy and space sciences)
        Includes bibliographical references
        ISBN 1-85233-164-X (alk. paper)
            1. Space robotics. I. Title. II. Series.

TL1097 .E45 2000
629.47--dc21   00-037375

Cover design: Jim Wilkie
Cover picture: Produced by Vince Shaw-Morton
Typesetting: Heather FitzGibbon, Christchurch, Dorset, UK

Printed on paper supplied by Precision Publishing Papers Ltd, UK

For my father, who died before he had the chance to write his book on the history of English common law that he admired so much as a product of human problem-solving ingenuity

Contents

Foreword

Space and robotics have sprung our attention as vital frontiers that now confront us as we march boldly into the new millennium. The phenomenal technological advances of the twentieth century firstly gave us manned flight and then sent us even further into our space backyard. These advances also provided us with robot manipulators and computers to make the robots think. With these forces combined we are now poised to push things further. For all the technological advances of the last century, this new century will see its own incredible innovations and space robotics will appear as the centre spread.

There are indeed many who feel that, because of the limitations presently placed on humans as they travel through, and perform in, space, it will be robots who themselves inherit space. In reality this does not (most likely) mean that a bunch of robots will be sent off with back-packs, into space, and told, 'OK it's all yours, over to you, get on with it'. Rather it means that robots will be the most important entities, for us, in space—they will be our eyes, ears, hands and legs.

It is therefore vital that we are able to get a good idea of what this all amounts to, not only philosophically but also technically. If robots will be operating way out there, what does it mean from an engineering perspective? What constraints are imposed, how do time and gravitational forces affect things? Will robots be robust enough to operate for long periods in outer space? What effect does planetary operation have on robot sensors and grasping mechanisms? How are humans going to interact with such robots, whether we be based on Earth or close by? Having a good understanding of these questions is vital for those interested in the field. Having some appropriate answers is even more vital.

Without robots, apart from those keen on long-distance picnics, space exploration is just about pointless. It is a critical technology in a central position. When it comes to the possibility of commercial exploitation this is even more true. Robotics is clearly a part of all space exploration. As further planets are explored this will become more and more the case due to the distances involved, the hostile environments encountered and the need for more autonomy and reliability in operation.

Having stressed the need for and importance of a book in this area, it is probably worth saying a word or two about this book in particular, before the Foreword becomes an afterthought. Quite simply when I first saw the draft I was amazed at the extent of the coverage and even more nonplussed at the information that Alex has amassed. He appears to have extracted much technical and fiscal information that I thought was all very hush-hush from a variety of people in the know.

Important though the subject area is, the pages which you will shortly unfold, present, to my knowledge, the first integrated text in the field of space robotics. But the text is much more than that. Whilst giving a very good overview of a complex field, the book contains much original material. The freeflyer analysis explained in Chapter 8 stems from Alex's own Ph.D. work, and this considerably simplifies the computations required by the servocontroller on board the spacecraft.

In its entirety, the book provides students with the background knowledge necessary for robotic analysis and design, in the targeted area of space travel. Whilst being directed towards spacecraft design engineering, however, the book in fact makes a significant contribution to the field of robotics in its own right, and students of robotics *per se* will find this an extremely useful addition to their shelves.

Kevin Warwick
Reading University
March 2000

Preface

This textbook is designed for students and professionals engaged in the analysis and design of robotic space missions. Such missions are becoming central to the success of near-Earth orbit space activities pertaining to in-orbit operations, both commercial and in support of International Space Station (ISS) activities. Indeed, they will fundamentally define in-orbit capabilities. They are also central to exploratory missions throughout our solar system, particularly Mars as the focus for exploration of the solar system. Planetary robotic rovers feature as a central component of all *in situ* planetary exploration missions. The recent loss of two US Mars landers highlights the non-trivial issues and difficulties involved in such missions. Indeed, the history of Mars exploration has been chequered with catastrophic failures interspersed with spectacular successes. This book is designed as a text to both instruct students in and present to the professional engaged in the design of such robotic missions a state-of-the-art overview of the field of space robotics. The central purpose of the book is to provide a reasonably comprehensive coverage of the field of space robotics, to present the essential concepts and to integrate into a single text the necessary knowledge of both spacecraft design and robotics technology.

The book introduces the reader to both spacecraft engineering and robotics engineering unified framework which integrates a mechatronics approach within the spacecraft systems design and mission analysis format. The mechatronics eye-view sees control as the central process in providing the interface between sensing of the environment and actuation on the environment. This view is appropriate to robotic spacecraft design as it focusses on the issues of performance of the space mission—whereas most robotic spacecraft to date have been almost entirely platforms for sensors, robotic actuation introduces much greater complexities in terms of the performance of tasks which physically affect their environments. The robotic control system and its analysis forms the central pillar of the robotics component of a robotic spacecraft while spacecraft orbit and attitude control issues define its mission profile. Such a mechatronics approach is a truly integrative one, sweeping across traditional engineering boundaries and traditional spacecraft subsystem specialisations. As will be seen throughout the text, there are pronounced interactions between both the robotic payload and the spacecraft bus subsystems which go beyond traditional subsystems trade-offs. There are intense trade-offs in the onboard control system division of labour across the robotic spacecraft payload and bus, between orbit actuation capabilities and mission profile performance, between onboard computational capabilities and communications bandwidth requirements to the ground control station, and the need for significant power generation to obtain the required performance levels of robotic actuation. Furthermore, with the advent of smart materials, control capabilities are imposed on the structural subsystem.

The overall structure of the book follows that broadly typical of any spacecraft systems design text. After consideration of the general purpose, aims and nature of space robotics, a systems design example of a robotic spacecraft (ATLAS) is introduced to provide a focus for the rest of the text. Subsequently, a detailed consideration of each robotic spacecraft subsystem is expounded. The robotic payload subsystems drive the design of any robotic spacecraft and these are considered first, followed by consideration of the spacecraft bus subsystems which are constrained by the robotic payload. The required tools of analysis are introduced as required at the appropriate points. It is important for the reader to keep in mind the ATLAS design example as the book ranges across a variety of disciplines, issues, analyses and technologies, all of which contribute to a single robotic spacecraft with a demanding mission to perform. The robotic spacecraft with manipulation capabilities introduces a much greater complexity of mission task performance that hitherto has been characteristic only of manned missions.

Chapters 1 to 4 introduce the nature of space robotics, the type of missions that robotic spacecraft are required to undertake, and issues in remote control of such spacecraft. Chapter 1 sets the scene, describing the variety of hostile environments encountered by robotic spacecraft for both in-orbit and exploratory missions. Chapter 2 provides a survey of robotic explorer missions, emphasising the uniqueness of such missions with regard to planetary rovers (concentrating on Mars exploration). Chapter 3 is devoted to in-orbit servicing missions emphasising the high degree of complexity of servicing operations and their nature. Chapter 4 begins the coverage of generic technologies in which we cover the variety of man–machine interface approaches, including virtual reality technologies and survey ground station architectures. Chapter 5 introduces our systems design example robotic spacecraft, ATLAS, which is a generic in-orbit servicer spacecraft with manipulator arms, to provide a concrete focus for the rest of the book.

Chapters 6 to 13 consider the robotic payload subsystems which will drive the design of the ATLAS robotic spacecraft. This is where the analysis of the robotic control system is covered which forms the central component in the design of the robotic payload. Chapter 6 begins the robotic analysis coverage, beginning with manipulator kinematics as manipulators are the key to in-orbit servicing and will define and extend the capabilities of future planetary explorers. Chapter 7 extends this analysis to the dynamics of manipulators, emphasising the recursive Newton–Euler approach for computational efficiency. Chapter 8 introduces the modifications required to the kinematics and dynamics equations of the two earlier chapters to apply to in-orbit servicing freeflyer spacecraft such as ATLAS. Chapter 9 covers manipulator trajectory interpolation and obstacle avoidance strategies so critical for successful robotic task completion. Chapter 10 provides a survey a various manipulator control strategies, emphasising computed torque control and covering force control strategies which are essential in in-orbit servicing. Chapter 11 provides the coverage of the onboard robotic payload support system, namely sensors and actuators (including a survey of MEMS technologies), through which the control system interacts with external world. Chapter 12 is devoted to onboard robotic vision sensors and vision processing which will be the primary sensory modality in monitoring the perform- ance of the robotic tasks. Chapter 13 introduces artificial intelligence technologies for both onboard reflexive behaviours and ground station task planning for both planetary exploration and in-orbit servicing.

Chapters 14 to 19 cover the spacecraft bus segment of the ATLAS robotic spacecraft design upon which the robotic payload impacts and constrains. Chapter 14 begins cover- age of the spacecraft bus subsystems with orbital analysis, orbit actuation and mission profiles which define the robotic mission constraints. Chapter 15 extends this coverage to attitude control which is the subsystem with which the manipulator control system inter- acts most strongly (for an in-orbit servicing freeflyer such as ATLAS), highlighting the need for the adoption of control moment gyroscopes.

Chapter 16 covers onboard avionics which is a critically scarce resource in terms of the requirement for real-time manipulator servocontrol. Chapter 17 covers the communi- cations subsystem of the spacecraft which implements tracking, telemetry and command functions, and emphasises bandwidth as another scarce resource which defines the division of labour between the ground and space segments of the robotic algorithm implementation architecture. Chapter 18 covers issues related to power raising, storage and dissipation which are critical as robotic motors are typically power-hungry components. Chapter 19 considers the structure of robotic spacecraft with an emphasis on advanced materials and the potential offered by smart materials in vibration suppression.

Chapter 20 changes pace by introducing commercial and legal aspects of robotic missions, particularly those related to in-orbit servicing missions such as ATLAS. The final chapter, Chapter 21, concludes the book by considering a broad sweep across future space exploration scenarios within which space robotics will be set and contribute, be they automated or manned lunar or Mars missions beyond the ISS programme.

Space robotics as a discipline is fundamental to the success of all space missions, and it will increasingly affect the spacecraft design process as it restructures the traditional divisions between spacecraft subsystem specialisations by placing control implementa- tion and control architecture at the centre of the robotic spacecraft design process.

ACKNOWLEDGEMENTS

Firstly, I would like to thank Joe Parrish at the University of Maryland Space Systems Laboratory (SSL) for providing helpful feedback in the review process, gently pointing out when I had strayed off the beaten track and providing much-valued advice. I'd like to thank Clive Horwood of Praxis Publishing Ltd who kept me on track and offered encour- agement when my enthusiasm was waning on the long, arduous and lonely task of writing a textbook such as this. I'd like to thank Jim and Rachael Wilkie and Vincent Shaw- Morton for skilfully managing to turn my abstract ideas into accurate concrete visual representations for both the diagrams and the front cover. I'd like to thank Llewelyn De Souza who unwittingly volunteered for and helped with the tedious task of looking up references in libraries and photocopying journal papers. Others who contributed advice and technical feedback include: Craig Carignan and David Akin of SSL at the University of Maryland; Hal Aldridge and Wendell Mendell at NASA Johnson Space Centre; Mark

Maimone, Paul Backes, Bruce Bon, Alberto Behar and H. Seraji at the NASA Jet Propulsion Laboratory; David Miller of the KISS Institute for Practical Robotics; Elaine Hinman-Sweeney at Oceaneering Inc; Lynne Vanin, Rob Leitch and Lawrence Reeves at MacDonald-Dettwiler & Associates Ltd (formerly Spar Aerospace Ltd); Louis Freidman and Jim Burke at the Planetary Society; Philip Davies, Jon Williams, Gary Lay, Nick Shave, Robert Harper and Roger Dewell at Space Division, Logica UK Ltd; Sean Hardacre of Analyticon; James Keravala at Surrey Satellites Ltd; Nicholas Watkins of the British Antarctic Survey; Karl Deutsch at the International Space University; Andrew Ball at ISSI; Maurizio Ricciardi at Carlo Gavazzi Space; Marcel Schoonmade at NLR (Holland), and many others including Alan Pritchard. There are others who helped in many different ways who are too numerous to name. All errors are of course exclusively my own. If any reader finds errors, omissions or mistakes, or even has general comments concerning course suitability and so on, I'd be grateful for feedback for future editions.

1

Introduction

Welcome to the field of space robotics. Recently, space robotics has become a burgeoning field of study unifying the disciplines of astronautics and robotics mainly due to the focus provided by the International Space Station programme. In many respects, this is only proper. Quite apart from the historical coincidences of both disciplines' development, both can benefit each other in a uniquely symbiotic relationship. The year 1957 saw the launch of the world's first spacecraft, Sputnik 1 into Earth orbit, while 1961 saw the launch of the first industrial production robot Unimate by Unimation; 1969 saw both the first landing of men on the Moon marking the height of manned space exploration, and the development of the Stanford robot arm which evolved into the Unimate PUMA robot which became the workhorse standard of robotics. Notwithstanding these parallel developments, robotics provides a means to effectively explore and commercialise space, while the space environment provides a unique impetus and applications arena for pioneering robotics and automation research.

There is little dispute that manned spaceflight affords a flexibility of space operations that is unattainable in unmanned missions. However, a robotic capability would enable some of this flexibility to be retained in unmanned missions without the human endurance limitations of bone decalcification and muscle atrophy. This is not to say that such robotic capabilities could replace men in space—the issue is one of complementation. There has been a longstanding (and often highly emotive) debate concerning the use of robots or humans in space exploration. Although most machines act as energy transformers to extend human capabilities, the robot, is a machine that is designed ultimately to reproduce human capabilities, particularly for strenuous or dull tasks in potentially harmful environments. Consistent, repetitive and routine tasks requiring precision are particularly suited to the machine since machines do not tire or become inattentive. They are ideally suited to tasks like monitoring multiple complex systems, and for fault detection and correction. Humans are suited to flexible pattern recognition in noisy and uncertain environments involving multiple sense stimuli and for reacting in response to unexpected occurrences. Humans can make inductive decisions and utilise generalisations from past experience. However, automated fault monitoring and diagnosis can provide valuable real-time aid in such human functions. Manual tasks that can be performed prescriptively by detailed procedures (algorithms) are suitable for automation to reduce the human workload. Tasks which exceed human capabilities due to a required high rate, high accuracy, repeatability and consistency over long periods of time should also be automated. The reasons for autonomy are multitudinous. The majority of operational errors that occur are usually resultant from human misinterpretation of information, though this can be minimised by effective displays in man–machine interfaces. The need is for information management to avoid information overload.

The next stage in the evolution of space infrastructure lies in the development of a permanent human presence in space together with robotic manipulators which will be essential both for in-orbit operations, and any future proposed attempt to return to the Moon, Mars or any of the other planets will rely extensively on unmanned robotic precursor missions. Humans are required to install and maintain complex equipment and to conduct field exploration. These tasks require the flexibility, skill and judgement of the human being. Very sensitive instruments cannot tolerate robotic deployment. Furthermore, when complex equipment fails, humans can effectively save a mission by repairing the equipment by virtue of their greater dexterity. So far in space missions, it has been human ingenuity and flexibility that has saved missions from unforeseen events. Exploration comprises two phases—global survey and field study [Spudis 1999]. Survey is best achieved by automated spacecraft, landers and rovers to give a broad overview of the environment under study. Field study is required to understand detailed planetary processes. Such field study requires human intelligence to collect and interpret data, formulate and test hypotheses and direct further data collection in a focussed manner in an open-ended process of investigation [Crawford 1998]. Fieldwork always involves unexpected discoveries. The Apollo astronauts were able to recognise geologically significant rock samples from multiple locations returning 382 kg of lunar material to Earth compared with the Russian Luna landers which selected single core samples indiscriminately returning 321 g of lunar material to Earth.

The term 'space robotics' is open to interpretation—indeed, "robotic spacecraft" refers to a generic description of deep space probes of all types such as robotic observatories or planetary probes implying their unmanned and autonomous nature, though we will adopt a more restricted view. Space robotics applications may be divided into three categories: extravehicular servicers (freeflyers), intravehicular science payload servicers (telescience), and planetary probes (including planetary rovers) [Lavery 1994]. We shall consider first, generic robotic space missions with an emphasis on landers and penetrators as modes of delivery for robotic payloads (Chapter 2), second, planetary rovers for terrestrial type planets as a specialised application of space robotics (Chapter 2), and then finally in-orbit servicer missions as a more traditional focus for space robotics which emphasises robotic manipulation (Chapter 3). This book will concentrate on in-orbit servicing missions through a design example ATLAS (Advanced TeLerobotic Actuation System) introduced later (Chapter 5), but our coverage of the technologies will be directly relevant to planetary missions.

Before we embark on space robotics *per se*, we shall first consider the variety of environments encountered by robotic spacecraft to outline the hostile environments that will constrain the design of the space robot. The extraterrestrial environment represents a wide diversity of challenging environments which must be catered for in the design of the space robot. The first environment encountered is that within the launch vehicle. The next environment encountered is that of the space environment after launch, and each is considered in turn. We then consider the space debris environment characteristic of near Earth space that in-orbit servicers inhabit, and finally we shall consider the Martian surface environment for which most rover missions are designed.

1.1 THE LAUNCH VEHICLE ENVIRONMENT

The launch vehicle is required to place the spacecraft payload into a ~200 km altitude orbit from Earth and so must supply orbit injection. The launch vehicle thus imposes an environment on the spacecraft to be launched that differs from its normal operational environment. Launch vehicles typically use cryogenic liquid oxygen/liquid hydrogen propellant with regenerative cooling engines. Such combustion reactions yield low specific volume products are advantageous, e.g.

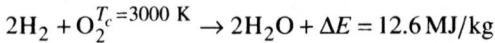

$$2H_2 + O_2^{T_c = 3000\ K} \rightarrow 2H_2O + \Delta E = 12.6\,MJ/kg$$

with $M_r = 18$. Such LOX/LOH combinations offer high $I_{sp} \sim 450$–475 s and tend to be used for launchers where the requirement is for ~10 + kN thrusts for ~10 minute flights. For instance, the NASA advanced LOX/LOH engine provides 90 kN of thrust at an I_{sp} of 470 s in a total mass of 245 kg. They can provide the required $\Delta v > 9.5$km/s to LEO (Low Earth Orbit) and >10.3 km/s to GEO (Geosynchronous Equatorial Orbit) with 9×10^3 kJ/kg specific energy. However, they impose severe mass penalties due to their need for cryogenic storage. Regenerative cooling involves cooling the nozzle and inner chamber by propellant flow in surrounding jackets prior to injection into the thrust chamber. The energy extracted in the cooling process then contributes to increasing the thermal energy of the combustion gases. Since structural mass increases rapidly with tank size and pressure, closed cycle turbopump feeds are used rather than pressure feeding (suitable only for low-moderate thrust levels <100 N due to the penalty for thick-walled pressure tanks). For high thrusts utilising turbopumps, the hot gas drives the turbine which drives the pumps to supply propellant at high pressure to the engine. These systems give ~2% higher I_{sp}. Boot-strapping extends this technique by bleeding off a small fraction of the propellant at the pump outlet which is burned in a gas generator to drive the turbine consuming ~3–5% of the propellant. Launch requires a Δv of ~9.5 km/s thrust to reach LEO plus 800 m/s to recover guidance constraints, 120 m/s to compensate atmospheric drag, 830 m/s to overcome gravity, and 1 m/s for lift (1751 m/s in losses) to generate a trajectory given by:

$$m\frac{dv}{dt} = T\cos(\alpha + \delta) - mg\sin\psi - D \text{ parallel to the flight direction}$$

$$m\frac{d\psi}{dt} = T\sin(\alpha + \delta) - mg\cos\psi + L \text{ normal to the flight direction}$$

where

T = thrust
D = drag

L = lift
v = vehicle velocity tangent
ψ = angle of trajectory tangent to the local horizontal
α = pitch angle of attack of the vehicle axis to the trajectory tangent
δ = angle of thrust from the vehicle axis

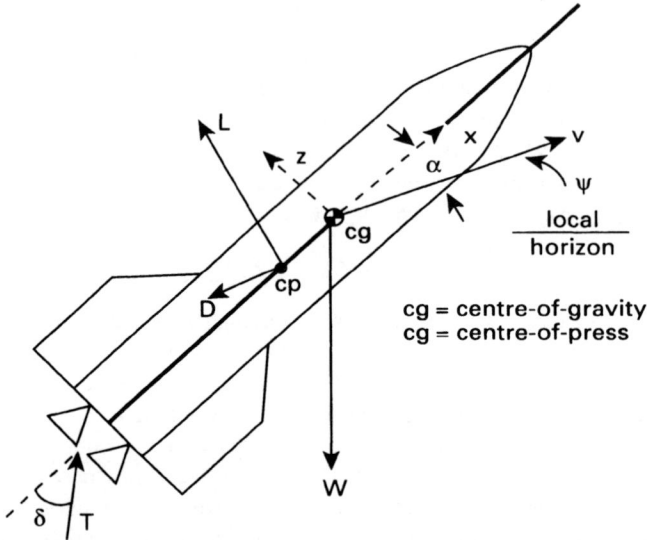

Fig. 1.1. Launch vehicle forces geometry (from Wertz & Larson 1999).

Vehicles are launched vertically but to minimise gravity loss the flight trajectory is rapidly deflected from the vertical by pitching through a gravity turn. For the launcher using modern-day propellants, this would require a typical mass ratio of initial to final mass of ~10. Hence, all expendable launch vehicles are three stage rockets as there is an asymptotic limit to the benefits of staging:

$$\Delta v = n v_{ex} \ln\left[\frac{1+\pi}{p^{1/n}+\pi}\right] \rightarrow \frac{\Delta v}{v_{ex}} = \ln\left(\frac{1}{p}\right) \quad \text{for} \quad n = \text{number of stages}$$

$$v_{ex} = \text{exhaust velocity}$$

The mounting configuration of stages is variable: in series (Saturn V), in parallel (Ariane) or piggyback (Shuttle). The launch profile typically uses the Earth's eastwards rotational velocity. Launcher second-stage cutoff occurs once the initial parking orbit has been achieved at ~200 km altitude. Prior to the 1980s, the USA was the only provider of launchers in the Western world until 1979 when Europe's Ariane 1 launcher became operational. It was later superseded by Arianes 2 and 3 in 1984 which has since been superseded by Ariane 4. It now commands 60% of the world market share. In 1984, the US Congress passed the Commercial Space Launch Act to promote economic growth in the private launch services market but after the Challenger disaster in 1986 the Act was amended in 1988 prohibiting the US Space Shuttle from routine commercial or foreign launches. Both Ariane and the Shuttle have had over 50 successful launches to date. The US Titan is the main competitor within the USA offering a payload capability of 18 tonnes to LEO and 4.5 tonnes to GEO. The US expendable launcher fleet (Atlas, Titan and Delta) represents a very old technology today. Indeed, Atlas is a modified US Air Force missile which was used to launch the manned Mercury missions and Titan is a modified US Air Force missile which was used to launch the manned Gemini missions. However, they have high reliabilities over large numbers of launches. The US Delta, Atlas and Titan launchers have had over ~400, ~230, and ~160 launches to date.

ESA's Ariane 4 (still operational with the advent of Ariane 5) is a three-stage, liquid-propelled expendable launcher with a lift-off mass of ~210 tonnes. The third stage burn at equatorial crossing takes the payload into an elliptical transfer orbit with its perigee at the parking orbit and apogee at the higher orbit. The first stage uses four Viking V engines and each produces a thrust of 550 kN while the second stage uses a single Viking IV engine with a thrust of 760 kN. The HM7 third stage produces a thrust of 62 kN. Ariane 4 can deliver 1.75 tonnes to GTO, 4.6 tonnes to low inclination LEO, and 2.7 tonnes to sun-synchronous polar LEO, but two solid-propellant strap-on boosters (42P configuration) increases this GTO payload to 2.3 tonnes. The two solid-propellant boosters increase the take-off thrust from 2490 to 4040 kN and the lift-off mass to 230 tonnes. This can be expanded to four boosters either all liquid (44L configuration) or in combination with solid boosters (44LP configuration). The 44P configuration yields a GEO payload capacity of 3 tonnes, the 42LP increasing this to 3.7 tonnes. The most powerful combination is four liquid boosters (44L) which increases the GTO payload capability to 4.2 tonnes. Ariane 4 is launched from Kourou in French Guiana at a latitude of 5°N. The equatorial rotational velocity of 0.47 km/s favours an eastward launch. Ariane offers dual launch facilities where two satellites are injected into the same orbit and oriented and spun up independently. SYLDA comprises a carbon-fibre-reinforced shell which encapsulates the lower spacecraft and supports the upper spacecraft in the payload fairing. An extended version of SYLDA is SPELDA which incorporates a longer fairing. This dual launch facility could be exploited by an in-orbit servicer if an additional space platform were launched as a 'warehouse/workshop' facility. Ariane 5 is an expanded version of Ariane 4 offering a payload capability of 7.4 tonnes to GEO or 15 tonnes to LEO. It utilises a more powerful second-stage motor to provide a thrust of 10^6N. This HM60 engine is similar to the Saturn V second stage. The first stage and strap-on boosters are recovered for re-use offering potential savings in launch costs. Each vehicle costs around 140M ECU similar to the cost of an Airbus 340.

The US Space Shuttle Transport System (STS) is a manned, mostly reusable launcher with a lift-off mass of 2040 tonnes. The central engines are the three Space Shuttle Main Engines (SSME). They are fed with LO_2/LH_2 propellant stored in the External Tank (ET) and fire in parallel with two strap-on solid rocket fuel boosters (SRB) for 120 s. Each SSME delivers 2×10^6N thrust with a specific impulse of 455 s. The SRBs are recovered downrange for re-use. The ET is jettisoned before final orbit injection. Following ET jettison, the Shuttle orbiter's orbital manoeuvring system (OMS) provides thrust for final orbit insertion, circularisation, rendezvous manoeuvres and de-orbiting. The OMS engines use MMH/N_2O_4 fuel-oxidiser with a specific impulse of 313 s and a thrust of 27 kN. The OMS tankage provides a total Δv capability of 350 m/s. The Space Shuttle launches from eastern Kennedy Space Centre at a latitude of 28.5°N to an altitude of 200–400 km (nominally 200 km). The maximum payload capability is 29.5 tonnes to 200 km LEO at an inclination of 28.5°. Increases in parking orbit altitude can be achieved by trading off payload mass with additional Shuttle OMS propellant. Indeed OMS kits are available to provide extra propellant mounted in the payload bay for the OMS engines to allow the Shuttle to reach altitudes of ~1000 km. However moving the 68 000 kg Orbiter is not efficient. It is more economic to employ dedicated auxiliary perigee stages for the particular mission spacecraft payload. The Space Shuttle places satellites into LEO parking orbit only and so perigee kick motor and an apogee kick motor is required. Ariane places satellites into GTO transfer orbit so that only the apogee kick motor is required. Solid-propellant upper stage payload modules provide a GTO injection capability. The Inertial Upper Stage (IUS) is a Shuttle compatible two-stage system capable of boosting 2.27 tonnes into GTO. The Payload Assist Module (PAM) comes in two varieties: the Delta-sized (D) and Atlas-sized (A) versions offering payload capabilities of 1.1 tonnes and 2.1 tonnes respectively. The Transfer Orbit Stage (TOS) is a single-stage perigee burn-only stage with a payload capability of 3.1 tonnes. Ulysses was launched to Jupiter utilising two upper-stage boosters—the IUS and PAM-S boosters. The Jupiter encounter was necessary to provide the gravity assist in 1992 to direct it out of the ecliptic towards the solar south pole. A second launch site at the Western Test Range at Vandenberg AFB in California at 35°N was to provide accessibility to polar orbits for Earth Observation spacecraft at reduced payload capability ~10 tonnes but this idea has been shelved. Only expendable launchers are operated from this site. On deployment, the spacecraft payload is spun up to 50 rpm and a retaining clamp is released pyrotechnically and the spacecraft is ejected by springs at ~1 m/s. Alternatively, the Remote Manipulator System (RMS) may be used for deployment but is usually used for retrieval of large payloads.

Larger expendable launch vehicles tend to offer lower costs per kilogram: the Titan 3 offers heavier payload capability than the Atlas-Centaur and offers a cost of $25 000/kg compared with Atlas-Centaur with $40 000/kg to GEO (1995 prices). Present launch costs are averaging ~$8000–$12 000/kg to LEO. The Shuttle prices are heavily subsidised by NASA as around 50% of Shuttle launch costs are associated with manpower and only 10% for fuel and 40% for booster and ET costs. Another consideration in selecting launchers is injection accuracy. Ariane offers superior apogee injection accuracy while the Shuttle offers superior perigee and inclination accuracies. Launch vehicle reliability is now estimated at ~0.95 generally: STS ~0.97, Ariane 4 ~0.88, Titan 3 ~0.94, Atlas-Centaur ~0.87 (1992 figures). Generally, the spacecraft launcher payload must be designed to be compatible with more than one launcher to enhance launcher availability.

1.2 THE SPACE ENVIRONMENT

Once the launch vehicle payload has been launched into space, it experiences a very different environment. The space environment is characterised by high vacuum ~10^{-14} Pa, extreme thermal differentials, ionising radiations, micrometeoroids and orbital debris.

The Earth's atmosphere exhibits an exponential decrease in pressure with altitude: $p(h) = p(h_0)e^{-\mu gh/RT}$ where μ = mean molecular weight of gas, R = gas constant = 8.31 J/kmol, T = temperature, h = height. Another useful factor is scale height $H = RT/\mu g$ at which atmospheric pressure decreases by $1/e \sim 8$ km for Earth. Atmospheric density variation is described similarly:

$$\rho(h) = \rho_0 e^{-\beta/(r-r_0)}$$

where

ρ_0 = reference altitude density

$$\beta = \frac{(\mu g/R + dT/dr)}{T} = 5\text{--}15 \text{ km (average 7 km)}$$

T = temperature

$$\frac{dT}{dr} \sim 10 \text{ K/m} = \text{lapse rate}$$

μ = molecular weight of air

$$\frac{1}{\beta} = H = \text{scale height}$$

r_0 = reference altitude (usually 100 km)

g = acceleration due to gravity (varies by 4% from sealevel to 120 km)

Atmospheric density decreases from 10^3 ng/m^3 at a height of 0–10 km to around 10^{-3} ng/m^3 at 1000 km. Indeed, in orbits of interest it can vary between 4×10^{-12}–3×10^{-15} kg/m^3. The pressure variation equation is valid, however, only up to 90 km since the mean molecular weight of air is constant only up to that altitude. The atmospheric composition is fully mixed to this height but thereafter nitrogen dominates up to 170 km, atomic oxygen up to 500 km, helium up to 900 km, and hydrogen beyond.

The troposphere extends from 0 to 15 km which exhibits temperature decreases from 290 K at sea level to 200 K at 15 km, the stratosphere from 15 to 50 km which exhibits a temperature increase from 200 to 280 K. The atmosphere is heated from below by the surface of the Earth which radiates in the infrared which is absorbed by water and carbon dioxide in the lower atmosphere (the greenhouse effect). This causes convection in the troposphere which is confined by the stratosphere. The stratosphere absorbs high-energy solar ultraviolet radiation through the resident ozone layer exhibiting a temperature increase with height, so convection cannot occur in the stratosphere forming an inversion layer. The mesosphere from 50 to 90 km which exhibits temperature increases to a maximum of 280 K near 50 km due to ozone absorption of ultraviolet radiation which subsequently decreases to a minimum of 180 K at 85 km. From 90 to 800 km lies the thermosphere which exhibits a temperature increase to a maximum value of 1500 K at 350 km and is maintained at that value for higher altitudes. This temperature increase is due to extreme solar ultraviolet and X-ray absorption. Repeated large thermal changes can cause thermal fatigue and in particular large temperature changes can disrupt the repeatability of robot arm positioning accuracy.

For altitudes beyond 90 km, the US Standard Atmosphere is adopted and is given in Fig. 1.2.

Altitude (km)	Temperature (K)	Pressure (N/m^2)	Density (kg/m^3)
0	288.15	1.01×10^5	1.23
100	195.08	3.20×10^{-2}	5.60×10^{-7}
200	854.56	8.47×10^{-5}	2.08×10^{-9}
300	976.01	8.77×10^{-6}	1.92×10^{-11}
400	995.83	1.45×10^{-6}	2.80×10^{-12}

Fig. 1.2. US standard atmosphere.

The ionosphere is created by solar ultraviolet and X-ray flux from the sun ionising the atmospheric components forming regions of positive ions and free electrons (plasma comprising a gas of electrically charged particles that are strongly influenced by electromagnetic forces). These ionised layers reflect HF radio transmission at wavelengths less than 15 km for long-range terrestrial communications. The Earth's geomagnetic field strongly influences the ionosphere's large-scale structure especially its variation with latitude. The ionosphere extends upwards from 50 km to 600 km altitude where free ions and electrons exist and from which radio waves in the band 3–30 MHz are reflected (scattering occurs for 30–100 MHz frequencies making this band useless for communications). Within the ionosphere exist the D, E and F layers. The D region, from 50 to 90 km, has a density of ~10^2–10^3 electrons/cm^3 and is caused by the solar photoionisation of NO by Lyman α-radiation at wavelengths of 1216 Å during daylight and absorbs radio waves; the E region (Heaviside layer), from 90 to 120 km, has a density of ~10^4 electrons/cm^3 peaking at 100 km caused by photoionisation by solar ultraviolet and X-rays during daylight and reflects daytime HF radio to provide single hop paths of less than 200 km; the F region (Appleton layer) from 120 to 600 km has a density of ~10^5–10^6 electrons/cm^3 peaking at 250 km.

The particles of the ionosphere move under the influence of the Earth's magnetic field described by the guiding centre approximation of three component motions constituting electric current within the plasma. Magnetic fields govern the motion of individual charged particles by forcing them on spiral paths along the magnetic field lines. The first component is a circular motion perpendicular to the magnetic field lines. The second component is motion along the field lines in the direction of increasing flux with Lorenz force reflection at the pole mirror points. The period of oscillation is much larger than the cyclotron frequency (which gives a gyration radius of the order of a kilometre). The third component is a slow drift in longitude with electrons drifting east and protons drifting west generating a magnetic shell. The magnetic poles are linked electrically to the magnetosphere through magnetic reconnection evidenced by the polar aurorae at altitudes of around 90–130 km. Tidal winds in the Earth's atmosphere are driven by solar heating in the infrared and ultraviolet regions of the electromagnetic spectrum. Such winds reach

50 m/s in the ionosphere at 100 km altitude. This induces electric fields across the geomagnetic field causing dynamo-driven electric currents in the dayside E layer of $\sim 10^5$ A which are detectable as geomagnetic fluctuations ~ 20 nT on the ground. The ionospheric current density depends on the relative motion of electrons and ions: $j = Ne(v_i - v_e) = \sigma E$ where v_i, v_e = drift velocities of ions and electrons and σ = ionospheric conductivity.

The number of electrons and ions of the plasma N must be large for significant conductivity and this occurs in the E-layer. In near-Earth space, the Larmour gyrofrequency is 1 MHz for an electron and 500 Hz for a proton typically. The faster and heavier the particle, the larger the radius of gyration. A 1-MeV electron has a gyroradius of 30 km while a 1-eV electron has a gyroradius of 10 cm. At a given height, the magnetic gyrofrequency-to-particle collision frequency ω/v is different for ions and electrons. In the E-layer, $\omega/v \gg 1$ for electrons and ~ 1 for ions, so current flows due to the relative motion between ions and electrons. In the F-layer, $\omega/v \gg 1$ for all charged particles so they drift together with a speed E/B and no current flows. The F-layer is the most important for LEO spacecraft as it ranges from 120 to 600 km. During the day, it has two divisions F_1 and F_2. The F_1 layer from 140 to 200 km is associated with an ion population of oxygen ions caused by solar ionisation and is the main daylight reflective medium providing radio signal paths of 2000 km, but it disappears at night. The F_2 layer from 200 to 400 km is a layer associated with peak electron density that exists at night and depends on locality, season and solar cycle providing radio signal paths of 4000 km. The ambient temperature varies from 800 to 2000 K between 50 and 300 km but above 300 km the temperature slowly increases to 10 000 K while the density falls exponentially. Similarly the ion composition changes from oxygen ions to hydrogen ions.

The ionosphere affects the spacecraft through spacecraft charging which occurs owing to the generation of electric potentials on the spacecraft with respect to the ambient 15–20 keV plasma and the formation of wakes and plasma sheaths from the passage of the spacecraft through the medium. Having different ion and electron mobilities causes charge separation and the generation of electric fields. Charge is built up on a spacecraft as a consequence of different sources generating a finite potential, introducing the possibility of electrostatic arc discharges. Transients generated could cause the spacecraft to fail. There is a day/night variation as photoemissive currents do not occur during eclipse but do during illumination. When a body exists in a plasma, it assumes a floating potential different to that of the plasma itself. Since electrons are more mobile than ions this potential is negative with respect to the plasma causing a space charge sheath of high electric field strength around the spacecraft. The spacecraft effectively moves through the plasma at velocities of Mach 4 to 8. The thermal velocity of electrons is ~ 200 km/s while that of ions is ~ 1 km/s compared with the orbital speed of the spacecraft of ~ 8 km/s. The ambient magnetic field induces electric fields causing a plasma wake. At low altitudes, this causes the wake to be depleted of ions and dominated by electrons. This can produce drag through ohmic dissipation. The plasma upstream is ram-compressed which can cause significant attenuation of radio signals. In LEO, the plasma environment is cool with a temperature of ~ 2000–3000 K but is dense at $\sim 10^8$–10^{12} particles/m^3. The Debye length defines the distance over which the satellite perturbs the ambient plasma and so defines the space charge sheath of the satellite. Over such short scales imbalances of charge can exist over short distances defined by the Debye length $\lambda_D = 6.9$ cm$\sqrt{T/n}$. In the ionosphere, the Debye length is around 1 cm. The Debye length is a function of spacecraft altitude and varies from $\sim 10^{-3}$ m at LEO to ~ 30 m at GEO and high latitudes. Actual charge separations will tend to oscillate with a characteristic frequency $f_p = 9$ kHz$\sqrt{n}$, which may be used to probe the plasma density directly. Large structures allow the space charge limited assumption that Debye length is much less than the satellite dimension [Martin 1994]. For smaller structures, the thick sheath limit that Debye length is much larger than the satellite size impling that the charge current is a function of the plasma and the spacecraft geometry. Spacecraft charging is typically low above 300 km and the typical satellite potential is +1–10 V with respect to the ambient plasma. In the polar auroral regions, however, spacecraft charging is high ~-1 kV to -20 kV due to the higher electron fluxes and energies ~ 5–10 keV. This could be problematic if two spacecraft with different potentials are in proximity, such as in the case of robotic servicing. An EVA (extravehicular activity) astronaut at a potential of -170 V within 10 m of the Shuttle would be severely affected by the Shuttle's sheath at -850 V which generates rapid beams of ions attracted to the Shuttle. Eclipse charging is particularly important at GEO generating negative differential voltages ~ 1–10 kV and arc discharging and electrical interference. Satellites comprise highly irregular surfaces and often with different properties and this can cause differential charging generating current flow from one region to another. This may be reduced by good grounding design such as the avoidance of cavities and the metallisation of dielectric surfaces to provide conductive paths. Active control of space charging using electron/ion emitters are possible but unusual. Such emitters may be based on hollow cathode technology developed for ion engine neutralisation.

The space radiation environment is composed of a solar contribution and a galactic contribution [Stassinopoulos & Raymond 1988]. Galactic cosmic rays are omnidirectional originating from outside the solar system and comprises $\sim 85\%$ protons with energies dominantly ~ 10–200 MeV (average 60 MeV) but with low fluxes ~ 4 particles/cm^{2}s. The rest of the cosmic rays are composed of alpha particles $\sim 14\%$ and heavier nuclei from Li to Fe $\sim 1\%$ with energies ~ 15 MeV. Heavy ions are strongly ionising. The solar contribution to the space radiation environment is significant. The Sun exhibits a solar dynamo effect generated by differential rotation and convection creating a magnetic field. The Sun is a source of plasma and solar wind is emitted as expansion of the solar corona. Charged particles of mostly electrons and protons from the Sun's coronal holes (temperature $\sim 10^6$ K) are emitted from the Sun travelling at hypersonic velocities of ~ 200–700 km/s and energies of ~ 1 keV generating a solar wind of density ~ 1–10 protons/cm^3. The coronal plasma ejected from the coronal holes expand supersonically. The solar wind expansion transports magnetic fields and the rotation of the Sun creates an Archimedes spiral structure to the interplanetary magnetic field which extends out to the heliopause at 5–100 AU. The solar wind thus has an associated magnetic field of ~ 5 nT. The solar magnetic field possesses an 11-year period cycle of activity indicated by sunspot activity. Sunspots are regions of magnetised plasma where the magnetic field suppresses convective energy transport beneath the photosphere, so generating the cool dark regions characteristic of sunspots. At solar maximum, about 200 sunspots and 12 solar flares occur per year whilst at solar minimum about 10 sunspots and 2 solar flares occur per year. The quiet sun X-ray flux is ~0.1–1.0 erg/m^{2}s but solar flares are also a major contributor to the solar xX-rays from the corona. A typical flare will emit ~10^4 protons/cm^{2}s (plus ~5–10% alpha particles and a few heavy nuclei and electrons) with energies exceeding 20 MeV (up to ~1–10 GeV) for a duration of ~48 h. Solar flare activity depends on the 11-y solar cycle. Solar cycle 19 (1953–64) provided an active solar flare environment and has been used as the model for spacecraft operations. Geomagnetic storms can occur at any time but typically result from solar coronal hole emissions which are more frequent during sunspot maximum [Beech et al 1995]. The resultant coronal mass ejections are much more powerful than the solar winds generating vast disturbances in space and causing large electric currents in the upper atmosphere and ionosphere. As the solar plasma impacts the outer boundary of the magnetosphere, the energy flux enters the ionosphere at high latitudes through the intervening magnetosphere to produce the auroral phenomena. This can cause electric power grids particularly sensitive transformers to break down and disrupt short-wave radio communications [Risbeth 1991], e.g. in 1989, geomagnetic storm activity left 6 million people in Quebec without electricity for 9 hours. The effect on spacecraft is also significant. The upper atmospheric density dramatically increases drag—Skylab in a 435-km orbit was dragged into re-entry in 1979 by high solar activity. The Canadian TV satellites Anik E1 and E2 were affected by a storm in 1994. Spurious electric currents caused their solar arrays to be commanded to point away from the Sun. Although E1's backup systems overrode this, E2 ran down its battery power and flew out of control for 6 months until its orbit brought its arrays back into sunlight.

The solar wind impinging on a solid planetary body generates an interaction dependent on the nature of the planetary obstacle. An atmosphere-less solid body will generate a wake of plasma behind the body. If a body has a planetary atmosphere, the plasma will pile up on the sunward side and generate a long plasma tail to the rear of the object. If a body possesses a planetary magnetic field like Earth, the planet is shielded from the solar wind by a well-defined magnetospheric surface around the body (magnetopause is formed by the boundary where solar wind slows from supersonic to subsonic speeds) forming a magnetic bow shock wave compression on the forward side at 10–12 Earth radii distance and a magnetic tail of 40 Earth radii in diameter extending out to ~100 Earth radii on the leeward side. The magnetosphere is defined as the region of completely ionised plasma and is constrained by the Earth's magnetic field. A region of trapped charged particles of density n_e ~10^{-15}/cm^3 which extends upwards from ~200 km to ~10 000 km altitude is dependent on solar activity. At the magnetopause, solar wind pressure on the outside is balanced by magnetic pressure on the inside of magnitude $p = B_m^2/2\mu_0$. The energy stored in the sheared magnetic fields ~ $B_\perp^2/\mu_0$ may be released in thin current sheets and transferred to particles with high energies. This is the cause of the aurorae borealis in the upper atmosphere.

Trapping of charged particles in the Van Allen radiation belts around the Earth were first discovered by Explorer 1 in 1958 within a magnetic bottle as utilised in nuclear fusion tokamaks. Centrifugal force causes positive ions to drift westwards and negative particles to drift eastwards generating the plasma constituents of the outer Van Allen radiation belts to form a large ring current around the Earth at a distance of 3–5R_E. This inner magnetosheath comprises the toroidal Van Allen radiation belts of >30 keV charged

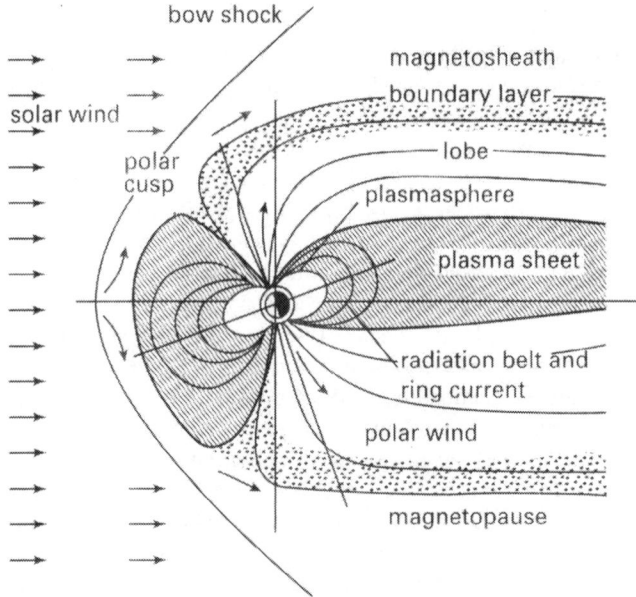

Fig. 1.3. The Earth's magnetosphere (from Wertz & Larson 1999).

particles which constitute a westward-flowing ring current at the magnetopause forming an inner belt from 7000–13 000 km and an outer belt from 19 000 to 41 000 km. The electron energy spectrum varies from 20 keV to 6 MeV. The electron population is divided into an inner belt from one Earth radius to 2.4 Earth radii of <5 MeV electrons and an outer belt from 2.8 Earth radii to 12 Earth radii of ~7 MeV electrons. The maximum electron population is ~10^8 electrons/cm²s. There is a minimum electron population between the inner and outer belts called the slot which disappears during high geomagnetic activity due to increased electron density and this geomagnetic activity is directly related to the proton flux incident on the Earth's atmosphere. The electron flux does not contribute significantly to radiation effects on the spacecraft. The proton population extends from one Earth radius to 3.8 Earth radii which varies with distance. Protons of energies <5 MeV are distributed throughout the magnetosphere at ~10^9 protons/cm²s. The maximum flux of protons occurs at 3000 km and 18 000 km. High energy protons from 10 to 600 MeV are confined to ~1000–40 000 km the maximum of which lies at 36 000 km where the number of protons with energies >40 MeV is ~2 × 10^4 protons/cm²s. Hence, GEO is in the upper regions of the radiation belt at 6.6 Earth radii.

The Earth's magnetic field is essentially a tilted dipole field at an inclination of 11° from the axis of rotation. It may be modelled by magnetic latitude λ as: $B = -(2\mu_0 M \sin \lambda)/4\pi r^3$, $B = -(\mu_0 M \cos \lambda)/4\pi r^3$ where M = magnetic moment of the dipole. It varies from $0.3G$ at the equator to $0.7G$ at the poles and is caused by convection in the Earth's molten Ni–Fe core. The South Atlantic Anomaly (SAA) is a high-radiation regional phenomenon that is caused by the displacement of the geomagnetic axis vector from the geographical axis vector of 500 km towards the North Pacific. This departure from the magnetic dipole model to <$0.25G$ causes the inner Van Allen radiation belt to be the manifestation of the southern mirror point which by virtue of geomagnetic field drift of 2.5 km/y generates a westward drift of the SAA by 0.3°/y. The high radiation region of the SAA in LEO may be avoided by choosing low inclination <45° orbits which give very low dose rates, even without radiation protection, up to 500 km altitude as the radiation is highly attenuated towards the magnetic poles by the magnetosphere (this is particularly important for EVA). Equatorial LEO is useful only for scientific observation and mobile communications missions. The high-altitude eccentric orbits (HEO) favoured by scientific missions have a perigee of 1000 km and an apogee of 70 000 km and this involves the spacecraft passing through the strongest radiation belts on each pass. However, because perigee passage is rapid only a small portion of the orbit lies within the belts as 70% of the orbit is above 40 000 km. This may be avoided altogether at the price of an active AOCS for rapid pointing at equatorial LEO to increase the observing from 30% to 75% (compared with 70% achievable from 24-h HEO) [Agonisti et al 1992].

Radiation can have significant effects on materials, particularly organics [Bouquet & Koprowski 1982]. The atomic oxygen environment in LEO is dominant in degrading the surfaces of coatings and material loss. The atomic oxygen arises from ultraviolet dissociation and ionisation of molecular oxygen in the upper atmosphere. It causes many materials to undergo oxidative degradation resulting in surface mass erosion and morphology changes [Reddy et al 1992]. The Solar Maximum Mission Kapton thermal blankets were found to be severely eroded during the Solar Maximum Repair Mission. The LDEF (Long Duration Exposure Facility) mission results indicate that for long-duration missions such as Eureca, consideration should be given to the use of oxygen-corrosion resistant materials and protective coatings especially on vulnerable components such as solar arrays which would otherwise suffer degradation. Oxygen atoms are typically travelling at ~8 km/s producing collision energies ~5 eV per atom. All metals except osmium and silver are stable to atomic oxygen erosion. Organic films such as Mylar and Kapton films, Teflon and any material containing CHON elements (e.g. carbon-reinforced plastic and fibreglass epoxy) are particularly susceptible. Kapton film will suffer surface erosion of 50–80 Å due to atomic oxygen sputtering. Different materials' susceptibility to damage is determined by their reactivity coefficients, R: Kapton ~3.0 ×10^{-24}cm^3/atom; silver ~10.5 × 10^{-24}cm^3/atom; carbon ~0.9–1.7 × 10^{-24}cm^3/atom; epoxy ~1.7–10^{-24}cm^3/atom. Atomic oxygen flux may be given by: $\varphi_0 = \rho v_{s/c}$, where ρ = atomic oxygen density = 10^9–10^{11} atoms/cm^3 at LEO and v_s/c = spacecraft ram velocity. The yearly surface loss is given by: $t = 365 \times 86\,400 \times \varphi_0 \times R$, where R = reactivity coefficient for the material.

The atomic oxygen flux increases as altitude is reduced from 1000 km to 200 km but increases at a given altitude at solar maximum periods due to atmospheric expansion. The thickness lost is 12 times greater during maximum solar activity than during solar minimum. The attack process is essentially one of oxidation since electronic excitation is caused by inelastic collisions increasing the material's reactivity. If the oxide is solid, volumetric expansion causes flaking and scaling by crack propagation. If the oxide is gaseous, the volatile material evaporates from the surface. Effective thin film coatings should be in their highest oxidative state and these include silicone polymers, siloxanes, aluminium oxide, silicon oxide and related compounds which are barely affected at all. Such coatings are absolutely necessary for Kapton and Mylar surfaces.

The meteoroid flux varies with the mass range and originates from cometary and asteroidal breakup and remnants from the proto-solar system formation process. They are mostly omnidirectional with average velocities ~20 km/s. They have an estimated accumulated mass of ~200 kg below 2000 km altitude. The impact cross-section of a meteoroid increases with decreasing size by virtue of increased population. A 25-g meteroid has an impact cross section of $\sim 5 \times 10^{-14}/m^2 s$. At 1.5-g masses, the impact cross-section increases to $\sim 1 \times 10^{-12}/m^2 s$ (this would require ~8-cm Al thick shielding at 28 km/s). At 0.015 g, the impact cross section has become $\sim 1 \times 10^{-9}/m^2 s$. A representative mass is 10^{-6} g for a 100-μm-sized micrometeoroid and their micrometeoroid flux is given by $\log_{10} N \sim 4$ with velocities ~20 km/s. More micrometeoroids occur at low altitudes due to the Earth's gravitational focussing effect. Their primary effect is to cause the degradation of thermal coatings and solar cells.

1.3 THE NEAR-EARTH SPACE DEBRIS ENVIRONMENT

There are currently around 500 spacecraft in Earth orbit of which only 6% are operational. Orbital debris is an ever-increasing problem [Lambert 1993]: in 1985, the USAF Space Command Surveillance Centre tracked with radar ~8000 space debris objects from 3800 launches since the dawn of the Space Age in 1957. The majority of these launches were from the USSR with 2400 and the USA with ~1000 followed a long way behind by ESA with around 50. The 8000 fragments vary in size from ~10 cm diameter in LEO to ~1 m diameter in GEO (limit of detection capability). Around 5900 of the 8000 objects reside in LEO at <1200 km altitude. The average relative velocity in LEO is ~8–14 km/s compared with ~100–500 m/s in GEO. It has been estimated that 40 000–80 000 nondetectable objects exist in LEO of size larger than 1 cm diameter and $\sim 10^9$ objects of size 0.01–0.5mm. The mass of the LEO debris was estimated at $\sim 2 \times 10^6$ kg with an average density of 2.7 g/cm^3. Particle impacts by ~0.3–0.5-mm objects will cause fatal ruptures in spacecraft structures if unshielded. A particle impact by a 4-mm piece of debris would be catastrophic for the Space Shuttle. Indeed, the STS 7 Challenger outer windscreen layer was cracked by a fleck of paint. There were almost 20 cockpit window replacements during the first 10 y of Shuttle operations due to micrometre-sized debris impacts at a cost of $40 000 a time. This is in spite of the intrinsic probability of collision being $\sim 10^{-9}/m^2/y$. The loss of the French microsatellite Cerise in 1996 was attributed to debris collision by virtue of the sudden discontinuous change in its moment of inertia due to the loss of its gravity gradient boom. A collision at more than 7 km/s in LEO with a debris fragment over 1 cm is size would impart the energy equivalent of a hand-grenade $\sim 10^5$ J to a spacecraft and lead to its catastrophic breakup, producing yet more debris. Shielding with bumper plates and meshes can only protect a spacecraft from objects up to 1 cm in size, but they involve considerable mass penalties so such methods are adopted only for manned space systems, but see Wilkinson et al (2000) for debris damage mitigation approaches. It has been estimated that an average satellite in LEO will survive ~50–100 y before collision with a trackable fragment

[Reynolds et al 1983]. There is a 4% risk that HST will be damaged by a collision with a fragment greater than 1 cm in size during its 17-year lifetime.

Only 6% of trackable objects are operational spacecraft; 45% are large fragments resulting from satellite breakups; 21% are inactive spacecraft (and so retrievable or disposable by robotic means); 16% are spent upper stages; and 12% are mission-related ejecta such as clamps, etc. Satellite explosions or breakups and launcher upper stages are the main contributors and it has been estimated that there have been ~90 satellite fragmentations since 1961 (mean failure rate ~3%). Around 18% of these fragmentation events were deliberate, 42% were propulsion-related malfunctions, and 40% of unknown origin (possibly hypervelocity debris collision). These breakups account for ~90% of all fragments. On average ~100–200 trackable fragments are generated by explosive fragmentation events which have occurred at an average rate of 4 per year since 1975. A large contribution was due to the explosions of Delta and Thor upper stages and from the Ariane 4 third stage carrying SPOT-1 in 1986 which itself generated 488 trackable fragments and up to an estimated 2000 untrackable fragment of size exceeding 1 mm. By June 1991, 110 of those trackable fragments remained in orbit. Arianespace has since modified its Ariane upper stage so that it retains its integrity and the US Delta has also been modified. Space debris is a considerable threat to reliability. The effect of debris impact is determined by the relative kinetic energy deposited: if the relative velocity of impact is 100 km/s, a 100-g fragment has a kinetic energy of impact equivalent to 1 kg of TNT. The collision probability/y for a satellite with a cross-section of 50–200 m^2 varies with altitude and debris fragment size. Large fragments can be detected and avoided, but smaller fragments are difficult to track. A 1 cm diameter fragment is the maximum shieldable size. For a fragment ~1 cm in diameter (with ~1 mg mass), the collision probability/y varies from ~10^{-4} at 300 km to ~10^{-3} at 400–500 km to ~10^{-2} at 600–1000 km, and ~10^{-3} beyond 1200 km. Hence, the worst-case probability is ~0.01/y. For a typical 10-y mission, the total collision probability is ~0.1.

GEO debris is persistent, whereas atmospheric drag in LEO acts as a natural cleansing agent below 600 km (the limit for realistic time scales of removal). Most GEO spacecraft are elevated to graveyard orbits after operational life for this reason. Most debris at <400 km altitude fall and burn up within less than 4 months. At 600 km atmospheric drag reduces debris persistence to ~10 y. Above 800 km the half-life of debris exceeds 30 y. The efficiency increases with smaller particles due to their higher ballistic coefficients. Solar maximum periods every 11 y also increases the rate and height of debris removal up to 800 km. Most debris ~99% is concentrated between 600 and 1500 km with peaks at 800, 1000 and 1500 km in near-circular orbits [Crowther 1994]. The collision probability in this band was estimated at $1.7 \times 10^{-5}/m^2/y$ for the 1–10 cm range. Polar orbits are particularly populous in orbital debris. Indeed as Sun-synchronous orbits are slightly retrograde with $i = 98$–$100°$ the spacecraft orbital velocity is higher than that at low inclination LEO. There are also peaks at $e = 0.6$ due to the 10.5-h period GTO and at $e = 0.8$ due to the 12-h period Molniya orbit. It is conceivable that the debris population at 1500 km is therefore virtually permanent. As the debris grows fragmentation due to hypervelocity impacts will increase dramatically.

The number of launches per year has been constant at 120/y but has recently been reduced to ~100 due to budgetary constraints (USSR ~80%, USA ~12%, others ~8%).

Around 60% of those launches are military. Furthermore, the average launch puts four spacecraft into orbit and an average mission produces three pieces of trackable debris. The expected growth of around 10% in Earth-orbiting satellite traffic for remote sensing in the future and the tendency towards lighter and smaller satellites acting to increase the number of launches/per year introduces the possibility of a runaway situation if the removal time for debris exceeds the average time between collisions. Computational modelling suggests that we could generate a catastrophic growth in debris at ~1000 km altitude by 2050 as a persistent structure forms presenting a permanent hazard to future operations severely restricting the utilisation of such orbits and even making space activities impossible [MacInnes 1993]. However, it has also been suggested that, even if all launches ceased, self-sustained collisions would continue to increase the number of fragments in orbit for a long period [Klinkrad & Jehn 1992, Rex et al 1989]. Above 800 km, even with solar cycle atmospheric altitude increases, atmospheric drag is too inefficient to clear the debris. The ESA Space Debris Working Group of 1988 specified LEO and GEO as particular areas of concern, the first due to the large traffic and characteristic high velocities of transit and the latter due to crowding of a limited resource. They stressed that steps must be taken to curb the space debris problem. This means essentially:

(i)   avoid explosive fragmentation;
(ii)   reduce the fragment debris population;
(iii) remove large objects from space.

The first point is being implemented by the major launch suppliers. The second is infeasible. The third may be provided by a freeflying robotic spacecraft.

Space debris is thus becoming a serious hazard to operational spacecraft in the near-Earth space environment. Fewer satellite launches would slow down the ever-increasing problem of crowding and consequent space debris generation. A robotic interceptor satellite such as ATLAS could alleviate this problem by salvaging the larger pieces of redundant hardware that pose such a threat to operational spacecraft, either for return to Earth (e.g. as in the case of the Long Duration Exposure Facility, LDEF, designed for retrieval and return to Earth), or for controlled re-entry disposal, or for putting into new graveyard orbits in a cost-effective and practical manner. For objects without grapple fixtures, highly compliant cushioned pads could be used [French & Boyce 1985]. The Soviet nuclear-reactor-powered Cosmos 954 incident, which reentered the Earth's atmosphere and spread nuclear debris over a remote area of Canada necessitating an expensive clean-up operation, illustrates the need for retrieval particularly where nuclear-powered spacecraft are concerned. The US Space Shuttle may provide the capability for LEO retrieval and return. Although satellite retrieval for salvage is unlikely to be cost-effective, propellant scavenging is a potentially useful operation.

1.4 THE MARS ENVIRONMENT

Mars is the most common target for rovers. The primary interest in Mars is in searching for evidence of life as might be suggested by the Martian meteorite ALH84001 found in Antarctica in conjunction with the recent discovery of extremophiles in hostile environments on Earth [Jakosky 1998]. Mars is about half the size of Earth with a radius of 3293 km and 1/10 its mass (and so a surface gravity of 3.72 m/s²). Its day is similar to Earth at 24 h 39 min. Its small size is probably due to the proximity of massive Jupiter accumulating much of the adjacent material during solar system formation. The solar radiation intensity is less than 50% that on Earth as it is further from the Sun at 1.52 AU, with a year of 687 days. It has two small moons, Phobos and Deimos. The 22 km diameter Phobos orbits closer to its primary than any other moon at 600 km altitude and its orbit is decreasing at 1.8 m/century so it will fragment into a ring in around 50 My. The 12.6 km diameter Deimos is the smallest moon in the solar system. Both are irregular C-type asteroids which were captured into Mars orbit after perturbations from Jupiter. Mars has a low density of 3.94 g/cm³ with a differentiated structure, this differentiation occurring early—the Martian mantle was virtually completely solidified by the end of the planetesimal accretion phase ~3.5 By ago. It has a dense central core ~1300 km in radius of 90% Fe/Ni and 10% FeS, now solid and a low density crust ~20 km thick. There is no evidence of global contraction but there is evidence of lithospheric expansion. Its deep interior is generally believed to be cold with a large fraction of its volume being at low temperature allowing only shallow interior heating during geologically recent times. Its present thermal flux from the surface of 0.04 J/m²/s suggests a uranium content of 0.03 ppm and most of its thermal activity would have ceased in its first 1 By.

Mars has a thin atmosphere of 95.3% carbon dioxide, 2.7% nitrogen, 1.6% argon plus variable traces of oxygen ~0.13%, carbon monoxide and water vapour ~0.03% giving a surface pressure of 6.1 mbar—it has lost much of its original atmosphere with its atmospheric pressure now only 0.08% that on Earth. The present atmosphere generates a negligible greenhouse effect from its thin carbon dioxide atmosphere which increases the global average surface temperature from 212 K to only 217 K. Only around mid-day at low latitudes will ice thaw but due to the low atmospheric pressure, it sublimes directly to water vapour as liquid water cannot exist on the Martian surface. A total surface pressure of 2 bar is required to allow melting of water—some 100 times its present pressure. The Martian surface solar ultraviolet flux averages 7×10^3 erg/cm²/s at the equator—some 43% that received by Earth. The Martian surface is cold, dry and highly oxidised with a mean surface temperature during the day of 220 K at low latitudes in the summer, some 53 K below the freezing point of water at that pressure, to a 150 K at night at the poles in winter.

Martian geology is dominated by iron- and magnesium-rich basaltic lava flows low in silica content originating from the upper mantle. Solar ultraviolet flux splits water into hydrogen and hydroxyl radicals to form highly oxidising hydrogen peroxide in the soil. Much of its surface rock is similar to andesite in which Fe and Mg minerals have precipitated out from the silicon-rich basalt melt. Typical minerals include 30% orthopyroxene, 30% plagioclase feldspars, 30% quartz and other minor minerals such as Fe/Ti oxides like magnetite Fe_3O_4. This makes the Martian crust similar to the Earth's basaltic oceanic crust. Its regolith is fine-grained and comprises 10% carbonates, 18% hydrates, 38% silicates, 15% Fe_2O_3, 5% Al_2O_3, 7% MgO and 7% CaO. There is some evidence of sedimentary rock in the form of pebbles and conglomerations in sandy and clay matrices rich in hydroxides such as kaolinite $Al_4(OH)_8Si_4O_8$ formed from feldspars in flowing water. Some 10–20% of the Martian regolith may be composed of salt minerals such as Mg sulphates. Martian geological history is divided into three eras: the Noachian in the first 1 By when its climate was warm and wet and still under-going heavy impactor flux from planetesimal gravitational sweep-up, the Hesperian from 3.5 to 2.5 By ago when the climate was cold interrupted with mild phases and extensive volcanism, and the Amazonian from 2.5 By ago to the present characterised by increasing geological quiescence and a permanent cold and dry climate.

Martian surface terrain is varied with impact craters, large canyons, large volcanoes and water-eroded channels. It exhibits a configuration due to meteor impacts, volcanic activity, and wind and water erosion particularly during its early history. Impact craters are widespread but dominate in the southern hemisphere while the northern hemisphere has been modified substantially by volcanism and erosion of older craters. Hence, the southern hemisphere is geologically ancient older than 3.5 By. Deep rift valley networks are found mostly in the southern uplands especially south of 30° latitude which were cut by running river water indicating that water flowed in the first 1 By of Martian history. The northern hemisphere is lower in elevation and covered with younger volcanic deposits which have re-surfaced the landscape. There is evidence of weathering and craters smaller than 15 km in diameter have been completely eroded. A huge east–west rift valley, the Valles Marineris has a width of 100–200 km, a depth of up to 5 km and a length of 4800 km near the Martian equator. The equator exhibits a cliff-like transitional zone of 3 km height between the two hemispheres. Shield volcanoes similar to terrestrial Hawaiian volcanoes but larger, are also distributed over the surface of the northern hemisphere up to 25 km above local elevation. The Tharsis region of the northern hemisphere, an elevated region of 10 km height and some 4000 km in diameter, holds four large shield volcanoes, the largest of which is Olympus Mons some 27 km high and 600 km in diameter with a caldera-type summit crater 80 km in diameter making it the largest volcano in the solar system. Extensive volcanic modification of the surface near the equator continued after the cessation of heavy meteoritic bombardment 3.8 By ago. Mars lacks tectonic plate mobility though it exhibits extensive rifting and faulting with evidence of lava flow from volcanic activity on the Martian surface perhaps as recent as 200–400 My ago. The Tharsis ridge represents the largest positive mass anomaly in the solar system. Such distortion causes a deviation from isostasy contributing a 6% alteration in Mars' moment of inertia and angular momentum. The eastern regions of Tharsis are characterised by vast canyons which stretch for 4000 km to the south of Chryse Basin representing stressed faults. The Chryse Basin appears to be an ancient river flow basin and was the landing site for both Viking 1 and Pathfinder. The uplift of the Tharsis region may have introduced a new, lower variability in planetary inclination and earlier geological events may have caused similar changes. Lithospheric uplift in conjunction with crustal fracturing and volcanic shield construction occurred possibly due doming and rifting of the lithosphere over a large stationary mantle plume or hot spot generating flood-basalts. The global lithosphere thickened over time to 25–150 km depth into a single plate over the planet exhibiting little tectonic activity. It has locally thin regions associated with volcanism causing local fracturing from tectonic stresses. These represent the easiest sites for magma access to the surface taking on a role similar to that of mid-oceanic ridges on Earth.

Liquid water appears to have been involved in its ancient surface morphology due to the existence of ancient dry river valleys and dendritic outflow channels suggesting that the Martian early atmosphere ~3.5–4.0 By ago was significantly thicker than the present one. Some highlands have extensive valleys draining into sediment-covered depression basins such as the southern and eastern canyons of Chryse. Such depression sites were possibly transient lakes for several My in the early geological history of Mars, e.g. the hypothetical ocean Oceanus Borealis that may have covered one-third of the planet in the northern hemisphere with as much as a global depth of 450 m of water, perhaps even reforming periodically [Baker et al 1991]. Water vapour from these liquid reservoirs could potentially generate a greenhouse effect with water vapour condensing into clouds with subsequent precipitation. Surface runoff would have percolated into the ground and frozen as subsurface permafrost and as polar ice or been lost to space. There is however no evidence of precipitation. The present Martian water inventory would provide a global coverage of only 50 μm [Kargel & Strom 1996]. Mars has not been thoroughly outgassed according to depleted Ar-36/Ar-38 ratios suggesting a global depth of 9.4 m of water, 0.14–0.53 bar of carbon dioxide and 20–30 mbar of nitrogen (presently 0.15 mbar) but these computations have assumed erroneously that the original Martian complement matched that of Earth—the discovery that Venus has a Ar-36 complement some 20–200 times higher than that of Earth shows that constant ratios cannot be assumed [Squyres & Kasting 1994]. Mars originated in cooler regions of the solar nebula suggesting a greater non-noble gas volatile complement while noble gases are adsorbed onto solar nebula grains which dominate in the inner solar system. Its D/H enrichment, some 5 times that of Earth, suggests significant outgassing of water onto its surface with subsequent loss of the lighter H fraction after the dissociation of water from solar ultraviolet flux [Jakosky and Jones 1994]. As much as 1–5 bar of water may have been outgassed, enough for a 100–500 m deep global ocean (compared with 3 km for the Earth). Around 3–10 bar of carbon dioxide and 0.1–0.3 bar of nitrogen may have been outgassed in addition. A maximum atmospheric pressure of 1–2 bars of CO_2 is sustainable at which saturation and condensation occurs, limiting the greenhouse warming below the temperature for the liquefaction of water. Most of the primordial atmosphere was subsequently lost to space mainly due to impact erosion and some hydrodynamic escape but, most of the original carbon dioxide may have subsequently been taken up as adsorbed gases into the regolith or incorporated as carbonates in the rock (which requires liquid water). There is evidence from the 12 SNC Martian meteorites found on Earth that veins of calcium carbonate were deposited less than 1.3 By ago. The primordial nitrogen could be deposited as nitriles and nitrates but ~99% would have escaped to space due to the high N–15/N–14 ratio. Much oxygen must be chemically bound in rock as nitrates, sulphates, carbonates and oxides. Water outgassed from the planet would have been less subject to impact erosion and hydrodynamic escape than the primary atmospheric gases. Much of the outgassed water could have subsequently been locked into permanent ice at the poles, subsurface permafrost ice, regolith adsorption and hydrated rock—it has been estimated that the amount of water that has escaped to space after dissociation to oxygen and hydrogen by solar ultraviolet is equivalent to a global depth of 1–2 m (its present rate of loss of water is $6 \times 10^7/cm^2/s$).

The higher atmospheric density in this early period would have yielded a greenhouse effect generating stable higher pressures and higher temperatures sufficient to allow permanent liquid water. It is doubtful if sufficient carbon dioxide was outgassed to raise the planetary temperature above 0°C given that the early Sun was some 25–30% less luminous than at present. If the primitive atmosphere were reducing however, the critical surface pressure required for liquid water is 0.15 bar, so that the required vertical column of carbon dioxide would be 4.5 times less with ammonia, methane, water providing an enhanced greenhouse effect. This greenhouse effect would depend critically on the ammonia (as the most powerful greenhouse gas) mixing ratios. Ammonia gas would subsequently be subject to photodissociation by solar ultraviolet radiation. However, volcanism could supply significant sources of methane. It appears that the atmosphere thinned around 3.8–3.5 By ago primarily due to impact erosion.

Some dry outflow channels appear to have originated during the later Hesperian era. Many outflow channels of 1000×100 km in size indicate water flow speeds of 60–75 m/s probably due to the sudden release of subsurface water usually present as permafrost in substantial quantities, possibly residing in up to 500 m thick layers. They are associated with volcanic regions such as the Tharsis region over several separate periods whereby volcanic activity heat freed subsurface permafrost. The largest outflow channels are located around the Chryse basin on the eastern margin of the Tharsis anomaly suggesting the flow of 5×10^6 km^3 of water. Hydrothermal vents may have been the sources of water heated at depth by active magma sources. Huge torrents of water were required to carve the outflow channels suggesting flow rates of 3×10^8 m^3/s [Carr 1986, 1987]. Tributaries indicate widening of channels below the inflow areas indicating downhill flow. Teardrop-shaped islands around 100 km long in alluvial plains indicate low viscosity flow patterns. Braided patterns indicate cycles of deposition of materials from liquid suspension. Networks of gullies exhibit dendritic patterns over large fractions of Mars, particularly over older regions, indicating melting of permafrost layers. Most such gullies formed at intermediate times between the formation of old cratered southern uplands and the volcanic plains of the northern hemisphere during a warmer, wetter period. Mars has been geologically inactive for $\sim 1 \times 10^9$ y or so, so volcanism ceased to replenish the atmosphere [Irwin et al 1999].

Many Martian craters exhibit mudflow features due to catastrophic flooding from melting of subsurface permafrost close to the surface. Certainly, volcanic activity or meteor impacts would have melted any ground ice from subsurface aquifers in catastrophic flooding events. Layered sediments in canyons indicate possible lakes and flash flooding. The nature and distribution of subsurface ice is not clear. Near the equator, craters over 4 km in diameter display muddy ejecta patterns while such features are associated with 1 km diameter craters near the poles suggesting that permafrost is deeper at the equator by 800 m than nearer the poles [Carr 1986, 1987]. The top of the permafrost layer may lie as close as 100–300 m from the surface. Some suggest that the water lies down to depths of 1–3 km. It is believed that at latitudes higher than 40°, subsurface ice lies at depths of ~30 m. Closer to the equator, the depth of the ice increases to an average of 500 m at the equator with most of the water incorporated into hydrated rock. Much of the water inventory at the equator has sublimed due to the higher surface temperature. Carr (1986) suggested that subsurface permafrost is stable at depths of 30 m – 3 km where the rock porosity is 10–20%, but below this porosity drops exponentially to zero at 10 km depth.

The whole Martian landscape is cold, dry desert with major formations windswept by sand and dust. Ravaging wind speeds are around 40–100 m/s which generate abrasive dust storms which can engulf the planet for several months at a time. The global dust storms occur at perihelion and such aeolian processes dominate erosion on Mars. Mars has frozen carbon dioxide ice caps at both hemispheres, but the two poles are different. The northern ice cap of ~1000 km in diameter and thickness of 4–6 km contains water while the southern ice cap of ~350 km in diameter and thickness of 1–2 km is purely carbon dioxide ice. The ice caps contain some 50 times as much carbon dioxide as resides in the atmosphere. They are both composed of very fine-grained dust particles and ice carried by winds and deposited at the polar regions from equatorial regions. Dust and ice act as nucleation centres for carbon dioxide condensation as ice particles which are preferentially deposited on the poles at winter. Successive layering indicate recent atmospheric erosion periods subject to fluctuations in net transport and deposition caused by cyclic variations. The zone of depletion stretches from 50° to 80° latitudes which show variable deposition of ice, the amount decreasing towards the equator. Some 30% of the Martian atmosphere freezes every winter and extends the coverage of the polar ice caps to be returned to the atmosphere every summer. The highest abundances of atmospheric water vapour occur at the edges of the North polar ice cap in midsummer but it drops towards the equator. It has been estimated that there may be 10^{19} kg of water ice locked in the polar caps and regolith.

Milankovitch cycles may be responsible for the present-day climate of carbon dioxide condensation as permafrost locking Mars into a self-perpetuating permanent ice age that exists now [Cockell 1995]. Martian obliquity, planetary precession and orbital eccentricity variations are much larger than those of Earth. Martian eccentricity has a short oscillation period of 95 000 y and a long period of 2×10^6 y with an amplitude of 0.14 (currently $e = 0.093$ on average). Eccentricity causes minor differences in mean annual insolation at the poles but large differences in insolation at subpolar regions is some 40% greater at perihelion than at aphelion. At perihelion, the Martian south pole experiences summer vaporising much of the carbon dioxide into the atmosphere. However, the summers are short and during the long winters, the carbon dioxide is re-deposited as ice, generating a drop in atmospheric pressure of 3 mbar. At perihelion with maximum $e = 1.22$, global dust storms are generated. The critical surface pressure for liquid water drops from 2 bar to 1 bar. Obliquity has periodicities of 1.2×10^5 y and 1.2×10^6 y with variations of 20° between 15° and 35° (currently 25°). Such changes lead to large variations in mean annual insolation at the poles which is enhanced quite dramatically at maximum obliquity of 35°. This causes a cycling of atmospheric pressure from 4 mbar to 0.7 mbar every 1.2 My with additional cycling every 100 000y. The mean atmospheric pressure is sensitive to solar insolation at polar regions, i.e. an inclination i indicating a positive feedback mechanism. The obliquity also varies chaotically rising abruptly to 60° every 10 My or so generating up to 3.5 times as much polar insolation during summer seasons. They may be responsible for cyclic variations in the Martian climate suggesting that the Martian volatile inventory is hidden in a reasonably accessible form. There may, however, have been relatively recent warm episodes ~500 000 y in duration, perhaps as recently as 10–20 My ago generating localised sapping of water. A thick atmosphere generated from sublimation at the poles due to higher temperatures.

Boston et al (1992) have suggested that hydrothermal processes below regions such as Valles Marineris and Olympus Mons could support a Martian ecology deep in the subsurface aquifers based on anaerobic chemolithographic processes. Such microbes could fix carbon dioxide into organic matter through energy released from oxidation using NO_3, SO_4, CO_2, CO or Fe as electron acceptors. Heat and reduced volcanic gases H_2, H_2S, SO_2, CH_4, CO_2 and CO would percolate upwards from subsurface magma sources, liquefying the water ice and/or clathrates. The biogenic species that are possible are the anaerobic methanogens, autogens and sulpur-reduction by denitrification (e.g. *Thiobacillus denitrificans*). The products of such microbial activity such as CH_4 and H_2S would diffuse and dissipate into the atmosphere through the surface.

1.5 SUMMARY

We have seen the variety of environments that will impose constraints on the design of a robotic spacecraft, be it an in-orbit servicer or a planetary rover for Mars. The extraterrestrial environment imposes strong constraints on the spacecraft and it is important for the spacecraft design to accommodate those features [Rycroft 1989]:

(i)   the spacecraft must be capable of functioning in a zero-g environment where there is no natural damping medium for the dissipation of vibrations;

(ii)   the spacecraft must be resistant to material outgassing and temperature extremes under high vacuum conditions;

(iii)   sensitive components and instruments must operate in a charged particle radiation environment so spot shielding and hardened electronics should be used;

(iv)   solar flares and charged particles can interfere with radio communication;

(v)   magnetospheric transients can induce large electric current flows induced in equipment by electromagnetic forces;

(vi)   the system must be of lightweight construction to comply with the severe mass restrictions for launch yet be robust enough to withstand launch and other loads;

(vii)   in-orbit spacecraft lifetimes are dependent on the atmospheric density which can vary by as much as 100 times as a function of solar activity;

(viii) the system must be reliable for a long mission and incorporate capabilities for self-repair.

The space environment imposes unique constraints to robotic systems in addition: highly variable illumination, remote observation of target objects, geometrically complex workspaces, handling of flexible extended objects (e.g. thermal blankets), limited knowledge of a potentially cluttered workspace, dynamic interactions in zero gravity, time-delayed communications and a lack of natural damping for manipulators.

2

Robotic explorer missions

In this chapter, we survey planetary missions. Initially, we look at some generic robotic spacecraft that have been developed and built to explore the solar system, but concentrating on delivery systems such as landers and penetrators. We then consider planetary rovers, particularly for Mars, and some of the esoteric methods of coping with diverse planetary conditions.

2.1 ROBOTIC EXPLORATION OF SPACE

Robotic spacecraft as a generic term is taken to refer to all robotic planetary exploration probes reflecting their unmanned, autonomous nature. They are regarded as complex missions characterised by extensive re-configurations and manoeuvres during the mission. Such automated robotic spacecraft are not new in planetary exploration. Planetary exploration is one of the most challenging and high-profile endeavours in science. The USSR has been active in robotic spacecraft missions and sent 23 robotic missions to the inner planets from 1965 to 1984. We shall briefly survey robotic explorers but with an emphasis on landers and penetrators for delivering payloads to the environments of interest as the lander generally provides essential support functions to planetary rovers such as store-and-forward communications relaying and may be regarded as part of the rover functionality.

All rovers have to be delivered to the surface of the planetary body in question, and this requires landers to convey the rover payload from space to the surface. Given that fuel is a critical mission commodity, any method that maximises the efficiency of fuel consumption or reduces the rate of fuel consumption during orbit manoeuvres will be a major enhancement to space missions generally, and planetary missions in particular. Single-pass aerocapture manoeuvres for planetary applications involve deep atmospheric penetration with a continuum flow environment and require high L/D (lift-to-drag ratios). The Magellan probe performed the first interplanetary aerobraking maneuvre through the upper atmosphere of Venus to circularise an eccentric orbit with limited ground support by a spacecraft not designed for aerobraking in a relatively unknown planetary atmosphere. The Mars exploration programme has extensively utilised aerocapture to provide controlled deceleration through atmospheric drag protected by an aeroshell and substantially reduce propellant requirements. The standard 140 kg Mars aeroshell design used by Viking and Pathfinder spacecraft was a 70° half-angle axisymmetric cone with an upper atmospheric entry velocity of 8.6 km/s. The lander also performs a multitude of support and complementary functions to the rover, the most obvious being acting as a relay station to communicate with Earth. Planetary surfaces vary tremendously but we can differentiate two types of surface—rock exemplified by asteroids and terrestrial planets or ice exemplified by comets and most gas giant moons. Small bodies such as cometary surfaces also require anchorage to the surface. Asteroid surfaces are much harder than cometary surfaces introducing greater impact energy absorption requirements. The landing system is an essential part of the rover delivery system and requires the use of shock alleviation mechanisms. Scientific instruments are usually highly sensitive to shock loading. The limit for sensitive payloads is 500g over 15 ms while less sensitive instruments can withstand 700g over 15 ms for hard landing. Semi-hard landing reduces impact loading to 100g over 5 ms and 200g over 0.5 ms [Doengi et al 1998]. Shock alleviation devices are based on frictional damping or irreversible deformation. Frictional damping includes spring-damper shock absorbers such as airbags but are difficult to model due to temperature variations affecting volume/pressure properties of the air. Irreversible deformation includes Al honeycomb materials which crush permanently in a predictable manner offering the highest specific energy absorption capacity. Al honeycomb has a low density of 16–150 kg/m^3 and can be compressed by 80% of their original axial length. The compression strength varies from 0.35–17 MPa prior to crushing increasing to 0.17–7 MPa after crushing. The specific energy absorption is 30 kJ/kg. Al foams are manufactured by injecting gas into the molten Al alloy. Their densities range from 70–550 kg/m^3 with compression strengths of 0.05–8 MPa. Foams constructed from Al powder have better homogeneity but higher densities of 300–1200 kg/m^3 and compression strengths of 2–25 MPa, but have lower specific energy absorption. Generally, Al honeycomb sandwich is used for hard landing and airbags for semi-hard landing.

Penetrators are beginning to become popular modes of payload delivery to solid surfaces and they impose strong design constraints on scientific instrument robustness. They have been used extensively by the Russians as drop zonds and are ideal for subsurface analysis such as the search for water on Mars. Penetrators are missile-shaped cylinders for delivering scientific payloads to planetary subsurfaces of typical velocities of 80–200 m/s. They require high specific energy to anchor onto the target and penetrate to 1–10 m depths typically. They typically comprise a forebody for subsurface operations and an aftbody to remain on the surface connected by a cable. Typical payloads include seismometer, heatflow probe, tiltmeter, accelerometer, magnetometer, conduction meter, gamma-ray spectrometer, neutron spectrometer, X-ray spectrometer, alpha particle spectrometer (typical of the sensors used in oil well borehole wireline logging) and rf transmitter powered usually by batteries. The front half comprises a percussive hammer in which a closed gas compresses a working gas which is rapidly released in the hammer chamber propelling the hammer into the soil. The science package is mounted on springs behind the percussion system. Behind the science package lies the communications and control subsystems. Penetrators require sufficient impact energy to embed themselves into the target and survive very high deceleration loads ~100–100 000g [Doengi et al 1998].

The first US planetary mission was the Mariner 2 Venus flyby in 1962. Venus has been visited by several other spacecraft—Venera 7 as the first spacecraft to land on the surface of another planet in 1970, Venera 9 as the first soft lander to relay images from the surface of Venus in 1975, which survived for 53 minutes, and Magellan in 1989, which radar mapped 98% of the surface at a high resolution of 300 m from orbit. Mercury has been explored only once, by Mariner 10 in 1973.

The Moon is the most-visited body in the solar system. Luna 1 flew by the Moon in 1959. Luna 2 was the first spacecraft to impact the lunar surface in 1959 and Luna 3 was the first to image the far side in 1959. The first US lunar landers were the Lunar Surveyor series which successfully landed on the Moon, beginning with Surveyor 1 in 1966 and ending with Surveyor 7 in 1968. They were essential for the success of the Apollo 11 lunar landing in 1969. While the US concentrated on its manned Apollo programme, the Russians adopted sophisticated robotic exploration. In particular, the Russians sent rovers capable of traversing wide areas of the lunar surface (Lunakhod) and for the return of lunar soil samples back to Earth (Luna 16 and 17) under ground-based control from Earth. Luna 16 and 17 performed some of the most complex tasks ever undertaken by unmanned spacecraft. Luna 16 landed on the Moon and returned samples of lunar soil in 1970. It soft-landed in the Sea of Fertility and its extensible arm employed a drilling rig to collect lunar soil core samples. It sealed 100 g of soil in a container payload on the ascent stage which returned to Earth successfully. Luna 20 soft-landed in 1972 in the Sea of Fertility to collect a further 30 g of lunar soil for return to Earth. The last of the Luna series, Luna 24 was the last Luna sample return mission in 1976 which returned 170 g of lunar soil from Mare Crisium. A total of 382 kg of rock samples from both the Russian Luna and US Apollo missions have been returned to Earth. Lunar Prospector (1998) was a lunar polar orbiter which carried a neutron spectrometer to detect the presence of water (as originally discovered by Clementine in 1994) in permanently shadowed regions of both polar craters in the top 50 cm layer of soil. Estimates of the amount of water are 6×10^9 tonnes with the northern pole having 30% more than the south pole.

Mariner 4 was the first spacecraft to image Mars in 1965 on a flyby, followed by Mariner 9 in 1971, which orbited Mars to provide mapping images of the surface with a resolution of 150–300 m. These missions finally laid to rest Schiaparelli's "canali" and Lowell's artificial irrigation canals built by a dying race of intelligent Martians. The Viking 1 and 2 landers landed on the Martian surface in 1976 in the Chryse Planitia and Utopia Planitia regions respectively on opposing sides of the planet. The primary purpose of Viking was to search for life. The landers ran three microbiological experiments of its complement of four to detect microbial life on the surface soil to a depth of 30 cm but failed to find such evidence. Mars Pathfinder launched in 1996 successfully landed the Sojourner rover on the Martian surface. The Mars Global Surveyor was the first of an aggressive NASA Mars Surveyor exploration programme which involves launching a spacecraft to Mars at every launch opportunity (every 26 months when Mars is in opposition to Earth). Unfortunately, the last two missions, the Mars Climate Orbiter and the Mars Polar Lander (launched in 1998) like their predecessor Mars Observer (1993) failed on arrival to Mars in 1999. The currently operational Mars Global Surveyor launched in

1997 to replace the Mars Observer mission carried five of its seven instruments. This spacecraft laid to rest the mythology that had built up regarding the "face" in the Cydonia region of Mars revealing a natural hillock feature.

The Mars Polar Lander was to be deployed at the Martian southern polar ice cap equipped with cameras, a robotic arm and scientific instruments to probe the Martian soil. Part of the failed Mars Polar Lander payload included an example of the NASA New Millennium programme concerned with testing of new technologies. The Deep Space 2 microprobe mission which was piggybacked on the Mars Polar Lander mission were to penetrate up to 2 m into the Martian soil in search of water. Two DS2 microprobes were attached to the spacecraft's cruise stage under the lander solar panels and were to separate on atmospheric entry from the lander to land some 50–100 km from the main lander. Each microprobe weighed 3 kg and was covered in a non-ablating aeroshell which could withstand heating rates of 300 W/cm^2. They did not employ parachutes, airbags or retro-rockets. The aeroshell was to shatter on impact on the surface with a velocity of 190–210 m/s. The DS2 microprobe penetrators were 18 cm and which were to separate into a 900-g fore and 1.05-kg aft body connected by a 2-m Kapton sheathed cable. The forebody penetrated to a depth of 0.3–1 m of soil and could withstand up to 30 000*g* shocks. The aftbody was to remain on the surface with its communications subsystem while the forebody was designed to measure soil composition and to determine if sub-surface water ice is present. The mission duration was to be 50 hours. The aftbody housed the power system comprising two 550 mAh $LiSOCl_2$ non-rechargeable batteries, the UHF communications system and antenna which provides a 500 kbps data rate with 640 mW transmit power to the Mars orbiter, a meteorological pressure sensor, descent accelerom-eter and sun detector. The forebody comprised an 8-bit 8051 microcontroller with 128 kB RAM and 128 kB EEPROM, ASIC (application-specific integrated circuits) based power electronics, subsurface water detector sensor, two thermistor-based temperature sensors and an impact accelerometer. The soil sample detector was to collect a 50-mg sample with a microdrill which was to be heated to release water which is detectable by a tunable laser diode spectrometer.

The standard deep space probe design is based on the Mariner Mark II spacecraft bus—a three-axis stabilised, radioisotope-thermal-generator-powered explorer spacecraft. The US Pioneer 10 launched in 1972 was the first spacecraft to transit the asteroid belt and flyby Jupiter in 1973. Pioneer 11, launched in 1973, went onto to flyby and observe Saturn in 1979. The identical 1-tonne Voyager 1 and 2 missions launched in 1977 both flew by Jupiter in 1979 and Saturn in 1980/81. Voyager 1's trajectory to the Saturn moon Titan (1980) sent the spacecraft's trajectory north out of the ecliptic plane while Voyager 2's trajectory was selected to maintain the option to encounter Uranus (1986) and Neptune (1989) after its encounter with Saturn by virtue of a rare geometric configuration of the outer planets which occurs once every 189 years, allowing a series of gravity assists at each planet to propel the spacecraft onto the next—the Grand Tour of the outer solar system. Voyager 2 flew past Uranus in 1986. Neptune had been visited by Pioneer 11 in 1979 and then Voyager 2 in 1989.

The joint NASA/ESA Ulysses mission was launched in 1990 and underwent a gravity assist at Jupiter to take it out of the ecliptic plane in 1992 to explore the Sun's poles. The NASA Galileo Jupiter spacecraft was launched in 1989 which followed a Venus–Earth–

Earth gravity assist (VEEGA). *En route* to Jupiter (arrival in 1995), the Galileo spacecraft performed a flyby of two asteroids (Gaspra and Ida) and provided a direct view of the of the comet Shoemaker–Levy's, nine fragments impacting Jupiter's atmosphere in 1994. Galileo included an atmospheric probe for Jovian atmosphere entry of mass 339 kg powered by $LiSO_2$ batteries of 580 W. The probe had its instruments housed in the descent module of titanium and aluminium protected by the fore and aft ablative heat shields of carbon phenolic/phenolic nylon respectively. The heat shield shell comprised the deceleration module. The probe had its own communication system of two parallel 128 bps L-band channels to send data to the orbiter's tape recorder and computer memory during atmospheric entry for later relay back to Earth. The probe remained attached to the orbiter until 5 months prior to arrival at Jupiter. The composite spacecraft was spun up to 10.5 rpm and the probe was separated from the orbiter. Jupiter's gravitational attraction was the sole source of propulsion as the probe had no rocket engines. There was no communication with the probe until after Jovian atmosphere entry. It remained in a dormant state with only the coast timer operating. The orbiter entered Jupiter orbit via a gravity assist at Io to explore the gas giant and fly by its large moons, Europa, Ganymede and Callisto. Meanwhile, at 6 hours prior to entry, the probe's coast timer woke up the probe by initiating calibrations. Three hours prior to entry, the Energetic Particles Experiment began to measure properties of Jupiter's inner radiation belts. The probe entered the Jovian atmosphere near the equator at an angle of 8.4° to the horizontal travelling at 47.4 km/s. The entry point was 450 km above the 1-bar pressure level in the Jupiter atmosphere (temperature –107°C at 1 bar). The Atmospheric Structure Instrument began recording the deceleration of the probe (initially $228g$) by an accelerometer to obtain the density, pressure and temperature profile of the Jovian upper atmosphere in conjunction with temperature and pressure sensors. The probe was decelerated to 0.5 km/s in 2 minutes by the heat shield from which the descent module was separated by parachute. The probe descended below the base of the ammonium ice and ammonium hydrosulphide ice cloud layer (pressure 1.6 bar, temperature –80°C)—this cloud layer was much less dense than expected as measured by the Net Flux Radiometer which measured the infrared and visible brightness of the sky in different directions and the Nephelometer which measured the light scattering of cloud particles by transmitting laser pulses observed by an optical receiver. Further confirmation of this came from the Neutral Mass Spectrometer which found less oxygen (in the form of water) than solar abundance but more carbon in the form of methane and sulphur in the form of hydrogen sulphide than solar abundance (for nitrogen also) probably due to comet and asteroid impacts. Radio signals from the probe were received by the orbiter up to 61.4 minutes after entry until the high temperatures (pressure 24 bar, temperature 153°C) caused failure of the probe at 146 km below the 1-bar pressure level. This was past the expected base of the water cloud height at –56 km (pressure 5 bar, temperature 0°C) indicating much drier conditions than expected. The ultimate fate of the probe was its melting as it descended into regions of higher temperature (aluminium melts at 660°C and 280 bar pressure). The Doppler Wind experiment measured the variation in frequency of the radio signals emitted from the probe to measure wind speeds found to be ~700 km/h at all levels below the upper cloud layer powered by the heat emission from the planet's interior. Lightning discharges were observed at radio frequencies within clouds, but were about 10% as numerous as those on

Earth, but ten times as powerful. This could account for the low measurement of organic compounds. It was concluded that the probe had entered at the edge of a 5 μm infrared hot spot characterised by reduced cloud cover and high dryness.

The Galileo Europa mission was a two-year extended follow-on mission to study the fourth largest Jovian moon, Europa, and characterise its properties via eight close encounters utilising the orbiter's high-resolution visual imaging cameras ~50-m resolution and its infrared/ultraviolet spectrometers. The main aim was to determine the presence or absence of a subsurface water ocean below its water ice surface. It used the radar sounder to bounce radio waves through the surface ice in order to attempt to determine its thickness (estimated at 100 km). NASA's Europa Ice Clipper orbiter is a potential future mission which may be launched in 2003 to arrive in 2008 at a 200-km orbit to use radar to determine the depth of the ice/water interface below Europa's surface. If a subsurface ocean does exist, the surface of Europa should rise and fall 30 m every 3.6 days.

The joint NASA/ESA Cassini–Huygens probe is an exemplar of a deep space robotic mission [Kohlhase & Peterson 1997]—indeed, it represents the last of the traditional big, sophisticated, expensive deep space autonomous robotic explorer missions. The 5.5-tonne, 6.8 m tall, 4 m diameter Cassini–Huygens probe was launched in 1997 (it is twice the size of Galileo). It is one of the heaviest and most sophisticated robotic spacecraft ever built (only the two Russian Phobos spacecraft were heavier)—it has 1630 circuits, 22 000 wire connections and 14 km of cabling. It is similar in configuration to the Voyager probe and is based on the three-axis stabilised Mariner Mark II spacecraft design with a dry mass of 2200 kg at launch (5548 kg fuelled), including 200 kg of scientific instruments. Cassini was originally designed with articulating platforms and arms but this approach was discarded in favour of having instruments fixed to the spacecraft body so that pointing must be done by slewing the vehicle in a similar way to Magellan. Cassini–Huygens will reach the Saturn system in 2004 via gravity assist manoeuvres twice at Venus, once at Earth and once at Jupiter. The ESA-designed Huygens component is due to explore Saturn's largest moon Titan while the larger NASA-designed Cassini component explores Saturn [Lambreton et al 1995, Hassan et al 1994, Hassan & Jones 1997, Raulin et al 1990]. The Huygens probe is similar in design to the Galileo probe descent module. The Huygens probe has a mass of 318 kg, including 40 kg of body-mounted scientific instruments, within a diameter of 1.3 m. Prior to the probe's separation, a final health check is initiated by uploading a signal to Huygens via Cassini's umbilical to wake up the mission timer unit. On approach to Saturn, the Cassini orbiter fires its pyrotechnic bolts to detach the Huygen probe from the Cassini carrier at 0.3 m/s. Huygens will coast for 22 days towards Titan. During the coasting phase, 24 minutes prior to Titan atmospheric entry, the mission timer triggers a sequence of commands to generate power to the probe's instruments and subsystems. This represents the initiation of a pre-programmed automatic sequence of instructions throughout the mission. Huygens will be completely autonomous as the single-way communications channel cannot accept commands and is dedicated to telemetry. The round-trip time delay of 160 minutes would make real time control impossible. Hence, it is controlled entirely by automated, time-sequenced software stored on its MIL-STD 1750A microprocessor architecture. Huygens will enter Titan's atmosphere at 1270 km altitude at a speed of 6 km/s protected by thermal tiles of the aeroshell rated to 1200 K. It will decelerate to 500 m/s at an altitude of 170 km at which point it will deploy its main parachute. The front heat shield of thermal tiles is then released. The controlled floating descent will take 2.5 h with a further parachute deployment, and it will impact on the surface of Titan at 6 m/s which it is expected to survive for at least a few minutes—the probe's five LiSO$_2$ batteries rated at 1800 Wh are designed to provide power for 30 minutes on the surface. Data will be relayed to Cassini at 8 kbps on the S-band communications link via its high-gain antenna and stored in Cassini's two solid-state recorders (for redundancy) for transmission to Earth. Huygens must automatically initiate onboard experiments, collect data from them and control its transmission back to the Cassini orbiter which will relay the data back to Earth. All activities after detachment are autonomous with no provision for ground control.

The ESA Rosetta probe is a cornerstone mission of the ESA Horizon 2000 programme for space exploration based on the Eurostar telecommunications spacecraft bus. It is projected to be launched in 2003 with a Δv of 3.4 km/s. Its mission is to rendezvous with the near-Earth, short-period 46P/Wirtanen comet in 2011 after two asteroid flybys to emplace a multisensor surface science package to provide *in situ* investigation of a cometary nucleus to follow on from the highly successful Halley flyby by ESA's Giotto spacecraft [Schwenn & Hechler 1993, Verdant & Schwehnn 1998]. Rosetta's trajectory will follow a 700-day Mars (once) and Earth (twice) gravity assists. As Rosetta crosses the asteroid belt twice after each of the Earth gravity assists, fast flybys of the asteroids Mimistrobell and Siwa to within 600 and 1200 km respectively are planned at relative velocities of 6 and 13 km/s respectively. There will be long outage periods (up to six in number comprising 68% of the transit) during the cruise phase without ground contact, varying from 1 month to 2.5 years and including an 8-month outage due to spacecraft–Sun–Earth collineation. During such periods, Rosetta will enter hibernation mode with minimal onboard activity to minimise power consumption and increase its operational lifetime by up to ten times. The communications link is restored autonomously at the end of the hibernation periods. As it is, two-way transmission delays will be ~1.7 hours at 6 AU. All operations will be scheduled and uploaded to the spacecraft in advance to be performed autonomously onboard. Rosetta's trajectory will take the spacecraft to within 500 000 km of the comet at a relative velocity of 100 km/s. Scientific instruments will then be activated. Rosetta will approach to within 100 km of the comet reducing its relative velocity to 25–100 m/s via ground-based navigation prior to comet acquisition by the onboard cameras. Online observations will fine-tune data concerning the comet's orbit parameters to allow a rapid approach which will reduce the relative velocity to 2 m/s at 300 comet radii distance. Close approach occurs at cm/s relative velocity at 60 comet radii distance followed by hyperbolic orbit insertion at 25 comet radii. This interception will occur at 3.25 AU from the Sun in 2013 where the two-way communications time delay between the ground and Rosetta will be around 100 minutes. Once orbit around the comet is achieved, navigation will be by the sun sensor and star trackers. During the elliptical polar orbit about the nucleus, it will globally map the nucleus surface from 5 to 25 comet radii altitude. Mapping images in the visible and infrared regions of the spectrum will provide data on the nucleus size, rotation state and other data which will be used to determine the milli-g spacecraft orbit around the irregularly shaped comet at an altitude between 5 and 10 km. The cometary rotation period is expected to be between 6 and 13 h. Mapping of 80% of the surface will take 30 days. Several regions (nominally five in number) will be selected for close observation for a further 30 days until a final landing site for the Rosetta Lander (RoLand) is selected, over which Rosetta will eccentrically orbit at a pericentre altitude of 1 km.

RoLand with its surface science package payload has a mass of 87.5 kg with dimensions $1.13 \times 1.08 \times 1.04$ m. It comprises a carbon-fibre polygonal construction with articulated tripod landing gear which is unfolded after separation from the orbiter. Such deployable, three-legged configurations with a large footprint is a typical and robust design. RoLand will be released autonomously from the Rosetta orbiter mounting panel by four pyrotechnic bolts at 1.5 m/s separation velocity. Roland has a flywheel attitude control system with 0.02 Nm capability to ensure that an attitude tolerance of 0.5° nutation is maintained, and it will be delivered to the selected site via its descent system of cold gas thrusters. The overall impact Δv on landing will be 1.5 m/s. RoLand incorporates impact damping layers to absorb shocks, and rebounds will be prevented by incorporating a pushdown thruster for 30 Ns over 5 seconds to allow anchorage to the surface with a harpoon-type penetrator (plus redundant system) which fires pyrotechnically on impact with the comet's surface and subsequently tensions the tether within the 5 seconds with 1–30 N force with a winch mechanism to ensure that the lander impact does not cause rebound. The surface package has two compartments, a warm compartment at –40°C and a cold compartment. On the comet surface, 6 W of the 15 W electrical power capacity is dissipated in the warm compartment. Low-intensity, low-temperature (LILT) solar cells mounted on RoLand's surface panels provide 11 W of power at 3 AU and its primary batteries are rated at 960 Wh (plus 110 Wh for secondary batteries). After 60 hours on the surface, the drilling of a surface and subsurface sample initially from a depth of 20 cm will proceed. RoLand will perform *in situ* analysis and chemically analyse samples from up to 200 m depth. The total lander operational phase is projected to last for 4 months. The science package will comprise eight instruments—the alpha-proton-X-ray spectrometer, a pyrolytic gas chromatography instrument, a mass spectrometer, a set of visual and infrared camera (one panoramic camera, a down-looking camera and an infrared microscope spectrometer), a thermal conductivity sensor for chemical species characterisation, an integrated dust impact monitor/dielectric sensor/acoustic sensor, an integrated flux-gate magnetometer/Langmuir probe plasma monitor, and a radio experiment where the orbiter transmits signals through the nucleus to the lander which returns the signal. Roland's feet also incorporate accelerometers. No telecommand capability will exist and Roland will relay packaged telemetry data back to the orbiting Rosetta spacecraft, so it must operate completely autonomously from onboard software. Telemetry data storage buffering will be enabled by 4 Mbytes of mass memory. An emergency override capability from the ground is possible. Roland has two redundant TX/RX units attached to two sets of the TX and RX antennas (each of mass 100 g and diameter 80 mm) with a maximum EIRP of –50 dB to –120 dB for a 7 W received signal. The chief challenge for Rosetta is in achieving a close orbit around an irregular body with a weak, asymmetric, rapidly rotating gravity field surrounded by gas and dust.

The NASA Stardust mission is the fourth Discovery mission and was launched in 1999, the first US sample return mission to a comet. It offers an interesting contrast in terms of simplicity to the Rosetta comet rendezvous mission. It is designed to capture dust samples from the comet Wild-2 and interstellar dust. Wild-2 was a long-period comet until its close encounter with Jupiter in 1974, so will represent primordial material left over from the formation of the solar system. Stardust is a lightweight spacecraft utilising graphite fibre/polycyanate matrix facesheet and Al honeycomb core panels with off-the-shelf components. Its total mass is 380 kg, including its pure hydrazine monopropellant, and its overall length is 1.7 m. Its command and data handling subsystem is based on the 32-bit RAD6000 CPU and 128 Mbyte data storage capacity, of which 20% is utilised for the spacecraft's internal programs. It utilises a 0.6 m diameter, fixed, high-gain antenna, one 15-W solid-state amplifier and the transponder developed for Cassini. Two fixed solar arrays provide 6.6 m^2 of area for power generation. During eclipse periods, power is generated by one 16 Ah NiH_2 battery. A Whipple shield protects the spacecraft during its encounter with the comet. There are only two scientific experiments onboard: the aerogel dust collector and the cometary dust analyser to determine the composition of the dust. Stardust will undergo two gravity assists at the Earth and perform a slow approach flyby within 150 km of the comet nucleus in 2004 at a relative speed of 6.1 km/s. Within the coma, it will capture the comet dust in a highly porous silica glass foam called aerogel, the lowest density material in the world—1/1000th that of glass as it is 99.8% empty space. A retractable grid is coated on both sides with aerogel will be deployed during the encounter and retracted after the encounter. The sample will be delivered to Earth in a blunt body return reentry capsule in 2006. A similar mission is being proposed for Europa in 2001, the Europa Ice Clipper, to return samples of the Europan ice crust back to Earth.

The chief lesson learned in planetary exploration missions is that further exploration always yields surprises; it is a poignant example that the results of scientific exploration cannot be predicted and there is no better reason for its pursuit than that—the accumulation of knowledge about the broader environment in which we live.

2.2 PLANETARY ROVER MISSIONS

All manned missions to planetary bodies, be it the Moon or Mars, will require prior unmanned exploration of the surface for reconnaissance and surveying purposes by robotic rovers. They thus comprise a central plank in all planetary exploration missions both manned and unmanned from their ability to provide detailed *in situ* data not obtainable from orbital or flyby missions. Robotic rovers are uniquely suited to special applications such as seismic survey and local site preparation. They are capable of parallel operation if deployed in numbers with inherent redundancy and increased area coverage. Teams of rovers can be deployed as coordinate packs with individual rovers specialised to particular operations. Robotic rovers will be critical in searching for prebiotic and biotic evidence for life elsewhere in the solar system that has recently become a high priority for NASA—this would highlight Mars (potential subsurface chemolithotrophic life inspired by the debatable discoveries in the Martian meteorite ALH84001), Titan (potential prebiotic nitrogenous assemblages), Europa (potential hydrothermophilic life in the subsurface ocean) and comets (potential biochemical molecules similar to those such as amino acids discovered in carbonaceous meteorites) as the major targets for investigation. The discoveries of extremophiles on Earth under hostile conditions has spurred this new incentive to search for life elsewhere in our solar system—it appears that where there's water, there's the potential for life.

Planetary landers and rovers are a highly specialised application of robotic spacecraft. Planetary surfaces are highly unstructured and susceptible to uncertainties which must be accommodated in real-time. Similar arguments apply to in-orbit servicing scenarios. Longer-range missions will require a high degree of intelligent automation for autonomy since human intervention will be limited to higher control levels with longer time constants. Substantial local intelligence is required on the planetary surface—tasks such as the deployment of experiments and sample collection are complex and occur in unknown, sometimes variable environments. Pre-programmed sequences have already been used on spacecraft for autonomous execution, e.g. the Viking landers on Mars. The two Viking landers on Mars (1975) were the first space missions to utilise articulated remote manipulators to dig Martian soil up to 2.5 m from the landers to a depth of 10 cm.

As planetary exploration advances, the diversity of methods of delivery and types of potential rover design will reflect the diversity of planetary environments to be explored. The typical terrestrial environments advocated for rover exploration are Venus, the Moon and Mars; gas giant planets such as Jupiter lacking solid surfaces suggest the use of atmospheric penetrators and aerostats for atmospheric exploration; icy gas giant moons such as Titan and Europa suggest the use of cryobotic moles and hydrobot submarines. Generally, each rover is designed for a specific environment, typically highly hostile.

Planetary rovers must be able to traverse unknown ground terrain with its attendant problems of steep slopes, large rocks and loose soil. It must have a suite of sensors including gyroscopes, inclinometers, wheel encoders, accelerometers and rangefinders to achieve its mission. Adequate power generation is a major challenge, typically involving solar cells, batteries, fuel cells or RTGs (radio-isotope thermal generators). Local sensing based primarily on stereovision as the range sensor is generally adopted. Semi-autonomous planning and navigation with obstacle avoidance is essential. In a very real sense, the planetary rover is a small spacecraft in itself with all the functions and sub-systems of a spacecraft—onboard navigation, command and data handling, power generation, thermal control, communications and sensors.

It is obvious that an important aspect of planetary rover design for solid planetary surfaces concerns vehicle mechanics and traction/mobility system suitable for the planetary environment of operation. The vehicle engine torque produces tractive effort and so vehicle acceleration. Engine power is related to engine torque through engine speed: $P = \tau w$, where $w = 2\pi f$ and f = frequency of armature rotation.

The nominal ground pressure (NGP) is a measure of the resistance to sinkage: $NPG = W/nrb$, where W = vehicle weight, n = number of wheels, r = wheel radius and b = wheel width. A low NGP indicates low sinkage and resistance to motion: in terrestrial vehicles 75 kN/m^2 is adequate for off-road temperate soils but a lower NGP of 40 kN/m^2 is required for compliant soils such as peat. Many wheels with large radius supporting a light vehicle are required for good performance but for an NGP of less than 35 KN/m^2, tracks are generally required. The soil shear strength must exceed the NGP of the vehicle: Soil shear strength $S = c + \mu P$, where c = cohesive strength of fine-grained (clay) soil component, μ = coefficient of friction of coarse-grained (sandy) soil component = tan ϕ, ϕ = internal angle of friction and P = normal stress.

The maximum tractive effort available from the soil over a wheel contact area A: $F_{max(s)} = Ac + \mu W \cos \theta$, where θ = surface gradient (wheels are generally limited to gradients less than 25°). The maximum tractive effort available from the vehicle: $F_{max(v)} = \tau n \eta / r$, where n = gear ratio and η = transmission efficiency. Rolling resistance, R, occurs when soil compaction and bulldozing of the soil occurs and it can absorb between 5 and 35% of gross engine power depending on the soil and vehicle speed. When $F_{max(v)} > F_{max(s)} + R$, then transverse movement will occur. This translates to a vehicle power requirement of $P = Fv$, where v = vehicle speed.

The Moon's surface comprises a regolith of pulverised rock (dust and breccias) to depths of 2–3 m in the marias to 3–16 m in the highlands covering the bedrock [Utreja 1993]. It has been sampled from the Apollo missions to a depth of only 3 m in a small number of localities. Although large impacts are rare on the lunar surface today, there is a high meteoroid flux impinging on the surface, so lunar rocks are peppered with zap pits less than a millimetre in length. The mean grain size of the regolith varies from 40 to 268μm with a preponderance for smaller grain sizes. The regolith has a density of over 1.0 g/cm^3 at the surface which increases with depth to 2 g/cm^3 at 20 cm depth due to compaction. Its mean porosity is 40–43% by volume with a density-dependent cohesive bearing strength from 0.03 N/cm^2 at the surface to 0.3 N/cm^2 at 20 cm depth. Hence, traction for wheeled lunar roving vehicles does not present a problem [Johnson & Chua 1993]. It is a good thermal insulator with a thermal conductivity of 0.9–1.3 × 10^{-2} W/mK. At a depth of 30 cm, it maintains a near-constant temperature of 250 K. Lunar regolith is also a good electrical insulator with a conductivity of 10^{-14} Ω/m. Dust is a persistent problem for lunar activities [Nitta et al 1991, Ellery 1993, Johnson et al 1995]. Apollo astronauts have reported that lunar dust interferes with lunar activities and appears to stick to everything, probably due to sunlight electrostatically charging the dust causing it to levitate.

The first rovers to be placed onto the surface of another planetary body were Lunokhod 1 and 2 on the Moon which were wheeled, teleoperated lunar rover vehicle laboratories controlled from Earth by five-man teams. Luna 17 soft-landed in 1970 on the Moon at the Sea of Rains carrying the Lunokhod 1 rover which traversed the lunar surface on its eight independently powered wheels. It carried two antennas, four TV cameras and robotic devices to test the lunar soil's mechanical properties. It also included an X-ray spectrometer, X-ray telescope, cosmic ray detector and laser. Although designed to operate for three lunar days, it successfully operated for 11 lunar days. Luna 21 carried the Lunakhod 2 rover to the lunar surface in 1972. Lunokhod 2 was an eight-wheeled rover, 15 cm in height, 170 cm in length and 160 cm in width, with a mass of 840 kg. It had four TV cameras, one of which was a panoramic camera. It was powered by solar panels with a Po-210 heat source for thermal control. It included an X-ray spectrometer, a visible/uv photometer, a boom-deployed magnetometer and a radiometer. A five man team on Earth controlled the rover with its two speeds of 1 km/h and 2 km/h. It operated for 4 months covering 37 km of rugged terrain. It was the success of the Lunokhod missions that inspired the later Russian Marsokhod rover design that is currently favoured by several prospective European missions.

The most famous rover is the US Apollo lunar rover which although manually oper-ated has some specifically rover-oriented lessons in store, particularly concerning the mobility system. The vehicle also had the capability of being operated remotely by ground control for unmanned exploration similar to an unmanned rover. It was designed to certain specifications [Burkhalter & Sharpe 1995]:

(i) four-wheel individual and reversible drive power with dedicated individual wheel motors and power storage batteries;
(ii) 210 kg mass limitation;
(iii) payload capability of two astronauts and science experiments to a total of 450 kg;
(iv) range of 4×30 km traverses in a 78-h period;
(v) operating life of 78-h;
(vi) stowable into one compartment bay of the ascent stage of the lunar module and deployable by one astronaut;
(vii) maximum speed of 18 km/h;
(viii) obstacle negotation of 30 cm height and crevasse capability of 70 cm;
(ix) slope negotiation of $25°$ and 35 cm chassis ground clearance;
(x) turn radius of one vehicle length of 3 m;
(xi) overall power generation of 150 W by two Ag-Zn batteries of 36 V each.

The Apollo rover comprised eight subsystems: locomotive mobility, navigation and control, power, data handling and communications, thermal protection, crew station, sensors and display, and vehicle structure and deployment. The locomotive mobility system, unlike the other subsystems, is peculiar to rovers rather than being part of general spacecraft design. This subsystem is concerned with ride, handling and braking, which is dependent on the interfaces between the ground and vehicle, i.e. wheels and suspension. The basic chassis of the lunar rover comprised Al alloy tubing similar to that used for aircraft wing tips. The stowage/deployment condition implied the need for four wheels mounted on the chassis by suspension arms attached to a torsion bar with damper suspension. Each 5.4-kg, 2.3-m diameter wheel comprised a non-inflatable woven wire mesh tyre. Each wheel had independent traction drive powered by a 0.25 hp series-wound dc motor with 17 000 rpm capability and a harmonic drive gear with an 80:1 stepdown ratio. Steering was based on a modified Ackerman geometry as used in terrestrial auto-mobiles so that the inner wheel pivots slightly more than the outer one. All steering, speed control and braking was performed using a single T-shaped handle. As the Moon had no magnetic field, compasses were useless so the navigation system comprised a directional gyro and odometers on each wheel measuring distance increments. Odometry tracks wheel rotation but its measurement is subject to uncertainty due to slippage which grows in proportion to the distance travelled. The communications subsystem comprised a high-gain antenna and a low-gain antenna.

Nomad is a 550-kg lunar rover prototype capable of traversing hundreds of kilometres at speeds of 0.5 m/s with a power requirement of 2 kW developed at a cost of $3.5M [Whittaker et al 1997]. It underwent successful field tests in the Atacama desert, Chile in 1997. It represents the traditional rover design style of large size, complexity and capability. It has a transforming chassis which can expand and contract as it moves to improve its stability via two pairs of four-bar linkage mechanisms. It has four cleated Al wheels with front and back independent steering and can negotiate slopes of up to $38°$ and obstacles up to 56 cm in height. It is powered by a petrol-driven electric generator that supplies power for its computers, communications, instruments, cameras and propulsion systems. It is equipped with 12 cameras including two pairs of 640×480 pixel stereo science cameras with pan/tilt capability, a 1024×1024 primary panoramic full-surround camera. It also uses a pair of DGPS (differential global positioning system) receivers of 20-cm resolution for terrestrial navigation (to be replaced by star trackers for lunar operation) supplemented by odometric wheel encoder data. A gyroscope-based inertial navigation unit provides angular data for active pointing of a high-gain directional antenna which has a 100 kbps data rate and range of less than 1 km. The primary non-visual scientific sensor is a three-axis magnetometer. However, it has a robotic manipulator arm for mounting of scientific instruments, currently an optical reflection spectrometer and metal detector for rock identification. It is an autonomous rover controlled by two onboard computers, a 50 MHz 68040 microprocessor and a 40 MHz 68030 microprocessor. Its vision processing is performed by an Earth-based 200 MHz dual-chip processor—images are acquired at a frame rate of 6–8 Hz. Its navigation computer is a 133-MHz PC which performs sensory data fusion. The dedicated science computer is also a 133-MHz PC. It has been successfully desert-tested. This type of planetary rover is required to traverse ~10–20 km over rough terrain over a period of weeks or months. It is currently in service in the Antarctic (Elephant Moraine rock-field) to search for meteorites in the ice autonomously, classifying them according to colour, lustre and texture and send the GPS-derived location back for field scientists.

2.2.1 Mars rovers

More rover missions have been proposed for Mars rather than any other celestial body. Planetary rover missions to Mars will be a necessary tentative step towards the search for possible fossilised biota—sedimentary rock deposits will be the obvious candidates such as clays, carbonates, phosphates, evaporites and cherts. Other possibilities include inactive volcanic vents and fumaroles. The distances involved in any planetary mission bar the Moon are so large that the communication time delay precludes any form of control by teleoperative or telepresence methods: Mars involves a 30-minute round trip delay in communication. Some form of autonomy on the rover is required—the Mars Pathfinder had only two 5-minute communication windows to Earth per day. Highly autonomous intelligent capability is beyond state-of-the-art AI (artificial intelligence) but degrees of autonomy can be provided [Wilcox et al 1992, Miller 1990b, Miller & Varsi 1993]. Mobility introduces uncertainty in that the environment constantly changes so that the rover must always undergo continuous calibration and parameter adjustment. Semi-autonomous navigation is feasible and has been demonstrated on the Sojourner Mars rover. The ideal capability is that of a field geologist. This implies local navigation, selection of sites of scientific interest and dealing with onboard experiments—an intelligent, mobile science station. Local routes may be planned onboard autonomously while global routes could be planned on Earth using maps obtained from images produced by a Mars orbiting probe. The orbital images may be used by a human operator to select an approximate corridor to avoid large obstacles, etc. in the desired direction. The rover then analyses the local scene using local landmarks and derives a local map of the terrain using the onboard sensors which is then mapped to the local part of the global map statistically. The rover may drive a few tens of metres at a speed of 10 cm/s and the process repeats. This is waypoint navigation. This was the initial approach suggested for the 1-tonne Mars Rover Sample Return rover with a 40-kg scientific payload (now cancelled because of its \$10B price tag). This was based on traditional rover approaches such as DARPA's ALV (autonomous land vehicle), a 4-tonne, eight-wheeled rover for autonomous off-road navigation and traversal. The Mars rover was to traverse over kilometre distances to survey and sample some 5 kg of geological objects which would be launched back to Earth using a return launcher 1 y later. Difficulties in rover missions are exemplified by the Dante rover, an 800-kg, eight-legged robotic rover that descended using a tether cable into an active volcanic crater in the Antarctic in 1993. Its role was to minimise dangers to field geologists after eight geologists were killed in 1993 exploring volcanoes. Dante's progression into the crater was slow, due to the limitations of the control system; then its fibre optic link between the rover and the remote station snagged and broke.

Due to the uncertainties in NASA's budget, emphasis has since been placed on 'fast, cheap and small' missions. There is a trade-off between the communications and computation subsystems of rovers as both are power-hungry, requiring non-trivial levels of RTG (radio-isotope thermal generator) or solar cell/battery power generation [Miller 1992, Miller 1990b]. There are multiple advantages to using microrovers of small mass and size: they reduce mission timelines, reduce launch mass and the mass of lander support hardware (which scales at 1.6 times the rover mass), making them cheaper to launch, provide greater robustness to shock, and provide greater reliability and flexibility of operations. Although small rovers can negotiate only smaller obstacles, they can circumvent larger obstacles more easily. Large obstacles can be avoided with little or no loss in operational capability [Miller 1990a,b]. A biological analogy is that small animals can go most places that a large animal can go. The communications subsystem is one function that does not scale down in size—hence the importance of autonomy. Furthermore, as CMOS power is determined by clock speed, the power required for computation can only be reduced by reducing the computational load. Microrovers have high strength-to-weight ratios and are capable of withstanding high landing shocks from aeroshell/parachute delivery. Microrovers require only ~10^4 bytes of memory for both programs and data for autonomous operation through behaviour control. Sensors ~$100g$ are available for use on such microrovers. Although microrovers cannot carry large payloads, system redundancy is enhanced with lower cost by using multiple autonomous units. Teams of microrovers may be specialised with different scientific payloads built into the rover as an integrated part of its structure such as multispectral imaging, gas chromatographs, seismic sensing and core sampling.

Microrovers cannot maintain detailed world models but can accomplish most tasks that a larger, more massive rover can due to their capability of reacting to unexpected eventualities with limited computational power through their stimulus–response behaviours. In using behaviour control, a goal location may be determined by a human operator which is transmitted to the main vehicle which passes the locations to each microrover which use their behavioural suites to robustly pursue the goals using their sensors. Behaviour control is the key that allows useful tasks to be performed using small programs for robot autonomy in an unstructured environment. In behaviour control, sets of simple behaviours can be activated or de-activated according to the situation. The basic capability is autonomous local navigation and obstacle avoidance. Complex problems are decomposed into simple tasks rather than functions such as the perception–plan–act cycle. Each behaviour is independent and communicates via a well-defined input–output protocol with no intervening world model by tightly coupling sensors directly to actuators with simple computations. This enables such rovers to react quickly and robustly to changes encountered in the environment. Simultaneous and temporal activation generate complex behaviours with real-time performance using only 0.1–1 Mips of computing power with programs of 10 kB or so easily stored in ROM. Behavioural methods require much less computation than traditional AI planning methods [Miller 1990a].

Zubrin (1992) considered modes of long-range mobility on planetary bodies with atmospheres such as Mars. Ground vehicle range is directly proportional to its speed which is directly proportional to its power. The internal combustion engine is power unlimited effectively with superior power–weight ratios. Such vehicles are fuel-hungry, however, necessitating the use of indigenous materials. As most terrestrial planets like Mars possess CO_2-dominated atmospheres, H_2/CO_2 offer 2.1×10^4 kJ/kg and CH_4/CO_2 offer 1.0×10^4 kJ/kg as fuel with the CO_2 being provided by the planetary atmosphere. Cryogenic hydrogen storage is difficult so CH_4/CO_2 is the generally preferred option for a ground vehicle.

The JPL Rocky series is an ambitious progressive programme of robotic rover innovation. Rocky III is a small six-wheeled, 15-kg mobile microrover (of $48 \times 64 \times 32$ cm dimensions) capable of autonomously navigating over rough terrain. Six wheels provide greater stability and obstacle crossing capability than four wheels. It used an 8-bit processor with 10 kbyte of memory for implementation of its control system [Miller et al 1992]. It carried its own onboard battery power, a 9600 baud digital radio modem for communication and a single Motorola 6811 microcontroller. It had a springless 'rocker-bogie' suspension system which comprised of two arms either side of the vehicle [Gat et al 1994]. Each rocker-bogie arm comprised a main rocker link and a secondary link which was pivoted at the front end of the main arm. The front and middle wheels were rigidly attached to each end of the forward link, while the rear wheel was attached to the rear end of the rear link. The forward end of the rear link was attached to the middle of the forward link. The rear link was attached to the vehicle chassis. Mechanical stops prevented excessive rotation of the secondary arm. The two arm assemblies were connected through a differential gear at the centre of gravity of the rover. The main arms pivoted relative to the vehicle body giving it a full wheel diameter (13 cm) clearance of obstacles by following the terrain contour (see Fig. 2.1). Each of the six wheels were independently driven and the front and rear wheels were steered independently with an Ackermann steering geometry such that the normals from each wheel intersected at a common point. A 3-degree-of-freedom arm with a soil scoop end effector was mounted on the front of the rover with a 5-cm soil depth capability. The rover's sensory suite was simple, including a mast-mounted flux gate compass, wheel encoders at the central wheels, magnetic reed switches on the front bogie pivot as limit detectors, arm joint encoders for arm control, and an infrared beacon detector (to detect a remote infrared beacon at the lander) mounted on top of the chassis with a 5 m range. It had a contact sensor at the front of its chassis to detect collisions (see Fig. 2.2).

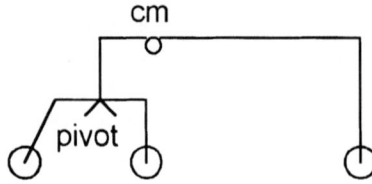

Fig. 2.1. Rocker-bogie suspension (adapted from Miller et al 1992). Reproduced with permission from David Miller.

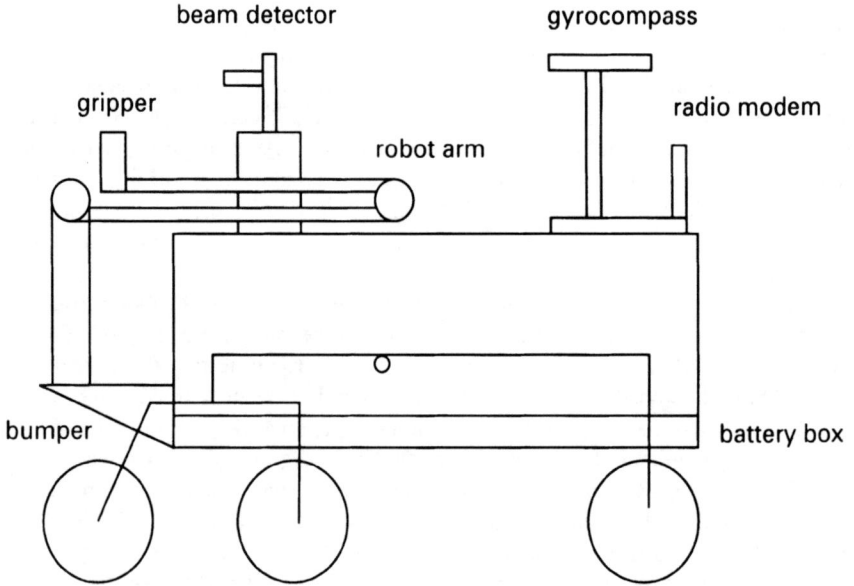

Fig. 2.2. Rocky III (adapted from Miller et al 1992). Reproduced with permission from David Miller.

Rocky III had three layers to its software control system (see Fig. 2.3) with each layer passing information down the control hierarchy, minimising the computation between sensory input and actuator output. The lowest layer controlled the rover's motors' speed and direction, i.e. servo-level control. The second layer instigated obstacle avoidance. Its basic obstacle avoidance strategy involved backing up and turning to one side. The third layer was concerned with high-level control in the form of a path sequencer to interpolate between waypoint goal locations and for manipulator grasping. It was programmed in ALFA to provide networks of computational modules which communicated with each other in a dataflow architecture. Rocky III did not require real-time communications and could house onboard power, communications and computation subsystems.

Rocky IV was a more sophisticated but functionally similar version of Rocky III. It was a 7.5-kg microrover which used six wheels in a similar 'rocker-bogie' configuration with Ackermann steering to enable it to climb obstacles up to 1.3 times its wheel

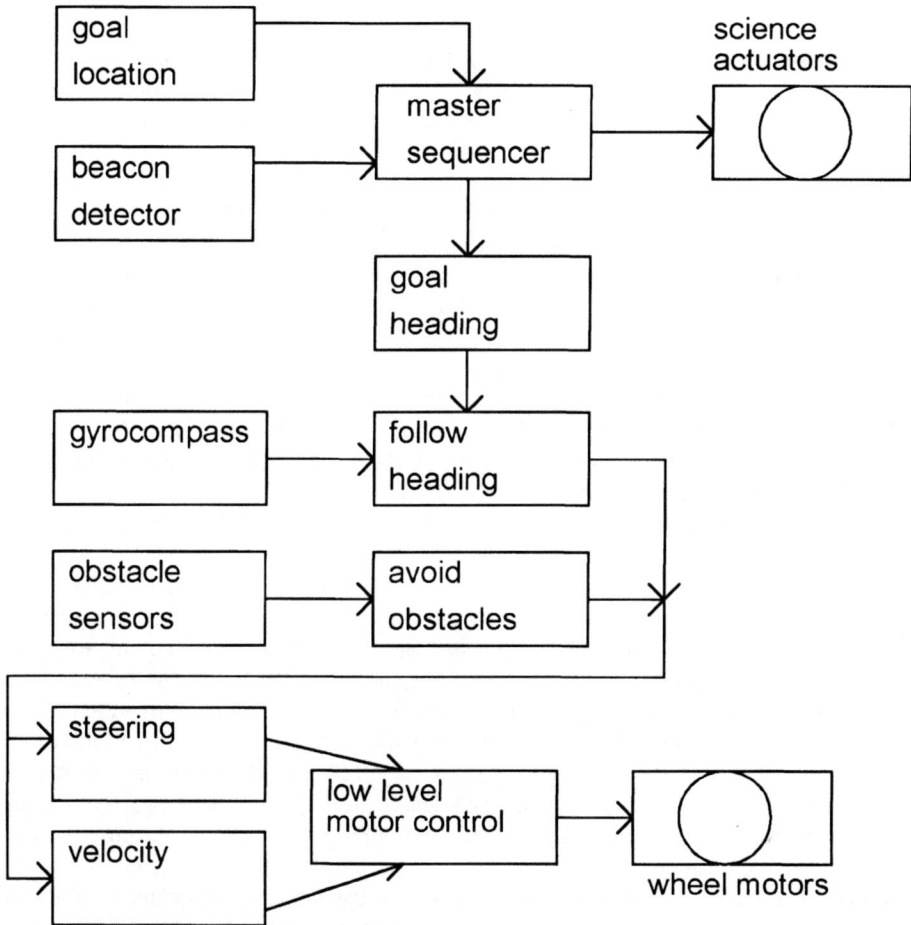

Fig. 2.3. Rocky III control architecture (adapted from Gat et al 1992). Reproduced with permission from David Miller.

diameter. It had tyres constructed from steel mesh with helical cleat patterns to provide superior traction. It had a main forward section for the scientific payload and a rear section for the battery and computer connected an Al cylinder. Its main processor was a 6811 microprocessor, as in Rocky III, with 20 kB of memory connected in series to three Intel 8751 slave microprocessors to provide additional input/output pins. Infrared proximity sensors were added to the obstacle detection sensor suite to provide remote obstacle detection. Its science payload added a CCD camera, a visible spectrometer and a microseismometer.

Rocky 7 employed a mast-mounted CCD stereovision imaging camera, 2-DOF stowable arm capable of digging into the soil with a 2-DOF end-effector for digging and grasping, an onboard visual/infrared spectrometer (360–8500 nm range) integrated into the rover chassis but with a fibre-optic cable running along the arm to enable placement of the sensor head by the arm or mast to the target. The 1.4-m 3-DOF mast with its stereo imaging camera gave panoramic views of the surrounding terrain. The fibre-optic cable also linked to a Fe-57 Mossbauer spectrometer designed to detect the oxidation state of Fe in the soil and rock, and to a Fe NMR spectrometer designed to detect magnetic minerals.

The MESUR (Mars Environmental SURveyor) microrover Pathfinder mission was the second of the NASA Discovery programme estimated to cost $200M [Pivirotto 1993]. This was a Mars direct mission launched in 1996 that did not manoeuvre into Mars orbit. The Mars Pathfinder spacecraft had a mass of 850 kg. The cruise stage of the spacecraft was jettisoned some 8500 km from Mars. The spin-stabilised Martian lander entered the Martian atmosphere at 125 km altitude at 7.5 km/s at an angle of 14.8°. Peak heating during atmospheric entry occurred at an altitude of 40 km. The heat shield aeroshell slowed the spacecraft down to 400 m/s at 8.6 km altitude when a 12.5 m diameter parachute was deployed, slowing the lander further to 60 m/s. At 6.5 km altitude, the heat shield was released and the lander was separated from its rear cover by a tether. The radar altimeter was activated at 1.5 km altitude and at 300 m altitude (8 seconds before landing), four airbags inflated into a 5.2 m diameter protective ball. Three solid rocket motors fired for 2 s at 80–100 m altitude to halt the descent at 12 m altitude. The parachute and rockets were jettisoned and the lander impacted on the Martian surface at 25 m/s. It bounced 15 m into the air, bounced another 15 times before coming to rest 2.5 minutes after the initial impact some 1 km from the initial impact site in the Ares Vallis region of Chryse Planitia (a large outflow channel formed 1.8–3.5 By ago) within 850 km of Viking 1 [Stone 1998, Matijevic 1998, Morrison & Nguyen 1998]. After landing, the airbags were deflated. The 250 kg lander, the Sagan Memorial Station (named after the deceased Professor Carl Sagan of Cornell University), housed an 11-kg, six-wheeled surface microrover Sojourner in a stowed configuration while mounted on one of the three enclosing triangular solar panels. The solar panels unfolded to right the lander, allowing the rover to drive off the lander ramp onto the Martian surface under the control of Earth-based operators with a time delay of 10–15 minutes. A Silicon Graphics work-station hosted the graphical user interface at the terrestrial ground station. The teleoperators wore special goggles that allowed 3D stereo imaging with commanding based on the Spaceball joystick tracker to move a rover icon on the screen. Sojourner was based on JPL's Rocky series (Rocky IV specifically). Its dimensions were 630 mm in length by 480 mm in width and a nominal height of 280 mm (including 130 mm ground clearance), and it had a top speed of 0.6 cm/s with a wheel torque of 1 Nm in motion and 12 Nm stall torque. It had 6×13 cm diameter wheels with a four-wheel steerable rocker-bogie configuration—the independent front and rear steerability provided the means to perform wheel–soil interaction experiments capable of surmounting 17 cm sized obstacles (though the hazard avoidance control system invokes avoidance of obstacles of 10-cm size). Rather than using typical space-rated brushless motors with integrated electronics, the rover used commercial brush motors due to their tolerance of low temperatures.

The lander's control computer was an IBM RAD6000, a radiation-hardened, single-board computer with a 32-bit architecture for 22 Mips of processing power in conjunction with 128 Mbyte of solid state memory. The lander was mounted with a variety of instruments for meteorological measurements—temperature, pressure, capacity and wind velocity. The rover's control system was based on the 2 MHz Intel 80C85 processor with an 8-bit architecture with 0.1 Mips of processing power plus 16 kB of PROM boot memory, 64 kB of RAM main memory, 176 kB of EEPROM program storage and 512 kB of temporary RAM storage. The rover included a telecommunications system, structural support, deployment mechanisms and a payload of science and technology experiments such as black-and-white/colour forward mounted CCD stereo imaging cameras with laser striping projectors, and an alpha-proton-X-ray backscattering absorption (APXS) spectrometer to analyse the soil and rock.

Sojourner's power was generated by 0.22 m^2 of GaAs on Ge solar array with 18% efficiency on the rover upper surface for 16 W peak power at mid-day backed up with 150 Wh non-rechargeable LiSOCl$_2$ D-cell battery power storage conditioned and distributed by a power electronics board (the batteries were capable of supplying power for the nominal 7-day mission). Peak power production was 30 W (of which 16 W were supplied by the solar arrays) with batteries being used to power experiments such as the APXS. In terms of power utilisation, computation drew 3.5 W plus 2 W for gyroscopic sensors and 3 W for wheel-driving motors. Imaging required an additional 0.8 W for the CCD camera. There was insufficient power to run the rf transmitter (2.5 W) with the wheel motors simultaneously as power was a major design constraint on the rover design. The power distribution system has several characteristic features with its 58 power switches: a hardware current limiter circuit ensured that the electrical bus maintained at least 13.5 V; a hardware load shedding circuit protected the CPU from excessive power surges; latching switches enabled batteries to power motors and the APXS; a hardware power monitor provided daily graceful shutdown; and an alarm clock controlled power switch that woke the rover up at daybreak.

All the electronics and batteries were housed in the warm electronics box (WEB) heated by an internally mounted heating unit which warms the WEB during the day but at night the temperature drops. The box was constructed from fibre-glass/epoxy resin face sheets sandwiching solid silica aerogel insulation (of density 20 mg/cm^3 originally designed to accumulate cometary dust grains for the NASA Stardust mission). This enabled maintenance of stable temperatures within the WEB between −40°C and +40°C insulated from the Martian environment. Only the radio modem which required an operating temperature above −15°C required a dedicated heater to raise its ambient day-time temperature to −10°C.

UHF communication to and from the rover was relayed through the lander's whip antenna with a BER (bit error rate) of 10^{-5}. The rover used a 2-W Motorola radio modem operating at 459 MHz with 2 kbps capability coaxed to the rover's omnidirectional whip antenna for 500 m line-of-sight communications. The lander communicated with Earth twice per day and telemetry/telecommands were stored on the lander for forwarding to and from Earth (store-and-forward communications strategy). The rover telemetry was transmitted as formatted packages with several error-detection and error-correction protocols.

The Earth-based teleoperator set waypoints and the rover employed a behaviour-control reactive control software architecture using 90 kbyte of EEPROM to navigate

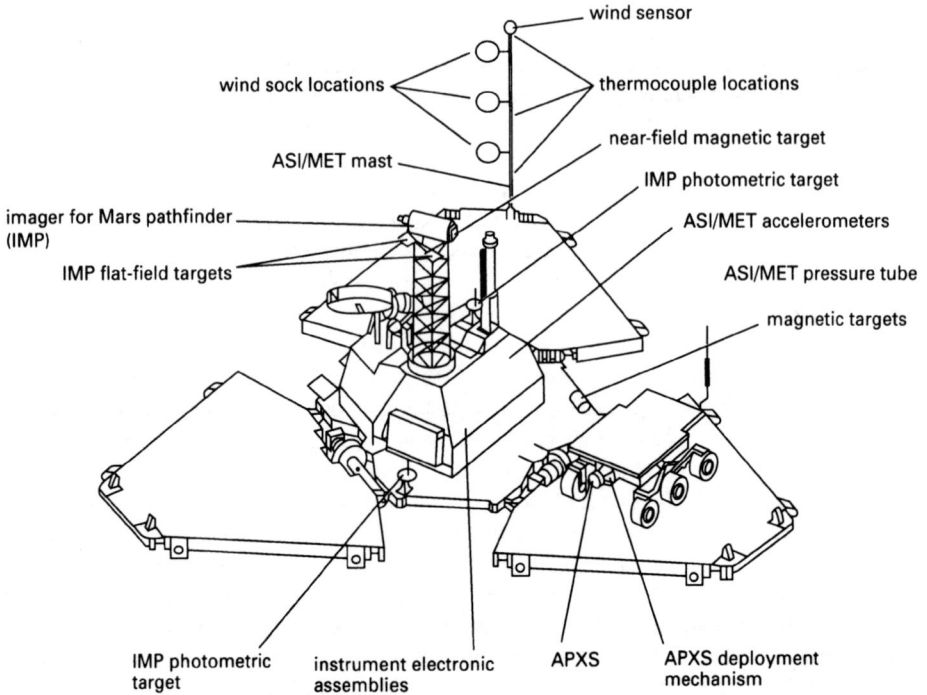

Fig. 2.4. (a) Mars Pathfinder lander; (from Surkov 1997). Reproduced with permission from Praxis Publishing Ltd.

autonomously to each waypoint. The operator selected the waypoints based on the data from the 1.5-m mast-mounted panoramic stereo cameras. This formed the basis of the Rover Activity Sequence File for telecommanding. The GoTo_XY command was the principal autonomous command for traversing between waypoints. Dead-reckoning navigation employed wheel odometry, potentiometers, gyroscopes and accelerometers to generate steering direction for the wheels. The vehicle steered autonomously to avoid obstacles using its odometry and gyroscopic sensors to update its progress in executing the commanded goal locations. Hazard detection was based on proximity sensors such the frontal stereocameras, the five laser striping projectors and the contact sensors on the front of the rover which invoked behavioural routines to negotiate the obstacles. Each CCD camera was a 767×484 pixel imager with images scanned slowly due to the computational limitations of the 80C85 processor. The laser striping projectors generated vertical planes of light to create stripes on obstacles making their identification easier. Each sensor was monitored and on failure was to be excised from the control system and marked with an error state flag. Interrupt handling provided the means for reflex events within the software control loop which implemented its various functions such as house-keeping and health checks invoked by software timers, command handlers, etc. Sojourner was programmed to stop periodically to scan the route for obstacles and then continue. It was also programmed to stop and wait if it encountered tilting or wheel slippage causing

Fig. 2.4. (b) Sojourner Rover; (from Surkov 1997). Reproduced with permission from Praxis
Publishing Ltd.

problems during its passage through the 'rock garden'. The vehicle traversed a total of
52 m via 114 commanded movements.

With the loss of the Mars Climate Orbiter and the Mars Polar Lander, the future of
currently planned Mars missions is uncertain. At the time of writing, the Mars Surveyor
2001 lander will include a 10-km-range rover (Marie Curie) almost identical to Sojourner
with a robotic arm, panoramic cameras and scientific instruments (such as Mossbauer
spectrometer and APXS for mineral identification and chemical analysis) in its search for
ancient fluvial or hydrothermal regions on the Martian surface. It will analyse the distribution of dust grain sizes as grain sizes less than 3 μm are dangerous to humans, causing
silicosis of the lung. The Lander will include an *in situ* demonstrator for rocket-propellant
production (specifically oxygen) using gases of the Martian atmosphere. Core drilling
typically based on the all-percussive demolition hammer is also likely to be a desirable
capability to a depth of at least 3 m for addition to the end effector suite—vibration
suppression is the most significant challenge, e.g. Venera 13 and 14 soft landers. A
motor-driven flywheel drives a crankshaft hammer onto a demolition spike repeatedly at
high speed to penetrate the rock. After Surveyor 2001, Mars missions will be devoted to
the sample return mission in 2003 and 2005 based on the Athena rover series which will
carry Mossbauer and Raman spectrometers in conjunction with the APXS. The miniaturised Raman spectrometer will be able to detect mineral and organic material. The Athena

rover will take 1 kg of rock core samples with a 30-W microdrill. The samples will be stored in a hermetically sealed canister which is inserted into the Mars Ascent Vehicle (MAV) for return to Earth. Contamination of Earth from potentially alien biogenic material is a priority issue, though realistically it is unlikely that any alien biota would interact with the Earth's biosphere.

ESA is proposing its €150M, 1.07-tonne Mars Express mission for Mars interception in 2003 [Schmidt et al 1999]. It will carry some of the instruments from the failed Russian Mars-96 mission, and it will carry a 60-kg lander, Beagle 2 (the original *Beagle* was the Royal Navy ship that carried Charles Darwin as the resident naturalist to the Galapagos Islands) with 25 kg of scientific instruments to the surface to search for subsurface water and evidence for life. Mars–Earth opposition occurs once every 780 days, but Mars Express will take advantage of a once-every-17-year opposition at Mars perihelion, minimising the distance between Earth and Mars at 55×10^6 km. The basic cost-cutting philosophy is to re-use Rosetta mission technology. The lander, Beagle 2, will be released 5 days before Mars Express enters an elliptical and then near-polar orbit at 250 km altitude. An orbital sounding radar onboard Mars Express will probe below the ground for water to depths of several kilometres. Beagle 2 will autonomously deploy its parachutes 1 km above the surface to reduce its velocity of descent, and gas-filled balloons will deploy around the lander to cushion it as it impacts the surface. It will then configure itself, deploying its solar panels and erect its antenna mast for surface operations. Beagle 2's payload includes a microscope, a gas analyser/mass spectrometer, a small robotic arm with a reach of 1 m and a tethered mole capable of crawling across the Martian surface at 6 m/h and burrowing beneath large boulders up to 5 m from the lander. The robotic arm will use a grinder to remove weathered oxidised rock from boulders and a microscope to examine the interior of rock to seek for mineral evidence of past life. It will also carry Mossbauer and X-ray spectrometers built into the arm to analyse the rock. The arm will then use a drill to core the rock which is taken back to the lander for stepped pyrolysis in oxygen between 200 and 500°C to generate carbon dioxide from any carbon compounds present. As different carbon materials combust at different temperatures, a gas chromatograph–mass spectrometer will measure the ratios of different carbon isotopes in the evolved carbon dioxide. The robotic arm will also deploy the mole Pluto (Planetary Undersurface Tool) to take subsurface samples from 1m depth back to the lander for analysis. An electrical motor compresses a spring between two masses; on release, the smaller mass strikes the front end of the tube causing the larger mass to recoil at the tail end, enabling it to burrow into the soil.

A dextrous planetary manipulator may be considered an important part of any surface rover payload for surface science experiments and soil acquisition. It is an essential part of any *in situ* analysis or sample return mission. The minimum requirement is for a 3 degree of freedom telescopic arm with a scoop for trench digging, but the addition of wrist pitch with a gripper gives much greater flexibility of deployment. Science experiments usually include analytical probing such as point spectrometry, instrument placement (like APXS, NMR or Mossbauer spectrometry), material sampling such as trench digging, core sampling and sample acquisition and containment.

Angle & Brooks (1990) suggested that legs rather than wheels would be appropriate for microrovers in that they act as sensors and offer greater locomotive capability over rough terrain—legs make use of separate distributed footholds, while wheels require unbroken support that cannot be guaranteed in slippery, steep, soft or uneven terrain. Their six-legged prototype Attila, based on Genghis had a mass of <2 kg with 24 actuators and 150 sensors connected via a LAN to 11 onboard computers. Legged walking robots are generally based on insect-type morphologies which in turn are based on the 3D pantograph mechanism of four links providing three degrees of freedom. Two frames each with three legs ensures that the rover is inherently stable. Twin-frame hexapod walking machines offer greater mobility over rugged terrain than wheels but are simpler to control than coordinated multiple-degree-of-freedom leg motion [Amati et al 1999]. The orthogonality of the leg configuration provided by prismatic joints allows the decomposition of the two-degrees-of-freedom horizontal and vertical motions for each leg. The legs of one frame are lifted, the frame moves forward relative to the other frame and then the legs are set down. The legs on the second frame follow suit, thereby completing the walk cycle. The robot can negotiate slopes of up to 30° retaining a level body orientation. The orthogonal SCARA type leg is generally chosen, comprising a vertical link activated by a prismatic joint and two more degrees of freedom to produce horizontal motion. An actuated foot may comprise a universal joint with force sensor and joint orientation sensors. Wave gaits like the tripod gait are used by insects which move the body at constant speed while legs perform a leg motion sequence. The front and back legs on each side of the body swing in phase with the middle leg on the opposite side of the body. Support legs move backwards while the transfer legs move forward, all at the same time. Discontinuous gaits are characterised by sequential motion of the legs and body. One transfer leg moves downwards until the foot touches the ground while the others remain still. Body motion is performed by simultaneous backward motion of all legs at the same time and speed in the support role. These actions are easier to implement than wave gaits and have better static stability. However, wave gaits can achieve higher speeds.

A Martian network of microrover landers ~5–10 kg each using behaviour-based control methods linked to a central lander computer to emplace seismometers, take samples and perform spectral analysis has become much more attractive an alternative [Chicarro et al 1993, 1994]. Alternatively, a dedicated low-orbital microsat system ~50 kg may provide a relay from the microrovers to Earth for data analysis [Kondo et al 1993]. Such a system would be characterised by small antennas, low-power transponders and low propagation losses. As low orbit imposes limited visibility from the ground, multiple units in constellation would be required for complete coverage. High data rates are not required and ~1–10 kbps would be sufficient. Parallel operation allows a wide area coverage with reliability through redundancy. A more thorough survey of the Martian surface could be made per unit mass of launched payload by using a buckshot spread of 100 microrovers. The parallel operation of multiple rovers offers reduced mission time, increased area coverage and a high degree of redundancy. They provide maximum utilisation of limited resources and the possibility of flexible exploration by waves of such swarms dispatched with specialised payloads as information is obtained over time. Several small rovers can duplicate all the scientific payloads of a larger rover. Integrated penetrometer payloads can be precisely positioned at the correct locations and driven into the soil with a precise distribution [Ellery & MacGuire 1993]. Oliviera et al (1991) suggested the use of the distributed AI paradigm for multiagent cooperation of microrovers. An example of such multi-agent cooperation is the construction of a termite nest with a multitude of ants exhibiting purposeful behaviour as a single entity. This activity emerges from a multitude of disorderly movements as termites randomly transport and drop small quantities of earth mixed with hormones. They move randomly until a small initial but sufficiently critical quantity of earth is accumulated at a number of points. The termites are then attracted over longer distances by the stronger concentration of hormones at these points and the probability of more earth and hormones being dropped there is reinforced. The application to construction activities such as digging, levelling or piling of regolith by microrovers is obvious. Little communication between Earth and the microrover would be required. To enable the microrovers to swarm as teams they would require infrared beacons mounted onto their structures. Each rover activates behaviours that move it toward the greatest concentration of beacons but at an offset distance from all other rovers. To operate as a coordinated pack one microrover disables its swarm behaviour and activates its leader behaviour while the other microrovers activate their swarm behaviours. In this way they can act as a colony which behaves as if under purposeful control. Each rover is then a semi-autonomous unit in that a community is created by the communication of messages. Each agent is a computational unit with its own problem-solving capabilities. This mode offers an alternative means for multi-agent control when the level of intelligence in each agent is higher.

2.2.2 The future of planetary rovers

A book such as this cannot cover all types of rover mission, given that the rover designs will be as varied as the environments in which they explore. However, brief mention will be made of planetary aero-vehicles which are lightweight, flying robots (aerovers or aerobots) for atmospheric environments. Typically, their missions are mapping of the local terrain at greater resolution than orbiters and the deployment of microrovers or dropzonds at different locations. There are a variety of modes of aerial flight but bouyancy control by balloon offers the most flexible concept [Smith & Cutts 1999]. Indeed, many spacecraft scientific instruments are flight-qualified on balloon flights. There are two types of balloon. The zero pressure balloon is launched filled with sufficient He that its lift exceeds the weight of the balloon and its payload. The balloon rises and, as it rises, atmospheric density decreases. The volume of gas expands until the balloon is completely filled, at which point the pressure inside matches the ambient atmospheric pressure. During the flight, the He gas cools and contracts, so ballast must be shed to maintain altitude. The time that the balloon can remain aloft is determined by the amount of ballast. Super-pressure balloons are not vented; during the day the internal pressure exceeds the atmospheric pressure but at night the differential pressure drops so it maintains a stable altitude. The mission concept of the aerobot is in delivering balloons from the planetary orbiter in entry capsules to the planetary atmosphere and on re-entry, the balloon opens and inflates during the descent by parachute. Flight planning requires global wind models that describe wind patterns at different altitudes supplemented by local wind-speed sensor measurements. There are several suitable targets for such exploration—Mars, Venus, Titan and Jupiter are the most attractive. The most critical technology is the balloon fabric that can maintain integrity between the inside and outside of the balloon. The balloon material must be strong, flexible, light and resistant. A typical balloon fabric is composite material. A polyester film may act as the barrier to He leaching, a polyester fabric for strength, and a polyethylene film to contain the gas, all of which must be bonded together by a soft adhesive. The composite has a density of 55 g/cm^2 and a strength of 2600 N/m. Evidently, each balloon fabric must be designed to cope with the planetary atmosphere in question. Pumpkin shapes are preferred over spherical designs which place loads on the meridional tendons and reduce the stresses on the fabric. The Martian atmosphere being thin represents a challenging environment for balloons—its pressure and density are similar to that of the Earth's stratosphere. Future developments include a reversible fluid aerobot which repeatedly drops to the surface and relaunches back to the heights of the atmosphere. Such a reversible fluid balloon is ideal for Titan where the temperature increases closer to the surface. A primary gas of He and a secondary gas of Ar would be suitable. At high altitudes, Ar condenses and the aerobot descends until it reaches the surface at which point it relaunches. Zubrin (1992) has suggested that winged atmospheric vehicles propelled by internal combustion piston engine or jet engine could be powered by H_2/CO_2 combustion generating a specific impulse of 600 s. The wing area required decreases inversely with the square of the aircraft velocity and such vehicles can provide long-range mobility of 500 km.

Probably the most significant challenge that is likely to arise in the foreseeable future is a robotic mission to Europa's hypothesised subsurface ocean [Rothery 1994, Pappalardo et al 1999]. There is a potential terrestrial testbed for the subsurface exploration of Europa on Earth—Lake Vostock in Antarctica, 1000 km from the South Pole, lies 3.7 km below the ice, 700 m below the nominal sealevel cut off from the surface for ~1 My [Kapitsa et al 1996, Maimone 1999]. It was discovered in 1974 by airborne radar survey and mapped extensively by ERS-1 in 1993. It is the size of Lake Ontario at 48×224 km in extent with a depth of up 500 m plus 50–300 m of sediment (average depth 125 m), and is the largest of several such freshwater subsurface lakes beneath Antarctica. It may be a tectonic rift valley like the Red Sea. Geothermal flux from the Earth's interior ~50 mW keeps it liquid and if is a tectonic rift valley, there exists the possibility of hydrothermal vents. The presence of methane gas hydrates in the overlying ice suggests that methane clathrates may exist in the lake which could provide a source of organic material to the lake. Microbes may reside in such subsurface lakes but the deepest core samples of ice down to 3623 m has reached 120 m above the liquid/ice surface in 1998 covering the past four glacial-interglacial cycles (over 400 000 y). The ice cores taken show evidence of extensive biota. It is due to be explored directly by a probe and extreme measures are being discussed to ensure that it is not contaminated under the provisions of the Antarctic Treaty.

A Europan planetary lander would require an ice penetrator cryobot ~3 m in length and 15 cm in diameter to melt through the ~100-km ice surface at a rate of 1–2 m/h with its electrically heated tip and deliver a hydrobot submersible to explore the subsurface ocean. This Europan cryobot will be fundamentally limited in power (nominally RTG-generated) and a minimum of 4 kW is required to penetrate the ice [Klaes 1999]. A cable unspools from the surface delivering power and data to the cryobot. Once through the ice, the cryobot splits leaving an ice/water interface station and releasing the sediment exploration station deployed by another cable until it reaches the subsurface ocean bottom. Once the sediment station is secured at the bottom, a submarine hydrobot is released from its rear section. The hydrobot will use sonar and cameras to obtain data from the subsurface ocean. The total length of the probe is determined by the amount of cable spooled within it—a 0.5-m length of penetrator can accommodate 500 m of spooled cable. As 0.3 W/cm^2 of heat is required to prevent the probe from freezing in the ice, 400 W of power per 0.5 m of probe length are required. Such a mission would represent a highly challenging problem given the inaccessiblity of the subsurface ocean and the multiple environments to be endured and traversed, most of which are presently unknown. However, it is likely that, given the potential for hydrothermal vents and the possibility of life, such a mission may be forthcoming within the next 20 years.

The logical extension to microrover technology is the nanorover with a mass range of ~100 g with a power requirement of 1 W—the chief problem would be the provision of backup battery power within such small masses. Behaviour-based techniques offer promising technologies for planetary exploration but are insufficient by themselves for an in-orbit servicer. They are ideal for reactive capabilities but task-oriented knowledge is required in addition and that means planning capabilities.

3

In-orbit servicing missions

The most complex space application of robotics is in-orbit servicing and we will use this type of mission as a design example to focus on space robotic technologies later. Furthermore, in-orbit servicing is based on using the more traditional anthropomorphic robotic manipulator of general utility rather than focussing on rovers designed for specific environments. That said, much will be relevant to planetary rover missions.

The commercial motive for robotic spacecraft and their future central role in space activities is comparatively new, particularly for general Earth orbit operations which are presently dominated by manned missions. There is little doubt that robotic and automated systems in space will contribute considerably to the future commercialisation of the space environment. Indeed, it may be considered that the true commercial utilisation of space relies on two major capabilities: advances in inexpensive re-usable space transportation from Earth to reduce launch costs, presently ~$20 000/kg, and the development of remote complex manipulation for in-orbit activities. The time is indeed ripe for the introduction of remote manipulation technology into space. Modern spacecraft are now being designed to be serviceable in-orbit. Advances in CCD sensor technology motivated by remote earth observation satellite and astronomical satellite applications may be utilised for advanced robotic vision. The presently available communications infrastructure, e.g. EDRS/TDRSS (European Data Relay System/Telemetry & Data Relay Satellite System), may be flexibly exploited by robotic spacecraft in Earth orbit.

In 1985, the NASA Advanced Technology Advisory Committee made recommendations to the US Congress Committee on Science & Technology to develop a permanently manned Space Station and highlighted the critical role to be played by robotics and automation. The reasoning behind their report and the emphasis on robotics and automation was that such capabilities would considerably reduce the cost of future space programmes by providing the basis for the development of a spaceflight infrastructure. The development of a robotics and automation capability was seen as critical for all areas of space exploration, global information technology, space utilisation and space industrialisation. It was consequently expected that robotics and automation would play a central role in the Space Station Freedom programme as a small step towards the expansion of a highly underdeveloped technology applications area ripe for exploitation. Greater autonomy would increase flexibility of space operations and so increase responsiveness to innovation and reduce human exposure to hazard, reduce reliance on ground support and so reduce operational costs and extend the capabilities of unmanned missions; and improved reliability and improved productivity would increase the probability of mission success and mission efficiency. The report recommended that the robotics and automation development budget should make-up no less than 10% of the total Space Station cost. Furthermore, the robotics and automation programme would benefit US technology in general, and consequently the US economy. The following recommendations were made [Cohen & Erickson 1985]:

1. Automation & Robotics should be a significant element of the Space Station programme.
2. The initial Space Station should be designed to accommodate the evolution and growth in Automation & Robotics technology.
3. The initial Space Station should utilise significant elements of Automation & Robotics technology.
4. The criteria for the incorporation of Automation & Robotics technology should be developed and promulgated.
5. Verification of the performances of automated equipment should be stressed, including terrestrial and space demonstrations to validate technology for Space Station use.
6. Maximum use should be made of technology developed for industry and the Government.
7. Automation & Robotics techniques should be used to enhance NASA's management capability.
8. NASA should provide measures and assessments to verify the inclusion of Automation & Robotics in the Space Station.
9. The initial Space Station should utilise as much Automation & Robotics technology as time and resources permit.
10. An evolutionary Space Station would achieve in stages a very high level of advanced automation.
11. An aggressive programme of long-range technology advancement should be pursued, recognising areas in which NASA must lead, provide leverage for, or exploit developments.
12. A vigorous programme of technology transfer to US industries and R&D communities should be pursued.
13. Satellites and their payloads accessible from the Space Station should be designed as to be serviced and repaired by robots.

The introduction of robotics to the space environment was seen as a means to increase productivity in space for both commerce and research and, through technology transfer to earth-based industries, increasing productivity on Earth. The ability to deploy, construct, inspect, repair and retrieve space systems may be considered to be a primary requisite for future expansion into the space environment. To this end, NASA had devoted ~$25m/y for the development of automation and robotics technology for space missions in the mid-1980's. It had been estimated that NASA could save $5b/y by AD 2000 if such research was pursued with vigour [Freitas & Gilbreath 1980]. Specifically, NASA began work on two specific proposals to support the Space Station Freedom programme: the Orbital Manoeuvering Vehicle (OMV) and the Flight Telerobotic Servicer (FTS). The OMV was an off-the-shelf spacecraft bus module designed for docking operations for bolt-on payloads such as robot manipulators: it had dimensions of 94 cm thickness and 4.52 m diameter, i.e. disk-shaped for manoeuverability, with a mass of 4761 kg of which 3039 kg was N_2O_4/MMH (monomethyl hydrazine) propellant. FTS was a dedicated freeflyer robotic spacecraft with two 7-degree-of-freedom manipulator arms and a third simpler arm designed for spacecraft retrieval in low Earth orbit (LEO). The spacecraft bus was hexagonal in shape with two cameras for stereoscopic imaging. Although both programmes were cancelled early in the Space Station programme due to budgetary constraints, research into robotics and automation is still a priority for NASA [Weisbin & Montemerlo 1992]. Indeed, NASA's new Ranger project (similar in concept to the FTS) presently under development is a dual-arm freeflying telerobotics flight experiment for in-orbit validation of robotic technologies relating to satellite servicing to be operated in conjunction with an under-water unit in a ground-based neutral bouyancy environment.

3.1 IN-ORBIT ROBOTIC SYSTEMS

Much of the US and elsewhere have used the International Space Station (ISS formerly known as Freedom and then Alpha) programme as a focus for their robotics and automation research particularly concerning in-orbit servicing repair and assembly. It is likely that the ISS despite its problems will be forthcoming in some form or other and it will require robotic capabilities.

The most well-known existing robotic manipulation capability in space is the US Space Shuttle Remote Manipulator System (RMS) which first flew in November 1981 has had around 50 missions to its credit including the capture of the Solar Maximum satellite, the recovery of LDEF (Long Duration Exposure Facility) and Eureca (EUopean REtrievable CArrier), and the capture of Intelsat VI and the Hubble Space Telescope (HST) amongst others. It was designed by the Canadian Space Agency with Spar Aerospace as the prime contractor and is at present the only operational space manipulator. It has been designed as fail-safe with redundant systems and built-in test circuitry such that crew safety would not be jeopardised [Hedley 1986]. Closed-circuit TV cameras with pan, tilt and zoom capabilities and light assemblies are mounted one each at the elbow and at the end effector. Its mass is 480 kg with a reach of 15 m, and a payload capacity in excess of 30 tonnes (recently upgraded to 266 tonnes). It is constructed from graphite carbon composite booms with Ti and Al alloy joints for high stiffness. Each of the upper and lower arm booms have a mass of 33 kg and length of 5 m by 33 cm cross-section diameter giving a boom stiffness of $\sim 10^9$ kg/cm^2. The shoulder has two degrees of freedom, the elbow one degree of freedom and the wrist three degrees of freedom. It is protected from impact by kevlar and has passive multilayer thermal blanket insulation. Furthermore, it is controlled by the Space Shuttle General Purpose Computer. Although it has over 15 000 parts it is designed for a lifetime of 100 shuttle missions without maintenance, i.e. 10y. It is capable of ~ 5 cm and 1° position/orientation accuracy with an unloaded tip velocity of 0.6 m/s (0.06 m/s when loaded) and a maximum end effector

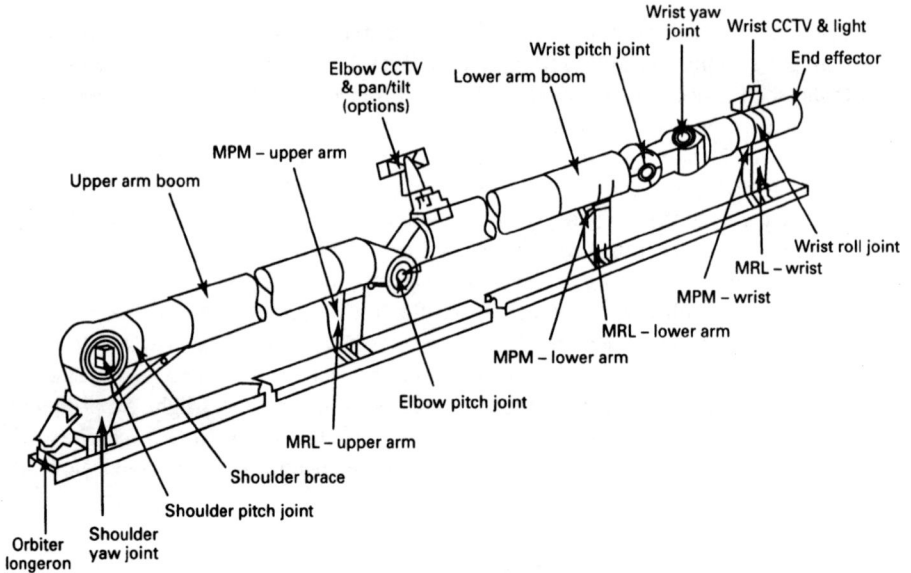

Technical details

Fig. 3.1. US space shuttle semote sanipulator system (reproduced with permission from MacDonald Dettwiler Robotics Ltd:http://www.mdrobotics.ca).

torque of 300 Nm. It is, however, based on 1970s technology and is capable only of simple positioning and retrieval tasks ('fly-swatting'). However, it has demonstrated the potential of robot manipulation in space.

The USA and Canada have concentrated their robotics research effort for the ISS on their Mobile Servicing System (MSS) designed by the Canadian Space Agency with Spar Aerospace as the prime contractor. The Canadian robotics contribution amounts to ~$1.3b. The Mobile Base System (MBS) is a mounting platform to store payloads and provide a work platform for both astronauts and robotic manipulators. It sits on the Mobile Transporter (MT) which is attached to the ISS truss which can be driven along truss guide rails to relocate the MBS. The MBS provides a portable work platform with extra-vehicular activity (EVA) support equipment, ORU pallets, propulsion module attachment system, tool carrier, cameras and lights. It also provides power and data links to the mounted system. It has a mass of ~2 tonnes and an average power requirement of 740 W. The MSS is the basis for the Space Station Remote Manipulator System (SSRMS) for handling large payloads including the US Space Shuttle for berthing and deberthing. It is based on the Shuttle RMS but is three times stronger. The SSRMS has 7 degrees of freedom with a reach of 17.6 m (similar to the Shuttle RMS) and a positioning accuracy of 0.5 m. Its mass is 1.7 tonnes and has a payload capacity of ~160 tonnes. It has end effectors at both ends to enable it to move around hand-over-hand along any truss struc-ture. It has four TV cameras, one at each end effector, and one on each side of the elbow joint. The smaller Special Purpose Dextrous Manipulator (SPDM) serves as a 'bolt-on' front end for more intricate EVA-type (extra-vehicular activity) operations such as ORU (orbital replacement unit) changeout. The SPDM has a mass of ~1.2 tonnes with an average operational power requirement of 1.4kW. It comprises a body with two 7 degree-of-freedom dextrous arms. The body can be attached to the SSRMS end effector and all joints are offset for maximum dexterity. Cameras with lights are mounted on each fore-arm boom and within each end effector. It has a reach of 2 m and a tip-positioning accuracy of 0.024 m and 1° and maximum velocities of 0.075 m/s and 2.5°/s. Both manipulators are designed to be ORU compatible.

The Autonomous EVA Robotic Camera AERCam SPRINT is a small (16 kg mass and 35 cm diameter) teleoperated freeflyer which acts as a mobile camera platform for remote inspection of the ISS. Its design was derived from the Simplified Aid for EVA Rescue (SAFER) backpack. It comprises two hemispherical aluminium shells covered with insulation mounted with a handle and attachment points. Three UHF radio antennas for communications are integrated into the shell. It has two colour cameras for stereo imaging and an infrared obstacle detection system. Vision processing occurred 8 frames/s implemented on a universal image processing chip, the Texas TMS320C80. It uses three rate gyroscopes for attitude control in conjunction with an inertial measurement unit and GPS system and 12 nitrogen gas thrusters for translation control powered by six lithium batteries capable of supplying power for a 7-h EVA. It is controlled from a 16-bit microcontroller with 64 kB of RAM and 64 kB of EEPROM. It was flown on STS-87 in 1987 for technology demonstration of free flying mode at a speed of 7.5 cm/s .

The US Ranger Telerobotic Flight Experiment (RTFX), now the Ranger Telerobotic Shuttle Experiment (RTSX) involves the development of a neutral buoyancy and flight prototypes for a number of low-cost, freeflying telerobots for in-orbit satellite servicing (particularly ISS orbital replacement unit exchange) to be controlled from the ground [Parrish 1998, Parrish & Akin 1995, Graves 1995]. It is administered by the University of Maryland Space Systems Laboratory for NASA's Johnson Space Centre (JSC). Ranger will comprise four manipulators—two 7-degree-of-freedom arms for EVA-equivalence, a 7-degree-of-freedom manipulator leg for securing the robotic spacecraft at the worksite, and a 6-degree-of-freedom manipulator to independently position a pair of stereo cameras. A second pair of stereo cameras are mounted on the vehicle centreline. The Ranger NBV (neutral buoyancy vehicle) is of identical kinematic configuration to, and has the same control and dynamics architecture as, the RTSX space system as a tech-nology demonstrator and simulator. An interactive computer graphics and telepresence control interface will be implemented using virtual reality techniques to aid real-time visualisation of the worksite. The original RTFX mission was to be of 30 days launched on an expendable booster and remaining attached to the upper stage. The upper stage would serve as the target for in-orbit operations. It was designed to fly in a 1700-km circular orbit at 45° inclination to ensure sufficient line of sight between the ground and the spacecraft. The communications link was to be by direct line of sight with the ground station with six overhead passes per day of 22 minutes each. It was to demonstrate simple and complex robotic servicing tasks, and finally to detach from the target vehicle to demonstrate freeflying proximity manoeuvres such as repositioning of the spacecraft base and hand-over-hand manoeuvring, flyaround visual inspection, rendezvous and grappling. The RTSX experiment will comprise a taskboard with three orbital replacement units mounted onto a Spacelab pallet in the Shuttle orbiter payload bay. The taskboard units include the ISS remote power control module, Hubble Space Telescope electronics orbital replacement unit, and the ISS battery box. The most complex task is the Articulating Portable Foot Restraint setup and disassembly requiring four different end effectors to complete. Other tasks include contour-following, hinged access door opening, EVA handrail usage, pin insertion and extraction, ORU insertion and removal and end effector interchange. Ranger's body with its manipulator payloads will be mounted onto the pallet. Interchangeable end effectors each mounted with wrist force/torque sensors will be stowed on Ranger's main body. The mission is to be 48 hours in duration conducted in twelve 4-hour blocks. The Ranger NBV will simulate the space operations in parallel to correlate ground and flight operations. The communications link to the ground will be via the Shuttle orbiter Flight Control Station (FCS) housed in a mid-deck locker and the Ground Control Station (GCS) at JSC. A TDRSS time delay will be simulated. The FCS includes one Silicon Graphics (SGI) O2 computer with keyboard, trackball, flat panel graphics displays. The GCS will include two operator consoles, one primary and the other supplementary. There are four SGI computers, multiple monitors, stereo imaging glasses, two pairs of three-axis hand controllers and two 6-degree-of-freedom head trackers.

The US Robonaut (robotic astronaut) project is an anthropomorphic design for EVA-equivalence to overcome the limitations on the ISS SSRMS and SPDM which require special alignment and grapple fixtures. They are also too large to fit through EVA access doors and suffer from limited dexterity for complex manipulations [Aldridge 1999]. Robonaut has a similar size to an EVA-suited astronaut comprising of three 7-degree-of-freedom arms, each with 12-degree-of-freedom multi-fingered hands, a 6 degree-of-freedom stinger tail, a 4-degree-of-freedom binocular stereo camera platform, all mounted onto a torso with a 3-degree-of-freedom head (i.e. pan, tilt and verge). The robotic hands are designed to handle EVA tools, including power tools, with human-like dexterity. The stinger tail fixes into worksite interface sockets located around the ISS for stabilisation. It can also be fitted as an end tool to the the SSRMS. Robonaut will be teleoperated by an ISS astronaut (intravehicular activity, IVA) utilising telepresence technology such as head-mounted display helmet and virtual reality gloves specifically developed (Dextrous Anthropomorphic Robot Testbed, DART). The dextrous hand has four fingers and a thumb arranged in a human opposition arrangement. The thumb, index and middle fingers have three degrees of freedom while the other two fingers each have one degree of freedom. The hand also has a palm degree of freedom to enable flexing of the whole hand for grasping. Six axis force/torque sensors will be placed at the wrist and the shoulder. The hand is connected to the arm with a pitch–yaw wrist. The hand/wrist module contains 14 motors and 42 sensors. The arm has a 0.7-m reach equivalent to that of the human arm with a roll–pitch–roll–pitch–roll–pitch–yaw configuration with elbow self-motion for collision avoidance. The total system has 48 degrees of freedom. Robonaut has some 150 or so sensors per arm. All actuation will be by brushless dc motors. The design philosophy of Robonaut is EVA-equivalence and the ability to use human tooling like the EVA toolkit. To minimise the wiring harness, electronics have been embedded into the mechanisms.

The Japanese experimental module (JEM) of the Space Station is also to have its own dedicated manipulators with six degrees of freedom each. A single main arm of 9.7 m

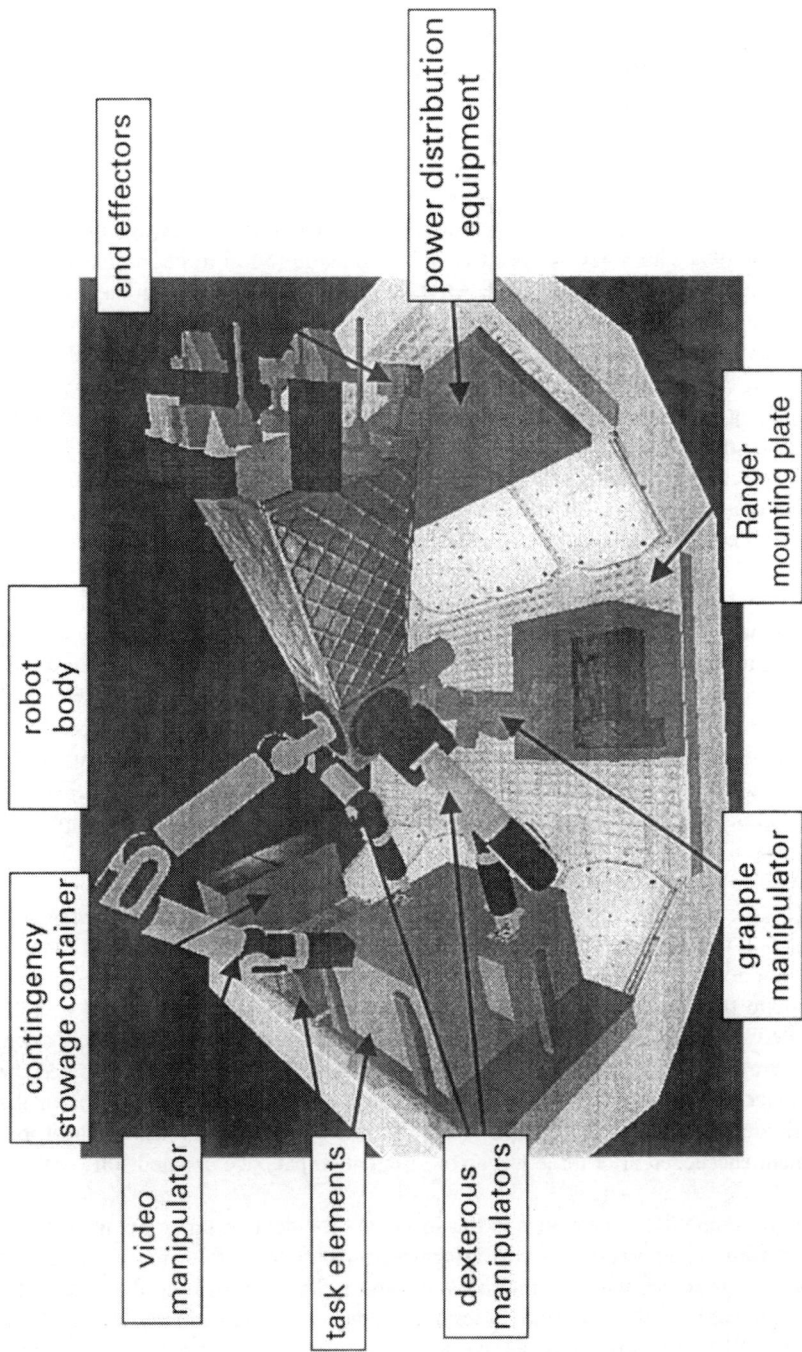

end effectors

power distribution equipment

robot body

Ranger mounting plate

contingency stowage container

grapple manipulator

video manipulator

task elements

dexterous manipulators

Fig. 3.2. Ranger RTSX configuration (from Parrish 1998). Reproduced with permission from Joe Parrish.

length and 370 kg mass, with a payload capacity of 7 tonnes and a maximum speed of 0.05 m/s, would be used for large payloads. It represents a central component of the JEM's Exposed Facility experiment platform. A smaller arm of 1.6 m length and 120 kg mass, with a payload capacity of 300 kg and a maximum tip speed of 0.1 m/s will be used for smaller payloads.

In support of their space robotics programme, Japan's NASDA (National Space Development Agency) designed and constructed their Experimental Test Satellite, ETS-VII launched in 1997 into a 550 km orbit at 35° inclination to test a robotic manipulator (particularly the Advanced Robotic Hand, ARH) in space teleoperated from the ground to demonstrate automated rendezvous and docking tasks on a small subsatellite such as fuel transfer, battery exchange and grappling of objects [Visentori & Didot 1999]. ETS-VII consisted of two satellites, the chaser of 248 kg and the target of 380 kg. During rendezvous and docking experiments, the chaser performed rendezvous with the target using the manipulator mounted onto the chaser. The ETS-VII robotic unit of mass 45 kg had a stowed configuration of dimensions 50 × 48 × 48 cm. Its power consumption was 80 W. It included a manipulator arm of six degrees of freedom with a reach of 2 m mounted with a three-fingered hand, a hand exchange platform, taskboard and latch mechanism. Its sensors included a hand–eye camera, three proximity rangefinders, a pair of grip force sensors, a compliance sensor and a wrist-mounted force-torque sensor. Computer vision was used to close the feedback control loop as a teleoperated system. Two wrist-mounted cameras generated the video signals which were JPEG-compressed. Each robot task had an operational duration of 20 minutes. The spacecraft-ground communication was performed through NASA's TDRSS (tracking and data relay satellite system) located in geosynchronous equatorial orbit (GEO). ESA provided an Autonomous Executive for high-level commanding as the front end to the ETS-VII control system. In teleoperative mode, telecommands are uploaded at a rate of 800 bps while telemetry and compressed video images at a frame rate of 2 Hz were downloaded at a rate of 12 kbps and 1.5 Mbps respectively. The total time delay between ground and space communication was 4–6 s, dominated by ground computer network delays. The taskboard on the chaser included a connector, a bolt and a tethered ball for capture. The telerobotic mode adopted was that of interactive autonomy in which tasks were split into typical subtasks of of medium complexity. Ground operators parametrised, monitored and re-scheduled the subtasks which were executed autonomously. ETS-VII successfully demonstrated automated docking from 2 m apart using three grappling claws mounted onto the target which grappled the chaser. The second rendezvous and docking experiment failed due to an error in the chaser's attitude thrusting which placed it in safe mode generating a misalignment, but the experiment succeeded after three weeks of failed attempts once updated software was uploaded.

ESA are also committed to robotics in the space environment as a major component in Europe's thrusting general research into information technology. ESA has a number of space robotic programmes with the International Space Station providing the focus. The European contribution will to be the Columbus Attached Module. Many studies have been done: the BIAS (Bi-Arm Servicer) was conceived as a smart front end to a larger manipulator for fine manipulation tasks; MTSU (Man-Tended Servicing Unit) was a manned small pressurised manoeuvering vehicle equipped with manipulator arms to be teleoperated from inside the chamber (now cancelled), IRAS (Interactive Remote Automation & Robotic Servicing) system for investigation into task-oriented telerobotics; EMATS (Experiment Manipulation And Transportation System) comprised two 6-degree-of-freedom robot arms mounted onto a 2-degree-of-freedom mobile gantry carriage to provide the IVA system for servicing and experimentation within the Columbus Attached Laboratory; ERA (European Robotic Arm) is 630 kg 11.3 m long 7-degree-of-freedom manipulator with 3-degree-of-freedom wrists and end effectors at both ends linked by a single elbow joint to enable it to walk along truss structures using the two end effectors as legs—it evolved from HERA (HErmes Robot Arm) after the cancellation of the ESA Hermes spaceplane and with a payload capacity of 8 tonnes now has a role in external servicing the Columbus laboratory of the European space station module, as well as for general equipment exchange, outer surface inspection and to provide support for EVA astronauts.

ROTEX (RObot Technology EXperiment) was the most extensive robotic experiment in space by ESA [Hirzinger 1987, 1993]. It flew on the Spacelab D2 mission in 1993 as one of the rack experiments to provide basic knowledge on microgravity robotics. ESA's Spacelab is a manned laboratory that flies in the US Space Shuttle payload bay. It has been used extensively to perform microgravity experiments of various kinds including materials and life science. For ROTEX, a 20-kHz AC power supply connected all sensors and actuator systems. In addition, D2 experiments included telescience nodes for microgravity investigations of single crystal growth of various semiconductor materials using zone melting techniques. The ROTEX comprised a 6-degree-of-freedom multi-sensory gripper with 15 sensory components sealed inside a working cell of an experiment rack. With 400 mechanical components, it represented one of the most complex robot grippers ever built. The sensory suite included two lockable force-torque wrist sensor, nine triangulation-based laser rangefinders for 0–50 cm ranges, a 32-element conductive rubber tactile array in each finger, and a pair of CCD cameras for stereo images mounted into the gripper. Its joint actuators could deliver 120 Nm of torque. The design philosophy was fundamentally based on using minimally sized modularly exchangeable sensors with the integration of the electronics into the sensors. ROTEX could carry its own mass of 10 kg as payload compared with most industrial manipulators which are restricted to ~20:1 robot mass-to-payload mass ratios. One of the tasks performed successfully by the manipulator was an ORU-exchange-type operation with electrical bayonet connectors as well as grasping a free-floating object. These operations were performed by both astronauts and from the ground in teleoperative modes and automated modes—the TeleSensor Programming approach. Fast switching between different operational modes was demonstrated to be a powerful tool for remote manipulation.

De Peuter et al (1993) proposed a dedicated robotic satellite system for operation in GEO (geostationary equatorial orbit) as a service for comsats. Such a system would provide commercial services such as satellite inspection, mechanical assistance to faulty deployment mechanisms and re-orbiting satellites into graveyard orbits at EOL (end of life). The fuel penalty of a GEO satellite to boost into a graveyard orbit is equivalent to 6 months further operational life. With the GSV (Geosynchronous Servicing Vehicle—the basis for the Experimental Servicing Satellite ESS) operating, no such fuel penalty is necessary. This would provide the 'bread-and-butter' revenue for the GSV for its 15-y lifetime. During such a lifetime, it was envisaged that there was a potential market for 25 re-orbiting missions, ten inspections, three mechanical interventions and two dead satellite removals. The GSV would have a mass of 4.2 tonnes (dry mass 1.2 tonnes) giving a Δv of 3.7 km/s. It would be operated from a dedicated portable ground station colocated with the customer's main ground station via a low-gain S-band antenna with wide lobes to maintain communications links even in the event of obstruction.

Even with the replacement of the original dual-keel Space Station Freedom with the slimmed-down International Space Station (formerly Alpha), robotic systems will continue to have a vital role to play in the space environment, and indeed such systems may be adapted comparatively straightforwardly to other projects if required (e.g. the Hermes Robotic Arm HERA became the External Robotic Arm ERA for the European Columbus module of the Space Station). The cancellation of FTS in the early stages of the original Space Station Freedom programme was an indication of short-sightedness on the part of NASA when under budgetary pressure, since its role was seen as in support of Space Station Freedom alone, rather than as a generic capability in space. The ISS will provide a test bed for new technologies and will require extensive in-orbit rendezvous, docking and assembly operations capability. Parts of the ISS have been launched to become fully operational in 2004 with a permanent crew of six. It will require 93 flights into orbit assuming no catastrophic launch or payload failures.

3.2 IN-ORBIT OPERATIONS

In this section, we will be examining in-orbit servicing operations. Our design example ATLAS (Advanced TeLerobotic Actuation System) will be introduced later in the form of a telerobotic interceptor spacecraft dedicated to in-orbit servicing, inspection, repair and maintenance of operational satellites in Earth orbit, be they Earth observation, telecommunications, global navigation, scientific or meteorological satellites. A freeflying telerobotic spacecraft is a natural evolution from astronaut EVA and the Shuttle-attached manipulator. In EVA, the astronaut acts directly as the manual worker for small activities while the Shuttle RMS operates for heavier duties with the Shuttle providing a stable platform (due to its high mass—the reaction control system is switched off when the RMS is deployed). A freeflying teleoperated system can be remotely controlled from the ground to provide great flexibility. It would provide a natural testbed for general automated facilities in space for construction, processing and experimentation. The manipulator system represents the spacecraft payload mounted onto the spacecraft bus platform. The robotic interceptor is likely to become the workhorse for such activities in space in the future. Such a capability has considerable immediate commercial application by opening up hitherto under-utilised markets. Indeed, the US Civil Space Policy of 1978 stated categorically that increased participation for private enterprises should be stimulated. There is little doubt that certain space activities have generated commercial successes: Intelsat/Inmarsat (telecommunications), BSkyB (television), Spot Image (remote sensing), Arianespace (launchers). Robotic in-orbit servicing will be no exception. Furthermore, robotic in-orbit servicing would free the astronaut from repetitive tasks and fully utilise astronauts for more appropriate tasks requiring human flexibility and ingenuity.

Spacecraft reliability may in some ways be regarded as the defining feature of a spectacular mission and a nominal one—the Voyager spacecraft are generally regarded as the flagships of NASA post-Apollo era. The identical Voyagers 1 and 2, launched in 1977, each comprised 65 000 individual components and were designed for encounters with Jupiter (1979) and Saturn (1980/1981) and their larger moons with a design lifetime of 4 years. Remote-controlled re-programming allowed them to be adapted to new tasks and missions, though their onboard memories are limited to 1700 words. The ultraviolet spectrometer on Voyager 1 is the only instrument on the scan platform still functioning but it will be switched off around 2000 once electrical power levels become insufficient to power the instruments' heaters. The current decay rate of available power is 5 W/year. The Voyager Interstellar Mission is an extension to the Voyager missions to utilise the cosmic ray detector, charged particle detector, magnetometer, plasma analysers as they cross the heliopause into interstellar space until 2020 when available electrical power becomes insufficient to support the instruments. Hence, the Voyager spacecraft designed for 5 years is still functioning more than 20 years later. A similar success story is apparent with ESA's Giotto spacecraft launched in 1985 which was designed to study Halley's Comet but its mission was subsequently extended to study comet Grigg–Skjellerup. It encountered Halley in 1986 at a distance of 0.89 AU from the sun within 540 km of Halley's nucleus with a flyby speed of 70 km/s. It was the European contribution to an armada which included two Russian Vega probes, two Japanese probes and the US International Cometary Explorer—this last was formerly the International Sun–Earth Explorer 3 which was re-tasked from studying the Earth's magnetosphere. Despite Giotto's dust shield of aluminium/Kevlar, it suffered damage from high speed dust particles to some its sensors. However, it was re-activated in 1990 and encountered P/Grigg–Skjellerup in 1992 within 200 km at a distance of 1.01 AU from the Sun, and eight of the ten scientific experiments onboard functioned correctly. Other highly successful missions include: the International Ultraviolet Explorer Observatory launched in 1978 which after over 15 y is still operational despite a breakdown in the ACS; Mars Viking Lander 2 in 1975 was designed to function for 90 days but was sending data for 6 y.

Spacecraft system redundancy is no longer sufficient to ensure reliability in spacecraft due to cost constraints. Redundancy imposes launch mass penalties and trade-off studies for individual systems indicate that there is a limit to which redundancy is cost-effective, at which point maintainability should be implemented. Particularly for long-life Earth-oriented missions, in-orbit exchange of parts is a necessary precursor to success. No small number of cases have been documented whereby satellites have failed to function properly in orbit—between 1962 and 1983, there have been ~2500 spacecraft failures of one kind or another. The majority of failures occur within the first few weeks or months of operational service due to infant mortality, and in the latter years of operational service due to propellant or battery lifetime limitations.

There have also been cases where failures have been recovered, however. The European EXOSAT launched in 1983 malfunctioned in 1986 with a permanent loss of attitude control. EXOSAT had been designed with a degree of onboard reprogramming flexibility. The AOCS (attitude and orbit control system) sensor processing and control algorithms were implemented in a 16-k microprocessor. Software was parametrised such that the code was contained in ROM (read-only memory) with parameter values held in RAM (random access memory). This allowed the updating of parameters. Some of the 40-k applications software which was not in constant use was stored in a ground library which could be uploaded as required into the spaceborne computer for operational execution allowing newly developed programs to be uploaded if required. Regardless of this operational flexibility, EXOSAT was plagued with problems concerning its AOCS. The drift bias circuitry of the gyros which was used to compensate for the inherent gyro drift regularly experienced jumps during passage through the van Allen radiation belts due to electrostatic discharge of accumulated charge. Unfortunately, during radiation belt passage the star tracker was closed down and so was unable to provide closed-loop control. An onboard program was developed and uploaded to reset the drift bias levels when jumping occurred. The hardwired safety mode circuits of the AOCS was also problematic. One of the two redundant safety modes started to trigger spuriously during perigee and was disabled from the ground. This was followed by one of the four gyros intermittently registering failure incorrectly and was automatically deselected by the remaining safety mode. Unfortunately, later a second gyro actually failed catastrophically and was switched out automatically by the failure mode leaving only two gyros operational. The only way to reactivate the incorrectly diagnosed failed gyro was to disable the remaining safety mode circuit and replace it in software. This was used successfully for the rest of the mission. Another AOCS problem was a spurious thruster-on condition which resulted in the automatic activation of the opposing thruster to counteract the rotation. However, due to a thruster imbalance the satellite spun up until the threshold rate should have triggered one of the safety mode circuits. This would have resulted in automatic switching to a redundant propulsion system but with the disabling of both hardwired safety mode circuits this did not occur. A means was developed, exploiting the sensors used to detect pressure changes in the propellant tanks, to detect the thruster-on condition and so invoke automatic switching to the redundant propulsion system but not until after much fuel loss. These events show the flexibility of augmenting ROM-based systems with RAM which can replace ROM processes if necessary through applications software uploaded into RAM.

The joint NASA/ESA SOHO mission after two years in a nominal halo orbit, the L1 Langrangian point, failed in 1998 when the AOCS switched into Emergency Sun Re-acquisition (ESR) safe mode and contact was lost [Vandenbusshe 1999]. The loss of contact was preceded by a routine calibration of the spacecraft's three gyroscopes and a momentum management manoeuvre which dumps excessive angular momentum from the momentum wheels by firing the thrusters. SOHO entered into the ESR mode which is a hardwired analogue control mode forming part of the Failure Detection Electronics. Two of SOHO's three gyroscope were damaged so, on firing the thrusters to initiate ESR, the spacecraft spun out of control with an increased coning motion, eventually tumbling. It was initially recovered and then the third gyroscope failed so no attitude information was available. The spacecraft kept its thrusters constantly firing to keep its arrays pointing towards the Sun using its precious fuel reserves. The NASA 70-m Deep Space Network Goldstone receiver was used to track the spacecraft and determine its spin condition. The batteries were periodically recharged when the solar arrays were illuminated by the Sun.

The hydrazine tanks had to be thawed as 48 kg of the 200 kg of hydrazine in the tank had frozen up. This had to be achieved by draining power from two batteries which had to be interleaved with battery recharging. This enabled a despin manoeuvre to be completed. The ground crew at ESTEC developed a program to use a star tracker for positional information to be fed to its three momentum wheels to continue its mission until 2003. The upgraded software was delivered by the 305-m Arecibo radiotelescope in Puerto Rico.

OAO-A2 lost its star tracker due to debris spoofing it; OAO-C lost attitude control causing excess momentum build up and excess spin rates; ATS-6 lost its star sensor due to debris spoofing and lost a momentum wheel actuator; NOAA-6 started tumbling after hydrazine venting. The failure of Galileo's main 4.8 m diameter X-band high gain antenna (134.4 kbps channel capacity) to open fully meant that all communication with Galileo had to proceed via its low data rate S-band low-gain antenna (8–16 bps channel capacity). New software was uploaded to the spacecraft to enable it to compress the data collected by the spacecraft to extract greater performance from the limited capability ~160 bps such that 70% of the original objectives would be achieved. Two data compression methods were used: lossless compression as used in PC modems and lossy compression through mathematical approximations. In all these failures, software corrections allowed re-establishment of the spacecraft mission. Attitude control failure corrections were possible due to the existence of redundant thrusters. Some faults may be corrected in this way either fully or partially from the ground by uploading such software changes (workarounds), but hardware faults cannot be corrected except by re-routing, often leading to suboptimal operation and degraded performance. Such workarounds and reconfiguring require extensive ground support with dynamic simulations for software update validation. Validated software is uploaded to the spacecraft via the communications link (assuming that it hasn't been compromised—not an uncommon scenario, e.g. Mars Observer in 1993). Although these software corrections often work quite well, they are not inexpensive solutions to the problems. The Hipparcus satellite was launched into the wrong orbit due to a fault in the apogee kick motor requiring extensive changes to the mission. Galileo which was launched to Jupiter in 1990 had a high-gain antenna which failed to deploy fully and the science mission return had consequently degraded by 30%. Some workarounds have been quite spectacular: Olympus, an ESA experimental telecommunications satellite of 2.6 tonnes was launched in 1989 to test EDRS (European Data Relay Satellite) technologies. In 1991, it went out of control and its fuel froze but the ground team was able to retrieve it and put it back into operation. Olympus I in May 1991 suffered control loss and recovered 2 months later when assisted and reoriented by astronauts; the UARS satellite required solar array replacement. However, workarounds can correct only ~17% of all failures in terms of track record. Loss of attitude sensing or control appears to be a ubiquitous problem. For a satellite operating a valuable service or function, this may be unacceptable (particularly for military satellites). Most failures occur due to environmental conditions or excess load beyond design tolerance, or through random failures. Design failures account for ~25% of all faults, environmental failures for ~21%, random failures for ~30%, software failures for ~5% while ~19% of all failures result from unknown origin. A dedicated robotic interceptor satellite for in-orbit servicing would alleviate these eventualities through maintenance and servicing and effectively increase the overall reliability of all accessible space systems.

The cost of space hardware has grown to the extent that spacecraft are being designed in modular fashion to be serviceable in-orbit despite the mass penalties. ORUs (orbital replacement units) are modules which serve as containment packages for equipment which may be replaced. For low levels of repair at component level, a high number of interfaces need to be broken and rejoined implying long time periods to perform as well as simulate prior to operation. On the other hand, higher standardisation of replaceable modules at subsystem level implies that higher masses have to be transported to the worksite for replacement. However, standardisation minimises the number and variety of tools that are required to be transported to orbit, minimisation of training and simulation requirements for different scenarios and minimisation of task execution. Availability of spare standard modules reduces unplanned down-time. Finally, modularity substantially reduces costs during integration and test phases prior to launch. If standardisation can be made at a sufficiently low level such as for common PCBs and electronics boxes (e.g. power supplies and conditioning) further advantages are gained. However, at no point should standardisation resolve down to component level as this impairs flexibility of design and probably escalates costs in the long run. Hence, although interface standardisation is at the modular level, it would be desirable to extend it to electronics box level. Such ORUs, since they are usually of low mass (~50–500 kg) can be carried as piggyback payloads at low cost on scheduled launches. Reliability and maintainability are related concepts—system operational availability (a measure of system performance) and cost are both cubic functions of reliability and maintainability such that they initially increase very rapidly before tailing off due to the decreasing return on investment [Sepehri 1987]. It is desirable to increase operational availability within cost restraints. This requires a mixed approach since:

$$\text{Availability, } A = (\text{MTBF}/(\text{MTBF} + \text{MTTR} + \text{MTFS})$$

where

MTBF = mean time between failures and reflects reliability

MTTR = mean time to repair and reflects maintainability

MTFS = mean time for supply and reflects logistic capability

The optimal cost is attained when an incremental change in reliability equates to an incremental change in maintainability for a given total investment. At present, MTFS is very large since it is subject to long delays due to long launch preparation times, advanced launch bookings, astronaut training, launch delays, turnaround times, etc. The HST had to wait three-and-a-half years after its launch before it was repaired. Only a dedicated in-orbit robotic system can reduce this parameter. Furthermore, since one of the purposes of robotics is to remove the human from the hostile environment, the system itself must have a high availablity and be capable of self-repair and self-reconfiguration— sending a human to a remote hostile environment to repair a robotic system defeats the object of the system. The system should therefore be highly modular, highly redundant, highly maintainable and accessible.

As well as repair, ageing satellites could be upgraded with more modern systems as technology proceeds, re-orbited from orbit degradation and resupplied with consumables such as propellant, batteries or cryogens to function beyond their design lifetimes (possibly up to 2–3 decades for comsats and 5–10 y for scientific and Earth observation satellites), enhancing their profitability and usefulness by eliminating these restrictions. Warships have a 10 y development schedule typical of spacecraft, but every 7–10 y such ships undergo a major overhaul lasting ~18 months for upgradings.

We now look to the in-orbit servicing market to define the kinds of spacecraft that may require such facilities [De Aragon et al 1998]. Space science is a new science involving Earth observation, planetary exploration and astronomical telescope observations which have all been enabled by the development of space technology. Space-based astronomical observations outside the Earth's atmosphere have opened up the infrared, ultraviolet, gamma-ray and X-ray spectrum to probe the skies. Infrared sensors are limited to 0–120 K for their functioning and so require refrigeration to very low temperatures. As passive radiant cooling is limited to above 50–100 K temperatures cryogenic systems have been employed in space particularly for photoconductive infrared sensors such as the infrared telescopes on IRAS and ISO. Cryostats which expand high-pressure gas through an orifice to achieve cooling through a Stirling cycle of compression–expansion. However, these materials are subject to boil off and so represent consumables. The UK/US/NL IRAS satellite launched in 1983 was fundamentally limited by its He cryogen coolant supply—it operated for 11 months until the He depleted even though all other subsystems were functioning. Its ESA successor ISO supplemented the sky survey of IRAS through detailed observations of selected sources between 2.5 and 200μm using its infrared Richey–Chretien telescope and four scientific instruments (infrared camera, infrared photopolarimeter, and short-wave and long-wave spectrometers) but was similarly limited to 18 months operational life. Indeed, there are several potential applications of cryogenic detection on space missions [Jewell et al 1992]:

(i)   microcalorimetry would provide highly sensitive, high-resolution thermometry at 0.1–2-K ranges, e.g. the COBE satellite;
(ii) photoconductive infrared detection at 1.5–3.5-K ranges, e.g. IRAS, ISO infrared astronomical spacecraft;
(iii) superconducting tunnel junctions for high-resolution X-ray and ultraviolet spectroscopy.

Cryogenically cooled telescopes operating at less than 1 K based on single photon detecting superconducting tunnel junctions such as ESA's S-Cam will become increasingly adopted for optical, near-infrared and soft X-ray astronomy [Rando et al 1999]. Hence, astronomy missions will have massively enhanced value with the potential for cryogenic refuelling. Furthermore, liquid He is expected to become increasingly used on comsats for cooling to improve their signal-to-noise ratios by reducing thermal dark currents [Eaton et al 1994, Morgan 1994]. The primary sources of waste heat on a spacecraft are dc to rf conversion losses between power amplifier output and the output of the antenna feeds: only 17% of the total dc power is actually radiated as rf energy although it is the most power-intensive subsystem—around 60% of the total power is waste heat due to these losses. Superconductivity can reduce these sources of internal heat by reducing the dc power requirement and further reducing waste heat in a positive feedback manner.

New broadband services by satellites are expected to yield around $5b/y. Lower orbits than GEO require a constellation of satellites for worldwide coverage [Williamson 1998, 1999a, Camponella 1986, Wakeling 1999]. The three major investors in satellite-based 2-MB data rate mobile telephony are ICO, Iridium and Globalstar who have invested in small, handheld terminals with satellite-based worldwide coverage. Iridium proposed a 77 LEO satellite constellation of eleven satellites each in seven different LEO orbits at 700–1000 km altitude. LEO offers lower link attenuation (~15–20 dB) and so allows smaller user units. This is a major step from the conventional GEO satellite system which essentially serves the existing communications network infrastructure in relaying international telephone calls and direct broadcast TV. Handheld telephone receivers require higher satellite antenna powers to compensate for lower receiver powers. ICO proposing a constellation of ten satellites plus two in-orbit spares in intermediate Earth orbit of 10 390 km altitude arranged in two diametrically opposed planes of five satellites plus one in-orbit spare in each to give global coverage. Iridium are providing a constellation of 66 satellites (originally 77) plus six in-orbit spares in low polar Earth orbit of 780 km altitude arranged in six orbital planes of eleven satellites each for global coverage. Although Iridium experienced no launch failures in its fifteen launches, seven satellites failed, of which one expended all its fuel accidentally and three had attitude control problems, highlighting the utility of an in-orbit servicing capability. INMARSAT's Project 21 (i.e. twenty-first century), however, is expected to utilise a mixture of such LEO systems with ever larger and more powerful GEO satellites. Globalstar is utilising a constellation of 48 spacecraft, six in each of eight orbital planes plus eight in-orbit spares. At the time of writing, it has 36 satellites successfully launched to LEO surpassing its minimal requirement of 32 satellites to initiate services. There are in addition, a host of smaller mobile communication satellite initiatives. It has been estimated that 62 million subscribers to satellite-based mobile communications will be reached by 2010. However, both Iridium and ICO are suffering financial difficulties at present having filed for Chapter 11 bankruptcy protection in 1999, and it not clear how the future of satellite-based mobile communications will fare. Internet services are the most rapidly growing sector of satcoms application. Teledesic is constructing a global, high-capacity, broadband 'internet in the sky' system using a constellation of 288 LEO satellites at an altitude of 696 km divided into 12 planes of 24 satellites each to provide multimedia access. It is due to become operational in 2003. Teledesic's main competitor for internet services, Skybridge, will operate 80 LEO satellites to become operational in 2001. The new constellation approach to telecommunications services has important repercussions for in-orbit servicing. Iridium's experience in satellite failures suggests a strong potential market to increase the reliability of constellations, reduction in space junk due to dysfunctional spacecraft and reduced costs to satellite service providers. EO (Earth Observation) satellites reside in Sun-synchronous polar orbit at 800–1000 km altitude, allowing a satellite to scan the entire planet's surface passing over the same point at the same local time of day for constant illumination. NASA's Mission to Planet Earth programme is an environmental monitoring scheme to solve problems of pollution, land management and land use. It is predicted to involve 240 Gb of information per day being generated and transmitted to Earth for the spatially integrated GIS (geographical information systems) database to provide a tool for decision-making concerning environmental issues. Remote sensing from space is the only effective way of studying the Earth as a unified system. The market distribution of French SPOTIMAGE data is instructive: 30% cartography, 20% vegetation (forestry and agriculture), 18% geological prospecting, 8% public civil engineering, 4% urban town planning, 3% coastal studies, 2% news media, 2% general public and 13% other. The demand for ESA's ERS-1 satellite data of global sea state, sea surface winds, sea surface temperature, ocean circulation, ocean and ice levels has been growing at a rate of 20–30%/y. This increasing demand for EO data together with the political and public support for environmental protection indicates that EO satellites will become increasingly populous in low-altitude polar orbits. These satellite-based telecommunications, EO, navigation, internet and e-commerce services represent a major market for in-orbit servicing.

Consumable replenishment (e.g. fuel, cryogenic fluid, batteries, etc) essentially spreads fixed cost overheads over longer operational periods. In fact, the cost parameter for a communications satellite is given by [Bargellini 1978]:

(Launch cost + spacecraft cost)/(number years of service × number telephone circuits)

Hence, increasing the lifetime directly reduces the cost parameter in linear fashion. Russel & Price (1990) declared that the largest costs for GEO comsats are hardware and transport (GEO is inaccessible to astronauts). To reduce these costs, it is desirable to maximise the payloads fraction, but this is subject to a limit. This limit can be curtailed by refuelling propellant and replacing worn-out batteries since when this is needed, the other systems are usually still functional. This effectively converts the fuel mass which has no direct earning capacity to payload mass and higher revenues without increasing spacecraft mass. Refuelling in particular increases the spacecraft net present value by up to $24.4m and increases the internal rate of return up to 8.3%. In addition, only one resupply launch is required to transport consumables to many client spacecraft via a dedicated servicer rather than each satellite operator resupplying its own satellites individually. Furthermore, the present constraint of introducing redundancy into spacecraft systems increases costs. The advent of in-orbit servicing capability effectively renders the redundancy of certain mission subsystems as obsolete, thereby significantly reducing satellite construction and design costs. Indeed, redundancy of some subsystems is virtually impossible (e.g. the propulsion subsystem). Module replacement is much cheaper than satellite replacement which now have cost ranges of ~$200–$600m.

Finally, in-orbit satellite insurance premium rates may be reduced by the existence of such a robotic servicer. Satellite compatibility for robotic servicing such as number and ease of access to replacement modules, accessibility and number of grappling and docking fixtures, existence of detailed and accurate CAD databases for subsystems and components, would in effect increase spacecraft reliability. Indeed, it has been estimated that serviceability could reduce the in-orbit post-launch operational fraction of insurance costs by ~50% [Russel & Price 1990].

There is a long lead time between failure and the contingency repair operation due to extensive preparations required and launch schedule contracts—the Hubble Space Telescope (HST) repair mission occurred 3.5y after its launch. Shuttle flights are not scheduled to coincide with the time constraints of any particular servicing requirement or worksite. The cost of EVA in-orbit servicing is also phenomenal. It has been quoted that the cost of EVA alone is ~$70 000/man-hour, but actual costs suggest that additional and hidden factors dominate, the cost of launch itself being probably the largest single factor and estimates of a single Shuttle flight have been estimated variously at costing $245m, $420m and other figures up to $1b—the true cost is probably difficult to gauge involving so many hidden factors such as training of the astronauts, etc., but $500m would seem a reasonable ballpark figure. The Intelsat VI repair operation in 1992 was estimated to have cost the insurers around $150m [Williamson 1992]. The cost of the more complex repair of the $2b Hubble Space Telescope in 1993 was much greater at $700m, almost three-quarters the cost of the $1b Mars Observer probe lost in 1993. Errors are costly in the space business. Remote robotic execution of these activities would greatly relieve astro-nauts from hazardous and repetitive tasks and reduce their workload allowing them to concentrate their valuable resources on tasks requiring beyond state-of-the-art machine intelligence such as scientific investigation and experiments. Furthermore, a dedicated robotic system capable of reasonably complex manipulation tasks would provide greater flexibility in task execution and enhanced performance, safety, cost-effectiveness and greater reliabilty for mission success. Finally, only low-inclination LEO is accessible to astronauts as the Eastern Test Range at Vandenberg AFB will no longer support Shuttle operation for polar orbit. LEO satellites are estimated to fail at a rate ~40–50% higher than those in GEO. High-inclination polar orbits (the preferred orbit for earth observation satellites) are inaccessible as are high orbits >500 km altitude. High-inclination (polar) orbit is the operational orbit for Earth observation and this orbit is likely to become increasingly populated with such satellites. Such satellites are limited in resolution to ~10 m at altitudes of 700–900 km. A 300-km operational altitude would allow higher resolution of ~2 m, but air drag limits the lifetime of such orbits—cheap smallsats may be perceived to fill this orbit and they would not require servicing. Although there is a recent trend to place communications satellites into LEO for mobile communications, the majority of comsats still reside at the geostationary altitudes ~36 000 km which are presently inaccessible to astronauts and also pose higher radiation hazard than LEO.

In summary, the need for robotic servicing in space has been outlined. We now examine extravehicular activity used in many in-orbit servicing missions by NASA astro-nauts, and then use one example of EVA-based servicing missions as a lesson for use in in-orbit servicing by robot.

3.3 EXTRAVEHICULAR ACTIVITY (EVA)

At present, all in-orbit operations are performed by astronauts in EVA (extravehicular activity) with a limited degree of assistance from teleoperated manipulators, if at all. This is both costly and hazardous. EVA tasks are restricted by access, risk and complexity, and trade studies comparing EVA against other alternative techniques are usually performed to assess payload mass limitations, reaction time requirements, performance time, astro-naut workload and reliability. Other constraints include translation to the worksite (usually the cargo bay), equipment transfer to and from the cargo bay, lighting and tool logistics. Nearly all tasks require foot restraints and handholds, and loose objects require tethers to attach points. All EVA tasks exceeding 30 minutes must be tested in the neutral buoyancy tank which involve the design and construction of mockups and extensive task testing followed by astronaut crew training. Crew training includes KC-135 simulated-zero-*g* parabolic flight (30 s weightlessness) for donning the EMU. NASA has three underwater neutral buoyancy training simulators, the largest and most used are at NASA Marshall Spaceflight Centre in Huntsville, Alabama, and NASA Johnson Space Centre in Houston, Texas.

EVA is defined as any space operation or activity whereby a crewmember leaves the protective environment of the spacecraft pressurised cabin and ventures out into the vacuum of space, requiring independent life-support equipment (a spacesuit). They are undertaken for a variety of reasons including contingency repairs, experiments and testing, spacecraft servicing and space structure construction. The first EVA was performed by Alexei Leonov in 1965 from the Russian Voskhod 2 (a modified Vostock spacecraft as flown by Yuri Gagarin in 1961 but with an airlock), marking the feasibility of survival of astronauts outside the spacecraft. He was followed by Ed White in the same year from Gemini 4. EVA techniques were evolved during the US Gemini, Apollo and Skylab programmes. All Gemini and Apollo EVAs were open-hatch affairs such that the spacecraft cabins were depressurised and all crewmembers inside were exposed to vacuum conditions necessitating their wearing protective pressure suits with life support and communications facilities. The purpose of the Gemini programme was to establish EVA feasibility for in-orbit operations amongst others and the lessons learned during that programme included the need for restraints, the need for one-handed task operations, the need for maintaining energy reserves due to physiological limits and the need for thorough training. One-handed operations allow the astronaut to use his other hand for the frequently required readjustments for stability. This requirement advocated the need for power tools and with the need for restraints dominating all in-orbit activities. To date, the majority of US EVAs were conducted during the Apollo programme and include lunar excursions. EVA was demonstrated as a valuable generic capability on the US Skylab. The US Skylab space station was built from surplus Apollo parts around an empty Apollo Saturn IVB tank and was launched in 1973 into 275 km circular orbit at 50° inclination. An airleak during launch in the booster shroud caused damage to the spacecraft when the micrometeoroid shield tore loose and jammed the deployment of one solar panel and partially deployed the other prematurely. The exhaust plume of the second stage tore the partially deployed solar array away completely. Occupation by the first crew was delayed by 11 days for repair plans to be made. This required the repair and deployment of the jammed solar panel by EVA (SL-2) to restore the 20 kW of electric power (including 7 kW for the three-man life-support system) and the construction of a parasol sunshade for thermal control (SL-1). This was achieved successfully. Skylab, due to its low orbit, re-entered the Earth's atmosphere and burned up in 1979. The longest US spaceflight was 84 days on Skylab in 1973/74. The Russians by way of contrast have had their Soyuz–Salyut space stations permanently in orbit since 1971 and the longest space mission was 237 days on Salyut 7. Since then, several spectacular in-orbit servicing missions via EVA have been undertaken by Shuttle crews. So far, >400 EVA man-hours have been logged (1995 figures).

Normal human breathing rate is 10–18 times per minute, each breath drawing 0.5-0.9 l of air into the lungs—deep breathing increases this by another 1.5 l. There is a residual 1.5 l of capacity in the lungs. A typical lung ventilation rate is 7.5 l/minute which can increase by an order of magnitude with hard labour. Around 20% of the oxygen inhaled is absorbed by the lungs. EVA typically involves leaks of spacecraft cabin atmosphere of around 2 kg/man-day. Altitude sickness is caused by anoxia due to decreased oxygen partial pressure at high altitude from the sealevel value of 21 kN/m^2. Above 3 km, symptoms begin to appear. Oxygen starvation at low pressure below 159 mm Hg affects the central nervous system generating headaches, drowsiness, sensory deterioration, apathy and eventually loss of consciousness. At extremely low pressure, below 6.3 kN (at 20 km altitude), bodily water which makes up 70% by mass of the human body, boils. This causes violent swelling, bleeding and death within seconds. Low partial pressure of oxygen can be increased to counter the effects of anoxia by the use of pressurised environment suits. Above 8 km altitude, spacesuits are required for the supply of oxygen for breathing and to maintain pressure around the human body to keep the body fluids in the liquid state. Similarly, oxygen partial pressure excess above 176 mm Hg must also be considered as this causes hypoxia, generating oxygen intoxication, sensory impairment, and eventually carbon dioxide retention and lung collapse. Rapid pressure changes must be avoided, especially large pressure drops, as this will cause damage to the diaphragm of the ear, gas volume expansion in the gastro-intestinal tract and aeroembolism of nitrogen gas in the tissues of the body ('the bends'). Decompression sickness occurs when people working in high ambient pressures are suddenly decompressed and nitrogen bubbles rapidly form out of blood solution, causing pain in the muscles and joints. Such a sudden decompression killed the crew of the Russian Soyuz 11 when a valve failed in the spacecraft in 1971.

The Space Shuttle EMU spacesuit (extravehicular mobility unit) of mass 131 kg provides environmental protection, life support and communications for EVA in LEO. EMUs are maintained at 4.3 psi (29.6 kPa) with near-100% oxygen compared with the Shuttle crew compartment which is kept at sealevel pressure (1033 g/cm^2) with 80% nitrogen and 20% oxgen atmosphere (similar to Earth's atmosphere of 78% nitrogen, 21% oxygen, and 1% argon, neon, etc.). The need for a similar environment to Earth in manned space missions was a hard-learned lesson—a flashfire in Apollo 1 killed three astronauts (Virgil Grissom, Ed White and Roger Chaffee) during a ground test, due to its low-pressure pure oxygen atmosphere. The EMU comprises several parts: a liquid-cooling and ventilation garment which can remove body heat up to 500 kW/h; a lower torso assembly, a hard upper torso assembly, two arms, two gloves with adjustable palm straps, communications carrier assembly with earphones, helmet with extravehicular visor assembly, adjustable visor and sunshades, a portable life support system including 0.55 kg of pressurised oxygen for 7 h and a display control module. Bearings in the shoulder, arm, wrist and waist joints allow freedom of movement. The difficulty with joints is that the major body joints can provide motions about two or more axes. If joint movements are restricted, the number of muscle groups involved in a given motion increases resulting in significant increases in metabolic rate and so decreases productivity. The life-support system is mounted at the back of the hard upper torso above which is mounted the radio communication system. The control and display unit is mounted on the chest at the front of the upper torso. The helmet has 55° up/65° down and 100° left/right angle of view through the visor. It also has light-mounting brackets for illumination, since every other 45 minutes are spent in eclipse, and the gloves have mobility to 90° wrist yaw, 120° wrist pitch, and 180° wrist roll. Nominal Shuttle EVA support hardware includes the airlock, two EMUs, a portable foot restraint, an RMS-mounted foot restraint and an MMU (manned manoeuvering unit) as well as the EVA tools and aids in the cargo bay. The MMU is a nitrogen gas propulsion unit with 6-degrees-of-freedom manoeuvrability for up to 6 hours and an automatic attitude hold capability and propulsion Δv of 22 m/s fully laden. Cargo bay aids include handrails on the starboard and port cargo bay sides, six floodlights, EVA winch, two bulkhead cameras on forward and aft ends of the cargo bay, an EVA safety tether, and several portable floodlights. Trembley (1994) outlined possible hazards to the hardware reliability of the EMU:

(i)   loss of carbon dioxide removal and cooling capability;
(ii)   external leakage of oxygen;
(iii)   suit overpressurisation;
(iv)   rupture of pressurised oxygen bottles;
(v)   loss of primary oxygen supply;
(vi)   oxygen fire;
(vii)   LiOH dust in ventilation loop;
(viii) helmet fogging from dehumidification failure;
(ix)   free water in suit;
(x)   decompression sickness.

Hazards are imposed by a hostile, radiation-filled vacuum and the possibility of spacesuit rupture due to snagging, micrometeoroid or debris impacts, exposure to cryogenic fuels such as LOX/LOH or corrosive fuels such as hydrazine or ammonia if operating near fuel tanks or lines or near cryogenically cooled sensors. The probability of micrometeroid or debris penetration of the EMU during a single two-man 6-h EVA has been estimated to be 0.0006, which, although small, is finite. Accidental RCS thruster activation is another remote possibility which could cause unexpected impact on the astronaut and/or exposure to hot gases.

Over 84% of the possible hardware failures would compromise an EVA mission, while 52% of those failures could result in compromising the crew safety. Several failures have been encountered: a puncture of the glove bladder occurred during an EVA on STS-37; both helmet fogging and free liquid in the suit occurred on EVA during the STS-41C mission. These occasions were not critical, but during the STS-5 EVA preparation, it was found that both EMUs had unrelated failures preventing the execution of the scheduled EVA. This is in fact more of a safety-critical issue since it is necessary to have two operational EMUs on board the spacecraft in case there is a need to close the payload bay doors manually prior to re-entry.

EVA is a major limiting factor for in-orbit activities, involving long preparation periods, a 7-h limit on human EVA operational duration, great expense and the exposure of astronauts to hazard [Anderson & Rockoff 1990, Rockoff & Anderson 1990]. Furthermore, Shuttle missions are limited to 7-day missions nominally though they can be extended to 21 days exceptionally. The 7-h limit on EVA operations includes 15 min for egress, 6 h useful work, 15 min for ingress and 30 min reserve [Weaver & Wickman 1993]. EVA is medically restricted to three shifts per week per astronaut. Indeed, one of the major criticisms of some of the original designs of the Space Station Freedom was that they required excessive EVA times for its construction beyond safety limits, necessitating its further redesign [Colucci 1990]. Further, for any EVA operation, only two astronauts are normally permitted outside the Shuttle at any one time, though this restriction was relaxed for the Intelsat VI repair operation when three astronauts were simultaneously in EVA mode. Pre-breathing pure oxygen reduces the amount of nitrogen in the body prior to EVA. Astronauts must spend several hours breathing pure oxygen before egressing the airlock to remove nitrogen gas dissolved in the body fluids and prevent the formation of nitrogen bubbles in the body fluids and so cause decompression sickness [McBarron 1994, McBarron et al 1994]. Pre-EVA includes equipment preparation, EMU checkout, EVA task preparation, airlock preparation prior to depressurisation and egress, donning the EMU and pre-breathe [Mickle 1993]. For 14.7 psi cabin pressure, 4 h of suit pre-breathe is required, but with 10.2 psi exposure for 24 h, a 40-min suit pre-breathe is required; with 10.2 psi cabin pressure exposure for 12 h, a 75-min pre-breathe is required. Prior to the cabin depressurisation to 10.2 psi, a 1-h mask pre-breathe is required. Usually, one of these two protocols (12 h or 24 h cabin depressurisation to 10.2 psi) is adopted. Depressurisation below this level would require an increased oxygen percentage. Normally, only two EMU's are carried on each flight, so personal rescue enclosures are available to the remaining crew in case emergency egress becomes necessary. These are 83.4 cm diameter pressurised spheres each with life support and communications equipment. Prompt evacuation would be required if life-threatening conditions arose, but it had been estimated that it could take 15–45 days before astronauts could be evacuated from the Space Station in LEO without a dedicated crew return vehicle. The cost of utilising a space shuttle as an ambulance is ~$125–250m in addition to the abandonment of a costly mission.

Post-EVA operations include EMU doffing, and EMU maintenance and recharge. The implication is that long access times render EVA unsuitable for some emergency operations where time may be at a premium. Suits with higher operating pressures would reduce the pre-breathe time but would restrict astronaut dexterity and mobility. As it is, the astronaut is restricted by hand mobility in performing fine operations which are limited to 1–2 h of continuous work due to fatigue. The tactile sensitivity of the gloved hand is reduced by the increased amount of free space between the hand and the suit at higher pressures as well as imposing a decreased maximum force capability of being exerted by the gloved hand. For the present Shuttle EMU, tactile sensitivity is reduced by ~4 to ~2 mm resolution and maximum force is limited to ~225–300 N, but it is desirable to restrict the required force to <110 N. Arm movements which use the larger muscle groups are preferred over small wrist movements to reduce fatigue. Furthermore, functional loading on the hands is very high since the astronaut is constantly concerned with using handholds and rehooking of safety tethers since all objects are required to be tethered (tethers are designed with a load limit of 334 N). Average EVA activities are comparable to cycling at 15 km/h while peak activities are equivalent to heavy physical activity. Gene Cernan on Gemini 9 was so exhausted after putting on his manoeuvring pack that he was unable to test it. The use of hand and foot restraints do alleviate the fatigue problem somewhat. Foot restraints in particular allow the astronaut to bend his/ her body to provide a proper working posture with adequate arm freedom and normal reach envelope in the frontal and sagittal planes without excessive exertion. EVA astronaut productivity is dominated by several factors [Barer & Filipenkov 1994]:

(i)   spacesuit microclimate (gas composition, pressure and temperature);
(ii)   limitation of motion activity, visibility and touch;
(iii) astronaut training for zero-g environment;
(iv)   general physical capabilities;
(v)   accumulated individual EVA experience;
(vi)   EVA duration and work rate;
(vii) tool selection.

All in all, human performance is limited by strength, vigilance, fatigue and reaction speed. Human error can lead to accidents—the US Navy mistakenly shot down a civil airline Iran Air 655 killing all 290 passengers as a direct result of human error in the interpretation of radar data. Accidents are not only unacceptable to the public, they also represent considerable costs. Attestable to the limits of human performance is that some 65–70% of civil airline accidents are caused by human pilot error and non-pilot personnel were generally involved in 44% of all air-related accidents in 1980. The accident rate is around 3.25 accidents per 100 000 hours. Most of these errors are due to limitations in vigilance and decision-making judgement, but this does not account for the avoidance of accidents through human reaction to and judgement of unexpected events [Chambers & Nagel 1985, Rouse & Cody 1987]. Fatigue alone in astronauts can cause up to ~30% of operational errors in completed operations in space. Astronauts in particular are susceptible to hiding fatigue or difficulties due their perceived self-image. The astronaut's metabolic rate is limited to ensure that excessive fatigue does not occur: average rate <1600 BTU/h in any hour and overall average for the duration of EVA <1000 BTU/h; peak metabolic rates are restricted to <2000 BTU/h. All these factors limit EVA as the only form of in-orbit servicing at present—the NASA Space Telerobotics Programme is seeking to implement 50% of EVA operations telerobotically by 2004.

3.4 IN-ORBIT ROBOTIC MANIPULATION

Space operations occur in an environment that is not well characterised in that it is subject to unexpected events which must be dealt with. General-purpose robotic manipulation represents the most varied and diverse sets of tasks that may be performed in space.

Nevins (1985) classified tasks as deterministic or non-deterministic: deterministic tasks are those driven by geometry, e.g. peg-in-hole task; non-deterministic tasks are those which require skill and involve not-well-understood process models, e.g. data interpretation. Non-deterministic tasks require either human intervention or a high degree of machine intelligence with adaptive learning capability. Deterministic tasks may be pre-programmed but may also require adaptive capabilities. The Robot Institute of America defines a robot as: 'A reprogrammable, multifunctional manipulator designed to move materials, parts, tools or specialised devices through variable programmed motions for the performance of a variety of tasks'. This definition is a little narrow and refers specifically to industrial robots which have limited capabilities. The Japanese Industrial Robot Association went further and classified robots into a series of classes based on their capabilities:

Class 1—teleoperated robots
Class 2—fixed sequence robots
Class 3—variable sequence robots
Class 4—teach-playback robots
Class 5—computerised numerically controlled robots
Class 6—adaptive/intelligent robots

Class 6 is in fact more properly subdivided into two categories:

(a) adaptively controlled robots
(b) autonomous robots employing artificial intelligence methods

A robot may be defined as [Nitzan 1985]: 'A general purpose machine manipulator that can perform a variety of difficult tasks in remote hostile conditions not necessarily known *a priori* by using an intelligent control system that observes the environment and takes appropriate autonomous actions to suit that environment and those task objectives'. This corresponds to Class 6 of the Japanese classification of robots and is partially realisable to a certain degree using today's technology and it is to this definition that space robotics should be working to attain. In-orbit servicing and planetary exploration are not highly structured tasks: they require effective reactive responses to unexpected events.

In-orbit manipulation will provide a good indicator of the feasibility of in-orbit manufacturing platforms in general—space-based manufacturing techniques lie in the future but their development will ultimately lie in the lessons learned in general-purpose task performance in zero gravity. Nitzan (1985) divided robotic manipulation tasks into four categories:

(i)   material handling;
(ii)   parts fabrication;
(iii) inspection;
(iv)   assembly.

Material handling (i.e. transport) represents the simplest application. Parts fabrication represents the commonest use of robots in industry with half being used for welding, mostly spot welding. Assembly represents the most complex application of robot manipulators and it is this that most space-based robotic devices will be concerned with. Here, we are specifically concerned with in-orbit servicing, maintenance and repair of spacecraft, with truss assembly and construction and astronaut rescue and retrieval being regarded as secondary functions. Retrieval of spacecraft is unlikely to be economic, but complex major failures to expensive satellites may require return to Earth for refurbishment. Broadly, in-orbit operations may be defined as:

(i)   construction operations involving in-orbit assembly;
(ii)   maintenance operations including routine calibration of equipment;
(iii) repair functions to meet mission objectives in the face of failure or damage to
       equipment;
(iv)   routine servicing by replenishment of expendables or installation of upgraded equip-
       ment [Siedman 1992].

Elfving (1990a,b) and Schroer (1988) highlighted a series of reference tasks:

(i)   RT1—assembly of trusses;
(ii)   RT2—mating of connectors;
(iii) RT3—exchange of ORU equipment box;
(iv)   RT4—removal of thermal blanket/panels;
(v)   RT5—operation of power tools;
(vi)   RT6—local inspection by sensors for faults.

Observation and inspection of spacecraft, subsystems or components for flaws, defects, etc. due to thermal stresses, micrometeoroid damage, gas leakages, fatigue cracking, etc.—such inspection constitutes the first phase of any servicing activity. Inspection is particularly man-intensive and monotonous so it is well suited for automation. Terrestrial NDT (non-destructive testing) is usually performed using ultrasonic techniques because they provide good depth capabilities, but space inspection would have to use other techniques: eddy current systems have little depth penetration and are limited to surface defects; the other alternative is X-ray radiography which has good depth capabilities, but has a high initial cost. In-orbit transportation of spacecraft for retrieval is also a critical capability and the in-orbit servicing process starts with grappling objects and docking fixtures on spacecraft. This may involve passivating tumbling about three rotational axes by momentum nullification through momentum exchange via the dual-spin turn for instance [Kaplan 1976]. All the more complex manipulation tasks will require certain basic capabilities: opening and closing access panels, doors and covers; operating mechanical connections such as latches, bolts, cranks, screws, plugs etc. and operating electrical connections such as connectors, soldering, replacing faulty PCB's, etc. 'Legged' locomotion on space truss structures is another potential capability. Dual grippers enable the robot to attach itself to threaded holes in the truss with walking accomplished by alternate grasping and releasing of the nodes by the grippers and swinging the arms from one node to the next. This requires sufficient span between the two grippers. The end effector must be able to grip to the nodes firmly as the robot platform shifts from one end effector to the other for support.

Two fundamental tasks for in-orbit servicing are: (i) fluid connector interchange; (ii) ORU interchange. The mating elements in both cases must be within the FOV (field of view) of any camera mounted on the arm so that guiding targets are visible. The ORU exchange task is treated later (as part of the Solar Maximum Repair Mission example).

Replenishment of consumables such as hazardous fluids like hydrazine to extend mission durations, eg. fuels, cryogenic fluids by fluid umbilicals or battery unit exchange, are critical capabilities. Fluid connector interchange represents a complex manipulation problem by virtue of the fact that it involves fluids and flexible elements [Abidi &

Gonzalez 1990, Abidi et al 1991]. The requirement is to mate and demate a fluid connector. The alignment/locking mechanism comprises a nozzle to be inserted into a receptacle mounted on the target satellite. The nozzle and receptacle are essentially male and female BNC connectors. The nozzle is cylindrical with two steel pins located 180° apart extending 0.5 in. It has a flat parallelepiped end acting as an attachment point for the robot end effector. The nozzle is 4.75 in long with inner and outer diameters of 1.25 in and 1.75 in respectively. The receptacle has a flared rim with inner and outer diameters of 2 in and 2.9 in respectively. It has two V-notches 180° apart in its rim leading into two grooves that lock the nozzle into position once it is inserted. The robot picks up the nozzle and locates the receptacle. The robot arm inserts the nozzle with the flared rim guiding the receptacle by generating corrective forces on the nozzle. As the nozzle is inserted, the pins enter the V-notches which align the nozzle. When the nozzle reaches the bottom of the V-notches, the robot then rotates the nozzle 15° clockwise which allows the pins to enter the grooves in the receptacle.

Now a real example of in-orbit servicing missions is considered. Although this is one of the earliest sophisticated repair missions, it illustrates many aspects of in-orbit servicing. In manual assembly sequences, line balance is achieved according to defining manual worker division of labour with similar timescales. A similar approach is valid for EVA tasks. In contrast, for robotic machines whose performance is different from humans, the assembly sequence must be defined differently. However, initially in-orbit tasks will be performed by human operators through teleoperation so EVA sequences in terms of human line balance are of direct relevance to robotic assembly until full automation has been achieved.

3.4.1 Solar Maximum repair mission (1984)
The STS 41-C Solar Maximum mission in April 1984 has become a standard for space robotics and *in situ* servicing. The 2.3-tonne astronomical satellite had been launched in 1980 into a 570×563 km orbit, but soon after orbit injection, three fuses blew in its attitude control system. Simpler EVAs had been performed such as the STS-37 mission when the two astronauts successfully manually freed a stuck antenna on the Gamma Ray Observatory satellite. The satellite would have had to have been returned to Earth otherwise. However, the Solar Maximum repair mission was significantly more complex. The intention was to demonstrate the feasibility of in-orbit repair of satellites to justify the manned space station concept as a maintenance base. The original space station architecture was to have a central set of habitation and laboratory modules with several unmanned co-orbiting platforms to be periodically visited by astronauts. The platforms would enable microgravity experiments to be performed without interference from crew movements. Such an architecture would require extensive EVA.

The Solar Maximum satellite was based on NASA's standard Multimission Modular Spacecraft (MMS) designed for in-orbit servicing. In general, the MMS comprises four separate modules weighing ~225 kg each: the Attitude Control Subsystem (ACS), the Communications & Data Handling (C&DH) Module, the Modular Power System, and the Propulsion Module. Each is externally a $1.4 \times 1.4 \times 0.5$ m box with the exception of the Propulsion Module which being optional was not installed on the Solar Max satellite. Each module is fastened to the spacecraft with two 2 cm hex bolts requiring 135 Nm torque for removal/installation. The Attitude Control Subsystem of the Solar Maximum satellite failed after 9 months of its 2-y operational life suffering a loss of yaw and pitch control rendering the $235m spacecraft unable to point to solar flare locations on the Sun. Curiously, one of the astronauts, attempting to capture the satellite manually using an MMU, caught it by a solar array, setting it tumbling. Only after the ground crew uploaded software corrections was the spacecraft detumbled to ~0.1 m/s and retrieved by the RMS. This illustrates the difficulties in capturing bodies under irregular motion even by human action. The total mission took two EVAs of two astronauts each. This operation has all the major features characteristic of satellite repair and it cost $77m (nominally—in reality, the cost of the Space Shuttle mission launch and the one year of neutral bouyancy training for the astronauts would far exceed this). Two major repairs were required: replacement of the satellite's attitude control ORU module—this ORU exchange was a designed-for-servicing task [Davis 1987]; the electronics box exchange, however, was not designed for servicing [Adams et al 1987]. The third operation was to install a Be-Cu deflector baffle near the propane exhaust port to eliminate plasma interference of the X-ray polychromator. This operation took only a few minutes to complete and is not considered further.

(1) MMS ACS exchange
Two astronauts, one riding the RMS foot restraint operated by a third astronaut replaced the MMS ACS using the EVA MST (Modular Servicing Tool) powered by an Ag–Zn MMU battery to clamp to the ORU and allow tightening and loosening of the bolts. The MST is a power tool comprising a 2-cm hex socket with 135 Nm torque capability to turn the two MMS bolts on two opposite vertices of the MMS outer face. Prior to this, the removable panel covering the MMS module had to be removed. The single blind-mate connector for electrical interface on the module mates perfectly when the two bolts are fastened. This procedure took 35 minutes for EVA, 15 minutes by laboratory teleoperation and 40 minutes by laboratory automation.

ORU module exchange is highly structured since all objects will probably have known dimensions, locations, masses, moments of inertia and positions of grapple fixtures. The procedure has been performed both teleoperatively and autonomously in the laboratory. Bronez et al (1986) characterised the robotic in-orbit servicing task of the ORU exchange:

1. Isolate the old module.
2. Open and disconnect the thermal interface blankets and secure.
3. Acquire the power tool.
4. Latch the power tool to the old module captive bolts and loosen the bolts with the power tool while holding the cover: (a) stow power tool; (b) remove and stow the cover; (c) release the old module retention clamps.
5. Detach the old module.
6. Stow the old module to the robotic spacecraft bus (for controlled re-entry disposal or for return to Earth for refurbishment and subsequent re-use).
7. Find the replacement module.
8. Orient the new module for insertion.

9. Install the new module to the spacecraft: (a) tighten the module retention clamps; (b) acquire the cover; (c) acquire the power tool; (d) tighten the captive bolts with the power tool; (e) test the bolts; (f) stow the power tool; (g) release and grasp thermal blanket, then drape and fasten the blanket.
10. Inspect and wait for the new module checkout.

Backes & Tso (1990) subdivided the single-arm autonomous ORU changeout sequence into subtasks. One of these subtasks, the dual-pin insertion/removal subtask required a decision-making strategy with compliant control to prevent jamming. They used a stiffness-type outer force control loop which modified the position trajectory. The subtasks used the RCCL (Robot Control C Library) and all were used more than once during the ORU changeout. The bolt was assumed to be located at the centre of the lug. The ORU exchange programme used subtask components to first unscrew the two bolts on the two opposite vertices of the outer face of the ORU. Then the ORU was changed out on unlatching them.

The Bolt sequence to turn the bolts with repeated 60° turns:

1. Grapple-lug-acquisition subtask to locate the lug position.
2. Angled-bolt-seating subtask to set the socket onto the bolt.
3. Bolt-turning subtask about tool z-axis by 60°.
4. Compliant-move subtask along tool z-axis to position above bolt to reduce contact forces.
5. Guarded-move subtask rotation of −60° about tool z-axis to return tool twist.
6. Vertical-bolt-seating subtask to seat the socket onto the bolt.
7. Repeat 3–6 until desired tool turn is attained.
8. Compliant-move subtask along tool z-axis to position above bolt and reduce contact forces.

The ORU changeout sequence was as follows with ORU replacement following a similar pattern:

1. Grapple-lug-acquisition subtask to locate lug position.
2. Guarded-move subtask to mounting region.
3. Move-to-touch subtask at 4 mm/s to contact between fingers and lug until the forces lie between the backoff force of 1 N and the threshold force of 5 N.
4. Level subtask to match finger surfaces on lug surface by reducing moments.
5. Compliant-grasp subtask to grapple ORU by closing the fingers to 60 psi.
6. Dual-pin insertion/removal subtask to remove the two-pin ORU from two-holed mount interface.
7. Guarded-move subtask to insertion point on stow platform.
8. Move-to-touch subtask at 4 mm/s to contact between ORU pins and stowage holes until the forces lie between the backoff force of 1 N and the threshold force of 5 N.
9. Dual-pin-insertion/removal subtask to insert the two-pin ORU into the two-holed stowage interface.
10. Compliant-ungrasp subtask to release ORU by opening fingers.
11. Compliant-move subtask to move away from ORU by reducing the contact forces.

As can be seen, the ORU exchange task is a fairly complicated, though well character-ised, task. Bruhm (1987) declared that the sensitivity of the single arm to disturbing forces due to the lever arm between the ORU centre of mass to the grapple points is high. Positioning/orientation errors arising from this may be as much as ~3–5 times higher than for the unloaded arm. The use of two arms significantly reduces the maximum compliance values.

(2) Main electronics box (MEB) exchange
The Solar Maximum repair mission astronauts followed a 31-page document to replace the chronograph/polarimeter main elecronics box using EVA hand tools and three Ni–Cd battery-powered screwdrivers, one with an Allen drive for retaining-cap screws, one with an Allen drive with a screw capture shroud, and one with a slotted driver blade. Other EVA tools included the MEB hinge assembly, panel support bracket, electrical connector removal tool, electrical connector installation tool, Essex ratchet, and an assortment of simple tools such as cutters and tape. The MEB hinge assembly and panel support bracket were both unique tools designed specifically for the SMM repair. The task involved 13 steps:

(i)   cutting the plastic Kapton tape holding the thermal blankets;
(ii)   folding, unfolding and taping the thermal blankets;
(iii)   installing the MEB panel hinge assembly;
(iv)   removing 14x MEB panel 10-32 non-capture screws and re-installing 4×10–32 captive cap crews;
(v)   installing the panel support bracket;
(vi)   removing 22×4–40 slotted head connector screws;
(vii)   removing 11 subminiature electrical connectors;
(viii) cutting the plastic tie wraps holding the wiring harness;
(ix)   removing the MEB from the hinge assembly and reinstalling the new MEB onto the hinge assembly;
(x)   reinserting the connectors using a three sided guide and a pair of spring clips at each end of each connector;
(xi)   removing the velcro straps from the thermal blankets;
(xii)   using the velcro straps as cable ties;
(xiii) replacing the thermal blankets and sealing using tape.

This procedure took 2h of EVA compared with the neutral buoyancy simulation time of 3.5 h. The remote MEB repair sequence in the lab was only possible using dual manipulators as some operations required parallel execution while others required sequential execution using both arms. Laboratory dual arm teleoperation took 3 h—automation of the task has yet to be attempted and it is unlikely to be feasible for the foreseeable future, although task sequence planning has been generated and simulated [Sanderson et al 1988]. This was the most difficult operation. In remote laboratory teleoperation, handling the thermal blanket and cutting tape were the most difficult processes while the removal of the connector crews and reinstalling the electrical connectors were much quicker in teleoperation than in EVA (38 min versus 60 min, and 27 min versus 55 min respectively). Hence, the handling of flexible extended objects appears to require complex manipulation capabilities. All in all, the total EVA time was 7 h.

Since the Solar Maximum repair mission, several major repair or retrieval missions have been undertaken. Palapa B2 and Westar 6 C band communications satellites were both retrieved with two EVAs and taken aboard the STS-51A Shuttle after their upper launch stages failed in 1984. Both satellites were outside rendezvous capability necessitating their use of onboard hydrazine to circularise their orbits and alter their orbital parameters. Similarly a series of de-spin attitude manoeuvres enabled the astronauts to berth them and return them to Earth for reconditioning and relaunch in November 1984. The Syncom satellite repair in September 1985 required two EVAs (STS-51D) which failed to activate the satellite time sequencer. The Challenger disaster effectively stopped all Shuttle operations for several years. The STS-49 4.5 tonne dual-spin Intelsat VI reboost in May 1992 required four EVAs with three astronauts, the longest EVA being 8.5 h in duration during the Intelsat VI capture. This satellite was launched in 1990 and could not achieve operational orbit because the Titan second stage failed to separate the satellite. The satellite perigee kick motor separated the satellite from the Titan by ground control command and was boosted to a stable orbit. This necessitated the addition of a new propulsion unit to place the satellite into its GEO orbit. STS-49 was also the mission on which two EVA astronauts demonstrated the feasibility of constructing a 15×15 ft truss structure. The Eureca satellite was retrieved with one EVA in June 1993 (STS-57).

The STS-61 Hubble Space Telescope (HST) repair mission in December 1993 required five in-orbit EVAs, 3.5 years after its launch to enable it to continue its 15-y life. The 11-tonne HST was designed to be serviced from the Space Station with two 2-hour EVAs every 22 months. The scientific instrument modules had been designed so that they could be removed and replaced by EVA as technology advanced but unfortunately not the mirrors of the telescope. HST has a stiff pointing requirement of 0.01 arcsec accomplished by its combination of six rate sensing gyro assemblies, three fixed-head star trackers, four gas-bearing reaction wheels for attitude control and magnetorquers for momentum dumping of gravity gradient torque accumulation. Details of its design are outlined by Wojtalik (1987). It was found that the HST 2.4-m main mirror was incorrectly shaped causing spherical aberration. Rather than focussing 70% of the incident light to its 0.1 arcsec radius resolution limit, it was focussing only 15%. This 70% was specified as the minimum scientific requirement set by the point spread function of a typical star. NASA's Space Shuttle Endeavour was selected as the newest shuttle with its superior fuel reserves to reach the high orbit of 310×297 km imposed by the requirement for minimal interference from the Earth. The astronauts were prepared for the mission by 400 h training in the neutral buoyancy tank at MSFC (Marshall Spaceflight Centre at Huntsville) and JSC (Johnson Space Centre at Florida). The mission lasted 10 days with five EVA excursions averaging over 7 h each. Besides installing the corrective optics, other replacements were installed: two of the six Goddard/Lockheed/Fairchild DF-224 coprocessor memory units had failed and three of the six rate sensing gyros had failed, all of which required replacement. The solar arrays also suffered thermal shock during the day–night terminator crossing causing spacecraft vibrations and so required replacement. Prior to EVA, the cabin pressure in the Shuttle was reduced from 14.7 psi to 10.2 psi to reduce pre-breathe periods. The HST was grappled by the RMS and berthed in the payload bay. The first EVA lasted 7 h 54 min and the EVA team successfully replaced three rate sensing gyros. The second EVA involved replacing the two solar arrays. The third EVA also involved the RMS for the installation of the corrective optics and DF-224 coprocessors. Finally, the fourth and fifth EVAs had the EVA team replace the solar array drive electronics. All in all, the whole mission involved a total of 35 h 28 min EVA. The total cost of the mission was $700m. HST was corrected and upgraded to achieve its original performance.

The need for in-orbit servicing has been demonstrated and the complexity of potential tasks that are required has been outlined. Performance of such tasks robotically rather than by EVA would be a considerable asset to space missions in the next millennium.

4

Man–machine interface

In this chapter, we look at the interface between man and the machine with reference to the ground station. It is perceived that all in-orbit servicing and planetary rover exploration will be telerobotically controlled from an Earth ground station—this may be regarded as an extension to standard tracking telemetry and commanding functions provided by a standard ground station. This is a critical component in any space robotic system as it effectively defines the task performance of the remote robotic system in space or on a planetary surface.

4.1 GROUND STATION MISSION OPERATIONS

Mission operations strongly drive the spacecraft mission lifecycle costs, especially over a number of years. It includes spacecraft operations and ground support from prelaunch, launch, early orbit checkout and normal operations and finally decommissioning. It also includes personnel training (e.g. simulators for teleoperative crews) and planning of orbit manoeuvres (e.g. simulations of scenarios), flight software maintenance and general engineering support. The ground system controls and monitors the spacecraft and its payload by transmitting command instructions and data using the received spacecraft telemetry and mission data. The ground station is responsible for the maintenance of the communications link, tracking functions and dealing with anomalies through workarounds. The ground segment usually comprises a series of ground stations and a mission control centre (MCC). The LEO pass time is limited since the maximum viewing angle of any LEO spacecraft from any point on the Earth's surface even with steerable tracking <20% of maximum of the orbital path (this increases to ~30% of maximum for inclinations above 30°), i.e. LEO spacecraft have a maximum viewing time of ~12 min per 95-min orbit. One way to overcome this is to use packet tracking, telemetry and command (TT&C) a few times a day and adopt increased autonomy to reduce the need for ground stations—this simplifies the ground system component complexity and increases the reliability since performance is not dependent on the Earth link, but the communications uplink may be lost at a critical time. Such packet telemetry was adopted for Eureca with its 7-min coverage from its 90-min orbit. Telemetry is recorded onboard and dumped at high speed during the ground station pass. Loss of signal is an emergency event—indeed, this occurred to Mars Observer in 1993 while it was approaching Mars and was lost. Cassini, however, has two radio receivers and its onboard software invokes emergency procedures when commands have not been received for a defined period. It automatically switches to the backup receiver. Generally if high coverage is required, auxiliary stations are required to provide full tracking, telemetry and command. Such geographical dispersion of ground stations requires synchronised timing systems of great complexity. The maintenance of a large number of fixed ground stations is costly, though most are in fact shared with other satellite systems, e.g. ESA operates a network of ground stations in Sweden, French Guiana, Germany, Belgium and Spain as well as ESOC (European Space Operations Centre) in Darmstadt, Germany. Sharing facilities introduces problems of coordination and priorities because it is desirable to use an existing facility rather than building a new one.

The basic ground station comprises an antenna with steerable mount and autotracking. The antenna operates in both receive mode to accept downloaded telemetry data for phase demodulation and in transmit mode to upload carrier modulated commands. The antenna is the primary active part of the ground station. The antenna gain determines its efficiency in detecting the messages being received. Much of the process is computer-controlled. Mission data is recovered from phase demodulated telemetry through buffers and synchronisers. On the basis of this data, tracking information is generated and

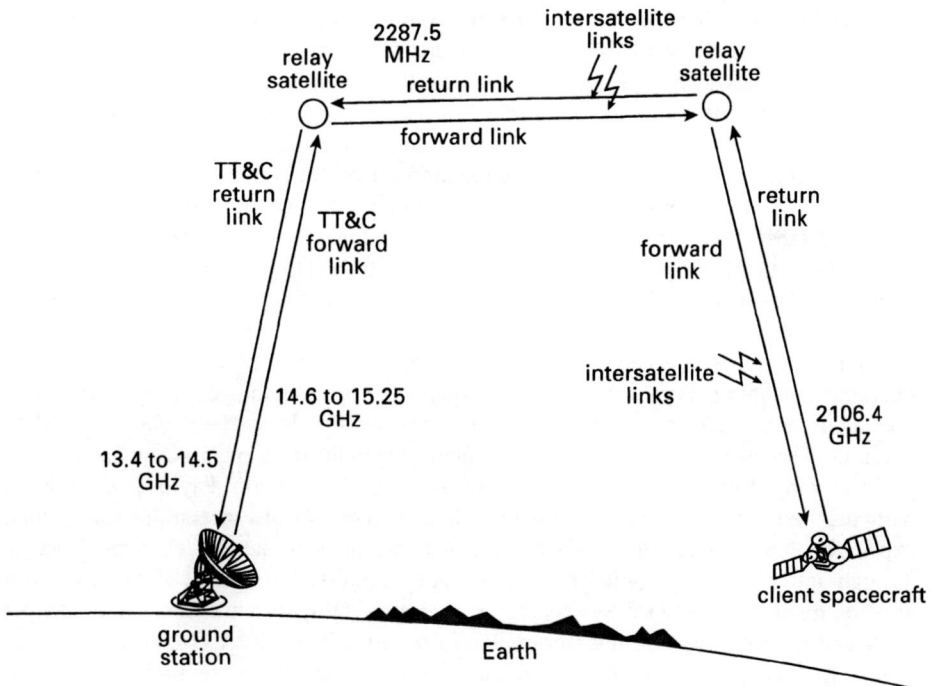

Fig. 4.1. Spacecraft/ground communications architecture (adapted from Barter (1999)). Reproduced with permission from TRW Space & Electronics Group.

commands issued for transmission. The ground station simulates tasks to be executed on its flight dynamics software. It implements a real-time simulator of spacecraft hardware, its environment and mission. The payload operations control centre (POCC) is responsible for payload operations while the spacecraft operations control centre (SOCC) is responsible for the spacecraft bus operations and usually some form of coordination is required between the two. The mission control centre (MCC) is responsible for monitoring, operating, configuring and scheduling the space mission for both ground and space segments. It receives telemetry on the health of the spacecraft and transmits telecommands to the spacecraft to control the mission. Typical MCC software amounts to 650 000 lines of code for mission sequencing, 1 615 000 lines of code for telemetry interpretation and 550 000 lines of code for GNC (for the Galileo example). It is important that for the period of critical spacecraft operations that the MCC is effectively dedicated in that 100% of the resources required to support the mission are available. Single ground stations with a single antenna limits the system to only one spacecraft at a time but deployment of several antennas enables support of more than one spacecraft link simultaneously. Employing some small mobile ground station terminals provides for directly interfacing with the customer ground station at low cost. Claros (1992) described a mobile off-the-shelf integrated TT&C system CLEO which requires colocation via LAN with the operations control centre. It requires six personnel per CLEO system. SCOS II is a next-generation spacecraft mission control and operations system developed for installation at ESOC to enhance the flexibility of such systems at reduced maintenance costs [Howard 1995]. It can support telemetry rates of 2–3 Mbps and video rates at 10 frames/s. It is designed to support the anticipated increased complexity of long-term spacecraft operations until 2010. It uses packetisation of telemetry and command and is event-driven. It has a reactive component of alarm raising in the event of limit violations and the generation of automated corrective actions. The Envisat Centre Monitoring & Control system uses AI (artificial intelligence) techniques such as constraint satisfaction for mission planning [Grew & Jones 1999]. It runs on two standard workstations. A plan typically covers 48 h with 5000 directives generated automatically which can be modified by the human operator. A directive is a high-level schedule constrained by the other directives. Search by constraint satisfaction ensures that all schedules are consistent. The mission planning segment uses a knowledge base search for operations planning. The flight dynamics component predicts the geometry of the flight operations and manoeuvre requirements to generate subplans. Heuristics are used to guide the knowledge base searches, e.g. battery reconditioning must occur before solar panel stowing. Remote access is becoming increasingly common. Payload operators in particular require access to onboard experiments. Traditionally accomplished through expensive POCCs with dedicated communications links to the MCC, remote access through internet technology offers a far cheaper alternative [Hoag et al 1999]. There is a growing trend towards Open Systems Interconnection (OSI) systems such as the Ethernet LAN and the internet. The internet evolved from the US DARPA (Defence Advanced Research Projects Agency) for exchanging military research information between researchers at different sites. In 1984, it became NSFNET providing access to educational facilities. No single group controls the internet as a loose organisation of computer networks across the world. The internet is thus a network of networks. A network is a series of computers linked together to share the same data and software. In a LAN, all the PCs are connected to a central server computer workstation. This client/server model of networks enables workstations on the network to share common data and resources by using a common set of communications protocols. The server is capable of multi-tasking at high speeds and scheduling network resources. User workstations are connected to servers in LANs such as Ethernet. Component networks may be WAN-based by modem connections through the telephone system. Each network has a gateway to other networks. Most internet computers use UNIX or NT operating systems which transmit messages through an internet network communications protocol (TCP/IP). IP addresses and delivers data to the correct destination in the form of data packets by the most efficient routes. TCP divides the data into the packets for transmission. The Worldwide Web (WWW) is a global hypertext system, a network interlinked and cross-referenced via multimedia documents. WWW was developed by the European Laboratory for Particle Physics (CERN) in 1990 as an internet front-end. WWW can support text, voice, video and images. Such multimedia technology has important implications for communications in the future including in the context of spacecraft. The internet protocol (IP) provides a low-cost integrated global multimedia network for the delivery of data and video streams between the TDRSS (Tracking & Data Relay Satellite System) White Sands ground station in New Mexico, ESOC (European Space Operations Centre) in Germany, the Marshall Spaceflight Centre Payload Operations Centre in Alabama and science experiment users. However, although this may be suitable for packet communications with scientific experiments, this is unlikely to become the case for continuous TT&C envisaged for robotics/rover payloads which will be implemented from dedicated ground facilities in direct communication with the robotic system.

For a robotic freeflyer system such as ATLAS, the payload is tightly integrated with the spacecraft bus so there is no requirement for separate SOCC and POCC and these may be merged with the MCC into a single ground control station. This eliminates any coordination problems between multiple ground sites (as well as cost). A complex spacecraft is defined as one which exhibits some of the following properties: 300+ elemetry data points, multifunctional capability, three-axis stabilisation, operator interaction, onboard processing, precision pointing, real-time operation and distributed computing architecture. ATLAS fulfils all these criteria and spacecraft complexity generally determines the size of the operations support team. However, by using extensive ground-based automation, the operator workload and operator numbers can be reduced. This essentially means robotic automation using AI (artificial intelligence) methods implemented on the ground.

4.2 ROBOT PROGRAMMING LANGUAGES

A basic requirement for robot manipulators is the need to communicate with them to command them to perform the required tasks. Ideally, this would be in the form of verbal commands using natural languages such as English but automatic speech recognition is not yet well enough developed for colloquial English. So the means of communication is usually through a structured programming language for the man–machine interface. The programming language is defined by finite, discrete set of symbolic primitives. The program's symbolic primitives directly map into the structural switching primitives of the machine. A task is essentially defined by a computer program comprising a sequence of commands. Sensor-based manipulation requires such a textual language. Traditional methods of programming through teach-by-showing and replaying cannot cope with sensor feedback and error recovery. Sensing is absolutely essential in order to characterise a spatially complex 3D world and to control intricate mechanical manipulation which requires an attention to detail unknown in any other computer application. A typical robot sequence is:

(i)   identify grasp point;
(ii) move end effector to grasp point;
(iii) grasp;
(iv) test grip.

Bonner & Shin (1982), Gruver et al (1984) and Lozano-Perez (1983a) provided reviews on robotic programming languages and stressed the need for high-level task-oriented languages as opposed to existing lower-level robot-oriented languages. Robot-oriented languages are sequences of robot operation commands and at present include the majority of robot programming languages, e.g. VAL, AL. They are essentially extensions of traditional computer languages. They have a basic 'MOVE (object) TO (destination)' command with segments defining kinematic requirements. A language processor then translates the higher-level command into a lower-level one suitable for the trajectory generator and controller. AL is a Pascal-based language with similarities to ALGOL and was developed at the Stanford Research Institute (SRI). It is also capable of controlling multiple robot arms for parallel and cooperative manipulation. RPL was also developed at SRI based on LISP with similarities to FORTRAN and is implemented as subroutine calls. VAL was developed for the Unimation PUMA and is founded on BASIC. It employs its own operating system. These robot programming languages require extensive pre-planning.

Object- or task-oriented languages consist of a sequence of goals or subgoals as assembly sequence operations to be performed on the object with a construction of the form: OPERATION (object1, object2), e.g. PLACE (object1) ON (object2). These are automatically translated into robot level commands. AUTOPASS (AUTOmated Parts Assembly SyStem) is one such language [Lieberman & Wesley 1977]. AUTOPASS was developed by IBM for assembly operation and concentrates on the assembly sequence and uses formalised English-like command statements embedded in a subset of PL/I language. A planner (automated or human) plans the overall assembly operation as a sequence of high-level assembly operations. It requires a world model database of a geometric nature for the objects involved, their locations, physical data about them and their attachment relationships. A suitable format would be CAD/CAM data. At compile time, the compiler runs the world model with the program sequence to test it as well as translating the AUTOPASS source code into a robot-oriented target code. Example statements are OPERATE <tool>, ATTACH <fastener> TO <attach point>. Unfortunately AUTOPASS has not yet been implemented since it requires complex support facilities which have yet to be developed. Bonner & Shin (1982) assessed this language as superior

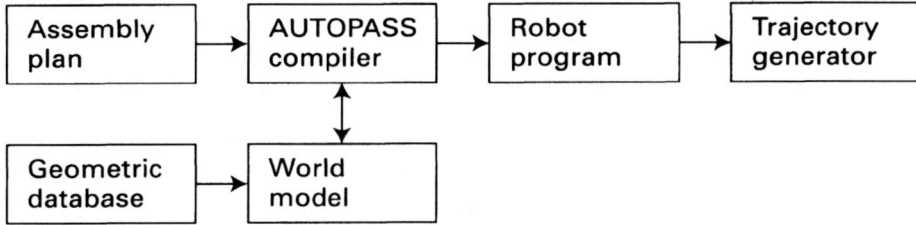

Fig. 4.2. AUTOPASS sequence.

to the others evaluated due to its decision-making capabilities and its interactive ability with world model objects for use in trajectory planning.

Bonner & Shin (1982) defined certain desirable characteristics of a good programming language:

(i)   clarity, simplicity and unity;
(ii)   clarity of program structure;
(iii) naturalness of application;
(iv)   ease of extension;
(v)   debug and support facilities;
(vi)   efficiency.

Additional desirable characteristics for specifically robot programming languages were:

(vii) decision-making capability;
(ix)   interaction with external devices and sensors;
(x)   interpreter code for greater interactivity;
(xi)   concurrent operation of devices;
(xii) interaction with world modelling systems.

The disadvantage of robot programming languages is that they are restrictive. Alternatively a general-purpose, high-level computer language such as FORTRAN or C may be used to develop hierarchies of robot function subroutines to provide robot programming capabilities. This would reduce programming costs and enhance portability: RCCL (Robot Control C Library) is a major step in this direction though it is applicable only to servocontrol levels. RPL is another such segment of special-purpose subroutines that are called by a general-purpose program. There are two types of supervisory control—supervised autonomy and shared control [Backes 1993]. Supervised autonomy allows the simulation and modification of task commands prior to being sent for actual execution. However, as the remote environment cannot be known exactly *a priori*, bounded behaviour execution is necessary to autonomously cope with unexpected perturbations through reflex action. Command sequences are generated and uploaded to the remote robot. The operator interface allows the operator to specify those commands for task planning. Such commands include a library of generic movement skills—Status Request, Cartesian Guarded Motion, Joint Guarded Motion, Move-to-Touch, Single-Arm Generalised Compliant Motion, Dual-Arm Generalised Compliant Motion, and Grasp. Task description and sequence generation is provided by the User Macro-Interface (UMI) which adopts an 3D iconic graphical interface which may be wireframe, transparent or solid with a hierarchical menu system for command selection. A sequence can be simulated with graphics overlay to verify the geometric model of the environment. A typical task would be ORU exchange. Backes et al (1994, 1998) introduced MOTES (MOdular Telerobot task Execution System) as a remote site robot control system to implement supervisory control on a Silicon Graphics VGX Power Series workstation. It used a command interpreter which is a limited robot command language. This allows a fixed flight software system to exhibit a wide range of robot behaviours. It also provided a stereo graphics overlay on a video monitor with an interactive update of a model of the remote environment. This approach was used on Galileo. MOTES supports both supervised autonomy and teleoperation. Its functionality is similar to the PRIM level and some of the servo-level functions of the NASREM architecture (see later) with a shared memory accessible to all other modules which have specific input–output parameters.

4.3 AUTOMATED SPEECH RECOGNITION (ASR)

Speech recognition is one mode of communication that is presently being developed to enhance man–machine interfaces. Speech recognition systems match a transform of speech input to a set of stored representations. Speech is one of the fastest ways to communicate with the advantage that it leaves the hands and eyes free for other tasks and can be used in the absence of visual input [White 1976, 1990, Rudnicky et al 1994]. Consequently, this mode of communication is highly desirable for adaptive man–machine interfaces which adapt to the user rather than the user having to adapt to the interface, e.g. in telerobotics [Norci & Stanley 1989]. Natural language as an information-processing activity of great complexity is the most flexible form of dialogue between the user and the system (for the user). A natural language interface must be more robust than a command language. The computer system should be able to adapt to the user by compensating for weaknesses and offering help to decrease the user's mental and physical workload. Natural languages have the advantage over structured command languages in that no learning of special vocabulary or artificial command syntax is required for communicating with the system and the computer system itself transforms the natural language into a formal language for interpretation, i.e. using a compiler/interpreter to associate a valid function template with natural language pattern input.

Natural language processing involves multiple processing tasks—phonetic analysis processes sounds, syntactic analysis examines how words combine to form sentences through grammatical rules, semantic analysis processes meaning of words for interpretation, and pragmatics is concerned with the contextual effects of language on agents. This is a hierarchical process. Natural language processing is similar to pattern recognition but is less well-defined and relies on statistical models of the regularities of syntax of the language. The essential requirement for realistic automatic speech recognition is to translate acoustic information into computer language commands.

Speech is produced in the vocal tract which may be modelled as a resonating cylinder open at one end which emits variable amplitude and frequency. Manipulation of this airstream by the larynx and the tongue provide the basis for distinguishing vowels and consonants. Information is encoded in the temporal patterns of the sound. The initial stage in speech recognition is the use of transducers to convert sound waves into an electric signal. Speech processing signals are typically filtered in the bandwidth 100–3200 Hz and then digitised at 16 kHz. Frames are constructed ~20 ms in length separated by 10 ms. The measured acoustic pattern represented as spectral properties is then matched against internal spectral models stored in memory, e.g. word templates and/or sets of grammatical rules. This matching is usually a Markov process. The hidden Markov model is a finite set of states each associated with a probability distribution. English would require some one thousand mapping rules to cope with pronunciation variations. All speech sounds may be represented as a linear superposition of sine waves of different phases. Consonants in speech carry most of the intelligibility while vowels possess most of the energy as they tend to be louder. Consonants also tend to be at higher frequencies than vowels by virtue of the constriction of the airstream. The most and least intense speech sounds can differ by as much as 50 dB. A power spectrum analyser (equivalent to the cochlea) processes the signal by measuring the relative intensities of each component sine wave (since phase is used for direction-finding rather than providing acoustic information). This provides the parametric reformulation of the speech sounds. Speech is a wideband signal (particularly in the 1–3 kHz range) modulated in three ways: (i) vocal cord frequency; (ii) pitch duration; and (iii) frequency modulation of the sound spectrum. Speech may contain redundant information which may be removed through data compression to reduce the computational load. One technique of data compression is formant tracking whereby formants are acoustic waveforms generated by the human vocal system: the major peaks in the multi-harmonic power spectrum (formances) are produced in the vocal cavity as resonances and most of the information content in speech is contained in three principal formants and these can be interpolated. Consonant/vowel combinations contain the most rapid formant transitions. These are the acoustic cues which mark specific phonetic features—second and third formant transitions are the acoustic cues to the position of articulation of the initial plosives. The speech amplitude spectrum over time contains peaks and valleys. Such fluctuations arise from the response of the vocal tract to excitation. Vowels have the greatest amplitude forming the core of syllables whereas amplitude is minimal ~15 dB in the consonant region between vowels due to restriction of the vocal tract. Each contrast produced by consonants gives a number of different patterns of spectral shape as a major cue to vowel identity. The identification functions for consonant contrasts are S-shaped with a steep gradient at the boundary point indicating that perception divides the acoustic continuum for discrimination. Amplitude variations occur at the syllabic rate of 200–300 ms (dominated by vowel duration). Stop consonants (p, t, k, d, d, g) and nasal consonants (m, n) exhibit rapid spectrum changes over 10–30 ms due to abrupt vocal tract changes though nasal consonants have greater low-frequency energy and lower high-frequency energy than stop consonants. The evolution over time of the formant patterns are tracked for frequency, pitch and amplitude to give all the information necessary for regenerating speech with compression. A major problem is in the recognition of a set of words of a large predefined vocabulary (more than a thousand words) in unconstrained speech independent of the speaker. The speech input is continuous but the words are discrete and their boundaries are difficult to delineate. Content words (nouns, verbs, adjectives, adverbs) are often emphasised but function words (articles, prepositions, pronouns, etc.) are often poorly articulated. Function words comprise only 4% of the vocabulary but comprise around 30% of speech by frequency. The approach used most commonly is hidden Markov models which utilise a parametric statistical approach. The input signal is modelled as a sequence of vocabulary words with a background signal. The background signal includes extraneous speech which the hidden Markov model represents. Another technique for data compression that is popular is linear predictive coding (LPC), a form of Wiener autocorrelation filtering. It represents speech in highly compressed form and is computationally efficient. It imposes a model of a linear acoustic tube on speech with cross-sectional area and length determined by the LPC coefficients. By normalising random speakers' characteristic tube length and cross-section to standard values, some small measure of speaker independence can be achieved.

Next, the parametric representation is grouped into words for matching against the word prototypes. Spoken words are typically ~10 ms long. These segments are compared with corresponding segments in word prototypes defined by their average pass band energies. A Euclidean or Hamming distance function is used as the similarity measure: the shorter the geometric distance in n-dimensional space, the greater the similarity. Linear predictive residuals may be used as another similarity measure such that the residual is the error between the LP filter to the speech form. Using word prototypes is inefficient, so words are represented as sequences of smaller phonemes as the basic unit. These represent syllables which form the digital units of sound composition into words, phrases, sentences, etc. Phonemes reduce the number of prototypes which need to be processed and reduces the memory requirement for word prototypes. Next morphological analysis distinguishes the word, its present participle root and its derived inflected form (e.g. study-ing) and reduces the number of lexical entries in the stored dictionary. The morpheme is the basic unit that carries meaning. Phonemes are strongly affected by neighbouring phonemes causing difficulties in phoneme recognition. Triphone models take into account neighbouring phonemes. Phonological rules that describe legal phoneme sequences provide some relief from this. The process then continues from words to higher hierarchical patterns. The recognition process must be performed at many hierarchical levels in interactive fashion using hypothesise-and-test methods to overcome many ambiguities. Models are required at all levels with lower-level models calling on higher models to resolve local ambiguities through high-level expectations. Words are characterised by acting as single discrete units which are meaningful on a stand-alone basis. They are searated by pauses and cannot be embedded within each other. The output of the morphological analysis is a sequence of words.

The next stage is semantic analysis which requires inferencing procedures. Stanford Research Institute have developed a system called TDUS (Task-oriented Dialogue Understanding System) for communicating with a human about repairs on electro-mechanical equipment [Hendrix & Sacerdoti 1982]. Information about the repairs was recorded in procedural network data structures. The network divides tasks into a number of subtasks such that each action is associated with a number of subactions which are partially ordered. The structure of the task-oriented dialogue closely follows the division of the task in the network into increasing detail of subactions. Furthermore, referential expressions refer to objects in the current or higher subtasks rather than sibling subtasks, i.e. referential expression in natural language follow similar conventions to variable references in block-structural programming languages though not explicitly (see Fig. 4.3).

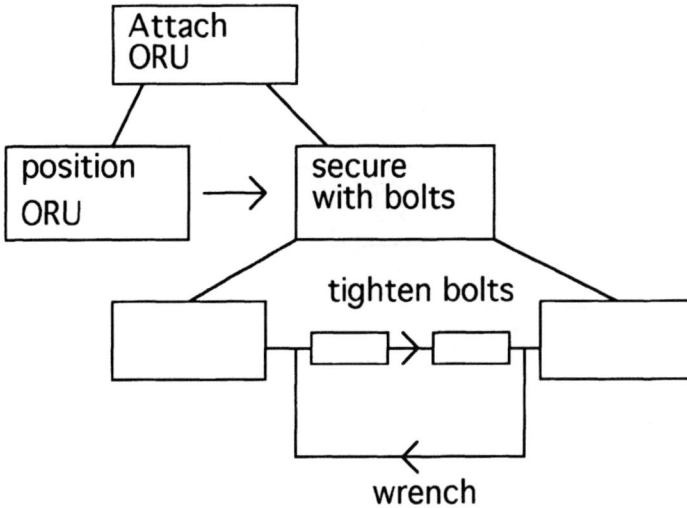

Fig. 4.3. Task-oriented Dialogue Understanding System.

Many automatic speech recognition (ASR) systems require careful pronunciation of a limited number of words (<1000) from a small set of known speakers and require ~0.1 s pauses between words to isolate them. Some ASRs, however, are capable of continuous speech recognition, but again have limited vocabulary (<100) words and a small set of speakers. Generally, isolated word recognition systems have larger vocabularies than continuous systems. Many of these systems can give word recognition accuracies >90% for vocabularies of a few hundred words. Commercial standards require a word error rate <5% which is still some five times greater than for natural speech error rate (~0.8%). Some ASR systems can have vocabularies up to 10 000+ but only under optimal conditions. Speaker-adaptive systems are initially speaker independent but are modified to become speaker-dependent. However, ASR is computationally expensive. The need is for increases in data dictionary compression and directed dictionary retrieval to reduce the recognition time. Hearsay II speech recognition and language understanding system is a blackboard system that integrates information from multiple independent knowledge bases and employs hypothesise-and-test paradigms such that phoneme strings predicted by top-down analysis are compared to strings of phonemes produced by acoustic bottom-up analysis. Arbib & Caplan (1979) provided the conceptual framework for Hearsay II (see Fig. 4.4). Each linguistic task is performed by several functional knowledge-based components both in parallel and in sequence. Because the knowledge sources act in parallel, they exchange results through the common blackboard memory to provide ccoperative computation. This provides a means of interconnecting different sources of knowledge. Hearsay II provided different levels of representation with each knowledge source representing a different functional module of the human brain. The blackboard

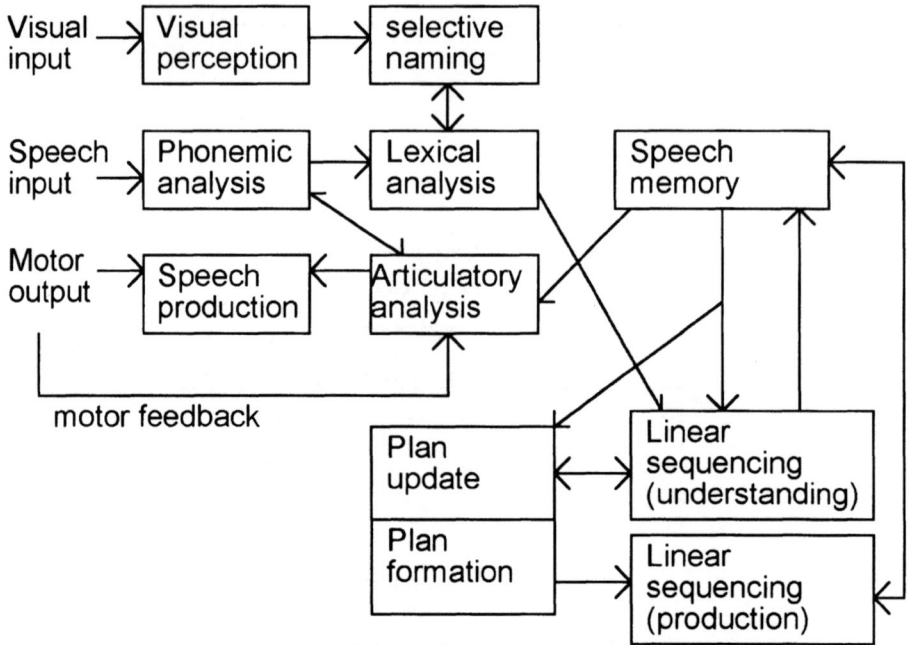

Fig. 4.4 Hearsay II blackboard (from Arbib & Caplan 1979). Reproduced with permission from
Behavioural and Brain Sciences by Cambridge University Press.

linking the knowledge sources represented working memory. Solutions were found by
iterative hypothesise-and-test processes. Natural language understanding should follow
ASR to achieve recognition of continuous speech because continuous speech is filled
with ambiguity which can only be resolved with additional semantic knowledge—this is
language understanding as opposed to speech recognition.

Although the present state-of-the-art ASR is limited, it is sufficiently advanced to be
of great importance to telerobotics in space where the environment is limited and the
number of possible scenarios can be accommodated with a limited number of trained
operators using discrete word recognition systems. Voice synthesisers may be used to
convey critical messages verbally.

4.4 TELEROBOTIC MODE OF CONTROL

Telerobotics may be regarded as a specialised form of man–machine interfacing. The
human–computer interface (HCI) is fundamentally concerned with human interaction
with computers and the need to maximise the efficiency of that interaction. HCI should
resemble human communication as closely as possible. Human communication is a transactional process in which the speaker and the listener alternate roles. Such social transaction has certain characteristics: similar modalities of representation, shared knowledge
about the topic under discussion, social etiquette defining the transactional procedure, and

corrective actions based on feedback on previous communications. There are certain limitations, however. Language is inherently ambiguous and vague. There are certain desirable properties of an HCI system. The user must utilise simple commands which launch unique actions which are in turn tightly related to the command, i.e. there must be a good mapping between the commands that are available and the goals that must be achieved. This is one aspect of 'user-friendliness'. Humans are primarily visual information processing animals—the ratio of visual to auditory pathway fibres to the brain is ~30:1. Graphical representation of data provides the most useful representation of data. Even text can be read at 15 000 words/h compared with voice which can convey 5000 words/h. Visualisation gives pictorial format to complex data using graphics and imaging techniques in conjunction with simulation and data compression techniques. The human retina comprises ~10^7 rod cell receptors distributed over the retinal surface for grey level discrimination and ~10^6 cone cell receptors clustered near the fovea for colour discrimination. The human eye is capable of receiving and interpreting ~7×10^8 bits with a resolution equivalent to 6000×6000 pixels (~4×10^7) with ~1×10^6 colour levels, while the average screen can yield ~6×10^6 bits with ~7×10^5 pixels and 256 colour levels. For sound, the machine is capable of analysing more data and with greater frequency discrimination and at faster speeds than the human ear. Speech recognition as a computer interface is only recently advancing, though presently sound is usually used for alarms. The other senses of taste, touch and smell are not well developed in machines though tactile research is progressing rapidly. Hence, visualisation is the best tool for delivering large volumes of complex multidimensional numerical data from diverse sources. Visual information is presented to the brain as a two-dimensional spatial array of intensity levels modulated over time, while auditory input is non-spatial one-dimensional information modulated over time. Hence, the value of utilising visual modes of HCI. The first visual mode of information presentation adopted in computing was text. Unlike spoken words, visually presented text is all ready broken down into words. Then came the Windows-based GUI (graphical user interface) philosophy which was developed for Apple Macintosh computers providing a visual–spatial interface with manipulatable objects for rapid feedback to command execution. The key was to reduce the time between the cause and effect of command execution and make such command execution obvious. Multimedia applications extend this philosophy to graphics, video, text and sound. Current techniques adopted for the control of spacecraft operations are based on these concepts via Mimic (Synoptic) displays. Screens show the telemetry data through static and dynamic 'blackbox' representations of onboard hardware with a characteristic colour coding. Often, flow animation is used. Alphanumeric data can be accessed via the blackbox representations for detailed information. Critical blackboxes are presented on the screen for rapid access. Such Mimic displays are, however, limited in capability and flexibility.

The joint NASA/ESA Ulysses mission was teleoperated whereby control was exerted from the ground station. The out-of-the-ecliptic gravity assist manoeuvre at Jupiter was controlled essentially in real-time during the gravity assist [Angold et al 1993]. In general, for deep space probes, mission programs for sequencing operations are usually uploaded in advance to the onboard computer memory for subsequent execution via time-tagged command sequences. Prior to Ulysses, spacecraft passing through Jupiter's magnetosphere had experienced anomalies due to the high radiation levels, electrostatic discharges and single event upsets. These resulted in spurious command execution, spontaneous semiconductor state changes, instrument degradation and saturation, computer outage and spacecraft timing corruption precluding the use of autonomous control. Timing corruption was particularly detrimental as instrument deployment is usually 'time-tagged' such that commands are uploaded into a buffer in the computer subsystem which later releases the command according to the internal clock. Ulysses had to operate so that instrument activities were interleaved with spacecraft operations without conflict and as Ulysses was operating at 5.4 AU with continuous Deep Space Network coverage, near real-time commanding from the ground was possible because of the long time constant of spacecraft operations. Hence, near real-time telecommanding is feasible in specraft operations if the task duration is long enough.

Bejczy (1979) defined teleoperation thus: 'A teleoperator is a robotic device having video and/or other sensors, manipulator arms with some mobility capability which is remotely controlled over a telecommunications channel by a human operator. This human operator can be a direct man-in-the-loop controller who observes a video display of the teleoperator and with a joystick or analogous device continuously controls the position of the teleoperator vehicle, its arm or its sensor orientation. Alternatively the teleoperator can employ a computer endowed with a modicum of artificial intelligence capable of executing simple control functions automatically through local force or proximity sensing; in this case the remote human operator shares and trades control with the computer'. The aim of teleoperation is to use a master/slave robot system such that a human operator (HO) uses the master device to control the slave manipulator within a remote environment (usually hostile) to perform work, i.e. teleoperation is a cybernetic man–machine interface system designed to augment and project human senses and dexterity across physical distances [Bejczy 1980]. Indeed, Arbib (1976) advocated that the goal of artificial intelligence is to enable the machine to cooperate with humans in a symbiotic fashion such that the machine enables humans to perform more tasks more intelligently.

The master environment is that of the human operator; the slave environment contains the remote worksite (including the manipulator) joined by a communications link between them. Teleoperation involves reflecting at a distance the physical motions of the HO. This is a manual dexterity skill but it is desirable to shift the interface to a more mental (symbolic) level, i.e. to shift the interface to higher levels of control. Telerobotics involves varying degrees of sophistication according to the degree of supervision at higher levels of control given by the HO. Essentially, control is shared between man and machine. This involves using commands at motion segment or even task level—this is desirable since it overcomes any communications time delay problems that may arise due to the distance separating the master and the slave devices. Higher-level commands have characteristic times of execution greater than that of motion reflection and often exceed the communication delay [Varsi 1990]. Complex tasks may be performed remotely, but work efficiency is less than that for direct manual task performance. Sensory feedback is absolutely essential between the master and the remote slave. A human operator workstation provides the interface between the human operator and the remote manipulator and closes the control loop—it provides multiple inputs to the human operator which must be interpreted. The most fundamental form of teleoperative feedback is vision where visual feedback guides the human operator in the conduct of the task. Enhancement of the man–machine interface can be pursued in three ways:

(i)   graphic display of the sensor information;
(ii)   voice communications;
(iii) kinaesthetic coupling between the HO and the remote manipulator.

This multimodal approach offers great versatility.

Teleoperation for space missions are characterised by the remoteness of the operations environment. A lunar mission has a round-trip delay of 2.6 seconds and a Mars mission involves round-trip delays of 40 minutes. This generates time delays in the round-trip signal transmission and so in the control loop. The maximum time delay for efficient teleoperation is ~0.2 s [Korf 1982]. When longer transmission delays occur, move-and-wait strategies are usually adopted by the operator. More than 1-s delay seriously degrades performance to the extent that move-and-wait strategies must be adopted automatically. For in-orbit servicing, some of these problems may be alleviated by using space-based HOs with a direct communications link to the remote robot, i.e. astronauts (e.g. the Shuttle mission specialist operates the RMS). However, astronauts 'on-site' are expensive and represent a sub-optimal use of astronauts who are better employed in other activities. For maximum flexibility, the robotic spacecraft must be operated from the ground.

Teleoperation to GEO involves ~0.5-s time delay due to the time for the command to reach and return from the spacecraft, a ~0.8-s delay for the modulation–demodulation of the signal (typically), plus ~0.2-s delay due to limited human reaction speed, i.e. all in all, the communication delay will be at least as high as ~1.5 s for GEO, though, in practice, time delays will be much higher. The problem of eclipse for LEO operations in terms of limited view windows from the ground station suggests the necessity of a network of ground stations for continuous communication (which is prohibitively costly). However, the requirement for continuous communication during manipulative operation with the ground may be maintained using data relay satellites in GEO. Since this introduces a GEO time delay for the communications link of the robotic freeflyer spacecraft in LEO, this may be regarded as a standard time delay [Kalaycroglu & Jaifu 1992]. Generally, for all space missions, the communication time delay increase beyond tolerable levels unless compensated for. It is expected that in reality, the time delay between low earth orbit remote sites such as the Space Station and the ground will be of the order of 2–3 seconds, even up to 8 seconds due to processing and transmission through ground-based computer networks which increase task execution times by up to 500%. This makes the capture of a moving target virtually impossible and prevents the operator from avoiding possible damage to the target or the manipulator [ESA 1986]. Furthermore, as the time delay increases, so the time for task completion increases—an adequate robot technology requires at least 80% usage time to be generally economic in the industrial environment and this provides a useful yardstick when applying robotics to other environments such as space: move-and-wait strategies are highly inefficient in terms of manipulator operation. Teleoperation can be performed in time scales ~3–5 times those of human manipulation on Earth using EVA gloves [Andre et al 1990].

Predictive computer graphics pre-views can considerably improve tracking performance under conditions of transmission time delays between a remote site and the operator station [Stark et al 1987]. Predictive displays involve heuristic synthetic models represented graphically to simulate the robot and the target motions in real-time. They operate by simulation of the task environment and simulation of the effects of operator actions. The predicted effects are displayed to the operator and are continuously updated to correspond to the delayed actual sensory data at the remote site. Hence, they provide both previews of action and predictions of the environment. The 3D graphics models of the manipulators and objects are overlayed over the 2D scene TV images. The overlap may be wire-frame or solid-polygonal models with varying degrees of transparency. A CAD/CAM model is often used for such superposition. Calibrated synthetic viewing involves overlaying graphics onto images manually—an 'operator-interactive' mode. This is time-consuming, so automation of this procedure is preferred. Calibration between the images is performed using the manipulator geometry as the calibration fixture and the cameras are modelled as ideal pin-hole image formation processes. The least squares fit vertex- and edge-matching algorithm computes the correspondence between the graphic overlay and the live video from multiple cameras. The human operator can correlate the model to the video image and bring the two into alignment by superimposing the vertices of the model to their equivalent vertices on the image. Calibrated 3D graphics reduces manipulation time by 50% enhancing the viewability of low-visibility regions. Several prediction algorithms are available and Smith's predictor algorithm is a popular one and assumes a stationary target [Delpech & Marrette 1985]. Transfer function is given by $G(s) = G_R(s) \exp [-sT]$, where $G_R(s) = $ rational transfer function and $T = $ time delay. A compensator $K(s)$ stabilises the rational part of the transfer function and an auxiliary loop around the compensator stabilises the irrational part of the transfer function (see Fig. 4.5). The closed-loop transfer function is given by: $G_{CLS}(s) = G(s)K(s)[1 + G_R(s)K(s)]^{-1}$. The compensator $K(s)$ is usually designed according to LQG optimal control theory. The Smith predictor feeds back predicted responses immediately after the prediction error after the real response has been measured. For stability the prediction error should be negligible. These predictive displays reduce the effects of time delays dramatically and reduce the total teleoperation time, but they do limit the manipulator speed of operation to a degree. Predictive displays reduce teleoperation time delays by 50% for a 1.5-s time delay, 60% for a 3-s time delay and 70% for a 6-s time delay. Sato et al (1993) suggested using automatic velocity reduction from the operator to the robot controller such that the simulator performs like a slow-motion action film to slow down the trajectory transmitted to the robot spacecraft. Another possibility to compensate for the signal delay times through the use of a Kalman filter to estimate the expected configuration of objects in a scene. A significant side-effect of teleoperation is that mismatches between the display representation of the remote scene motion and the motion actually experienced by the operator can cause simulator sickness caused by mismatches between visual input by the display and vestibular motion input similar to space adaptation syndrome experienced by astronauts.

Shared control for advanced telerobotic operation with a mixture of automation and teleoperation is desirable, i.e. supervised autonomy. The well-structured nature of many robot motion segments such as pick-and-place, tool exchange and subsystem control (e.g.

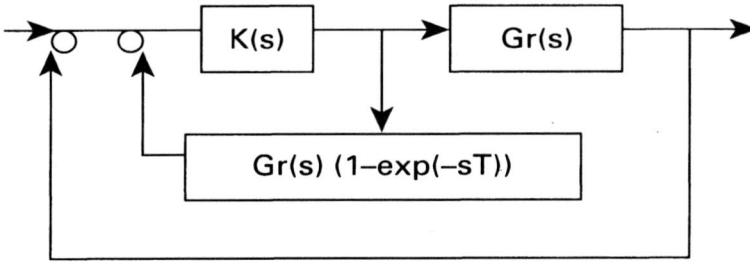

Fig. 4.5. Smith predictor.

TV tracking of the task) are executed repeatedly in all teleoperations and impose a burden on the human operator (HO). This introduces the opportunity for high degrees of autonomy for such tasks through pre-progamming, automatic control and object-level languages. This may be enabled by the fact that target spacecraft are man-made and deterministic in structure and CAD/CAM models of high accuracy should be available. Automatic control of camera position and orientations based on manipulator and camera positions would significantly reduce the operator's workload since continuous work on the task is possible without interruption for camera adjustment. For override to alter requirements, the teleoperator may use object-level languages (possibly spoken) to specify the aim of the task or camera requirements which is automatically decomposed into motion-level procedures and automated control routines [Sato & Hirai 1987]. Supervisory telerobotic control may be maintained whereby remote devices operate in autonomous mode and only interacts with the HO for high-level instructions or when it encounters a problem that it cannot handle. High-level open-loop planning tasks are performed by the human operator while low-level path planning is performed autonomously by the robot locally through closed-loop, sensor-based control [Landzettal et al 1995] (see Fig. 4.6). Gross commands are refined autonomously by the robot.

Pin et al (1992) suggested that the ultimate goal of the intelligent machine is to support humans in the efficient accomplishment of increasingly complex tasks, and that former man–machine interfaces were characterised by a static allocation of tasks to the human and to the machine based on rigid, a priori capability characteristics. They regarded this as limiting. Only through dynamic allocation of tasks based on changing skills, intellectual capabilities and performance of the human such as including the effects of fatigue will man–machine interfaces become effective. This forms the basis of a dynamic symbiotic relationship between humans and the machine or synergistic man–machine interface. This dynamic process allows the man-machine symbiont to cope with a changing environment, causing the resources most appropriate for performing a subtask to be assigned that subtask while the other monitors it. This essentially optimises the division of labour between man and machine. This requires the system to:

(i)   acquire new knowledge over time through experience and observation;
(ii) evaluate current capabilities based on the current state of the environment and the skills of the system's components;
(iii) dynamically optimise the allocation of tasks between the human and the machine based on current capabilities.

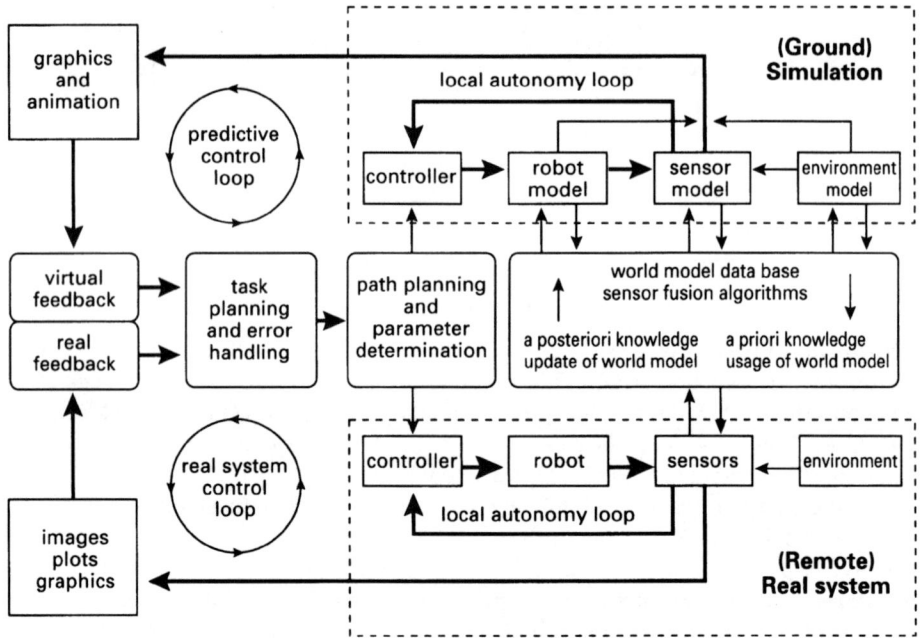

Fig. 4.6. TeleSensor Programming concept. (Reproduced with permission from Landzettal K © 1995).

A closed-circuit TV system provides visual feedback of the remote site. Clarke (1985) suggested that camera positions should be placed in human-like relationships to the remote slave arms, i.e. midway between the arms and near the horizontal plane. Illumination is also a considerable problem, particularly for the space environment: factory assembly is recommended in a ~1000–6000 lux environment. To minimise glare, the environment should be less than 3000–4000 lux, suggesting a baseline illumination of ~3000 lux. Control of the camera and lighting functions occupy a significant fraction of a teleoperator's time justifying the use of automatic tracking to relieve the operator's workload [Brooks 1992]. Appropriate lighting arrangements should avoid shadows. The end effector is kept in the centre of the FOV (field of view) and a 3D graphical image of the kinematic configuration of the arm is superimposed on the reference data. A graphical representation of the current state of cameras and lighting with computer-aided camera/ lighting optimisation would aid the operator.

Computer-generated three-dimensional visual representation is preferable to any other mode. A skeleton image may be constructed to provide the most useful information in salient form to the operator. Zone displays may be used to highlight regions in the visual field that place restrictions on the human operator, e.g. imminent collision points may be false-coloured or lost, or blurred information may be filled in from a CAD/CAM database. Depth information is fundamental and may be incorporated through virtual views. The simulated view is generated from a virtual camera placed at any angle or location required. Alternatively, perspective grids may be superimposed onto the image with markers indicating the end effector position in relation to the target. Visual cues such as high-contrast stripes or patterned markings may be used to identify distance and orientation of targets to aid performance. Furthermore, visual cues can be generated graphically and superimposed on the monitor without the need for mounting or painting special target cues on the satellite. Stereo allows the display of a scene in three dimensions from different viewpoints from stereo cameras at the remote site. LCD sequential shutter glasses in which each lens becomes opaque in sequence with the display of left and right images on the VDU screen can generate a stereoscopic image. Such binocular vision allows depth perception of single objects.

Bandwidth constraints impose a low bit rate capacity for telemetry and command so limiting the transmission of large numbers of frame-by-frame video camera images. For both a robotic freeflyer and a planetary rover, video transmission would dominate the data transmission bit-rate due to the frequency of frame changes possibly requiring dedicated facilities for communication of this data. It would certainly require extensive data compression onboard the robot spacecraft to allow digital transmission down-link in real-time (e.g. Galileo compression). Uttal (1989) suggested that low-resolution colour images combined with high-resolution black-and-white images from two remote binocular cameras can be combined into a functional quality colour stereo image, reducing the bandwidth requirements. However, it is only necessary to represent essential parameters only and update only corrections and standard resolution black-and-white TV may be sufficient for task execution—this reduces the required bandwidth further. Space tends to limit colour distinguishibility so this is not likely to produce difficulties. Indeed, false colour may be added at the ground station to enhance distinguishability. Furthermore, a distributed approach should be used such that processing appropriate to the spacraft should be performed onboard whilst processing appropriate to the ground station should be performed at the ground station with its superior computational resources. The ground station may comprise sensor processors, simulators for world modelling and task planning, with the HO executing the remote manipulation aided by the simulators.

The Shuttle RMS is controlled from two joysticks onboard the Space Shuttle with the astronaut closing the control loop visually. One joystick has a three-axis handle for controlling translation and the other has a three-axis rotational grip for controlling orientation. The JPL MOTES ground telerobotic control system has interface devices including a 6-axis spaceball and LCD shutter glasses. However, fundamental limitations on the performance of vision systems invoke the need for force information through force sensing. Bilateral force-reflecting servo feedback via a joystick alleviates many of the limitations of video feedback by providing kinaesthetic coupling between the HO and the manipulator. The joystick senses the operator's hand position and provides backdriveable forces and torques in all six dimensions to the HO in response to those sensed by the slave. Alternatively, the slave and master may both be manipulators such that the master arm is a kinematic replica of the slave manipulator (though smaller). Force reflection greatly enhances the HO's feeling of remote presence through direct duplication to the HO and so increasing the transparency of the interface while reducing the visual workload on the HO. However, transmission delays between local and remote sites can make force feedback unstable more rapidly than in visual feedback: $F_{master}(t) = F_{slave}(t - T)$, where T = time delay. Without force feedback, assembly is difficult, time-consuming and error-prone. Force feedback allows micro-displacements to be made and felt. Under certain circumstances, making certain displacements difficult and recommended displacements easy allows improved performance—this may be achieved by artificially increasing or reducing the forces required to execute a manoeuvre according to requirements [Schenker 1988]. Bilateral systems provide this through 6 degree-of-freedom force reflection to the hand controller. JPL (NASA's Jet Propulsion Laboratory) have developed a 6-degree-of-freedom, force-reflecting, universal hand controller joystick. It transforms the HO's translational and rotational motion into equivalent 6-degree-of-freedom motion of the teleoperated manipulator and it reflects the forces acting at the slave hand to the operator's hand. It can operate any slave arm system geometry by means of real-time transformation of the kinematic dissimilarities between the slave and the master. Reduced force feedback on the hand controller reduces operator fatigue and allows manipulation of loads beyond that of direct manipulation. Alarms may be used to alert the human operator to important phenomena. The human operator can develop a high degree of skill for a wide variety of tasks.

4.4.1 Virtual reality systems

Telerobotic feedback data may be gathered at a rate higher than is manageable by the human analyst. Large amounts of information are made available to humans in visual form to allow them to make judgements about the data more quickly than if it was in numerical form. This process of visualisation provides a means of displaying symbolic information in the form of concrete objects and physical relations. Virtual reality (VR) is a natural evolutionary development in man–machine interface based on realistic 3D computer graphics which aspire towards ultimate transparency between the computer and the human sensorimotor system. It is expected to overcome the limitations of simple teleoperated movements that currently plague telerobotics to allow more complex operations such as blanket handling, dextrous manipulation of variable-mass and variable-sized objects, and even tie knots. VR is the integration of computer graphics with the human sensor–control–actuation cycle to create the illusion of 'telepresence' that the operator is physically present at the remote worksite in a computer-generated environment through this transparency. It requires real-time feedback of a similar quality and form as that received by the eyes, ears and skin. It increases the situation awareness and reduces the operator's physical and mental workload. Indeed, telepresence projects human capabilities to remote locations. It allows the computer to simulate the particular 3D environment with dynamic objects controlled only in part by the human and allows human actions to be translated within that simulated world (cybernetic simulation or 'cyberspace'). This cyberspace comprises the storage medium for all the interactive information accessible by the operator. Cyberspace in turn generates sensations to make the illusion of participation and interaction in that world. This property of immersion gives the observer the experience of reality rather than the mere observation of reality from outside. Networked Silicon Graphics workstations through an Ethernet backbone should be able to provide the required computing power to generate this cyberspace.

Virtual reality has its origins in flight simulation, head-up displays and telerobotic master–slave developments. Advanced telerobotics where the operator manoeuvres a remote manipulator which the operator observes through the VR 3D graphic display provides a real-time telepresence capability. Real-time interactions between humans and machines through perception, feedback and control was required. For such real-time operation, a minimum of ~30 sensor readings/second are ideally required though lower rates ~10–15 Hz have been used effectively. Indeed for delicate manipulation with tactile and force feedback, ~1000 samples/s may be required. As long as the feedback from the robot correlates with that expected by the simulator, the robot proceeds with its operation, but if the feedback is unexpected, the robot freezes until the operator determines the course of action. This requires extensive parallel processing facilities. Presently, VR applications are performed on Silicon Graphics workstations such as the Iris 380 GT/VGX system which has a series of 8 Mips R3000 processors giving 37 Mflops performance.

Current implementations of virtual reality requires special headgear and special gloves. TV offers a severely reduced visual field in comparison with the human FOV of 50° up, 80° down, and 90° sideways. Furthermore, a 2D representation of a 3D world is not the ideal representation. The helmet-mounted display is a spin-off from military aircraft cockpit headup displays and allows stereo pair imaging and telepresence [NASA Telerobotics Unit 1988]. The head-mounted display generates two stereoscopic images one for each eye of the 3D scene and allows head motion coupling. It detects the operator head motions and controls remote cameras accordingly to cover the simulated task environment. Three axis-aligned Helmholtz coils detect the head movements along the three rotational axes. Two viewfinders mounted onto the helmet each comprising a CRT screen and converging lens give the operator a view of the scene. A computer-generated stereo picture pair is displayed onto the screens, one for each eye. Each eye is projected with a slightly offset view of the same image. Further depth information is provided by motion parallax through sensory feedback from head position. A lens forms a virtual image of the stereogram pair behind the display screen. With the correct geometric conditions, left and right images overlay to form a single 3D image. Lasers could be used to detect and track eye motion to produce a continuous virtual world. Such displays by spanning the visual field give the observer the sense of immersion rather than looking in from outside. Stereo imaging gives the sensation of depth and gives faster task performance, particularly for more complex tasks. A counterweight on the helmet provides balance for the helmet. Such perspective and 3D displays of simulated environments are computationally intensive and should be performed at the ground station using transputer technology. The chief disadvantage of head-mounted displays are that they are tiring for the eyes, are heavy, causing neckache, and can induce nausea.

Data gloves allow tactile and force feedback to be incorporated. The virtual hand representation is slaved to the finger and hand sensors of the data glove. Fibre-optic cables between the layers of cloth run along each finger. An LED at the wrist sends light down the cables to phototransistors at each of the finger ends and measure the bending of the fingers. Alternatively, piezoelectric materials may be used as strain gauges. Both the helmet and the gloves are studded with position and orientation sensors to allow the head and hands to be referenced to point in the 3D graphical cyberspace. For complete VR immersion, a body suit which senses the orientation of the operator's arms and legs within the 3D space may be used. Exoskeletons may be donned which allow or resist any movement of the body through feedback from the computer. Further developments have incorporated a virtual workstation where specified hand signals generate a graphic control console to appear within the internal world allowing the operator to issue commands without exiting from the virtual world. The critical requirement is to maximise the transparency of the man–machine interface by tight coupling through the removal of the distinction between the system and the user's environment.

The point behind VR is to exploit human natural skills to explore the virtual world. For use in environments where factors such as nonoptimal illumination, smoke, turbid water, etc. render TV visual images almost useless, infrared laser scanners, strip laser lighting and radar can be used to reconstruct the remote environment visually with pre-existing CAD representations [Stone 1992]. However, the simulation process is inter-active and time-consuming. Rule-based expert systems for guidance may be able to improve the effectiveness of VR systems to assist the HO. VR may be used to simulate space robotic operations. The VR system could use the operator's arm as the manipulator model constrained kinematically to move like the manipulator while the data gloves use force feedback to allow the operator to 'feel' the workspace through piezoelectric vibration. Caldwell et al (1994) succeeded in replicating the four basic human skin sensations of pressure, texture, temperature and pain. Force sensing provided information on pressure for shape and hardness determination, piezoelectric dynamic vibration sensors provided information on texture and slip, while fast response thermocouples provided information on thermal conductivity. Pain was recorded as an overload in all three sensors. Feedback was provided through an operator's glove and sleeve. Thermal feedback was provided through a Peltier effect heat pump, texture/slip and pressure feedback was provided by increasing temperatures towards the thermal design limit of 50°C. The system revealed that operators gave 90% correct identification of surfaces after only 10 minutes of training.

Virtual reality does not have to be immersive. It may be accessed through conventional screen-based interfaces eliminating the need for a headset. This provides a mechanism for visualisation of complex designs, particularly to evaluate assembly and repair of structures. If a sufficiently complex CAD model of the target is available with its parts and their integration, a virtual world could be created to test servicing scenarios prior to their actual execution, possibly offering opportunities to automate part or all of the servicing process. Only if the execution departed from the virtual world simulation would human intervention be required. Immersive VR requires visual aids of one form or other. Holographic techniques can dispense with viewing devices. Picture volume data generated from wide and deep viewing angles can be obtained using lasers projected onto a photoconductive film. Individual volume elements (voxels) may be created by the intersection of the laser beams. An array of small lenses can direct the incident wavefront each with a unique view of the scene. The full 3D VR image is produced as the integration of all intersecting beams.

A closely related technology to telerobotics is that of telescience in which the human operator conducts scientific experiments on an experimental platform remotely using real-time data feedback. Eureca (EUropean REtrievable CArrier) is a re-usable space platform supplying power, thermal control, ground communication and data processing to a variety of independent, individual experimental payloads. Eureca requires packet telemetry and telecommand since its nominal altitude ~500 km entails a number of short-duration passes over the ground station. Eureca may be regarded as a precursor mission to a re-usable robotic freeflying servicer. Its primary function is as a follow-on to Spacelab as an automated retrievable freeflying experiment platform for long-duration microgravity experiments at $\sim 10^{-5}$ g [Nellesson 1992]. It is re-usable (with a maximum turn-around time of two years) and can make up to five flights from the Space Shuttle over a 10-y period with each flight operating for 6 months (extendable to 12 months) under the control of ESOC (European Space Operations Centre) at Darmstadt, Germany. This gives it the capability of experimental durations beyond the capabilities of present manned laboratories without the potential degradation to experiment integrity induced by manned activities of $\geqslant 10^{-4}$ g. Its re-useability offers reduced costs through a space-proven platform with existing ground infrastructure. Its cost efficiency to client experiments is such that it is 20 times cheaper than the traditional approach of developing and building new satellites for onboard experimentation. Indeed, it is half the price of developing smallsats with their short development time, low cost and reduced launch costs. The US Space Shuttle is at present the only launcher capable of retrieving and and returning payloads to Earth. Eureca's first flight was in 1992 when it was deployed by Space Shuttle Atlantis (STS-46) to 300 km altitude. It's own propulsion system manoeuvred it up to 500 km altitude to minimise air drag effects and then returned it to its 300 km parking orbit for retrieval by the Shuttle at the end of its mission. It has a total mass of 4.5 tonnes of which 1 tonne is payload and 750 kg is fuel. The payload experiments are mounted onto the carrier's upper deck in plug in/plug out fashion. Its solar array output is 5 kW, of which 1 kW is available to the payload. It can transmit data at 256 kbps downlink and can receive commands at 2 kbps uplink at S band frequencies. The main ground station is at Gran Canaria with a second station at Kourou, French Guiana. A backup station is at Perth in Australia. As the time between ground station contacts with Eureca can be as long as 18 h with visibility comprising only 3% of the mission duration, Eureca is designed to function autonomously for up to 48 h. Onboard operations are programmed 60 hours ahead in its master schedule with autonomous conduct of all experiments in that schedule. Its data handling system interface is industrial bus standard compatible (IEEE 488). All data including housekeeping telemetry and scientific results are held in memory onboard which is transmitted during passes of visibility. Hence, Eureca must operate autonomously most of the time. Most such LEO platforms will be subject to this restriction. However, the EDRSS/TDRSS relay offers the possibility for teleoperated manipulators and the opportunity for interactive and interdictive experimentation and for the retrieval, return and replacement of experiments without returning the whole platform. ESA is introducing a telescience facility for its Columbus Orbital Facility of the ISS based on the World Wide Web infrastructure [Christ et al 1998].

4.5 GROUND STATION COMPUTER HARDWARE ARCHITECTURES

The ground station will need to employ powerful information processing architectures incorporating parallel and distributed hardware in multiple computing elements, particularly for man–machine interfacing through telerobotics. Computer systems with several users executing concurrent programs nominally operate whereby programs are executed without interference with each other until a critical level of memory is occupied. More requests for memory will cause the system to move to a different mode of stable operation whereby the space for other programs shrinks such that they occupy space for longer, creating an increase in the number of programs in need of memory space. This amplifying effect is called thrashing. Similar effects can occur in operational modes of communications satellites which utilise multiaccess broadcast shared by a large number of ground stations. Hence, it is critical to size the hardware to cope with the potential computational loads. Furthermore, a major cost to operating spacecraft arises from flight operations in the form of support personnel for day-to-day spacecraft routine. Flight operations may be run from multiple ground station sites increasing costs. Multiple computing facilities may be linked by satellite through a wide area wireless network. The wireless star topology assumes a static channel allocation with separate frequency allocations. This is rigid and inefficient, while a wireless bus topology allows sharing of the same communications medium for multiple users.

Computer RAMs have been growing faster than CPU processing speed so that the fraction of time that the computer spends performing useful computations has been relatively reduced. The von Neumann architecture with its single centralised control processor imposes a processor/memory bottleneck due to its sequential data processing. Programs operate through three phases. The memory register address of the instructions is determined in a time t_s, the instructions are then fetched from the memory register in time t_m, the execution of the instructions on the data takes time t_e, and the result is stored in another register. This instruction sequencing process is then a sequence of memory changes which illustrate the sequential manipulation of data, each of which represents a potential bottleneck. The only way to overcome this bottleneck is to use explicit VLSI technology and to adopt supercomputer architectures enabling ~10^2–10^3 Mips compared with 5–10 Mips for large mainframes. They are often used for numerical computations in science and engineering and are similarly useful for complex dynamics solutions in real time as well as for real-time simulations and virtual worlds.

The ground station will need to employ supercomputing hardware particularly for real-time control and real-time support capabilities to avoid system instabilities inherent in signal processing and data transmission delays. AI (artificial intelligence) techniques such as production systems as data directed control structures, which may be invoked in parallel, are an ideal application for parallelisation as they are computation bound, amenable to parallelism and are based on functional programming languages which are also suited to parallel execution. The choice of hardware architecture should logically follow the software architectures for efficiency and compatibility. AI requires intensive irregularly patterned memory accesses and input–output activities. Symbolic processing involves knowledge representation as the encoding of information into symbolic structures and knowledge processing as the manipulation of those structures. AND-parallelism involves concurrently executing a set of necessary and independent tasks while OR-parallelism involves evaluating alternative paths.

There exists a communications overhead which increases with architectural complexity of supercomputing multiple processor hardware. The problem with multi-processor architectures is that concurrency is limited by the communications network which itself imposes an overhead between processors which can devour any speedup advantages from parallelism. Processor interconnection is limited by physical feasibility such that each node must be connected to a fixed number of other nodes independent of the number of nodes. The asymptotic optimal speedup of $O(\log N)$ in theory but the ratio of communication to processing time increases. A task may be represented as parallel components and serial components such that $P + S = 1$. Each processor experiences an overhead of H due to the extra steps required for parallel execution so the requirement for N processors is $P + Hn + S$ so the total speedup advantage is

$$SU = \frac{1}{1/n(P + H + S)}.$$

Hence, a small overhead can limit the speedup factor. This speedup factor will vary according to the class of problem from $O(P \log P)$ to only $O(\log P)$. However, performance is still dependent on component switching speds. When n processors share access to m memories, the fraction of processors blocked at each memory cycle is given by:

$$f = \sqrt{1 + (n/m)^2} = n/m$$

For $n = m$, 40% of processors are blocked. Some means for synchronisation is necessary which is a parallel programming problem. The present mode of synchronisation is dataflow computation. By replacing control flow with dataflow, high parallelism with limited cost of synchronisation can be achieved. In control flow, each processor has an instruction pointer which determines the instructions to be executed. With dataflow, the arrival of all data dictates which instructions are executed. A dataflow program comprises of a set of instructions in memory. Instructions are disabled until all the required data has arrived. Enabled instructions are passed across the network to the processor array for execution and the results distributed back to the next set of instructions waiting for them.

Special-purpose parallel architectures designed to solve specific problems may be employed to overcome computational bottlenecks. Parallel architectures exploit VLSI technology and may be classified into four Flynn categories [Graham 1989, Hwang 1987, Denning & Tichy 1990]:

(i)   SISD (single instruction stream, single data stream)—this is the conventional von Neumann single microprocessor architecture that operates sequentially with one control unit and one processor combined on one chip as a microprocessor. The control unit fetches instructions from program memory and decodes them. It then sends data operations such as add, multiply etc. to the processor unit. These memory access operations are slow—the fastest workstations operate at 20 Mflops. Pipelining is a method for speeding up the operation of a single processor similar to an assembly line. A single operation is performed on multiple data sets in sequence, particularly useful for floating-point operations. Pipelines are well-suited to vector and matrix processing of ordered arrays. Similar operations are performed on the data elements of the matrices, e.g. matrix multiplication.

(ii) SIMD (single instruction stream, multiple data stream)—a single control unit is implemented within an array of processors. The central control processor broadcasts the same instructions through an interconnected network to subsidiary processors which execute instructions in parallel on different data sets—such systems are highly coupled. It is well suited to vector and matrix operations.

(iii)   MIMD (multiple instruction stream, multiple data stream)—multiple processors each controlled by its own control unit. Each processor fetches and executes its own instructions on multiple data sets in parallel overseen by a central scheduler— these systems may be either loosely coupled (each processor has its own dedicated memory distributed through the network) or tightly coupled (shared memory across all processors is accessible to all processors).

(iv)   MISD (multiple instruction/single data stream)—separate instructions are executed on a common data stream—this is not common.

The Connection Machine is a SIMD computer of 2^{16} processors connected as a Boolean hypercube. Each processor is a 1-bit processor with 8 bits of internal state memory and 4096 bits of local memory. It communicates by passing single bit markers and each cell applies a simple Boolean operation to the marker bits.

Dataflow computers use a message passing approach which is a radical departure from the traditional von Neumann architecture. It involves a form of overlap processing so that one part of the CPU fetches instructions whilst the other part executes instructions. Instructions are ready for execution when their data operands arrive rather than being coordinated by a scheduling control unit. This data-activated instruction execution allows high concurrency with independent computations proceeding in parallel. The data flow program comprises activity units each of which has a unique address. When the required data for an activity unit is available the unique address of the activity is enqueued, assigned a processor and executed. Conventional programming languages which are based on von Neumann sequentialism are ill-suited to data flow architectures but several data flow languages have been developed such as VAL. If staggered, processing speeds can be doubled but only if no program branches occur, limiting their usefulness.

The rest of the approaches considered are all MIMD approaches whereby multiple processors are adopted with each processor being fully programmable with its own memory and capable of executing its own program. This approach is more flexible than SIMD approaches but involves more complex interconnection control structures and data communications between processors are critical. Interconnection approaches for parallel processing between processors may be critical. The most common local interconnection scheme is NEWS (north-east-west-south) across a 2D array of processors. Global interconnections schemes switch to all other processors. For optimal time effectiveness, the number of interconnections required equals the number of nodes (e.g. crossbar switch) but is costly in terms of hardware. The crossbar permits random point-to-point communications but requires n^2 switching points. Alternatively, few connections may be made with messages routing through intermediate nodes (e.g. ring network). Such broadcast interconnects connect all processors through one bus but can only do so sequentially due to traffic. There do exist compromise networks between these extremes (e.g. shuffle exchange network). The n-cube interconnection scheme connects each processor with the same topology as the corners of an n-dimensional hypercube. An n-dimensional cube can connect 2^n processors. The longest path between two processors is $\log_2 n$. Asynchronous parallelism can achieve linear speedup with the number of processors. However if crossbar switching is used which increases as N^2, then this linear speedup is not feasible for large N. Any form of coordination and synchronisation will impose an overhead [Haynes et al 1982].

Crossbar switching involves full interconnection with every processor connected to every other in a kind of common shared memory system. It is the most general and flexible interconnection scheme and N^2 switches are required to interconnect N processors/memories. Hence, for large arrays >30 processors, it is infeasible due to the high hardware complexity. The Banyan network is a tradeoff which provides complete interconnection of N processors with a complexity in switching circuitry that grows linearly with $N \log N$. The network is divided into distinct levels of processing nodes and forms a loosely coupled distributed set of local memory machines. Each node has a set spread number of connections fanning into it and a set fanout number of connections fanning out from it. The binary k-hypercube, perfect shuffle and tree networks are homomorphic images of the Banyan network. A binary k-cube connects $n = 2^k$ processors which are at the corners of the k-dimensional hypercube, i.e. nodes m and n are connected only if $n = m + 2^j$ for $0 \leqslant j \leqslant k - 1$. Hence, there are $k = \log N$ connections per processor. Cube-connected cycles and perfect shuffle networks are versions of the k-cube but with a constant number of connections per processor. They use a divide-and-conquer strategy by recursively breaking down complex problems into smaller problems and recombining them after solution. A typical interconnection scheme for the cube connected cycle structure comprises 2^k processors arranged in rings of size 2^r at the corners of a $(k - r)$ cube. As r increases or decreases the number of connections per processor decreases towards a ring or decreases towards a cube. To retain cube performance but with minimised interconnection complexity, $(r - 1) + 2^{r-1} \leqslant k \leqslant r + 2^r$. The shuffle exchange network connects $n = 2^k$ multistage processors with each node having only four connections. The composition of the k shuffle exchange networks (omega networks) is equivalent to the k-cube. Essentially, sets of processors are interleaved. Finally, binary tree networks involve processors connected to two children and a single parent. They use a minimal number of connections per processor but create a bottleneck near the top of the tree if the tree is large. Essentially, it uses a divide-and-conquer strategy with a time complexity of $O(n)$ and is applicable to a wide variety of problems including NP-complete problems. They have been used for relational database operations.

The CHiP (Configurable Hardware Processor) scheme is essentially a polymorphic switchable systolic array of specialised machines [Snyder 1982]. Algorithmically specialised parallel processors are hardware specific and cannot be programmed to perform functions other than those for which they were designed and constructed. Based on the notion that the power of computers lie in their general-purpose capabilities endowed by changeable programs, CHiP adopts programmable switches which allow changes to the processor configuration through different interconnection structures between processing elements from the two-dimensional array (e.g. for dynamic programming) to tree networks (e.g. for sorting) without cost in computation time, e.g. Cedar. The CHiP computer architecture comprised three components: a set of homogeneous microprocessing elements, a switch lattice and a controller. The switch lattice is programmable to connect processing elements in different structural configurations. Each switch contains local memory to store the configuration setting. The controller is responsible for loading the switch memories in parallel by broadcasting the command to all switches to invoke a particular configuration. Switches at the intersection of vertical and horizontal switch corridors perform most of the routing while those between two adjacent processors act like ports for selecting data paths from the corridor busses. Increasing the number of data paths/unit area improve processor utilisation when complex interconnection patterns are required—a corridor of width $N/\log N$ was optimal where N = number of processors but a width of $w \leqslant 4$ provides a good cost/benefit tradeoff. This approach offers a high degree of fault tolerance.

The need to have 2^k processors communicate is problematic. The interconnection structure can be optimised for a particular class of problems when the data pattern movements are known in advance. However, this does not solve the dynamic problem of moving data in unpredictable patterns such as in memory requests from a large parallel array. Furthermore many tasks are input–output bounded (particularly real-time operations). The MIMD paracomputer is one way to alleviate this difficulty [Gottlieb & Schwartz 1982]. All processors communicate through a single, shared, common or public memory (e.g. Cray series). This is the tightly coupled multiprocessor technique with shared memory compared with the loosely coupled multicomputer technque which uses distributed local memories. The passing of messages of arbitrary size and performing complex operations on these messages demand powerful node processors.

The hypercube topology has the advantage of being uniformly structured with its $\log N$ diameter but still suffers from communications traffic congestion. Both may be combined by retaining the shared global memory with the individual processors also

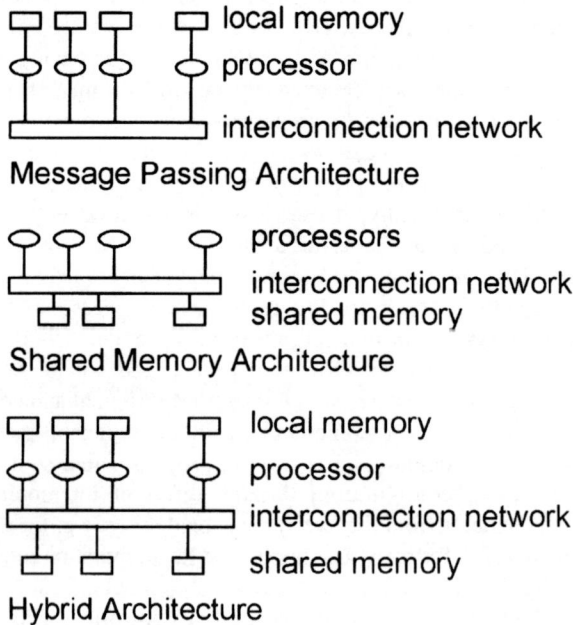

Fig. 4.7. Hardware architectures.

having private local memories which communicate through message passing. Shared memory across multiple processors suffers from the latency of memory requests as only one memory request can be processed at a time. The addition of cache memories to each processor which is private and local to each processor reduces the average memory access time. Cache memories may perform the majority ~95% of all memory requests. Shared memories, on the other hand, offer simplicity of code and data structures for application processes which can be perfomed in parallel. These hybrid architectures combine shared memory and message passing offering flexibility (e.g. Cedar project). The need for memory beyond physical central memory space local to the CPU implies the need for large amounts of virtual memory space which may be used as the shared global memory with a large proportion of the operating system and all applications programs running in this space. The communications buffer for synchronising functions is, however, part of the central physical memory. The global address space is distributed across the processors. In addition, part of each local memory may be allocated to or from the global memory. Dynamic partitioning of memory allows a pure shared memory through a crossbar network, a pure local memory system using message passing or an intermediate mixed mode. Shared memory applications may be allocated to private local memory for improved efficiency or message-oriented applications allocated to global memory to balance the workload. An important issue is dynamic load balancing so that processes are efficiently allocated to processors and that processes migrate automatically from heavily loaded to lightly loaded processors. The load may include CPU time, communication time, memory utilisation and the number of concurrent processes. Receiver-initiated load balancing involves sending requests to neighbouring nodes for work when lightly loaded. Sender-initiated load balancing involves the busy node exporting processes to neighbouring nodes. Hybrid load balancing uses receiver-initiated loading when the system becomes excessively loaded as it favours heavily loaded systems while using sender-initiated load balancing when the system load becomes light since that favours lightly loaded systems.

Amdahl's law states that the relative performance, P, of parallel processing over sequential processing is given by:

$$P = \frac{1}{(1-f) + f/r}$$

where

f = parallelisation ratio = percentage of code capable of being parallelised

r = relative speedup ratio

Software must be designed and decomposed for parallel execution and must be coordinated [Hwang 1987]. This requires complex scheduling to exploit parallelism. Scheduling requires assignment of processors and memory to nodes of a computation graph—this is an NP type problem. Physically, synchronisation can be achieved to 10 μs accuracy determined by clock frequency. Intelligent compilers should be able to detect parallelism in sequential source programs and convert them into parallel object machine code. Less than 4% of most scientific and engineering programs account for

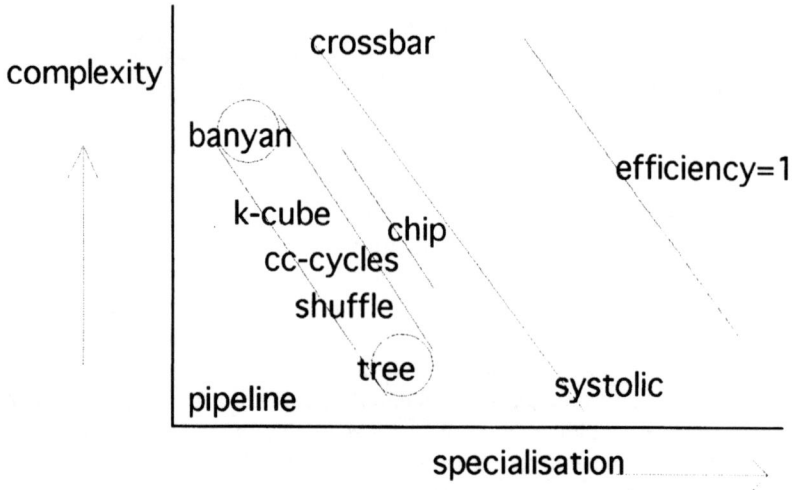

Fig. 4.8. Hardware architecture types (adapted from Gottlieb & Schwartz 1982). Reproduced with permission from © 1982 IEEE.

more than 50% program running time, mostly due to DO loops. Hence, the basic unit of parallelisation is the DO loop which relies heavily on data independence in the loop at each iteration, though concurrent processing of loops with data dependency can be achieved. This requires re-entrancy such that a copy of the program module is used by more than one task in parallel and this may be implemented using a stack mechanism. Some languages have special constructs for specifying parallel execution, e.g. ADA, OCCAM, VAL. Algorithms have traditionally had a sequential bias but parallelism offers the potential for 'pre-algorithmic' computation. One problem with massively parallel computer networks that has yet to be well-characterised is the potential for chaotic dynamics with thrashing. One future enabling technology for parallelism is fibre optics, which offer large data bandwidths for point-to-point communication $\sim 10^{15}$ bps.

4.6 THE NASREM SOFTWARE COMPUTING ARCHITECTURE

Autonomy in space has three main advantages over manual task execution by EVA and teleoperation [Erickson 1987]: firstly, automated robots generate lower costs and improved productivity; secondly, increased flexibility and availability due to higher precision and quality; thirdly and most importantly, there is greater safety due to reduced hazards to humans. Robots with powerful onboard machine intelligence will be able to do many operations in space autonomously, and have the advantage that they can be given perceptual abilities beyond those of humans (e.g. infrared vision for inspection tasks). Modern technology has endowed mankind with the tools to use a multitude of senses beyond our natural five. The central point is that autonomy imposes a heavy reliance on sensory data to deal with an uncertain and changing environment. The general conclusion is: 'Remote systems require the ability to do autonomous operations to be effective' [Heer 1978].

NASA has developed an evolutionary strategy of direct teleoperated systems gradually being replaced by increasing autonomy with human supervision as new technologies become available. During the late 1980s, NASA had developed a baseline computational architecture to support space robotic systems and this envisaged gradual evolution from teleoperative mode towards autonomy (NASREM) [Albus et al 1987]. The first phase of evolution is that of the 'intelligent aid' whereby automatic control of individual subsystems would be achieved. The 'intelligent apprentice' stage follows when automatic control of multiple subsystems has been achieved. Next comes the 'intelligent assistant' characterised by hierarchical control of multiple subsystems and finally the 'intelligent accomplice' will be characterised by automated distributed control of multiple subsystems [Decker 1987]. At all stages, humans will have top-level authority, responsibility and control. This evolution towards autonomy will result in major reductions in ground segment costs which can be as high as ~60% of space mission lifecycle costs [Larson & Wertz 1992]. The communication time delay, bandwidth limitations and human performance limits will contribute to this evolution. Teleoperation requires high information flows and will force the use of computational techniques to be employed remotely.

Most mobile robot systems have implemented a common blackboard database/world model type intelligent control system (see Chapter 13) and NASREM is no exception [Levi 1987]. The NASA/NBS Standard Reference Model architecture (NASREM) for telerobotic control was developed as a logical and functional computing architecture for telerobotics in support of the Space Station programme with particular reference to the FTS (now cancelled) [Albus et al 1987, 1988]. It has since been adopted by ESA as their standard architecture for space robot systems. It may be implemented with the lower levels of the hierarchy performed onboard the spacecraft and the higher levels executed on the ground due to their greater temporal windows. It defines a set of computing modules and standard interfaces to allow the integration of telerobotic software without specification of each module's mode of operation. It provides the basic requirements for flexible control strategies involving both cognitive and reflexive behaviour. The central component is the world model blackboard since such knowledge representation will affect planning, control and sensing. It implies the manipulation of knowledge to provide context by integrating *a priori* sensory data with *a posteriori* knowledge base data. It is hierarchically structured into multiple layers: layer 1 is concerned with kinematic coordinate transformations; layer 2 performs the dynamic analysis and trajectory interpolation; layer 3 is concerned with path generation and obstacle avoidance; layer 4 transforms object tasks into specified sequences of elementary movements; layer 5 decomposes multiple actions and sequences and schedules individual tasks; layer 6 involves allocation of resources between distributed individual telerobots at different worksites according to mission priorities. Levels 1 to 3 are part of the low-level control system (the operations performed by the cerebellum) which tend to be fixed while levels 4 to 6 are concerned with high-level planning and are task specific and highly variable (the operations performed by the frontal cortex). NASREM is also horizontally partitioned into three processing legs comprising a distributed heterarchical structure (i.e. multiple units at the same level of abstraction): the task decomposition (TD) leg performs planning operations; the world modelling (WM) leg represents the internal representation estimate of the external world and provides a simulation capability and includes knowledge sources

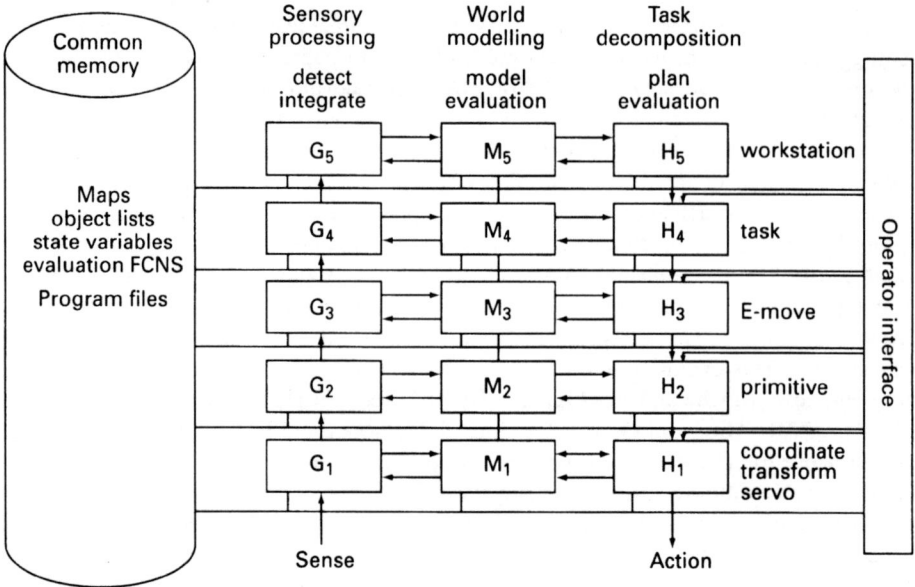

Fig. 4.9. Hierarchical software control system architecture (NASREM) [from Albus et al 1988].
Reproduced with permission from © 1988 IEEE.

pertinent to the task requirement; the sensory processing (SP) leg involves extracting information from the external world and processing sensory input data to update the world model and comparing observations with expectations. The output is provided by actuators and input by sensors. Albus (1990) later expanded the task decomposition leg into two systems elements called the behaviour generation modules and the value judgement modules. The value judgement system determined the reward/punishment values of the world and of predictions and made decisions on that basis. This included considerations of priorities, beliefs and uncertainties. The behaviour generation system selected goals and plans and executed tasks. The goals provided the objective and the task execution provided the behaviour.

The sharing of data is horizontal such that more information flows horizontally at the same level than flows between levels. The flow of commands and status data is vertical: sensory processing is bottom-up (data-driven) and task decomposition is top-down (goal-directed), though the world model can act as an expectation generator for goal directed sensory processing and as a feedback mechanism for task decomposition to allow monitoring and provide data-driven control. The hierarchical levels are defined by temporal and spatial decomposition of tasks into finer levels of resolution through a 'chunking' process—the control bandwidth decreases by an order of magnitude at each higher level. The sensory processing modules filter sensory data from multiple sources, detect and recognise patterns in the environment. Data from multiple sources must be unified into a single representation scheme in the world model (sensor fusion). Perception comprises

comparing sensory observations with expectations generated by the internal world model which is a unified estimate of the state of the external world. The world model includes a database and a simulation capability to generate expectations and predictions. Task decomposition selects and analyses goals and plans and executes tasks. Tasks are decomposed recursively into subtasks and sequenced to achieve the desired goals. The world model predicts the result of plans and the task decomposition modules select plans on the basis of the state of the world and the results of the hypothetical plans. Each SP module comprises four submodules: a comparison matcher to match observed data with the world model predictions, temporal integrator to integrate differences and similarities of observations and predictions with time, a spatial integrator to integrate data over space and a recognition thresholder which defines that when a correlation function exceeds a threshold recognition occurs. Hence the SP and WM modules are coupled to form a feedback loop and the merging of SP and WM data forms an adaptive resonance process. Level 1 comprises brightness/colour intensities and their derivatives at each pixel; level 2 comprises connected edges and vertices (Marr's primal sketch); level 3 consists of textures, orientations, shape, size and boundaries (Marr's 2.5D sketch); level 4 comprises 3D volumes and objects (Marr's 3D model); level 5 comprises groups of objects and their relationships to each other; level 6
comprises the total environment.

The task decomposition modules comprise one job assignment module and a set of planner and executor modules. They decompose the task goals both spatially and temporally into finer resolutions into plans of sequential and concurrent actions (possibly represented by AND/OR graphs). The job assignment manager partitions tasks into spatially distinct jobs to the planners and the planners decompose each job into temporal sequences of subtasks. The optimal plan is then selected according to its evaluation function. For each planner the executor executes and monitors the plan execution through feedback to provide reactive control. Reactive control required the SP modules to be directly linked to the TD modules to bypass the WM. The WM modules provide the interface between sensory processing and task decomposition such that the SP modules update the world model which in turn provides the context for the TD modules to plan. The WM comprises a hierarchical short-term memory (STM) which effectively implements a 'chunking' process. The WM makes predictions and evaluates current states and provides expectations of future states as a simulator. Knowledge in knowledge bases is used to generate task dependent knowledge and may include CAD models. The WM essentially comprises a 'blackboard' such that it updates the knowledge base and provides data accessible to all other modules. The task information in the knowledge bases may be represented as frame data structures. The WM uses *a posteriori* knowledge (already known) from the knowledge bases and *a priori* knowledge (learned information) from the sensors. The global memory is a virtual database where knowledge about the internal state and current state of the world model is maintained. It is accessible to the world model and maintains the current best estimates of the state of the world. It is kept current by the world model. It includes cognitive maps to represent distributions of bodies in space at different resolutions, lists of objects, features, relationships (CAD models, topological trees, semantic nets, schemas, production rules). Global memory may be physically distributed but appears globally transparent.

The operator interface allows the human operator to interact either online or offline with the control system at any level via joystick, keyboard, mouse, voice, etc. from the operator terminal. The operator terminal must support graphical displays, direct TV images and graphical overlays. At levels 1, 2 and 3 control is teleoperative; at levels 4, 5 and 6 control is supervisory telerobotic; otherwise control is autonomous.

The communications system routes data and messages between the modules to exchange information via the global memory either via bus or LAN. Traffic is clock-controlled such that there is a data-transfer/compute cycle whereby computed output variables at $t = i$ become available as input variables at time $t = i + 1$. Data is conveyed from all output buffers mediated by flags to signify readiness to global memory/world model at the end of the compute cycle and thence to the input buffers representing information at the next cycle. The communication system should be aided by protocols to allow for event-driven access to data and information. Although NASREM implements data exchange through the blackboard it is possible to extend the architecture to allow modules to exchange data directly amongst themselves also in a heterarchically distributed fashion, e.g. contract net protocol. At different levels of the hierarchy different temporal synchronisations are required and perceptual resolution decreases an order of magnitude at each higher level: at level 1, < 1 ms; at level 2, a few ms; at level 3, ~1/10 s; at level 4, a few s; at level 5, minutes; at level 6, minutes to hours. Similarly higher bandwidth communication is required across hierarchical levels than for command/status traffic between levels. The control bandwidth decreases an order of magnitude at each higher level of the hierarchy. NASREM is completely open regarding its mode of implementation, allowing connectionist architectures [Albus 1991].

5

Systems design example (ATLAS)

Consideration is now given to a design example of a dual manipulator robotic freeflying space robot dedicated to satellite servicing (ATLAS—Advanced TeLerobotic Actuation System). This is an example of a potential robotic spacecraft and comparisons will be made with NASA's Ranger spacecraft being designed at the University of Maryland. This concept has been selected for a number of reasons. Firstly, planetary rovers are designed for particular environments that vary tremendously. Secondly, the technologies applicable to planetary rover design are applicable to in-orbit servicing robots. Thirdly, in-orbit servicing involves more traditional robotics in the form of manipulators, and finally, in-orbit servicing represents an extremely challenging environment that would have significant impact on the performance of planetary robots, adding complex manipulation capabilities to rovers. The detailed design considerations will be elucidated in subsequent chapters from the payload to the spacecraft bus. This chapter serves to focus attention on an example system which should be borne in mind when reading the following chapters. The starting point is to assess the mission goals and constraints to generate the functional requirements to define the system and subsystems required to meet those goals within the constraints of the control system [Larson & Wertz 1992]. The spacecraft configuration as a whole is dominated by the requirements of the mission payload, the selection of the launch vehicle and the mode of attitude stabilisation of which the payload is the most significant driver.

Mission statement: Because in-orbit failures have an increasing impact on the efficiency and cost of Earth-orbiting spacecraft operations there is a need for an effective system to cope with these failures.

Primary objectives:
(i)   To detect, identify and characterise the major types of spacecraft in-orbit failures and repair those failures in near real-time at profit.
(ii) To develop a safe and effective space infastructure support system for future missions.

Secondary objectives:
(i)   To demonstrate the feasibility and actuality of large scale-space debris removal.
(ii) To provide support for Space Station operations at profit or with equity.

(iii) **Hidden agenda:** To provide a focus for Information Technology research and development and demonstrate an effective partnership between long-term government investment and short term private commercial exploitation.

Mission characterisation:
Alternative mission concepts: In-orbit servicing by EVA astronauts from the US Space Shuttle—this is costly, hazardous and suffers from scheduling difficulties as well as being limited in orbital parameters (i.e. near equatorial LEO only with $i = 28.5°$ and $e = 0$).
Hidden agenda: As the US manned space programme has been justified primarily through its in-orbit servicing capabilities it is important that the proposed system is not seen as a threat to this justification. Negotiations with NASA should be undertaken to reach an equitable arrangement without jeopardising their marketing strategy, e.g. the proposed system should be Shuttle RMS compatible as a bolt-on capability to extend aid to EVA astronauts for complex in-orbit servicing in equatorial LEO and/or the system should also operate independently outside of near equatorial LEO.

Mission architecture:
Mission orbit
— testbed LEO between 200 and 1200 km altitude at near equatorial or high inclination (polar) orbit
— evolution to a small fleet to cover GEO and other Earth applications orbits
— variable orbital requirements, variable eclipse, variable thermal environments
— variable thrusting, i.e. liquid propellant
Mission attitude
— variable attitude slewing with high pointing accuracy, i.e. 3-axis stabilisation
— flexible solar paddles stowed during manipulation
— high-stiffness structure without booms
Mission reliability
— hardened electronics with distributed and functionally redundant computers
— distributed processing between ground and spacecraft
— evolution towards autonomy
— modular design with self-repair capabilities
— logistic resupply of fuel tanks

Ground operations
— minimal ground personnel for teleoperation and support in single Earth station
— extensive use of expert systems on the ground
— limited ground visibility
— TDRSS/EDRS communications link to support multiple video data rates
Mission duration
— 15–20 y with resupply of consumables and upgraded modules
Payload
— EVA equivalence requires two 6-degree-of-freedom arms with a reach >1 m
— extensive sensor suite including vision and tactile capabilities
Design budgets
— total loaded mass ~1.5 tonnes including fuel

— manipulation phase peak power by regenerative fuel cells
— sufficient fuel for Δv orbit changes over 200–1000 km several times

The proposed dual-arm robotic freeflyer ATLAS has as its primary objective the maintenance, servicing and repair of Earth applications satellites or other satellites. Its secondary objectives are to provide assistance to astronauts in EVA and in retrieval and rescue of spacecraft and to provide a means for large-scale debris removal. Like Robonaut, it should provide a functionality equivalent to a suited astronaut. A spacecraft comprises two basic elements: the mission payload and the spacecraft bus. The payload performs the mission function while the spacecraft bus supports the payload and provides housekeeping.

Typically, the payload mass comprises from ~15% to 50% of the spacecraft dry mass. The payload classifies the satellite and drives the design of the spacecraft bus: the basic payload of the communications satellite is the active repeater/transponder and such satellites tend to reside in the 24-h circular geosynchronous orbit at 36 000 km altitude or in the highly eccentric 12-h Molniya orbit inclined at 63.4° where perigee rotation is nullified allowing a fixed latitude apogee for ground coverage; the basic payload of the meteorological satellite is the high-resolution visual/infrared radiometer and such satellites may reside in high inclination LEO or in GEO; the basic payload of the Earth observation satellite is the multispectral scanner and such satellites reside in near polar Sun-synchronous orbits between ~700 and 1200 km altitude for a repeating ground track with orbit plane precession of 1°/day to match the Earth's revolution around the Sun; the basic payload of the astronomical satellite is the imaging telescope and such satellites tend to reside in high-altitude, highly eccentric orbits ~800–1400 km above the Earth's atmosphere. All these satellites except the astronomical satellite are Earth-oriented. Usually, satellites are designed for a single specific purpose but multimission satellites are not unknown: INSAT-1 of ISRO (Indian Space Research Organisation) is an example which was launched in 1982 for telecommunications and meteorological observation. Its payload comprised 12 C-band transponders for long-distance telephone communication, two C-band transponders for TV broadcast, a very-high-resolution radiometer (VHRR) with visible/infrared sensitivity and a data channel for relaying meteorological, hydrological and oceanographic data to unmanned ground stations. It was 3-axis stabilised providing a field of view to deep space for the radiative cooling of the Earth-pointing VHRR, Sun-pointing for the solar arrays and Earth-pointing for the communications antenna [Agrawal 1980].

ATLAS also has a multifunctional role. It comprises many elements from more traditional spacecraft: optical and infrared imaging sensors (including star trackers) similar to those used on Earth observation and astronomical satellites, a high data rate communications antenna similar to those on telecommunications satellites, and possibly microwave tracking pulse radar similar to those used on Earth observation satellites. However, it is characterised by very complex operational requirements typical of some military satelllites and resembling the variable operational environments like some interplanetary spacecraft. Indeed, only interplanetary spacecraft to date have been exposed to such multiple design criteria. There are other similarities regarding mode of operation: real-time ground control versus autonomous onboard control. Only 12% of spacecraft launches are scientific exploratory probes or manned missions so such criteria are unusual in spacecraft missions generally. The majority of space missions are Earth-oriented applications tools which do not exhibit complex requirements though Earth Observation satellites tend to exhibit complex sensors due to our extensive knowledge of Earth compared with other planets.

ATLAS is essentially a tradeoff between human, technical, financial and political risk. It reduces the human risk by minimising EVA requirements for in-orbit servicing. The high technical risk may be reduced by using off-the-shelf technologies and no new major terrestrial technological innovations are required for its implementation. The financial risk may be offset by launch and in-orbit insurance. Political risk is minimised since the political motive is process-oriented (i.e. technology R&D) rather than results-oriented. In the first instance, it is a spacecraft which may be required to perform orbital transfers of payloads and equipment such as an OTV/OMV (Orbit Transfer/Manoeuvring Vehicle) between limited orbits. The ATLAS payload comprises its robotic manipulators and robotic support subsystems. This drives its size, its mass, its power and fuel requirements. The spacecraft bus should be modular to enable multiple payload capabilities to enhance future spacecraft operational flexibility by providing multi-orbit multi-attitude capabilities to a basic spacecraft bus. It should be designed for operations in all orbits from LEO to GEO and from equatorial to polar orbits so a single spacecraft design can serve in all Earth-applications orbits. Hence, due consideration must be given to the variable environments that the spacecraft will experience.

5.1 SPACECRAFT SIZING

The first concern is that of size—the operational Space Shuttle RMS has a capability for large payloads up to 30 tonnes but has limited repeatability and dexterity. This suggests that servicing should involve a smaller-sized manipulator about the same size as the human arm to replace or augment EVA operations. Assembly has been found to involve ~80% of small part (<4 kg) manipulation [Nevins & Whitney 1980], so it is likely that assembly of small objects forms a major part of the servicing process.

There has been a recent tendency for some space missions to become smaller, simpler and less costly. It provides a complementary approach to missions which are getting larger, more complex and more expensive. The evolution of increasing complexity and size is well characterised by the evolution of GEO comsats which, since SYNCOM in 1963, have been increasing in size and complexity. The spin-stabilised Intelsat 1 launched in 1965 had a mass of 38 kg with a power output of 40 W and bandwidth of 50 MHz compared with the 3-axis stabilised Intelsat 5 launched in 1980 with a mass of 970 kg and a power output of 1200 W and bandwidth of 2300 MHz. More powerful high-directional antennas, increased use of spacecraft onboard processing, longer mission lifetimes and tighter tolerances on attitude and orbital drift have been responsible allowing increased simplicity of fixed ground-based receivers for lower costs to the customer. However, the same drive to lower costs and smaller ground receivers is also a key component in the trend to mobile communications by LEO smallsat constellations. Smallsats are most suited to specialised activities such as mobile communications, small-scale space science, well focussed quick-response Earth observation and in-orbit technology demonstration.

The UK's Space Technology Research Vehicle (STRV) is a minisatellite programme for in-orbit technology demonstration by a series of 52-kg minisatellites developed by UK's DERA (Defence Evaluation & Research Agency). They offer short design timescales of ~12–18 months, low costs of <$1m, low launch mass of <200 kg and small design teams. Small teams imply minimum bureaucracy and streamlined management structures, which can cut development costs by ~40%—a useful lesson for any space project. Micro-satellites exploit miniaturisation of electronic components, emplace all complexity in software and employ redundancy in only critical systems of limited reliability such as the control system and input–output devices. Operations can often be supported from a user's PC as the ground station. Examples include the highly successful UoSAT series designed by the UK's Surrey Satellite Technology Ltd and the OSCAR series designed by AMSAT for radio amateurs. The major requirement for ATLAS is EVA equivalence and this dictates human-level capabilities. This offers the advantage of relatively low mass and lower costs and according to the US DoD's classification would not rank as a 'major' mission (less than $1b costs).

The next question concerns the orbit at which ATLAS may reside. Equatorial LEO at altitudes less than 500 km is accessible to astronauts and the orbit of the ISS at ~500 km altitude is likely to require extensive robotic support. GEO is another potential orbital regime that may benefit from robotic support as it is inaccessible to astronauts (it also imposes a significant radiation environment). High-altitude equatorial orbit is favoured by science payloads which represent high cost, limited life payloads which may significantly benefit from servicing, replenishment and upgrading. The present NASA emphasis on 'Mission to Planet Earth' programmes for high inclination LEO remote sensing satellites favours operation in polar LEO. As the Eastern Test Range at Vandenburg AFB will not support Shuttle operations, this orbit is inaccessible to astronauts. Near-equatorial LEO operation could also serve as a testbed for any further proposals for operation of similar spacecraft in GEO or polar LEO. Hence, at the outset, the freeflyer should be designed to operate in all orbital environments but the initial test unit should serve in near-equatorial LEO in support of Space Station activities though not necessarily exclusively. A major consideration that must be attended to is the projected long lifetime of the 15–20-y operational period and the reliability of onboard equipment and systems. The effects of radiation, thermal, vacuum and other environmental phenomena have a detrimental impact on survivability.

5.2 ROBOTIC PAYLOAD

The robot servicer subsystems constitute the payload of the spacecraft. The primary aim of the ATLAS freeflyer robot payload is to eliminate or at least reduce the requirement for EVA in-orbit servicing by providing general-purpose Earth orbital operations (deployment, maintenance, repair and retrieval of spacecraft). This implies the replacement of the *in situ* human astronaut with an Earth-based human teleoperator for EVA equivalence (perhaps evolving autonomy over time). Schenker (1988) and Elfving (1990a) both stipulated that space robots for in-orbit servicing must exhibit EVA equivalence. This requirement presupposes a minimum of two approximately human-dimensioned arms, side-mounted onto the spacecraft platform. The two arms offer greater efficiency in task execution than a single arm, generate larger workspaces and increase the number of configuration possibilities [Grossman et al 1985]. This is particularly critical in space since fixtures and jigs are not always available in the highly demanding zero-g environment. The kinematic configuration suggested is a generic one based on the PUMA 560/600 geometry suggestive of the human arm—Ranger adopts an 8-degree-of-freedom redundant configuration of R-P-R-P-R-P-Y-R design [Parrish 1998]. The standard task for space robotics is generally considered to be the ORU (orbital replacement unit) exchange. The multi-modular spacecraft (MMS) attitude control module used for the Solar Maximum Mission of 1984 had a mass of 225 kg (and dimensions $1.4 \times 1.4 \times 0.5$ m) and this represents the upper limit to the ORU (orbital replacement unit) changeout mass expected to be undertaken by ATLAS during satellite servicing. There are certain tasks that cannot be performed by a single arm by virtue of its nature. Legged motion along truss structures requires two arms for a freeflyer. Handling of extended and flexible objects requires two-arm manipulation. For instance, a basic requirement is the need to cut through and subsequently repair thermal insulation to get to components. Satellite grappling also implies the necessity of using dual arms since two arms can provide up to 2–5 times the force-torque capability of a single arm [Bruhm 1987]. Indeed, two may be regarded as the absolute minimum number of arms required for parallel and coordinated task execution. Side-mounting generates larger workspaces than overhead mounting, better manoeuvrability, ease of stowage and anthropomorphic configuration [Hemami 1985, 1986a,b]. A modified Zambesi bridge configuration would eliminate the problems of inter-arm collision [Mowforth & Bratko 1987]. For two-arm control several modes of operation are possible providing great flexibility of operation [Bruhm 1987]:

(i)   single arm function—in this function, the second arm may be used as a stabiliser or a leg by attaching the spacecraft to the worksite;

(ii)   coordinated arm function—parallel execution of similar multiple tasks to reduce the task completion time;

(iii) simultaneous cooperative or auxiliary actions on a single task—examples include the rigid transfer of extended or flexible objects (closed chain configuration).

In 1988, the UKAEA, under contract to the DTI as part of the Advanced Robotics Initiative, studied the feasibility of developing space robotic systems. They considered three separate systems:

(a)   Internal Experiment Manipulator for use within a space station environment (this has now seen development as ESA's EMATS).

(b)   Local External Manipulator fixed externally to a space station (such as RMS-derived manipulators for use on ISS).

(c)   Satellite Servicing Vehicle (SSV) with a mobile (freeflying) capability.

They highlighted the SSV as being considerably more sophisticated than the other proposals in that it possessed two specific properties of complexity: the variable platform reference and its bi-arm dexterity. Both are considered in this text. Schenker (1988) devised an evolutionary programme of task capabilities for such a concept:

(i)   stationary multi-arm function for simple satellite servicing tasks;
(ii) mobile multi-arm function for simple servicing tasks (e.g. grappling);
(iii) mobile multi-arm function for complex servicing tasks (e.g. ORU exchange);
(iv) mobile multi-arm function for unplanned repair involving fabrication (e.g. thermal blanket handling).

A symmetric control scheme should be used rather than a master/slave configuration to provide operational flexibility, allowing role exchange between the arms, and provide for simpler operation, e.g. panel removal, tool replacement, camera positioning. Hence, the two arms should have similar dimensions, masses and structures to retain symmetry. This is consistent with the Ranger design which utilises two identical dextrous manipulator arms. Furthermore, each manipulator should be 6-degree-of-freedom (minimum) revolute-jointed structures. Revolute joints offer larger workspaces and higher flexibility of operation over prismatic joints, which exhibit large inertias from the slide and require higher motive power. The prismatic structure also has a tendency to interfere with the workspace. Since the freeflyer system already possesses 6 degrees of freedom through the platform translation and rotation, 6-degree-of-freedom manipulators should be sufficient. The adoption of 7+-degree-of-freedom arms merely adds unnecessary mass to the space-craft with little improvement in functional performance and a great increase in computational complexity of control—Ranger opts for the functional performance criterion by adopting 8 degrees of freedom for its manipulators.

The baseline rendezvous operation is taken to occur from below and behind the target so the workspace must extend both in front and above the spacecraft bus mount (up being defined as the vector opposed to the thrust vector). The baseline servicing task is the ORU exchange of standard modules for logistic resupply and inspection activities. Over time the learning curve may allow more sophisticated capabilities to be included in the manipulation inventory such as construction and assembly and thermal blanket handling. Like BIAS, ATLAS should be compatible with the Shuttle RMS as a smart front end so it should incorporate a grapple fixture on the upper torso opposite to direction of the workspace. A carousel mounted between the lower propulsion system assembly and the upper main torso housing provides a dynamic tool rack for easy access. If the upper assembly is also rotatable this provides an additional degree of freedom to the system enhancing its flexibility. This is a different configuration to Ranger which adopts a 30.5 cm^3 cubic manipulator module for mounting its four manipulators on each side, separated from the main electronics housing which comprises a truncated pyramid 76.2 cm long and 50.8 cm^2 at its base. The novel parallel configuration of Ranger with its workspace directly forward of the manipulator module and the Ranger spacecraft bus directly aft of the workspace ensures that the Ranger spacecraft bus structure does not impinge on the workspace—ATLAS has an orthogonal configuration with the workspace lying orthogonal to the spacecraft bus with the potential for interference, but this was opted for in providing easy access to the carousel tool-rack between the upper and lower assemblies. To provide a telerobotic capability, sensing and actuation is required with man–machine interfaces to the human operator on the ground. Force feedback and stereo vision should be supported for the human operator, i.e. bilateral force reflecting hand controllers and virtual reality representation of visual information.

Following Deptovich & Houghton (1989), a design process methodology has been adopted based on robot task definition and requirements. A set of suitable performance requirements have been derived based on the functional requirements of peak torques/forces, velocities, etc. and workspace. On the basis of these performance requirements the manipulator criteria may be defined. The primary task requirement centres around the assembly process. The control system should be part of an integrated architecture of sensors and actuators. The design parameters selected for ATLAS are in agreement with other analyses performed particularly ESA's BIAS system [Andre et al 1990].

Position repeatability: 0.1mm
                         0.1°.
Payload capability: 250 kg (maximum ORU mass)
Workspace reach radius: 1.2 m/1 mm resolution
Maximum end effector velocity: 1 m/s (1 mm/s resolution)
                                30°/s (1°/s resolution)
Force/torque capability: 100 N with 0.5 N resolution
                          (cf. Ranger 30 lbf, Robonaut 5 lbf)
                          100 Nm with 1 Nm resolution
                          (cf. Ranger 30 ftlbf, Robonaut 5 ftlbf)
Arm stiffness: 2×105 N/m (double the lateral stiffness of the remote centre compliance device)
Power requirement: 500 W peak

The inclusion of power tools would provide a force/torque capability exceeding that of the ATLAS general-purpose end effector and such tools may be useful for capture operations of large satellites with the manipulator arms adopting degrees of limpness/stiffness. As the attitude control system is the most error-prone spacecraft subsystem, it would not be acceptable to rely on the target manoeuvring capability. The system should carry a basic tool set of power tools designed to cope with the majority of servicing tasks that are likely to be encountered and augment any force limitations of the servicer. The primary task is the ORU exchange. Much of it could be automated with the human operator providing supervisory control and dealing with more difficult tasks such as the need to cut through thermal insulation to get to components. If the manipulator joints are stiffened and locked, ATLAS may use its onboard control moment gyro attitude control system to passivate the target on capture. The inclusion of a telescopic stabiliser with an EVA handhold gripper would provide a docking fixture to allow the use of both arms for all parallel tasks. The stabiliser should be more robust than the arms with higher stiffness and force/torque capabilities ~1 kN/1 kNm. This would provide the ability to withstand large reaction forces from the target particularly during force control. An additional telescopic capability for the arms may extend the arm extension to near 2 m by lengthening the upper and lower arm links would increase its operational capabilities but compliance could become a problem. The stabiliser is not adopted here for simplicity and mass constraints (the second arm can act as a stabiliser) but it is included in the Ranger design as a dedicated 2.7 m long, 7-degree-of-freedom grapple manipulator of Y-P-R-P-P-Y-R design capable of outputting 25 lbf force and 225 ftlbf torque at its endpoint to ensure the transmission of reaction forces back to the target structure [Parrish 1998]. During the RTSX, it will be permanently grappled to the Spacelab pallet.

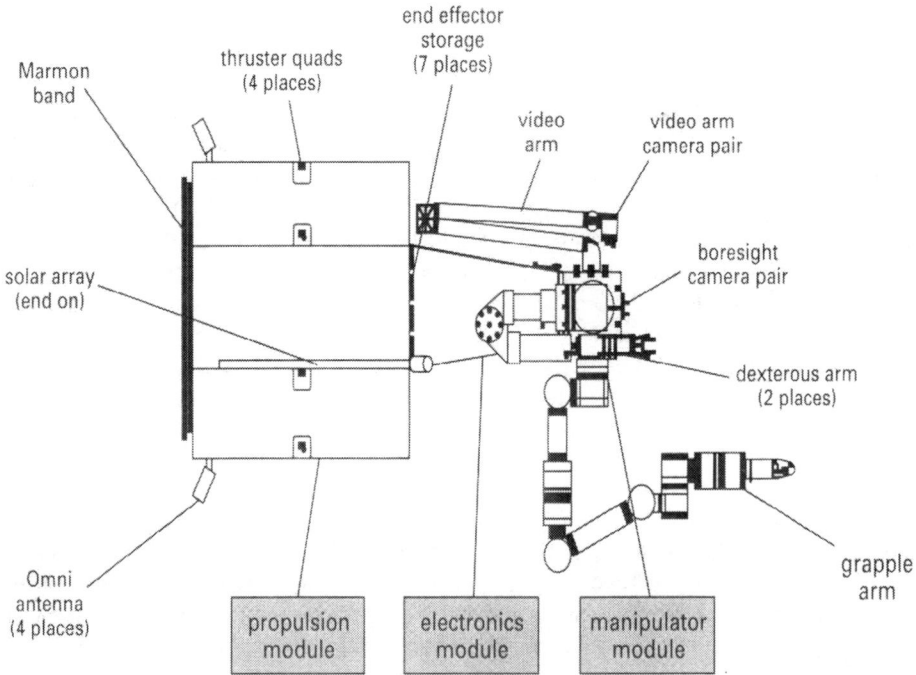

Fig. 5.1. Ranger TFX Vehicle Configuration (from Parrish & Akin 1995). Reproduced with permission by Joe Parish.

5.2.1 The sensing subsystem

Teleoperation incorporating a man in the control loop requires a suite of sensors for sensor feedback. Sensing enables tasks to be performed in the presence of environmental uncertainty and sensory feedback provides information for coping with that uncertainty. Such sensors should be smart with dedicated onboard processing to buffer and compress information to reduce the requirement for data rate transmission. The most important sensor is the visual imaging system. This requires closed-circuit TV cameras with pan, tilt and zoom capability—CCD cameras offer better fidelity than vidicon cameras. Vision also requires active illumination as there is little ambient lighting. For real-time tracking of a dynamically changing environment, 30 frames/s are required. Typical resolution is 512×512 pixels. Two black and white cameras are superior to a single colour camera. The Lunar Rover camera had a mass of 4 kg and colour CCD cameras are available with masses of less than 1 kg (excluding actuator drives) although typically they are ~10–30 kg including drives. The baseline chosen here is 8 kg. Two stereoscopic vision cameras with illumination mounted between the arms provides a general vision capability similar to Ranger. Each wrist should also have camera mounts with lighting units for close work—this is consistent with Ranger. An overhead panoramic camera with a lighting unit on an extendible boom could provide an overall workspace view at different angles but not adopted here—this is the approach adopted by Ranger with a dedicated 1.4 m long 7-degree-of-freedom manipulator of R-P-R-P-R-P-R design mounted with a camera to provide orthogonal views of the worksite for general tasks [Parrish 1998]. Vision processing involves low-level feature extraction (segmentation, edge detection, edge clustering and chain coding) which may be implemented as hardware using special-purpose pipelined processing for real-time operation to aid the human operator. Such low-level processing can provide the information necessary on centroids, principal moments of inertia, perimeter and area of the outlined object. Onboard low-level processing allows the use of reactive control by the robot control system. CAD database models as a source of geometric information should be available to ATLAS to enable it to perform pattern recognition by template matching and to the human operator. Distance information can be derived from the stereoscopic cameras through triangulation and this could be combined redundantly with a laser radar range finder to ease the registration problem characteristic of stereovision.

Microwave radar is well established for rendezvous and docking operations in space and offer accuracies ~3 m and are well suited to highly reflective metal objects. They offer the ability for coarse target acquisition and tracking. The laser rangefinder offers greater accuracy but shorter ranges. A CO_2 laser radar requires ~50–150W of power. Radar ranges for noncooperative targets (without beacons) are typically ~70 km for microwave and ~20 km for optical signals [Korf 1982]. Radar is not included in the baseline design as stereoscopic vision provides range finding for docking distances without the additional incurrred mass penalties of rangefinder radar. For closer distances proximity detectors may be used at the end effector and pressure pads for tactile information. Proximity sensors are usually light sources and photocells between the fingers of the end effector. Tactile sensors at the fingers (possibly at the elbows and distributed over the gripper itself) are usually doped silicone rubber skins whose conductance changes with pressure. Resolutions of 250 points/cm^2 are available. Compliance is an absolutely essential requirement for effective manipulation. Six-degree-of-freedom force/torque sensors may be implemented at the wrist as piezoelectric transducers with a range of 0.5–1000 N. Position and velocity sensors at the joints provide proprioceptive internal servocontrol information.

5.2.2 The actuation subsystem
Joining tasks require at least two manipulators unless a jig is available. Each manipulator has six degrees of freedom with revolute joints as a minimum for complete workspace reachability as the spacecraft offers redundant degrees of freedom of translation and attitude. The end effector should comprise of exchangeable general-purpose three-fingered grasping hands and a suite of special-purpose tools. The end effector should use snare wires to close the end effector evenly for soft grappling. All exchangeable components are stored in the carousel storage racks. An additional standardised ball-and-socket mechanism may be advantageous for docking with the target spacecraft as a separate specialised telescopic stabiliser arm to maintain the ATLAS platform in a rigid attitude with respect to the target. The manipulator links should be constructed from carbon-fibre composite for high stiffness and strength in a truss-patterned structure to minimise mass

with impact protection rubber and insulated with multilayer thermal blankets to protect the interior wiring and motors. The ratio of Young's modulus to density for graphite is 7.2 compared with 2.45 for Ti. The ROTEX manipulator was constructed in this way with grid-like carbon fibre links each of 0.2 kg. Brushless direct drive dc electric motors for each joint would eliminate the need for gearing and pulleys as well as offering high torques. The joint structure should be constructed from Ti alloy for high strength. ROTEX used Ti motors of mass 1 kg. DC direct drive motors may be used with the following characteristics (similar to the direct drive robot motor): link 1 motor of 6 kg has a diameter of 50 cm and a 200 Nm torque capability; link 2 and 3 motors of 4 kg mass have a diameter of 30 cm with 140 Nm capability; the three wrist motors of 3 kg have diameters of 20 cm with 50 Nm capabilities. Ranger uses brushless dc motors with harmonic drive gear transmission. Each link has the following lengths: $d_2 = 0.25$ m, $a_2 = d_4 = 0.5$ m and $d_6 = 0.2$ m+ (dependent on the tool or end effector—the remote centre compliance has a length of 0.2 m while the HST power tool has a length of 0.3 m). This is almost identical in reach to the Ranger dextrous manipulator. The motors contribute most to the link mass. With Ti motors and carbon fibre truss links and thermal insulation an additional 30% above motor mass is added so that link 1 has a mass of 8 kg, links 2 and 3 have a mass of 6 kg and the wrist has a mass of 12 kg. The end effector is projected to have a mass of 6 kg, most of which is due to a 4-kg, 30-cm diameter motor to apply 100 N applied forces and 100 Nm applied torques. These figures are consistent with the ⩽2 kg EVA HST power tool applied torques and the Japanese Space Station Module Arm capabilities. Ranger carries seven interchangeable end effectors with six different tools. Inclusion of the remote centre compliance adds a negligible mass at 0.45 kg. Because the joints cannot be thermally controlled through insulation alone, resistance heaters may be incorporated at the interior of each joint and inside the end effector to maintain sufficiently high temperatures for operation of the lubricant. The joint torques required should be ~100 Nm.

5.2.3 The control system and man–machine interface

The man–machine interface connects the operator on the ground to the sensors and the manipulators. As a minimum, the interface includes graphic displays for sensor data, stereoscopic visual imaging and bilateral force reflecting joysticks for controlling the manipulator. Sensors must also be pointed in the desired direction automatically by default with supervisory override. To cope with the time delays of several seconds due to the GEO round trip and network transit times, predictive graphics are required which may be updated with sensory feedback to avoid move-and-wait strategies. The addition of speech recognition and auditory feedback would provide an additional modality for the interface. Overall control is shared between the human operator, ground-based expert systems and the spacecraft onboard processors. Low-level control includes the formulations outlined later for dynamic control of the manipulator up to trajectory generation level which should be implemented by the spacecraft. Task planning and other higher-level functions may be implemented on the ground and shared between the human and the AI (artificial intelligence) expert systems.

The robot control system will require dedicated processors for each arm plus a higher level processor for co-ordination functions totalling a capability ~10 Mips. The RMS is controlled by the Space Shuttle General Purpose Computer which is limited to a 12.5 Hz update rate. The best computational architecture for various spacecraft and robotic functions are based around a distributed set of dedicated microprocessors for each subsystem or function. On the lunar probe Clementine, general spacecraft processing was performed by a 1.7 Mips MIL-STD 1750A, while image processing was performed by a 18 Mips 32-bit RISC processor. Each arm has its own decentralised controller with which to perform dynamic hybrid position/force control functions. Dedicated processors have the advantage of localisation of software changes, simple interfaces and error isolation. Embedded real-time microcontrollers such as the Intel MCS-96 and the Motorola 68HC11 offer input–output intensive capabilities. Transputers offer ~10 Mips and, with on-chip RAM disabled, offer ~50 krad radiation hardness. This is sufficient for LEO which is characterised by ~1 krad/y environment but probably not sufficient for other orbits such as GEO. The currently space-rated microprocessor is the ESA MA31750 8086 16-bit processor based on silicon-on-sapphire technology, though the SPARC processor has recently become available in radiation hardened form. The MA31750 offers ~1–3 Mips processing power with a radiation hardness of 500 krad in a mass of 20 kg and power requirement of 50 W. Bubble memories offer robust performance. Their lack of flexibility can be overcome by using the shared central processors for dynamic resource allocation. A centralised spacecraft bus processor coordinates spacecraft bus functions and a coordinating centralised robot processor provides for manipulator arm coordination particularly in the closed chain configuration. Both centralised processors communicate with each other through an overall spacecraft-wide processor (a hierarchical architecture). Such a hardware architecture is compatible with the NASREM software architecture which is distributed between the spacecraft and the ground. This is based on the hierarchical/distributed control architecture paradigm. All electronic systems require shielding from radiation, temperature extremes, mechanical shocks and vibration. Fault avoidance through the use of highly reliable components should be augmented with fault tolerance through protective redundancy from component to subsystem level. Three processors offers functional redundancy. The implementation of computation through software rather than hardware allows reprogrammability. Fault tolerance implies the necessity of cross-strapping all arm control units and employing redundant power switch units for arm power switching to reinforce the cross-strapping redundancy. Backup joint electronics units and redundant force-torque sensor cross-strap switch ensure safety-critical redundancy for joint control. Ranger data management is to be implemented by two primary R4400 RISC cpu with 32 local 80386 processors at each manipulator link distributed across a MIL-STD 1553 databus [Parrish & Akin 1995].

Around 25% of NASA's manpower is devoted to ground support and computing for ground support imposing heavy costs. The ATLAS system must support an evolutionary framework of the gradual adoption of spacecraft automation to reduce mission support costs. This implies the need for the implementation of robust artificial intelligence techniques for reliable autonomy. The boundary between autonomy and teleoperation is vague with many different degrees of supervisory control (telerobotics) in between. ATLAS provides excellent suitability to this evolutionary trend from teleoperation to autonomy through computer-aided telerobotics by virtue of the task environment. The task environment possesses a degree of structure yet it is subject to changeability. Machine intelligence requires large processing and memory capabilities so these functions would be implemented on the ground with their greater resources. Autonomy requires task level software to convert goal specifications into robot level commands. For this, the planner requires a description of the objects to be manipulated, the present state of the environment, the desired state of the environment and procedures on how to achieve goals. Central to this capability is a world model which must provide geometric knowledge and dynamic knowledge of the objects involved (e.g. from CAD/CAM models) and a simulation of the succession of environmental states between the initial and the goal situation in which the task is completed. Finally, it must also monitor the plan execution and react to changes. These techniques are particularly applicable to repetitive and well-characterised tasks such as ORU exchange. Advanced man–machine interface technologies such as telepresence should be the approach for more complex operations such as thermal blanket handling.

The parameters given in Table 5.1 have been adapted from the BIAS arm servicer unit to suit the ATLAS robot servicer [Andre et al 1990]. This represents our baseline. The payload support system of robotic control and video power of 450 W is identical to Ranger's average power requirement. The manipulator drive power of 500 W is comparable to Ranger's power requirement of 400 W for arm motors.

Table 5.1. Robotic payload characteristics

	Mass (kg)	Power (W)
Manipulator system		
Link 1	8	50
Link 2	6	50
Link 3	6	50
Wrist	12	90
End effector	8	
Subtotal	40	250
Dual arms	80	500
Payload support system		
MA31750 ($\times 3$)	20	50
Offline memory	15	
Camera ($\times 5$)	8	30
(illumination) + drive		30
Subtotal	115	450
Tool kit of 10 items of 5 kg	50	
Total	245	950

5.3 SPACECRAFT BUS DESIGN BUDGET

There are three major budgets for a spacecraft [Larson & Wertz 1992]: the propellant budget, the power budget and the mass budget. There is usually a tradeoff between these budgets for any Earth applications programme though the attitude control system is particularly important in determining the spacecraft configuration and complexity. ATLAS adds the orbit control system to this tradeoff for its requirement for variable orbits and this introduces an extremely demanding set of requirements. In the past, it has been assumed rather glibly that the Space Station could be used as a warehouse for fuel, power raising and hardware, thereby solving the problems of logistics. Furthermore, the cost of transporting such items on a regular basis to the ISS and usage of valuable ISS assets were not considered. Since ATLAS is a stand-alone system, all these considerations must be explicitly examined. Most workers to date have considered the dynamic control problem due to the coupling of the spacecraft platform to the robot manipulators to be the major stumbling block to the realisation of a robotic freeflyer, e.g. Andre et al (1990). This particular problem is tackled in the text later but it appears in retrospect that the design tradeoffs and spacecraft budgets may be just as difficult to resolve.

The propellant budget is determined by the Δv requirement plus a 25% margin (10% margin, 10% contingency and 5% off-nominal performance and trapped residual fuel overhead):

$$\text{Propellant mass, } m_p = m_i\left(1 - e^{-\Delta v/gI_{sp}}\right)$$

where m_i = initial spacecraft mass, Δv = manoeuvring requirement, g = acceleration due to gravity = 9.8 m/s^2 and I_{sp} = specific impulse. Typically, the dry-to-propellant mass varies from 2:1 to 7:1. The power budget is determined by the payload and spacecraft by energy requirements plus a 5–25% margin. For large spacecraft this will exceed 500 W and is usually apportioned: payload ~40–80%; propulsion ~0–5%; guidance, navigation and control ~5–40%; communications including telemetry, tracking and command ~5–50%; thermal control ~0–5%; power system ~5–25%; structure ~0%. The mass budget comprises the payload mass, the spacecraft bus dry mass, and the propellant mass, plus a margin of 4% and 2% for electrical and mechanical integration respectively and a 10% miscellaneous margin for design growth. The spacecraft dry mass comprises: propulsion system mass ~0.1 m_p; guidance, navigation and control mass ~65 + 0.022(m_i – 700); mass of communication system; thermal control system mass ~0.02 m_i, structural mass ~0.09 m_i; and mass of the power system plus a 25% margin.

The mission may be regarded as having two modes: the dormant mode and the operational mode. The spacecraft mission is divided into three phases:

(i)   launch and early orbit checkout
(ii)   operational phase
        — elliptical transfer orbit
        — manipulation task in operational orbit
(iii)   dormant phase—housekeeping or other assigned duties

The design approach adopted here is a conventional approach. Interfaces which provide functional boundaries joining separate subsystems across which energy, matter or information are transmitted are critical. Standardisation allows interdependability and cost savings. The packaging of all subsystems should provide good access for self-servicing and self-repair and maintain proximity between functional interfaces without interference.

The spacecraft bus can be divided into six subsystems:

(i)   attitude control subsystem
(ii)   propulsion subsystem
(iii)   electric power subsystem
(iv)   thermal control subsystem
(v)   structural subsystem
(vi)   telemetry, command and communications subsystem

5.3.1 The propulsion subsystem

The initial assumption is that ATLAS has been placed into the required orbital family by the launcher and/or by dedicated kick motors. The propulsion subsystem injects the spacecraft into the desired orbit (nominally LEO) and maintains the orbital parameters within the required limits by providing thrust. ATLAS must employ a dedicated propulsion system to perform two-burn Hohmann transfers and rendezvous and dock with targets. The manipulator arms must grapple each target and the attitude control system must passivate the complete system. ATLAS requires fuel storage and rocket engine thrusters to expel the fuel. Electric propulsion consumes a significant fraction of the spacecraft power and relates to power plant mass by $m_p = \alpha P$, where $P = \frac{1}{2}\dot{m}v_{ex}^2$ and α =specific power density = 20 kg/kW. High power requirements usually imply either very large solar arrays which are undesirable during slewing manoeuvres or the adoption of nuclear sources which are politically undesirable. Hence, as power may be regarded as a scarce resource, the propulsion system for ATLAS will be conventional liquid bipropellant fuel-oxidiser (N_2O_4/MMH) for multiple thrusts. Some comparisons are given in Tables 5.2 and 5.3.

Resistance heaters which increase the enthalpy of combustion impose a 5% power requirement during thrusting. The propulsion system comprises the lower section of the spacecraft. A single translational thruster provides the Δv capability. The spherical propellant fuel tanks are mounted symmetrically around the lower assembly and are detachable for refuelling. The total mass of the propulsion system includes propellant,

Table 5.2. Thruster types

System	Thrust (N)	I_{sp}(s)	Propellant	Life (s)	Mass (kg)
RL-10A	7.3×10^4	446	LO_2/LH_2	400	138.4
R-40B	4.00×10^3	309	N_2O_4/MMH	25 000	7.26

Table 5.3. Propellant types (adapted from Larson & Wertz 1992)

Type	Propellant	Energy	I_{sp}(s)	Thrust range (N)	Bulk density (g/cm^3)
Solid	Al/NHCl$_2$	Chemical	210–300	50–5 × 10^6	1.8
Cold gas	N$_2$	Pressure	50–75	0.05–200	0.28
Liq (mono)	N$_2$H$_4$.	Catalytic	150–225	0.05–0.5	1.0
Liq (bi)	O$_2$/H$_2$	Chemical	450	5–5 × 10^6	1.14/0.7
	N$_2$O$_4$/MMH	Chemical	300–340	5–5 × 10^6	1.43/0.86
	F$_2$/N$_2$H$_4$	Chemical	425	5–5 × 10^6	1.5/1.0
Resistojet	N$_2$, N$_2$H$_4$.	Electric	150–799	0.005–0.5	0.28, 1.0
Ion	Hg	Electric	2000–6000	5 × 10^{-6} – 0.05	13.5

tankage, propellant feed lines and thrusters. Tankage and feed lines are the heaviest components other than the propellant itself and usually comprise ~10% of the propellant mass. Mounting hardware comprises ~20% of the mass of the propulsion system mass. For intercept and rendezvous, a guidance radar is usually required but Global Positioning System (GPS) is preferred to provide a backup system for the baseline spacecraft adopting stereovision. A two-way 1000-km transfer between 200 km and 1200 km allows access to LEO spacecraft and HEO scientific spacecraft and this is the most demanding manoeuvre that ATLAS may be required to perform as dedicated spacecraft for polar orbital rendezvous and GEO rendezvous require only a very narrow range of altitudes. This two-way manoeuvre requires Δv = 1.06 km/s. One such transfer plus a 50% margin implies a Δv of 1.59 km/s. Hence ATLAS will need to be refuelled either incrementally through piggy-back launches of fuel tanks for every worst-case operation (smaller transfer changes will reduce the required refuelling rate). Further, by remaining in the target orbit after operation completion and making transfers between operational orbits, this further reduces the refuelling rate.

$$\text{Total fuel fraction,} \quad \frac{m_p}{m_i} = 1 - e^{-\Delta v/gI_{sp}}$$

where

$$\Delta v = 1.59 \text{ km/s}$$

$$I_{sp} = 425 \text{ s}$$

for hydrazine resistojet systems

$$\rightarrow \frac{m_p}{m_i} = 0.317$$

The total spacecraft mass is 1.5 tonnes so the propellant mass is 475.6 kg, allowing 1024.4 kg of useful equipment.

5.3.2 The attitude control subsystem

The attitude control subsystem maintains the spacecraft attitude in space particularly for pointing requirements. It requires sensors for attitude measurement, actuators to effect changes of attitude and a control system. Three-axis stabilisation is the only viable attitude control mode using control moment gyro actuators (CMG) since reaction wheels are limited to <11 Nm torques, and magnetorquers for momentum desaturation. Sensors include Earth and Sun sensors for coarse attitude determination and star sensors for accurate attitude determination in conjunction with gyroscopes. The Earth sensor is not strictly necessary for ATLAS. Star sensors may include filters to enable utilisation as sun sensors. Directional antennas, solar arrays and the robotic payload sensors all require gimballing. The antenna and the robotic sensors require two-axis gimballing while the solar array requires only one axis of freedom with the orientational axis being provided by body attitude for sun tracking. Slewing and feedforward compensation of manipulator movements are the dominant requirements.

Table 5.4. Typical double-gimballed CMGs

	Mass (kg)	rpm (10^3)	Ang.mom. capy (Nms)	Max. output torque (Nm)	Max gimbal rate (°/s)	Size	Power req. (W)
Bendix DGCMG	253	4–12	1400–4000	237	5–30	1.1 m diameter	15–30 W (standby)
Skylab DGCMG	200	—	—	500	—	—	0.2 W/Nm (torque)

Torque,

$$\tau = \tfrac{1}{2}\frac{w^2 I}{\theta}$$

provides an attitude change of angle θ where w = angular velocity. Reaction wheel size is determined by the amount of angular momentum to be absorbed but the CMG size is determined by its torque capability $\tau = h \times w$, where h = angular momentum, w = angular rate limited by the gimbal drive ~2 rad/s usually [Heimel & Schultz 1985].

The mass of a reaction wheel including housing and control electronics is given by: $m = 3.2h^{0.4}$ (kg), where $h = I_w w_w$ (Nms), w_w = wheel angular velocity, I_w = wheel inertia [Dougherty et al 1971, Huddleston 1991].

The rotor mass of a CMG may be chosen optimally with the required reaction torques, rotor speed, moment of inertia and gimbal ring radii [Heimel & Schultz 1985]:

$$\text{Rotor mass } m = \frac{2I_w}{r_a^2 + v_i^2}$$

where

$$l_w = \frac{h}{w_w}$$

and r_i, r_a = inner/outer gimbal ring radii.

Peak power required by motor depends on peak torque: $P_{max}(W) = 38\,\tau_{max}$.

The average maximum torque exerted by a CMG is 500 Nm, but peak torques are much higher. Typically, $m > 40$ kg with a power requirement of ~90–150 W for such torques. Usually, the total power includes 10% losses due to friction, etc.

Table 5.5. Typical onboard attitude control component properties

	Mass (kg)	Power (W)
Earth sensor	2–3.5	2–10
Sun sensor	0.2–2	0–3
Magnetometer	0.2–1.5	0.2–1
Gyroscope	0.8–3.5	5–200
Star sensor	5–7	2–20
Processors	5–25	5–25
Reaction wheel	$5 + 0.1h$	300 W/Nm
CMG	$35 + 0.05h$	0.1 W/Nm
Single axis actuators	$4 + 0.03\tau$	3 W/Nm

The Honeywell GG1320 ring laser gyro package of three laser gyros and three accelerometers (the 150×125 mm Miniaturised Inertial Measurement Unit) has a mass of 2.5 kg and a power requirement of 25 W. A strapdown system requires in addition a star tracker and other scanners which are available at 500g. The attitude control system will utilise two gimballed star trackers which may also be used as sun sensors with filters, two CMGs for functional redundancy and three magnetometers. A typical onboard spacecraft computer has 150 kword memory (with a 25% margin) and a 400 kbps processor speed (with a 30% margin). The computational power was limited until recently to the space qualified 8086 processor. The MA31750A (MAS281) computer is an 8086 CMOS/SOS architecture with 1.5μm feature sizes offering 3 Mips and 128 k ROM and a maskable 48 k ROM with 8087 or 6800 coprocessors. It has 0.5 Mrad radiation hardness for a 10-y life in GEO. Its mass is 20 kg and has a 50 W power requirement. A 10^9 bit magnetic bubble memory offers offline storage in a mass of 12 kg. The Raytheon fault tolerant computer offers 0.25 Mips and 600 kbyte memory in a package of 23 kg mass and a power requirement of 25 W. The SARDS (Spacecraft Attitude Real-time Determination System) is a multiprocessor distributed architecture and is a MIMO system of 16-bit CPUs arranged as six microprocessor boards with eight coprocessors each in a $50.3 \times 48.3 \times 18.0$ cm box for recursive Kalman filtering which can be performed on just three of the processors [Gary 1987]. Ranger has adopted Sun sensors, a magnetometer, a GPS receiver as GNC (guidance, navigation and control) sensors. It also adopts reaction wheels, magnetorquers and cold gas thrusters for attitude actuation rather than CMGs adopted here. It also uses the R4600 RISC processor for spacecraft control.

5.3.3 The electrical power subsystem

The electrical power subsystem provides and regulates the energy raising for all spacecraft functions. It comprises a power source, power storage and power conversion units. ERS-1 (Earth Resources Satellite) was ESA's 2.4 tonne Earth Observation flagship for oceanic observation based on the French SPOT spacecraft design and was launched into a Sun-synchronous slightly retrograde polar orbit ($i = 98.5°$) in 1991 at a mean altitude of 785 km with a design lifetime of 2–3 y. It was the first European spacecraft to use active microwave devices for Earth observation. Such microwave techniques had only been used for very short missions such as Seasat in 1987 which operated for only 90 days–a massive electrical short circuit destroyed its operational capability. ERS-1's major payload, the C-band Active Microwave Instrument operated in two modes: the SAR (synthetic aperture radar) mode with 30-m resolution and the wind scatterometer mode to measure oceanic wind velocity. While the wind scatterometer required only 55 W of transmit power, the SAR mode required 4.8 kW of power but the spacecraft solar arrays could supply only 1.8 kW. ERS-1, however, used 3 kW of battery power to supplement the array power to operate the SAR for short durations of maximum 10 minutes per orbit with battery recharging during sunlit periods of non-usage. This is the general approach adopted here.

The power system uses solar arrays and batteries to cope with nominal dormant loads. Typical housekeeping requires ~150 W. Operational loads are handled by nonregenerative or regenerative fuel cells (with electrolyser at 60% efficiency). Fuel cells have an operational life of 10 000 h with a specific power of 0.5 kW/kg when run at ~1 kW power levels. Shuttle fuel cells have power outputs of 12 kW. The operational lifetime of 10 000 h gives over a year's continuous operation. Peak loads are expected to last less than 10 d at a time, offering a potential for up to 40 operational peak loads. The solar array is flexible and retractable for deployment during nominal dormant phases so lightweight cells are used. Conventional solar cells have a specific mass of 35 g/W while lightweight cells have a specific mass of 15 g/W. Solar array areal power densities are typically 0.007 m^2/W. The power budget needs to account for degradation of solar arrays from beginning of life (BOL) to end of life (EOL).

Nominal power (dormant) = 500 W

Peak power (operational) = 2.0 kW

Beginning of life power, $P_{BOL} = \dfrac{P_{EOL}}{(1 - L_d)\cos\theta(1 - T_e)}$

where L_d = life degradation = 0.22, T_e = operating temperature effect = $(T_o - T_n)\,\alpha$ = 0.08, T_o = operating temperature = 67°C, T_n = nominal temperature = 28°C, α = temperature coefficient = –0.2%, θ = sun off normal angle = 0.

This gives a $P_{BOL} = P_{EOL}/0.72 = 695.5$ W

$$\text{End of life power } P_{EOL} = P_{LOAD} + \frac{C_B \times V}{15h}$$

where V = array voltage = $1.2C_B$ and C_B = battery capacity.

This gives P_{EOL} of 500.05 W. A planar array panel has a power output in proportion to its projection area to the Sun.

$$\text{Array area, } A_a = \frac{P_{BOL}}{\eta \times S}$$

where η = efficiency ~18% for GaAs at 28°C and S = solar intensity constant = 1351 W/m².

For an EOL power output of 500 W, array area is 2.86 m². Ranger's solar arrays generate 750 W. Two symmetric arrays of 1.43 m² may be implemented as square arrays 1.2 m on a side. Array mass $m_a = P/\alpha$ where α = specific performance which varies from 70 W/kg at BOL to 15–45 W/kg (usually 25 W/kg) at EOL due to degradation from high-energy particles. The German ultralow mass panel offers a specific power of 45 W/kg (almost the same specific power of muscle at 50 W/kg). Each panel has a mass of 6.3 kg with a power output of 280 W. If α = 70 W/kg then m_a = 0.07 P_{BOL}. Equivalently, for 500 W of EOL power m_a = 9.9 kg. The adoption of deployable flexible arrays was favoured over stiffened arrays due to mass considerations. The Aristoteles mission to generate models of the Earth's gravitational and magnetic fields by remote sensing adopted the opposite approach. It required orbit raising manoeuvres between mission phases [Schuyer et al 1992]. All solar arrays and booms were stiffened to prevent the buildup of undesirable vibrations. The arrays were non-rotatable being fixed with respect to the main spacecraft. The arrays were composed of two deployable paddles and a fixed body-mounted panel. The TT&C and GPS antennas were mounted at the edges of the two deployable solar arrays paddles. The approach adopted for ATLAS in using flexible deployable arrays was constrained by the need for multiple orientations of the spacecraft. Such flexible arrays may be constructed from thin film ~7μm silicon photovoltaic cells deposited on a lightweight substrate such as Kapton polyimide.

For batteries the critical parameters are battery capacity and specific performance.

$$\text{Battery capacity } C_B(\text{Wh}) = \frac{P_{LOAD} \times t}{DNV\eta}$$

where η = charge efficiency = 0.96, D = maximum depth of discharge = 40–60% for NiH batteries at LEO, N = number of cells = bus voltage/cell voltage = 28/1.25 = 23, V = bus voltage = 28V dc and t = load duration (h) = 0.5 at LEO. For a load of 500 W, the battery capacity is 0.81 Wh.

Battery specific performance of battery mass to battery capacity is 25–40 Wh/kg for NiH$_2$ batteries with cycle limits of 10 000. Ranger adopted the use of AgZn batteries with a capacity of 3500 W h. Battery mass is given by $m_B = C_B/25 = 0.035$ kg. The power control system has a mass of ~0.02 kg/W of power throughput. Electronic equipment requires dc voltage regulation from which ~20% is dissipated, so power conversion units have masses ~0.025 kg/W. Ranger uses two main dc power buses—28V dc for control electronics and 48 V dc for actuator motors. Wiring and switching dissipates ~2-5% of operating power and the wiring harness comprises ~1-4% of dry mass. The batteries provide eclipse storage during dormant phases while the solar arrays slowly trickle charge the fuel cells used for operational periods with excess capacity from EOL array sizing.

5.3.4 The thermal control cubsystem

The thermal control subsystem maintains the temperature of the spacecraft equipment within its specified ranges. It comprises thermal insulation and usually heat pipes to redistribute thermal energy. It may also include radiating louvres for heat rejection. The thermal control system is dominated by passive control methods and typically comprises ~2% of the spacecraft dry mass and consumes around 20 W in a medium-sized spacecraft which employs active thermal control.

5.3.5 The structural subsystem

The structural subsystem provides stiff mechanical support for all the spacecraft sub-systems. It must be capable of sustaining launch loads. The maximum shell radius for the Ariane standard fairing is 1.9 m and the minimum spacing for the robot arm separation to eliminate elbow collision is 1.5 m for the PUMA-type 1.2 m robot arms. The robotic manipulators will be constructed from graphite fibre composite trusses which offer high strengths $\sigma \cong 4.5 \times 109$ N/m^2 with densities of 1.59 g/cm^3 giving $\sigma/\rho \cong 3 \times 10^6$ J/kg. A spacecraft primary Al structure of cylindrical or polygonal design would have a radius of 0.75 m to give a maximum spacecraft diameter and arm separation of 1.5 m. For a monocoque cylindrical shell under compression the critical buckling stress is:

$$\sigma_{cr} = \frac{0.68Et}{R}$$

where R = shell radius = 1.6 m, t = shell thickness and E = Young's modulus = 71×10^9 N/m^2 for 7075 Al alloy.

Critical buckling load $F_{cr} = \sigma_{cr}A = \sigma_{cr}2\pi Rt$

The critical load must exceed accelerations in excess of $8g$ steady-state load and $20g$ transient load. Ultimate design loads should be 1.25-1.5 times the load limit. Hence the ultimate design load is $F_{cr} = m_i a = 4.4 \times 10^5$ N assuming 50% margin onto a load of 20 g. This gives a shell thickness of 1.2×10^{-3} m or 2 mm. Additional mass is imposed by shielding etc. Similarly for a monocoque shell the natural frequency is given by:

$$f_n = \frac{1}{2\pi}\sqrt{\frac{EI}{\mu L^3}}$$

where μ = mass/unit length and ρ = density = 2.8×10^3 kg/m^3 for Al.

The principal driver to the structural subsystem is the requirement for minimum mass yet providing structural stiffness for the payload and spacecraft subsystems. The primary structure typically comprises ~10–20% of the spacecraft dry mass with fasteners and fittings increasing the primary structure mass by around 10%. The electric wiring harness alone can account for as much as 10% of the spacecraft dry mass. A design margin of 25% should be included for structural growth. Equivalently the spacecraft structure typically comprises ~8–12% of the spacecraft launch mass. The dynamic properties of the spacecraft is determined by the following heuristic rule of thumb:

$$\text{Moment of inertia (kgm}^2) \quad I_s = 0.01 M_s^{5/3}$$

External arrays greatly increase the moment of inertia of the spacecraft by $\sim l_a^2 m_a$ where m_a = array mass, l_a = array offset from spacecraft mass centre $= 0.5l + 0.5\sqrt{A_a/2}$, where l = spacecraft bus diameter. The array axis moment of inertia increases, however, by $\sim A_a^2 m_a$. As the arrays are not deployed during manipulation the arrays will not affect the dynamic properties of the system in this phase.

5.3.6 The communications subsystem

The communications subsystem maintains a two-way data link (uplink for command and downlink for telemetry) between the spacecraft and the ground. It requires a receive/transmit antenna and a modulation system. The communication system comprises a transmitter/receiver, and additional small omnidirectional transmitter for functional redundancy and an rf diplexer for the single transmit/receive antenna. The command decoder and two telemetry multiplexers for redundancy use the basic units of the command and data handling subsystem and a dedicated digital computer should be included for coordinating complex data processing. The waveband would be allotted and controlled by the ITU within the S band for TT&C services (2–4 GHz). The requirement for a clear and continuous field of view (FOV) to the ground during real-time operations implies the need for a data relay to avoid the need for a distributed network of ground stations. The S-band allows compatibility with TDRSS/EDRSS, the Ground Space Tracking & Data Network and the Space Shuttle Orbiter. Although data relay introduces a two-way time delay, this is not considered problematic as it is negligible compared with network transit times and predictive displays at the ground station can compensate for this. The S band offers 300 kbps uplink at 2.025–2.120 GHz and 1.2 Mbps for the downlink at 2.2–2.3 GHz. EDRS offers 150 kbps uplink and 3 Mbps downlink and this is the baseline relay. EDRS requires the user spacecraft to provide an EIRP (product of transmitter power and gain) of 45 dBW and a G/T of 8 dB/K.

A typical spacecraft generates 870 bits of telemetry of which 330 is from the AOCS and 280 is from the payload. Similarly, a typical spacecraft receives 540 commands of which 270 is for the AOCS and 130 for the payload. Generally, housekeeping telemetry varies from 100–1000 bps whilst housekeeping for the payload can generate up to 10 kbps to 500 Mbps. The command rates rarely exceed 1 kbps though they have been known to reach up to 100 kbps. Most spacecraft require <1 kbps for command but it is anticipated that for teleoperation ~100 kbps is more appropriate—Ranger adopts 128 kbps real-time command uplink. Most of the data rate requirement, however, is for the downlink of telemetry data. Data rates in excess of 1 kbps require the use of high-gain directional antenna. It is anticipated that data rates for the downlink will be ~36 Mbps since a 512×512 pixel TV image requires ~63 Mbps with 8-bit words at 30 frames per second (assuming no preprocessing). The frame rate may be reduced to 2 frames/s with marginal loss in quality. The use of 8-bit words allows 256 colour levels. The use of orthogonal polarisation (doubles bandwidth capacity), adaptive DPCM (reduces data rate by orders of magnitude), QPSK (which offers 1.7 bs/Hz) and other data compression techniques can reduce this the data rate to 400 kbps (the RMS has a data link of 600 kbps and all ESA video services are compressed to 384 kbps). Ranger offers a downlink capability of 4 Mbps for its video transmission to the ground. The advantage of voice commanding is that it requires only ~64 kbps (48 dBs/s)–a typical transmitter power for an audio band is 1 W. A bandwidth allocation of 36 MHz (typical of a comsat) should easily cope with five cameras and numerous sensor data through the use on onboard processing. Diameter of receiver $d = 100\lambda/\delta$ where δ = beamwidth ~1° gives a ground-based antenna diameter of 13.6 m at 2.2 GHz. The maximum power received on Earth is limited to -148 dBW/m^2/4 kHz. If the satellite EIRP is 38 dBW (typical of a comsat) with a transmitter gain of 28 dBW the transmitter power required is 10 dBW = 10 W (Voyager 2 transmitter power was 30 W). A 28-dB high-gain antenna can communicate via the S band and steer through a solid cone of $\pm110°$. It has a mass of 40 kg including drive and support assembly of which 5 kg is the antenna. It requires a nominal power of 10 W. The mass of a standard S band parabolic antenna is 10 kg (with a 28-dB gain and 150 cm diameter). A typical 10-W solid state transmitter requires ~20 W at the S band due to 50% efficiency. With a gain of 28 dB and a bandwidth of 36 MHz (typical of a comsat) such a transmitter would have a mass of 2 kg and occupy 3400 cm^3. High-directional antennas range in size from 0.3 to 3 m in diameter. The diplexer and filters for the S band antenna have masses of 2 kg in total. Receivers have a size and mass of 150 cm^3 and 1.8 kg typically with a power consumption of 3 W. A TDRSS-compatible S band trans-ponder has a mass of 7 kg and a power requirement of 4.5 W and dimensions of $14 \times 33 \times 14$cm. A typical TT&C central microcomputer will have a mass of 7.5 kg and a power requirement of 27 W and dimensions of $12 \times 23 \times 40$ cm.

The link budget is determined as follows with $S(\text{dBW}) = 10 \log_{10} P(\text{W})$:

$$\frac{E_b}{N_0} = \frac{P_t G_t G_r}{L_a R} \left(\frac{\lambda}{4\pi r} \right)^2 \left(\frac{1}{k} \right) \left(\frac{G_r}{T_s} \right)$$

where $L_s = (\lambda/4\pi r)^2$ = free space losses = 195–215 dB, k = Boltzmann's constant = 1.38×10^{-23} J/K = 228.6 dBW/K/Hz, $G_r = 4\pi A_e/\lambda^2$ = receiver gain =15–65 dB for a parabolic reflector, A_e = effective aperture area = $\eta A_r \sim 0.55 A_r$ for a parabolic reflector, $A_r = \pi D^2/4$, L_a = transmission losses due to rain, etc. = 0 dB, $N_0 = kT_s$ = system noise, T_s = system temperature, P_t = transmitter power = 10 W (identical to Ranger's BPSK transmitter) G_t = transmitter gain = 28 dB for a 150 cm antenna, R = data rate = 1.2 Mbps and B = bandwidth = 36 MHz.

Receiver performance and quality is dictated by the G/T ratio: $G/T \sim 23$ dB typically assuming an Earth station temperature of 200 K (23 dBK) and an omnidirectional receiver. System noise is typically $N = 30$ dBW and GaAs device have noise temperatures of 70 K. Carrier to noise ratio may be given by:

$$\frac{C}{N} = P_t G_t \left(\frac{\lambda}{4\pi r}\right)^2 \left(\frac{G_r}{T_s}\right)\left(\frac{1}{kB}\right)$$

For a 2.2-GHz waveband and antenna efficiency of 0.55, a 1.5 m diameter transmitter (28 dB) the link budget for a spacecraft–Earth link:

Transmitter power	10 dBW
Transmitter gain	28 dB
EIRP	38 dBW
Free space loss	−206 dB
Receiver G/T	23 dB/K
Boltzmann contribution	228.6 dB
Data rate	−60.8 dB

Hence

$$\frac{E_b}{N_0} = EIRP + L_s + \frac{G}{T} + 228.6 - 10 \log R = 22.8 \text{ dB}$$

$$\frac{C}{N} = \frac{E_b}{N_0} + 10 \log R - 10 \log B = 22.8 + 60.8 - 75.6 = 8 \text{ dB}$$

This margin for a 10^{-6} bit error rate is more than sufficient for the Earth–spacecraft link budget.

Table 5.6. Telecommunications component properties

	Mass (kg)	Power (W)
TDRSS transponder	7	4.5
TT&C PCM computer	7.5	27
Antenna (incl. drive)	40	0
Diplexer	2	0
Transmitter	2	20
Receiver	2	3

Ranger adopted four omnidirectional antennas with an uplink capabity of 0.5 Mbps and downlink capability of 4 Mbps (including imaging downlink) to communicate with a high-gain receiver ground station with six 22-minute overhead passes per day. The need for lengthy passes restricts the orbit operations to a minimum of 100 km altitude. The intermittent nature of conducting manipulations was regarded as undesirable here, so the baseline ATLAS design adopts a continuous EDRS relay with a directional antenna.

5.4 SPACE MISSION RELIABILITY

Spacecraft are long-life systems which cannot be easily or cheaply maintained manually over the operational lifetime. Reliability is critical in situations where malfunctions can have disastrous consequences. ORUs are also subject to failure of which the component failure rates are outlined in MIL-HNBK-217C [Winchell 1987]. Their MTBF (mean-time-before-failure) rating is 20 y. Industrial class robots currently have MTBF ratings of ~10 000 h of continuous operation—this is acceptable for space application since the manipulators will be operational for limited periods of time during the mission. Standby failure rate while unpowered is 10% that at 100% duty cycle. Hence, shutting down systems into standby or dormant mode should be adopted when systems are not required to reduce their operating times, though some systems will be vital throughout the mission lifetime.

All components and subsystems are subject to the possibility of failure which generally increases with the mission elapsed time. Failures occur when the actual performance of a system or subsystem deviates from the required performance. Faults are manifested as errors. The larger and more complex the system (i.e. the more components), the greater the probability of failure. Several kinds of failure are possible: random component failures are characteristic of electronic equipment, wearout failures are characteristic of the power subsystem, design faults are caused by unanticipated stresses, and manufacturing faults may be prevented by effective quality assurance. Faults can be classified in a number of ways [Siewjorek 1991]. Transient faults occur over a limited duration and tend to subside but do not have well-defined manifestations and occur only under certain unstable or intermittent conditions (e.g. power surges, static discharges). Permanent faults are caused by hardware failures and recovery involves the use of duplicate components. Single faults are limited to a simple logic variable while distributed faults affect more than one variable. Distributed faults are often caused by single failures of some critical element such as clock, power system, databases, reconfiguration switches, etc. and they then propagate through the system. Hence, some form of reliability analysis is required to attempt to reduce the impact and incidence of failures.

Reliability may be defined as the conditional probability that the system has survived the interval $[0, t]$ given that it was operational at time $t = 0$. Basic reliability is the probability that all elements will survive while mission reliability is the probability that essential mission elements will survive and this is the most important type of reliability as far as spacecraft are concerned. This usually involves functional redundancy concerned with meeting mission requirements from dissimilar elements, e.g. use of thrusters if reaction wheels fail. The reliability of a typical comsat is ~0.8, i.e. an average of 20% downtime. Expected cost of failure may be given by [Hecht 1973]:

$$V_f = V_s \int_0^W \left(1 - \frac{t}{W}\right) R(t) \frac{dF}{dt}$$

$$= cV_s \int_0^W \left(1 - \frac{t}{W}\right) e^{-(c+s)} dt$$

$$= \left(\frac{cV_s}{c+s}\right)\left[1 + \frac{e^{-W(c+s)}}{W(c+s)}\right] \approx F_s \times V_s$$

where W = specific wearout time, $R(t) = e^{st}$ = reliability for a single spacecraft, t = mission elapsed time, $dF/dt = ce^{-ct}$ = failure rate, c = potential hazard in mission life/y ~0.1/y typically, $s = -4\lambda$, $\lambda = 1/MTBF$ where MTBF = mean time between failures, V_s = cost of loss of spacecraft = cost of replacement spacecraft at launch and F_s = probability of failure.

Reliability improvement is beneficial as long as the decrease in failure cost exceeds the cost of reliability improvement:

$$F_0^u = \frac{F_s}{F_0}$$

where F_0 = failure probability, F_s = failure probability after improvement, $u = V_r/V_0$, V_r = fractional cost of reliability and V_0 = baseline spacecraft cost.

If $u = 0$, $V_r = 0$ and $F_s = F_0$ (no redundancy); if $u = 1$, $V_r = V_s$ and $F_s = F_0^2$ (full redundancy). The reliability benefit is determined by the added mass fraction $F_s = F_0^{u+1}$ where $u = m_r/m_0$. To a first approximation, redundancy doubles or triples costs.

Fault avoidance is the initial procedure in increasing reliability by using high-reliability components which have been extensively tested to reduce the possibility of failure. The failure rate of high quality components are defined in MIL-STD 1543. Class S electronic components have one-quarter the failure rate of general military specification and one-tenth the failure rate of high-grade commercial specifications. Indeed, the cost of failure for space systems justifies the use of only the highest-grade components, materials and parts. All critical spacecraft components and system must be self-testing and removable for repair and replacement, i.e. be modular and easily accessible. The mean-time-to-repair (MTTR) should also be as minimal as possible. However, fault-avoidance procedures are not sufficient for fault tolerance since they still imply high-cost delays from failures when they do occur.

Fault tolerance is the ability to continue operations after a failure and usually involves redundancy and redundancy management should be performed automatically to reconfigure the system to minimise the effects of failure and limit damage. First and foremost is the requirement to eliminate the possibility of single point failures which are catastrophic. It is also desirable to allow non-catastrophic graceful degradation in performance in the event of increasing failures. It is also desirable to avoid failure propagation by using 'stand-alone' systems. Redundant parts are present in standby mode ready to be invoked automatically. The two approaches of fault-tolerance and fault-avoidance are complementary such that resources are divided between fault-avoidance approaches and fault-tolerant approaches to yield the highest reliability [Aviziennis 1976].

Fault tolerance is difficult for structures and parts of the propulsion system. However, these are designed for well-modelled stresses. Fault tolerance involves detecting failures on which control is switched to redundant elements. Parallel redundancy is preferred over series redundancy since if one component fails in series redundant configuration the system fails: for series redundancy: $R_s = R_1 R_2 R_3$; for parallel redundancy: $R_p = 1 - (1 - R_1)(1 - R_2)(1 - R_3)$ such that $R_p > R_s$.

Systems are composed of a hierarchy of levels and errors may be generated at any level. As errors propagate up the hierarchy they affect an increasing fraction of the system and have longer response times so it is essential to detect errors at the lowest levels of the hierarchy. Fault tolerance may be at the component level or subsystem level and both should be employed complementarily. Components with high initial probability of failure F_c and low unit cost V_s (high failure value ratio F_c/V_s) are good candidates for redundancy:

$$\text{Component reliability, } R = \left[1 - \left(1 - e^{-\lambda t} \right)^2 \right]^4$$

For system-level reliability the best approach is to reduce the number of components selecting components with low failure rates and introduce redundancy:

$$\text{System level reliability, } R = 1 - \left(1 - e^{-4\lambda t} \right)^2$$

Component redundancy offers greater reliability since the cause of most failures is due to component degradation [Hecht 1973]. Standby redundancy (where the redundant units are in dormant mode) is preferred over active redundancy (where the redundant units are operational). This is because standby failure rates are ~10% that of active redundancy. The reduced operating time enhances reliability. Redundancy is especially effective when a function requires several identical components. This is k-out-of-n redundancy where k = number of active elements required and n = total number of components initially provided where $k < n$. One shareable spare will make redundant each of the identical units required for the function, e.g. a single shared spare gives $R = 1 - (1 - R_i)^4$. Functions are divided so they can be served by identical units, e.g. processors.

Active dual redundancy,

$$R = 2e^{-\lambda t} - e^{-2\lambda t}$$

with failure rate

$$\frac{dF}{dt} = 2\lambda \left(e^{-\lambda t} - e^{-2\lambda t} \right)$$

Triple modular redundancy,

$$R = 3e^{-2\lambda t} - 2e^{-3\lambda t}$$

with failure rate

$$\frac{dF}{dt} = 6\lambda \left(e^{-2\lambda t} - e^{-3\lambda t} \right)$$

Most subsystem failures are 74% attributable to the spacecraft bus and 26% to the payload in the following distribution: TT&C ~25%, ACS ~14%, power ~13%, avionics ~9%, thermal control ~6%, propulsion ~4%, structure ~3%, sensors ~13%, communications ~5%, navigation ~3%, and dedicated payload ~3%.

There is a certain amount of risk inherent in technological innovation based on the uncertainty in the success of applying new technologies. The newer the technology the greater the risk of failure. NASA has devised a technology classification and relative cost risk listing (see Table 5.7). The most likely estimate (MLE) of costs are subject to uncertainty distributions due to the technology impacts quantified as standard deviation from standard cost estimates. Hence, the cost of ATLAS should have a probability distribution with a standard deviation of 25%.

Table 5.7. Technological risk

Technology readiness level:	Definition	Relative risk level	SD about MLE (%)
1	Basic principles validated	High	>25
2	Conceptual design formulated	High	>25
3	Conceptual design tested analytically or experimentally	Medium	20–25
4	Critical characteristics demonstrated	Medium	15–20
5	Breadboard tested in relevant environment	Medium	10–15
6	Prototype engineering model tested	Low	<10
7	Engineering model test in space	Low	<10
8	Full operational capability	Low	<10

Static redundancy is employed extensively in space computers and involves using permanently connected and powered redundant components to mask the effects of hardware failures. Static redundancy relies on the assumption that failures of redundant copies are independent. One typical form of this type of redundancy is the triple modular replication of individual hardware modules and components for three identical copies of a computation. A separate processor is dedicated to each copy of the computational software which are operated concurrently. This offers independence of hardware operations from software. A majority vote acts in redundancy management and decides the correct output automatically neutralising the effect of the fault. This is fault masking which hides the effects of errors. This is often used in conjunction with a Hamming or Golay error correcting code in primary memory which adopt code redundancy methods for critical computations (which varies from 10 to 40% depending on the data bit-length). Indeed, RAM failures typically comprise ~60% of all system failures and computer failure rates are typically ~75 per 10^6 h. Functional redundancy is provided by a computer EX-OR gate across the driver of each transistor output pin. The failure of one component will cause a mismatch in comparators. Used in self-healing, multiply paired processors, software can remove the failed pair from the system and proceed with operations. Pairs of pairs involve one pair in processing with the other pair shadowing. The shadow executes in step with the primary but does not drive the outputs. If both are connected via an error-reporting channel then the primary will send an error report and disconnect itself without software interruption at the occurrence of failure. The shadow then proceeds with the operation.

Dynamic redundancy involves dynamically changing the system configuration automatically in response to a fault to effect online repair. This also requires backup duplication of the system with the spare standby unit operating in parallel to the active unit. When comparison detects a failure, the good unit is switched in and the faulty unit is switched off line. Dual ports in all units allow access by alternative paths in case of unit failures controlled by switches. The addition of a redundant system and switching involves 2.5 times the hardware cost of non-redundant systems. Masking through static redundancy is the method that is usually adopted for critical computations and long-life applications since error detection, recovery and correction are all part of the same process. Error detection occurs in hardware while recovery is performed by software. Usually a combination of fault avoidance, fault detection, static redundancy and dynamic redundancy are used. A two-channel system is fail-safe since a failure can be detected. A three-channel system is fail-operational since tasks can be completed with a failure. A four-channel system is fail-operational/fail-operational as it can tolerate two failures and complete the task. It is usually recommended that a minimum of dual redundancy is implemented for computer systems in unmanned spacecraft.

Voting logic itself has a probability of failure and itself is a single-point failure mode. Hence, it is preferable to let each channel remain independent and use cross-strapping whilst utilising voting as an auxiliary process for backup. Cross-strapping systems to each redundant unit such that all instruments are available to all processors increases reliability substantially [Brodie & Giardina 1975]. Cross-strapping is used extensively with banks of sensors and sensor data processors. Any failure in one unit effects a replacement of the failed component with a redundant unit permitting continuous operation in the presence of failure. If all systems are functional either voting may be employed or all data can be combined to provide a best estimate. Such cross-strapped redundant systems are more reliable than isolated redundant systems in that they are minimally fail-safe. No single unit failure affects all channels and each processor receives all the data which may be averaged. The only loss incurred by unit failure is a measure of the statistical independence between channels. A computer failure does not cause cessation of operations as no unit is dedicated to any particular channel. The data from n sensors is filtered to decrease the output variance of the instrument errors:

Output covariance matrix, $R = \sigma^2 (H^T H)^{-1}$ which is to be minimised

where H = geometric coordinate design matrix referred to body coordinates and σ^2 = noise variance

Best linear estimate of X is $Y = HX + U$ where U = zero-mean noise vector

$$\text{Gauss–Markov theorem states: } \hat{X} = \left(H^T H\right)^{-1} H^T Y$$

Failure detection and isolation (FDI) specifically deals with partial failure but can also handle hard failures. FDI detects the error vector from a sensor before it propagates through the system. If a unit fails its output is replaced by a composite output from other similar units by restructuring the design matrix H. Sensor failure is observed where the failure vector $e = X_i - X_j$ from two different sources is large. Usually e will have a small finite value without source failure since no two sensors are identical. If the threshold is set too low then false alarms will become problematic while if the threshold is set too high missed failures will occur. Furthermore, the error vector can yield a unique signature pattern characteristic of the failure mode. Cross-strapping implies dynamically variable interconnections which may themselves be subject to failure. Voting logic may be used in backup mode.

Functional testing provides maximum assurance that units will function when installed. At the instant the equipment passes its acceptance tests its reliability is 100% after which reliability falls linearly. Failure mode effects and criticality analysis (FMECA) is usually used to identify all possible failure modes, their effects and their probabilities of occurrence including single point failures. Guerin & Lane (1991) proposed that ground assembly, integration and testing is a similar problem to in-orbit control as they involve similar functions. Testing of flight systems and mission control involve the same basic functions: mission/test preparation, mission/test implementation and mission/test evaluation. The major difference is that testing involves operating all satellite subsystems in all possible modes while in-orbit control involves operating all satellite subsystems in in-orbit modes only. Garner & Ross (1991) extended this notion to pre-launch testing which involves the use of telemetry and telecommand as in satellite control.

5.5 DISCUSSION

The ATLAS total power requirement of 1742 W is comparable to that of the average power during operations for Ranger of 1650 W (though it is considerably less than the Ranger peak power requirement of 2700 W which includes power supplied to the pallet subsystems). The ATLAS total mass of 1425 kg is comparable to the the total mass of the Ranger system, task equipment and support structure of 1055 kg. The Ranger spacecraft itself has a mass of 773 kg compared with ATLAS of 1425 kg. However, ATLAS is a stand-alone system with its own propulsion system and 475 kg of fuel so this is expected.

The high structural mass was estimated using a heuristic formula hence the discrepancy from the structural mass based on a structural thickness of 1.2 mm as calculated from the Euler buckling formula without consideration of spot radiation shielding. The figure of 225 kg for a spacecraft offers a potential 1 cm thick Al structure which is greatly in excess of requirements for an unmanned spacecraft defined earlier for a structural thickness of 1.2 mm. As composites may be used (at least for the secondary structure), this mass discrepancy effectively offers a substantial margin for other tradeoffs. This is especially the case as spacecraft structural subsystems have negligible failure rate compared with active spacecraft systems. Note that the high power requirement of the

	MASS	POWER
Fuel	475.6057	75
Propulsion	57.07268	0
Communications	60.5	54.5
Power	81.2274	1115.012
Thermal	20.48789	25.00262
Structure	225.3668	0
Attitude control	47.5	103
Payload	457.5633	1369.322
Spacecraft bus	967.7604	372.5147
Total	1425.324	1741.836
Payload	Includes fuel cells	Includes CMG

Fig. 5.2. ATLAS design budget parameters (from Ellery 1996).

payload includes the power requirements of the attitude control actuators which are operated for reaction moment compensation during manipulation.

A systems design proposal for a robotic in-orbit servicer, ATLAS, has been presented and it was found that if logistic factors were taken into account, the spacecraft subsystem tradeoffs were far from trivial. The specific design of the dual-arm robotic interceptor presented was based on conventional technologies. We have indicated a feasible systems design for the ATLAS spacecraft on the basis of standard spacecraft technologies and standard robotic technologies both of which may advance over time. The logistics problems are still relevant but have been reduced to manageable proportions and it is this aspect on which new technologies will have an impact. The technology is thus in place to build such a system perhaps using off-the-shelf hardware and software which require very little major new technological innovation. ATLAS has been proposed for service in LEO—LEO operation would provide a test bed for any such spacecraft to operate in other orbits such as GEO or polar orbits. The tradeoffs yielded a 1.5-tonne spacecraft with EVA-equivalence capable of two-way 1000-km orbit changes. As always, propellant usage limits the lifetime of the spacecraft and periodic replenishment of consumables launched as piggy-back payloads would be necessary to extend the operational lifetime of the system to realistic timescales. Onboard power was another problem which required a flexible approach of using fuel cells during manipulation for peak power loads and deployable flexible arrays during non-operational phases. The desirability of autonomy within the limits of space-rated microprocessor technology suggests distribution of the computational workload through a communications link sufficient to provide video data transmission. This allows a facility for teleoperation from a ground station using EDRS for a continuous communications link during manipulation phases of the mission. The control system automatically compensates for the manipulator reactions on the spacecraft so that ground operators would not need to be retrained from terrestrial manipulators (this is intoduced later). In fact, this control compensation may be regarded as the first step towards supervised autonomy of space robotic systems. The market with an average charge of $150m per operation is estimated to potentially yield around $2b over the following 15 year operational period, representing up to a 250% discounted return on investment (see Chapter 20).

6

Robotic manipulator kinematics

Over the next few chapters, we will be examining some of the analytical methods used in robotics that are required for the robot manipulator control system of ATLAS or any other spacecraft employing robotic manipulators.

Most machines (defined as transducer devices used to apply mechanical power) are kinematic chains such that they comprise paired links between which there is relative motion at the joint. Indeed, most are reducible to one of two kinematic chains: the four-bar chain and the slider-crank chain. The four-bar chain comprises four rigid bodies connected in a closed chain by pivots, e.g. the human knee. The slider-crank is in fact a special case of the four-bar chain with one of the four members replaced by the slider. A complete revolution of the crank causes a reciprocating motion of the slider, e.g. the single slider-crank chain forms the basis of the simple reciprocating engine (see Fig. 6.1)

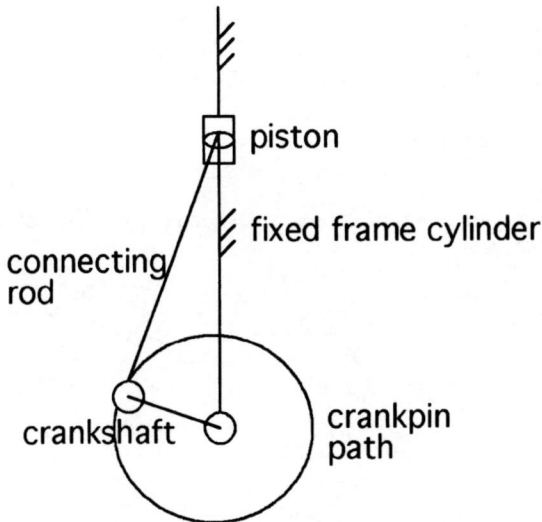

Fig. 6.1. Slider crank chain

If the mating elements are in contact across a surface then the joint is a lower pair such as those characterising linkages; if the mating elements contact at a point or line then the joint is a higher pair such as those characterising gears. Robotic manipulators are serial multiple-link kinematic chains that are powered by motors at the lower-pair joints between the links. The robot manipulator is unique in all of Man's technological creations in that it is a general-purpose tool. When used with advanced sensors and an information processing computer, the resultant system embodies all the characteristics that mark Man from the rest of the animal kingdom: binocular vision for high-resolution data processing, complex symbol manipulation and information processing, and opposable fingered gripping hand with multiple tool use for complex manipulation to change the state of the environment.

A robot manipulator must have six joints to provide the six orthogonal degrees of freedom (DOF) in position and orientation required to produce any arbitrary position and orientation within a given workspace. For a general configuration 6 DOF robot, there are $6^6 = 46\,656$ different kinematic chain configurations possible. By decoupling the first 3 DOF into arm-positioning variables and the final 3 DOF into an orienting spherical wrist where the 3 intersecting wrist axes are oriented at $0°$ or $90°$ to each other, there exist 20 arm and 12 wrist configurations which allow closed-form analytic solutions to the inverse kinematics problem. This effectively allows the decomposition of a 6 DOF problem into two 3 DOF computations, one for the arm position and the other for wrist orientation. By eliminating degenerate, equivalent and planar configurations, 12 arm configurations and 5 wrist configurations are possible [Wang & Lien 1988]. There six different lower pair joints (revolute, prismatic, cylindrical, spherical, screw and planar), but only revolute and prismatic joints are used in robotic manipulators whereby the two surfaces slide over each other while remaining in contact, so that each joint has one degree of freedom. The other joint types are merely implementations to achieve the same function (e.g. the screw joint to provide translational motion through rotary motion) or provide additional degrees of freedom in their function (e.g. the spherical joint which allows three degrees of freedom). From these two types of joint there are four common arm configurations and two common wrist configurations when neighbouring joint axes are either parallel or perpendicular to each other. The spherical wrist enables derivation of its position from the position of the end effector directly.

(1) $R\bot R\|R$ (revolute configuration)—this comprises ~25% of industrial robots, e.g.
       PUMA 560/600

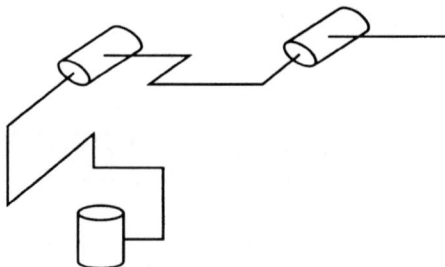

(2) $R\perp R\perp L$ (spherical configuration)—this corresponds to ~13% of industrial robots, e.g. Stanford arm

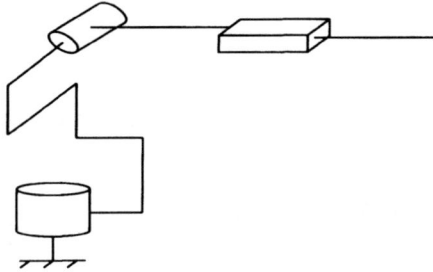

(3) $R\|L\perp L$ (cylindrical configuration)—this comprises ~47% of industrial robots

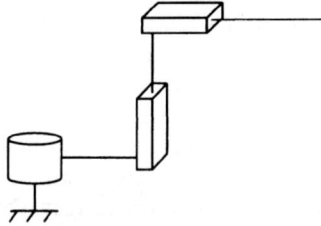

(4) $L\perp L\perp L$ (cartesian configuration)—this comprises ~14% of industrial robots

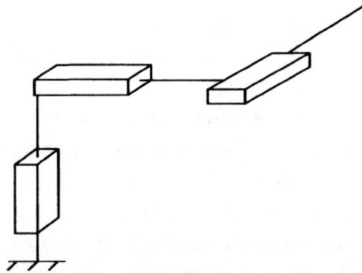

The $R\perp R\|R$ geometric configuration with all revolute joints yields the maximum workspace. It also allows a simple sequence of motions to perform the basic human arm motions of lift in the vertical direction, sweep along the horizontal direction and reach in the line of sight as well as motion in a spherical workspace. The wrist configuration requires all three neighbouring wrist axes which are perpendicular or parallel to each other to intersect at a point [Horak 1984]. Each wrist joint is a revolute joint and the two forms are: ZYZ Euler angle sequence or YXZ roll–pitch–yaw sequence, e.g. Unimation PUMA 560/600 has RRR arm and Euler configuration wrist while the Cincinnati Milacron T^3 has RRR arm and a RPY configuration wrist. The Euler angle sequence bears direct relations with the twist–turn–tilt motions commonly exhibited by human wrist actions and is the commonest convention. The human wrist very rarely engages in yaw motions and the ZYZ wrist double roll provides great flexibility in anthromorphic tasks.

The roll axis also provides redundancy for rotary tools. Furthermore, the *ZYZ* wrist tends to be more compact than the *RPY* wrist.

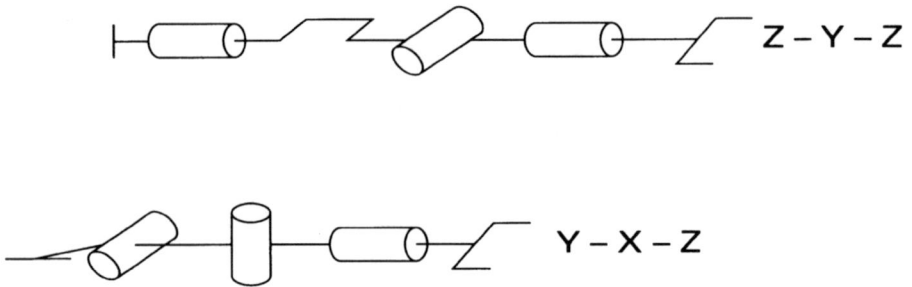

Fig. 6.2. Wrist kinematic configurations.

In terms of the kinematic formulation, a 6 DOF robot comprises six sequential links and six joints with the last two links having zero length and mass—the wrist parameters characterising the last three joints are lumped into the last link. This spherical wrist then enables the derivation of the end effector position directly from the wrist position/orientation. The end effector/tool parameters are usually lumped with the wrist into the final link. The end effector is usually a dextrous multifunctional gripper (either two parallel jaw gripper or three-fingered hand) which in fact offer additional degrees of freedom beyond the manipulators' normal 6 DOF (degrees of freedom). In addition, specialised tools, e.g. capture tools, ORU replacement tools, connector engagement tools, etc. may be replaced in substitution to the end effector. Such tools could be stored in a carousel mounted onto the spacecraft lower assembly [Elfving 1990a,b].

The robot control problem may be characterised in the statement that task requirements are normally specified in terms of cartesian world coordinates describing the motion and trajectory of the end effector, while control and commanding of the robot actuators is performed at the joints as motor torque commands. The desired motion of the end effector is specified as a trajectory in cartesian coordinates whilst the control system requires inputs in joint coordinates. Furthermore, proprioceptive sensors generate data concerning the actual state of the environment at joint level. This implies a requirement for expressing the kinematic variables of position, velocity and acceleration in end effector cartesian coordinates to be transformed into their equivalent joint coordinates. These transformations are highly dependent on the kinematic geometry of the manipulator. Hence, the robot control problem comprises three separate computational problems:

(i)   the determination of the desired trajectory in cartesian world coordinates;
(ii)   the transformation of the cartesian trajectory into equivalent body (joint) coordinates;
(iii) the generation of the motor torque commands to realise the trajectory.

For smooth straight line trajectories, several joint motors must be driven simultaneously at different rates in a coordinated fashion to generate steady hand (end effector) motion. The relationship between joint kinematic variables and cartesian end effector kinematic variables is given by complex trigonometric transformations. This defines resolved motion control whereby given the end effector cartesian kinematic variables, we need to find the equivalent inverse joint kinematic variables in order to drive the joint actuators.

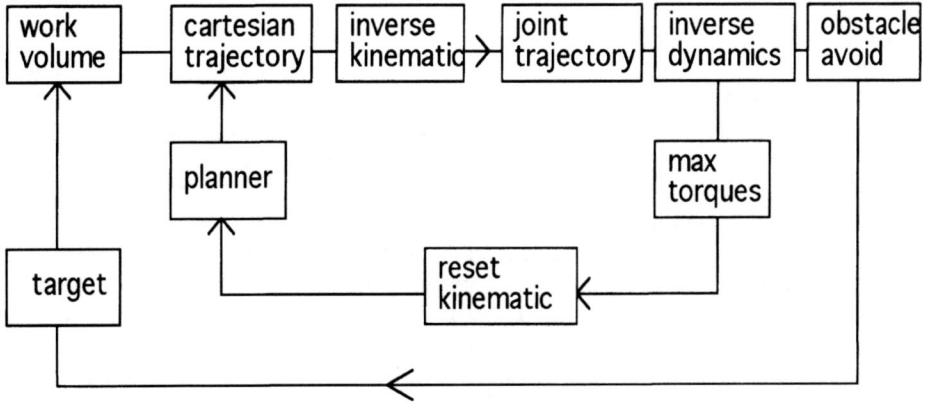

work volume	cartesian trajectory	inverse kinematic	joint trajectory	inverse dynamics	obstacle avoid

planner

max torques

target

reset kinematic

Fig. 6.3. Path construction algorithm.

Kawato et al (1988) suggested that these are precisely the computations that are carried out neurally in the human brain in voluntary movement and this provides an illustrative example of the robot control algorithm as a whole. Areas 2, 5 and 7 of the sensory-association cortex perform coordinate transformation from the desired joint angles θ^d and the generation of motor commands τ with the actual movements θ fed back to the motor cortex by proprioception. Sensorimotor integration shifts positional frames of reference in spatial coordinates to derive the muscular trajectories. Area 2 produces motor torque commands while areas 5 and 7 provide the cartesian trajectory and their transformation into joint coordinates. The parietal-association cortex receives visual and sensory information about the world. Neural circuits are capable of computing the non-linear transformation and the inverse dynamics model of the musculoskeletal system through motor learning of inverse dynamics reference models in the cerebellum, i.e. stimulus–response associations. This motor learning provides a nonlinear mapping between the torque input τ and the trajectory output θ^d by cross-correlation. Now, we consider the transformation between cartesian and joint coordinates of kinematic variables which is central to robotics.

6.1 RESOLUTION OF POSITION COORDINATES

The direct kinematics problem is concerned with the transformation of joint coordinate angles (angular rates and angular accelerations) into the equivalent cartesian coordinates (velocities and accelerations) of the end effector. The inverse kinematics problem is to find the joint coordinate angles (angular rates and angular accelerations) that correspond to a given set of cartesian coordinates (velocities and accelerations) of the end effector. It is the inverse kinematics relation that is required by the control system. It is assumed until we consider freeflyer dynamics later that the manipulator base coincides with world coordinates from which all tasks are referred (a possible alternative world coordinates system may be defined at the workpiece). It is considered that such actor-oriented reference coordinate frames are natural for describing events relative to the robot.

6.1.1 Forward position kinematics solution

Robotic links are characterised by two parameters ('i' subscript signifies joint number counted consecutively from the base of the manipulator): link length a_i (common normal distance between joint axes z_{i-1} and z_i) and offset link twist angle α_i (angle between axes z_{i-1} and z_i perpendicular to a_i). These two parameters determine the link structure. Joints are characterised by two parameters: offset distance d_i (distance between two joint axis normals x_{i-1} and x_i) and joint angle θ_i (angle between normals perpendicular to joint axes z_{i-1} and z_i). These two parameters determine the relative positioning of each link. For prismatic joints, a_i is variable and θ_i is constant while for revolute joints, θ_i is variable and a_i is constant. For both joints d_i and α_i are constant parameters of the link and joint and define the geometry of the robot manipulator.

Coordinate frames are assigned to each link at the intersection of the common normal at the joint axis. Base coordinates represent the coordinate frame at the origin of link 1. Hand coordinates represent the coordinate frame at the end effector. By convention, coordinate frames are designated by [Paul 1981, Paul et al 1981b]:

(i) z_{i-1} axis of joint i lies parallel to the rotation axis of joint i;
(ii) x_i axis of joint i is perpendicular to the z_{i-1} axis and points away from it;
(iii) y_i axis of joint i satisfies a right-hand coordinate system contraint.

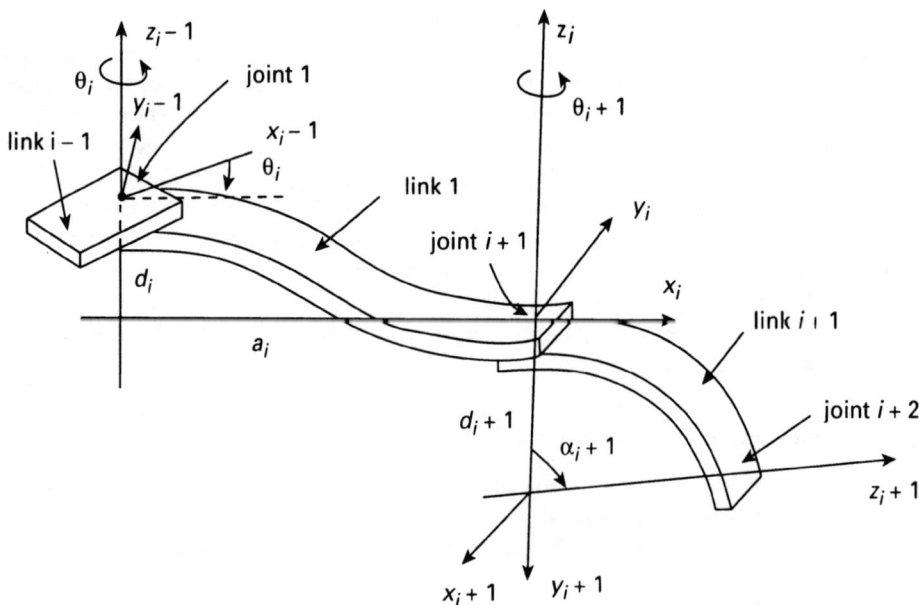

Fig. 6.4. Robot link and joint kinematic variables (from Paul 1981). Reproduced with permission from MIT Press ©1981.

For virtually all control operations, kinematic transformation has to be performed before any other transformations. The Denavit–Hartenberg (DH) 4×4 homogeneous 'A' matrix transforms between each coordinate reference frame assigned to each link at each joint sequentially from the base of the manipulator $(x_0 y_0 z_0)$ to the end effector $(x_n y_n z_n)$ through each link [Denavit & Hartenberg 1955, Paul et al 1981a,b]. The 4×4 DH matrix incorporates all link parameters in the position vector submatrix and contains a 3×3 direction cosine matrix and floating row. Successive multiplication of these coordinate frames will yield the required coordinates of the manipulator end effector with respect to base (world) coordinates.

The DH matrix that represents this sequence of events is constructed thus:

(i)   rotate through θ_i about the z_{i-1} axis to align the x_{i-1} axis with the x_i axis: $\mathrm{Rot}(z_{i-1}, \theta_i)$
(ii) translate by d_i along the z_{i-1} axis to coincide x_{i-1} and x_i axes: $\mathrm{Trans}(0, 0, d_i)$
(iii) translate by a_i along the x_i axis to coincide the two coordinate frame origins: Trans $(a_i, 0, 0)$
(iv) rotate through α_i about the x_i axis to coincide the two coordinate frames completely: $\mathrm{Rot}(x_i, \alpha_i)$

The DH 'A'-matrix that represents this sequence of events has the form:

$$A_i^{i-1} = \mathrm{Rot}(z_{i-1}, \theta_i)\,\mathrm{Trans}(0, 0, d_i)\,\mathrm{Trans}(a_i, 0, 0)\,\mathrm{Rot}(x_i, \alpha_i)$$

$$= \begin{pmatrix} \cos\theta_i & -\sin\theta_i & 0 & 0 \\ \sin\theta_i & \cos\theta_i & 0 & 0 \\ 0 & 0 & 1 & 0 \\ 0 & 0 & 0 & 1 \end{pmatrix} \begin{pmatrix} 1 & 0 & 0 & 0 \\ 0 & 1 & 0 & 0 \\ 0 & 0 & 1 & d_i \\ 0 & 0 & 0 & 1 \end{pmatrix}$$

$$\begin{pmatrix} 1 & 0 & 0 & a_i \\ 0 & 1 & 0 & 0 \\ 0 & 0 & 1 & 0 \\ 0 & 0 & 0 & 1 \end{pmatrix} \begin{pmatrix} 1 & 0 & 0 & 0 \\ 0 & \cos\alpha_i & -\sin\alpha_i & 0 \\ 0 & \sin\alpha_i & \cos\alpha_i & 0 \\ 0 & 0 & 0 & 1 \end{pmatrix}$$

$$= \begin{pmatrix} \cos\theta_i & -\cos\alpha_i \sin\theta_i & \sin\alpha_i \sin\theta_i & a_i \cos\theta_i \\ \sin\theta_i & \cos\alpha_i \cos\theta_i & -\sin\alpha_i \cos\theta_i & a_i \sin\theta_i \\ 0 & \sin\alpha_i & \cos\alpha_i & d_i \\ 0 & 0 & 0 & 1 \end{pmatrix}$$

This is the DH matrix for a revolute or prismatic joint. In both cases, the link structure parameters ai and α_i are constant. For a prismatic joint θ_i is constant and d_i is variable; for a revolute joint d_i is constant and θ_i is variable. This matrix transforms a point P_{i-1} to point P_i: $P_i = A_i^{i-1} P_{i-1}$. This 4×4 matrix may be partitioned into two submatrices which represent the rotational and translational components:

$$A_i^{i-1} = \begin{pmatrix} R_i & p_i \\ 0 & 1 \end{pmatrix}$$

where

$$R_i = \begin{pmatrix} \cos\theta_i & -\sin\theta_i\cos\alpha_i & \sin\theta_i\sin\alpha_i \\ \sin\theta_i & \cos\theta_i\cos\alpha_i & -\cos\theta_i\sin\alpha_i \\ 0 & \sin\alpha_i & \cos\alpha_i \end{pmatrix} \quad \text{and} \quad p_i = \begin{pmatrix} a_i\cos\theta_i \\ a_i\sin\theta_i \\ d_i \end{pmatrix}$$

The DH formulation provides a means of relating joint coordinates for the servo control system to cartesian coordinates for task characterisation, and this relation is nonlinear. The elements of these matrices are complex trigonometric functions. For a 6 DOF robot manipulator, 6 DH 'A' matrices, one for each link, are required to transform from base coordinates to hand coordinates (direct kinematics problem). Base coordinates are the body (world) coordinates lying at the base of the manipulator.

The final matrix (sometimes called the T-matrix) has the form (n defines the number of degrees of freedom, in this case $n = 6$) with generalised position coordinates:

$$q = T_{0n} = f(\theta) = \sum_{i=1}^{n} A_i(\theta) = A_1^0 \dots A_n^{n-1}$$

$$= \begin{pmatrix} x_n & y_n & z_n & p_n \\ 0 & 0 & 0 & 1 \end{pmatrix} = \begin{pmatrix} n & s & a & p \\ 0 & 0 & 0 & 1 \end{pmatrix}$$

where

$$n = \left(n_x n_y n_z\right)^{\mathrm{T}}$$

$$s = \left(s_x s_y s_z\right)^{\mathrm{T}}$$

$$a = \left(a_x a_y a_z\right)^{\mathrm{T}}$$

$$p = \left(p_x p_y p_z\right)^{\mathrm{T}}$$

(6.1)

where $\mathbf{n}$ = normal vector perpendicular to the fingers at the end effector (tilt); $\mathbf{s}$ = slide vector parallel to the finger grip direction of the end effector (twist); $\mathbf{a}$ = approach vector perpendicular to the palm of the end effector (turn); $\mathbf{p}$ = position vector of the hand with respect to base coordinates (sweep, reach, lift).

An alternative form represents the rotational 3×3 and translational 3×1 components separately—it is this representation that is used in Freeflyer Dynamics chapter for application to the space environment as the translational and rotational components are treated separately:

Generalised position coordinates:

$$q = T_{0n} = \sum_{i=1}^{n} R_i p_i \qquad (6.2)$$

Note that $R_n^0 = (x_6, y_6, z_6) = (\mathbf{n\ s\ a})$.

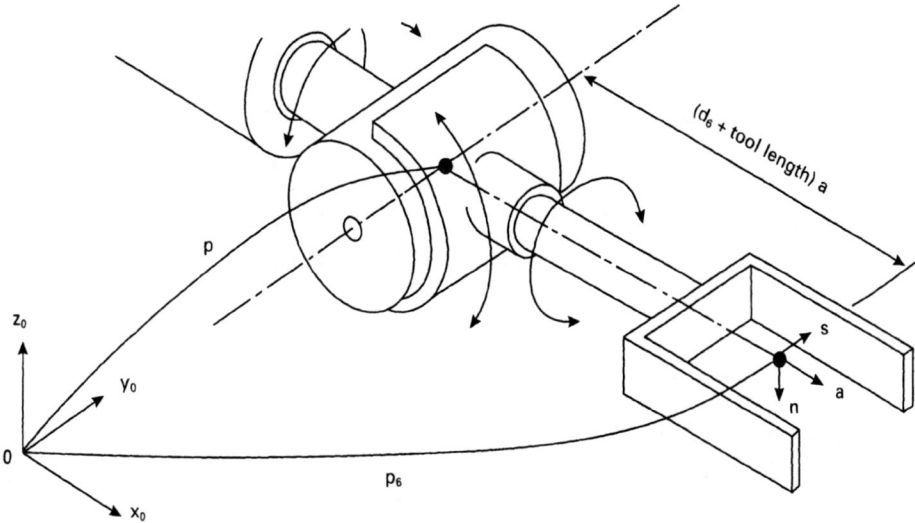

Fig. 6.5. Hand coordinates with respect to base coordinates [from Fu et al 1987]. Reproduced with permission from Fu, Gonzalez & Lee (1987), *Robotics: Control, Sensing, Vision and Intelligence*, McGraw-Hill.

A robot manipulator geometry derived from the Unimate PUMA 560/600 has been selected as the baseline case because:

(i)   it represents a generalised 6 DOF configuration of revolute joints;
(ii)   it yields closed form analytic solutions to the inverse kinematics problem;
(iii) the PUMA is widespread and has been the workhorse for robotics research for many years (though this place is now being eroded by the Cincinnati Milacron T^3);
(iv)   it is anthromorphic being based on the human arm geometry facilitating manoeuvrability and flexibility of operation (though the human upper arm has 3 degrees of freedom at the shoulder through a ball joint).

The author also suggests that a 6 DOF manipulator system is more appropriate for space application than a redundant design since a spacecraft mounting already possesses inherent redundancy by virtue of the 6 DOF of the mounting platform. Furthermore, additional links on the manipulator arm imposes additional and unnecessary mass penalties. However, this is a minority view with the majority of space manipulator designs opting for redundant configurations.

The matrix representation of rotation in the DH matrix has nine elements in the 3×3 R-submatrix representation. They are not generalised independent coordinates. However, an Euler angle set can provide such a representation describing the orientation of a rigid body target with respect to a world reference frame. Several possibilities exist according to the rotation sequence, but the *RPY* sequence is popular and is intuitively straight-forward. It is used for aerospace vehicles to describe their orientations about reference principal coordinate axes. For roll *R* about the x axis, pitch *P* about the *y* axis and yaw *Y* about the *z* axis:

$$R_{ypr} = \begin{pmatrix} cR & -sR & 0 \\ sR & cR & 0 \\ 0 & 0 & 1 \end{pmatrix} \begin{pmatrix} cP & 0 & sP \\ 0 & 1 & 0 \\ -sP & 0 & cP \end{pmatrix} \begin{pmatrix} 1 & 0 & 0 \\ 0 & cY & -cY \\ 0 & sY & cY \end{pmatrix}$$

$$= \begin{pmatrix} cRcP & cRsPsY - sRcY & cRsPcY + sRsY \\ sRcP & sRsPsY + cRcY & sRsPcY - cRsY \\ -sP & cPsY & cPcY \end{pmatrix}$$

Equating these to the (*n s a*) rotational submatrix:

$$n_x = \cos R \cos P$$

$$n_y = \sin R \cos P$$

$$n_z = -\sin P$$

$$s_x = \cos R \sin P \sin Y - \sin R \cos Y$$

$$s_y = \sin R \sin P \sin Y + \cos R \cos Y$$

$$s_z = \cos P \sin Y$$

$$a_x = \cos R \sin P \cos Y + \sin R \sin Y$$

$$a_y = \sin R \sin P \cos Y - \cos R \sin Y$$

$$a_z = \cos P \cos Y$$

Solutions are given by:

$$P = -\sin^{-1}(n_z)$$

$$R = \cos^{-1}(n_x / \cos P)$$

$$Y = \cos^{-1}(a_z / \cos P)$$

However, these solutions are ill-conditioned because $\cos \theta = \cos(-\theta)$ and $\sin^{-1} \theta = 0$ when $\theta = 0$ or 180. To use the $\tan^{-1}$ function, rearrange the equation equating (*n s a*) to *RPY* angles:

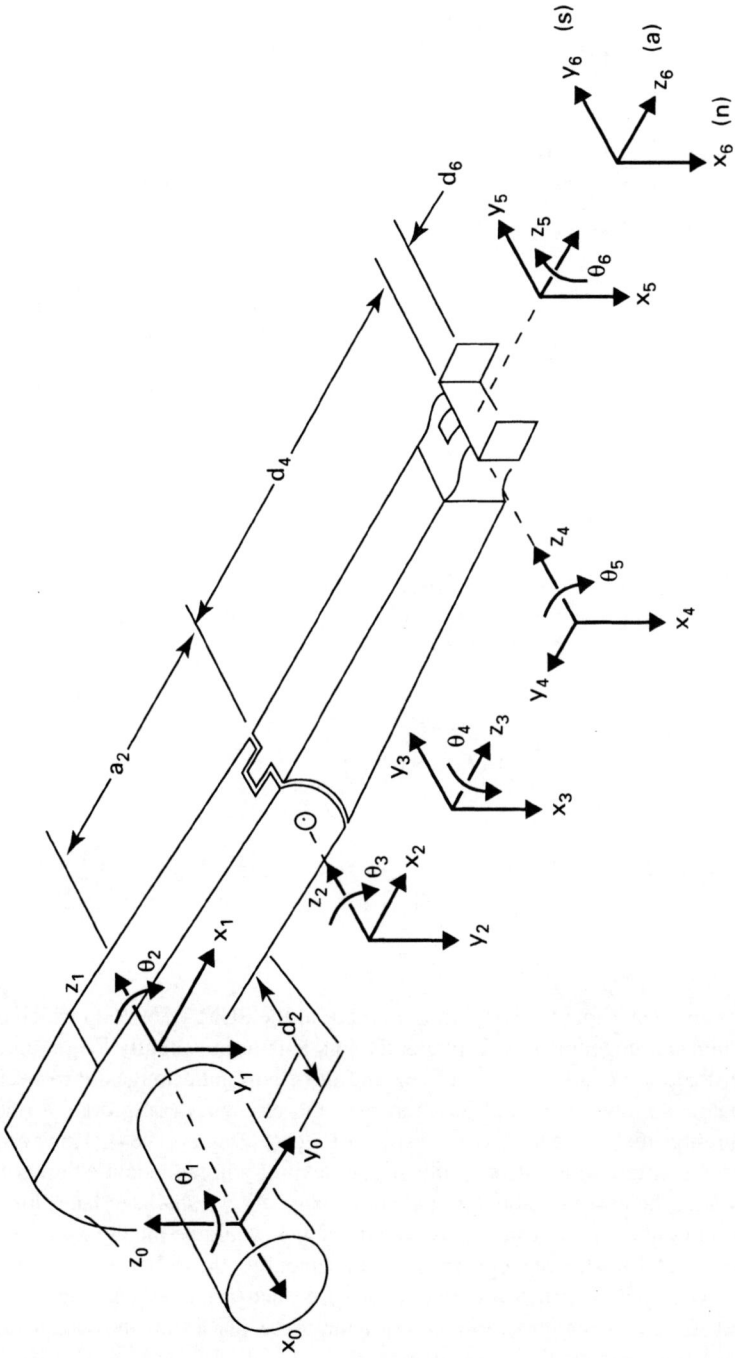

Fig. 6.6. PUMA 560/600 kinematic configuration (adapted from Lee 1982). Reproduced with permission from IEEE ©1982 IEEE.

$$
\begin{pmatrix} n_x & s_x & a_x \\ n_y & s_y & a_y \\ n_z & s_z & a_z \end{pmatrix} = \begin{pmatrix} cR & -sR & 0 \\ sR & cR & 0 \\ 0 & 0 & 1 \end{pmatrix} \begin{pmatrix} cP & 0 & sP \\ 0 & 1 & 0 \\ -sP & 0 & cP \end{pmatrix} \begin{pmatrix} 1 & 0 & 0 \\ 0 & cY & -sY \\ 0 & sY & cY \end{pmatrix}
$$

$$
\begin{pmatrix} cR & sR & 0 \\ -sR & cR & 0 \\ 0 & 0 & 1 \end{pmatrix} \begin{pmatrix} n_x & s_x & a_x \\ n_y & s_y & a_y \\ n_z & s_z & a_z \end{pmatrix} = \begin{pmatrix} cP & 0 & sP \\ 0 & 1 & 0 \\ -sP & 0 & cP \end{pmatrix} \begin{pmatrix} 1 & 0 & 0 \\ 0 & cY & -sY \\ 0 & sY & cY \end{pmatrix}
$$

$$
\begin{pmatrix} cRn_x + sRn_y & cRs_x + sRs_y & cRa_x + cRa_y \\ -sRn_x + cRn_y & -sRs_x + cRs_y & -sRa_x + cRa_y \\ n_z & s_z & a_z \end{pmatrix} = \begin{pmatrix} cP & sPsY & sPcY \\ 0 & cY & -sY \\ -sP & cPsY & cPcY \end{pmatrix}
$$

Hence,

$$
-sRn_x + cRn_y = 0 \rightarrow -sRn_x = -cRn_y
$$

$$
\rightarrow sR/cR = \left(n_y/n_x \right) \rightarrow R = \tan^{-1}\left(n_y/n_x \right)
$$

Similarly,

$$
\tan P = sP/cP \rightarrow P = \tan^{-1}\left(\frac{-n_z}{cRn_x + sRn_y} \right)
$$

and

$$
\tan Y = Y \rightarrow Y = \tan^{-1}\left(\frac{sRa_x - cRa_y}{-sRs_x + cRs_y} \right)
$$

The parameters given in Table 6.1 are similar to those of the PUMA 560/600 which is modelled on the human arm geometry except that the link lengths are slightly longer for a larger workspace, there is no offset at the elbow and the elbow joint ranges have been reversed to favour the elbow-down configuration over the elbow-up configuration (the elbow-up configuration may obscure the workspace) [Lee & Ziegler 1984]. However, elbow-up and -down configurations afford some of the flexibility in redundant elbow-roll configurations without the cost in computational complexity. As we shall see later, for a freeflyer space robot system, these restrictions on joint angles would be much less severe due to the requirement for flexible operations. Conceptually, the robot manipulator workspaces must cover both the forward region of space and the overhead region of space. This coverage of two regions allows constraints to be placed on the spacecraft design which would otherwise be difficult to constrain.

Table 6.1

Joint	Variable	α_i (°)	a_i (m)	d_i (m)	Range (°)	Range midpoint, β_i (°)
1	θ_1	−90	0	0	−160 to 160	0
2	θ_2	0	0.5	0.25	−45 to 225	90
3	θ_3	90	0	0	−225 to −45	−90
4	θ_4	−90	0	0.5	−110 to 170	30
5	θ_5	90	0	0	−100 to 100	0
6	θ_6	0	0	0.2	−266 to 266	0

Using these parameters in the DH matrix scheme for a PUMA 560/600 configuration:

$$A_1^0 = \begin{pmatrix} c_1 & 0 & -s_1 & 0 \\ s_1 & 0 & c_1 & 0 \\ 0 & -1 & 0 & 0 \\ 0 & 0 & 0 & 1 \end{pmatrix}, \quad A_2^1 = \begin{pmatrix} c_2 & -s_2 & 0 & a_2 c_2 \\ s_2 & c_2 & 0 & a_2 s_2 \\ 0 & 0 & 1 & d_2 \\ 0 & 0 & 0 & 1 \end{pmatrix},$$

$$A_3^2 = \begin{pmatrix} c_3 & 0 & s_3 & 0 \\ s_3 & 0 & -c_3 & 0 \\ 0 & 1 & 0 & 0 \\ 0 & 0 & 0 & 1 \end{pmatrix}, \quad A_4^3 = \begin{pmatrix} c_4 & 0 & -s_4 & 0 \\ s_4 & 0 & c_4 & 0 \\ 0 & -1 & 0 & d_4 \\ 0 & 0 & 0 & 1 \end{pmatrix}$$

$$A_5^4 = \begin{pmatrix} c_5 & 0 & s_5 & 0 \\ s_5 & 0 & -c_5 & 0 \\ 0 & 1 & 0 & 0 \\ 0 & 0 & 0 & 1 \end{pmatrix}, \quad A_6^5 = \begin{pmatrix} c_6 & -s_6 & 0 & 0 \\ s_6 & c_6 & 0 & 0 \\ 0 & 0 & 1 & d_6 \\ 0 & 0 & 0 & 1 \end{pmatrix} \tag{6.3}$$

where $s_i = \sin\theta_i$ and $c_i = \cos\theta_i$.

The parameters are measured from the principal axes of the links through their respective centres of mass. Link 0 is the stationary pillar link bolted to a platform, typically the ground for a terrestrial manipulator. Coordinates (x_i, y_i, z_i) are associated with joint $i - 1$ so that z_{i-1} is the axis of rotation for joint i. If an end effector or tool is used whose origin or axes do not coincide with the coordinate system of link 6, the tool must be related by a constant transform E to link 6. It is assumed generally that the tool origin does coincide with the link origin and its dimensions are included in the d_6 end effector term in A_6^5.

To exploit the structure of the equations and avoid the extraneous computations of '0's and '1's in the floating row [Roth 1985], the usual procedure is to partition the computation into two parts, each of which is performed by hand, leaving the computer to do the final number-crunching in multiplying the two resultant matrices:

For the first three matrix multiplications:

$$T_1 = A_3^0 = \begin{pmatrix} c_1c_{23} & -s_1 & c_1s_{23} & a_2c_1c_2 - s_1d_2 \\ s_1c_{23} & c_1 & s_1s_{23} & a_2s_1c_2 + c_1d_2 \\ -s_{23} & 0 & c_{23} & -a_2s_2 \\ 0 & 0 & 0 & 1 \end{pmatrix}$$

where $s_{ij} = c_js_i + c_is_j$ and $c_{ij} = c_ic_j - s_is_j$.

The last three matrix multiplications:

$$T_2 = A_6^4 = \begin{pmatrix} c_4c_5c_6 - s_4s_6 & -c_4c_5s_6 - s_4c_6 & c_4s_5 & c_4s_5d_6 \\ s_4c_5c_6 + c_4s_6 & -s_4c_5s_6 + c_4c_6 & s_4s_5 & s_4s_5d_6 \\ -c_6s_5 & s_5s_6 & c_5 & c_5d_6 + d_4 \\ 0 & 0 & 0 \end{pmatrix}$$

Hence,

$$T_{0n} = T_1T_2 = A_n^0 = \begin{pmatrix} n_x & s_x & a_x & p_x \\ n_y & s_y & a_y & p_y \\ n_z & s_z & a_z & p_z \\ 0 & 0 & 0 & 1 \end{pmatrix}$$

where

$$p_x = c_1[d_6(c_{23}c_4s_5 + s_{23}c_5) + s_{23}d_4 + a_2c_2] - s_1(s_4s_5d_6 + d_2)$$

$$p_y = s_1[d_6(c_{23}c_4s_5 + s_{23}c_5) + s_{23}d_4 + a_2c_2] + s_1(s_4s_5d_6 + d_2)$$

$$p_z = d_6[c_{23}c_5 - s_{23}c_4s_5] + c_{23}d_4 - a_2s_2$$

$$n_x = c_1[c_{23}(c_4c_5c_6 - s_4s_6) - s_{23}c_6s_5] - s_1(s_4c_5c_6 + c_4s_6)$$

$$n_y = s_1[c_{23}(c_4c_5c_6 - s_4s_6) - s_{23}s_5s_6] + c_1(s_4c_5c_6 + c_4s_6)$$

$$n_z = -s_{23}(c_4c_5c_6 - s_4s_6) - c_{23}c_6s_5$$

$$s_x = -c_1[c_{23}(c_4c_5s_6 + s_4c_6) - s_{23}s_5s_6] + s_1(s_4c_5c_6 - c_4c_6)$$

$$s_y = -s_1[c_{23}(c_4c_5s_6 + s_4c_6) - s_{23}s_5s_6] + c_1(c_4c_6 - s_4c_5s_6)$$

$$s_z = s_{23}(c_4c_5s_6 + s_4c_6) + c_{23}s_5s_6$$

$$a_x = c_1(c_{23}c_4s_5 + s_{23}c_5) - s_1s_4s_5$$

$$a_y = s_1(c_{23}c_4s_5 + c_5s_{23}) + s_1s_4s_5$$

$$a_z = -s_{23}c_4s_5 + c_{23}c_5 \qquad\qquad (6.4)$$

To check that the forward kinematics solution is correct, the stowed configuration has joint angles

$$\theta_1 = 90, \theta_2 = 90, \theta_3 = 90, \theta_4 = \theta_5 = \theta_6 = 0$$

yielding $p_x = -d_2$, $p_y = d_4 + d_6$, $p_z = -a_2$ and $R_6^0 = I_3$ as expected.

6.1.2 Inverse position kinematics solution

The inverse solution is to find the joint angles for given cartesian variables q, i.e. $\theta = f^{-1}(q)$. This is the information required for the robot control system. To find the inverse solution of joint angles corresponding to a particular end effector configuration in cartesian space it is necessary to decouple the angular components of the wrist from the position components of the arm. For a 6 DOF manipulator with six joints employing a spherical wrist where the last three orienting degrees of freedom intersect, closed form analytic solutions are possible. The position of the wrist subassembly may be found to separate the wrist orientation from the arm positioning assembly since the wrist contributes a distance d_6 in the direction of the approach vector to the final position of the end effector:

$$
\begin{pmatrix} p_x \\ p_y \\ p_z \end{pmatrix} = \begin{pmatrix} p_x^{arm} \\ p_y^{arm} \\ p_y^{arm} \end{pmatrix} + \begin{pmatrix} p_x^{wrist} \\ p_y^{wrist} \\ p_z^{wrist} \end{pmatrix} = \begin{pmatrix} p_x^{arm} \\ p_y^{arm} \\ p_y^{arm} \end{pmatrix} + d_6 \begin{pmatrix} a_x \\ a_y \\ a_z \end{pmatrix}
$$

$$
\begin{pmatrix} p_x^{arm} \\ p_y^{arm} \\ p_y^{arm} \end{pmatrix} = \begin{pmatrix} p_{4x}^0 \\ p_{4y}^0 \\ p_{4z}^0 \end{pmatrix} = \begin{pmatrix} p_x \\ p_y \\ p_z \end{pmatrix} - \begin{pmatrix} p_x^{wrist} \\ p_y^{wrist} \\ p_z^{wrist} \end{pmatrix} = \begin{pmatrix} p_x \\ p_y \\ p_z \end{pmatrix} - d_6 \begin{pmatrix} a_x \\ a_y \\ a_z \end{pmatrix}
$$

$$
= \begin{pmatrix} c_1(a_2 c_2 + d_4 s_{23}) - d_2 s_1 \\ s_1(a_2 c_2 + d_4 s_{23}) + d_2 c_1 \\ d_4 c_{23} - a_2 s_2 \end{pmatrix} \tag{6.5}
$$

For solutions to the inverse kinematic problem, a well-balanced trigonometric function is required: cosines and sines are often undefined or multivalued, so the arctangent is used with proper quadrants.

$$
\begin{aligned}
\theta = \tan^{-1}(y/x) \text{ for } \quad & 0 < \theta < 90 \ \text{ for } x > 0 \text{ and } y > 0 \\
& 90 < \theta < 180 \ \text{ for } x < 0 \text{ and } y > 0 \\
& -180 < \theta < -90 \ \text{ for } x < 0 \text{ and } y < 0 \\
& -90 < \theta < 0 \ \text{ for } x > 0 \text{ and } y < 0
\end{aligned}
$$

The position of the arm is determined by the first three positioning links d_2, a_2, d_4 and the first three joint angles θ_1, θ_2, θ_3 [Lee & Ziegler 1984; Lee 1982; Fu et al 1987].

To find θ_1:

$$-s_1 p_x^{arm} + c_1 p_y^{arm} = -s_1\{c_1(a_2c_2 + d_4s_{23}) - d_2s_1\}$$

$$+c_1\{s_1(a_2c_2 + d_4s_{23}) + d_2c_1\} = d_2$$

Let

$$p_x^{arm} = r\cos\varphi, \quad p_y^{arm} = r\sin\varphi$$

where

$$r = \sqrt{\left(p_x^{arm}\right)^2 + \left(p_y^{arm}\right)^2} \quad \text{and} \quad \varphi = \tan^{-1}\left(p_y^{arm}/p_x^{arm}\right)$$

Substitute:

$$-r(\cos\varphi)s_1 + (r\sin\varphi)c_1 = d_2$$

$$\sin\varphi c_1 - \cos\varphi s_1 = d_2/r \rightarrow \sin(\varphi - \theta_1) = d_2/r$$

Now,

$$\cos(\varphi - \theta_1) = \sqrt{1 - \sin^2(\varphi - \theta_1)} = \sqrt{1 - \left(d_2/r\right)^2}$$

i.e.,

$$\tan(\varphi - \theta_1) = \frac{d_2/r}{\sqrt{1 - \left(d_2/r\right)^2}} \rightarrow \theta_1 = \varphi - \tan^{-1}\left[\frac{d_2/r}{\sqrt{1 - \left(d_2/r\right)^2}}\right]$$

$$= \varphi - \tan^{-1}\left[\frac{d_2}{\sqrt{r^2 - d_2^2}}\right]$$

Now,

$$\tan\varphi = r\sin\varphi/r\cos\varphi = p_y^{arm}/p_x^{arm}$$

Hence,

$$\theta_1 = \tan^{-1}\left(\frac{p_y^{arm}}{p_x^{arm}}\right) - \tan^{-1}\left[\frac{d_2}{\pm\sqrt{(p_x^{arm})^2 + (p_y^{arm})^2 - d_2^2}}\right] \tag{6.6}$$

The '+' indicates the right arm/shoulder configuration and the '−' indicates the left arm/shoulder configuration.

To find θ_2:

$$c_1 p_x^{arm} + s_1 p_y^{arm} = c_1\{c_1(a_2c_2 + d_4s_{23}) - d_2s_1\}$$

$$+ s_1\{s_1(a_2c_2 + d_4s_{23}) + d_2c_1\} = a_2c_2 + d_4s_{23}$$

Let

$$A = c_1 p_x^{arm} + s_1 p_y^{arm} = a_2 c_2 + d_4 s_{23} \rightarrow s_{23} = \left(\frac{A - a_2 c_2}{d_4} \right)$$

Similarly,

$$p_z^{arm} = d_4 c_{23} - a_2 s_2 \rightarrow c_{23} = \left(\frac{p_z^{arm} + a_2 s_2}{d_4} \right)$$

Now,

$$c_{23}^2 + s_{23}^2 = 1, \quad \text{so} \quad \left(p_z^{arm} + a_2 s_2 \right)^2 + \left(A - a_2 c_2 \right) = d_4^2$$

This yields,

$$-s_2 p_z^{arm} + c_2 A = \frac{A^2 + (p_z^{arm})^2 + a_2^2 - d_4^2}{2 a_2} = B$$

Let

$$p_z^{arm} = r \cos \varphi, \quad A = r \sin \varphi$$

where

$$r = \sqrt{A^2 + (p_z^{arm})^2} \quad \text{and} \quad \varphi = \tan^{-1}\left(\frac{A}{p_z^{arm}} \right)$$

Substitute:

$$(r \sin \varphi) c_2 - (r \cos \phi) s_2 = B \rightarrow \sin (\varphi - \theta_2) = B/r$$

Now,

$$\cos(\varphi - \theta_2) = \sqrt{1 - (B/r)^2}$$

i.e.

$$\tan(\varphi - \theta_2) = \frac{B/r}{\sqrt{1 - (B/r)^2}} \rightarrow \theta_2 = \varphi - \tan^{-1}\left[\frac{B}{\sqrt{r^2 - B^2}} \right]$$

Now,

$$\varphi = \tan^{-1}\left(\frac{A}{p_z^{arm}} \right)$$

Hence,

$$\theta_2 = \tan^{-1}\left(\frac{A}{p_z^{arm}} \right) - \tan^{-1}\left[\frac{B}{\pm \sqrt{A^2 + (p_z^{arm})^2 - B^2}} \right] \tag{6.7}$$

The '+' corresponds to the elbow-down configuration $(\theta_3 - \theta_2 > 0)$ such that the elbow lies below the wrist, and the '−' corresponds to the elbow 'up' configuration $(\theta_3 - \theta_2 < 0)$ such that the elbow is above the wrist. The elbow-down configuration is preferred since the elbow-up configuration could obscure the workspace.

To find θ_3:

$$\tan(\theta_2 + \theta_3) = s_{23}/c_{23} = \left[\frac{A - a_2 c_2}{p_z^{arm} + a_2 s_2}\right]$$

$$\theta_3 = \left\{\tan^{-1}\left[\frac{A - a_2 c_2}{p_z^{arm} + a_2 s_2}\right]\right\} - \theta_2 \tag{6.8}$$

To find $\theta_4, \theta_5, \theta_6$ for the wrist joints, the rotational components of the DH matrix formulation are used [Hollerbach & Sahar 1983, Wang & Lien 1988]:

$$R_6^0 = R_3^0 R_6^3 = R_{arm}^0 R_{wrist}^{arm}$$

$$R_6^3 = \left(R_3^0\right)^{-1} R_6^0 = \left(R_3^0\right)^T R_6^0 \quad \text{where} \quad R_6^0 = (nsa)$$

Now,

$$\left(R_3^0\right)^{-1} = \left(R_3^0\right)^T = \begin{pmatrix} c_1 c_{23} & s_1 c_{23} & -s_{23} \\ -s_1 & c_1 & 0 \\ c_1 s_{23} & s_1 s_{23} & c_{23} \end{pmatrix}$$

Hence,

$$R_{wrist}^{arm} = \begin{pmatrix} c_1 c_{23} & s_1 c_{23} & -s_{23} \\ -s_1 & c_1 & 0 \\ c_1 s_{23} & s_1 s_{23} & c_{23} \end{pmatrix} \begin{pmatrix} n_x & s_x & a_x \\ n_y & s_y & a_y \\ n_z & s_z & a_z \end{pmatrix} = \begin{pmatrix} n_x^w & s_x^w & a_x^w \\ n_y^w & s_y^w & a_y^w \\ n_z^w & s_z^w & a_z^w \end{pmatrix}$$

where

$$n_x^w = c_1 c_{23} n_x + s_1 c_{23} n_y - s_{23} n_z$$

$$n_y^w = -s_1 n_x + c_1 n_y$$

$$n_z^w = c_1 s_{23} n_x + s_1 s_{23} n_y + c_{23} n_z$$

$$s_x^w = c_1 c_{23} s_x + s_1 c_{23} s_y - s_{23} s_z$$

$$s_y^w = -s_1 s_x + c_1 s_y$$

$$s_z^w = c_1 s_{23} s_x + s_1 s_{23} s_y + c_{23} s_z$$

$$a_x^w = c_1 c_{23} a_x + s_1 c_{23} a_y - s_{23} a_z$$

$$a_y^w = -s_1 a_x + c_1 a_y$$

$$a_z^w = c_1 s_{23} a_x + s_1 s_{23} a_y + c_{23} a_z$$

Equating these values to R_6^3 :

$$R_6^3 = \begin{pmatrix} c_4 c_5 c_6 - s_4 s_6 & -c_4 c_5 s_6 - s_4 c_6 & c_4 s_5 \\ s_4 c_5 c_6 + c_4 s_6 & -s_4 c_5 s_6 + c_4 c_6 & s_4 s_5 \\ -c_6 s_5 & s_5 s_6 & c_5 \end{pmatrix} = \begin{pmatrix} n_x^w & s_x^w & a_x^w \\ n_y^w & s_y^w & a_y^w \\ n_z^w & s_z^w & a_z^w \end{pmatrix}$$

Now,

$$\tan \theta_4 = \frac{s_4 s_5}{c_4 s_5} = \frac{a_y^w}{a_x^w} \rightarrow \theta_4 = \tan^{-1} \left(\frac{-s_1 a_x + c_1 a_y}{c_{23}(c_1 a_x + s_1 a_y) - s_{23} a_z} \right) \qquad (6.9)$$

Now,

$$(c_4 s_5)^2 + (s_4 s_5)^2 = s_5^2 \rightarrow \sin \theta_5 = \sqrt{\left(a_x^w\right)^2 + \left(a_y^w\right)^2} \quad \text{and} \quad \cos \theta_5 = a_z^w$$

i.e.

$$\theta_5 = \tan^{-1} \left[\frac{\sqrt{(c_1 c_{23} a_x + s_1 c_{23} a_y - s_{23} a_z)^2 + (-s_1 a_x + c_1 a_y)^2}}{s_{23}(c_1 a_x + s_1 a_y) + c_{23} a_z} \right] \qquad (6.10)$$

Now,

$$\tan \theta_6 = \frac{s_5 s_6}{s_5 c_6} = -\frac{s_z^w}{n_z^w} \rightarrow \theta_6 = \tan^{-1} - \left(\frac{s_{23}(c_1 s_x + s_1 s_y) + c_{23} s_z}{s_{23}(c_1 n_x + s_1 n_y) + c_{23} n_z} \right) \qquad (6.11)$$

This completes the analytic solution to the inverse kinematics problem. Alternatively, numerical techniques are available to solve the inverse kinematics, e.g. the Newton–Raphson iteration method:

$$q_{k+1} = q_k - f(q_k)/f'(q_k) \quad \text{or} \quad \theta_{k+1} = \theta_k - J^{-1}(\theta_k) q_k$$

where $q_k = f(\theta_k)$.

This method requires the selection of an initial seed estimate and the number of iterations depends on the accuracy of this initial estimate. Usually four iterations are sufficient to reach an accuracy of <0.1 mm. Numerical methods are much more complex than the closed-form analytic solution since they involve inversion of the Jacobian matrix and a close estimate for the initial seed. Being an interpolation approximation, it is also subject to divergence if the initial seed is not sufficiently close to the solution. To numerically differentiate position for velocity and acceleration is highly undesirable as it amplifies noise.

6.1.3 Manipulator workspace

The robot workspace is defined as the region which may be reached by the centre of the manipulator hand [Tsin & Sani 1983]. The dextrous primary workspace is defined as the volume within which every point is reachable by the hand at any orientation, i.e. the arm must be able to deliver the wrist subassembly to any point within that sphere. The secondary workspace is reachable from limited directions due to singularities in that workspace: $W_T^{primary}(H) \subset W_T^{secondary}(H)$ [Kumar & Waldron 1981]. At singular points, the end effector loses one or more degrees of freedom since the kinematic equations become linearly dependent or certain solutions become undefined [Featherstone 1983]. One example is when the wrist lies above or below the shoulder such that it is parallel to the z_0 axis of base coordinates. Singular configurations must be avoided as the velocities required to move the end effector become effectively infinite. The solution is to ensure that singular configurations are forbidden within the controller. Lee & Yang (1983) and Yang & Lee (1983) derived a recursive formula for determining the manipulator work-space volume $W_k(H)$ with $H(x_k, y_k, z_k)$, where H denotes the centre of the manipulator hand and $W_k(H)$ denotes the reachable workspace generated by point H by turning all revolute joints $k \dots n$ within their permitted ranges whilst holding the axis k as fixed. This represents a circle with its centre at the origin of the coordinate frame k:

$$
\begin{aligned}
x_k &= r_k \cos\theta_k \\
y_k &= r_k \sin\theta_k \quad \text{and} \quad \begin{pmatrix} x \\ y \\ z \\ 1 \end{pmatrix}_k^* = A_{k+1} \begin{pmatrix} x \\ y \\ z \\ 1 \end{pmatrix}_{k+1} \\
z_k &= z_k^*
\end{aligned}
$$

with θ_k = total permitted range of revolution of joint k

$$
r_k = \sqrt{\left(x_k^*\right)^2 + \left(y_k^*\right)^2}
$$

β_k = location angle of midrange point of joint k

$$
A_k = \text{Rot}(x_k, \alpha_k)\, \text{Trans}\,(a_k, 0, 0)\, \text{Trans}\,(0, 0, d_k)\, \text{Rot}(z_{k-1}, \beta_k)
$$

$$
= \begin{pmatrix} 1 & 0 & 0 & a_k \\ 0 & c\alpha_k & s\alpha_k & d_k s\alpha_k \\ 0 & -s\alpha_k & c\alpha_k & d_k c\alpha_k \\ 0 & 0 & 0 & 1 \end{pmatrix} \begin{pmatrix} c\beta_k & -s\beta_k & 0 & 0 \\ s\beta_k & c\beta_k & 0 & 0 \\ 0 & 0 & 1 & 0 \\ 0 & 0 & 0 & 1 \end{pmatrix}
$$

$$
= \begin{pmatrix} c\beta_k & -s\beta_k & 0 & a_k \\ c\alpha_k s\beta_k & c\alpha_k c\beta_k & s\alpha_k & d_k s\alpha_k \\ -s\alpha_k s\beta_k & -s\alpha_k c\beta_k & c\alpha_k & d_k c\alpha_k \\ 0 & 0 & 0 & 1 \end{pmatrix}
$$

Initial conditions for joint n are given by:

$$\left.\begin{array}{l} x_n = r_n \cos\theta_n \\ y_n = r_n \sin\theta_n \\ z_n = d_n \cos\alpha_n \end{array}\right\} \quad \text{with} \quad r_n = \sqrt{a_n^2 + (d_n \sin\alpha_n)^2}$$

The total workspace of an n-jointed manipulator [Gupta & Roth 1982].

$$W_T(H) = \sum_{i=1}^{n} A_i^{i+1}\{W_n(H)\}$$

For the PUMA 560/600, the workspace volume $W_T(H) = 3.94$ m^3. For any given manipulator, a constant volume index may be defined: $VI = W_T/(l_1 + \ldots + l_n)^3$. For the PUMA 560/600, $VI = 3.944/2.744 = 1.44$. There exists a theoretical maximum such that the manipulator workspace is a sphere of radius L^3 where $L = (l_1 + \ldots + l_n)$ at the first joint, i.e. $VI_{max} = 4\pi L^3/L^3 \approx 4.2$. This generates a normalised index, $NVI = VI/VI_{max}$. For a PUMA 560/600, $NVI = 1.44/4.2 = 0.34$. Joint limits reduce the ideal unlimited workspace by 34%. For most industrial robots, at least two of the positioning links are coplanar. For maximum workspace, the coplanar links should have the same dimensions: $a_2/d_4 = 1$. Offsets do not contribute to the workspace volume and so should be reduced—in this formulation, the elbow offset of the industrial PUMA 560/600 model has been eliminated. Furthermore, increasing the wrist size d_6 increases the total workspace but reduces the primary workspace, so a small wrist enhances dexterity. The workspace is bounded by concentric spheres of radii $R_i - d_6$ and $R_0 - d_6$ where $R_i = a_2 - d_4$ and $R_0 = a_2 + d_4$. The concentric spheres of radii R_i and R_0 define the workspace of the wrist point. The surface of the sphere of radius d_6 within the region between the concentric spheres defines all possible approach angles to that point [Hanson et al 1983].

6.2 RESOLUTION OF VELOCITY COORDINATES

The Jacobian matrix is important and is the fundamental component of resolved rate, resolved acceleration and resolved force control schemes and itself relies on the joint solution given by the inverse position kinematics solution.

6.2.1 Forward differential kinematics solution
The differential kinematics relating joint velocities to cartesian end effector velocities is determined by a square matrix of partial differential terms—the Jacobian matrix. The $\mathbf{n} \times \mathbf{n}$ Jacobian matrix $J(\theta)$ specifies how the end effector velocities may be linearly transformed into joint rates [Whitney 1969].

From the kinematic analysis:

$$q = f(\theta) = \sum_{i=1}^{n} R_i p_i$$

By differentiating:

$$\dot{q} = J(\theta)\dot{\theta} \quad \text{where} \quad J(\theta) = \frac{\partial f(\theta)}{\partial \theta}$$

Hence,

$$\dot{q} = \frac{d}{dt}\sum_{i=1}^{n} R_i p_i = \sum_{i=1}^{n} \dot{R}_i p_i \quad \text{since} \quad \dot{p}_i = 0 \tag{6.12}$$

Now,

$$\dot{R}_i^{i-1} = \frac{\partial R_i^{i-1}}{\partial \theta_i}\dot{\theta}_i$$

So,

$$\dot{R}_i^0 = \frac{d}{dt}\sum_{k=1}^{i} R_i^{i-1} = \sum_{k=1}^{i} \frac{\partial R_i^0}{\partial \theta_k}\dot{\theta}_k$$

Substitute,

$$\dot{q} = \left(\sum_{i=1}^{n}\sum_{k=1}^{i} \frac{\partial R_i^0}{\partial \theta_k} p_i\right)\dot{\theta}_k \rightarrow \dot{q} = J(\theta)\dot{\theta}_k \quad \text{where} \quad J(\theta) = \sum_{i=1}^{n}\sum_{k=1}^{i} \frac{\partial R_i^0}{\partial \theta_k} p_i \tag{6.13}$$

A more convenient and efficient formulation of the Jacobian matrix may be derived from the Newton–Euler dynamics equations as cross product relations [Whitney 1972, Orin & Schrader 1984]. From the forward Newton–Euler dynamics formulation (see next chapter), the linear and angular velocity of the end effector, $\dot{q} = \begin{pmatrix} v_i & w_i \end{pmatrix}$ are given by:

$$w_i = w_{i-1} + z_{i-1}\dot{\theta}_i = \sum_{j}^{i} z_{j-2}\dot{\theta}_{j-1} + z_{i-1}\dot{\theta}_i = \sum_{j=1}^{i} z_{j-1}\dot{\theta}_j$$

$$v_i = v_{i-1} + w_i \times l_i = \left(\sum_{i=1}^{i} z_{j-2}\dot{\theta}_{j-1}\right) \times l_{j-1} + z_{i-1}\dot{\theta}_i \times l_i = \sum_{j=1}^{i} (z_{j-1} \times l_j)\dot{\theta}_j$$

where z_{i-1} = unit vector along axis of motion of joint i and $l_i = (p_i - p_{i-1})$ = position of link i with respect to base coordinates.

Hence, at the end effector:

$$\begin{aligned} w_n &= \sum_{i=1}^{n} z_{i-1}\dot{\theta}_i \\ v_n &= \sum_{i=1}^{n} (z_{i-1} \times l_i)\dot{\theta}_i \end{aligned} \quad \text{or, alternatively,} \quad \begin{pmatrix} v_n \\ w_n \end{pmatrix} = \begin{pmatrix} \sum_{i=1}^{n} (z_{i-1} \times l_i) \\ \sum_{i=1}^{n} z_{i-1} \end{pmatrix} \dot{\theta}_i$$

such that

$$J(\theta) = \begin{pmatrix} z_{i-1} \times p_i^{i-1} \\ z_{i-1} \end{pmatrix} \tag{6.14}$$

To apply and calculate the Jacobian matrix from the kinematic equations, the kinematics may be reformulated from the end effector to the base—the normal formulation from the base to the end effector requires the calculation of the Jacobian at each joint in terms of all preceding links and this is wasteful [Orin & Schrader 1984]. For resolved rate control, only the Jacobian for the end effector coordinate system is required and allows advantage to be taken of the kinematic simplicity of the wrist.

Hence, $J = (J_1 \ldots J_n)$ so

$$J = \begin{pmatrix} z_0 \times p_6^0 & z_1 \times p_6^1 & z_2 \times p_6^2 & z_3 \times p_6^3 & z_4 \times p_6^4 & z_5 \times p_6^5 \\ z_0 & z_1 & z_2 & z_3 & z_4 & z_5 \end{pmatrix}$$

for a 6-link manipulator. For the PUMA 560/600-type structure adopted here:

$$p_6^0 = \begin{pmatrix} c_1\big[d_6(c_{23}c_4s_5 + s_{23}c_5) + s_{23}d_4 + a_2c_2\big] - s_1(s_4s_5d_6 + d_2) \\ s_1\big[d_6(c_{23}c_4s_5 + s_{23}c_5) + s_{23}d_4 + a_2c_2\big] + c_1(s_4s_5d_6 + d_2) \\ d_6(c_{23}c_5 - s_{23}c_5) + c_{23}d_4 - a_2s_2 \end{pmatrix} \quad z_0 = \begin{pmatrix} 0 \\ 0 \\ 1 \end{pmatrix}$$

$$p_6^1 = \begin{pmatrix} c_{23}(c_4s_5d_6) + s_{23}(c_5d_6 + d_4) + a_2c_2 \\ s_{23}(c_4s_5d_6) - c_{23}(c_5d_6 + d_4) + a_2s_2 \\ s_4s_5d_6 + d_2 \end{pmatrix} \quad z_1 = \begin{pmatrix} -s_1 \\ c_1 \\ 0 \end{pmatrix}$$

$$p_6^2 = \begin{pmatrix} c_3(c_4s_5d_6) + s_3(c_5d_6 + d_4) \\ s_3(c_4s_5d_6) - c_3(c_5d_6 + d_4) \\ s_4s_5d_6 \end{pmatrix} \quad z_2 = \begin{pmatrix} -s_1 \\ c_1 \\ 0 \end{pmatrix}$$

$$p_6^3 = \begin{pmatrix} c_4s_5d_6 \\ s_4s_5d_6 + d_4 \\ c_5d_6 \end{pmatrix} \quad z_3 = \begin{pmatrix} c_1s_{23} \\ s_1s_{23} \\ c_2 \end{pmatrix}$$

$$p_6^4 = \begin{pmatrix} s_5d_6 \\ -c_5d_6 \\ 0 \end{pmatrix} \quad z_4 = \begin{pmatrix} -(c_1c_{23}s_4 + s_1c_4) \\ (c_1c_4 - s_1c_{23}c_4) \\ s_{23}s_4 \end{pmatrix}$$

$$p_6^5 = \begin{pmatrix} 0 \\ 0 \\ d_6 \end{pmatrix} \quad z_5 = \begin{pmatrix} c_1(c_{23}c_4s_5 + s_{23}c_5) - s_1s_4s_5 \\ s_1(c_{23}c_4s_5 + s_{23}c_5) + c_1s_4s_5 \\ -s_{23}c_4s_5 + c_{23}c_5 \end{pmatrix}$$

Hence,

$$J = \begin{pmatrix} J_{11} & J_{21} & J_{31} & J_{41} & J_{51} & J_{61} \\ J_{12} & J_{22} & J_{32} & J_{42} & J_{52} & J_{62} \\ J_{13} & J_{23} & J_{33} & J_{43} & J_{53} & J_{63} \\ J_{14} & J_{24} & J_{34} & J_{44} & J_{54} & J_{64} \\ J_{15} & J_{25} & J_{35} & J_{45} & J_{55} & J_{65} \\ J_{16} & J_{26} & J_{36} & J_{46} & J_{56} & Jl66 \end{pmatrix}$$

where

$$J_{11} = -s_1[d_6(c_{23}c_4s_5 + s_{23}c_5) + s_{23}d_4 + a_2c_2] - c_1(s_4s_5d_6 + d_2)$$

$$J_{12} = c_1[d_6(c_{23}c_4s_5 + s_{23}c_5) + s_{23}d_4 + a_2c_2] - s_1(s_4s_5d_6 + d_2)$$

$$J_{13} = 0, \quad J_{14} = 0, \quad J_{15} = 0, \quad J_{16} = 1$$

$$J_{21} = c_1(s_4s_5d_6 + d_2)$$

$$J_{22} = s_1(s_4s_5d_6 + d_2)$$

$$J_{23} = -s_1[s_{23}c_4s_5d_6 - c_{23}(c_5d_6 + d_4) + a_2s_2]$$
$$\qquad -c_1[c_{23}c_4s_5d_6 + s_{23}(c_5d_6 + d_4) + a_2c_2]$$

$$J_{24} = -s_1, \quad J_{25} = c_1, \quad J_{26} = 0$$

$$J_{31} = c_1s_4s_5d_6$$

$$J_{32} = s_1s_4s_5d_6$$

$$J_{33} = -s_1[s_3c_4s_5d_6 - c_3(c_5d_6 + d_4)] - c_1[c_3c_4s_5d_6 + s_3(d_4 + c_5d_6)]$$

$$J_{34} = -s_1, \quad J_{35} = c_1, \quad J_{36} = 0$$

$$J_{41} = s_1s_{23}(d_6c_5 + d_4) - d_6c_{23}s_4s_5$$

$$J_{42} = d_6c_{23}c_4s_5 - c_1s_{23}(d_6c_5 + d_4)$$

$$J_{43} = d_6s_{23}(c_1s_4s_5 - s_1c_4s_5)$$

$$J_{44} = c_1s_{23}, \quad J_{45} = s_1s_{23}, \quad J_{46} = c_{23}$$

$$J_{51} = s_{23}s_4c_5d_6$$

$$J_{52} = s_{23}s_4s_5d_6$$

$$J_{53} = d_6[c_5(c_1c_{23}s_4 + s_1c_4) + s_5(s_1c_{23}s_4 - c_1c_4)]$$

$$J_{54} = -(c_1c_{23}s_4 + s_1c_4)$$

$$J_{55} = c_1c_4 - s_1c_{23}s_4$$

$$J_{56} = s_{23}s_4$$

$$J_{61} = d_6 \left[s_1 (c_{23}c_4s_5 + c_5s_{23}) + c_1s_4s_5 \right]$$

$$J_{62} = -d_6 \left[c_1 (c_{23}c_4s_5 + s_{23}c_5) - s_1s_4s_5 \right]$$

$$J_{63} = 0$$

$$J_{64} = c_1 (c_{23}c_4s_5 + s_{23}c_5) - s_1s_4s_5$$

$$J_{65} = s_1 (c_{23}c_4s_5 + s_{23}c_5) + c_1s_4s_5$$

$$J_{66} = -s_{23}c_4s_5 + c_{23}c_5 \tag{6.15}$$

Hence, we have resolution of the velocity coordinates between the end effector cartesian coordinates and the joint coordinates:

$$\begin{pmatrix} v_x \\ v_y \\ v_z \\ w_x \\ w_y \\ w_z \end{pmatrix} = J \begin{pmatrix} \dot{\theta}_1 \\ \dot{\theta}_2 \\ \dot{\theta}_3 \\ \dot{\theta}_4 \\ \dot{\theta}_5 \\ \dot{\theta}_6 \end{pmatrix} \tag{6.16}$$

This is a statement of the forward differential kinematics problem. Paul (1981) used a differential translation and rotation method derived directly from the DH kinematic equations to compute the Jacobian as:

$$J_i = \begin{pmatrix} p_x^i n_y^i - p_y^i n_x^i \\ p_x^i s_y^i - p_y^i s_x^i \\ p_x^i a_y^i - p_y^i a_x^i \\ n_z^i \\ s_z^i \\ a_z^i \end{pmatrix} \qquad \text{for revolute joints}$$

$$J_i = \begin{pmatrix} n_z^i \\ s_z^i \\ a_z^i \\ 0 \\ 0 \\ 0 \end{pmatrix} d_i \qquad \text{for prismatic joints}$$

where i defines the Jacobian matrix column.

However, the method is marginally computationally more expensive than the Whitney's cross product method [Hollerbach & Sahar 1983, Orin & Schrader 1984]:

DH formulation: $30n - 25$ multiplications, $15n - 25$ additions, $2n$ trigonometric functions.

Cross product formulation: $30n - 55$ multiplications, $15n - 38$ additions, $2n - 2$ trigonometric functions.

The advantage of Paul's method lies in that if the joint angles are not available, but the cartesian coordinates of each link are available, then the Jacobian can be calculated. However, although the end effector coordinates will be available, it is unlikely that the cartesian coordinates of the other joints will be available without the joint solutions.

6.2.2 Inverse differential kinematics solution

For the inverse velocity kinematics, we require the differential change in joint angles expressed in terms of the cartesian linear and angular velocities of the end effector. The unique inverse solution may be found by inverting the Jacobian matrix: $\dot{\theta} = J^{-1}\dot{q}$. The Jacobian matrix is difficult to invert analytically so numerical techniques are usually used. Such numerical techniques usually involve Gaussian elimination. The numerical inversion is generally regarded as a computational bottleneck and the Jacobian can become singular at certain configurations. At this point the matrix loses its full rank, i.e. the manipulator loses a degree of freedom and cannot contribute to moving the end effector position. At these unstable singular positions, infinite joint velocities are required to move the end effector through infinitesimal motions and no inverse exists. Physically, the end effector velocity is parallel to the motion of two separate joints and the joints become degenerate. Singular positions may be found by solving det$|J| = 0$. These points may then be avoided by storing lookup tables within the arm control system, or alternatively, additional joints in a kinematically redundant configuration may be introduced to give the extra degrees of freedom to overcome the singularities [Yang & Lee 1983]. Paul et al (1981b) avoided using the Jacobian altogether by directly differentiating the inverse joint angle kinematics solution to give differential joint coordinates in terms of the end effector velocity, thereby avoiding the Jacobian singularity problem. The method involves differentiating the inverse position kinematics solution directly to give for the modified PUMA 560/600 model which are quoted without proof:

$$\dot{\theta}_1 = \frac{c_1 v_y^{wrist} - s_1 v_x^{wrist}}{c_1 p_x^{wrist} + s_1 p_y^{wrist}}, \qquad \dot{\theta}_2 = \frac{c_2 \dot{A} - s_2 v_z^{wrist}}{c_2 p_2^{wrist} + s_2 A} \qquad (6.17)$$

where

$$A = c_1 p_x^{wrist} + s_1 p_y^{wrist}$$

$$\dot{A} = c_1(v_x^{wrist} + p_y^{wrist}) + s_1(v_y^{wrist} - p_x^{wrist})$$

$$\dot{\theta}_3 = \frac{a_2^2 \dot{\theta}_2 + a_2 c_2 \left(v_z^{wrist} - A\dot{\theta}_2\right) + a_2 s_2 \left(p_z^{wrist}\dot{\theta}_2 + \dot{A}\right) + p_z^{wrist}\dot{A} - Av_z^{wrist}}{d_4^2} - \dot{\theta}_2$$

$$\qquad (6.18)$$

$$\dot{\theta}_4 = \frac{c4sd4 - s4cd4}{(c4)^2 + (s4)^2} \qquad (6.19)$$

where

$$c4 = c_{23}\left[(c_1 a_x + s_1 a_y) - s_{23} a_z\right]$$

$$s4 = -s_1 a_x + c_1 a_y$$

$$cd4 = c_{23}\left[s_1(-a_x + \dot{a}_y) + c_1(\dot{a}_x + a_y) - \dot{a}_z\right] - s_{23}\left[c_1 a_x + s_1 a_y - \dot{a}_z\right]$$

$$sd4 = -c_1 a_x - s_1 a_y - s_1 \dot{a}_x + c_1 \dot{a}_y$$

$$\dot{\theta}_5 = \frac{c5 sd5 - s5 cd5}{(c5)^2 + (s5)^2} \tag{6.20}$$

where

$$c5 = s_{23}(c_1 a_x + s_1 a_y) + c_{23} a_z$$

$$s5 = \sqrt{\left[c_{23}(c_1 a_x + s_1 a_y) - s_{23} a_z\right]^2 + [-s_1 a_x + c_1 a_y]^2} = 1$$

$$cd5 = s_{23}\left[s_1(-a_x + \dot{a}_y) + c_1(a_y + \dot{a}_x) - a_z\right] = c_{23}[c_1 a_x + s_1 a_y + \dot{a}_z]$$

$$sd5 = 0$$

$$\dot{\theta}_6 = \frac{c6 sd6 - s6 cd6}{(c6)^2 + (s6)^2} \tag{6.21}$$

where

$$c6 = s_{23}(c_1 n_x + s_1 n_y) + c_{23} n_z$$

$$s6 = s_{23}(c_1 s_x + s_1 s_y) + c_{23} s_z$$

$$cd6 = s_{23}\left[s_1(-n_x + \dot{n}_y) + c_1(n_y + \dot{n}_x) - n_z\right] - c_{23}[c_1 n_x + s_1 n_y - \dot{n}_z]$$

$$sd6 = s_{23}\left[s_1(-s_x + \dot{s}_y) + c_1(s_y + \dot{s}_x) - s_z\right] - c_{23}[c_1 s_x + s_1 s_y - \dot{s}_z]$$

with

$$\dot{R} = (\dot{n}\,\dot{s}\,\dot{a}) = (nsa)\begin{pmatrix} 0 & -w_z & w_y \\ w_z & 0 & -w_x \\ -w_y & w_x & 0 \end{pmatrix}$$

so that:

$$\dot{n}_x = s_x w_x - a_y w_y \quad \dot{s}_x = -n_x w_z + a_x w_x \quad \dot{a}_x = n_x w_y - s_x w_x$$

$$\dot{n}_y = s_y w_z - a_y w_y \quad \dot{s}_y = -n_y w_z + a_y w_x \quad \dot{a}_y = n_y w_y - s_y w_x$$

$$\dot{n}_z = s_z w_x - a_z w_y \quad \dot{s}_z = -n_z w_z + a_z w_x \quad \dot{a}_z = n_z w_y - s_z w_x$$

This technique offers the inverse joint solution and a similar technique may be applied for resolved acceleration control, but the formulations (particularly for resolved acceleration control) are so complex that little is gained in terms of computational advantage. The technique is also manipulator-specific, restricting its application. The Jacobian need be inverted only once for both resolved rate and resolved acceleration control. Furthermore, the Jacobian itself is explicitly required for force control. One possible utilisation of the analytic solution to the inverse kinematics is for overcoming points when the Jacobian becomes singular.

6.2.3 Manipulator redundancy

Although manipulator redundancy has not been employed in this study due to the additional mass penalty incurred by additional links, actuators and drive mechanisms, manipulators that are nonredundant with respect to certain tasks may become redundant for other tasks—this situation can occur in dual arm manipulation. Redundancy increases the reachability of points in the workspace particularly if environmental constraints are present and enables points to be reached with smaller joint movements than nonredundant manipulators. Furthermore, the redundant degrees of freedom may be used to generate joint motion to overcome possible singular positions. The NASA dextrous Robotics Research K-1207 7 DOF arm (Fig. 6.7) is a revolute anthropomorphic design which adds an upper arm roll as the seventh joint to the traditional 6 DOF design.

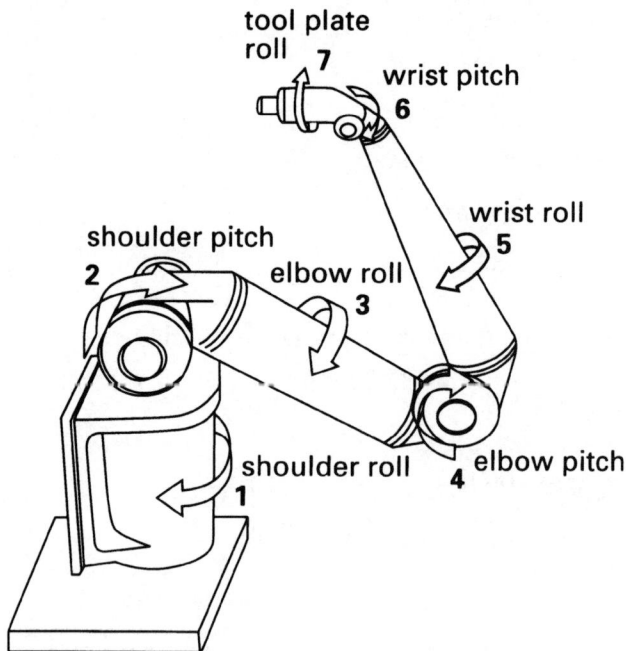

Fig. 6.7. 7 DOF Robotics Research manipulator arm (from Seraji & Long 1993). Reproduced with permission from IEEE © 1993 IEEE.

Resolved motion control uses a Jacobian matrix. In general J is an $m \times n$ matrix. If $m = n$, then the rank of J is n, generating a square matrix which is invertible, giving a unique solution for joint rates. However, if $m < n$, then m generates an infinite number of solutions: more degrees of freedom exist than are required for the task. This under-determination arises from the fact that there are fewer rows than columns. Since J is no longer square, J^{-1} does not exist. Similarly, the transformation from cartesian coordinates to joint coordinates have no closed form solution. However, it can be expressed as roots of a polynomial of degree 2^n with constraints to reduce the multivalue problem. Iterative root finding may be time-consuming and is approximate only. The way to solve the Jacobian is to make the Jacobian square by adding constraints to the joint variables. There exists a generalised pseudo-inverse Jacobian which provides a useful least squares solution such that [Whitney 1969]:

$$\dot{\theta} = J^+ \dot{q}$$

The least-squares estimator can be used to find the generalised inverse. The error to be minimised: $e = \lambda^T (\dot{q} - J\dot{\theta})$ where λ = Lagrange multiplier vector. This is minimised with respect to $\dot{\theta}$:

$$\begin{matrix} \dot{\theta} = J^T \lambda \\ \dot{q} = J\dot{\theta} \end{matrix} \rightarrow \dot{q} = JJ^T \lambda \rightarrow \lambda = (JJ^T)^{-1}\dot{q}$$

Now,

$$\dot{\theta} = J^T \lambda \rightarrow \dot{\theta} = J^T (JJ^T)^{-1}\dot{q} \quad \text{where} \quad J^+ = J^T (JJ^T)^{-1}$$

This is the Moore–Penrose pseudoinverse which is optimal in terms of the minimum covariance of $\dot{\theta}$ and satisfies the following properties [Liegeois 1977, Klein & Huang 1983]:

(i) $JJ^+ J = J^+$;
(ii) $J^+ JJ^+ = J$;
(iii) $(J^+ J)^T = J^+ J$;
(iv) $(JJ^+)^T = JJ^+$

This pseudoinverse is complex to compute being of complexity order n^3 and yields a minimum Euclidean norm least-squares solution. This minimum norm property mini-mises the product $\dot{\theta}^T \dot{\theta}$ which is closely related to the kinetic energy function and associated actuator power requirements. Similarly, singularities at which high joint velocities occur are avoided due to the tendency of the function to reduce such joint velocities. It is possible to modify the basic Moore–Penrose pseudoinverse to optimise additional position-dependent performance criteria in a least squares manner (the 'null space vector'):

$$\dot{\theta} = J^+ \dot{q} + \alpha(I - J^+ J)\frac{\partial g(\theta)}{\partial \theta}$$

where α = scalar gain constant, and $g(\theta)$ = minimised performance index.

Different workers have proposed different schemes for $g(\theta)$ whereby secondary criteria are satisfied as well as the primary criterion of trajectory tracking:

(a) $g(\theta)$ represents a position-dependent restriction to actuator angle limits and the extra degrees of freedom are used to optimise joint angle availability [Liegeois 1977]:

$$g = \sum_{i=1}^{n} \left(\frac{\theta_i - \theta_{ci}}{\delta\theta_i} \right)$$

where

$$\theta_{ci} = \frac{\beta}{2} = \left(\theta_i^{\min} + \theta_i^{\max} \right)\Big/2 = \text{midpoint of joint angle, and}$$

$\delta\theta_i$ = maximum allowable drift.

(b) $g(\theta)$ represents a manipulability index such that g is maximised and singularites where $w = 0$ are avoided [Klein & Huang 1983]:

$$g(\theta) = -w(\theta) = -\sqrt{\det JJ^T}$$

Unfortunately, any scheme involving the pseudoinverse generates a relation between end effector velocity and joint angle velocity which is not integrable. The path generated becomes nonconservative such that a closed path in x does not yield a closed path in θ. After a number of cycles around a closed path in x, a drift will occur generating unpredictable states for θ. Baillieul (1985) introduced an extended Jacobian for redundant formulations to introduce conservative motion by adding an n-row vector to the $m \times n$ Jacobian to make the extended Jacobian square. This imposed limitations on the number of possible configurations of the manipulator corresponding to any set of cartesian co-ordinates. Hollerbach & Suh (1987) examined the effects of various optimisation techniques used in redundant formulations and they found that only the kinematic unweighted Moore–Penrose pseudoinverse is globally stable and that the extended Jacobian technique experienced instabilities at trajectory extremes and for longer trajectories. Stability problems occurred for dynamic formulations such as torque constraints and inertia weightings due to whiplash effects generated from conflicting requirements. This suggests that attempts to modify the Moore–Penrose pseudoinverse dynamically are detrimental and that kinematic formulations with the pseudoinverse only should be used.

Configuration control utilises manipulator joint redundancy for additional tasks beyond end effector placement such as elbow rotation, joint limit avoidance, singularity avoidance, collision avoidance and base placement [Seraji & Long 1993, Lim & Seraji 1996]. The position kinematics relation is given by:

$$q' = \begin{pmatrix} q_e \\ \phi \end{pmatrix}$$

where $q_e = (p_x, p_y, p_z, \alpha, \beta, \gamma)^T$ = six cartesian end effector coordinates, and $\phi = \phi_1,...,\phi_r$ = user-defined cartesian kinematic task constraint assigned to r redundant degrees of freedom such that $6 + r = n$.

To enable elbow self-motion without motion of the end effector in a 7 DOF manipulator, $\phi = 180 + \theta_4$. To implement collision avoidance,

$$\phi(\theta) = d_c(\theta) - r_0 \geq 0$$

where d_c = critical distance between arm and obstacles, and r_0 = radius of object space volume of influence.

We define the $n \times n$ augmented Jacobian by:

$$\dot{q}' = J^+(\theta)\dot{\theta} \quad \text{where} \quad J^+(\theta) = \begin{pmatrix} J_e(\theta) \\ J_c(\theta) \end{pmatrix} = \text{augmented Jacobian}$$

and where $J_e(\theta) = 6 \times n$ end effector Jacobian which relates 6×1 hand velocity vector to the $n \times 1$ joint velocities, and $J_c(\theta) = \partial\phi/\partial\theta = r \times n$ task constraint Jacobian.

This provides an over-determined set of $6 + r$ equations in seven unknowns for a 7 DOF manipulator. The optimal approximate solution is given by:

$$\dot{\theta} = \left(J_+^T W_t J^+ + W_u \right)^{-1} J_+^T W_t \dot{q}_d'$$

where $W_t = (6 + r) \times (6 + r)$ task space error weights = diag$\{W_e, W_c\}$, and $W_v = 7 \times 7$ user-defined joint velocity damping weights.

This scheme is a generalised version of the extended Jacobian above which allows additional task constraints to be applied beyond end effector placement.

Some of the advantages of redundant joints may be retained without the large increase in computational control complexity. An additional rotary joint may be included near the elbow joint on link a_2 of the PUMA 560/600 manipulator configuration. This joint may be locked at $0°$ to yield the elbow-up/down configuration or locked at $90°$ about the x_2 axis to rotate the z_2 axis to produce a sideways elbows-out/in configuration (with elbow-out being utilised rather than elbow-in which might obstruct the workspace). This provides some of the flexibility of the redundant joint by introducing two possible configurations of the elbow (similar to a human arm) such as for obstacle avoidance but by restricting the manipulator to one or other configuration (elbow down or elbow out), the pseudoinverse is not required. The kinematic formulation need only be changed by replacing the link rotation matrix R_3 with

$$R_3' = \begin{pmatrix} c_3 & -s_3 & 0 \\ s_3 & c_3 & 0 \\ 0 & 0 & -1 \end{pmatrix} \text{ to rotate } (x_2, y_2, z_2) \text{ into } (x_3, -z_3, -y_3)$$

6.3 RESOLUTION OF ACCELERATION COORDINATES

Resolved acceleration control specifies how end effector accelerations are transformed to joint coordinates [Luh et al 1980b]:

$$\ddot{q} = J\ddot{\theta} + \dot{J}\dot{\theta} \quad \text{or} \quad \ddot{\theta} = J^{-1}(\ddot{q} - \dot{J}\dot{\theta}) \tag{6.22}$$

Although this has always in the past been performed numerically, analytic solutions to $\dot{J}$, which represents the nonlinear components to acceleration, are derived for the first time here using the Newton–Euler dynamics formulation for $\dot{w}_i, \dot{v}_i$ similar to Whitney's cross product dynamics method of finding the Jacobian (see next chapter where the variables are defined):

$$\dot{w}_i = \dot{w}_{i-1} + z_{i-1}\ddot{\theta}_i + w_{i-1} \times (z_{i-1}\dot{\theta}_i)$$

$$= \sum_{j=1}^{i} z_{j-1}\ddot{\theta}_j + \sum_{j=1}^{j-1}\left(\sum_{k=1}^{j-1} z_{k-1}\dot{\theta}_k\right) \times \sum_{j=1}^{i} z_{j-1}\dot{\theta}_j$$

$$\dot{v}_i = \dot{v}_{i-1} + \dot{w}_i \times l_i + w_i \times (w_i \times l_i)$$

$$= \sum_{j=1}^{i} \dot{w}_j \times l_j + \sum_{j=1}^{i} w_j \times (w_j \times l_j)$$

$$= \left[\sum_{j=1}^{i} z_{j-1}\ddot{\theta}_j + \sum_{j=1}^{i} w_{j-1} \times z_{j-1}\dot{\theta}_j\right] \times l_j + \left[\sum_{j=1}^{i}\sum_{k=1}^{j} z_{k-1}\dot{\theta}_k \times \left(\sum_{k=1}^{j} z_{k-1}\dot{\theta}_k \times l_j\right)\right]$$

$$= \sum_{j=1}^{i} (z_{j-1} \times l_j)\ddot{\theta}_j + \sum_{j}\left[\sum_{k=1}^{j-1} z_{k-1}\dot{\theta}_k \times z_{j-1}\dot{\theta}_j\right] \times l_j$$

$$+ \sum_{j=1}^{i}\left[\sum_{k=1}^{j} z_{k-1}\dot{\theta}_k \times \left(\sum_{k=1}^{i} z_{k-1}\dot{\theta}_k \times l_j\right)\right]$$

At the end effector, these equations become:

$$\dot{w}_n = \sum_{i=1}^{n} z_{i-1}\ddot{\theta}_i + \sum_{i=1}^{n}\left(\sum_{k=1}^{i} z_{k-2}\dot{\theta}_{k-1}\right) \times z_{i-1}\dot{\theta}_i$$

$$\dot{v}_n = \sum_{i=1}^{n} (z_{i-1} \times l_i)\ddot{\theta}_i + \sum_{i=1}^{n}\left(\sum_{k=1}^{i} z_{k-2} \times z_{i-1} \times l_i\right)\dot{\theta}_k\dot{\theta}_i$$

$$+ \sum_{i=1}^{n}\left[\sum_{k=1}^{i} z_{k-1}\dot{\theta}_k \times \left(\sum_{k=1}^{i} z_{k-1}\dot{\theta}_{k-1} \times l_i\right)\right]$$

or, an alternative formulation:

$$\begin{pmatrix} \dot{v}_n \\ \dot{w}_n \end{pmatrix} = \begin{pmatrix} \sum\limits_{i=1}^{n} z_{i-1} \times l_i \\ \sum\limits_{i=1}^{n} z_{i-1} \end{pmatrix} \ddot{\theta}_i$$

$$+ \begin{pmatrix} \sum\limits_{i=1}^{n}\left[\sum\limits_{k=1}^{i} z_{k-2} \times z_{i-1} \times l_i\right]\dot{\theta}_{k-1} + \left[\sum\limits_{k=1}^{i} z_{k-1}\dot{\theta}_k \times \left(\sum\limits_{k=1}^{i} z_{k-1}\dot{\theta}_k \times l_i\right)\right] \\ \sum\limits_{i=1}^{n}\left[\sum\limits_{k=1}^{i} z_{k-2} \times z_{i-1}\right]\dot{\theta}_{k-1} \end{pmatrix}\dot{\theta}_i$$

(6.23)

This is consistent with $\ddot{q} = J\ddot{\theta} + \dot{J}\dot{\theta}$ and represents the problem statement of the forward acceleration kinematics solution in analytic form. For the PUMA-type configuration being adopted:

$$\dot{J} = \begin{pmatrix} \dot{J}_1 & \dot{J}_2 & \dot{J}_3 & \dot{J}_4 & \dot{J}_5 & \dot{J}_6 \end{pmatrix}$$

where

$$\dot{J}_1 = \begin{pmatrix} z_0\dot{\theta}_1 \times \left(z_0\dot{\theta}_1 \times p_6^0\right) \\ 0 \end{pmatrix}$$

$$\dot{J}_2 = \begin{pmatrix} z_0\dot{\theta}_1 \times z_1 \times p_6^1 + (z_0\dot{\theta}_1 + z_1\dot{\theta}_2) \times \left[(z_0\dot{\theta}_1 + z_1\dot{\theta}_2) \times p_6^1\right] \\ z_0\dot{\theta}_1 \times z_1 \end{pmatrix}$$

$$\dot{J}_3 = \begin{pmatrix} (z_0\dot{\theta}_1 \times z_1\dot{\theta}_2) \times z_2 \times p_6^2 + (z_0\dot{\theta}_1 + z_1\dot{\theta}_2 + z_2\dot{\theta}_3) \times \left[(z_0\dot{\theta}_1 + z_1\dot{\theta}_2 + z_2\dot{\theta}_3) \times p_6^2\right] \\ (z_0\dot{\theta}_1 + z_1\dot{\theta}_2) \times z_2 \end{pmatrix}$$

The rest of the differential Jacobian columns are derived similarly. This may be resolved into the differential Jacobian matrix elements as:

$$\dot{J} = \begin{pmatrix} \dot{J}_{11} & \dot{J}_{21} & \dot{J}_{31} & \dot{J}_{41} & \dot{J}_{51} & \dot{J}_{61} \\ \dot{J}_{12} & \dot{J}_{22} & \dot{J}_{32} & \dot{J}_{42} & \dot{J}_{52} & \dot{J}_{62} \\ \dot{J}_{13} & \dot{J}_{23} & \dot{J}_{33} & \dot{J}_{43} & \dot{J}_{53} & \dot{J}_{63} \\ \dot{J}_{14} & \dot{J}_{24} & \dot{J}_{34} & \dot{J}_{44} & \dot{J}_{54} & \dot{J}_{64} \\ \dot{J}_{15} & \dot{J}_{25} & \dot{J}_{35} & \dot{J}_{45} & \dot{J}_{55} & \dot{J}_{65} \\ \dot{J}_{16} & \dot{J}_{26} & \dot{J}_{36} & \dot{J}_{46} & \dot{J}_{56} & \dot{J}_{66} \end{pmatrix}$$

$$\dot{J}_{11} = -p_{6x}^0\dot{\theta}_1; \quad \dot{J}_{12} = -p_{6x}^0\dot{\theta}_1; \quad \dot{J}_{13} = \dot{J}_{14} = \dot{J}_{15} = \dot{J}_{16} = 0;$$

$$\dot{J}_{21} = -s_1\dot{\theta}_1(s_4s_5d_6 + d_2) - c_1\dot{\theta}_2(s_1\dot{\theta}_2 p_{6y}^1 + c_1\dot{\theta}_2 p_{6x}^1) - \dot{\theta}_1(s_1\dot{\theta}_2 p_{6z}^1 + \dot{\theta}_1 p_{6x}^1)$$

$$\dot{J}_{22} = c_1\dot{\theta}_1(s_4s_5d_6 + d_2) - s_1\dot{\theta}_2(s_1\dot{\theta}_2 p_{6y}^1 + c_1\dot{\theta}_2 p_{6x}^1) + \dot{\theta}_1(c_1\dot{\theta}_2 p_{6z}^1 - \dot{\theta}_1 p_{6y}^1)$$

$$\dot{J}_{23} = s_1\dot{\theta}_1 p_{6x}^1 - x_1\dot{\theta}_1 p_{6y}^1 - s_1\dot{\theta}_2(s_1\dot{\theta}_2 p_{6z}^1 + \dot{\theta}_1 p_{6x}^1) - c_1\dot{\theta}_2(c_1\dot{\theta}_2 p_{6z}^1 - \dot{\theta}_1 p_{6y}^1)$$

$$\dot{J}_{24} = -c_1\dot{\theta}_1; \quad \dot{J}_{25} = -s_1\dot{\theta}_1; \quad \dot{J}_{26} = 0$$

$$\dot{J}_{31} = -s_1\dot{\theta}_1(s_4s_5d_6) + c_1(\dot{\theta}_2 + \dot{\theta}_3)\left[-s_1(\dot{\theta}_2 + \dot{\theta}_3)p_{6y}^2 - c_1(\dot{\theta}_2 + \dot{\theta}_3)p_{6x}^2\right]$$
$$- \dot{\theta}_1\left[s_1(\dot{\theta}_2 + \dot{\theta}_3)p_{6z}^2 + \dot{\theta}_1 p_{6x}^2\right]$$

$$\dot{J}_{32} = c_1\dot{\theta}_1(s_4s_5d_6) + s_1(\dot{\theta}_2 + \dot{\theta}_3)\left[-s_1(\dot{\theta}_2 + \dot{\theta}_3)p_{6y}^2 - c_1(\dot{\theta}_2 + \dot{\theta}_3)p_{6x}^2\right]$$
$$+ \dot{\theta}_1\left[c_1(\dot{\theta}_2 + \dot{\theta}_3)p_{6z}^2 - \dot{\theta}_1 p_{6y}^2\right]$$

$$\dot{J}_{33} = -c_1\dot{\theta}_1 p_{6y}^2 - s_1\dot{\theta}_1 p_{6x}^2 - s_1(\dot{\theta}_2 + \dot{\theta}_3)\left[s_1(\dot{\theta}_2 + \dot{\theta}_3)p_{6z}^2 + \dot{\theta}_1)p_{6x}^2\right]$$
$$- c_1(\dot{\theta}_2 + \dot{\theta}_3)\left[c_1(\dot{\theta}_2 + \dot{\theta}_3)p_{6z}^2 - \dot{\theta}_1 p_{6y}^2\right]$$

$$\dot{J}_{34} = -c_1\dot{\theta}_1; \quad \dot{J}_{35} = -s_1\dot{\theta}_1; \quad \dot{J}_{36} = 0$$

$$\dot{J}_{41} = c_{23}\left[s_1(\dot{\theta}_2 + \dot{\theta}_3) + c_1\dot{\theta}_1\right]c_5d_6 + c_{23}(\dot{\theta}_2 + \dot{\theta}_3)(s_4s_5d_6 + d_4)$$
$$+ a_{3y}\left[a_{3x}(s_4s_5d_6 + d_4) - a_{3y}(c_4s_5d_6)\right]$$
$$- a_{3z}\left[a_{3z}(c_4s_5d_6) - a_{3x}(c_5d_6)\right]$$

$$\dot{J}_{42} = -c_{23}\left[c_1(\dot{\theta}_2 + \dot{\theta}_3) - s_1\dot{\theta}_1\right]c_5d_6 - c_{23}(\dot{\theta}_2 + \dot{\theta}_3)(c_4s_5d_6)$$
$$+ a_{3z}\left[a_{3y}(c_5d_6) - a_{3z}(s_4s_5d_6 + d_4)\right]$$
$$- a_{3x}\left[a_{3x}(s_4s_5d_6 + d_4) - a_{3y}(c_4s_5d_6)\right]$$

$$\dot{J}_{43} = c_{23}\left[c_1(\dot{\theta}_2 + \dot{\theta}_3) - s_1\dot{\theta}_1\right](s_4s_5d_6 + d_4) - c_{23}\left[s_1(\dot{\theta}_2 + \dot{\theta}_3) + c_1\dot{\theta}_1\right](c_4s_5d_6)$$
$$+ a_{3x}\left[a_{3z}(c_4s_5d_6) - a_{3x}(c_5d_6)\right] - a_{3y}\left[a_{3y}(c_5d_6) - a_{3z}(s_4s_5d_6 + d_4)\right]$$

$$\dot{J}_{44} = c_1c_{23}(\dot{\theta}_2 + \dot{\theta}_3) - s_1c_{23}\dot{\theta}_1; \quad \dot{J}_{45} = s_1c_{23}(\dot{\theta}_2 + \dot{\theta}_3) + c_1c_{23}\dot{\theta}_1;$$

$$\dot{J}_{46} = -c_{23}(\dot{\theta}_2 + \dot{\theta}_3)$$

$$\dot{J}_{51} = c_5 d_6 \left\{ z_{4y} \left[-s_1(\dot{\theta}_2 + \dot{\theta}_3) + c_1 c_{23} \dot{\theta}_4 \right] - z_{4x} \left[c_1(\dot{\theta}_2 + \dot{\theta}_3) + s_1 c_{23} \dot{\theta}_4 \right] \right\}$$
$$+ d_6 \left[a_{4y}(a_{4y} s_5 - a_{4x} c_5) - a_{4z} a_{4z} s_5 \right]$$

$$\dot{J}_{52} = s_5 d_6 \left\{ z_{4y} \left[-s_1(\dot{\theta}_2 + \dot{\theta}_3) + c_1 c_{23} \dot{\theta}_4 \right] - z_{4x} \left[c_1(\dot{\theta}_2 + \dot{\theta}_3) + s_1 c_{23} \dot{\theta}_4 \right] \right\}$$
$$+ d_6 \left[a_{4z} a_{4z} c_5 - a_{4x}(a_{4y} s_5 - a_{4x} c_5) \right]$$

$$\dot{J}_{53} = -c_5 d_6 \left\{ z_{4z} \left[c_1(\dot{\theta}_2 + \dot{\theta}_3) + s_1 c_{23} \dot{\theta}_4 \right] - z_{4y} \left[\dot{\theta}_1 + c_{23} \dot{\theta}_4 \right] \right\}$$
$$- s_5 d_6 \left\{ z_{4z} \left[s_1(\dot{\theta}_2 + \dot{\theta}_3) - c_1 c_{23} \dot{\theta}_4 \right] + z_{4x} \left[-s_1(\dot{\theta}_2 + \dot{\theta}_3) + c_1 c_{23} \dot{\theta}_4 \right] \right\}$$
$$+ d_6 \left[a_{4z}(a_{4x} s_5 - a_{4y} c_5) \right]$$

$$\dot{J}_{54} = z_{4z} \left[c_1(\dot{\theta}_2 + \dot{\theta}_3) + s_1 c_{23} \dot{\theta}_4 \right] - z_{4y} \left[\dot{\theta}_1 + c_{23} \dot{\theta}_4 \right]$$

$$\dot{J}_{55} = z_{4z} \left[s_1(\dot{\theta}_2 + \dot{\theta}_3) - c_1 c_{23} \dot{\theta}_4 \right] + z_{4x} \left[-s_1(\dot{\theta}_2 + \dot{\theta}_3) + c_1 c_{23} \dot{\theta}_4 \right]$$

$$\dot{J}_{56} = z_{4y} \left[-s_1(\dot{\theta}_2 + \dot{\theta}_3) + c_1 c_{23} \dot{\theta}_4 \right] - z_{4x} \left[c_1(\dot{\theta}_2 + \dot{\theta}_3) + s_1 c_{23} \dot{\theta}_4 \right]$$

$$\dot{J}_{61} = d_6 \left\{ -z_{5z} \left\{ -s_1(\dot{\theta}_2 + \dot{\theta}_3) + c_1 c_{23} \dot{\theta}_4 + z_{4x} \dot{\theta}_5 \right] \right.$$
$$\left. + z_{5x} \left[\dot{\theta}_1 + c_{23} \dot{\theta}_4 + z_{4z} \dot{\theta}_5 \right] \right\} + d_6 a_{5x} a_{5z}$$

$$\dot{J}_{62} = -d_6 \left\{ z_{5z} \left[c_1(\dot{\theta}_2 + \dot{\theta}_3) + s_1 c_{23} \dot{\theta}_4 + z_{4y} \dot{\theta}_5 \right] \right.$$
$$\left. - z_{5y} \left[\dot{\theta}_1 + c_{23} \dot{\theta}_4 + z_{4z} \dot{\theta}_5 \right] \right\} + d_6 a_{5y} a_{5z}$$

$$\dot{J}_{63} = -d_6 (a_{5x} a_{5x} + a_{5y} a_{5y})$$

$$\dot{J}_{64} = z_{5z} \left[c_1(\dot{\theta}_2 + \dot{\theta}_3) + s_1 c_{23} \dot{\theta}_4 + z_{4y} \dot{\theta}_5 \right] - z_{5y} \left[\dot{\theta}_1 + c_{23} \dot{\theta}_4 + z_{4z} \dot{\theta}_5 \right]$$

$$\dot{J}_{65} = -z_{5z} \left[-s_1(\dot{\theta}_2 + \dot{\theta}_3) + c_1 c_{23} \dot{\theta}_4 + z_{4x} \dot{\theta}_5 \right] + z_{5x} \left[\dot{\theta}_1 + c_{23} \dot{\theta}_4 + z_{4z} \dot{\theta}_5 \right]$$

$$\dot{J}_{66} = z_{5y} \left[-s_1(\dot{\theta}_2 + \dot{\theta}_3) + c_1 c_{23} \dot{\theta}_4 + z_{4x} \dot{\theta}_5 \right]$$
$$- z_{5x} \left[c_1(\dot{\theta}_2 + \dot{\theta}_3) + s_1 c_{23} \dot{\theta}_4 + z_{4z} \dot{\theta}_5 \right] \tag{6.24}$$

with

$$\begin{pmatrix} a_{3x} \\ a_{3y} \\ a_{3z} \end{pmatrix} = \begin{pmatrix} -s_1(\dot{\theta}_2 + \dot{\theta}_3) + c_1 s_{23} \dot{\theta}_4 \\ c_1(\dot{\theta}_2 + \dot{\theta}_3) + s_1 s_{23} \dot{\theta}_4 \\ \dot{\theta}_1 + c_{23} \dot{\theta}_4 \end{pmatrix}; \quad \begin{pmatrix} a_{4x} \\ a_{4y} \\ a_{4z} \end{pmatrix} = \begin{pmatrix} a_{3x} + z_{4x} \dot{\theta}_5 \\ a_{3y} + z_{4y} \dot{\theta}_5 \\ a_{3z} + z_{4z} \dot{\theta}_5 \end{pmatrix};$$

$$\begin{pmatrix} a_{5x} \\ a_{5y} \\ a_{5z} \end{pmatrix} = \begin{pmatrix} a_{4x} + z_{5x}\dot{\theta}_6 \\ a_{4y} + z_{5y}\dot{\theta}_6 \\ a_{4z} + z_{5z}\dot{\theta}_6 \end{pmatrix}$$

Hence $\dot{J}$ may be computed analytically rather than numerically. The inverse solution which is required by the servocontroller is found by numerically inverting the Jacobian:

$$\ddot{\theta} = J^{-1}\left(\ddot{q} - \dot{J}\dot{\theta}\right) = J^{-1}(\ddot{q} - \alpha)$$

where

$$\begin{pmatrix} \alpha_x \\ \alpha_y \\ \alpha_z \\ \alpha_\alpha \\ \alpha_\beta \\ \alpha_\gamma \end{pmatrix} = \dot{J}\begin{pmatrix} \dot{\theta}_1 \\ \dot{\theta}_2 \\ \dot{\theta}_3 \\ \dot{\theta}_4 \\ \dot{\theta}_5 \\ \dot{\theta}_6 \end{pmatrix} \qquad (6.25)$$

6.4 RESOLUTION OF STATIC FORCE COORDINATES

Although force is not a kinematic variable, cartesian static external forces and torques at the end effector may be transformed directly into their joint torque equivalents by the Jacobian transpose [Paul 1981; Whitney 1977]. This stipulates a requirement for the explicit calculation of the Jacobian matrix. Force feedback is a critical part of the robot control structure and is required to correct errors in position to allow the manipulator to accommodate the physical constraints of the task which usually require high degrees of positioning accuracy between mating parts, e.g. insertion tasks. Force control effectively reduces the required positioning accuracy to perform a task: there will always be random fluctuations in final positioning which need to be accounted for. Small variations in relative position will generate large contact forces when parts interact providing a natural mechanism of amplifying small position errors at hand level. Much manipulation involves contact forces encountered at the end effector rather than just positioning and so force sensing and control is a fundamental requirement for any robot working in a variable structure environment. Indeed, Paul (1987) stated that manipulation comprises of a series of collisions between object and manipulator. The three types of manipulator motion are therefore: motion in freespace, contact, and force exertion. Force sensors may be mounted on the joints or the wrist [Shimano & Roth 1975]. Joint sensors measure motor torque currents directly at the joints but they limit the accuracy of the force coefficients at the end effector due to uncertainties in joint friction and damping making this method unsuitable when small hand force resolution <1 N are required. Such inaccuracies are intolerable for precise assembly tasks. Furthermore, it is time-consuming to convert from joint torques to cartesian hand forces and moments since it involves inversion of the J^T matrix—a major problem with force feedback is that even short delays in computation can cause instability [Whitney 1987]. Hence, wrist sensors such as the Scheinmann wrist should be used since they eliminate motor torque and dynamic link effects by measuring forces directly at hand level to high resolution. The wrist sensor should be mounted as close to the hand as possible. Mounting at the gripper fingers itself is not practicable as the sensor will experience noise problems and wear-and-tear. The application of signal processing, filtering and estimation for prediction is required for noise handling, and this is too time-consuming for the servocontrol loop level.

The resolution of force coordinates is based on virtual work arguments. Virtual work is performed when a real force moves through a virtual displacement. The principle of virtual work is one of the two cornerstones underlying structural analysis (the other being the principle of equilibrium). It states that a rigid body in equilibrium will have zero net work done by a system of forces acting on it through an arbitrary displacement. Conservation of energy states that the external energy derived from the work done by external forces entering the closed strucure of such a body will be stored as internal strain energy in the structure. Hence, between any two coordinate frames, the virtual work done by any applied force is the same [Paul 1981]: $dW_i = dW_{ext}$ or equivalently $F_i^T dx_i = F_{ext}^T dx_{ext}$, where F = generalised force vector. Since the generalised force vector at the joints are of revolute form they represent the joint torques: $\tau_i^T dx_i = F_{eff}^T dx_{eff}$ for revolute jointed manipulator where $F_i = \tau_i$.

The external forces acting on the manipulator (subscript 'ext') are assumed to be precisely coincident with those forces measured by the wrist force sensor (subscript 'eff').

Now,

$$dx_{eff} = v_{eff} = J\theta_i = Jv_i :$$

Hence,

$$\tau_i^T v_i = F_{eff}^T Jv_i$$

i.e.

$$\tau = J^T F_{ext}$$

where

$$F_{ext} = \left(f_x^{ext} \, f_y^{ext} \, f_z^{ext} \, n_x^{ext} \, n_y^{ext} \, n_z^{ext} \right)^T$$

$$\tau = (\tau_1 \tau_2 \tau_3 \tau_4 \tau_5 \tau_6)^T$$

$$J^T = \begin{pmatrix} J_{11} & J_{12} & J_{13} & J_{14} & J_{15} & J_{16} \\ J_{21} & J_{22} & J_{23} & J_{24} & J_{25} & J_{26} \\ J_{31} & J_{32} & J_{33} & J_{34} & J_{35} & J_{36} \\ J_{41} & J_{42} & J_{43} & J_{44} & J_{45} & J_{46} \\ J_{51} & J_{52} & J_{53} & J_{54} & J_{55} & J_{56} \\ J_{61} & J_{62} & J_{63} & J_{64} & J_{65} & J_{66} \end{pmatrix} \tag{6.26}$$

This enables force control to be implemented explicitly at hand level within the servo-controller.

6.5 DUAL MANIPULATOR KINEMATICS

Dual robot manipulators offer several advantages over the use of a single robot manipulator:

(i)   reliability is enhanced as two arms offer redundancy;
(ii)   parallel operation decreases task completion time;
(iii) some tasks explicitly require more than one manipulator.

Indeed, it is conceivable that the space environment will provide the first deployment of multiple arm robotic systems. Indeed, Adams et al (1987) noted that for the Solar Maximum main electronics box exchange, two-arm-parallel manipulation was essential, even if only to carry a camera. Hence any robotic freeflyer such as ATLAS must have a minimum of two arms (plus possibly a telescopic docking device). To this end, this chapter covers methods for dealing with dual arm operation, particularly for coordinated motion, and develops techniques that are directly utilisable by the space-based dual arm kinematics formulation of Chapter 8.

6.5.1 Inter-arm collision avoidance
Maimon & Nof (1985) proposed that dual arm control requires some form of coordination and introduces an Activity Controller which could be embedded in a hierarchical control structure above the level of individual robot arm servocontrollers. It could be implemented by either a central spacecraft processor or more likely a dedicated processor for dual arm coordination. This Activity Controller was divided into two parts, AC1 and AC2. AC1 provided planning of the coordination of tasks for each robot arm with respect to spatial constraints. AC2 provided planning of the synchronisation of tasks for each robot arm with respect to temporal constraints to effect collision avoidance. This highlights one of the chief difficulties in dual-arm operation—that of inter-arm collision. Coordinated motion of two arms may be loose such that the two arms execute independently controlled sequences of actions, or tight when both arms have a fixed relation over the motion to execute a common task. Both types of operation can generate collisions, but it is particularly susceptible in loose coordination since the arms are not maintained in fixed positions relative to each other. Furthermore, Chapter 9 gives an indication of the difficulties inherent in obstacle avoidance particularly if the obstacle is moving.

Lee & Lee (1987) attempted to tackle the problem of inter-arm collision by representing each robot wrist as a sphere centred on the hand coordinate frame, thereby restricting their formulation to wrist collisions only (admittedly the most likely, however). The use of a spherical model was motivated by the desire for minimal computation for modelling and yet allowing for rotations. They used a collision map of path length against time to generate straight line path segments and locate potential collision regions. Using a time scheduling algorithm to modify the travelling speed of the planned trajectories rather than path modification, they incorporated time delays in the trajectory generation algorithm. The main problem was that task execution time was considerably increased due to the time-consuming scheduling operations. The difficulty of the planning approaches is that they are complex and cannot be performed in real-time—collision avoidance is a real-time reflexive operation and not a cognitive reasoning problem.

Collision between arms is always a possibility when the work envelopes of the two arms overlap. Mowforth & Bratko (1987) eliminated the possibility of collision using the Zambesi bridge configuration such that the manipulators were positioned so that they could not touch yet were capable of passing extended components, i.e. their separation distance (taken to be in the x-direction) was the maximum extent of their combined workspaces: $h = 2 \times p_n^{max}$ assuming both manipulators to be of the same geometry. Since the workspaces of the two robots intersect only at a point, this configuration is highly limited.

Collision may be avoided by using safety routines which halt the manipulator under certain conditions. Collision can occur between any two points on the respective manipulators but certain kinematic constraints can be imposed to effect collision avoidance. Hemami (1986a) introduced the notion of software monitoring of the distances between the central axes of each link of each arm at every update step as part of an intercept-driven algorithm. If the separation distance between any two links on separate manipulators falls below a prescribed value (i.e. the dimensions of each arm) then the motion is halted and replanning is invoked as necessary. The number of repetitions of this algorithm depends on the number of link segments, i.e. for three link segments (i.e. shoulder/elbow/end effector) there must be nine repetitions at each update point.

Let t = diameter of a manipulator link, l_2 = length of link 2 from shoulder to elbow, p = position vector of the specified joint of specified arm with respect to own base coordinates:

(i)   No collision occurs if $p_{shoulder}^{right} - p_{shoulder}^{left} > t$—this condition is trivially true since $p_{shoulder}^{right} - p_{shoulder}^{left} = h > t$, so this does not require computation.

(ii) No collision occurs if $p_{elbow}^{right} - p_{elbow}^{left} > t$—this condition will always be true if $l_2 < h/2$.

(iii)
$$\left. \begin{array}{l} p_{elbow}^{right} - p_{shoulder}^{left} > t \\[2mm] p_{shoulder}^{right} - p_{elbow}^{left} > t \end{array} \right\} \text{—these conditions will always be true if } l_2 < h.$$

Hence, by fixing $l_2 \leqslant h/2$, elbow/shoulder collisions can be avoided. For the PUMA 560/600 variant considered here, $l_2 = 0.5$ m $\rightarrow h \geqslant 1.0$ m.

A modification of the Zambesi bridge concept will overcome wrist collisions by allowing the robot workspaces to intersect generating a plane of intersection equidistant between the two arms at $h/2$. This allows a wide area for component swapping. However, the motions of each arm are restricted to their own workspace quadrants with no arm being permitted to cross the plane of workspace intersection. As no arm overlapping can occur, so neither can any collisions. The end effector positions must be repeatedly evaluated at each update step to ensure that the following condition holds at all times during loose cooperation: $p_{n(x)}^{right} \leqslant h/2$ and $p_{n(x)}^{left} \leqslant h/2$ for each end effector in the x-coordinate. This modified Zambesi bridge constraint may be relaxed during closed chain maneouvres as the maintenance of fixed position/orientation between the arms precludes the possibility of collision—in this way the Zambesi bridge configuration will not restrict large-scale payload transfers by the manipulator arms. Furthermore, for a space freeflyer, adjustment of the orientation and attitude of the spacecraft with respect to the target can provide additional flexibility in redefining the plane of workspace intersection.

Certain other kinematic constraints may be applied for other reasons:

(i)   to enable self-repair—upper arm and forearm lengths to exceed the half-separation distance between the arms: $a_2 + d_4 \geqslant h/2$ [Wang 1987];

(ii)   to enable truss walking between truss nodes: $a_2 + d_4 \geqslant d/2$ where d = truss node interval = 1.67 m [Xu et al 1992]—this is true for $a_2 = d_4 = 0.5$ m. For the PUMA 560/600 variant considered here, this gives $1 \text{ m} \leqslant h \leqslant 2 \text{ m}$ constraint on the separation distance between the two arms.

6.5.2 Dual manipulator kinematic configuration

Hemami (1985) considered that side-mounting of two manipulators onto a platform such that the elbow or shoulder joints moved in opposition (i.e. left/right configurations) offered not only an anthropomorphic configuration but better manoeuvrability than conventional mounting whereby the identical arms are related by a rotation and a simple translation of separation distance h. This is the configuration adopted here: a left shoulder is mounted onto the left side of the spacecraft platform mount (master arm—superscript/subscript 'm') and the right shoulder configuration is mounted onto the right side of the spacecraft mount (slave arm—superscript/subscript 's'). The relationship between the two arms is the matrix H which translates h in the $+x$-direction and rotates 180° about y axis of the master arm to the slave arm. H is also its own inverse [Hemami 1985, 1986b].

$$H = \begin{pmatrix} -1 & 0 & 0 & h \\ 0 & 1 & 0 & 0 \\ 0 & 0 & -1 & 0 \\ 0 & 0 & 0 & 1 \end{pmatrix}$$

where h = separation distance between the two manipulators.

This allows specification of the kinematics of the slave arm in terms of the coordinates of the master arm, assuming that the kinematic structures of both robot manipulators are the same (except for handedness). The reference coordinate system is the base of the master arm (arbitrarily taken as the left arm). Hence, $T_s = HT_m$ denotes the same position/orientation of the slave arm with respect to base reference coordinates where

$$T_m = \begin{pmatrix} n & s & a & p \\ 0 & 0 & 0 & 1 \end{pmatrix}$$

Two types of control for dual manipulator configurations are possible:

(i)   tasks requiring independent programming for each arm—although independent unrelated tasks may be performed by each manipulator in parallel they must be coordinated (loose coordination);

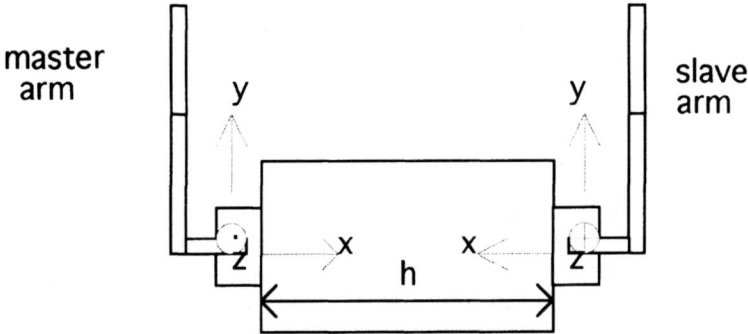

Fig. 6.8. Dual manipulator configuration.

(ii)   tasks requiring the two manipulators to cooperate in some fashion and require
       programming with respect to each other (tight coordination)—this is typical of
       most assembly type tasks:

   (a)   open chain formulation where no closed kinematic loops exist—generally the
         tasks will be of the main task/subsidiary role for each manipulator, but two
         special cases exist whereby the relative actions of the two end effectors are
         constrained in one or more directions [Hemami 1986b; Tao et al 1987]:

      (1)   copying motion whereby each arm performs the same operation;
      (2)   symmetric motion through a plane whereby each arm mirrors the other;

   (b)   closed chain formulation in which synchronisation is necessary—the arms
         hold a common object for spatial transfer making compliant motion neces-
         sary due to the kinematic constrants for the maintenance of constaint position
         and orientation.

6.5.3 Open chain kinematic configuration

The coordination of two arms in the open kinematic chain mode is necessary for two
special cases of important types of task characteristic of assembly type operations [Tao,
et al 1987; Hemami 1986b].

(1)   For copied motion, only the tool point position differs between the master and slave
      arms—this difference is constant and serves as the task constraint:

$$\begin{pmatrix} R_s & p_s \\ 0 & 1 \end{pmatrix} = H \begin{pmatrix} R_m & p_m + b \\ 0 & 1 \end{pmatrix} = \begin{pmatrix} -n_x^m & -s_x^m & -a_x^m & -(p_x^m + b_x) + h \\ n_y^m & s_y^m & a_y^m & p_y^m + b_y \\ -n_z^m & -s_z^m & -a_z^m & -(p_z^m + b_z) \\ 0 & 0 & 0 & 1 \end{pmatrix}$$

      where

$$b = \begin{pmatrix} b_x \\ b_y \\ b_z \end{pmatrix} = \text{tool point bias offset} \tag{6.27}$$

An example of where this formulation may be used is in ORU exchange by removing two bolts at once at different positions on the module or spacecraft.

(2)   For symmetric motion, the position of the tool point differs only in one coordinate if the reflection occurs in the zy plane at $x = d = h/2$. Furthermore the z components of the **a** and **s** rotation vectors are opposed and since $\mathbf{n} = \mathbf{s} \times \mathbf{a}$, the x and y components of the **n** vector are also opposed for symmetric motion.

$$\begin{pmatrix} R_s & p_s \\ 0 & 1 \end{pmatrix} = H \begin{pmatrix} -n_x^m & s_x^m & a_x^m & p_x^m + 2d \\ -n_y^m & s_y^m & a_y^m & p_y^m \\ n_z^m & -s_z^m & -a_z^m & -p_z^m \\ 0 & 0 & 0 & 1 \end{pmatrix} = \begin{pmatrix} n_x^m & -s_x^m & -a_x^m & -p_x^m \\ -n_y^m & s_y^m & a_y^m & p_y^m \\ -n_z^m & s_z^m & n_z^m & -p_z^m \\ 0 & 0 & 0 & 1 \end{pmatrix}$$

$$\tag{6.28}$$

This would represent a task such as grappling—the capture operation for a space manipulator is regarded as an important elementary task for space manipulators [Umetani & Yoshida 1989]. Dual insertion tasks such as peg-in-hole operations also have this form since **n**, **s**, and p_z are variable with p_x, p_y and **a** fixed to allow compliance. Screwing operations are a subsidiary of insertion operations whereby the gripper orientation must change but the direction of screw turn does not, i.e. $\theta_6^s = \theta_6^m$ saving a 180° rotation at the joint [Hemami 1986b]:

$$\begin{pmatrix} R_s & p_s \\ 0 & 1 \end{pmatrix} = H \begin{pmatrix} n_x^m & s_x^m & a_x^m & p_x^m \\ n_y^m & s_y^m & a_y^m & p_y^m \\ n_z^m & s_z^m & -a_z^m & p_z^m \\ 0 & 0 & 0 & 1 \end{pmatrix} = \begin{pmatrix} -n_x^m & -s_x^m & -a_x^m & -p_x^m \\ -n_y^m & s_y^m & a_y^m & p_y^m \\ -n_z^m & -s_z^m & a_z^m & -p_z^m \\ 0 & 0 & 0 & 1 \end{pmatrix} \tag{6.29}$$

Both insertion and screwing operations are generic assembly tasks so these conversion are critical.

6.5.4 Closed chain kinematic configuration

The closed chain configuration is a critical mode of dual arm cooperation on a single object as this is the primary purpose of using dual manipulators. For any kind of cooperative motion, an object must be accessible to both manipulators, i.e. the object must extend into the workspace of both robots and/or the object must lie within the intersection of their common workspace [Uchiyama et al 1987]. The chain closure is effected by the two manipulators gripping the same object and the end effectors are separated by an offset bias. Both assembly operations and object transfer through the closed chain configuration determine that the behaviour of each robot manipulator is no longer independent of the other. The kinematic constraints alter for an open-loop tree configuration which suddenly changes to a closed-loop topology on target acquisition. In the closed loop configuration, the manipulators exert forces and torques on the object to cause tension, compression or torsion in the object without affecting the motion trajectory of the object. It is desirable to minimise these forces and torques.

Tarn et al (1987, 1988) considered the closed kinematic chain case and allowed rotation between the fingers of the end effectors and the object. The closed chain formed is essentially the addition of the object as a link of a longer chain. These two extra degrees of freedom contributed to a total of eight degrees of freedom to the closed chain system as a whole as defined by the Kutsbach–Grubler criterion: $n = 6(m - 1) - 5p = 8$, where n = number of degrees of freedom controllable, m = number of links = 14, and p = number of arbitrary joints = 14.

Hence, eight coordinates are required to specify this closed chain. The closed chain imposes nonholonomic constraints to the manipulator configurations implying actuator redundancy since the number of outputs (8) was less than the number of controllable inputs (12 independent torque-generating actuators), i.e. the configuration of the closed chain was determined by $\theta = (\theta_1^m,...,\theta_6^m,\theta_7,\theta_8)$ where θ_7 and θ_8 represented the rotation angles between the contact surfaces of the object and the two end effectors (assuming no relative translation). The slave joint angles $\theta_1^s,...,\theta_6^s$ are determined by $\theta_1^m,...,\theta_6^m,\theta_7,\theta_8$.

However, if the object is rigidly grasped, the Kutzbach–Grubler criterion is altered. In this case, $m = 12$ since the last link of each arm and the payload now represent one link. They are fixed relative to each other such that θ_7 and θ_8 are invariant. The other five links of each arm and the mounting generate $m = 12$. There are also 12 actuators, one for each manipulator joint, hence $p = 12$ giving six as the number of controllable degrees of freedom. This is effectively reduces the problem of dual arm, closed chain control to a holonomic problem similar to that of controlling a single manipulator. By adopting rigid grasp to form a closed kinematic chain of 12 joints and 12 links, the relative positions and orientations of the arms with respect to each other remains invariant—this is a statement of the task constraint as defined by Lim & Chyung (1987) in their resolved position control formulation for closed chain kinematics. Although the object coordinate frame is centred at the load's centre of mass and the reference coordinates are centred on the mounting centre of mass, the end effector coordinates of each arm may be found enabling inverse kinematics solution to be applied to find the joint angles. The end effector 4×4 coordinates of arm l (master or slave) with respect to its own base is given by:

$$T^l = \left(T_0^l\right)^{-1} T_{0n}^{ref} T_{n+1}^l$$

where $l = 1$ for master arm (superscript 'm'), or 2 for slave arm (superscript 's'), T_{n+1}^l = constant task constraint 4×4 DH matrix, T_{0n}^{ref} = object trajectory defined by the object coordinate frame 4×4 DH matrix, and $T_0^l = 4 \times 4$ DH matrix that maps from the centre of mass of the mount to the base of arm l.

Uchiyama & Dauchez (1988) characterised T_{n+1}^l as virtual sticks fixed at each hand pointing towards the object centre of mass. This interpretation allows the 4×4 matrix to

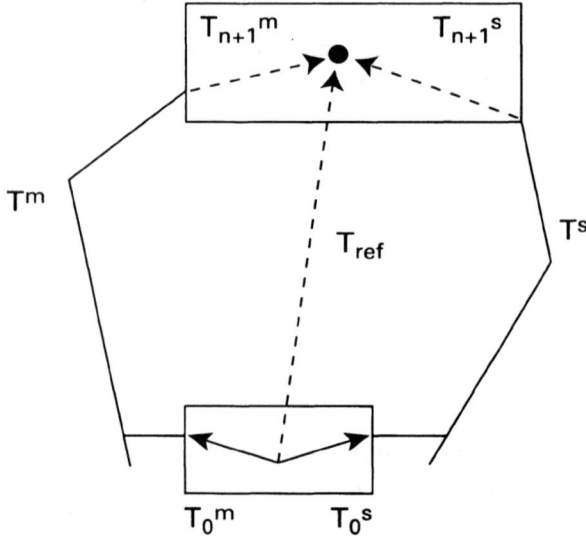

Fig. 6.9. Dual manipulator closed chain configuration.

be represented as a relative position vector r_{n+1}^l without the rotational component R_{n+1}^l, particularly as the relative orientation with respect to the end effectors will not change. Similarly, the 4×4 matrix T_0^l may be represented as position vectors vectors s_0^l with no platform attitude component R_0^l. The arm vectors are related by: $s_0^{l=1} = H s_0^{l=1}$.

Tao et al (1987) and Zheng & Luh (1986) went a step further by deriving a set of holonomic constraints on position and orientation of the slave arm (superscript 's') to constrain the motion of the kinematic chain uniquely. The motion of the master arm (superscript 'm') is planned to generate the required motion of the payload. The motion of the slave arm is then calculated to have a coordinated relation with respect to the master arm. They defined the relative position/orientation of the end effectors to be invariant whilst holding a rigid body such that [Tao et al 1987]:

$$T^s = H^{-1} T^m O$$

where

$$O = \begin{pmatrix} n_{obj} & s_{obj} & a_{obj} & l_{obj} \\ 0 & 0 & 0 & 1 \end{pmatrix} = \left(T_{n+1}^s\right)^{-1} T_{n+1}^m = \text{object } 4 \times 4 \text{ DH matrix}$$

$$H = \left(T_0^m\right)^{-1} T_0^s$$

Since the relative positions and orientations of the end effectors are invariant in the closed chain, object matrix **O** is constant. Equivalently, the position and orientations of **O** may be decoupled

$$O = \begin{pmatrix} R_{obj} & l_{obj} \\ 0 & 1 \end{pmatrix}$$

to yield two sets of holonomic equality constraints [Luh & Zheng 1986, Zheng 1989].

Holonomic angular constraint:

$$R_{obj} = \left(R_n^m\right)^T R_n^s = (n^m s^m a^m)^T (n^s s^m a^m) = (n_{obj} s_{obj} a_{obj}) \tag{6.30}$$

Holonomic linear constraint:

$$p_n^m + T_{n+1}^m - \left(p_n^s + T_{n+1}^s\right) = 0$$

$$\rightarrow p_n^m + R_{obj} l_{obj} - p_n^s = 0 \tag{6.31}$$

Hence, the slave arm end effector coordinates are dependent on the end effector coordinates of the master arm and the matrix R_{obj} which is constant:

$$p_n^s = p_n^m + R_{obj} l_{obj}$$

$$R_n^s = H^{-1} R_n^m R_{obj} \tag{6.32}$$

This then allows the inverse kinematic joint angles to be found for the control system.

These constraints may be extended to velocity and acceleration variables [Luh & Zheng 1986, Zheng 1989]. Both linear and angular velocities of both end effectors must be the same at all times so that no relative motion occurs between the end effectors:

$$\begin{pmatrix} v^s \\ w^s \end{pmatrix} = \begin{pmatrix} v^m \\ w^m \end{pmatrix} \rightarrow \dot{\theta}^s = (J^s)^{-1} \begin{pmatrix} v^m \\ w^m \end{pmatrix} = (J^s)^{-1} J^m \dot{\theta}^m$$

and

$$\dot{\theta}^m = (J^m)^{-1} \begin{pmatrix} v^m \\ w^m \end{pmatrix} \tag{6.33}$$

Note that in general the slave and master Jacobians will be different since they will have different configurations. Similarly, there is a similar acceleration constraint:

$$\begin{pmatrix} \dot{v}^s \\ \dot{w}^s \end{pmatrix} = \begin{pmatrix} \dot{v}^m \\ \dot{w}^m \end{pmatrix} \rightarrow J^s \ddot{\theta}^s + j^s \dot{\theta}^s = J^m \ddot{\theta}^m + j^m \dot{\theta}^m$$

$$\rightarrow \ddot{\theta}^m = (J^m)^{-1} \left[\begin{pmatrix} \dot{v}^m \\ \dot{w}^m \end{pmatrix} - j^m \dot{\theta}^m \right] \quad \text{and} \quad \ddot{\theta}^s = (J^s)^{-1} \left[\begin{pmatrix} \dot{v}^m \\ \dot{w}^m \end{pmatrix} - j^s \dot{\theta}^s \right] \tag{6.34}$$

This completes the dual manipulator kinematic configuration which may be regarded as a straightforward extension of the single manipulator case.

6.5.5 Dual arm configuration summary

This section has considered both open chain and closed chain kinematics for dual manipulator control while utilising the techniques of single manipulator control. The central philosophy was to enable individual robot arm servocontrollers to perform their functions independently of each other yet observing certain constraints. This makes the schemes proposed as suitable for space application as indicated in Chapter 8. However, for the implementation of global schemes such as load distribution (outlined in Chapter 10), a higher-level controller must be implemented to perform such tasks. This implies a hierarchical distribution of control. Suffice to state that an Activity Controller of one description or another could be used to implement this in a control framework compatible with NASREM. The approach above ensures that each arm's servocontrol is executed as low down the control hierarchy as possible to avoid computational bottlenecks and problems of differing temporal horizons higher up the control hierarchy. This section shows that the methods applicable to single arm control are applicable to dual arm control with the addition of straightforward constraints.

7

Robotic manipulator dynamics

Next, we must derive the joint torques of the manipulator required to realise the end effector kinematic variables of position, velocity and acceleration. Although space robotics per se has only recently been considered in the literature, multibody dynamics in space has attracted interest for some years due to its highly complex nature and applicability to orbiting spacecraft. Hooker & Margulies (1965) first considered the dynamics formulation of multibody systems in space. They found that for n rigid bodies connected by rotational joints in an arbitrary open topological tree configuration (i.e. no closed loops), any rotational motion of the bodies produced a relative translation with respect to each other. They introduced the notion of the connection barycentre which they defined as the composite centre of mass of each body obtained by loading each joint of the body with the total residual mass of the system connected to that joint, i.e. the mass distribution is represented by the total mass of inboard links Mu_i acting on the inboard joint i and the total mass of outboard links $M(1 - u_{i+1})$ acting on the outboard joint $i + 1$ to form an augmented body.

The barycentric vector from the ith barycentre to ith body centre of mass:

$$c_i = r_i u_i + s_i(1 - u_{i+1})$$

where

$$u_i = \sum_{j=0}^{i-1} \frac{m_j}{M_T} \quad \text{for} \quad i = 0, \ldots, n$$

and where M_T = total mass of all links, r_i = link vector from joint $i - 1$ to link i centre of mass and s_i = link vector from link i centre of mass to joint i.

They used the barycentric concept to derive the inertia dyadics for the Newton–Euler dynamics equations to describe the attitude equations of the system. Hooker (1970) extended the barycentric approach by eliminating the constraint torques for the n body attitude equations using the Lagrangian method, reducing the number of equations from $3n$ to r where r = number of rotational degrees of freedom of the multibody system. This method is not really suitable for robotic application since it requires recomputation of the barycentres whenever any payload is acquired or released, generating high computational overheads and is restricted to open chain configurations.

Ho (1977) stated that Hooker later abandoned the barycentric approach in favour of the direct path method advocated by Ho. This method involves determining the contribution of the motion of each body singly to coefficient matrices of the overall dynamic equations describing the system. The direct path is a vectorial path from the main reference body 0 to each separate body j centre of mass comprising the system. The motion of each body influences the motion of the whole system. Since linear and angular positions, velocities and accelerations are related to the other bodies and to the reference coordinates kinematically, these direct path vectors transform the influence of the motion of one body to the other bodies. However, for revolute robots, this method is inappropriate since almost inevitably, more than one link motion is required to move to attain any particular point in space. Jerkovsky (1978) adapted the direct path method by introducing the concept of path matrices and reference matrices which describe the particular topology of the system.

$$\text{Path matrix, } \pi_{ij} = \begin{cases} 1 \\ 0 \end{cases}$$

1 signifies body j between body 1 and body i, 0 otherwise

$$\text{Reference matrix, } \rho_{ij} = \begin{cases} 1 \\ -1 \\ 0 \end{cases}$$

1 if $i = j$, -1 if body i referred to body j, 0 otherwise and $\rho\pi = \pi\rho = I$

He derived the primitive equations of motion for each body separately, considering them to be free bodies. By converting inertial velocities to relative velocities using linear operators, he was able then to apply the constraints of motion to the system equations. Hughes (1979) suggested that the direct path method is superior to the barycentric approach. However, the method is unsuitable for closed chain configurations whose inertial velocities cannot be expressed as independent sets of relative velocities for each body. These methods are very general and may be applied to multiple degree-of-freedom joints as well as for forward dynamic analysis for simulation. However, generality is not an issue for space robotic control which only require inverse dynamics solutions where each actuator has only a single degree of freedom, and efficient equations have been developed for these problems. Robot dynamics is concerned with relating joint motions (position, velocity, acceleration) to the required forces and torques to achieve those motions. The required actuator joint torques are computed to enforce tracking of the prescribed trajectory. The torques are fed forward to the joint servos to provide fast response without instability.

We consider initially terrestrial manipulators before tackling space-based manipulators. Generally speaking, two main approaches exist for the description of the dynamics of robots: the Lagrange–Euler technique and the Newton–Euler technique (excluding more exotic techniques such as characterising robot dynamics as a Kalman filter problem [Rodriguez 1987, Johnson & Hill 1985]) and these two formulations are in fact equivalent [Silver 1982]. Both generate a set of six coupled, second-order, nonlinear differential equations in position, velocity and acceleration. The Lagrange–Euler technique yields highly structured equations highly suited to control analysis in the state space form by decomposing the complex system into a set of smaller systems [Bejczy & Tarn 1986].

Generalised force,

$$\tau_i = \frac{d}{dt}\left(\frac{\partial L}{\partial \dot{\theta}_i}\right) - \left(\frac{\partial L}{\partial \theta_i}\right)$$

where $L = T - V = $ Lagrangian, $T = T = \frac{1}{2}\dot{\theta}^T D(\theta)\dot{\theta} = \frac{1}{2}mvv + \frac{1}{2}wIw$ and $V = $ potential energy $= 0$ in zero-gravity, i.e.

$$\tau = \frac{d}{dt}\left(\frac{\partial T}{\partial \dot{\theta}}\right) - \left(\frac{\partial T}{\partial \theta}\right)$$

The kinetic energy term T forms the basis of the Routhian and Hamiltonian functions where $D(\theta) = $ inertia matrix.

Since potential energy is effectively zero in space, the concept of energy specifically entails kinetic energy and kinetic energy is used for investigating the stability of manipulators, e.g. in the Lyapunov function. Its usefulness may be illustrated by the following relations for force F and moment N:

$$\left.\begin{aligned}\frac{d}{dt}\left(\frac{\partial K}{\partial v}\right) + w \times \left(\frac{\partial K}{\partial v}\right) &= F \\[2mm] \frac{d}{dt}\left(\frac{\partial K}{\partial w}\right) + w \times \left(\frac{\partial K}{\partial v}\right) + v \times \frac{\partial K}{\partial v} &= N\end{aligned}\right\}$$

$$\text{where} \quad \left(\frac{\partial K}{\partial w}\right) = Iw = L \quad \text{and} \quad \left(\frac{\partial K}{\partial v}\right) = mv = P$$

where $L = $ angular momentum and $P = $ linear momentum.

The kinetic energy of a manipulator is given by:

$$dK_i = \frac{1}{2}\dot{\theta}_i D(\theta)\dot{\theta}_i = \frac{1}{2}tr\left(v_i v_i^T\right)dm$$

For any point fixed on link i:

$$r_i^0 = A_i^0 r_i^i$$

where

$$A_i^0 = A_i^0 \dots A_i^{i-1}$$

$$r_i^i = (x_i y_i z_i l)^T$$

Differentiate:

$$v_i^0 = v_i = \frac{d}{dt}\left(A_i^0 r_i^i\right) = \left(\sum_{j=1}^{i}\frac{\partial A_i^0}{\partial \theta_j}\dot{\theta}_j\right)r_i^i$$

Now,

$$\frac{\partial A_i^{i-1}}{\partial \theta_i} = Q_i A_i^{i-1}$$

where

$$Q_i = \begin{pmatrix} 0 & -1 & 0 & 0 \\ 1 & 0 & 0 & 0 \\ 0 & 0 & 0 & 0 \\ 0 & 0 & 0 & 0 \end{pmatrix} \quad \text{for a revolute joint}$$

For $i = 1, \ldots, n$: let

$$U_{ij} = \frac{\partial A_i^0}{\partial \theta_j} = \begin{cases} A_{j-1}^0 Q_j A_i^{j-1} & \text{if} \quad j \leqslant i \\ 0 & \text{if} \quad j > i \end{cases}$$

Hence,

$$v_i = \left(\sum_{j=1}^{i} U_{ij} \dot{\theta}_j \right) r_i^i$$

Substitute:

$$dK_i = \frac{1}{2} tr \left[\left(\sum_{j=1}^{i} U_{ij} \dot{\theta}_{ij} \right) r_i^i \right] \left[\left(\sum_{j=1}^{i} U_{ij} \dot{\theta}_{ij} \right) r_i^i \right]^T dm$$

$$= \frac{1}{2} tr \sum_{j=1}^{i} U_{ij} \left(r_i^i \ dm \ r_i^{iT} \right) U_{ij}^T \dot{\theta}_j \dot{\theta}_j$$

The matrix U_{ij} represents the rate of change of points on link i relative to base coordinates as the joint angle changes.

Integrate:

$$K_i = \int dK_i = \frac{1}{2} tr \sum_{j=1}^{i} U_{ij} \left(\int r_i^i r_i^{iT} \ dm \right) U_{ij}^T \dot{\theta}_j \dot{\theta}_j$$

where

$$I_i = \int r_i^i r_i^{iT} \ dm$$

$$= \begin{pmatrix} \frac{1}{2}(-I_{ixx} + I_{iyy} + I_{izz}) & I_{ixy} & I_{ixz} & m_i \bar{x}_i \\ I_{ixy} & \frac{1}{2}(I_{ixx} - I_{iyy} + I_{izz}) & I_{iyz} & m_i \bar{y}_i \\ I_{ixz} & I_{iyz} & \frac{1}{2}(I_{ixx} + I_{ijj} - I_{izz}) & m_i \bar{z}_i \\ m_i \bar{x}_i & I_{iyz} & m_i \bar{z}_i & m_i \end{pmatrix}$$

where I_i = inertia tensor with respect to own link coordinates and $\bar{r}_i = (\bar{x}_i \bar{y}_i \bar{z}_i)^T$ = centre of mass of link i with respect to link i coordinates.

In principal axis coordinates, the off-diagonals are zero.

Total kinetic energy:

$$K = \sum_{i=1}^{n} K_i = \frac{1}{2} \sum_{i=1}^{n} \sum_{j=1}^{i} tr \, U_{ij} I_j U_{ij}^T \dot{\theta}_j \dot{\theta}_j$$

For a space manipulator, the KE function also equates to the Langrangian and Hamiltonian functions as potential energy is zero. This enables the manipulator mass matrix to be computed with a complexity $O(n^2)$:

$$D_{ik} = \sum_{j=i}^{n} \sum_{k=1}^{j} tr \left(U_{jk} I_j U_{ji}^T \right)$$

This is the acceleration of the joint and is symmetric such that $D_{ik} = D_{ki}$. This allows computation of reaction moments N_r on the spacecraft from the generalised Jacobian formulation [Umetani & Yoshida 1989]:

$$N_r = -D_{ik}(J^*)^{-1} v^*$$

where J^* = generalised Jacobian and v^* = inertial cartesian velocity of end effector.

Although the mass matrix represents the simplest part of the Lagrange–Euler formulation to compute, it is still complex and a more efficient means of calculating the kinetic energy can be obtained from the Newton–Euler recursive method from which v_{ci}^0 is available in cartesian coordinates:

$$K_i = \frac{1}{2} m_i \left(v_i^0 v_i^{0T} \right) + \frac{1}{2} w_i^{0T} I_i w_i^0$$

These may be differentiated to yield:

$$\dot{K}_i = \frac{1}{2} \begin{pmatrix} m(v^T \dot{v} + v \dot{v}^T) \\ \dot{w} I w^T + w I \dot{w}^T \end{pmatrix} = \begin{pmatrix} w \dot{v} v \\ \dot{w} I w \end{pmatrix} = \begin{pmatrix} F^T v \\ N^T w \end{pmatrix}$$

Generally, the Lagrange–Euler method yields the general form of the dynamics [Walker & Orin 1982]:

$$\tau_i \sum_{k=1}^{n} D_{ik}(\theta) \ddot{\theta}_k + \sum_{k=1}^{n} \sum_{m=1}^{n} H_{ikm}(\theta, \dot{\theta}) + G_i(\theta) + B_i \dot{\theta}_i \tag{7.1}$$

$D(\theta)$ represents the symmetric positive definite $n \times n$ mass/inertia matrix whereby the diagonals represent joint inertias and off-diagonals the inertial coupling between joints, $(\theta, \dot{\theta})$ represents the $n \times 1$ vector of coriolis and centrifugal forces, $G(\theta)$ represents the $n \times 1$ gravity loading vector, and B represents the actuator viscous damping coefficient. H and G represent the nonlinear terms, and viscous damping is usually neglected to a first approximation. Terms involving $\dot{\theta}_i^2$ are the centrifugal components while terms involving $\dot{\theta}_i\dot{\theta}_j$ are the coriolis components. The Lagrangian equations are thus coupled and highly nonlinear. Hamman (1985) considered that the space environment exerts special conditions on manipulator dynamics. Based on ESA contract report, ESA CR(P) 2048, he outlined low gravity, low end effector speed (and so low centrifugal and coriolis forces) but strong inertial cross coupling to be peculiar to space robots. The advantage of microgravity entails that actuator torques are available for robot motion without the need to support the weight of the manipulator or payload [Lumia & Wavering 1989]. Motion trajectories comprise an acceleration phase, a constant velocity phase and a deceleration phase. Inertial cross-coupling occurs when the manipulator changes configuration. If the constant velocity phase is short compared with the acceleration/deceleration phases (e.g. for near time-optimal bang-bang manoeuvres) then the inertial cross coupling will be significant over most of the trajectory. The net moment of inertia opposes the accelera-tion at one joint due to the effective stiffness of the other joints and loads. Coupling is especially dominant for the positioning links over those of the wrist due to their higher moments of inertia. Coriolis and centrifugal terms are significant and dominate at high-velocity phases, becoming insignificant near the goal point when joint velocities are decreasing [Kosha & Kanade 1988]. These terms become small only when the joints are driven slowly and generally their neglect will generate significant errors in joint torque generation. However, the gravity term even for terrestrial manipulators is small. The average execution time for the Lagrangian formulation is ~8 sec for the Stanford manipu-lator on a PDP11/45—clearly this is not suitable for real-time control and is suitable only for modelling analysis. It is conceivable that future control systems may be implemented on parallel systolic computer hardware architectures, allowing real-time control, but this is not likely for space systems in the near future.

Khatib (1987) suggested casting all the Lagrange–Euler dynamics in end effector (operational space) coordinates:

Generalised cartesian force,

$$F = \lambda(q)\ddot{q} + \mu(q,\dot{q}) + p(q)$$

Since

$$\tau = J^T F = D(\theta)\ddot{\theta} + C(\theta,\dot{\theta}) + G(\theta)$$

$$F = \lambda(\theta)\ddot{\theta} + J^{-T}C(\theta,\dot{\theta}) + J^{-T}G(\theta)$$

where $\lambda(\theta) = J^{-T}(\theta)D(\theta)$.

According to this formulation:

$$\lambda(x) = J^{-T}(\theta)D(\theta)J^{-1}(\theta)$$

$$\mu(x,\dot{x}) = J^{-T}(\theta)C(\theta,\dot{\theta}) - \lambda(\theta)\dot{J}(\theta)\dot{\theta}$$

$$p(x) = J^{-T}(\theta)G(\theta)$$

$$F = J^{-T}(\theta)\tau$$

Although this formulation allows the specification of the control law in end effector and task coordinates, e.g. $\ddot{q} = \ddot{q}^d + K_p(q^d - q) + K_v(\dot{q}^d - \dot{q})$, its extreme complexity exceeds that of the Lagrange–Euler formulation in joint coordinates (K_p and K_v are position and velocity control gains respectively; superscript 'd' indicates desired variable value, and unsuperscripted variable indicates actual variable value as measured by sensors).

Kane & Levinson (1980) compared several dynamics formulations for multibody systems in terms of equation simplicity and difficulty of derivation. They found that the Newton–Euler methods are much more efficient computationally than Langrangian approaches. They introduced their own formulation of generating partial velocities which linearise the kinematics to give generalised velocities and enabled the derivation of generalised forces by serial approximation. They later adapted those equations specifically for robotics applications [Kane & Levinson 1983] which formed the basis of Konigstein et al's (1989) approach to the Newton–Euler type of computed torque control of a two-armed space robot using Kane's equations for rate and acceleration resolution with momentum constraints. This approach exploited the purely closed chain kinematic configuration without considering the contact interaction forces. Kane's equations involve serial approximations whereas the classical Newton–Euler method is exact. Nagashima & Nakaruma (1992) suggested that the Newton–Euler dynamics method of computation offers the most efficient technique for space-based robotics due to the recursive nature of those equations.

Luh et al (1980a) derived a Newton–Euler formulation for robot mechanisms where D'Alembert's principle of virtual work applied to constraint forces is applied to each link sequentially. This method was found to be extremely efficient with a linear computational complexity of $O(n)$ where n = number of links. They further reduced the computational requirement by referring all link parameters to their own coordinate systems rather than in base reference coordinates. However, for space-based manipulators, the total reaction moment on the mounting platform is required for control purposes which necessitates the Newton–Euler computation to be performed in base reference coordinates as well but this increases the computational overhead only marginally. Hollerbach (1980) derived a recursive formulation for the Lagrangian dynamics method to reduce the computational overheads associated with the formulation. Since Coriolis and centrifugal forces are significant at moderate speeds of motion, they cannot be ignored. He derived a series of forward recursive equations of velocities and accelerations from the base to the end effector and backward recursive equations of forces from the end effector to the base for real-time Lagrange–Euler computation but this destroyed the structure of the Lagrange–Euler approach. He also used 3×3 rotation matrices and vectors rather than the usual 4×4 homogeneous matrices to reduce unnecessary computation. This reduced the complexity of the Lagrangian method from $O(n^4)$ to $O(n)$. Silver (1982) later showed through tensor methods that the Newton–Euler and Lagrangian recursive formulations were equivalent since the nine element time derivative of the 3×3 rotational matrix of the Lagrangian equates to the three element angular velocity components of the Newton–Euler formulation. However, the Newton–Euler method remains 60% more efficient than the recursive Lagrangian method. Hence this is the algorithm of choice for deriving the robot dynamics.

The Newton–Euler technique explicitly states all constraint forces and moments and generates a set of recursive equations [Orin et al 1979, Luh et al 1980a]. It is stipulated here that all dynamics equations and variables are referred to base coordinates in addition to their own link coordinates for reasons that are elaborated in Chapter 8. Although this means that the moments of inertia of each link changes as the arm configuration changes, this does not disadvantage the scheme with excessive computation. Kinematic information is propagated from the base to the hand and dynamic information is propagated back from the hand to the base. Such a recursive structure lends itself to implementation using a pipelined microprocessor architecture reducing the number of wires inside the robot arm. Newton's laws of motion are applied through D'Alembert's principle to each link in turn: 'For any body, the algebraic sum of imaginary external forces and forces resisting motion in any direction for equilibrium is zero'.

The forward recursive equations derive the velocities and accelerations of each link at its joint

$$\begin{pmatrix} v_i \\ w_i \end{pmatrix} \quad \text{and} \quad \begin{pmatrix} \dot{v}_i \\ \dot{w}_i \end{pmatrix},$$

the velocities of each link centre of mass (subscript ci), and the forces at each link centre of mass

$$\begin{pmatrix} F_{ci} \\ N_{ci} \end{pmatrix}:$$

$$w_i = w_{i-1} + z_{i-1}\dot{\theta}_i$$

$$\dot{w}_i = \dot{w}_{i-1} + z_{i-1}\ddot{\theta}_i + w_{i-1} \times (z_{i-1}\dot{\theta}_i)$$

$$v_i = v_{i-1} + w_i \times l_i$$

$$\dot{v}_i = \dot{v}_{i-1} + \dot{w}_{i-1} \times l_i + w_i \times (w_i \times l_i)$$

$$v_{ci} = v_i + w_i \times (-s_i)$$

$$\dot{v}_{ci} = \dot{v}_i + w_i \times (-s_i) + w_i \times \left(w_i \times (-s_i)\right)$$

$$F_{ci} = m_i \dot{v}_{ci}$$

$$N_{ci} = I_i \dot{w}_i + w_i \times (I_i w_i) \tag{7.2}$$

The backward recursive equations are:

$$f_i = f_{i+1} + F_{ci} - m_i g$$

$$n_i = n_{i+1} + N_{ci} + l_i \times f_{i+1} + (l_i - s_i) \times F_{ci}$$

$$\tau_i = n_i^T z_{i-1} + b_i \dot{\theta}_i \tag{7.3}$$

where

$$z_{i-1} = \text{rotational axis of joint } i$$

$$l_i = \text{length of link } i$$

s_i = position of link i centre of mass from joint i

m_i = mass of link i

F_{ci} = total force on link i centre of mass

N_{ci} = total moment about link i centre of mass

I_i = inertia matrix of link i

f_i = force on linke i due to link $i-1$

n_i = moment on link i due to link $i-1$

w_i = angular velocity of joint i

$\dot{w}_i$ = angular acceleration of joint i

v_i = linear velocity of joint i

$\dot{v}_i$ = linear acceleration of joint i

v_{ci} = linear velocity of link i centre of mass

$\dot{v}_{ci}$ = linear acceleration of link i centre of mass

τ_i = actuator torque on joint i

b_i = viscous friction on joint i

This formulation generates the joint torques required to provide the stipulated end effector positions, velocities, and accelerations.

7.1 DERIVATION OF NEWTON–EULER DYNAMICS

Consider a fixed inertial frame of reference with origin O_{XYZ} coincident with the origin of a rotating frame O_{xyz} with a vector r to point P [Fu et al 1987] (Fig. 7.1):

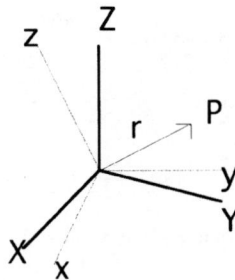

Fig. 7.1. Rotating reference frames.

$r = XI + YJ + ZK$ from the fixed frame and

$r = xi + yj + zk$ from the rotating frame

From the fixed frame:

$$\frac{dr}{dt} = \dot{X}I + \dot{Y}J + \dot{Z}K$$

this is the velocity of P relative to O_{XYZ}. From the rotating frame:

$$\frac{\partial r}{\partial t} = \dot{x}i + \dot{y}j + \dot{z}k$$

this is the velocity of P relative to O_{xyz}. Now,

$$\frac{di}{dt} = w_z j - w_y k = w \times i; \quad \frac{dj}{dt} = w_x k - w_z i = w \times j; \quad \frac{dk}{dt} = w_y i - w_x j = w \times k$$

Hence,

$$\frac{dr}{dt} = \frac{\delta r}{\delta t} + x(w \times i) + y(w \times j) + z(w \times k)$$

If r is fixed in the rotating frame,

$$\frac{\delta r}{\delta t} = 0 \rightarrow \frac{dr}{dt} = w \times r$$

Acceleration with respect to the fixed reference frame is found by differentiation:

$$\frac{d^2 r}{dt^2} = \frac{d}{dt}\left(\frac{\delta r}{\delta t} + w \times r\right) = \frac{d}{dt}\left(\frac{\delta r}{\delta t}\right) + \frac{d}{dt}(w \times r)$$

$$= \left(\frac{\delta^2 r}{\delta t^2} + w \times \frac{\delta r}{\delta t}\right) + w \times \frac{dr}{dt} + \frac{dw}{dt} \times r$$

$$= \frac{\delta^2 r}{\delta t^2} + w \times \frac{\delta r}{\delta t} + w \times \left(\frac{\delta r}{\delta t} + w \times r\right) + \frac{dw}{dt} \times r$$

$$= \frac{\delta^2 r}{\delta t^2} + 2w \times \frac{\delta r}{\delta t} + w \times (w \times r) + \frac{dw}{dt} \times r$$

Now,

$$\dot{w} = \frac{\delta w}{\delta t} + w \times w = \frac{\delta w}{\delta t}$$

$$\frac{d^2 r}{dt^2} = \frac{\delta^2 r}{\delta t^2} + 2w \times \frac{\delta r}{\delta t} + \dot{w} \times r + w \times (w \times r)$$

This is the coriolis theorem: the first term represents the acceleration of the point with respect to the rotating frame O_{xyz}; the second term represents the coriolis acceleration perpendicular to the trajectory path; the third term represents the rate of change of angular velocity; and the fourth term represents the centrifugal acceleration towards the axis of revolution.

Now, consider translating the rotating coordinate system such that the origin of O_{xyz} is displaced by a vector h with respect to O_{XYZ} (Fig. 7.2). Now,

$$r = s + h$$

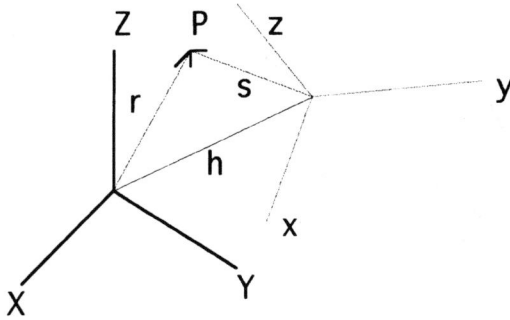

Fig. 7.2. Rotating and translating reference frames.

From a fixed frame of reference O_{XYZ}, the velocity of P is

$$v = \frac{dr}{dt} = \frac{ds}{dt} + \frac{dh}{dt} = v_s + v_h$$

where v_s = velocity of P relative to O_{xyz}, and v_h = velocity of O_{xyz} relative to O_{XYZ}. Now,

$$\frac{ds}{dt} = \frac{\delta s}{\delta t} + w \times s$$

so relative to the fixed coordinate system O_{XYZ}:

$$v = \left(\frac{\delta s}{\delta t} + w \times s \right) + \frac{dh}{dt} \quad \text{and}$$

$$\dot{v} = \frac{d^2 r}{dt^2} = \frac{d^2 s}{dt^2} + \frac{d^2 h}{dt^2} = \frac{\delta^2 s}{\delta t^2} + 2w \times \frac{\delta s}{\delta t} + \dot{w} \times s + w \times (w \times s) + \frac{d^2 h}{dt^2}$$

7.1.1 Link kinematics

These results are now applied to the link kinematics [Luh et al 1980a, MacInnes & Lin 1986, de Silva 1991]. Link kinematics define the required information for the Newton–Euler dynamics formulation with respect to base reference coordinates $(x_0 y_0 z_0)$. The centroid of each link i is assumed to be located along the line segment connecting joint $i - 1$ and i, i.e. each link is a body of revolution about its central axis of symmetry.

Some preliminary definitions follow:

$(x_i y_i z_i)$ defines the coordinate system at joint i where z_{i-1} is the axis of rotation of link i; $(x_0 y_0 z_0)$ defines the base reference coordinate system; $(x_6 y_6 z_6) = (nsa)$ defines the hand coordinate system.;

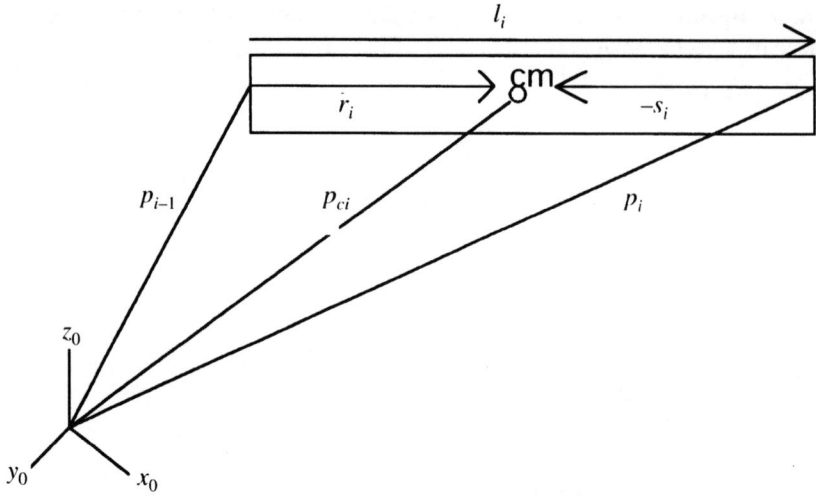

Fig. 7.3. Link kinematics.

v_i^0 = linear velocity of moving coordinate system $(x_i y_i z_i)$ with respect to base reference coordinates $(x_0 y_0 z_0)$;

w_i^0 = angular velocity of moving coordinate system $(x_i y_i z_i)$ with respect to base reference coordinates $(x_0 y_0 z_0)$;

v_i^{i-1} = linear velocity of coordinate system $(x_i y_i z_i)$ with respect to coordinate system $(x_{i-1} y_{i-1} z_{i-1})$;

w_i^{i-1} = angular velocity of coordinate system $(x_i y_i z_i)$ with respect to coordinate system $(x_{i-1} y_{i-1} z_{i-1})$.

By definition: $l_i = p_i - p_{i-1} \rightarrow p_i = l_i + p_{i-1}$ where $l_i =$ position from the origin of the i frame with respect to the $i-1$ frame. Now,

$$v_i^0 = v_{i-1}^0 + \frac{\delta l_i}{\delta t} + w_{i-1}^0 \times l_i$$

$$\dot{v}_i^0 = \dot{v}_{i-1}^0 + \frac{\delta^2 l_i}{\delta t^2} + 2w_{i-1}^0 \times \frac{\delta l_i}{\delta t} + \dot{w}_{i-1}^0 \times l_i + w_{i-1}^0 \times (w_{i-1}^0 \times l_i)$$

Furthermore,

$$w_i^0 = w_{i-1}^0 + w_i^{i-1}$$

$$\dot{w}_i^0 = \dot{w}_{i-1}^0 + \dot{w}_i^{i-1}$$

$$\rightarrow \dot{w}_i^0 = \dot{w}_{i-1}^0 + \frac{\delta w_i^{i-1}}{\delta t} + w_{i-1}^0 \times w_i^{i-1} \text{ since } \dot{w}_i^{i-1} = \frac{\delta w_i^0}{\delta t} + w_{i-1}^0 \times w_i^0$$

For a rotational joint, the coordinate system $(x_i y_i z_i)$ has angular velocity w_i^{i-1} about the z_{i-1} axis in the $(x_{i-1} y_{i-1} z_{i-1})$ coordinate system. Hence, relative to link $i - 1$ coordinates $(x_{i-1} y_{i-1} z_{i-1})$:

$$\left. \begin{aligned} w_i^{i-1} &= z_{i-1}\dot{\theta}_i \\[6pt] \frac{\delta w_i^{i-1}}{\delta t} &= z_{i-1}\ddot{\theta}_i \end{aligned} \right\} \text{ for revolute joints, and } \left. \begin{aligned} w_i^{i-1} &= 0 \\[6pt] \frac{\delta w_i^{i-1}}{\delta t} &= 0 \end{aligned} \right\} \text{ for prismatic joints}$$

Hence, referred to base coordinates $(x_0 y_0 z_0)$:

$$w_i^0 = w_{i-1}^0 + z_{i-1}\dot{\theta}_i \quad \text{for a revolute joint}$$

$$\quad\;\; = w_{i-1}^0 \qquad\qquad\quad \text{for a prismatic joint}$$

$$\dot{w}_i^0 = \dot{w}_{i-1}^0 + z_{i-1}\ddot{\theta}_i + w_{i-1}^0 \times (z_{i-1}\dot{\theta}_i) \quad \text{for a revolute joint}$$

$$\quad\;\; = \dot{w}_{i-1}^0 \qquad\qquad\qquad\qquad\qquad \text{for a prismatic joint}$$

Similarly, relative to link $i - 1$ coordinates $(x_{i-1} y_{i-1} z_{i-1})$:

$$\left. \begin{aligned} \frac{\delta l_i}{\delta t} &= w_i^{i-1} \times l_i \\[8pt] \frac{\delta^2 l_i}{\delta t^2} &= \frac{\delta w_i^{i-1}}{\delta t} \times l_i + w_i^{i-1} \times \frac{\delta l_i}{\delta t} \\[8pt] &= \frac{\delta w_i^{i-1}}{\delta t} \times l_i + w_i^{i-1} \times (w_i^{i-1} \times l_i) \end{aligned} \right\} \text{ for revolute joints}$$

$$\left. \begin{aligned} \frac{\delta l_i}{\delta t} &= z_{i-1}\dot{\theta}_i \\[8pt] \frac{\delta^2 l_i}{\delta t^2} &= z_{i-1}\ddot{\theta}_i \end{aligned} \right\} \text{ for prismatic joints}$$

Hence referred to base coordinates $(x_0 y_0 z_0)$:

$$v_i^0 = v_{i-1}^0 + w_{i-1}^0 \times l_i + w_i^{i-1} \times l_i = v_{i-1}^0 + w_i^0 \times l_i \quad \text{for a revolute joint}$$

$$= v_{i-1}^0 + z_{i-1}\dot{\theta}_i + w_i^0 \times l_i \quad \text{for a prismatic joint}$$

For the centre of mass of a link: $v_{ci} = v_i^0 + w_i^0 \times (-s_i)$ since $p_{ci} = p_i - s_i$. Now,

$$\dot{v}_i^0 = \dot{v}_{i-1}^0 + \frac{\delta^2 l_i}{\delta t^2} + \dot{w}_{i-1}^0 \times l_i + 2w_{i-1}^0 \times \frac{\delta l_i}{\delta t} + w_{i-1}^0 \times (w_{i-1}^0 \times l_i)$$

Using the following relations:

$$(a \times b) \times c = b(a.c) - a(b.c)$$

$$a \times (b \times c) = b(a.c) - c(a.b)$$

and

$$\frac{\delta^2 l_i}{\delta t^2} = \frac{\delta w_i^{i-1}}{\delta t} \times l_i + w_i^{i-1} \times (w_i^{i-1} \times l_i);$$

$$\frac{\delta w_i^{i-1}}{\delta t} = z_{i-1} \ddot{\theta}_i \times l_i; \quad \frac{\delta l_i}{\delta t} = w_i^{i-1} \times l_i$$

we have:

$$\frac{\delta^2 l_i}{\delta t^2} + w_i^0 \times l_i = w_{i-1}^0 \times (w_{i-1}^0 \times l_i) + w_i^0 \times l_i - (w_{i-1}^0 \times w_i^{i-1}) \times l_i = \dot{w}_i^0 \times l_i$$

$$2w_{i-1}^0 \times \frac{\delta l_i}{\delta t^2} = 2\left[w_i^{i-1} (w_{i-1}^0.l_i) - l_i(w_{i-1}^0.w_i^{i-1}) \right] = 0$$

$$w_{i-1}^0 \times (w_{i-1}^0 \times l_i) = w_i^0 \times (w_i^0 \times l_i) - w_i^{i-1}(w_i^{i-1}.l_i) - l_i(w_i^{i-1}.w_i^{i-1})$$

$$= w_i^0 \times (w_i^0 \times l_i)$$

Hence,

$$\dot{v}_i^0 = \dot{v}_{i-1}^0 + \dot{w}_i^0 \times l_i + w_i^0 \times (w_i^0 \times l_i) \quad \text{for a revolute joint}$$

$$\dot{v}_i^0 = \dot{v}_{i-1}^0 + z_{i-1} \ddot{\theta}_i + \dot{w}_i^0 \times l_i + 2w_i^0 \times (z_{i-1} \dot{\theta}_i)$$

$$+ w_i^0 \times (w_i^0 \times l_i) \quad \text{for a prismatic joint}$$

$$\dot{v}_{ci}^0 = \dot{v}_i^0 + \dot{w}_i^0 \times (-s_i) + w_i^0 \times (w_i^0 \times -s_i)$$

7.1.2 Link dynamics

Next, the forces and torques need to be calculated (Fig. 7.4)

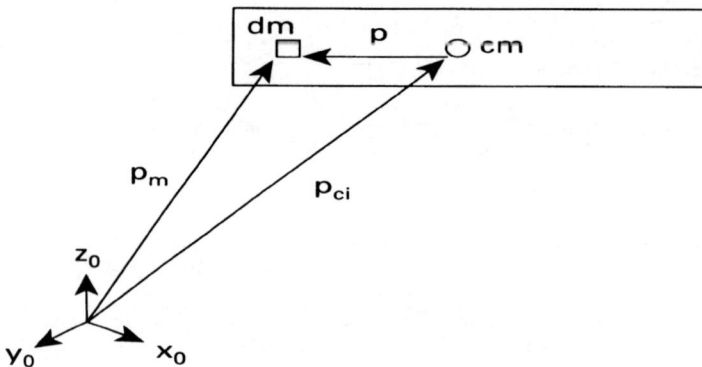

Fig. 7.4. Robotic link mass distribution.

The total angular momentum about the base reference point is the sum of angular moments of all rigid bodies composing the manipulator (assuming zero or negligible gravity):

$$H_0 = \int_L p_m \times \frac{dp_m}{dt}\,dm = \sum_{i=1}^n \int \left(p_m \times \frac{dp_m}{dt} \right) dm \text{ at any mass element } dm.$$

Now,

$$p_m = p_{ci} + p: \ H_0 = \sum_{i=1}^n \int \left(p_{ci} \times \frac{dp_{ci}}{dt} \right) dm$$

$$+ \sum_{i=1}^n \int \left(p \times \frac{dp}{dt} \right) dm \text{ where } v_{ci} = \frac{dp_{ci}}{dt}$$

Now, $^0dp/dt$ in reference coordinates is related to $^idp/dt$ in link coordinates by:

$$\frac{^0dp}{dt} = \frac{^idp}{dt} + w_i \times p$$

where w_i = angular velocity of link i centre of mass from base reference coordinates, i.e.

$$H_0 = \sum_{i=1}^n \left[m_i p_{ci} \times \frac{dp_{ci}}{dt} + \int p \times \left(\frac{^idp}{dt} + w_i \times p \right) dm \right]$$

The position of the centre of mass of each link is fixed with respect to link coordinates so

$$\int p\,dm = 0$$

so:

$$H_0 = \sum_{i=1}^n m_i p_{ci} \times \frac{dp_{ci}}{dt} + \int p \times (w_i \times p)dm = \sum_{i=1}^n m_i p_{ci} \times \frac{dp_{ci}}{dt} + I_{ci} w_i$$

where

$$I_{ci} = \int_V rr\,dm = mk^2$$

is the moment of inertia of body i about its own centre of mass and k = radius of gyration, e.g.

moment of inertia of a rod about an axis through its mass centre, $I_{rod} = \frac{2}{3}ml^2$

moment of inertia of a disc about its central axis, $I_{disc} = \frac{1}{2}mr^2$

moment of inertia of a sphere about a diameter, $I_{sphere} = \frac{2}{5}mr^2$

moment of inertia of a uniform block about one central face-parallel axis,

$$I_{block} = \frac{1}{3}m\left[\left(\frac{l_1}{2} \right)^2 + \left(\frac{l_2}{2} \right)^2 \right]$$

Assuming that the links are bodies of revolution and are symmetric about their central axes and have a uniform mass distribution:

$$I_{ci} = diag(J) = \begin{pmatrix} I_{ix} & 0 & 0 \\ 0 & I_{ij} & 0 \\ 0 & 0 & I_{iz} \end{pmatrix} \text{ where } \begin{cases} I_{ix} = I_{iy} = 0.5 m_i r^2 \\ \\ I_{iz} = 0.67 m_i l^2 \end{cases}$$

Now for revolute joints, p_{ci} is constant:

$$H_0 = \sum_{i=1}^{n} I_{ci} w_i$$

By definition, moments are the rate of change of angular momentum and

$$N_0 = \sum_{i=1}^{n} N_{ci}$$

$$N_0 = \dot{H}_0 = \frac{d}{dt} \int_L p_m \times \frac{{}^0 dp_m}{dt} dm = \int_L p_m \times dF \text{ where } dF = \frac{{}^0 d^2 p_m}{dt^2} dm$$

Hence,

$$N_0 = \sum_{i=1}^{n} \left[m_i p_{ci} \times \frac{{}^0 d^2 p_{ci}}{dt^2} + \frac{{}^0 d(I_{ci} w_i)}{dt} + w_i \times (I_{ci} w_i) \right]$$

Now,

$$\frac{{}^0 d^2 p_{ci}}{dt^2} = 0$$

since p_{ci} is constant with respect to its own coordinates:

$$N_0 = \sum_{i=1}^{n} \left[I_{ci} \dot{w}_i + w_i \times (I_{ci} w_i) \right]$$

The first term is the rate of change of angular momentum relative to base reference coordinates and the second term is the rate of change of angular momentum relative to a body-corotational reference frame. Similarly,

$$F_0 = \sum_{i=1}^{n} F_{ci} = \sum_{i=1}^{n} m_i \frac{d^2 p_{ci}}{dt^2} \quad \text{where} \quad \dot{v}_{ci} = \frac{d^2 p_{ci}}{dt^2}$$

At each link:

$$F_{ci} = m_i \dot{v}_{ci} \quad \text{and} \quad N_{ci} = I_{ci} \dot{w}_i + w_i \times (I_{ci} w_i)$$

For the backward recursion of forces and torques assuming that gravity is negligible, we have at the centre of mass of each link:

$$F_{ci} = f_i - f_{i+1}$$

where f_i = force due to link $i - 1$, f_{i+1} = force due to link $i + 1$

$$\rightarrow f_i = F_{ci} + f_{i+1} = m_i \dot{v}_{ci} + f_{i+1}$$

Similarly,

$$N_{ci} = n_i - n_{i+1} - r_i \times f_i - s_i \times f_{i+1}$$

Note that f_{n+1} and n_{n+1} represent external forces and torques on the manipulator hand. Now,

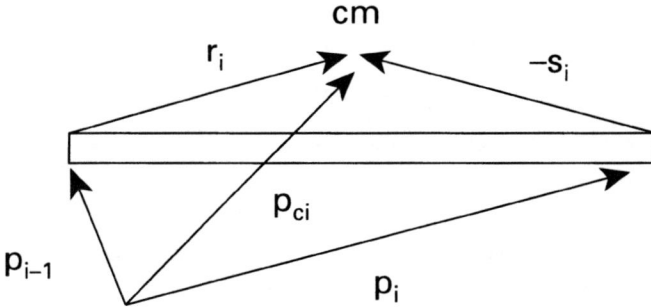

Fig. 7.5. Robotic link vectors.

$$r_i = p_{ci} - p_{i-1}$$
$$-s_i = p_{ci} - p_i$$
$$l_i = r_i - s_i = p_i - p_{i-1}$$

Hence,

$$N_{ci} = n_i - n_{i+1} - p_{ci} \times f_i + p_{i-1} \times l_i + p_{ci} \times f_{i+1} - p_i \times f_{i+1}$$

Now,

$$n_i - n_{i+1} + (p_{i-1} - p_{ci}) \times F_{ci} - l_i \times f_{i+1}$$
$$= n_i - n_{i+1} - p_{ci} \times f_i + p_{ci} \times f_{i+1} - p_i \times f_{i+1}$$

Hence,

$$N_{ci} = n_i - n_{i+1} + (p_{i-1} - p_{ci}) \times F_{ci} - l_i \times f_{i+1}$$

or

$$N_{ci} = n_i - n_{i+1} - (l_i - s_i) \times F_{ci} - l_i \times f_{i+1}$$
$$\rightarrow n_i = N_{ci} + n_{i+1} + (l_i - s_i) \times F_{ci} + l_i \times f_{i+1}$$

The torque at joint i produces a rotation of θ_i of coordinates $(x_i y_i z_i)$ about the z_{i-1} axis (for revolute joints), so $\tau_i = z_{i-1} n_i$ or $\tau_i = z_{i-1} n_i + b_i \dot{\theta}_i$ if viscous damping is included. Other slightly different formulations exist, but they offer no computational advantage, e.g. de Silva (1991). Hence, from the joint kinematic variables, the joint torques may be computed.

7.1.3 Link-referenced formulation

The drawback of these recursive equations is that all link inertia matrices are referenced to base coordinates, i.e. they will change as the robot arm configuration changes. By referring all variables to their own link coordinates, the moments of inertia of each link are referred with respect to their own principal axis coordinates, and so will be invariant: $I_{ci} = R_0^i (R_i^0 I_{ci} R_0^i) R_i^0$ since the inverse of the 3×3 R-submatrix of the 4×4 DH 'A' matrix is its transpose, i.e.

$$
\begin{pmatrix} N_{cix} \\ N_{ciy} \\ N_{ciz} \end{pmatrix} = \begin{pmatrix} I_{cix} & 0 & 0 \\ 0 & I_{ciy} & 0 \\ 0 & 0 & I_{ciz} \end{pmatrix} \begin{pmatrix} \dot{w}_{ix} \\ \dot{w}_{iy} \\ \dot{w}_{iz} \end{pmatrix} + \begin{pmatrix} w_{ix} \\ w_{iy} \\ w_{iz} \end{pmatrix} \times \begin{bmatrix} \begin{pmatrix} I_{cix} & 0 & 0 \\ 0 & I_{ciy} & 0 \\ 0 & 0 & I_{ciz} \end{pmatrix} \begin{pmatrix} w_{ix} \\ w_{iy} \\ w_{iz} \end{pmatrix} \end{bmatrix}
$$

$$
= \begin{pmatrix} I_{cix}\dot{w}_{ix} + (I_{ciz} - I_{ciy})w_{iy}w_{iz} \\ I_{ciy}\dot{w}_{iy} + (I_{ix} - _{iz})w_{ix}w_{iy} \\ I_{ciz}\dot{w}_{iz} + (I_{iy} - I_{iz})w_{ix}w_{iy} \end{pmatrix}
$$

To transform all dynamics variables back into their own link coordinates the inverse of the rotation matrix is required:

$$
R_{i-1}^i = \left(R_i^{i-1}\right)^T = \begin{pmatrix} c\theta_i & s\theta_i & 0 \\ -c\alpha s\theta_i & c\alpha c\theta_i & s\alpha \\ s\alpha s\theta_i & -s\alpha c\theta_i & c\alpha \end{pmatrix}
$$

which transforms $(x_i y_i z_i)$ to $(x_{i-1} y_{i-1} z_{i-1})$ so that the Newton–Euler equations now become:

$$
R_0^i w_i = R_0^i(w_{i-1} + z_{i-1}\dot{\theta}_i) = R_{i-1}^i \left(R_0^{i-1} w_i + R_0^{i-1} z_{i-1}\dot{\theta}_i \right)
$$

$$
= R_{i-1}^i \left(R_0^{i-1} w_i + z_0 \dot{\theta}_i \right)
$$

$$
R_0^i \dot{w}_i = R_{i-1}^i \left(R_0^{i-1} \dot{w}_{i-1} + R_0^{i-1} z_{i-1}\ddot{\theta}_i + R_0^{i-1} w_i \times R_0^{i-1} z_{i-1}\dot{\theta}_i \right)
$$

$$
= R_{i-1}^i \left(R_0^{i-1} \dot{w}_{i-1} + z_0 \ddot{\theta}_i + R_0^{i-1} w_{i-1} \times z_0 \dot{\theta}_i \right)
$$

$$
R_0^i \dot{v}_i = R_{i-1}^i R_0^{i-1} \dot{v}_{i-1} + \left(R_0^i \dot{w}_i \times R_0^i l_i \right) + R_0^i w_i \times \left[\left(R_0^i w_i \right) \times \left(R_0^i l_i \right) \right]
$$

$$
R_0^i \dot{v}_{ci} = R_0^i \dot{v}_i + R_0^i \dot{w}_i \times R_0^i s_i + R_0^i w_i \times \left[\left(R_0^i w_i \right) \times \left(R_0^i s_i \right) \right]
$$

$$
R_0^i F_{ci} = m_i R_0^i \dot{v}_{ci}
$$

$$
R_0^i N_{ci} = I_{ci} R_0^i \dot{w}_i + R_0^i w_i \times \left(I_{ci} R_0^i w_i \right)
$$

$$
R_0^i f_i = R_{i+1}^i R_0^{i+1} f_{i+1} + R_0^i F_{ci}
$$

$$R_0^i n_i = R_{i+1}^i \left[R_0^{i+1} n_{i+1} + R_0^i l_i \times R_0^{i+1} f_{i+1} \right] + \left[R_0^i l_i + R_0^i s_i \right] \times R_0^i F_{ci} + R_0^i N_{ci}$$

$$\tau_i = \left(R_{i-1}^i z_0 \right)^T \left(R_0^i n_i \right) \quad \text{where} \quad z_0 = (001)^T$$

These equations are necessary to derive link-referenced joint motor torques for the robot controller but the formulation for space manipulators followed in Chapter 8 requires the dynamics equations to be referred to base coordinates also.

7.2 NEWTON–EULER DYNAMICS OF A PUMA 560/600 SPACE MANIPULATOR

For the sake of simplicity, the recursive scheme is applied assuming that the link geometries are formed as bodies of revolution about their central axes of symmetry. The forward recursion is performed from link 1 to link n, whilst the backward recursion is performed from link n to link 1 via iterative loops. The following section applies the base-referenced Newton Euler method to the PUMA 560/600 geometry particularly concerning supplying initial conditions for the recursion.

7.2.1 Forward recursion

We illustrate the base-referenced formulation only as we will need to compute base-referenced forces and torques for the freeflyer dynamic formulation introduced later. Link-referenced forward recursion is required to compute the joint torques through backward recursion, but is trivially different to the base-referenced forward recursion, so it is not outlined here. The forward recursion is performed by a recursive loop algorithm from $i = 1$ to n. To find $w_i = w_{i-1} + z_{i-1} \dot{\theta}_i$ and $\dot{w}_i = \dot{w}_{i-1} + z_{i-1} \ddot{\theta}_i + w_{i-1} \times z_{i-1} \dot{\theta}_i$ with respect to base coordinates. From the DH matrix formulation:

$$\begin{pmatrix} z_{0x} \\ z_{0y} \\ z_{0z} \end{pmatrix} = \begin{pmatrix} 0 \\ 0 \\ 1 \end{pmatrix}; \quad \begin{pmatrix} z_{1x} \\ z_{1y} \\ z_{1z} \end{pmatrix} = \begin{pmatrix} -s_1 \\ c_1 \\ 0 \end{pmatrix}; \quad \begin{pmatrix} z_{2x} \\ z_{2y} \\ z_{2z} \end{pmatrix} = \begin{pmatrix} -s_1 \\ c_1 \\ 0 \end{pmatrix}; \quad \begin{pmatrix} z_{3x} \\ z_{3y} \\ z_{3z} \end{pmatrix} = \begin{pmatrix} c_1 s_{23} \\ s_1 s_{23} \\ c_{23} \end{pmatrix};$$

$$\begin{pmatrix} z_{4x} \\ z_{4y} \\ z_{4z} \end{pmatrix} = \begin{pmatrix} -(c_1 c_{23} s_4 + s_1 c_4) \\ c_1 c_4 - s_1 c_{23} s_4 \\ s_{23} s_4 \end{pmatrix}; \quad \begin{pmatrix} z_{5x} \\ z_{5y} \\ z_{5z} \end{pmatrix} = \begin{pmatrix} a_x \\ a_y \\ a_z \end{pmatrix}$$

The initial conditions for a stationary base are:

$$\begin{pmatrix} w_{0x} \\ w_{0y} \\ w_{0z} \end{pmatrix} = \begin{pmatrix} \dot{w}_{0x} \\ \dot{w}_{0x} \\ \dot{w}_{0z} \end{pmatrix} = \begin{pmatrix} 0 \\ 0 \\ 0 \end{pmatrix}$$

Whence:

$$
\begin{pmatrix} w_{ix} \\ w_{iy} \\ w_{iz} \end{pmatrix} = \begin{pmatrix} w_{i-1x} \\ w_{i-1y} \\ w_{i-1z} \end{pmatrix} + \begin{pmatrix} z_{i-1x} \\ z_{i-1y} \\ z_{i-1z} \end{pmatrix} \dot{\theta}_i
$$

$$
\begin{pmatrix} \dot{w}_{ix} \\ \dot{w}_{iy} \\ \dot{w}_{iz} \end{pmatrix} = \begin{pmatrix} \dot{w}_{i-1x} \\ \dot{w}_{i-1y} \\ \dot{w}_{i-1z} \end{pmatrix} + \begin{pmatrix} z_{i-1x} \\ z_{i-1y} \\ z_{i-1z} \end{pmatrix} \ddot{\theta}_i + \begin{pmatrix} w_{i-1y}z_{i-1z} - w_{i-1z}z_{i-1y} \\ w_{i-1z}z_{i-1x} - w_{i-1x}z_{i-1z} \\ w_{i-1z}z_{i-1y} - w_{i-1y}z_{i-1x} \end{pmatrix} \dot{\theta}_i
$$

To find $\dot{v}_i = \dot{v}_{i-1} + \dot{w}_i \times l_i + w_i \times (w_i \times l_i)$ and $\dot{v}_{ci} = \dot{v}_i + \dot{w}_i \times s_i + w_i \times (w_i \times s_i)$ with respect to base coordinates. From the DH matrix formulation, l_i is in base coordinates:

$$
\begin{pmatrix} l_{1x} \\ l_{1y} \\ l_{1z} \end{pmatrix} = \begin{pmatrix} p_{1x} \\ p_{1y} \\ p_{1z} \end{pmatrix} = \begin{pmatrix} 0 \\ 0 \\ 0 \end{pmatrix} \quad \text{and} \quad \begin{pmatrix} s_{1x} \\ s_{1y} \\ s_{1z} \end{pmatrix} = \begin{pmatrix} 0 \\ 0 \\ 0 \end{pmatrix}
$$

$$
\begin{pmatrix} l_{2x} \\ l_{2y} \\ l_{2z} \end{pmatrix} = \begin{pmatrix} p_{2x} - p_{1x} \\ p_{2y} - p_{1y} \\ p_{2z} - p_{1z} \end{pmatrix} = \begin{pmatrix} a_2c_1c_2 - s_1d_2 \\ a_2s_1c_2 + c_1d_2 \\ -a_2s_2 \end{pmatrix} \quad \text{and} \quad \begin{pmatrix} s_{2x} \\ s_{2y} \\ s_{2z} \end{pmatrix} = -\frac{1}{2} \begin{pmatrix} a_2c_1c_2 - s_1d_2 \\ a_2s_1c_2 + c_1d_2 \\ -a_2s_2 \end{pmatrix}
$$

$$
\begin{pmatrix} l_{3x} \\ l_{3y} \\ l_{3z} \end{pmatrix} = \begin{pmatrix} p_{3x} - p_{2x} \\ p_{3y} - p_{2y} \\ p_{3z} - p_{2z} \end{pmatrix} = \begin{pmatrix} 0 \\ 0 \\ 0 \end{pmatrix} \quad \text{and} \quad \begin{pmatrix} s_{3x} \\ s_{3y} \\ s_{3z} \end{pmatrix} = \begin{pmatrix} 0 \\ 0 \\ 0 \end{pmatrix} \quad \text{since} \quad p_2 = p_3
$$

$$
\begin{pmatrix} l_{4x} \\ l_{4y} \\ l_{4z} \end{pmatrix} = \begin{pmatrix} p_{4x} - p_{3x} \\ p_{4y} - p_{3y} \\ p_{4z} - p_{3z} \end{pmatrix} = \begin{pmatrix} a_2c_1c_2 - s_1d_2 + d_4c_1s_{23} \\ a_2s_1c_2 + c_1d_2 + d_4s_1s_{23} \\ -a_2s_2 + d_4c_{23} \end{pmatrix} - \begin{pmatrix} p_{3x} \\ p_{3y} \\ p_{3z} \end{pmatrix} = \begin{pmatrix} d_4c_1s_{23} \\ d_4s_1s_{23} \\ d_4c_{23} \end{pmatrix}
$$

$$
\text{and} \quad \begin{pmatrix} s_{4x} \\ s_{4y} \\ s_{4z} \end{pmatrix} = -\frac{1}{2} \begin{pmatrix} d_4c_1s_{23} \\ d_4s_1s_{23} \\ d_4c_{23} \end{pmatrix}
$$

$$
\begin{pmatrix} l_{5x} \\ l_{5y} \\ l_{5z} \end{pmatrix} = \begin{pmatrix} p_{5x} - p_{4x} \\ p_{5y} - p_{4y} \\ p_{5z} - p_{4z} \end{pmatrix} = \begin{pmatrix} s_{5x} \\ s_{5y} \\ s_{5z} \end{pmatrix} = \begin{pmatrix} 0 \\ 0 \\ 0 \end{pmatrix} \quad \text{since} \quad p_5 = p_4
$$

$$
\begin{pmatrix} l_{6x} \\ l_{6y} \\ l_{6z} \end{pmatrix} = \begin{pmatrix} p_{6x} - p_{5x} \\ p_{6y} - p_{5y} \\ p_{6z} - p_{5z} \end{pmatrix} = \begin{pmatrix} d_6a_x \\ d_6a_y \\ d_6a_z \end{pmatrix} \quad \text{and} \quad \begin{pmatrix} s_{6x} \\ s_{6y} \\ s_{6z} \end{pmatrix} = -\frac{1}{2} \begin{pmatrix} d_6a_x \\ d_6a_y \\ d_6a_z \end{pmatrix}
$$

The initial conditions for a stationary base are:

$$
\begin{pmatrix} \dot{v}_{0x} \\ \dot{v}_{0y} \\ \dot{v}_{0z} \end{pmatrix} = \begin{pmatrix} \dot{v}_{0x} \\ \dot{v}_{0y} \\ \dot{v}_{0z} \end{pmatrix} = \begin{pmatrix} \dot{v}_{c0x} \\ \dot{v}_{c0y} \\ \dot{v}_{c0z} \end{pmatrix} = \begin{pmatrix} 0 \\ 0 \\ 0 \end{pmatrix}
$$

Hence,

$$
\begin{pmatrix} \dot{v}_{ix} \\ \dot{v}_{iy} \\ \dot{v}_{iz} \end{pmatrix} = \begin{pmatrix} \dot{v}_{i-1x} \\ \dot{v}_{i-1y} \\ \dot{v}_{i-1z} \end{pmatrix} + \begin{pmatrix} \dot{w}_{iy}.l_{iz} - \dot{w}_{iz}.l_{iy} \\ \dot{w}_{iz}.l_{ix} - \dot{w}_{iz}.l_{iz} \\ \dot{w}_{ix}.l_{iy} - \dot{w}_{iy}.l_{ix} \end{pmatrix} + \begin{pmatrix} w_{iy}(w_{ix}.l_{iy} - w_{iy}.l_{ix}) - w_{iz}(w_{iz}.l_{ix} - w_{ix}.l_{iz}) \\ w_{iz}(w_{iy}.l_{iz} - w_{iz}.l_{iy}) - w_{ix}(w_{ix}.l_{iy} - w_{iy}.l_{ix}) \\ w_{ix}(w_{iz}.l_{ix} - w_{ix}.l_{iz}) - w_{iy}(w_{iy}.l_{iz} - w_{iz}.l_{iy}) \end{pmatrix}
$$

$$
\begin{pmatrix} \dot{v}_{cix} \\ \dot{v}_{ciy} \\ \dot{v}_{ciz} \end{pmatrix} = \begin{pmatrix} \dot{v}_{ix} \\ \dot{v}_{iy} \\ \dot{v}_{iz} \end{pmatrix} + \begin{pmatrix} \dot{w}_{iy}s_{iz} - \dot{w}_{iz}s_{iy} \\ \dot{w}_{iz}s_{ix} - \dot{w}_{iz}s_{iz} \\ \dot{w}_{ix}s_{iy} - \dot{w}_{iy}s_{ix} \end{pmatrix} + \begin{pmatrix} w_{iy}(w_{ix}s_{iy} - w_{iy}s_{ix}) - w_{iz}(w_{iz}s_{ix} - w_{ix}s_{iz}) \\ w_{iz}(w_{iy}s_{iz} - w_{iz}s_{iy}) - w_{ix}(w_{ix}s_{iy} - w_{iy}s_{ix}) \\ w_{ix}(w_{iz}s_{ix} - w_{ix}s_{iz}) - w_{iy}(w_{iy}s_{iz} - w_{iz}s_{iy}) \end{pmatrix}
$$

The inertia matrix must be referenced to base coordinates in the formulation used for space manipulators such that $I_i^0 = R_i^0 I_{ci} R_0^i$ where $R_0^i = (R_i^0)^T$ and $R_0^i = (x_i y_i z_i)$. From the DH formulation:

$$
\begin{pmatrix} x_{1x} \\ x_{1y} \\ x_{1z} \end{pmatrix} = \begin{pmatrix} c_1 \\ s_1 \\ 0 \end{pmatrix}; \quad \begin{pmatrix} y_{1x} \\ y_{1y} \\ y_{1z} \end{pmatrix} = \begin{pmatrix} 0 \\ 0 \\ -1 \end{pmatrix}; \quad \begin{pmatrix} z_{1x} \\ z_{1y} \\ z_{1z} \end{pmatrix} = \begin{pmatrix} -s_1 \\ c_1 \\ 0 \end{pmatrix}
$$

$$
\begin{pmatrix} x_{2x} \\ x_{2y} \\ x_{2z} \end{pmatrix} = \begin{pmatrix} c_1 c_2 \\ s_1 c_2 \\ -s_2 \end{pmatrix}; \quad \begin{pmatrix} y_{2x} \\ y_{2y} \\ y_{2z} \end{pmatrix} = \begin{pmatrix} -c_1 s_2 \\ -s_1 s_2 \\ -c_2 \end{pmatrix}; \quad \begin{pmatrix} z_{2x} \\ z_{2y} \\ z_{2z} \end{pmatrix} = \begin{pmatrix} -s_1 \\ c_1 \\ 0 \end{pmatrix}
$$

$$
\begin{pmatrix} x_{3x} \\ x_{3y} \\ x_{3z} \end{pmatrix} = \begin{pmatrix} c_1 c_{23} \\ s_1 s_{23} \\ -s_{23} \end{pmatrix}; \quad \begin{pmatrix} y_{3x} \\ y_{3y} \\ y_{3z} \end{pmatrix} = \begin{pmatrix} -s_2 \\ c_1 \\ 0 \end{pmatrix}; \quad \begin{pmatrix} z_{3x} \\ z_{3y} \\ z_{3z} \end{pmatrix} = \begin{pmatrix} c_1 s_{23} \\ s_1 s_{23} \\ c_{23} \end{pmatrix}
$$

$$
\begin{pmatrix} x_{4x} \\ x_{4y} \\ x_{4z} \end{pmatrix} = \begin{pmatrix} c_1 c_{23} c_4 - s_1 s_4 \\ s_1 c_{23} c_4 + c_1 s_4 \\ -s_{23} c_4 \end{pmatrix}; \quad \begin{pmatrix} y_{4x} \\ y_{4y} \\ y_{4z} \end{pmatrix} = \begin{pmatrix} -c_1 s_{23} \\ -s_1 s_{23} \\ -c_{23} \end{pmatrix}; \quad \begin{pmatrix} z_{4x} \\ z_{4y} \\ z_{4z} \end{pmatrix} = \begin{pmatrix} -c_1 c_{23} s_4 - s_1 c_4 \\ -s_1 c_{23} + c_1 c_4 \\ s_{23} s_4 \end{pmatrix}
$$

$$
\begin{pmatrix} x_{5x} \\ x_{5y} \\ x_{5z} \end{pmatrix} = \begin{pmatrix} c_5(c_1 c_{23} c_4 - s_1 s_4) - c_1 s_{23} s_5 \\ c_5(s_1 c_{23} c_4 + c_1 s_4) - s_1 s_{23} s_5 \\ -s_{23} c_4 c_5 - c_{23} s_5 \end{pmatrix}; \quad \begin{pmatrix} y_{5x} \\ y_{5y} \\ y_{5z} \end{pmatrix} = \begin{pmatrix} -c_1 c_{23} s_4 - s_1 c_4 \\ -s_1 c_{23} s_4 + c_1 c_4 \\ s_{23} s_4 \end{pmatrix}
$$

$$
\begin{pmatrix} z_{5x} \\ z_{5y} \\ z_{5z} \end{pmatrix} = \begin{pmatrix} s_5(c_1c_{23}c_4 - s_1s_4) + c_1s_5s_{23} \\ s_5(s_1c_{23}c_4 + c_1s_4) + s_1c_5s_{23} \\ -s_{23}c_4s_5 + c_{23}c_5 \end{pmatrix}
$$

$$
\begin{pmatrix} x_{6x} \\ x_{6y} \\ x_{6z} \end{pmatrix} = \begin{pmatrix} n_x \\ n_y \\ n_z \end{pmatrix}; \quad \begin{pmatrix} y_{6x} \\ y_{6y} \\ y_{6z} \end{pmatrix} = \begin{pmatrix} s_x \\ s_y \\ s_z \end{pmatrix}; \quad \begin{pmatrix} z_{6x} \\ z_{6y} \\ z_{6z} \end{pmatrix} = \begin{pmatrix} a_x \\ a_y \\ a_z \end{pmatrix}
$$

$$
\begin{pmatrix} I_{c3x} \\ I_{c3y} \\ I_{c3z} \end{pmatrix} = \begin{pmatrix} I_{c5x} \\ I_{c5y} \\ I_{cyz} \end{pmatrix} = \begin{pmatrix} 0 \\ 0 \\ 0 \end{pmatrix} \text{ since links 3 and 5 are fictitious links}
$$

Hence,

$$
\begin{pmatrix} I^0_{ixx} & I^0_{ixy} & I^0_{ixz} \\ I^0_{iyx} & I^0_{iyy} & I^0_{iyz} \\ I^0_{izx} & I^0_{izy} & I^0_{izz} \end{pmatrix} = \begin{pmatrix} x_{ix} & y_{ix} & z_{ix} \\ x_{iy} & y_{iy} & z_{iy} \\ x_{iz} & y_{iz} & z_{iz} \end{pmatrix} \begin{pmatrix} I_{cix} & 0 & 0 \\ 0 & I_{ciy} & 0 \\ 0 & 0 & I_{ciz} \end{pmatrix} \begin{pmatrix} x_{ix} & x_{iy} & x_{iz} \\ y_{ix} & y_{iy} & y_{iz} \\ z_{ix} & z_{iy} & z_{iz} \end{pmatrix}
$$

such that

$$
I^0_{ixx} = I_{cix}x^2_{ix} + I_{ciy}y^2_{ix} + I_{ciz}z^2_{ix};
$$

$$
I^0_{iyx} = I^0_{ixy} = I_{cix}x_{ix}x_{iy} + I_{ciy}y_{ix}y_{iy} + I_{ciz}z_{ix}z_{iy};
$$

$$
I^0_{izx} = I^0_{ixz} = I_{cix}x_{ix}x_{iz} + I_{ciy}y_{ix}y_{iz} + I_{ciz}z_{ix}z_{iz};
$$

$$
I^0_{iyy} = I_{cix}x^2_{iy} + I_{ciy}y^2_{iy} + I_{ciz}z^2_{iy};
$$

$$
I^0_{izy} = I^0_{iyz} = I_{cix}x_{iy}x_{iz} + I_{ciy}y_{iy}y_{iz} + I_{ciz}z_{iy}z_{iz};
$$

$$
I^0_{izz} = I_{cix}x^2_{iz} + I_{ciy}y^2_{iz} + I_{ciz}z^2_{iz}.
$$

From this, we can calculate the forward forces and moments about each link centre of mass in turn:

$$
N_{ci} = I^0_i \dot{w}_i + w_i \times I^0_i w_i
$$

$$
F_{ci} = m_i \dot{v}_{ci}
$$

$$
\begin{pmatrix} N_{cix} \\ N_{ciy} \\ N_{ciz} \end{pmatrix} = \begin{pmatrix} I^0_{ixx} & I^0_{ixy} & I^0_{ixz} \\ I^0_{iyx} & I^0_{iyy} & I^0_{iyz} \\ I^0_{izx} & I^0_{izy} & I^0_{izz} \end{pmatrix} \begin{pmatrix} \dot{w}_{ix} \\ \dot{w}_{iy} \\ \dot{w}_{iz} \end{pmatrix} + \begin{pmatrix} 0 & -w_{iz} & w_{iy} \\ w_{iz} & 0 & -w_{ix} \\ -w_{iy} & w_{ix} & 0 \end{pmatrix} \begin{pmatrix} I^0_{ixx} & I^0_{ixy} & I^0_{ixz} \\ I^0_{iyx} & I^0_{iyy} & I^0_{iyz} \\ I^0_{izx} & I^0_{izy} & I^0_{izz} \end{pmatrix} \begin{pmatrix} w_{ix} \\ w_{iy} \\ w_{iz} \end{pmatrix}
$$

$$
(7.4)
$$

such that

$$N_{cix} = I^0_{ixx}\dot{w}_{ix} + I^0_{ixy}\dot{w}_{iy} + I^0_{ixz}\dot{w}_{iz} - w_{iz}\left(I^0_{iyx}w_{ix} + I^0_{iyy}w_{iy} + I^0_{iyz}w_{iz}\right)$$

$$+ w_{iy}\left(I^0_{izx}w_{ix} + I^0_{izy}w_{iy} + I^0_{izz}w_{iz}\right)$$

$$N_{ciy} = I^0_{iyx}\dot{w}_{ix} + I^0_{iyy}\dot{w}_{iy} + I^0_{iyz}\dot{w}_{iz} + w_{iz}\left(I^0_{ixx}w_{ix} + I^0_{ixy}w_{iy} + I^0_{ixz}w_{iz}\right)$$

$$- w_{ix}\left(I^0_{izx}w_{ix} + I^0_{izy}w_{iy} + I^0_{izz}w_{iz}\right)$$

$$N_{ciz} = I^0_{izx}\dot{w}_{ix} + I^0_{izy}\dot{w}_{iy} + I^0_{izz}\dot{w}_{iz} - w_{iy}\left(I^0_{ixx}w_{ix} + I^0_{ixy}w_{iy} + I^0_{ixz}w_{iz}\right)$$

$$+ w_{ix}\left(I^0_{iyx}w_{ix} + I^0_{iyy}w_{iy} + I^0_{iyz}w_{iz}\right)$$

Similarly,

$$\begin{pmatrix} F_{cix} \\ F_{ciy} \\ F_{ciz} \end{pmatrix} = m_i \begin{pmatrix} \dot{v}_{cix} \\ \dot{v}_{ciy} \\ \dot{v}_{ciz} \end{pmatrix} \text{ where } m_3 = m_5 = 0 \text{ and } m_2 = m'_1 + m'_2 \text{ (composite link)}$$

$$(7.5)$$

since links 3 and 5 are fictitious and m_2 is composite.

If any payload for a space manipulator system is included, F_{c7} and N_{c7} must be added into the formulation of Chapter 8:

$$F_{c7} = m_7 \dot{v}_{c7} \quad \text{where} \quad \dot{v}_{c7} = \dot{v}_6 + w_6 \times r_7 + w_6(w_6 \times r_7)$$

$$N_{c7} = I^0_7 \dot{w}_6 + w_6 \times \left(I^0_7 w_6\right) \quad \text{where} \quad I^0_7 = R^0_6 I_7 R^6_0$$

The orientation of the object with respect to the end effector is invariant so $w^7_7 = \dot{w}^7_7 = 0$ so end effector variables are applicable with respect to the target object. Furthermore, the spacecraft moment of inertia will in general not have zero off-diagonals when referred to the end effector frame.

7.2.2 Backward recursion

Backward recursion to find the joint torques requires link-referenced forward recursion formulation. The backward recursion proceeds from the end effector towards the base to find the joint torques and is performed by a recursive loop algorithm from $i = n$ to 1.

$$f_i = F_{ci} + f_{i+1}$$

$$n_i = N_{ci} + n_{i+1} + (l_i + s_i) \times F_{ci} + l_i \times f_{i+1}$$

Let

$$\begin{pmatrix} f_{ext} \\ n_{ext} \end{pmatrix} = \begin{pmatrix} f_7 \\ n_7 \end{pmatrix} = \begin{pmatrix} 0 \\ 0 \end{pmatrix}$$

since external forces and moments are dealt with separately and explicitly through the Jacobian transpose.

$$\begin{pmatrix} f_{ix} \\ f_{iy} \\ f_{iz} \end{pmatrix} = \begin{pmatrix} F_{cix} \\ F_{ciy} \\ F_{ciz} \end{pmatrix} + \begin{pmatrix} f_{i+1x} \\ f_{i+1y} \\ f_{i+1z} \end{pmatrix} \tag{7.6}$$

$$\begin{pmatrix} n_{ix} \\ n_{iy} \\ n_{iz} \end{pmatrix} = \begin{pmatrix} N_{cix} \\ N_{ciy} \\ N_{ciz} \end{pmatrix} + \begin{pmatrix} n_{i+1x} \\ n_{i+1y} \\ n_{i+1z} \end{pmatrix} + \begin{pmatrix} 0 & -l_{iz} & l_{iy} \\ l_{iz} & 0 & -l_{ix} \\ -l_{iy} & l_{ix} & 0 \end{pmatrix} \begin{pmatrix} f_{i+1x} \\ f_{i+1y} \\ f_{i+1z} \end{pmatrix}$$

$$+ \begin{pmatrix} 0 & -(l_i+s_i)_z & (l_i+s_i)_y \\ (l_i+s_i)_z & 0 & -(l_i+s_i)_x \\ -(l_i+s_i)_y & (l_i+s_i)_x & 0 \end{pmatrix} \begin{pmatrix} F_{cix} \\ F_{ciy} \\ F_{ciz} \end{pmatrix}$$

such that

$$n_{ix} = N_{cix} + n_{i+1x} + f_{i+1z}l_{iy} - f_{i+1y}l_{iz} + F_{ciz}(l_{iy}+s_{iy}) - F_{ciy}(l_{iz}+s_{iz})$$

$$n_{iy} = N_{ciy} + n_{i+1y} + f_{i+1x}l_{iz} - f_{i+1z}l_{ix} + F_{cix}(l_{iz}+s_{iz}) - F_{ciz}(l_{ix}+s_{ix})$$

$$n_{iz} = N_{ciz} + n_{i+1z} + f_{i+1y}l_{ix} - f_{i+1x}l_{iy} + F_{ciy}(l_{ix}+s_{ix}) - F_{cix}(l_{iy}+s_{iy}) \tag{7.7}$$

To find the input torques, the input torque at joint i is the projection of n_i onto the z_{i-1} axis:

$$\tau_i = n_i^T . z_{i-1} = \begin{pmatrix} n_{ix} & n_{iy} & n_{iz} \end{pmatrix} \begin{pmatrix} z_{i-1x} \\ z_{i-1y} \\ z_{i-1z} \end{pmatrix} = (n_{ix}z_{i-1x} + n_{iy}z_{i-1y} + n_{iz}z_{i-1z})$$

This completes the recursive dynamics formulation for the PUMA 560/600 and it allows the computation of the joint torques given the kinematic joint variables in link-referenced coordinates.

7.3 COLLISION/CAPTURE DYNAMICS

Iwata et al (1991) and Yoshida et al (1992) have considered the problem of collisions of a manipulator end effector and a payload in space. The robot receives momentum from the object. A very large force acts on the collision points on the end effector for a very short time to cause a momentum distribution change between the colliding objects due to very rapid velocity changes through very short distances. The behaviour of the impacting body is dependent on the elasticity of the impacting objects. Newton's law of restitution states that when two bodies collide, their relative parting velocities in the direction of the common normal at the point of impact is $-e \times$ relative approach velocity whereby

$0 < e < 1$: $v_f - u_f = -e(v_i - u_i)$. For an inelastic collision, the bodies adhere and $e = 0$; for an elastic collision, the bodies rebound and $e = 1$. Assuming inelastic rigid-body collisions, the total system momentum is conserved before and after collision, i.e. momentum is lost from the payload to the spacecraft end effector (subscript '*obj*' refers to the target; subscript '*s/c*' refers to the spacecraft; subscript '*eff*' refers to the end effector; and subscript '*com*' refers to the composite body):

$$m_{obj}v_{obj} + m_{s/c}v_{eff} = (m_{obj} + m_{s/c})v$$

$$I_{obj}w_{obj} + I_{s/c}w_{s/c} = I_{com}w$$

Assume that the final velocity of the composite object in inertial space is zero, i.e. the manipulator brings the object to rest:

$$\begin{pmatrix} v \\ w \end{pmatrix} = 0$$

Impulse,

$$\int_{t_0}^{t_f} F\,dt = m\int_{t_0}^{t_f} \frac{dv}{dt}\,dt = m\delta v$$

$$\rightarrow \left.\begin{matrix} m_{obj}\delta v = f_{ext}\delta t \\ I_{obj}\delta w = n_{ext}\delta t \end{matrix}\right\} \quad \text{where} \quad \begin{cases} \delta v = v_{obj} - v = v_{obj} \\ \delta w = w_{obj} - w = w_{obj} \end{cases}$$

i.e.

$$f_{ext} = m_{obj}v_{obj}/\delta t$$
$$n_{ext} = I_{obj}w_{obj}/\delta t \tag{7.9}$$

External forces and moments introduce a problem for space manipulators—the need to express the external forces and moments at the end effector with respect to their reactive effects on the manipulator base point (see chapter 8). Now, the virtual work done by applied force is constant [Paul 1981]—subscript '0' signifies base-referenced variables and subscript '*eff*' signifies end effector-referenced variables:

$$\delta W_0 = \delta W_{eff} \rightarrow F_0^T dq_0 = F_{eff}^T dq_{eff} \rightarrow F_0^T v_0 = F_{eff}^T v_{eff}$$

The question is to find the relationship between v_0 and v_{eff} which is defined by a matrix N. The Jacobian J transforms the end effector velocities to each joint rate by propagating the end effector velocities to the required joint coordinates and summing the contributions to the end effector velocities by each joint's motion. A similar principle may be applied to the problem of determining the transform from the end effector velocities to the robot base. The total effect of the end effector velocities on the base may then be determined and so allow the explicit calculation of the external forces reaction effect on the base:

$$\begin{pmatrix} v_{eff} \\ w_{eff} \end{pmatrix} = N \begin{pmatrix} v_0 \\ w_0 \end{pmatrix} \quad \text{where} \quad N = \begin{pmatrix} z_n \times l_n & 0 \\ 0 & z_n \end{pmatrix}$$

using a cross product formulation. Now,

$$F_0^T v_0 = F_{eff}^T N v_0 \rightarrow F_0^T = F_{eff}^T N \rightarrow F_0^T = N^T F_{eff}$$

where

$$N^T = \begin{pmatrix} 0 & z_n \times l_n \\ z_n & 0 \end{pmatrix}$$

For a PUMA 560/600,

$$\begin{pmatrix} z_{6x} \\ z_{6y} \\ z_{6z} \end{pmatrix} = \begin{pmatrix} a_x \\ a_y \\ a_z \end{pmatrix} \quad \text{and} \quad \begin{pmatrix} l_{6x} \\ l_{6y} \\ l_{6z} \end{pmatrix} = \begin{pmatrix} p_x \\ p_y \\ p_z \end{pmatrix} \quad \text{so} \quad a \times p = \begin{pmatrix} 0 & -a_z & a_y \\ a_z & 0 & a_x \\ -a_z & a_x & 0 \end{pmatrix} \begin{pmatrix} p_x \\ p_y \\ p_z \end{pmatrix}$$

Explicitly, we have:

$$\begin{pmatrix} F_{cn+1} \\ N_{cn+1} \end{pmatrix} = \begin{pmatrix} f_x^0 \\ f_y^0 \\ f_z^0 \\ n_x^0 \\ n_y^0 \\ n_z^0 \end{pmatrix} = \begin{pmatrix} (a_y p_z - a_z p_y) n_x^{eff} \\ (a_z p_x - a_x p_z) n_y^{eff} \\ (a_x p_y - a_y p_x) n_z^{eff} \\ a_x f_x^{eff} \\ a_y f_y^{eff} \\ a_z f_z^{eff} \end{pmatrix} \qquad (7.10)$$

This represents

$$\begin{pmatrix} F_{c7} \\ N_{c7} \end{pmatrix}$$

referenced to base coordinates for the forward dynamics for modelling impacts with a 6 DOF manipulator.

8

Space freeflyer kinemo-dynamics formulation

The tools of both robotic and spacecraft dynamics have an unparalleled pedigree. Indeed, their development reflects the history of science and the scientific method itself. The fundamental science of mechanics was developed by those great minds who in the process gave birth to modern science as natural philosophy: Galileo Galilei (1564–1642), Johannes Kepler (1571–1630), Isaac Newton (1642–1729), Gottfried Leibnitz (1646–1716), Leonhard Euler (1707–1783), Jean d'Alembert (1717–1783), Joseph Lagrange (1736–1813), Pierre Laplace (1749–1827) and William Hamilton (1805–1865)—all made fundamental contributions that laid the foundations of mechanics and established a unique relationship between empirical science as a universal description of the world and mathematics as the language of science. Space robotics may be regarded as a specialised application of mechanics.

Space introduces a complicating factor to robotic systems that is not apparent on Earth—the manipulator base is not fixed in space and so reference coordinates are no longer fixed with respect to the spacecraft. This introduces a high degree of dynamic complexity. This effect will be significant as it has been suggested that the mass of a space manipulator payload may comprise up to one-third the mass of the spacecraft bus mounting [Vafa & Dubowsky 1990]. The motion of the manipulator will generate reaction forces and moments on the interceptor platform at the spacecraft mounting/manipulator coupling point. This will induce translational and rotational motion of the satellite platform in response to the manipulator movements. If no compensation is made for the motion of the interceptor mount, the robot end effector will not attain its target since the coupling has a significant effect on the manipulator kinematics, dynamics and control. Feedforward thruster control may be utilised to compensate for the motion of the mount, but this introduces undesirable and prohibitively excessive expenditure of fuel as well as generating exhaust plumes which may be detrimental to sensor operation. In addition, they are difficult to control proportionally since they operate in pulse mode.

This chapter utilises the techniques of robot kinematics and dynamics introduced in the previous chapters and modifies them for use in the control of space-based manipulators. The methodology is illustrated by applying the techniques to single manipulators initially and then expanded to apply to dual manipulator configurations as more realistic implementations of space-based freeflyer robots. It is assumed that no external forces or moments act on the total system other than those which are expressed explicitly, i.e. gravity, magnetic, aerodynamic and solar pressure torques are negligible. This allows the application of the laws of linear and angular momentum which hold in an inertial frame and these provide the constraints to solve the kinematic equations. Such conservation laws are the basis of orbit raising using tethers such that if the tether is sufficiently long, the system may be modelled as a rigid dumbbell of two point masses linked by a rigid massless bar suspended at the centre of mass of the system along the local vertical and lowering a mass and releasing it.

The robotic freeflyer spacecraft system under consideration here comprises a spacecraft bus mount and one or more manipulators as payload. Most solutions in the literature to the moving platform problem discussed above involve a considerable computational burden for the control of such systems in space. A 6 DOF robot mounted onto a 6 DOF platform generates a 12 DOF system controlled by only six joints' inputs. The system is redundant since both vehicle and manipulator have more controllable states than are necessary to specify the motion of the end effector. They do not generate closed-form solutions for position/orientation due to their inherent redundancy. Longman (1988) considered this form of 'kinetic' position control of a satellite with a 3 DOF manipulator model and found that an infinite number of solutions exist to the inverse kinematics due to this redundancy. The solutions are a function of the history of manipulator motion rather than joint angle configuration alone. Solutions can be obtained but usually involve the introduction of selective cyclic 'coning' motions superimposed on the desired manipulator trajectory to maintain constant spacecraft attitude without employing dedicated attitude control actuators [Vafa & Dubowsky 1990]. This 'coning' motion is equivalent to using spinning wheels within the spacecraft. The coning motions may also introduce possible collision hazards with target satellite appendages which would require complex path planning to avoid. Overcoming this by using small cyclic motions to eliminate non-negligible nonlinear terms introduces the requirement for many cycles for even small changes in spacecraft attitude. Vafa (1990) used a Virtual Manipulator approach to derive the conservation of angular momentum in relative joint coordinates. Only the end effector trajectory was controlled whilst the satellite attitude was allowed to be arbitrary. Clearly this is not desirable since the spacecraft bus will have components and subsystems that have their own specific pointing requirements. Another formulation is to find the generalised Jacobian matrix J^* by applying applied momentum constraints to a complete free-floating spacecraft/manipulator to account for platform translation and rotation by relating end effector velocities to joint and platform velocities [Umetani & Yoshida 1989, Papadopoulos & Dubowsky 1991a,b]. Such free-floating systems are defined by their lack of a dedicated spacecraft attitude control system. The generalised Jacobian due its dynamic nature is complex to compute with a complexity of $O(n^2)$ and so not conducive to real-time operation [Masutani et al 1989]. Xu (1993) introduced a dynamic coupling coefficient to quantify the relation between the end effector velocity (subscript 'eff') and base velocity (subscript '0'):

$$\begin{pmatrix} v_0 \\ w_0 \end{pmatrix} = P \begin{pmatrix} v_{eff} \\ w_{eff} \end{pmatrix}$$

This dynamic factor P is similar to the generalised Jacobian and may be reduced to a single value $w = \det(P^T P)$ to quantify the coupling effect. Evidently, dedicated attitude control would reduce this coupling effect by making $w_0 = 0$. However, this was only an analytical factor to characterise the coupling rather than having any application for control purposes. Masutani et al (1989) found that the generalised Jacobian comprises the conventional Jacobian of computational complexity of $O(n)$ with additional terms dependent on the masses, inertias and geometric structures of each link which generate the additional complexity. The conventional Jacobian could be used to derive a transposed Jacobian which yielded satisfactory results when the platform-manipulator mass ratios were in excess of ~5. This, however, limits the applicability of the formulation where it is conceivable that freeflyer robotic systems will have mass ratios <1.

Hence, these solutions are less than ideal since the computational power of space-rated microprocessors is a limited and valuable resource and the requirement for real-time operation is strict. Space system computer technology lags terrestrial computer technology by ~10–15 y. Present-day space-rated processors are of the 80386 variety while terrestrial technology has the Pentium III processor commercially available. Surprisingly, although the motivation for this time lag is reliability (as evidenced by a proven track record of Earth operation), this can and does often lead to reduced performance, increased design costs and delayed launches. Spacecraft now tend to be designed with subsystem-dedicated computers interconnected across a vehicle-wide databus. As well as providing a degree of functional redundancy this suggests that a piecewise approach to the problem is more suitable for a distributed computer system implementation. Finally, and most important, it has been found that freefloating systems are subject to unpredictable dynamic singularities in their manipulator workspaces due to the attitude motion of the platform at which point they become unstable [Papadopoulos & Dubowsky 1989]. Indeed, dynamic singularities are characteristic of freefloating systems and are functions of the mass and inertia of the composite spacecraft/manipulator system. These singular joint configurations cannot be mapped into unique points in the workspace since the generalised Jacobian is a dynamic function rather than being kinematic, and spacecraft attitude coordinates do not map uniquely to end effector coordinates. Hence, these singularities cannot be predicted from the kinematic configuration alone since they are functions of the history of the end effector path also, i.e. points in a workspace may become singular depending on the path taken to reach it. Any control system which uses inverse generalised Jacobian techniques for freefloating systems will encounter such singularities within the workspace. Although the bidirectional approach attempts to alleviate the problem of freefloating systems, this method is computationally intensive since the scheme relegates the problem to the path planning algorithms higher up the control heirarchy by using the extra degrees of freedom for path planning [Nakaruma & Mukherjee 1989]. They used a bidirectional search to map between the original state and final desired configuration using the generalised Jacobian to avoid excessive joint torques, for obstacle avoidance or for re-orientation of attitude. Nenchev et al (1992) used a Moore–Penrose pseudo-inverse version of the generalised Jacobian $(J^*)^+$ to overcome the dynamic singularities problem by using the redundant degrees of freedom. It allowed change of spacecraft attitude whilst keeping the position/orientation of the end effector fixed with respect to the inertial reference frame to provide greater flexibility of operation.

Joint angular velocity, $\dot{\theta} = (J^*)^+ \dot{q}$ where $\dot{q}$ = generalised cartesian velocity

However, this formulation is even more complex than the generalised Jacobian technique. Dubowsky et al (1989) noted that excessive spacecraft motions may occur which cannot be accounted for in path planning and so adapted standard robotic minimum-time optimal trajectory generation to limit joint torques and attitude rates to within specified bounds. Control schemes have been devised which switch between different coordinated modes of control. This involves control of the platform/manipulator being switched from the freefloating formulation to a redundancy formulation for the control of the platform alone whenever dynamic singularites are encountered [Papadopoluos & Dubowsky 1991a,b, Spofford & Akin 1988]. However, switching is characteristic of variable structure control schemes and these tend to suffer from 'chattering' induced by the switching. This is clearly undesirable for a space-borne manipulator which operates in an undamped medium. In conclusion then, freefloating systems are subject to certain constraints not encountered in terrestrial robotics (Papadopoulos & Dubowsky 1990):

(a)   spacecraft orientation is required to derive the generalised Jacobian;
(b)   dynamic properties affect the kinematics;
(c)   dynamic singularities occur in the workspace;
(d)   nonholonomic redundancy implies path dependency of joint angle configuration.

All such schemes which leave spacecraft attitude uncontrolled cannot cope with the input dynamics of target acquisition in real-time using present-day and near-future space-rated computational hardware.

Longman et al (1987) and Lindberg et al (1986) applied classical dynamics techniques to the Remote Manipulator Servicer mounted onto the Space Shuttle. They used two models of the manipulator: one with a prismatic elbow joint and one with a revolute elbow joint. These two papers marked the first detailed analysis of the space robot problem. They decoupled the translation and rotation components of the combined RMS/Shuttle system. They calculated the total reaction moment on the Space Shuttle due to the motion of the RMS using the Newton–Euler method referenced to inertial coordinates. This provided the basis for a feedforward compensation component to be input to the Shuttle's dedicated attitude control system (comprising of orthogonal reaction wheels) to compensate for the torques applied to the Shuttle at the base of the arm. Further, they reformulated the position kinematics to include dynamic variables by the application of equilibrium constraints derived fom the definition of the system centre of mass. They found that for the Shuttle/RMS system, reaction wheels were insufficient to effect reaction moment compensation of 34 N ms, suggesting that CMGs might be within the required performance bounds. It is important to note that these formulations were manipulator geometry specific rather than generalised. It was not evident how the scheme may be applicable beyond the manipulator geometries chosen. Walker & Wee (1991a,b) analysed a single-arm 15 DOF space robot system with orthogonal reaction wheels and found that such a system comprises a 9 DOF invertible portion including manipulator joint angles and base orientation and a 6 DOF component including the base position and reaction wheel positions. The reaction wheel dynamics were incorporated to eliminate reaction wheel position, and base translation was eliminated by the application of conservation of linear momentum. Once again, it was quite complex in its approach by globalising the dynamics and control of the whole system.

Some workers have considered two-arm systems since these offer more realistic models of future robotic systems. Yoshida et al (1991) extended their generalised Jacobian approach to consider dual arm coordination whilst mounted onto a space platform. Their generalised Jacobian was a composite 18×18 element matrix even larger than the original 6×6 generalised Jacobian required for a single arm. To provide coordinated control they suggested that whilst one arm was used for task operations, the other arm should be moved in compensatory mode to keep the satellite attitude stable and minimise the total torques applied to the spacecraft mount. Quite apart from the excessive computational overhead of the 18×18 Jacobian, the use of a manipulator as a dedicated attitude stabiliser is wasteful of costly hardware since it effectively reduces (operationally speaking) the dual arm system to a single arm system as well as imposing possible collision problems. The appropriateness of two arms derives from its operational capabilities rather than any other factor. Murphy et al (1991) also considered a two-armed robot mounted onto a satellite platform. They derived a set of closed form Newton–Euler equations for both manipulators. Once again, their formulation involved an 18×18 Jacobian transpose matrix precluding the formulation from real-time operation.

In conclusion, it appears that once the generalised Jacobian method had been developed, many workers in this field built on it to develop ever more fanciful control algorithms without considering the realities of the situation regarding limited computational resources in space and the inherent disadvantages of the generalised Jacobian. Although many of these techniques are elegant and yield interesting and useful insights to the problem, they are not implementable on available space-rated hardware. Here, we adopt a more pragmatic approach in seeking to achieve simplicity and achievability.

We consider an arbitrary robotic spacecraft with a single manipulator arm initially. Linear momentum conservation law states that:

$$P = \sum_{i=0}^{n+1} m_i \dot{r}_i = const = 0 \quad \left(\text{arbitrarily since } \dot{P} = \sum_{i=0}^{n+1} F_{ext} = 0 \right)$$

where P = linear momentum of the system

As a holonomic constraint, this is integrable to the equilibrium of moments principle:

$$\int_0^t \sum_{i=0}^{n+1} m_i \dot{r}_i \, dt = \left[\sum_{i=0}^{n+1} m_i r_i \right]_0^t = 0$$

Angular momentum conservation states that:

$$L = \sum_{i=0}^{n+1} I_i w_i + m_i \dot{r}_i \times r_i = const = 0$$

$$\left(\text{arbitrarily since } \dot{L} = \sum_{i=0}^{n+1} N_{ext} + \sum_{i=0}^{n+1} r \times F_{ext} = 0 \right)$$

where L = angular momentum of the system.

As a nonholonomic constraint, this is not integrable—integration of rotational velocity does not yield a unique vectorial representation of orientation since rotational motion is path-dependent due to the noncommutativity of rotations. There are thus an infinite number of paths yielding a particular orientation, i.e. orientation depends on the time history of motion [Nakaruma & Mukherjee 1989]. This derives from the definition of holonomy [Masutani et al 1989]: 'A system is holonomic if and only if its motion is constrained by a set of algebraic equations involving only general angular coordinates and time. This implies that a system is holonomic if and only if a set of vector fields defining the linear space of possible velocities is completely integrable everywhere in general coordinates. Otherwise, it is nonholonomic.' Hence, in this case there are more controllable states than are necessary to specify the motion of the end effector.

By decoupling the system such that linear and angular motions are treated separately, real-time control is made possible and straightforward by considerably reducing the complexity of the problem. The translation effects are holonomic constraints and so linear momentum constraints may be integrated and be applied to the system through equilibrium of moments conservation to account for the translational motion of the satellite in inertial space. The system centre of mass remains invariant in inertial space, providing unique closed-form solutions for translation motion control by incorporating linear compensation into the robot kinematics formulation. These translational kinematic equations have the same form as those for a terrestrial manipulator. Hence, the linear component of the reaction effect is compensated automatically within the linear portion of the controller without the use of fuel. Fuel is a non-renewable resource, expenditure of which must be minimised as a mission constraint to maximise the satellite's operational lifetime. In fact, it has been found that fuel use is critical in the performance and flexibility of task domains and that fuel expenditure during manipulation is usually prohibitive [Marcyk & Bellazzi 1989]. Typically, the fuel expenditure to compensate for the translational motion alone of the robotic satellite would exceed the capability of cold gas thrusters by several orders of magnitude.

The angular effects are nonholonomic constraints since the attitude of the platform and the orientation of the end effector depends on the history of joint displacements because the noncommutativity of finite rotation sequences. Different paths of the manipulator to the same point will result in different attitudes for the spacecraft. Hence, angular momentum conservation constraints are not integrable to spacecraft attitude as a function of manipulator joint angles, and so must be applied directly. It is possible to do this straightforwardly by employing active three-axis attitude stabilisation by non-fuel expending orthogonal reaction wheels or control moment gyros to compensate for the dynamic attitude reactions based on computations of the reaction forces and moments applied at the base of the robotic manipulator. The reaction forces and torques acting on the satellite bus due to the manipulator may be calculated directly from the manipulator dynamics and be fed forward to the attitude control system. The reaction wheels are then driven to counteract the reaction forces and moments on the spacecraft mounting in conjunction with standard attitude control algorithms to maintain spacecraft attitude. Maintenance of constant attitude is desirable to enable constant remote viewing conditions. Hence, all angular properties of the manipulator are identical to those of their terrestrial counterparts. The (n s a) rotational submatrix component of the 4×4 DH matrix remains unchanged from the terrestrial case since the spacecraft bus platform employs attitude control to maintain constant stabilisation of attitude. This method avoids the dynamic singularities associated with uncontrolled attitude and allows the end effector position and orientation to be formulated as a unique function of joint angles independent of the end effector path. Indeed, it is generally considered that explicit control of attitude is essential for any docking or grappling manoeuvres. All in all, this method will not exert excessive computational burdens on the spacecraft attitude and robot controllers, enabling real-time operation. Apart from on-board data-handling and power supply, the manipulator functions as a stand-alone system—this is more in line with the general developing preference for distributed computer systems on spacecraft. Furthermore, as well as being less computationally demanding on any single processor, this reduces the impact of any control equipment failure on the system as a whole. Finally, the method offers more favourable workspace dimensions over the free-floating mode [Vafa & Dubowsky 1987, 1990]. If satellite attitude is not controlled, the 'path-independent workspace' reachable by the manipulator by any path is much smaller than the 'constrained workspace' available to a spacecraft employing attitude control. Dynamic singularities are characteristic of a 'path-dependent workspace' (which is larger than the path-independent workspace) reachable by freefloating manipulators only via certain paths [Papadopoulos & Dubowsky 1989]. The path-independent workspace which does not suffer from dynamic singularities has its maximum extent when attitude control is employed, i.e. when it coincides with the constrained workspace. This mode of control represents a free-flying system (i.e. where a dedicated attitude control system is employed) which is a partial restriction of the more general free-floating system.

This approach is suggested by the results of Walker & Wee (1991a,b) and is a generalisation, extension and simplification of the classical approach originally formulated by Longman et al (1986) and Lindberg et al (1987) as applied to the Shuttle Remote Manipulator System. The Longman/Lindberg/Zedd method was highly specific and was not conducive to a straightforward extraction of the inverse kinematics solution of manipulator joint angles. Dubowsky et al (1989) suggested that attitude control actuator saturation may provide a problem but Spofford & Akin (1988) suggested that this may be unlikely since manipulator motions tend to be cyclic. Certainly, this will tend to be the case when employing dual manipulators. Either way, environmental torque sources may be used to desaturate wheels and gyroscopes.

8.1 SPACE FREEFLYER KINEMATICS

The central concepts introduced in this chapter are represented as three theorems and proofs concerning single-manipulator freeflyer spacecraft. Later sections extend those concepts to the dual-manipluator freeflyer case. All the concepts are based on the utilisation of the robotic manipulator control algorithms of the earlier chapters.

8.1.1 Resolution of inertial coordinates

Theorem (8.1) is the central core of the space robot control methodology outlined in this chapter. This is because the manipulator configuration dependency of the kinematics, dynamics and control algorithms is fundamental. The other theorems essentially follow from theorem (8.1).

Theorem 1: For a freeflying robotic manipulator employing dedicated attitude control of the spacecraft bus, the Denavit–Hartenberg formulation is given by [Ellery 1994, 1996]:

$$q = \begin{pmatrix} n & s & a & p^* \\ 0 & 0 & 0 & 1 \end{pmatrix}$$

where $R = (n\ s\ a)$ as for terrestrial manipulators, and

$$p^* = p_{cm}^* + \left(\frac{m_0}{m_T}\right)s_0 + \sum_{i=1}^{n} R_i \lambda_i - \left(\frac{m_{n+1}}{m_T}\right)R_{n+1}r_{n+1}$$

where

$$\lambda_i = \frac{1}{m_T}\left(\sum_{j=0}^{i} m_j l_i - m_i r_i\right) \tag{8.1}$$

m_0 = mass of the spacecraft bus and m_T = total mass of the system.

Proof: The position kinematics of a space manipulator with respect to inertial space may be given by:

$$p^* = r_{c0} + R_0 s_0 + \sum_{i=1}^{n} R_i^0 l_i \tag{8.2}$$

Walker & Wee (1991) used several frames between the inertial and manipulator base frames to model virtual links of zero mass, but no gain in generality is obtained by this. In general, the attitude of the spacecraft platform (for roll-pitch-yaw coordinates α, β, γ):

$$R_0 = Rot(z, \gamma)Rot(y, \beta)rot(x, \alpha)$$

$$= \begin{pmatrix} c\beta c\gamma & s\alpha s\beta c\gamma - c\gamma s\gamma & c\alpha s\beta c\gamma - s\alpha s\gamma \\ c\beta s\gamma & s\alpha s\beta s\gamma + c\alpha c\gamma & c\alpha s\beta s\gamma - s\alpha c\gamma \\ -s\beta & s\alpha c\beta & c\alpha c\beta \end{pmatrix}$$

For an attitude-controlled platform, $R_0 = I_3$. The centre of mass of the complete system

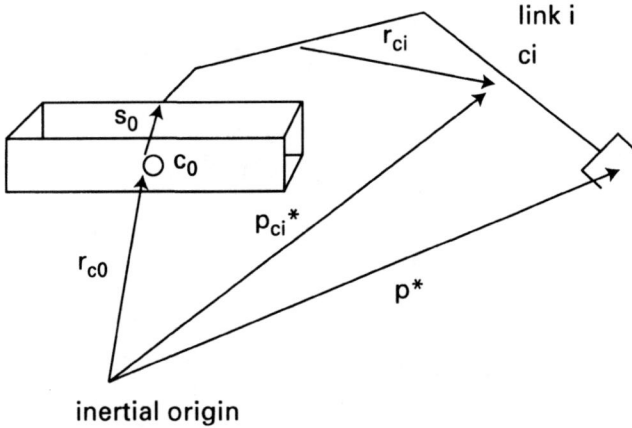

Fig. 8.1. Space freeflyer geometry.

(including the satellite bus mount, robotic manipulator and the payload) is given by [Longman et al 1987, Lindberg et al 1986]:

$$\left(\sum_{i=0}^{n+1} m_i\right) p_{cm}^* = \sum_{i=0}^{n+1} m_i p_{ci}^* \qquad \text{(equilibrium of moments)}$$

i.e.

$$p_{cm}^* = \sum_{i=0}^{n+1} m_i p_{ci}^* \bigg/ \sum_{i=0}^{n+1} m_i \tag{8.3}$$

The location of the system centre of mass will remain invariant in inertial space if no external forces act on the system

$$\sum_{i=0}^{n+1} F_i = \sum_{i=0}^{n+1} m_i \ddot{p}_{ci}^* = 0$$

such that $\ddot{p}_{cm}^* = 0$, i.e. this point corresponds to the 'virtual ground' defined by the virtual manipulator approach [Vafa & Dubowsky 1987, 1990]. The virtual ground is the point in inertial space at which an ideal virtual kinematic chain manipulator has its base when its end effector coincides with the end effector of the actual manipulator. This virtual ground point coincides with the centre of mass of the spacecraft–manipulator system. This is to be expected since the constraints applied were the equilibrium of moments for the derivation of the kinematics of the virtual manipulator. They did not, however, explicitly describe the kinematics and the virtual manipulator approach is limited to non-contact situations. The system centre of mass is the point at which all the mass may be considered to be concentrated (there is no corresponding centre of inertia).

Now,

$$p_{ci}^* = \sum_{j=0}^{i} r_{cj} \quad \text{and} \quad \sum_{i=0}^{n+1} m_i = m_T = \text{total mass of the system}$$

$$p_{cm}^* = \frac{1}{m_T}\left(\sum_{i=0}^{n+1} m_i \sum_{j=0}^{i} r_{cj}\right) = \frac{1}{m_T}\left(\sum_{i=0}^{n+1}\sum_{j=i}^{n+1} m_j r_{ci}\right)$$

$$= r_{c0} + \frac{1}{m_T}\sum_{i=1}^{n+1}\sum_{j=i}^{n+1} m_j r_{ci}$$

i.e.

$$r_{c0} = p_{cm}^* - \frac{1}{m_T}\sum_{i=1}^{n+1}\sum_{j=i}^{n+1} m_j r_{ci}$$

Substitute into (8.2):

$$p^* = p_{cm}^* + s_0 + \sum_{i=1}^{n} R_i l_i - \frac{1}{m_T}\sum_{i=1}^{n+1}\sum_{j=i}^{n+1} m_j r_{ci} \tag{8.4}$$

Now,

$$r_{ci} = R_i r_i + R_{i-1} s_{i-1}; \quad l_i = r_i + s_i$$

so substitute into (8.4):

$$p^* = p_{cm}^* + s_0 + \sum_{i=1}^{n} R_i(r_i + s_i) - \frac{1}{m_T}\sum_{i=1}^{n+1}\sum_{j=i}^{n+1} m_j(R_i r_i + R_{i-1} s_{i-1})$$

$$= p_{cm}^* + s_0 + \sum_{i=1}^{n} R_i(r_i + s_i) - \frac{1}{m_T}\left[\sum_{i=1}^{n+1}\sum_{j=i}^{n+1} m_j(R_{i-1}r_{i-1} + R_{i-1}s_{i-1}) + \sum_{i=1}^{n+1} m_i R_i r_i\right]$$

$$= p_{cm}^* + s_0 + \sum_{i=1}^{n} R_i(r_i + s_i) - \frac{1}{m_T}\left[\sum_{i=2}^{n+1}\sum_{j=i}^{n+1} m_j R_{i-1}(r_{i-1} + s_{i-1})\right]$$

$$- \frac{1}{m_T}\left[\sum_{i=1}^{n+1} m_i R_0(r_0 + s_0) + \sum_{i=1}^{n+1} m_i R_i r_i\right]$$

$$= p_{cm}^* + \frac{m_0}{m_T} s_0 + \sum_{i=1}^{n} R_i l_i - \frac{1}{m_T}\left[\sum_{i=1}^{n+1}\sum_{j=i+1}^{n+1} m_j R_i(r_i + s_i) + \sum_{i=1}^{n+1} m_i R_i r_i\right]$$

$$= p_{cm}^* + \frac{m_0}{m_T} s_0 + \sum_{i=1}^{n} R_i l_i - \frac{1}{m_T}\left[\sum_{i=1}^{n}\sum_{j=i+1}^{n+1} m_j R_i l_i + \sum_{i=1}^{n} m_i R_i r_i\right] - \left(\frac{m_{n+1}}{m_T} R_{n+1} r_{n+1}\right)$$

i.e.

$$p^* = p_{cm}^* + \frac{m_0}{m_T} s_0 + \frac{1}{m_T} \sum_{i=1}^{n} \left(\sum_{j=0}^{i} m_j l_i - m_i r_i \right) R_i - \frac{m_{n+1}}{m_T} R_{n+1} r_{n+1}$$

$$p^* = p_{cm}^* + \frac{m_0}{m_T} s_0 + \sum_{i=1}^{n} R_i \lambda_i - \frac{m_{n+1}}{m_T} R_{n+1} r_{n+1}$$

where

$$\lambda_i = \frac{1}{m_T} \sum_{j=0}^{i} (m_j l_i - m_i r_i) \qquad \text{QED (8.1)}$$

This equation has the same form as that for an earth-based manipulator (8.2) with additional constants. Note that p_{cm}^* is constant, λ_i is a lumped kinematic/dynamic parameter constant and that the payload term is constant since the payload remains fixed relative to the end effector. No loss of generality ensues by assuming that $R_{n+1} = I_3$. The r_{n+1} corresponds to a 'virtual stick' vector which points from the hand of the manipulator to the payload centre of mass which remains fixed in the target object [Uchiyama & Dauchez 1988]. This vector from the grasp point to the object's centre of mass will if large impose a large inertia matrix for the object to be manipulated. It is desirable therefore to minimise r_{n+1}. It is now necessary to find the location of the system centre of mass with respect to inertial coordinates where it remains invariant through an initialisation procedure [Longman et al 1987; Lindberg et al 1986].

$$p_{cm}^* = r_{c0} + \frac{1}{m_T} \sum_{i=1}^{n+1} \sum_{j=i}^{n+1} m_j r_{ci}$$

$$= r_{c0} + \frac{1}{m_T} \sum_{i=1}^{n+1} \sum_{j=i}^{n+1} m_j (R_i r_i + R_{i-1} s_{i-1})$$

$$= r_{c0} + \frac{1}{m_T} \sum_{i=1}^{n+1} \sum_{j=i}^{n+1} m_j R_{i-1} (r_{i-1} + s_{i-1}) + \frac{1}{m_T} \sum_{i=1}^{n+1} m_i R_i r_i$$

$$= r_{c0} + \frac{1}{m_T} \sum_{i=2}^{n+1} \sum_{j=i}^{n+1} m_j R_{i-1} l_{i-1} + \frac{1}{m_T} \left[\sum_{i=1}^{n+1} m_i l_0 + \sum_{i=1}^{n+1} m_i R_i r_i \right]$$

$$= r_{c0} + \frac{1}{m_T} \sum_{i=1}^{n+1} m_i s_0 + \frac{1}{m_T} \sum_{i=1}^{n} \sum_{j=i+1}^{n+1} m_j R_i l_i + \frac{1}{m_T} \sum_{i=1}^{n} m_i R_i r_i + \frac{m_{n+1}}{m_T} R_{n+1} r_{n+1}$$

$$= r_{c0} + \left(1 - \frac{m_0}{m_T} \right) s_0 + \frac{1}{m_T} \sum_{i=1}^{n} R_i \left(\sum_{j=i+1}^{n+1} m_j l_i + m_i r_i \right) + \frac{m_{n+1}}{m_T} R_{n+1} r_{n+1}$$

hence,

$$p_{cm}^* = r_{c0} + \left(1 - \frac{m_0}{m_T}\right)s_0 + \sum_{i=1}^{n} R_i L_i + \left(\frac{m_{n+1}}{m_T}\right)r_{n+1}$$

where

$$L_i = \frac{1}{m_T} \sum_{j=i+1}^{n+1} (m_j l_i + m_i r_i) \tag{8.5}$$

The robot/spacecraft system centre of mass is determined by:

(i)   the vector from the inertial origin to the spacecraft centre of mass r_{c0};
(ii)   the moment arm due to the fractional mass of the robot/payload to the total system centre of mass m_{17}/m_T (for a 6-DOF robot manipulator), and the fixed lever arm distance from the spacecraft centre of mass to the base of the robot arm s_0;
(iii) the kinematic structure of the manipulator and the link masses.

To find the system centre of mass with respect to inertial space, we assume arbitrarily that the local inertial reference frame initially coincides with the satellite bus centre of mass, i.e. $r_{c0} = 0$, since any point fixed in the interceptor body may be regarded as inertially fixed prior to any robotic manoeuvre. That point will remain fixed in inertial space as long as the task execution time is short compared with the satellite's orbital period and it may be regarded as an inertial reference for any particular payload of mass m_{n+1}. Although this choice of inertial coordinates is not strictly inertial since it moves along the orbital trajectory with the spacecraft at orbital velocity

$$w_{orb} = \sqrt{GM/R^3}$$

it is sufficiently inertial such that spin angular momentum and orbital angular momentum may be decoupled and considered separately, i.e. $L = L_{orbital} + L_{spin}$. The system centre of mass will then remain fixed in inertial space. Initialisation must be recomputed on payload acquisition or release to locate the new system centre of mass in inertial coordinates. This recomputation of the link parameters L_i involves simple recomputation of the total mass and fractional mass through adding or deleting the payload mass m_{n+1}. If the manipulator grasps an object attached to the target satellite and transfers it to the robotic spacecraft bus, the object mass is deleted from the target and added to the robotic spacecraft bus. In fact, for a zero payload or payload of known mass, the computation may be performed offline and stored in memory since m_i and l_i are all constants. Initialisation may be performed for any arbitrary robot configuration, e.g. stored ('tucked') position. Once p_{cm}^* is known, r_{c0} is defined with respect to the inertial origin for any further configuration:

$$r_{c0} = p_{cm}^* - \left(1 - \frac{m_0}{m_T}\right)s_0 - \sum_{i=1}^{n} R_i L_i - \frac{m_{n+1}}{m_T}r_{n+1}$$

though this is not specifically required for the kinematic formulation.

An equivalent form for the forward/orientation kinematics in terms of the 4×4 DH matrix representation is:

$$q = \begin{pmatrix} I_3 & \overset{*}{p}_{cm} \\ 0 & 1 \end{pmatrix} \begin{pmatrix} I_3 & m_0 s_0 / m_T \\ 0 & 1 \end{pmatrix} \begin{pmatrix} n & s & a & p(\lambda) \\ 0 & 0 & 0 & 1 \end{pmatrix}$$

$$\begin{pmatrix} -(m_{n+1} R_{n+1} / m_T) & -(m_{n+1} r_{n+1} / m_T) \\ 0 & 1 \end{pmatrix}$$

$$= \begin{pmatrix} n & s & a & \overset{*}{p} \\ 0 & 0 & 0 & 1 \end{pmatrix} \tag{8.6}$$

This form of the DH matrix allows the computation of the drive function D of the cartesian trajectory generator algorithm as outlined in Chapter 9. The advantage of this method over Longman et al's formulation is firstly that it is completely general and is not restricted by manipulator kinematics and that explicit calculation of r_{c0} as it varies with manipulator configuration is not required for the kinematics and terrestrial algorithms can be applied directly with little modification, i.e.

(i)   replacement of link parameters l_i by λ_i in the standard DH formulation; and
(ii)   addition of constants

$$\overset{*}{p}_{cm} + \frac{m_0}{m_T} s_0 - \frac{m_{n+1}}{m_T} r_{n+1}$$

to the DH formulation.

The inverse kinematics may be solved in the same way as that for an Earth-based manipulator to generate body-referenced joint angles which correspond to inertial cartesian coordinates by solving:

$$\overset{*}{p}_{arm} = \sum_{i=1}^{n-3} R_i \lambda_i = \overset{*}{p} - \overset{*}{p}_{cm} - \left(\frac{m_0}{m_T}\right) s_0 + \left(\frac{m_{n+1}}{m_T}\right) R_{n+1} r_{n+1} - \lambda_n a \tag{8.7}$$

to find positioning joint angles. The orienting joint angles of the wrist are identical to those for an Earth-based manipulator since attitude is stabilised.

Example: Here we consider a 6 DOF manipulator with an intersecting wrist and a payload (link 7) whereby all parameters are represented by assuming that their centres of mass lie on the links' longitudinal axes as bodies of revolution. The intersecting wrist means that links 4 and 5 are fictitious. Application of equation (8.2) gives:

$$\overset{*}{p} = r_{c0} + R_0 s_0 + R_1 l_1 + R_2 l_2 + R_3 l_3 + R_6 l_6 \qquad \text{since} \qquad l_4 = l_5 = 0$$

Now, application of equation (8.3):

$$m_T \overset{*}{p}_{cm} = m_0 \overset{*}{p}_{c0} + m_1 \overset{*}{p}_{c1} + m_2 \overset{*}{p}_{c2} + m_3 \overset{*}{p}_{c3}$$

$$+ m_6 \overset{*}{p}_{c6} + m_7 \overset{*}{p}_{c7} \qquad \text{since} \qquad m_4 = m_5 = 0$$

Hence,

$$
\begin{aligned}
p_{cm}^* &= \frac{1}{m_T}\left(m_0 p_{c0}^* + m_1 p_{c1}^* + m_2 p_{c02}^* + m_3 p_{c3}^* + m_6 p_{c6}^* + m_7 p_{c7}^*\right) \\
&= \frac{1}{m_T}\big[m_0 r_{c0} + m_1(r_{c0} + r_{c1}) + m_2(r_{c0} + r_{c1} + r_{c2}) \\
&\qquad\qquad + m_3(r_{c0} + r_{c1} + r_{c2} + r_{c3}) + \ldots + m_7(r_{c0} + \ldots + r_{c7})\big] \\
&= \frac{1}{m_T}\big[m_T r_{c0} + m_{17} r_{c1} + m_{27} r_{c2} + m_{37} r_{c3} + m_{67} r_{c6} + m_7 r_{c7}\big]
\end{aligned}
$$

where

$$
m_{ij} = \sum_{k=i}^{j} m_k
$$

Hence,

$$
r_{c0} = p_{cm}^* - \frac{1}{m_T}[m_{17} r_{c1} + m_{27} r_{c2} + m_{37} r_{c3} + m_{67} r_{c6} + m_7 r_{c7}]
$$

Substitute in equation (8.2):

$$
\begin{aligned}
p^* &= p_{cm}^* - \frac{1}{m_T}[m_{17} r_{c1} + m_{27} r_{c2} + m_{37} r_{c3} + m_{67} r_{c6} + m_7 r_{c7}] \\
&\quad + s_0 + R_1 l_1 + R_2 l_2 + R_3 l_3 + R_6 l_6
\end{aligned}
$$

Now,

$$
l_i = r_i + s_i \quad \text{and} \quad r_{ci} = R_{i-1} s_{i-1} + R_i r_i
$$

$$
\begin{aligned}
p^* &= p_{cm}^* + s_0 + R_1(r_1 + s_1) + R_2(r_2 + s_2) + R_3(r_3 + s_3) + R_6(r_6 + s_6) \\
&\quad - \frac{1}{m_T}\big[m_{17}(R_1 r_1 + R_0 s_0) + m_{27}(R_2 r_2 + R_1 s_1) + m_{37}(R_3 r_3 + R_2 s_2) \\
&\qquad\qquad + m_{67}(R_6 r_6 + R_3 s_3) + m_7(R_7 r_7 + R_6 s_6)\big]
\end{aligned}
$$

Multiply out and rearrange:

$$
\begin{aligned}
p^* &= p_{cm}^* + s_0\left(1 - \frac{m_{17}}{m_T}\right) + R_1\left[r_1\left(1 - \frac{m_{17}}{m_T}\right) + s_1\left(1 - \frac{m_{27}}{m_T}\right)\right] \\
&\quad + R_2\left[r_2\left(1 - \frac{m_{27}}{m_T}\right) + s_2\left(1 - \frac{m_{37}}{m_T}\right)\right] \\
&\quad + R_3\left[r_3\left(1 - \frac{m_{37}}{m_T}\right) + s_3\left(1 - \frac{m_{67}}{m_T}\right)\right] \\
&\quad + R_6\left[r_6\left(1 - \frac{m_{67}}{m_T}\right) + s_6\left(1 - \frac{m_7}{m_T}\right)\right] - \frac{m_7}{m_T} R_7 r_7
\end{aligned}
$$

$$= p_{cm}^* + \frac{m_0}{m_T} s_0 + R_1\left(\frac{m_0}{m_T} r_1 + \frac{m_{01}}{m_T} s_1\right) + R_2\left(\frac{m_{01}}{m_T} r_2 + \frac{m_{02}}{m_T} s_2\right)$$

$$+ R_2\left(\frac{m_{02}}{m_T} r_3 + \frac{m_{05}}{m_T} s_3\right) + R_6\left(\frac{m_{03}}{m_T} r_6 + \frac{m_{06}}{m_T} s_6\right) - \frac{m_7}{m_T} R_7 r_7$$

$$= p_{cm}^* + \frac{m_0}{m_T} s_0 + R_1\lambda_1 + R_2\lambda_2 + R_3\lambda_3 + R_6\lambda_6 - \frac{m_7}{m_T} R_7 r_7$$

where

$$\lambda_1 = \frac{1}{m_T}(m_{01}l_1 - m_1 r_1)$$

$$\lambda_2 = \frac{1}{m_T}(m_{02}l_2 - m_2 r_2)$$

$$\lambda_3 = \frac{1}{m_T}(m_{03}l_3 - m_3 r_3)$$

$$\lambda_6 = \frac{1}{m_T}(m_{06}l_6 - m_6 r_6)$$

These 'kinetic variables' replace the terrestrial algorithm kinematic variables by straight substitution and the other constants added. This is now applied to the the PUMA 560/600 configuration with zero elbow offset the characteristic equation for which is given by:

$$\begin{pmatrix} p_x^* \\ p_y^* \\ p_z^* \end{pmatrix} = \begin{pmatrix} r_{c0x} \\ r_{c0y} \\ r_{c0z} \end{pmatrix} + \begin{pmatrix} s_{0x} \\ s_{0y} \\ s_{0z} \end{pmatrix} + \begin{pmatrix} p_x \\ p_y \\ p_z \end{pmatrix}$$

where

$$\begin{pmatrix} p_x \\ p_y \\ p_z \end{pmatrix} = \begin{pmatrix} -s_1 d_2 + c_1(c_2 a_2 + s_{23} d_4) + a_x d_6 \\ c_1 d_2 + s_1(c_2 a_a + s_{23} d_4) + a_y d_6 \\ -s_2 a_2 + c_{23} d_4 + a_z d_6 \end{pmatrix}$$

By multiplying through, this can be reduced to:

$$\begin{pmatrix} p_x^* \\ p_y^* \\ p_z^* \end{pmatrix} = \begin{pmatrix} p_{cmx}^* \\ p_{cmy}^* \\ p_{cmz}^* \end{pmatrix} + \frac{m_0}{m_T}\begin{pmatrix} s_{0x} \\ s_{0y} \\ s_{0z} \end{pmatrix} + \begin{pmatrix} -s_1\delta_1 + c_1(c_2\alpha_2 + s_{23}\delta_4) + a_x\delta_6 \\ c_1\delta_2 + s_1(c_2\alpha_a + s_{23}\delta_4) + a_y\delta_6 \\ -s_2\alpha_2 + c_{23}\delta_4 + a_z\delta_6 \end{pmatrix} - \frac{m_7}{m_T}\begin{pmatrix} x_7 \\ y_7 \\ z_7 \end{pmatrix}$$

where

$$\delta_2 = \left(\frac{m_{01}d_2 - m_1r_1}{m_{07}}\right); \quad \alpha_2 = \left(\frac{m_{02}a_2 - m_2r_2}{m_{07}}\right);$$

$$\delta_4 = \left(\frac{m_{03}d_4 - m_3r_3}{m_{07}}\right); \quad \delta_6 = \left(\frac{m_{06}d_6 - m_6r_6}{m_{07}}\right)$$

This defines the forward position kinematics of a space-based PUMA-type manipulator
which describe the inertial position of the end effector in terms of their joint angles. The
orientation of the end effector is given by the terrestrial 3×3 rotation component of the
DH matrix (as the attitude is maintained as static by feedforward compensation):

$$(nsa) = \sum_{i=0}^{n} R_i = \sum_{i=1}^{n} R_i \quad \text{since} \quad R_0 = I_3$$

The initialisation process allows the computation of $\overset{*}{p}_{cm}$:

$$\overset{*}{p}_{cm} = r_{c0} + \frac{1}{m_T}(m_{17}r_{c1} + m_{27}r_{c2} + m_{37}r_{c3} + m_{67}r_{c6} + m_7r_{c7})$$

$$= r_{c0} + \frac{1}{m_T}\left[m_{17}(R_1r_1 + R_0s_0) + m_{27}(R_2r_2 + R_1s_1) + m_{37}(R_3r_3 + R_2s_2)\right.$$

$$\left. + m_{67}(R_6r_6 + R_3s_3) + m_7(R_7r_7 + R_6s_6)\right]$$

$$= r_{c0} + \frac{1}{m_T}\left[m_{17}R_0s_0 + R_1(m_{27}l_1 + m_1r_1) + R_2(m_{37}l_2 + m_2r_2)\right.$$

$$\left. + R_3(m_{67}l_3 + m_3r_3) + R_6(m_6l_6 + m_6r_6) + m_7R_7r_7\right]$$

$$= \left(1 - \frac{m_0}{m_T}\right)s_0 + R_1L_1 + R_2L_2 + R_3L_3 + R_6L_6 + \frac{m_7}{m_T}R_7r_7 \quad \text{with} \quad r_{c0} = 0$$

where

$$L_1 = \left(\frac{m_{27}l_1 + m_1r_1}{m_T}\right); \quad L_2 = \left(\frac{m_{37}l_2 + m_2r_2}{m_T}\right)$$

$$L_3 = \left(\frac{m_{67}l_3 + m_3r_3}{m_T}\right); \quad L_6 = \left(\frac{m_7l_6 + m_6r_6}{m_T}\right)$$

For the PUMA 560/600 configuration:

$$\begin{pmatrix} \overset{*}{p}_{cmx} \\ \overset{*}{p}_{cmy} \\ \overset{*}{p}_{cmz} \end{pmatrix} = \left(1 - \frac{m_0}{m_T}\right)\begin{pmatrix} s_{0x} \\ s_{0y} \\ s_{0z} \end{pmatrix} + \begin{pmatrix} -s_1D_2 + c_1(c_2A_2 + s_{23}D_4) + a_xD_6 \\ c_1D_2 + s_1(c_2A_2 + s_{23}D_4) + a_yD_6 \\ -s_2A_2 + c_{23}D_4 + a_zD_6 \end{pmatrix}$$

$$+ \frac{m_7}{m_T}\begin{pmatrix} x_7 \\ y_7 \\ z_7 \end{pmatrix}$$

where

$$D_2 = \left(\frac{m_{27}d_2 + m_1 r_1}{m_T}\right); \quad A_2 = \left(\frac{m_{37}d_2 + m_2 r_2}{m_T}\right);$$

$$D_4 = \left(\frac{m_{67}d_4 + m_3 r_3}{m_T}\right); \quad D_6 = \left(\frac{m_7 d_6 + m_6 r_6}{m_T}\right)$$

To find the inverse solution to the joint angles, a slightly modified version of the terrestrial inverse algorithm can be used. The inverse kinematics solution must generate body-referenced joint angles corresponding to the inertial position coordinates:

$$\sum_{i=1}^{n} R_i \lambda_i = p^* - p_{cm}^* - \frac{m_0}{m_T} s_0 + \frac{m_7}{m_T} R_7 r_7$$

$$\rightarrow p_{arm}^* = R_1 \lambda_1 + R_2 \lambda_2 + R_3 \lambda_3$$

$$= p^* - p_{cm}^* - \frac{m_0}{m_T} s_0 + \frac{m_7}{m_T} r_7 - R_6 \lambda_6 \quad \text{for} \quad R_7 = I_3.$$

$$\begin{pmatrix} p_{ax}^* \\ p_{ay}^* \\ p_{az}^* \end{pmatrix} = \begin{pmatrix} -s_1 \delta_2 + c_1(c_2\alpha_2 + s_{23}\delta_4) \\ c_1\delta_2 + s_1(c_2\alpha_2 + s_{23}\delta_4) \\ -s_2\alpha_2 + c_{23}\delta_4 \end{pmatrix}$$

$$= \begin{pmatrix} p_x^* \\ p_y^* \\ p_z^* \end{pmatrix} - \begin{pmatrix} p_{cmx}^* \\ p_{cmy}^* \\ p_{cmz}^* \end{pmatrix} - \frac{m_0}{m_T}\begin{pmatrix} s_{0x} \\ s_{0y} \\ s_{0z} \end{pmatrix} + \frac{m_7}{m_T}\begin{pmatrix} x_7 \\ y_7 \\ y_z \end{pmatrix} - \delta_6\begin{pmatrix} a_x \\ a_y \\ a_z \end{pmatrix}$$

The first three joint angles determine the position in space of the arm and the final three joint angles for the orientation of the arm are calculated as for the terrestrial case. The inverse calculations of Chapter 6 with the appropriate substitutions to equations (6.6–6.8) may be found to be:

$$\theta_1 = \tan^{-1}\left(\frac{p_{ay}^*}{p_{ax}^*}\right) - \tan^{-1}\left(\frac{\delta_2}{\sqrt{(p_{ax}^*)^2 + (p_{ay}^*)^2 - \delta_2^2}}\right)$$

$$\theta_2 = \tan^{-1}\left(\frac{A}{p_{az}^*}\right) - \tan^{-1}\left(\frac{B}{-\sqrt{A^2 + (p_{az}^*)^2 - B^2}}\right)$$

where $\quad A = c_1 p_{ax}^* + s_1 p_{ay}^* \quad$ and $\quad B = \dfrac{A^2 + (p_{az}^*)^2 + \alpha_2^2 - \delta_4^2}{2\alpha_2}$

$$\theta_3 = \tan^{-1}\left(-\frac{A - \alpha_2 c_2}{p_{az}^* + \alpha_2 s_2}\right) - \theta_2 \tag{8.8}$$

The inverse kinematics solution for the positioning joint angles have precisely the same form as for an Earth-based manipulator. The orienting joint angles are identical to the fixed base case since the attitude of the spacecraft is maintained as constant:

$$\theta_4 = \tan^{-1}\left(\frac{-s_1 a_x + c_1 a_y}{c_{23}(c_1 a_x + s_1 a_y) - s_{23} a_z}\right)$$

$$\theta_5 = \tan^{-1}\left(\frac{\sqrt{(c_1 c_{23} a_x + s_1 c_{23} a_y - s_{23} a_z)^2 + (-s_1 a_x + c_1 a_y)^2}}{s_{23}(c_1 a_x + s_1 a_y) + c_{23} a_z}\right)$$

$$\theta_6 = \tan^{-1}\left(-\frac{s_{23}(c_1 s_x + s_1 s_y) + c_{23} s_z}{s_{23}(c_1 n_x + s_1 n_y) + c_{23} n_z}\right) \tag{8.9}$$

Note that these equations are only marginally more complex than those for a terrestrial manipulator. Theorem 1 is the most fundamental of the theorems presented, the others being derivable from it by direct differentiation. Different positioning joint angles required to position the robotic hand in base coordinates (for a terrestrial manipulator) and in inertial coordinates (for a space-based manipulator). It is important to note that a cartesian trajectory defined in base coordinates will differ from the same cartesian trajectory defined in inertial space. Hence, any attempt to position a robotic manipulator in space without regard to the dynamics of the system will result in incorrect trajectories.

8.1.2 Resolution of inertial velocity, acceleration and static force

Theorem 2: For a freeflying robotic manipulator employing attitude control, the freeflyer Jacobian matrix is given by [Ellery 1994, 1996]:

$$\bar{J} = \sum_{i=1}^{n} \sum_{k=1}^{i} \frac{\partial R_i}{\partial \theta_k} \lambda_i \tag{8.10}$$

This freeflyer Jacobian, derived from the resolved rate kinematic control formulation, may be applied in the usual manner to resolved acceleration control and resolved force control problem since they derive directly from the resolved motion rate control problem.

Proof: Differentiate (8.2) with respect to time:

$$v^* = \dot{p}^* = \dot{r}_{c0} + \dot{R}_0 s_0 + \sum_{i=1}^{n} \dot{R}_i l_i = \dot{r}_{c0} + \sum_{i=1}^{n} \dot{R}_i l_i \quad \text{where} \quad \dot{R}_0 = 0 \tag{8.11}$$

Now,

$$\dot{r}_{c0} = -\frac{1}{m_T}\sum_{i=1}^{n+1}\sum_{j=1}^{n+1} m_j \dot{r}_{ci} \qquad \text{since} \qquad \dot{p}_{cm}^* = 0 \tag{8.12}$$

This represents the conservation of linear momentum to eliminate the platform velocity.

$$\dot{r}_{ci} = \dot{R}_i r_i + \dot{R}_{i-1} s_{i-1}$$

$$\dot{r}_{c0} = -\frac{1}{m_T}\sum_{i=1}^{n+1}\sum_{j=i}^{n+1} m_j (\dot{R}_i r_i + \dot{R}_{i-1} s_{i-1})$$

Substitute into (8.9):

$$v^* = \dot{R}_0 s_0 + \sum_{i=1}^{n} \dot{R}_i (r_i + s_i) - \frac{1}{m_T}\sum_{i=1}^{n+1}\sum_{j=i}^{n+1} m_j (\dot{R}_i r_i + \dot{R}_{i-1} s_{i-1})$$

Now,

$$\dot{R}_i^0 = \sum_{k=1}^{i} \frac{\partial R_i^0}{\partial \theta_k} \theta_k$$

and in general

$$\dot{R}_0 = \dot{\alpha}\frac{\partial R_0}{\partial \alpha} + \dot{\beta}\frac{\partial R_0}{\partial \beta} + \dot{\gamma}\frac{\partial R_0}{\partial \gamma} = \dot{\theta}_0 \frac{\partial R_0}{\partial \theta_0}$$

Substitute:

$$v^* = \dot{\theta}_0 \frac{\partial R_0}{\partial \theta_0} s_0 + \left\{ \sum_{i=1}^{n}\sum_{k=0}^{i} \frac{\partial R_i}{\partial \theta_k} l_i - \frac{1}{m_T}\sum_{i=1}^{n+1}\sum_{j=i}^{n+1} m_j \right.$$

$$\left. \left[\sum_{k=1}^{i} \frac{\partial R_i}{\partial \theta_k} r_i + \sum_{k=1}^{i-1} \frac{\partial R_{i-1}}{\partial \theta_k} s_{i-1} \right] \right\} \dot{\theta}_k$$

$$= \dot{\theta}_0 \frac{\partial R_0}{\partial \theta_0} s_0 + \sum_{i=1}^{n} \frac{\partial R_i}{\partial \theta_0} l_i - \frac{1}{m_T}\sum_{i=1}^{n+1} m_j \frac{\partial R_0}{\partial \theta_0} s_0$$

$$+ \left\{ \sum_{i=1}^{n}\sum_{k=1}^{i} \frac{\partial R_i}{\partial \theta_k} l_i - \frac{1}{m_T}\sum_{i=1}^{n+1}\sum_{j=1}^{n+1} m_j \left[\sum_{k=1}^{i} \frac{\partial R_i}{\partial \theta_k} r_i + \sum_{k=1}^{i-1} \frac{\partial R_{i-1}}{\partial \theta_k} s_{i-1} \right] \right\} \dot{\theta}_k$$

i.e.

$$\begin{pmatrix} v^* \\ w^* \end{pmatrix} = \bar{J}_0 \dot{\theta}_0 + \bar{J}_m \dot{\theta}_m$$

where

$$\bar{J}_0 = \frac{\partial R_0}{\partial \theta_0} s_0 + \sum_{i=1}^{n} \frac{\partial R_i}{\partial \theta_0} l_i - \frac{1}{m_T} \sum_{j=1}^{n+1} m_j \frac{\partial R_0}{\partial \theta_0} s_0 \quad \text{for} \quad \dot{\theta}_0 = (\dot{\alpha}\dot{\beta}\dot{\gamma})^T$$

$$\bar{J}_m = \sum_{i=1}^{n} \sum_{k=1}^{i} \frac{\partial R_i}{\partial \theta_k}(r_i + s_i) - \frac{1}{m_T} \sum_{i=1}^{n+1} \sum_{j=i}^{n+1} m_j \left[\sum_{k=1}^{i} \frac{\partial R_i}{\partial \theta_k} r_i + \sum_{k=1}^{i-1} \frac{\partial R_{i-1}}{\partial \theta_k} s_{i-1} \right]$$

for $\dot{\theta}_m = (\dot{\theta}_1 ... \dot{\theta}_n)^T$

These are the freeflyer Jacobians of the spacecraft and manipulator respectively such that

$$v^* = (\bar{J}_0 \ \bar{J}_m) \begin{pmatrix} \dot{\theta}_0 \\ \dot{\theta}_m \end{pmatrix}$$

Now, $(\bar{J}_0 \ \bar{J}_m)$ is not square and so non-invertible. However, it is possible to obtain an invertible square 'generalised' Jacobian matrix.

Conservation of linear and angular momentum state:

$$\sum_{i=0}^{n+1} m_i \dot{r}_i = 0$$

$$\sum_{i=0}^{n+1} I_i w_i + m_i r_i \times \dot{r}_i = \sum_{i=0}^{n+1} I_i w_i = 0$$

For the spacecraft and manipulator:

$$I_0 \dot{\theta}_0 + I_m \dot{\theta}_m = 0$$

where I_0 = inertia matrix of the spacecraft and I_m = mass matrix of the manipulator

$$\rightarrow \dot{\theta}_0 = -I_0^{-1} I_m \dot{\theta}_m$$

Now,

$$v^* = \bar{J}_0 \dot{\theta}_0 + \bar{J}_m \dot{\theta}_m = \bar{J}_0 \left(-I_0^{-1} I_m \dot{\theta}_m \right) + \bar{J}_m \dot{\theta}_m = \left(\bar{J}_m - \bar{J}_0 I_0^{-1} I_m \right) \dot{\theta}_m$$

This represents the generalised Jacobian J^* such that $\dot{\theta} = (J^*)^{-1} \dot{x}^*$ where $J^* = (\bar{J}_m - \bar{J}_0 I_0^{-1} I_m)$ [Umetani & Yoshida 1989, Nakaruma & Mukherjee 1989]. For an attitude-controlled spacecraft, the freeflyer manipulator Jacobian is its generalised Jacobian $J^* = (\bar{J}_m - \bar{J}_0 I_0^{-1} I_m) = \bar{J}_m$, specifying the relation between joint velocities and end effector velocities since $J_0 = 0$, $w_0 = 0$ and $\dot{R}_0 = 0$. It may be simplified to:

$$\bar{J} = \sum_{i=1}^{n}\sum_{k=1}^{i}\frac{\partial R_i}{\partial \theta_k}(r_i + s_i) - \frac{1}{m_T}\sum_{i=1}^{n+1}\sum_{j=i}^{n+1}m_j\left(\sum_{k=1}^{i-1}\frac{\partial R_{i-1}}{\partial \theta_k}r_{i-1} + \sum_{k=1}^{i-1}\frac{\partial R_{i-1}}{\partial \theta_k}s_{i-1}\right)$$

$$-\frac{1}{m_T}\sum_{i=1}^{n+1}\sum_{k=1}^{i}m_i\frac{\partial R_i}{\partial \theta_k}r_i$$

$$= \sum_{i=1}^{n}\sum_{k=1}^{i}\frac{\partial R_i}{\partial \theta_k}(r_i + s_i) - \frac{1}{m_T}\sum_{i=2}^{n+1}\sum_{j=i}^{n+1}m_j\left(\sum_{k=1}^{i-1}\frac{\partial R_{i-1}}{\partial \theta_k}r_{i-1} + \sum_{k=1}^{i-1}\frac{\partial R_{i-1}}{\partial \theta_k}s_{i-1}\right)$$

$$-\frac{1}{m_T}\sum_{i=1}^{n+1}\sum_{k=1}^{i}m_i\frac{\partial R_i}{\partial \theta_k}r_i - \frac{1}{m_T}\sum_{i=1}^{n+1}\sum_{k=1}^{i}m_i\frac{\partial R_0}{\partial \theta_k}(s_0 + r_0)\quad \text{where}\quad R_0 = \text{const.}$$

$$= \sum_{i=1}^{n}\sum_{k=1}^{i}\frac{\partial R_i}{\partial \theta_k}(r_i + s_i) - \frac{1}{m_T}\sum_{i=1}^{n+1}\sum_{j=i+1}^{n+1}m_j\left(\sum_{k=1}^{i}\frac{\partial R_i}{\partial \theta_k}(r_i + s_i)\right)$$

$$-\frac{1}{m_T}\sum_{i=1}^{n+1}\sum_{k=1}^{i}m_i\frac{\partial R_i}{\partial \theta_k}r_i$$

$$= \sum_{i=1}^{n}\sum_{k=1}^{i}\frac{\partial R_i}{\partial \theta_k}(r_i + s_i) - \frac{1}{m_T}\sum_{i=1}^{n}\sum_{j=i+1}^{n+1}m_j\left(\sum_{k=1}^{i}\frac{\partial R_i}{\partial \theta_k}(r_i + s_i)\right)$$

$$-\frac{1}{m_T}\sum_{i=1}^{n+1}m_i\frac{\partial R_i}{\partial \theta_k}r_i - \frac{m_{n+1}}{m_T}\sum_{i=1}^{n+1}\frac{\partial R_{n+1}}{\partial \theta_k}r_{n+1}$$

Now R_{n+1} is constant, so $\partial R_{n+1}/\partial \theta_k = 0$:

$$\bar{J} = \sum_{i=1}^{n}\sum_{k=1}^{i}\frac{\partial R_i}{\partial \theta_k}\left[1 - \frac{1}{m_T}\sum_{j=i+1}^{n+1}m_j l_i\right] - \frac{1}{m_T}\sum_{i=1}^{n}\sum_{k=1}^{i}m_i\frac{\partial R_i}{\partial \theta_k}r_i$$

$$= \sum_{i=1}^{n}\sum_{k=1}^{i}\frac{\partial R_i}{\partial \theta_k}\left(\frac{1}{m_T}\sum_{j=0}^{i}m_j l_i - m_i r_i\right)$$

i.e.

$$\bar{J} = \sum_{i=1}^{n}\sum_{k=1}^{i}\frac{\partial R_i}{\partial \theta_k}\lambda_i \quad \text{where} \quad \lambda_i = \frac{1}{m_T}\sum_{j=0}^{i}(m_j l_i - m_i r_i) \quad (8.10) \quad \text{QED}$$

The freeflyer Jacobian for a robotic spacecraft employing attitude control has the same form as that of an Earth-based manipulator with the replacement of kinematic constants l_i for dynamic constants λ_i. This is consistent with differentiating (8.1) directly:

$$v^* = \dot{p}^* = \dot{p}_{cm}^* + \frac{m_0}{m_T}\dot{s}_0 + \sum_{i=1}^{n}\dot{R}_i\lambda_i - \frac{m_{n+1}}{m_T}\dot{R}_{n+1}r_{n+1} = \sum_{i=1}^{n}\sum_{k=1}^{i}\frac{\partial R_i}{\partial \theta_k}\lambda_i\dot{\theta}_i$$

since $\dot{p}_{cm}^{*} = \dot{s}_0 = \dot{R}_{n+1} = 0$. Note how the freeflyer Jacobian requires only the replacement of kinematic link parameters with kinematic–dynamic parameters and the positional constants differentiate to zero. Hence, the Jacobian may be inverted normally as with terrestrial manipulators. This is far simpler than the immensely complex, fixed-attitude, restricted Jacobian given by: $\dot{\theta}_m = [J_m(I - I_m^{+} I_m)]^{+} v^{*}$ based on the Moore–Penrose pseudoinverse which used the inherent redundancy of the robotic manipulator spacecraft to maintain attitude [Nenchev et al 1992].

Similar arguments for deriving the freeflyer Jacobian above apply to other Jacobian formulations since the Jacobian is unique [Whitney 1969, 1972]. The freeflyer Jacobian is now derived for a PUMA 560/600 manipulator adopting the cross product formulation of the Jacobian [Whitney 1972]. The derivation begins with the dynamics formulation [Ellery 1994, 1996]:

$$v = v_0 + \sum_{i=1}^{n}(z_{i-1} \times l_i)\dot{\theta}_i$$

$$w = w_0 + \sum_{i=1}^{n}z_{i-1}\dot{\theta}_i = \sum_{i=1}^{n}z_{i-1}\dot{\theta}_i$$

since $w_0 = 0$ for an attitude controlled base.

Now,

$$\dot{r}_{c0} = -\frac{1}{m_T}\sum_{i=1}^{n+1}m_i\dot{p}_{ci}^{*} \quad \text{where} \quad \dot{p}_{ci}^{*} = v_{ci} = v_{i-1} + w_i \times r_i = \sum_{k=1}^{i}v_{k-1} + w_i \times r_i$$

Hence,

$$\dot{r}_{c0} = v_0 = -\frac{1}{m_T}\sum_{i=1}^{n+1}m_i\left[\sum_{k=1}^{i}v_{k-1} + w_i \times r_i\right]$$

$$= -\frac{1}{m_T}\sum_{i-1}^{n+1}\left[\sum_{i-j}^{n+1}m_j v_{i-1} - \sum_{i=1}^{n+1}m_i(w_i \times r_i)\right]$$

Now,

$$v_i = v_{i-1} + w_i \times l_i = \sum_{k=1}^{i}w_k \times l_k \quad \text{where} \quad w_i = \sum_{k=1}^{i}z_{k-1}\dot{\theta}_k$$

$$v_i = \sum_{k=1}^{n}(z_{k-1} \times l_k)\dot{\theta}_k \rightarrow v_n = \sum_{i=1}^{n}(z_{i-1} \times l_i)\dot{\theta}_i$$

Hence,

$$v_0 = -\frac{1}{m_T}\sum_{i=1}^{n+1}\left[\sum_{i=j}^{n+1}m_j\sum_{i=1}^{n-1}(z_{i-1}\times l_i)\dot\theta_i + \sum_{i=1}^{n+1}m_i\left(\sum_{i=1}^{n+1}z_{i-1}\times r_i\right)\dot\theta_i\right]$$

$$= -\frac{1}{m_T}\sum_{i=1}^{n+1}\left[\sum_{j=i+1}^{n+1}m_j\sum_{i=1}^{n}(z_{i-1}\times l_i)\dot\theta_i + m_i(z_{i-1}\times r_i)\dot\theta_i\right]$$

So,

$$v_n = \sum_{i=1}^{n}(z_{i-1}\times l_i)\dot\theta_i - \frac{1}{m_T}\left[\sum_{j=i+1}^{n+1}m_j\sum_{i=1}^{n}(z_{i-1}\times l_i)\dot\theta_i + \sum_{i=1}^{n+1}m_i(z_{i-1}\times r_i)\dot\theta_i\right]$$

$$= \sum_{i=1}^{n}(z_{i-1}\times l_i)\dot\theta_i - \frac{1}{m_T}\left[\sum_{j=i+1}^{n+1}m_j\sum_{i=1}^{n}(z_{i-1}\times l_i)\dot\theta_i + \sum_{i=1}^{n+1}m_i(z_{i-1}\times r_i)\dot\theta_i\right]$$

where

$$\frac{m_{n+1}}{m_T}(z_n\times r_{n+1})\dot\theta_{n+1} = 0$$

$$\to v_n = \sum_{i=1}^{n}\left(1-\frac{1}{m_T}\sum_{j=i+1}^{n+1}m_j\right)(z_{i-1}\times l_i)\dot\theta_i - \frac{1}{m_T}\sum_{i=1}^{n}m_i(z_{i-1}\times r_i)\dot\theta_i$$

$$= \sum_{i=1}^{n}(z_{i-1}\times \lambda_i)\dot\theta_i \quad \text{where} \quad \lambda_i = \frac{1}{m_T}\left(\sum_{j=0}^{i}m_j l_i - m_i r_i\right)$$

Hence, the cross product Jacobian is consistent with the concept of replacing kinematic link parameters with kinematic–dynamic link parameters.

For the PUMA 560/600 model:

$$\bar J = \begin{pmatrix} \bar J_{11} & \bar J_{21} & \bar J_{31} & \bar J_{41} & \bar J_{51} & \bar J_{61} \\ \bar J_{12} & \bar J_{22} & \bar J_{32} & \bar J_{42} & \bar J_{52} & \bar J_{62} \\ \bar J_{13} & \bar J_{23} & \bar J_{33} & \bar J_{43} & \bar J_{53} & \bar J_{63} \\ \bar J_{14} & \bar J_{24} & \bar J_{34} & \bar J_{44} & \bar J_{54} & \bar J_{64} \\ \bar J_{15} & \bar J_{25} & \bar J_{35} & \bar J_{45} & \bar J_{55} & \bar J_{65} \\ \bar J_{16} & \bar J_{26} & \bar J_{36} & \bar J_{46} & \bar J_{56} & \bar J_{66} \end{pmatrix}$$

where

$$\bar J_{11} = -s_1[\delta_6(c_{23}c_4 s_5 + s_{23}c_5) + \alpha_2 c_2] - c_1(s_4 s_5 \delta_6 + \delta_2)$$

$$\bar J_{12} = c_1[\delta_6(c_{23}c_4 c_5 + s_{23}c_5) + \alpha_2 c_2] - s_1(s_4 s_5 \delta_6 + \delta_2)$$

$$\bar J_{13} = 0; \quad \bar J_{14} = 0; \quad \bar J_{15} = 0; \quad \bar J_{16} = 1$$

$$\bar{J}_{21} = c_1(s_4 s_5 \delta_6 + \delta_2)$$

$$\bar{J}_{22} = s_1(s_4 s_5 \delta_6 + \delta_2)$$

$$\bar{J}_{23} = -s_1\left[\delta_6(s_{23}c_4c_5 - c_{23}c_5) + \alpha_2 s_2\right] - c_1\left[\delta_6(c_{23}c_4s_5 + s_{23}c_5) + \alpha_2 c_2\right]$$

$$\bar{J}_{24} = -s_1; \quad \bar{J}_{25} = c_1; \quad \bar{J}_{26} = 0$$

$$\bar{J}_{31} = c_1 s_4 s_5 \delta_6$$

$$\bar{J}_{32} = s_1 s_4 s_5 \delta_6$$

$$\bar{J}_{33} = -s_1\left[\delta_6(s_3 c_4 c_5 - c_3 c_5)\right] - c_1\left[\delta_6(c_3 c_4 s_5 + s_3 c_5)\right]$$

$$\bar{J}_{34} = -s_1; \quad \bar{J}_{35} = c_1; \quad \bar{J}_{36} = 0$$

$$\bar{J}_{41} = \delta_6(s_1 s_{23} c_5 - c_{23} s_4 s_5)$$

$$\bar{J}_{42} = \delta_6(c_{23} c_4 s_5 - c_1 s_{23} c_5)$$

$$\bar{J}_{43} = \delta_6 s_{23}(c_1 s_4 - s_1 c_4 s_5)$$

$$\bar{J}_{44} = c_1 s_{23}; \quad \bar{J}_{45} = s_1 s_{23}; \quad \bar{J}_{46} = c_{23}$$

$$\bar{J}_{51} = s_{23} s_4 c_5 \delta_6$$

$$\bar{J}_{52} = s_{23} s_4 s_5 \delta_6$$

$$\bar{J}_{53} = c_5 \delta_6(c_1 c_{23} s_4 + s_1 c_4) + s_5 \delta_6(s_1 c_{23} c_4 - c_1 c_4)$$

$$\bar{J}_{54} = -(c_1 c_{23} s_4 + s_1 c_4)$$

$$\bar{J}_{55} = c_1 c_4 - s_1 c_{23} c_4$$

$$\bar{J}_{56} = s_{23} s_4$$

$$\bar{J}_{61} = \delta_6\left[s_1(c_{23} c_4 s_5 + c_5 s_{23}) + c_1 s_4 s_5\right]$$

$$\bar{J}_{62} = \delta_6\left[c_1(c_{23} c_4 s_5 + s_{23} c_5) - s_1 s_4 s_5\right]$$

$$\bar{J}_{63} = 0$$

$$\bar{J}_{64} = c_1(c_{23} c_4 s_5 + s_{23} c_5) - s_1 s_4 s_5$$

$$\bar{J}_{65} = s_1(c_{23} c_4 s_5 + s_{23} c_5) + c_1 s_4 s_5$$

$$\bar{J}_{66} = -s_{23} c_4 s_5 + c_{23} c_5 \tag{8.13}$$

Hence resolved velocity coordinates are given by:

$$\begin{pmatrix} v^* \\ w \end{pmatrix} = \bar{J}\dot{\theta}_i \quad \text{with} \quad \dot{\theta}_i = \bar{J}^{-1}\begin{pmatrix} v^* \\ w \end{pmatrix}$$

Resolved acceleration control follows directly by differentiation [Ellery 1994, 1996]:

$$\begin{pmatrix} \dot{v}^* \\ \dot{w} \end{pmatrix} = \bar{J}\ddot{\theta}_i + \dot{\bar{J}}\dot{\theta}_i \quad \text{with} \quad \ddot{\theta}_i = \bar{J}^{-1}\begin{pmatrix} \dot{v}^* \\ \dot{w} \end{pmatrix} - \dot{\bar{J}}\dot{\theta}_i^*$$

Resolved force control also follows from resolved rate by virtual work arguments:

$$\tau = \bar{J}^T f_{ext}$$

In general, the forward kinematic solutions for a freeflyer employing dedicated attitude control have the same form as for Earth-based manipulators so that the same algorithms may be used for both forward and inverse kinematic solutions with only minor modifications. This is essentially a restatement of Papadopoulos & Dubowsky's (1990, 1991a,b) findings for the general case of freefloating manipulators that the dynamics formulation of any terrestrial manipulators have the same form as those of space manipulators and so terrestrial robotic control algorithms such as the computed torque control law [Konigstein et al 1989], are applicable to space robotics, though their formulation was much more complex due to the constraints. The differential kinematics of freefloating systems have the same structure as those for a fixed base manipulator. By applying attitude control, however, the position/orientation kinematics themselves also have this property. All these new 'lumped' dynamic parameters may be precalculated offline and stored in memory with the exception of the payload parameters. The change of payload involves negligible online calculation.

8.2 SPACE FREEFLYER DYNAMICS FEEDFORWARD COMPENSATION

For an attitude-controlled platform with a space manipulator, the dynamic formulation includes a moving platform with a finite translational velocity

$$v_0 = -\frac{1}{m_T}\sum_{j=1}^{n}\sum_{i=1}^{n}m_j v_{ci}$$

Walker and Wee (1991) used a Lagragian formulation for their characterisation of space robotic problems but suggested that a recursive formulation would be more efficient. The Newton-Euler recursive formulation is better suited to the problem when employing attitude control to calculate explicitly the reaction moments applied to the spacecraft mounting by the manipulator movements—such constraint forces are eliminated in the Lagrangian approach. Nagashima & Nakaruma (1992) derived a means for calculating $(v_0 w_0)^T$ explicitly using the Newton–Euler recursive equations which reduces to the formulation above when $w_0 = 0$. Longman et al (1987) and Lindberg et al (1986) proved a theorem specifying the feedforward component to the spacecraft attitude controller to compensate for the manipulator reaction forces on the spacecraft. When attitude control is employed, $w_0 = \dot{w}_0 = 0$, but v_0 and $\dot{v}_0$ are finite. This requires that the robot dynamics are calculated with respect to base coordinates. This is more efficient than calculating the robot dynamics with reference to link coordinates and then transforming them to inertial coordinates as has been proposed [Nagashima & Nakaruma 1992]. Outlined here is a slightly altered version of the theorem and its proof.

Theorem 3: The feedforward signal from the robot controller to the spacecraft attitude control system enables the attitude control system to compensate for the applied moments to the spacecraft such that the total moments about the satellite centre of mass sum to zero. The feedforward dynamics component with respect to local inertial coordinates is given by [Ellery 1994, 1996]:

$$N_r = N_T + (p_{cm}^* - r_{c0} - s_0) \times F_T$$

where

$$F_T = \sum_{i=1}^{n+1} F_{ci} = \sum_{i=1}^{n+1} m_i \ddot{v}_{ci} \quad \text{and} \quad N_T = \sum_{i=1}^{n+1} N_{ci} = \sum_{i=1}^{n+1} I_i \dot{w}_i + w_i \times I_i w_i \qquad (8.15)$$

Proof: Moment on spacecraft due to manipulator movements about the coupling point at the manipulator base is given by:

$$N_T = \sum_{i=1}^{n+1} \int p_{ci} \times \ddot{p}_{ci}\, dm$$

Moment about coupling point referred to inertial coordinates: $p_{ci} = p_{ci}^* - r_{c0} - s_0$, i.e.

$$N_0 = \sum_{i=1}^{n+1} \int p_{ci} \times \left(\ddot{p}_{ci}^* - \ddot{r}_{c0} - \ddot{s}_0 \right) dm$$

Now, $\ddot{s}_0 = 0$ since s_0 is invariant:

$$N_0 = \sum_{i=1}^{n+1} \int p_{ci} \times \ddot{p}_{ci}^*\, dm - \sum_{i=1}^{n+1} \int p_{ci} \times \ddot{r}_{c0}\, dm$$

For any inertially fixed frame of reference:

$$\int p_{ci} \times \ddot{p}_{ci}^*\, dm = p_{ci} \times dF \quad \text{i.e.} \quad N_0 = \sum_{i=1}^{n+1} p_{ci} \times dF - \sum_{i=1}^{n+1} \int p_{ci} \times \ddot{r}_{c0}\, dm$$

Now,

$$N_T = \sum_{i=1}^{n+1} N_{ci} = \sum_{i=1}^{n+1} p_{ci} \times dF \quad \text{i.e.} \quad N_0 = N_T - \sum_{i=1}^{n+1} m_i p_{ci} \times \ddot{r}_{c0}$$

Centre of mass is defined by

$$p_{cm}^* = \frac{1}{m_T} \sum_{i=0}^{n+1} m_i p_{ci}^* = \frac{1}{m_T} \sum_{i=0}^{n+1} \sum_{j=i}^{n+1} m_j r_{ci}$$

$$= \frac{1}{m_T} \left(\sum_{i=0}^{n+1} m_i r_{c0} + \sum_{i=1}^{n+1} m_i s_0 + \sum_{i=1}^{n+1} m_i p_{ci} \right)$$

$$= r_{c0} + \frac{1}{m_T} \left[(m_T - m_0) s_0 + \sum_{i=1}^{n+1} m_i p_{ci} \right]$$

$$\sum_{i=1}^{n+1} m_i p_{ci} = m_T \left(p_{cm}^* - r_{c0} \right) - (m_T - m_0)s_0$$

Substitute the summation:

$$N_0 = N_T - \left[m_T(p_{cm}^* - r_{c0}) - (m_T - m_0)s_0 \right] \times \ddot{r}_{c0} \qquad (8.16)$$

Similarly, since no external forces are acting:

$$F_T = \sum_{i=0}^{n+1} F_{ci} = \sum_{i=0}^{n+1} m_i \ddot{p}_{ci}^* = 0$$

$$= \sum_{i=0}^{n+1}\sum_{j=i}^{n+1} m_j \ddot{r}_{ci} = \sum_{i=0}^{n+1} m_i \ddot{r}_{c0} + \sum_{i=0}^{n+1} m_i \ddot{s}_0 + \sum_{i=0}^{n+1} m_i \ddot{p}_{ci}$$

$$= m_T \ddot{r}_{c0} + \sum_{i=1}^{n+1} m_i \ddot{p}_{ci} = 0$$

Hence,

$$\ddot{r}_{c0} = -\frac{1}{m_T} \sum_{i=1}^{n+1} m_i \ddot{p}_{ci}$$

Now, total reaction force on base of manipulator:

$$F_T = \sum_{i=1}^{n+1} m_i \ddot{p}_{ci} \rightarrow \ddot{r}_{c0} = -\frac{1}{m_T} F_T$$

Now,

$$F_0 = m_0 \ddot{r}_{c0} = -\left(\frac{m_0}{m_T} \right) F_T \qquad (8.17)$$

$$N_0 = N_T + \left[m_T(p_{cm}^* - r_{c0}) - (m_T - m_0)s_0 \right] \times F_T / m_T \qquad (8.18)$$

This gives the moments and forces on the spacecraft at the manipulator base with respect to inertial coordinates. The sum of moments about the spacecraft bus centre of mass which must be compensated by the attitude controller:

$$N_r = N_0 + s_0 \times F_0$$

$$= N_T + \left[m_T(p_{cm}^* - r_{c0}) - (m_T - m_0)s_0 \right]$$

$$\times F_T / m_T - s_0 \times \left(m_0 / m_T \right) \times F_T \qquad (8.15) \quad \text{QED}$$

$$= N_T + (p_{cm}^* - r_{c0} - s_0) \times F_T$$

This comprises the feedforward component to the attitude control system of the spacecraft that must be nullified to maintain attitude such that $w_0 = 0$. These results must be transformed to map into spacecraft body coordinates so that x, y, and z coordinates of the spacecraft and the manipulator base coincide. This is accomplished by the premultiplication matrix:

$$\begin{pmatrix} 0 & 1 & 0 \\ 1 & 0 & 0 \\ 0 & 0 & -1 \end{pmatrix}.$$

Base-referenced dynamic parameters, N_{ci}, F_{ci} are calculated by the robot controller, hence N_T and F_T as their summations is trivial. While calculation of the joint torques are usually referred to the respective joint coordinates, reaction moment calculation for the attitude control system requires the dynamic variables F_{ci} and N_{ci} to be calculated with respect to base coordinates for summation. Recalculation in base coordinates is a fairly trivial calculation as it required only a few additional calculations to be performed which does not burden the algorithm as the algorithmic complexity of the Newton–Euler recursive method is only $O(n)$. In fact, the only change in the dynamics algorithm for the dual computation of F_{ci} and N_{ci} lie through the moment of inertia terms of the links. The other parameters p_{cm}^*, r_{c0}, s_0 are known or calculated during the kinematics formulation. Reaction moment compensation commands are the same no matter where the mounting of the reaction wheels or control moment gyros are within the spacecraft since the moment of a couple is independent of the reference point. This feedforward compensation scheme improves the stability of the satellite attitude by more than ten times [Sato et al 1993]. Base reactions are exerted directly on the supporting space vehicle and de Silva (1991) proposed minimising the base reactions to reduce the accelerations of the base by applying a weighted quadratic function $Q = R^T W R$ where $R = (F_T N_T)^T$. However, the method is applicable to redundant manipulators only and requires finding a matrix of joint velocities and accelerations to be minimised during trajectory interpolation using the redundant degrees of freedom. This form of control is also computationally intense as well as not being applicable here. Another advantage of the scheme presented here is that spacecraft attitude control remains referenced to the vehicle centre of mass rather than the total system centre of mass allowing the formulation of vehicle moment of inertia as constant relative to body fixed coordinates to compensate for the moments on the spacecraft—the spacecraft is likely to have a somewhat complex asymmetric shape so this is significant (e.g. due to solar arrays). The reaction moments generated will depend on the payload mass and the maximum acceleration/deceleration profile of the joint trajectory.

We outline some simulation results of the joint torques and reaction moments experienced by a robotic spacecraft with the same physical properties as ATLAS [Ellery 1996]. The manipulator joint torques as calculated in the first phase of a simulated grasp manoeuvre utilising computed torque position control only indicate fairly low joint torques—~1 Nm, well within the capabilities of the PUMA 560/600 joint motors (Fig. 8.4).

The reaction moments are of a similar order of magnitude, but with a tendency to have greater absolute values <5 Nm, as expected from the reactive effects of simultaneously

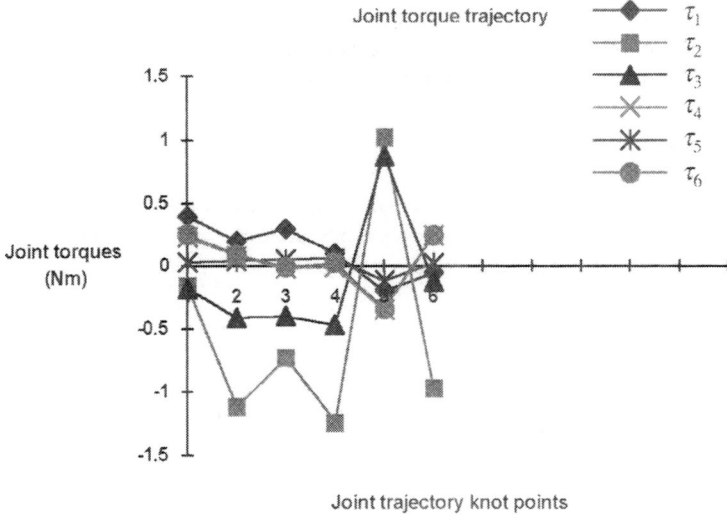

Fig. 8.4. Joint torque trajectory (from Ellery 1996).

driven multiple joints (Fig. 8.5). These are large however in comparison to the typical attitude disturbance torques experienced by spacecraft $\sim 10^{-6}$–10^{-3} g. Furthermore, they are barely within the bounds of the capabilities of reaction wheels which are limited to ~ 1–5 Nm torques typically. The reaction moments have similar orders of magnitude as the joint torques.

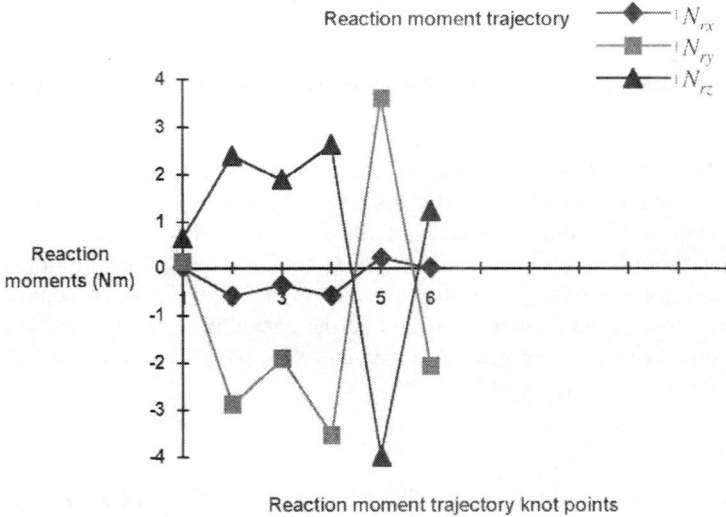

Fig. 8.5. Reaction moment trajectory (from Ellery 1996).

After the first phase of the trajectory, the end effector has collided with the target object and experienced impulsive type forces necessitating the use of force control schemes for the second phase of the capture manoeuvre. Force control has been a much neglected consideration in the literature on space robotics. The typical joint torques required for force control are very high ~1–10 kNm. These are well in excess of the capabilities of PUMA 560/600 joint torque motors, though space robotic motors offer much greater capabilities. Clearly, force control is a critical capability and this implies the need for high torque motors in conjunction with low force control gains (Fig. 8.6). The reaction moments have similar orders of magnitude to the joint torques ~kN (Fig. 8.7).

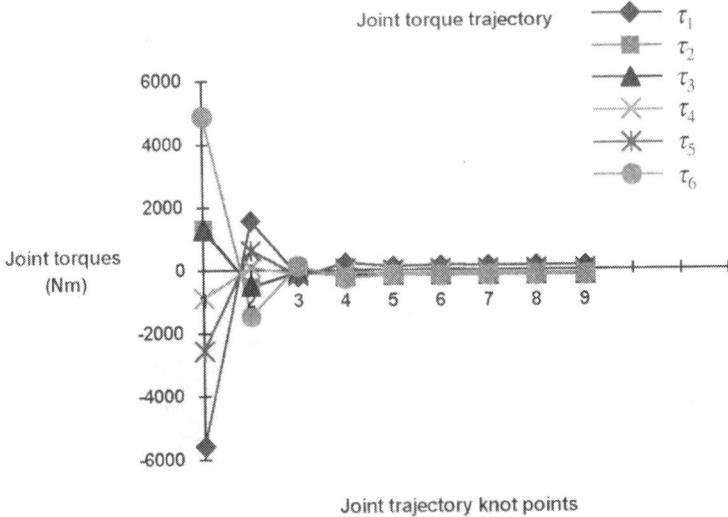

Fig. 8.6. Joint torque trajectory with relative velocity of collision of 0.1 m/s (from Ellery 1996).

These simulations illustrate the need for control moment gyro attitude control as reaction wheel control is not capable of reaction moment compensation during force control manoeuvres. Furthermore, the difference between force control and position control is evident. The implementation of force control yields very large motor torque requirements from both the manipulator joint motors and the spacecraft attitude control actuators, a factor of ~1000 greater than those imposed by pure position control. In other words, force control requirements will define the capabilities of any robotic system in space. This has been little appreciated in previous work which has been dominated by the space-based position control problem.

8.3 DUAL-MANIPULATOR ROBOTIC FREEFLYER SPACECRAFT

We have seen how dual-manipulator systems are essentially a straightforward extension of the single manipulator.

Reaction moment trajectory

N_{rx}
$N_{ry}\ (v = 1.0)$
N_{rz}
$N_{ry}\ (v = 0.1)$

Reaction moments (Nm)

2000
1000
0
-1000
-2000
-3000
-4000
-5000
-6000

2 3 4 5 6 7 8 9

Reaction moment trajectory knot points
(n)

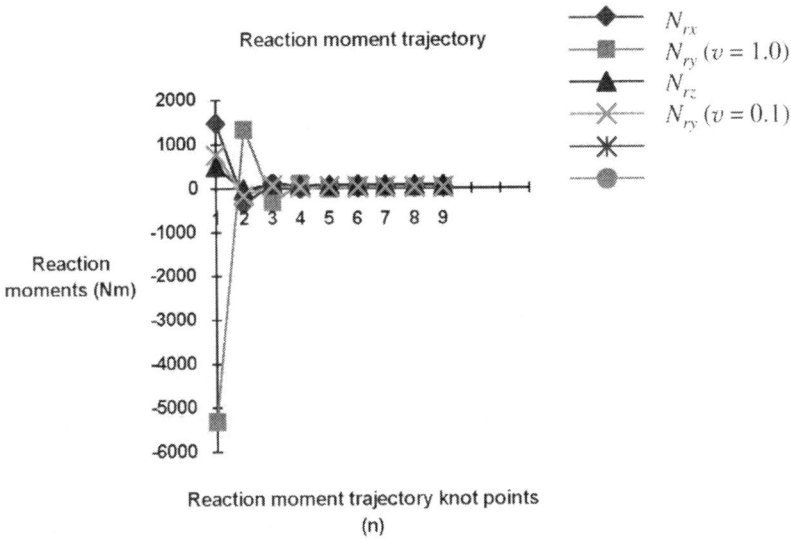

Fig. 8.7. Reaction moment trajectory during force control with relative velocity of 0.1 m/s and 1.0 m/s (from Ellery 1996)

8.3.1 Dual-manipulator position kinematics

The kinematic control equations may be adapted for two-arm control in a straightforward manner to allow the use of the control techniques outlined in Chapter 6, but the formulation presented here is more generally applicable. It assumes the case of $l = 2$ manipulator arms [Carignan & Akin 1988], but may be extended to more than two arms. The same assumptions are utilised: that linear momentum is conserved allowing the use of the equilibrium of moments and that 3-axis feedforward stabilisation is employed.

Equilibrium of moments:

$$\left(m_0 + \sum_{l=1}^{2} \sum_{i=1}^{n+1} m_i^l \right) p_{cm}^* = \left(m_0 + \sum_{l=1}^{2} \sum_{i=1}^{n+1} m_i^l \right) \sum_{j=0}^{n+1} p_{cj}^l$$

$$\rightarrow p_{cm}^* = r_{c0} + \frac{1}{m_T} \sum_{l=1}^{2} \sum_{i=1}^{n+1} \sum_{j=i}^{n+1} m_j^l r_{ci}^l$$

where

$$m_T = m_0 + \sum_{l=1}^{2} \sum_{i=1}^{n+1} m_i^l$$

Hence,

$$r_{c0} = p_{cm}^* - \frac{1}{m_T} \sum_{l=1}^{2} \sum_{i=1}^{n+1} \sum_{j=i}^{n+1} m_j^l r_{ci}^l \tag{8.19}$$

Now the inertial position of the end effector of arm l is given by:

$$p_l^* = r_{c0} + s_0^l + \sum_{i=1}^{n} R_i^l l_i^l$$

Substitute (8.19):

$$p_l^* = p_{cm}^* + s_0^l + \sum_{i=1}^{n} R_i^l l_i^l - \frac{1}{m_T} \sum_{l=1}^{2} \sum_{i=1}^{n+1} \sum_{j=i}^{n+1} m_j^l r_{ci}^l$$

$$= p_{cm}^* + s_0^l + \sum_{i=1}^{n} R_i^l \left(r_i^l + s_i^l \right) - \frac{1}{m_T} \sum_{l=1}^{2} \sum_{i=1}^{n+1} \sum_{j=i}^{n+1} m_j^l \left(R_i^l r_i^l + R_{i-1}^l s_{i-1}^l \right)$$

$$= p_{cm}^* + s_0^l + \sum_{i=1}^{n} R_i^l \left(r_i^l + s_i^l \right) - \frac{1}{m_T} \sum_{l=1}^{2} \sum_{i=1}^{n+1} \sum_{j=i}^{n+1} m_j^l \left(R_{i-1}^l r_{i-1}^l + R_{i-1}^l s_{i-1}^l \right)$$

$$- \frac{1}{m_T} \sum_{l=1}^{2} \sum_{i=1}^{n+1} m_i^l R_i^l r_i^l$$

Now, with $r_0^l = 0$:

$$p_l^* = p_{cm}^* + s_0^l + \sum_{i=1}^{n} R_i^l \left(r_i^l + s_i^l \right) - \frac{1}{m_T} \sum_{l=1}^{2} \sum_{i=1}^{n+1} \sum_{j=i}^{n+1} m_j^l R_i^l \left(r_i^l + s_i^l \right)$$

$$- \frac{1}{m_T} \sum_{l=1}^{2} \sum_{i=1}^{n+1} m_i^l R_i^l r_i^l - \frac{1}{m_T} \sum_{l=1}^{2} \sum_{i=1}^{n+1} m_i^l R_0^l \left(r_0^l + s_0^l \right)$$

$$= p_{cm}^* + \left(s_0^l - \frac{1}{m_T} \sum_{l-1}^{2} \sum_{i-1}^{n+1} m_i^l s_0^l \right)$$

$$+ \left(\sum_{i=1}^{n} R_i^l \left(r_i^l + s_i^l \right) - \frac{1}{m_T} \sum_{l=1}^{2} \sum_{i=1}^{n+1} \sum_{j=i+1}^{n+1} m_j^l R_i^l \left(r_i^l + s_i^l \right) \right)$$

$$- \frac{1}{m_T} \sum_{l=1}^{2} \sum_{i=1}^{n+1} m_i^l R_i^l r_i^l - \frac{m_{n+1}}{m_T} \sum_{l=1}^{2} R_{n+1}^l r_{n+1}^l \tag{8.20}$$

This equation now requires splitting into its component contributions from each arm l and l' (assuming two arms, but the formulation can be extended to any number of arms, but the complexity increases substantially):

$$p_l^* = p_{cm}^* + \left(1 - \frac{m_{1n+1}^l}{m_T}\right)s_0^l - \left(\frac{m_{1n+1}^l}{m_T}\right)s_0^{l'} + \left(1 - \frac{1}{m_T}\sum_{j=i+1}^{n+1}m_j^l\right)\sum_{i=1}^{n}R_i^l\left(r_i^l + s_i^l\right)$$

$$- \frac{1}{m_T}\sum_{i=1}^{n}\sum_{j=i+1}^{n+1}m_j^{l'}R_i^{l'}\left(r_i^{l'} + s_i^{l'}\right) - \frac{1}{m_T}\sum_{i=1}^{n}m_i^l R_i^l r_i^l$$

$$- \frac{1}{m_T}\sum_{i=1}^{n}m_i^{l'}R_i^{l'}r_i^{l'} - \frac{1}{m_T}\sum_{l=1}^{2}m_{n+1}^l R_{n+1}^l r_{n+1}^l$$

$$= p_{cm}^* + \frac{m_{0n+1}^{l'}}{m_T}s_0^l - \frac{m_{1n+1}^{l'}}{m_T}s_0^{l'} + \sum_{i=1}^{n}R_i^l\lambda_i - \sum_{i=1}^{n}R_i^{l'}L_i - \frac{1}{m_T}\sum_{l=1}^{2}m_{n+1}^l R_{n+1}^l r_{n+1}^l$$

where

$$\lambda_i = \left(\frac{1}{m_T}\sum_{i=1}^{n}\sum_{j=0}^{i}\left(m_j^l l_i^l + m_{1n+1}^{l'} - m_i^l r_i^l\right)\right)$$

and

$$L_i = \left(\frac{1}{m_T}\sum_{i=1}^{n}\sum_{j=i+1}^{i}\left(m_j^{l'}l_i^{l'} + m_i^{l'}r_i^{l'}\right)\right)$$

$$s_0^{l'} = Hs_0^l$$

Example: Now an illustrative example of a dual PUMA 560/600 manipulator system mounted onto a spacecraft platform is given. For generality, the open-loop configuration is analysed as the closed-loop configuration may be regarded as an open-loop system with the appropriate kinematic constraints. The formulation presented here is applied to the dual conguration of 6 DOF left/right-handed PUMA 560/600 manipulators with the left arm representing the master arm (superscript m) and the right arm representing the slave arm (superscript s). They are separated by a distance h in the x coordinate. The kinematic structures are assumed identical such that $m_i^m = m_i^s$ and $l_i^m = l_i^s$. Now,

$$p_l^* = p_{cm}^* + \left(s_0^l - \frac{1}{m_T}\sum_{l}^{2}\sum_{i=1}^{n+1}m_i^l s_0^l\right) + \sum_{i=1}^{n}R_i^l l_i^l - \frac{1}{m_T}\sum_{l=1}^{2}\sum_{i=1}^{n}\sum_{j=i+1}^{n+1}m_j^l R_i^l l_i^l$$

$$- \frac{1}{m_T}\sum_{l=1}^{2}\sum_{i=1}^{n}m_i^l R_i^l r_i^l - \frac{1}{m_T}\sum_{l=1}^{2}m_{n+1}^l R_{n+1}^l r_{n+1}^l$$

We assume for simplicity that the payload distribution is equal between the arms, though the formulation is not restricted to this assumption. For the master arm:

$$p_m^* = p_{cm}^* + \left(1 - \frac{m_{17}}{m_T}\right)s_0^m - \frac{m_{17}}{m_T}s_0^s$$

$$+ \left(1 - \frac{1}{m_T}\sum_{i}^{6}\sum_{j=i+1}^{7}m_j\right)\begin{pmatrix} c_1(a_2c_2 + d_4s_{23}) - d_2s_1 + d_6a_x \\ s_1(a_2c_2 + d_4s_{23}) + d_2c_1 + d_6a_y \\ d_4c_{23} - a_2s_2 + d_6a_z \end{pmatrix}_m$$

$$- \frac{1}{m_T}\sum_{i=1}^{6}m_i\begin{pmatrix} c_1(r_2c_2 + r_3s_{23}) - r_1s_1 + r_6a_x \\ s_1(r_2c_2 + r_3s_{23}) + r_1c_1 + r_6a_y \\ r_3c_{23} - r_2s_2 + r_6a_z \end{pmatrix}_m$$

$$- \frac{1}{m_T}\sum_{i=1}^{6}\sum_{j=i+1}^{7}m_j\begin{pmatrix} c_1(a_2c_2 + d_4s_{23}) - d_2s_{23} + d_6a_x \\ s_1(a_2c_2 + d_4s_{23}) + d_2c_1 + d_6a_y \\ d_4c_{23} - a_2s_2 + d_6a_z \end{pmatrix}_s$$

$$- \frac{1}{m_T}\sum_{i=1}^{n}m_i\begin{pmatrix} c_1(r_2c_2 + r_3s_{23}) - r_1s_1 + r_6a_x \\ s_1(r_2c_2 + r_3s_{23}) + r_1c_1 + r_6a_y \\ r_3c_{23} - r_2s_2 + r_6a_z \end{pmatrix}_s$$

$$- \frac{1}{m_T}\left(m_7^m R_7^m r_7^m + m_7^s R_7^s r_7^s\right)$$

Multiplying the mass terms out gives:

$$p_m^* = p_{cm}^* + \left(\frac{m_{07}}{m_T}\right)s_0^m + \left(\frac{m_{17}}{m_T}\right)s_0^s$$

$$+ \begin{pmatrix} c_1\left[c_2a_2\left(\dfrac{m_{02} + m_{17}}{m_T}\right) + s_{23}d_4\left(\dfrac{m_{03} + m_{17}}{m_T}\right)\right] - s_1d_2\left(\dfrac{m_{01} + m_{17}}{m_T}\right) + a_xd_6\left(\dfrac{m_{06} + m_{17}}{m_T}\right) \\ s_1\left[c_2a_2\left(\dfrac{m_{02} + m_{17}}{m_T}\right) + s_{23}d_4\left(\dfrac{m_{03} + m_{17}}{m_T}\right)\right] + c_1d_2\left(\dfrac{m_{01} + m_{17}}{m_T}\right) + a_yd_6\left(\dfrac{m_{06}^m + m_{17}^s}{m_T}\right) \\ c_{23}d_4\left(\dfrac{m_{03} + m_{17}}{m_T}\right) - s_2a_2\left(\dfrac{m_{02} + m_{17}}{m_T}\right) + a_zd_6\left(\dfrac{m_{06} + m_{17}}{m_T}\right) \end{pmatrix}_m$$

$$- \begin{pmatrix} c_1\left[c_2r_2\left(\dfrac{m_2}{m_T}\right) + s_{23}r_3\left(\dfrac{m_3}{m_T}\right)\right] - s_1r_1\left(\dfrac{m_1}{m_T}\right) + a_xr_6\left(\dfrac{m_6}{m_T}\right) \\ s_1\left[c_2r_2\left(\dfrac{m_2}{m_T}\right) + s_{23}r_3\left(\dfrac{m_3}{m_T}\right)\right] + c_1r_1\left(\dfrac{m_1}{m_T}\right) + a_yr_6\left(\dfrac{m_6}{m_T}\right) \\ c_{23}r_3\left(\dfrac{m_3}{m_T}\right) - s_2r_2\left(\dfrac{m_2}{m_T}\right) + a_zr_6\left(\dfrac{m_6}{m_T}\right) \end{pmatrix}_m$$

$$-\left(\begin{array}{c} c_1\left[c_2 a_2\left(\dfrac{m_{37}}{m_T}\right)+s_{23}d_4\left(\dfrac{m_{67}}{m_T}\right)\right]-s_1 d_2\left(\dfrac{m_{27}}{m_T}\right)+a_x d_6\left(\dfrac{m_7}{m_T}\right) \\[2mm] s_1\left[c_2 a_2\left(\dfrac{m_{37}^s}{m_T}\right)+s_{23}d_4\left(\dfrac{m_{67}}{m_T}\right)\right]+c_1 d_2\left(\dfrac{m_{27}}{m_T}\right)+a_y d_6\left(\dfrac{m_7}{m_T}\right) \\[2mm] c_{23}d_4\left(\dfrac{m_{67}}{m_T}\right)-s_2 a_2\left(\dfrac{m_{37}}{m_T}\right)+a_z d_6\left(\dfrac{m_7}{m_T}\right) \end{array}\right)_s$$

$$-\left(\begin{array}{c} c_1\left[c_2 r_2\left(\dfrac{m_2}{m_T}\right)+s_{23}r_3\left(\dfrac{m_3}{m_T}\right)\right]-s_1 r_1\left(\dfrac{m_1}{m_T}\right)+a_x r_6\left(\dfrac{m_6}{m_T}\right) \\[2mm] s_1\left[c_2 r_2\left(\dfrac{m_2}{m_T}\right)+s_{23}r_3\left(\dfrac{m_3}{m_T}\right)\right]+c_1 r_1\left(\dfrac{m_1}{m_T}\right)+a_y r_6\left(\dfrac{m_6}{m_T}\right) \\[2mm] c_{23}r_3\left(\dfrac{m_3}{m_T}\right)-s_2 r_2\left(\dfrac{m_2}{m_T}\right)+a_z r_6\left(\dfrac{m_6}{m_T}\right) \end{array}\right)_s$$

$$-\frac{1}{m_T}\left(m_7^m\begin{pmatrix}x_7\\y_7\\z_7\end{pmatrix}_m+m_7^s\begin{pmatrix}x_7\\y_7\\z_7\end{pmatrix}_s\right)$$

By grouping the common factors, the inertial position for the master arm may be found:

$$p_m^* = p_{cm}^* + \frac{m_{07}}{m_T}s_0^m + \frac{m_{17}}{m_T}s_0^s + \begin{pmatrix} c_1(c_2\alpha_2+s_{23}\delta_4)-s_1\delta_2+a_x\delta_6 \\ s_1(c_2\alpha_2+s_{23}\delta_4)+c_1\delta_2+a_y\delta_6 \\ c_{23}\delta_4-s_2\alpha_2+a_z\delta_6 \end{pmatrix}_m$$

$$-\begin{pmatrix} c_1(c_2 A_2+s_{23}D_4)-s_1 d_2+a_x D_6 \\ s_1(c_2\alpha_2+s_{23}D_4)+c_1 D_2+a_y D_6 \\ c_{23}D_4-s_2 A_2+a_z D_6 \end{pmatrix}_s$$

$$-\frac{1}{m_T}\left(m_7^m\begin{pmatrix}x_7\\y_7\\z_7\end{pmatrix}_m+m_7^s\begin{pmatrix}x_7\\y_7\\z_7\end{pmatrix}_s\right)$$

where

$$\delta_2 = \frac{(m_{01} + m_{17})d_2 - m_1 r_1}{m_T} \qquad\qquad D_2 = \frac{m_{27} + m_1 r_1}{m_T}$$

$$\alpha_2 = \frac{(m_{02} + m_{17})a_2 - m_2 r_2}{m_T} \qquad\qquad A_2 = \frac{m_{37} a_2 + m_2 r_2}{m_T}$$

and

$$\delta_4 = \frac{(m_{03} + m_{17})d_4 - m_3 r_3}{m_T} \qquad\qquad D_4 = \frac{m_{67} d_4 + m_3 r_3}{m_T}$$

$$\delta_6 = \frac{(m_{06} + m_{17})d_6 - m_6 r_6}{m_T} \qquad\qquad D_6 = \frac{m_7 d_6 + m_6 r_6}{m_T}$$

A similar expression may be derived for the slave arm:

$$p_s^* = p_{cm}^* + \frac{m_{07}}{m_T} s_0^s + \frac{m_{17}}{m_T} s_0^m + \begin{pmatrix} c_1(c_2 \alpha_2 + s_{23} \delta_4) - s_1 \delta_2 + a_x \delta_6 \\ s_1(c_2 \alpha_2 + s_{23} \delta_4) + c_1 \delta_2 + a_y \delta_6 \\ c_{23} \delta_4 - s_2 \alpha_2 + a_z \delta_6 \end{pmatrix}_s$$

$$- \begin{pmatrix} c_1(c_2 A_2 + s_{23} D_4) - s_1 d_2 + a_x D_6 \\ s_1(c_2 \alpha_2 + s_{23} D_4) + c_1 D_2 + a_y D_6 \\ c_{23} D_4 - s_2 A_2 + a_z D_6 \end{pmatrix}_m$$

$$- \frac{1}{m_T} \left(m_7^s \begin{pmatrix} x_7 \\ y_7 \\ z_7 \end{pmatrix}_s + m_7^m \begin{pmatrix} x_7 \\ y_7 \\ z_7 \end{pmatrix}_m \right)$$

This defines the forward inertial position kinematics of each end effector. Now, the geometric configuration for the dual arms was defined in Chapter 6:

$$s_0^s = H s_0^m = \begin{pmatrix} -s_{0x}^m + h \\ s_{0y}^m \\ -s_{0z}^m \end{pmatrix} \qquad \text{and} \qquad p_s = H p_m = \begin{pmatrix} -p_x^m \\ p_y^m \\ -p_z^m \end{pmatrix}$$

so these equations become:

$$p_m^* = p_{cm}^* + \begin{pmatrix} \dfrac{m_0}{m_T} s_{0x} + \dfrac{m_{17}}{m_T} h \\ s_{0y} \\ \dfrac{m_0}{m_T} s_{0z} \end{pmatrix} + \begin{pmatrix} c_1(c_2 \alpha_2 + s_{23} \delta_4) - s_1 \delta_2 + a_x \delta_6 \\ s_1(c_2 \alpha_2 + s_{23} \delta_4) + c_1 \delta_2 + a_y \delta_6 \\ c_{23} \delta_4 - s_2 \alpha_2 + a_z \delta_6 \end{pmatrix}_m$$

$$p_s^* = p_{cm}^* + \begin{bmatrix} -\begin{pmatrix} -c_1(c_2 A_2 + s_{23} D_4) + s_1 d_2 - a_x D_6 \\ s_1(c_2 A_2 + s_{23} D_4) + c_1 D_2 + a_y D_6 \\ -c_{23} D_4 + s_2 A_2 - a_z D_6 \end{pmatrix}_s \\ -\frac{1}{m_T}\begin{bmatrix} m_7^m \begin{pmatrix} x_7 \\ y_7 \\ z_7 \end{pmatrix}_m + m_7^s \begin{pmatrix} -x_7 \\ y_7 \\ -z_7 \end{pmatrix}_s \end{bmatrix} \end{bmatrix}$$

$$p_s^* = p_{cm}^* + \begin{pmatrix} -\frac{m_0}{m_T} s_{0x} + \frac{m_{07}}{m_T} h \\ s_{0y} \\ -\frac{m_0}{m_T} s_{0z} \end{pmatrix} + \begin{pmatrix} -c_1(c_2 \alpha_2 + s_{23} \delta_4) + s_1 \delta_2 - a_x \delta_6 \\ s_1(c_2 \alpha_2 + s_{23} \delta_4) + c_1 \delta_2 + a_y \delta_6 \\ -c_{23} \delta_4 - s_2 \alpha_2 - a_z \delta_6 \end{pmatrix}_s$$

$$\begin{aligned}&- \begin{pmatrix} c_1(c_2 A_2 + s_{23} D_4) - s_1 D_2 + a_x D_6 \\ s_1(c_2 A_2 + s_{23} D_4) + c_1 D_2 + a_y D_6 \\ c_{23} D_4 - s_2 A_2 - a_z D_6 \end{pmatrix}_s \\[6pt] &- \frac{1}{m_T}\begin{bmatrix} m_7^s \begin{pmatrix} -x_7 \\ y_7 \\ -z_7 \end{pmatrix}_m + m_7^m \begin{pmatrix} x_7 \\ y_7 \\ z_7 \end{pmatrix}_m \end{bmatrix}\end{aligned}$$

The complete solution to the inertial forward kinematics requires the calculation of p_{cm}^* which may be found from:

$$\begin{pmatrix} p_{cmx}^* \\ p_{cmy}^* \\ p_{cmz}^* \end{pmatrix} = \begin{pmatrix} c_1(c_2 A_2 + s_{23} D_4) - s_1 D_2 + a_x D_6 \\ s_1(c_2 A_2 + s_{23} D_4) + c_1 D_2 + a_y D_6 \\ c_{23} D_4 - s_2 D_2 + a_z D_6 \end{pmatrix}_m$$

$$+ \begin{pmatrix} -c_1(c_2 A_2 + s_{23} D_4) + s_1 D_2 - a_x D_6 \\ s_1(c_2 A_2 + s_{23} D_4) + c_1 D_2 + a_y D_6 \\ -c_{23} D_4 + s_2 A_2 - a_z D_6 \end{pmatrix}_s$$

$$+ \frac{1}{m_T}\begin{bmatrix} m_7^m \begin{pmatrix} x_7^m \\ y_7^m \\ z_7^m \end{pmatrix} + m_7^s \begin{pmatrix} -x_7^s \\ y_7^s \\ -z_7^s \end{pmatrix} \end{bmatrix} + \begin{pmatrix} \left(1 - \frac{m_0}{m_T}\right) s_{0x} - \frac{m_{17}}{m_T} h \\ s_{0y} \\ \left(1 - \frac{m_0}{m_T}\right) s_{0z} \end{pmatrix} \tag{8.21}$$

The inverse kinematics solution generates the joint angles $(\theta_1^m, \theta_2^m, \theta_3^m)$ for the master arm and $(\theta_1^s, \theta_2^s, \theta_3^s)$ for the slave arm respectively. These first three respective joint angles define the end effector positions for each arm. However, the kinematics of the two arms are now coupled and the equations must be solved simultaneously. Let

$$
\begin{pmatrix} X^m \\ Y^m \\ Z^m \end{pmatrix} = \begin{pmatrix} \overset{*}{p}_{mx} - \overset{*}{p}_{cmx} - \dfrac{m_0}{m_T}s_{0x} - \dfrac{m_{17}}{m_T}h + \dfrac{1}{m_T}\left(m_7^m x_7^m - m_7^s x_7^s\right) - a_x^m \delta_6 - a_x^s D_6 \\[2mm] \overset{*}{p}_{my} - \overset{*}{p}_{cmy} - s_{0y} + \dfrac{1}{m_T}\left(m_7^m y_7^m + m_7^s y_7^s\right) - a_y^m \delta_6 + a_y^s D_6 \\[2mm] \overset{*}{p}_{mz} - \overset{*}{p}_{cmz} - \dfrac{m_0}{m_T}s_{0z} + \dfrac{1}{m_T}\left(m_7^m y_7^m - m_7^s y_7^s\right) - a_z^m \delta_6 - a_z^s D_6 \end{pmatrix}
$$

$$
\begin{pmatrix} X^s \\ Y^s \\ Z^s \end{pmatrix} = \begin{pmatrix} \overset{*}{p}_{sx} - \overset{*}{p}_{cmx} + \dfrac{m_0}{m_T}s_{0x} - \dfrac{m_{07}}{m_T}h + \dfrac{1}{m_T}\left(-m_7^s x_7^s + m_7^m x_7^m\right) + a_x^s \delta_6 + a_x^m D_6 \\[2mm] \overset{*}{p}_{sy} - \overset{*}{p}_{cmy} - s_{0y} + \dfrac{1}{m_T}\left(m_7^s y_7^s + m_7^m y_7^m\right) + a_y^s \delta_6 + a_y^m D_6 \\[2mm] \overset{*}{p}_{sz} - \overset{*}{p}_{cmz} + \dfrac{m_0}{m_T}s_{0z} + \dfrac{1}{m_T}\left(-m_7^s y_7^s + m_7^m y_7^m\right) + a_z^s \delta_6 + a_z^m D_6 \end{pmatrix}
$$

Hence,

$$
\begin{pmatrix} X^m \\ Y^m \\ Z^m \end{pmatrix} = \begin{pmatrix} c_1(c_2\alpha_2 + s_{23}\delta_4) - s_1\delta_2 \\ s_1(c_2\alpha_2 + s_{23}\delta_4) + c_1\delta_2 \\ c_{23}\delta_4 - s_2\alpha_2 \end{pmatrix}_m - \begin{pmatrix} -c_1(c_2 A_2 + s_{23} D_4) + s_1 D_2 \\ s_1(c_2 A_2 + s_{23} D_4) + c_1 D_2 \\ -c_{23} D_4 + s_2 A_2 \end{pmatrix}_s
$$

$$
\begin{pmatrix} X^s \\ Y^s \\ Z^s \end{pmatrix} = \begin{pmatrix} -c_1(c_2\alpha_2 + s_{23}\delta_4) + s_1\delta_2 \\ s_1(c_2\alpha_2 + s_{23}\delta_4) + c_1\delta_2 \\ -c_{23}\delta_4 + s_2\alpha_2 \end{pmatrix}_s - \begin{pmatrix} c_1(c_2 A_2 + s_{23} D_4) - s_1 D_2 \\ s_1(c_2 A_2 + s_{23} D_4) + c_1 D_2 \\ -c_{23} D_4 - s_2 A_2 \end{pmatrix}_m
$$

These are six highly coupled, nonlinear, simultaneous equations in six unknowns $(\theta_1^m, \theta_2^m, \theta_3^m, \theta_1^s, \theta_2^s, \theta_3^s)$, so they may be solved numerically (e.g. by Newton's method) to yield unique solutions. The two manipulator motions are coupled together in such a way that it is not possible to control them independently without regard to each other. Note that to find the joint angles for the respective orienting wrists, terrestrial algorithms may applied to each arm independently due to the imposition of attitude control. As noted in Chapter 6, there are, however, special cases where the motion of the slave arm is restricted by the motion of the master arm, e.g. parallel copying motion, parallel symmetric motion and closed-chain configurations. For these cases, analytic solutions are available.

A particularly relevant example of an analytic solution for a space freeflyer is the situation where the master arm grapples the satellite and remains fixed such that $(\theta_1^m, \theta_2^m, \theta_3^m)$ are specified and locked:

$$p_s^* = p_{cm}^* + \begin{pmatrix} -\dfrac{m_0}{m_T}s_{0x} + \dfrac{m_{07}}{m_T}h \\ s_{0y} \\ -\dfrac{m_0}{m_T}s_{0z} \end{pmatrix} - \dfrac{1}{m_T}\left(m_7^s\begin{pmatrix} -x_7 \\ y_7 \\ -z_7 \end{pmatrix}_s + m_7^m\begin{pmatrix} x_7 \\ y_7 \\ z_7 \end{pmatrix}_m \right)$$

$$+ \begin{pmatrix} -c_1(c_2\alpha_2 + s_{23}\delta_4) + s_1\delta_2 - a_x\delta_6 \\ s_1(c_2\alpha_2 + s_{23}\delta_4) + c_1\delta_2 + a_y\delta_6 \\ -c_{23}\delta_4 + s_2\alpha_2 - a_z\delta_6 \end{pmatrix}_s$$

$$- \begin{pmatrix} c_1(c_2A_2 + s_{23}D_4) - s_1D_2 + a_xD_6 \\ s_1(c_2A_2 + s_{23}D_4) + c_1D_2 + a_yD_6 \\ c_{23}D_4 - s_2A_2 + a_zD_6 \end{pmatrix}_m$$

This gives the equation to be solved with known $(X^sY^sZ^s)$ as:

$$\begin{pmatrix} X^s \\ Y^s \\ Z^s \end{pmatrix} = \begin{pmatrix} -c_1(c_2\alpha_2 + s_{23}\delta_4) + s_1\delta_2 \\ s_1(c_2\alpha_2 + s_{23}\delta_4) + c_1\delta_2 \\ -c_{23}\delta_4 + s_2\alpha_2 \end{pmatrix}_s$$

where

$$\begin{pmatrix} X^s \\ Y^s \\ Z^s \end{pmatrix} = \begin{pmatrix} p_{sx}^* - p_{cmx}^* + \dfrac{m_0}{m_T}s_{0x} - \dfrac{m_{07}}{m_T}h \\ p_{sy}^* - p_{cmy}^* - s_{0y} \\ p_{sz}^* - p_{cmz}^* + \dfrac{m_0}{m_T}s_{0z} \end{pmatrix}$$

$$+ \dfrac{1}{m_T}\left(m_7^s\begin{pmatrix} -x_7^s \\ y_7^s \\ -z_7^s \end{pmatrix} + m_7^m\begin{pmatrix} x_7^m \\ y_7^m \\ z_7^m \end{pmatrix} \right) + \begin{pmatrix} a_x\delta_6 \\ -a_y\delta_6 \\ a_z\delta_6 \end{pmatrix}$$

$$+ \begin{pmatrix} c_1(c_2A_2 + s_{23}D_4) - s_1D_2 + a_xD_6 \\ s_1(c_2A_2 + s_{23}D_4) + c_1D_2 + a_yD_6 \\ c_{23}D_4 - s_2A_2 + a_zD_6 \end{pmatrix}_m$$

Unique solutions may be found for $(\theta_1^s, \theta_2^s, \theta_3^s)$ using the same inverse kinematics techniques as used for a single arm robot spacecraft:

$$\theta_1^s = \tan^{-1}\left(\frac{Y^s}{-X^s}\right) - \tan^{-1}\left(\frac{\delta_2}{\pm\sqrt{(-X^s)^2 + (Y^s)^2 - \delta_2^2}}\right)$$

$$\theta_2^s = \tan^{-1}\left(\frac{A}{-Z^s}\right) - \tan^{-1}\left(\frac{B}{\pm\sqrt{A^2 + (-Z^s)^2 - B^2}}\right)$$

where

$$A = -c_1 X^s + s_1 Y^s \qquad \text{and} \qquad B = \frac{A^2 + (-Z^s)^2 + \alpha_2^2 - \delta_4^2}{2\alpha_2}$$

$$\theta_s^s = \left[\tan^{-1}\left(\frac{A - \alpha_2 c_2}{-Z^s + \alpha_2 s_2}\right)\right] - \theta_2^s$$

The solutions to the wrist angles proceed directly from the terrestrial manipulator algorithms.

8.3.2 Dual manipulator velocity kinematics

Using the same techniques as before, for the inertial velocity of each arm l where $l = 2$:

$$v_l^* = \dot{p}_l^* = \dot{r}_{c0} + \sum_{i=1}^{n} \dot{R}_i^l l_i^l \tag{8.22}$$

Now

$$\dot{r}_{c0} = -\frac{1}{m_T}\sum_{l=1}^{2}\sum_{i=1}^{n+1}\sum_{j=i}^{n+1} m_j^i \dot{r}_{ci}^l$$

since $\dot{p}_{cm}^* = 0$ due to the invariance of the system centre of mass.

Now,

$$\dot{r}_{c0} = -\frac{1}{m_T}\sum_{l=1}^{2}\sum_{i=1}^{n+1}\sum_{j=i}^{n+1} m_j^l \left(\dot{R}_i^l r_i^l + \dot{R}_{i-1}^l s_{i-1}^l\right)$$

Now, substitute into (8.22):

$$v_l^* = \sum_{i=1}^{n} \dot{R}_i^l l_i^l - \frac{1}{m_T}\sum_{l=1}^{2}\sum_{i=1}^{n+1}\sum_{j=i}^{n+1} m_j^l \left(\dot{R}_i^l r_i^l + \dot{R}_{i-1}^l s_{i-1}^l\right)$$

$$= \left\{\sum_{i=1}^{n}\sum_{k=1}^{i} \frac{\partial R_i^l}{\partial \theta_k^l}\left(r_i^l + s_i^l\right) - \frac{1}{m_T}\sum_{i=1}^{n}\sum_{i=1}^{n+1}\sum_{j=i}^{n+1} m_j^l \left(\sum_{k=1}^{i} \frac{\partial R_i^l}{\partial \theta_k^l} r_i^l + \sum_{k=1}^{i-1} \frac{\partial R_{i-1}^l}{\partial \theta_k^l} s_{i-1}^l\right)\right\}\dot{\theta}_i^l$$

This has the form $v_l^* = \bar{J}^l \dot{\theta}_i^l$ with:

$$\bar{J}^l = \sum_{i=1}^{n}\sum_{k=0}^{i}\frac{\partial R_i^l}{\partial \theta_k}\left(r_i^l + s_i^l\right) - \frac{1}{m_T}\sum_{l=1}^{2}\sum_{i=1}^{n+1}\sum_{j=i}^{n+1}m_j^l\left(\sum_{k=1}^{i-1}\frac{\partial R_{i-1}^l}{\partial \theta_k^l}r_{i-1}^l + \sum_{k=1}^{i-1}\frac{\partial R_{i-1}^l}{\partial \theta_k^l}s_{i-1}^l\right)$$

$$-\frac{1}{m_T}\sum_{l=1}^{2}\sum_{i=1}^{n+1}\sum_{k=1}^{i}m_j^l\frac{\partial R_i^l}{\partial \theta_k^l}r_i^l$$

Now, with $\partial R_0/\partial \theta_k^l = 0$ for an attitude-controlled mounting platform:

$$\bar{J}^l = \sum_{i=1}^{n}\sum_{k=0}^{i}\frac{\partial R_i^l}{\partial \theta_k^l}\left(r_i^l + s_i^l\right) - \frac{1}{m_T}\sum_{l=1}^{2}\sum_{i=1}^{n+1}\sum_{j=i}^{n+1}m_j^l\left(\sum_{k=1}^{i-1}\frac{\partial R_{i-1}^l}{\partial \theta_k^l}r_{i-1}^l + \sum_{k=1}^{i-1}\frac{\partial R_{i-1}^l}{\partial \theta_k^l}s_{i-1}^l\right)$$

$$-\frac{1}{m_T}\sum_{l=1}^{2}\sum_{i=1}^{n+1}\sum_{k=1}^{i}m_j^l\frac{\partial R_i^l}{\partial \theta_k^l}r_i^l - \frac{1}{m_T}\sum_{i=1}^{n+1}\sum_{k=1}^{i}m_i\frac{\partial R_0}{\partial \theta_k^l}(r_0 + s_0)$$

$$= \sum_{i=1}^{n}\sum_{k=0}^{i}\frac{\partial R_i^l}{\partial \theta_k^l}\left(r_i^l + s_i^l\right) - \frac{1}{m_T}\sum_{l=1}^{2}\sum_{i=1}^{n+1}\sum_{j=i+1}^{n+1}m_j^l\left(\sum_{k=1}^{i}\frac{\partial R_i^l}{\partial \theta_k^l}r_i^l + \sum_{k=1}^{i}\frac{\partial R_i^l}{\partial \theta_k^l}s_i^l\right)$$

$$-\frac{1}{m_T}\sum_{l=1}^{2}\sum_{i=1}^{n+1}\sum_{k=1}^{i}m_i^l\frac{\partial R_i^l}{\partial \theta_k^l}r_i^l$$

Now, with $\partial R_{n+1}^l/\partial \theta_k^l = 0$ as the payload is invariant with respect to the end effector:

$$\bar{J}^l = \sum_{i=1}^{n}\sum_{k=0}^{i}\frac{\partial R_i^l}{\partial \theta_k^l}\left(r_i^l + s_i^l\right) - \frac{1}{m_T}\sum_{l=1}^{2}\sum_{i=1}^{n}\sum_{j=i+1}^{n+1}m_j^l\left(\sum_{k=1}^{i}\frac{\partial R_i^l}{\partial \theta_k^l}r_i^l + \sum_{k=1}^{i}\frac{\partial R_i^l}{\partial \theta_k^l}s_i^l\right)$$

$$-\frac{1}{m_T}\sum_{l=1}^{2}\sum_{i=1}^{n}\sum_{k=1}^{i}m_j^l\frac{\partial R_i^l}{\partial \theta_k^l}r_i^l - \frac{1}{m_T}\sum_{l=1}^{2}m_{n+1}^l\sum_{k=1}^{n+1}\frac{\partial R_{n+1}^l}{\partial \theta_k^l}r_{n+1}^l$$

This may now be decomposed into the respective arm contributions l and l':

$$\bar{J}^l = \sum_{i=1}^{n}\sum_{k=1}^{i}\frac{\partial R_i^l}{\partial \theta_k^l}\left(1 - \frac{1}{m_T}\sum_{i=1}^{n}\sum_{j=i+1}^{n+1}m_j^l\left(r_i^l + s_i^l\right) - \frac{1}{m_T}m_i^l r_i^l\right)$$

$$-\sum_{i=1}^{n}\sum_{k=1}^{i}\frac{\partial R_i^{l'}}{\partial \theta_k^{l'}}\left(\frac{1}{m_T}\sum_{j=i+1}^{n+1}m_j^{l'}\left(r_i^{l'} + s_i^{l'}\right) + \frac{1}{m_T}m_i^{l'}r_i^{l'}\right)$$

Hence,

$$\bar{J}^l = \sum_{i=1}^{n}\sum_{k=1}^{i}\frac{\partial R_i^l}{\partial \theta_k^l}\lambda_i - \sum_{i=1}^{n}\sum_{k=1}^{i}\frac{\partial R_i^{l'}}{\partial \theta_k^{l'}}L_i$$

where

$$\lambda_i = \left(\frac{1}{m_T} \left[\sum_{i=1}^{n} \sum_{j=0}^{i} m_j^l l_i^l + m_{1n+1}^{l'} - m_i^l r_i^l \right] \right)$$

$$L_i = \left(\frac{1}{m_T} \left[\sum_{j=i+1}^{n+1} m_j^{l'} l_i^{l'} + m_i^{l'} r_i^{l'} \right] \right) \tag{8.23}$$

Note that this formulation is applicable to dissimilar arms.

An equivalent derivation is given now for the cross product Jacobian with an illustrative example from the PUMA 560/600. From the cross product method for deriving the Jacobian matrix, we can calculate the Jacobian matrices of the respective manipulators for a dual arm configuration mounted onto a platform, i.e. $l = 2$ arms:

$$v_l^* = v_0 + \sum_{i=1}^{n} \left(z_{i-1}^l \times l_i^l \right) \dot{\theta}_i^l$$

$$w_l^* = \sum_{i=1}^{n} z_{i-1}^l \dot{\theta}_i^l$$

Now,

$$\dot{r}_{c0} = -\frac{1}{m_T} \sum_{l=1}^{2} \sum_{i=1}^{n+1} m_i^l \dot{p}_{ci}^l$$

and

$$\dot{p}_{ci}^l = v_{ci}^l = \sum_{k=1}^{i} v_k^l + w_i^l \times \left(-s_i^l \right) = \sum_{k=1}^{i} v_{k-1}^l + w_i^l \times r_i^l$$

$$\dot{r}_{c0} = v_0 = -\frac{1}{m_T} \sum_{l=1}^{2} \sum_{i=1}^{n+1} m_i^l \left(\sum_{k=1}^{i} v_{k-1}^l + w_i^l \times r_i^l \right)$$

$$= -\frac{1}{m_T} \sum_{l=1}^{2} \sum_{i=1}^{n+1} \left(\sum_{i=j}^{n+1} m_j^l v_{i-1}^l + m_i^l \left(w_i^l \times r_i^l \right) \right)$$

Now

$$v_i^l = \sum_{k=1}^{i} \left(z_{k-1}^l \times l_k^l \right) \dot{\theta}_k^l :$$

$$v_0 = -\frac{1}{m_T} \sum_{l=1}^{2} \sum_{i=1}^{n+1} \left(\sum_{i=j}^{n+1} m_j^l \sum_{i=1}^{n-1} \left(z_{i-1}^l \times l_i^l \right) \dot{\theta}_i^l + m_i \left(\sum_{i=1}^{n} z_{i-1}^l \times r_i^l \right) \dot{\theta}_i^l \right)$$

$$= -\frac{1}{m_T} \sum_{l=1}^{2} \sum_{i=1}^{n+1} \left(\sum_{j=i+1}^{n+1} m_j^l \sum_{i=1}^{n} \left(z_{i-1}^l \times l_i^l \right) \dot{\theta}_i^l + m_i^l \left(z_{i-1}^l \times l_i^l \right) \dot{\theta}_i^l \right)$$

Substitute:

$$v_i^* = \sum_{i=1}^{n} \left(z_{i-1}^l \times l_i^l \right) \dot{\theta}_i^l - \frac{1}{m_T} \sum_{l=1}^{2} \sum_{i=1}^{n+1} \left[\sum_{j=i+1}^{n+1} m_j^l \sum_{i=1}^{n} \left(z_{i-1}^l \times l_i^l \right) \dot{\theta}_i^l + m_i^l \left(z_{i-1}^l \times r_i^l \right) \dot{\theta}_i^l \right]$$

Now, the payload is invariant with respect to the end effector such that $\dot{\theta}_{n+1}^l = 0$:

$$v_i^* = \sum_{i=1}^{n} \left(z_{i-1}^l \times l_i^l \right) \dot{\theta}_i^l - \frac{1}{m_T} \sum_{l=1}^{2} \sum_{i=1}^{n} \left[\sum_{j=i+1}^{n+1} m_j^l \left(z_{i-1}^l \times l_i^l \right) + m_i^l \left(z_{i-1}^l \times r_i^l \right) \right] \dot{\theta}_i^l$$

$$- \frac{m_{n+1}}{m_T} \left(z_n^l \times r_{n+1}^l \right) \dot{\theta}_{n+1}^l$$

Separating the expression into its two separate arm contributions l and l':

$$v_l^* = \sum_{i=1}^{n} \left[\left(1 - \frac{1}{m_T} \sum_{i=1}^{n} \sum_{j=i+1}^{n+1} m_j^l \right) \left(z_{i-1}^l \times l_i^l \right) - \left(\frac{m_i^l}{m_T} \right) \left(z_{i-1}^l \times r_i^l \right) \right] \dot{\theta}_i^l$$

$$- \frac{1}{m_T} \sum_{i=1}^{n} \left[\left(\sum_{j=i+1}^{n+1} m_j^{l'} \left(z_{i-1}^l \times l_i^{l'} \right) + m_i^{l'} \left(z_{i-1}^l \times r_i^{l'} \right) \right) \dot{\theta}_i^l \right]$$

$$= \sum_{i=1}^{n} \left[\left(z_{i-1}^l \times \lambda_l \right) - \left(z_{i-1}^{l'} \times L_i \right) \right] \dot{\theta}_i^l$$

where

$$\lambda_i = \frac{1}{m_T} \sum_{i=1}^{n} \sum_{j=0}^{i} \left(m_j^l l_i^l - m_i^l r_i^l \right)$$

$$L_i = \frac{1}{m_T} \sum_{j=i+1}^{n+1} \left(m_j^{l'} l_i^{l'} + m_i^{l'} r_i^{l'} \right)$$

Hence, the dual configuration manipulator:

$$\bar{J}^l = \sum_{i=1}^{n} \left(\left(z_{i-1}^l \times \lambda_i \right) - \left(z_{i-1}^{l'} \times L_i \right) \right)$$

Example: For the PUMA 560/600 configuration of dual manipulators related by the *H*-matrix the following freeflyer Jacobian elements may be derived.

Master arm Jacobian:

$$\bar{J}_{11}^m = \left\{-s_1\left[\delta_6(c_{23}c_4s_5 + s_{23}c_5) + s_{23}\delta_4 + \alpha_2c_2\right] - c_1(s_4s_5\delta_6 + \delta_2)\right\}_m$$

$$+ \left\{-s_1\left[D_6(c_{23}c_4s_5 + s_{23}c_5) + s_{23}D_4 + A_2c_2\right] - c_1(s_4s_5D_6 + D_2)\right\}_s$$

$$\bar{J}_{12}^m = \left\{c_1\left[\delta_6(c_{23}c_4s_5 + s_{23}c_5) + s_{23}\delta_4 + \alpha_2c_2\right] - s_1(s_4s_5\delta_6 + \delta_2)\right\}_m$$

$$- \left\{c_1\left[D_6(c_{23}c_4s_5 + s_{23}c_5) + s_{23}D_4 + A_2c_2\right] - s_1(s_4s_5D_6 + D_2)\right\}_s$$

$$\bar{J}_{13}^m = 0; \quad \bar{J}_{14}^m = 0; \quad \bar{J}_{15}^m = 0; \quad \bar{J}_{16}^m = 1$$

$$\bar{J}_{21}^m = \left\{c_1(s_4s_5\delta_6 + \delta_2)\right\}_m + \left\{c_1(s_4s_5D_6 + D_2)\right\}_s$$

$$\bar{J}_{22}^m = \left\{s_1(s_4s_5\delta_6 + \delta_2)\right\}_m - \left\{s_1(s_4s_5D_6 + D_2)\right\}_s$$

$$\bar{J}_{23}^m = \left\{\begin{array}{c} -s_1\left[s_{23}c_4s_5\delta_6 - c_{23}(c_5\delta_6 + \delta_4) + \alpha_2s_2\right] \\ -c_1\left[c_{23}c_4s_5\delta_6 + s_{23}(c_5\delta_6 + \delta_4) + \alpha_2c_2\right] \end{array}\right\}_m$$

$$+ \left\{\begin{array}{c} -s_1\left[s_{23}c_4s_5D_6 - c_{23}(c_5D_6 + D_4) + A_2c_2\right] \\ -c_1\left[c_{23}c_4s_5D_6 + s_{23}(s_5D_6 + D_4) + A_2c_2\right] \end{array}\right\}_s$$

$$\bar{J}_{24}^m = \{-s_1\}_m; \quad \bar{J}_{25}^m = \{c_1\}_m; \quad \bar{J}_{26}^m = 0$$

$$\bar{J}_{31}^m = \{c_1s_4s_5\delta_6\}_m + \{c_1s_4s_5D_6\}_s$$

$$\bar{J}_{32}^m = \{s_1s_4s_5\delta_6\}_m - \{s_1s_4s_5D_6\}_s$$

$$\bar{J}_{33}^m = \left\{-s_1\left[s_3c_4s_5\delta_6 - c_3(c_5\delta_6 + \delta_4)\right] - c_1\left[c_3c_4s_5\delta_6 + s_3(\delta_4 + c_5\delta_6)\right]\right\}_m$$

$$+ \left\{-s_1\left[s_3c_4s_5D_6 - c_3(c_5D_6 + D_4)\right] - c_1\left[c_3c_4s_5D_6 + s_3(D_4 + c_5D_6)\right]\right\}_s$$

$$\bar{J}_{34}^m = \{-s_1\}_m; \quad \bar{J}_{35}^m = \{c_1\}_m; \quad \bar{J}_{36}^m = 0$$

$$\bar{J}_{41}^m = \left\{s_1s_{23}(\delta_6c_5 + \delta_4) - \delta_6c_{23}s_4s_5\right\}_m + \left\{s_1s_{23}(D_6c_5 + D_4) - D_6c_{23}s_4s_5\right\}_s$$

$$\bar{J}_{42}^m = \left\{\delta_6c_{23}c_4s_5 - c_1s_{23}(\delta_6c_5 + \delta_4)\right\}_m - \left\{D_6c_{23}c_4s_5 - c_1s_{23}(D_6c_5 + D_4)\right\}_s$$

$$\bar{J}_{43}^m = \left\{\delta_6s_{23}(c_1s_4s_5 - s_1c_4s_5)\right\}_m + \left\{D_6s_{23}(c_1s_4s_5 - s_1c_4s_5)\right\}_s$$

$$\bar{J}_{44}^m = \{c_1s_{23}\}_m; \quad \bar{J}_{45}^m = \{s_1s_{23}\}_m; \quad \bar{J}_{46}^m = \{c_{23}\}_m$$

$$\bar{J}_{51}^m = \{s_{23}s_4c_5\delta_6\}_m + \{s_{23}s_4c_5D_6\}_s$$

$$\bar{J}_{52}^m = \{s_{23}s_4s_5\delta_6\}_m - \{s_{23}s_4s_5D_6\}_s$$

$$\bar{J}_{53}^m = \left\{\delta_6\left[c_5(c_1c_{23}s_4 + s_1c_4) + s_5(s_1c_{23}s_4 - c_1c_4)\right]\right\}_m$$

$$+ \left\{D_6\left[c_5(c_1c_{23}s_4 + s_1c_4) + s_5(s_1c_{23}s_4 - c_1c_4)\right]\right\}_s$$

$$\bar{J}_{54}^m = \left\{-(c_1c_{23}s_4 + s_1c_4)\right\}_m; \quad \bar{J}_{55}^m = \left\{c_1c_4 - s_1c_{23}s_4\right\}_m; \quad \bar{J}_{56}^m = \{s_{23}s_4\}_m$$

$$\bar{J}_{61}^m = \left\{\delta_6\left[s_1(c_{23}c_4s_5 + c_5s_{23}) + c_1s_4s_5)\right]\right\}_m$$

$$+ \left\{D_6\left[s_1(c_{23}c_4s_5 + c_5s_{23}) + c_1s_4s_5)\right]\right\}_s$$

$$\bar{J}_{62}^m = \left\{-\delta_6\left[c_1(c_{23}c_4s_5 + s_{23}c_5) - s_1s_4s_5)\right]\right\}_m$$

$$- \left\{-D_6\left[c_1(c_{23}c_4s_5 + s_{23}c_5) - s_1s_4s_5)\right]\right\}_s$$

$$\bar{J}_{63}^m = 0; \quad \bar{J}_{64}^m = \left\{c_1(c_{23}c_4s_5 + s_{23}c_5) - s_1s_4s_5\right\}_m$$

$$\bar{J}_{65}^m = \left\{s_1(c_{23}c_4s_5 + s_{23}c_5) + c_1s_4s_5\right\}_m; \quad \bar{J}_{66}^m = \{-s_{23}c_4s_5 + c_{23}c_5\}_m$$

Slave arm Jacobian

$$\bar{J}_{11}^s = \left\{-s_1\left[\delta_6(c_{23}c_4s_5 + s_{23}\delta_4 + \alpha_2c_2\right] - c_1(s_4s_5\delta_6 + \delta_2)\right\}_s$$

$$+ \left\{-s_1\left[D_6(c_{23}c_4s_5 + s_{23}c_5) + s_{23}D_4 + A_2c_2\right] - c_1(s_4s_5D_6 + D_2)\right\}_m$$

$$\bar{J}_{12}^s = \left\{c_1\left[\delta_6(c_{23}c_4s_5 + s_{23}c_5 + s_{23}\delta_4 + \alpha_2c_2\right] - s_1(s_4s_5\delta_6 + \delta_2)\right\}_s$$

$$- \left\{c_1\left[D_6(c_{23}c_4s_5 + s_{23}c_5 + s_{23}D_4 + A_2c_2\right] - s_1(s_4s_5D_6 + D_2)\right\}_m$$

$$\bar{J}_{13}^s = 0; \quad \bar{J}_{14}^s = 0; \quad \bar{J}_{15}^s = 0; \quad \bar{J}_{16}^s = 1$$

$$\bar{J}_{21}^s = \left\{c_1(s_4s_5\delta_6 + \delta_2)\right\}_s + \left\{c_1(s_4s_5D_6 + D_2)\right\}_m$$

$$\bar{J}_{22}^s = \left\{s_1(s_4s_5\delta_6 + \delta_2)\right\}_s - \left\{s_1(s_4s_5D_6 + D_2)\right\}_m$$

$$\bar{J}_{23}^s = \left\{\begin{matrix} -s_1\left[s_{23}c_4s_5\delta_6 - c_{23}(c_5\delta_6 + \delta_4) + \alpha_2s_2\right] \\ -c_1\left[c_{23}c_4s_5\delta_6 + s_{23}(c_5\delta_6 + \delta_4) + \alpha_2c_2\right] \end{matrix}\right\}_s$$

$$+ \left\{\begin{matrix} -s_1\left[s_{23}c_4s_5D_6 - c_{23}(c_5D_6 + D_4) + A_2c_2\right] \\ -c_1\left[c_{23}c_4s_5D_6 + s_{23}(c_5D_6 + D_4) + A_2c_2\right] \end{matrix}\right\}_m$$

$$\bar{J}_{24}^s = \{-s_1\}_s; \quad \bar{J}_{25}^s = \{c_1\}_s; \quad \bar{J}_{26}^s = 0$$

$$\bar{J}_{31}^s = \{c_1s_4s_5\delta_6\}_s + \{c_1s_4s_5D_6\}_m$$

$$\bar{J}_{32}^s = \{s_1s_4s_5\delta_6\}_s - \{s_1s_4s_5D_6\}_m$$

$$\bar{J}_{33}^s = \left\{ -s_1 \left[s_3 c_4 s_5 \delta_6 - c_3 (c_5 \delta_6 + \delta_4) \right] - c_1 \left[c_3 c_4 s_5 \delta_6 + s_3 (\delta_4 + c_5 \delta_6) \right] \right\}_s$$

$$+ \left\{ -s_1 \left[s_3 c_4 s_5 D_6 - c_3 (c_5 D_6 + D_4) \right] - c_1 \left[c_3 c_4 s_5 D_6 + s_3 (D_4 + c_5 D_6) \right] \right\}_m$$

$$\bar{J}_{34}^s = \{ -s_1 \}_s; \qquad \bar{J}_{35}^s = \{ c_1 \}_s; \qquad \bar{J}_{36}^s = 0$$

$$\bar{J}_{41}^s = \left\{ s_1 s_{23} (\delta_6 c_5 + \delta_4) - \delta_6 c_{23} s_4 s_5 \right\}_s + \left\{ s_1 s_{23} (D_6 c_5 + D_4) - D_6 c_{23} s_4 s_5 \right\}_m$$

$$\bar{J}_{42}^s = \left\{ \delta_6 c_{23} c_4 s_5 - c_1 s_{23} (\delta_6 c_5 + \delta_4) \right\}_s - \left\{ D_6 c_{23} c_4 s_5 - c_1 s_{23} (D_6 c_5 + D_4) \right\}_m$$

$$\bar{J}_{43}^s = \left\{ \delta_6 s_{23} (c_1 s_4 s_5 - s_1 c_4 s_5) \right\}_s + \left\{ D_6 s_{23} (c_1 s_4 s_5 - s_1 c_4 s_5) \right\}_m$$

$$\bar{J}_{44}^s = \{ c_1 s_{23} \}_s; \qquad \bar{J}_{45}^s = \{ s_1 s_{23} \}_s; \qquad \bar{J}_{46}^s = \{ c_{23} \}_s$$

$$\bar{J}_{51}^s = \{ s_{23} s_1 c_5 \delta_6 \}_s + \{ s_{23} s_4 c_5 D_6 \}_m$$

$$\bar{J}_{52}^s = \{ s_{23} s_4 s_5 \delta_6 \}_s - \{ s_{23} s_4 s_5 D_6 \}_m$$

$$\bar{J}_{53}^s = \left\{ \delta_6 \left[c_5 (c_1 c_{23} s_4 + s_1 c_4) + s_5 (s_1 c_{23} s_4 - c_1 c_4) \right] \right\}_s$$

$$+ \left\{ D_6 \left[c_5 (c_1 c_{23} s_4 + s_1 c_4) + s_5 (s_1 c_{23} s_4 - c_1 c_4) D_4) \right] \right\}_m$$

$$\bar{J}_{54}^s = \left\{ -(c_1 c_{23} s_4 + s_1 c_4) \right\}_s$$

$$\bar{J}_{55}^s = \{ c_1 c_4 - s_1 c_{23} s_4 \}_s$$

$$\bar{J}_{56}^s = \{ s_{23} s_4 \}_s$$

$$\bar{J}_{61}^s = \left\{ \delta_6 \left[s_1 (c_{23} c_4 s_5 + c_5 s_{23}) + c_1 s_4 s_5 \right] \right\}_s$$

$$+ \left\{ D_6 \left[s_1 (c_{23} c_4 s_5 + c_5 s_{23}) + c_1 s_4 s_5 \right] \right\}_m$$

$$\bar{J}_{62}^s = \left\{ -\delta_6 \left[c_1 (c_{23} c_4 s_5 + c_5 s_{23}) - s_1 s_4 s_5 \right] \right\}_s$$

$$- \left\{ -D_6 \left[c_1 (c_{23} c_4 s_5 + c_5 s_{23}) - s_1 s_4 s_5 \right] \right\}_m$$

$$\bar{J}_{63}^s = 0; \qquad \bar{J}_{64}^s = \left\{ c_1 (c_{23} c_4 s_5 + s_{23} c_5) + c_1 s_4 s_5 \right\}_s;$$

$$\bar{J}_{65}^s = \left\{ s_1 (c_{23} c_4 s_5 + s_{23} c_5) + c_1 s_4 s_5 \right\}_s; \qquad \bar{J}_{66}^s = \{ -s_{23} c_4 s_5 + c_{23} c_5 \}_s$$

8.3.3 Dual-manipulator dynamic feedforward compensation

The dual manipulator case does not vary much from the single manipulator case since the reaction forces are merely summed at the mounting platform. Similarly if more than two manipulators are mounted. Indeed, this makes the procedure very flexible so that it may be applied to any mechanism (such as deployable arrays, extendable booms, gimballed sensors and antenna) which may be operational on the spacecraft. Such mechanisms may be treated (if they are rigid bodies) using robotics techniques such as the Newton–Euler recursive method to calculate the reaction forces and moments that they exert on the mounting spacecraft which may then be compensated by the dynamically feeding them forward to the attitude controller. For the dual arm case:

$$N_0^l = \sum_{i=1}^{n} N_{ci}^l - \sum_{i=1}^{n+1} m_i^l p_{ci}^l \times \ddot{r}_{c0}$$

Now,

$$N_{SUM} = \sum_{l=1}^{2} N_0^l = \sum_{l=1}^{2} \left(\sum_{i=1}^{n} N_{ci}^l - \sum_{i=1}^{n+1} m_i^l p_{ci}^l \times \ddot{r}_{c0} \right)$$

Now,

$$p_{cm}^* = \frac{1}{m_T} \sum_{l=1}^{2} \sum_{i=0}^{n+1} m_i^l p_{ci}^{*l} = \frac{1}{m_T} \sum_{l=1}^{2} \left(\sum_{i=0}^{n+1} m_i^l r_{c0}^l + \sum_{i=1}^{n+1} m_i^l s_0^l + \sum_{i=1}^{n+1} m_i^l p_{ci}^l \right)$$

$$= r_{c0} + \frac{1}{m_T} \left[\sum_{l=1}^{2} \left(m_{1n+1}^l s_0^l + \sum_{i=1}^{n+1} m_i^l p_{ci}^l \right) \right]$$

where

$$m_T = m_0 + \sum_{l=1}^{2} \sum_{i=1}^{n+1} m_i^l$$

Hence,

$$\sum_{l=1}^{2} \sum_{i=0}^{n+1} m_i^l p_{ci}^l = m_T \left(p_{cm}^* - r_{c0} \right) - \sum_{l=1}^{2} m_{1n+1}^l s_0^l$$

Substitute:

$$N_{SUM} = \sum_{l=1}^{2} \left(\sum_{i=1}^{n} N_{ci}^l - \left[m_T \left(p_{cm}^* - r_{c0} \right) - m_{1n+1}^l s_0^l \right] \times \ddot{r}_{c0} \right)$$

Similarly,

$$F_0^l = -\left(\frac{m_0}{m_T} \right) \sum_{i=1}^{n} F_{ci}^l, \qquad F_{SUM} = -\left(\frac{m_0}{m_T} \right) \sum_{l=1}^{2} \sum_{i=1}^{n+1} F_{ci}^l$$

and

$$\ddot{r}_{c0} = -\left(\frac{m_0}{m_T} \right) \sum_{l=1}^{2} \sum_{i=1}^{n+1} F_{ci}^l.$$

The sum of moments about the centre of mass of the mounting spacecraft to be compensated by the attitude-control system:

$$N_r = N_{SUM} + \sum_{l=1}^{2} s_0^l \times F_0^l$$

$$= \sum_{l=1}^{2} \left(\sum_{i=1}^{n} N_{ci}^l - \left[m_T \left(p_{cm}^* - r_{c0} \right) - \left(m_{1n+1}^l s_0^l \right) \right] \times \right.$$

$$\left. - \frac{1}{m_T} \sum_{i=1}^{n} F_{ci}^l + s_0^l \times - \left(\frac{m_0}{m_T} \right) \sum_{i}^{n} F_{ci}^l \right)$$

$$= \sum_{l=1}^{2} \left(\sum_{i=1}^{n} N_{ci}^l + \left[p_{cm}^* - r_{c0} - s_0^l \right] \times \sum_{i=1}^{n} F_{ci}^l \right) \qquad (8.24)$$

Notice that this formulation may easily be generalised to more than two manipulators and that, if $l = 1$, it is consistent with the single arm formulation.

8.4 SUMMARY

The decoupling of orientation and translation effects in the robotic spacecraft system trades off vast amounts of complexity with little loss of flexibility—for this reason the author refers to this methodology as the 'engineering' approach, as opposed to a strictly mathematical approach. The motion of the manipulator(s) will generate reaction forces and moments on the spacecraft platform at the spacecraft/manipulator coupling point(s). This will induce translational and rotational motion of the spacecraft platform in response to the manipulator movements. If no compensation is made for this effect, the robot end effector will miss its target. Feedforward thruster control introduces an unacceptable propulsive fuel overhead. The alternative solution in the literature involves the use of the generalised Jacobian and the inherent redundancy of the system does not permit closed-form solutions to the inverse position kinematics problem. The generalised Jacobian has a high complexity of $O(n^2)$ imposing a major computational burden on the limited processing capability of space-rated microprocessors (currently of the 80386 class). Furthermore, the generalised Jacobian suffers from dynamic singularities in the manipulator workspace and these singularities are not unique kinematic functions but are dependent on the dynamic properties of the system and are thereby unpredictable. This is a serious limitation. The approach adopted here was to decouple the translation and rotation components of the kinematics and dynamics of the system.

Initialisation occurs at a default configuration which could be set arbitrarily (for instance in the stowed configuration, or at zero joint angle values). The parameters D_2, A_2, D_4, and D_6 are calculated at this configuration. The computation of the system's centre of mass position with respect to local inertial coordinates. This is invariant for any given payload—evidently, this parameter must be recomputed on releasing and acquiring target objects of manipulation. The inverse kinematic/dynamic solution follows utilising equations (6.6–6.11) as modified above (equation (8.8)). The robot control system for the manipulator accounts for the translation effect of reaction forces by incorporating dynamic terms into the kinematic control algorithms as 'lumped' parameters.

The system utilises a dedicated attitude-control system to which the reaction moments on the spacecraft are fed forward for online compensation. This requires the computation of the reaction moments at the base of the manipulator which induce moments on the spacecraft bus. While computation of the joint torques are usually referred to the respective joint coordinates, reaction moment computation for the attitude-control system requires the dynamic variables F_{ci} and N_{ci} to be calculated with respect to base coordinates for summation. This is a fairly trivial calculation as it requires only a few additional computations to be performed. The computations of reaction moments are fed back to the spacecraft attitude control system for feedforward compensation.

The additional complexity of the formulation is small and utilises standard control methodologies for both spacecraft attitude and robotic manipulator control with small modifications. It has also been demonstrated that inertial control of position in space yields different joint requirements than for base control of position on Earth. Furthermore the formulation is readily extended to the dual-manipulator case.

The procedure outlined in this text is computationally straightforward and solves the space manipulator control problem simply—consequently it may be regarded as the 'engineering approach' to space robotic control since it trades off large computational overheads for a small loss of flexibility. The only advantage that the generalised Jacobian approach has over the methods presented here are that it allows the freeflyer to be manoeuvred such that the inertial positioning of the end effector remains invariant whilst altering the attitude of the mounting platform. This is not, it is conjectured, a highly useful capability and that its exclusion is justified by the computational savings of the outlined method given the limited computational resources available to spacecraft. It allows the use of standard spacecraft attitude and robot control algorithms with little modification and little extra computational effort due to the ease of incorporation of the modifications. Indeed, these algorithms are applicable not just to spacecraft employing manipulators, but to all spacecraft which employ communications antennae, rigid solar panels, orientable sensors or any spacecraft deployment mechanism (assuming no flexure) such as boom deployment and antenna deployment. Coriolis forces are generated during boom deployment as the boom moves relative to the body axes and reaction moments imposed on the spacecraft may be compensated by the feedforward compensation technique. Hence, given that the dynamic properties of the deployable are known the formulation introduced in this chapter provides a generalised methodology for the dynamics and control of virtually all rigid spacecraft regarding space-based actuation and deployment/pointing.

These algorithms are applicable to robotic submersibles. The underwater environment is used for simulation of weightlessness of space. Conceptually at least, sub-oceanic applications of robotics is similar to space applications. Indeed, ROVs have such strong similarities to robotic spacecraft that neutrally bouyant versions of a robotic spacecraft are used during the development, testing and operation of robotic spacecraft. The free-flying mode of control presented here is also applicable to oceanic ROVs, particularly as ROVs are operated under neutral bouyancy conditions for manoeuvrability. Furthermore, station-keeping alone occupies up to 35% of an operator's total workload. One example is the Ranger Neutral Buoyancy Vehicle (NBV) designed as a demonstrator for freeflying telerobotic control to perform operational tasks such as EVA-type activities, in-orbit package replacement, and deployment of failed mechanisms that will be employed in the Ranger spaceccraft. It has a propulsion system of eight ducted propeller thrusters to control the position and orientation of the vehicle. However, ROV dynamics are much more complex than those encountered by space robots due to uncertainties and non-linearities in high-density ocean currents and in the ducted propeller-driven electric thrusters. The pilot has insufficient capability to operate thrusters rapidly and correctly to maintain the vehicle on station and counter side-acting ocean currents. By their very nature, ROVs are multivariable dynamic systems characterised by nonlinearities through rigid body yaw–pitch coupling, significant hydrodynamic forces from rapid, high-density, multi-directional ocean currents on the vehicle with magnitudes ~4 knots at the surface and ~7 knots subsurface, and high drag effects in the ducted propeller-driven electric thrusters. These factors damage the capability of maintaining fine position accuracy for the robotic arms [Yuh 1990]. The hydrodynamic and propeller-driven thruster effects are absent in space robotic systems and accurate models are difficult to develop as ROVs do not have hydrodynamically shaped profiles and thrusters are highly nonlinear actuators. However, the space freeflyer control approach combined with online adaptation offers a potential solution. It provides a means for accounting for the robotic manipulator operation in the dynamics of the ROV allowing the implementation of force control—manipulator operation is rarely considered in ROV analysis and modelling. The global coordinate system is fixed to the ship on the ocean surface, and the local coordinate system is fixed to the ROV centre of mass. The global coordinates of the position and orientation of the vehicle may be specified by a vector and Euler angles both of which may be lumped into a DH matrix. The ROV is characterised by several additional forces and torques that are absent from those in space that may be included in the dynamic model: the forces and moments exerted on the vehicle by fluid currents; viscous drag on the motion of the vehicle is proportional to the square of the vehicle's velocity; thruster forces and torques may be modelled as [Yoerger et al 1990]: $N = FD$ where D = thruster diameter; $F = v\rho w D^3$, v = thruster axial speed, ρ = density of water, w = thruster shaft angular velocity, and D = thruster diameter. These terms may be added to the dynamic model of the ROV, but they are difficult to model accurately particularly due to phenomena such as cavitation, so adaptive control methods are essential to compensate for model inaccuracies.

9

Manipulator trajectory planning

We now consider the generation of end effector trajectories and for convenience the [*] superscript notation is dropped for simplicity. Trajectory planning is the process of specifying the desired time-dependent smooth path trajectory for the robot end effector to track from an initial location to a final desired location. It comprises four steps:

(i)   compute a series of points in cartesian space;
(ii) convert to joint space via the inverse kinematics;
(iii) quadratic spline fit to join the joint space points;
(iv) check the actuator input violations via the dynamics formulation.

The end effector is moved through a series of 'knot/interpolation points' on a straight line cartesian path, i.e. trajectory generation algorithm is given by:

$t = t_0$

loop: wait for next control interval
        $t = \delta t$ where δt = control sampling interval
        $H(t)$ = hand path position function at time t
        $J(t)$ = joint solution corresponding to $F(t)$
        if $t = t_f$ the exit
        return to loop.

The time to move between point j and $j + 1$ is defined as $T_j = t_f - t_0$. For accurate tracking, position, velocity and acceleration at hand level must be controlled so that no deviation from a staight line trajectory occurs. Paul (1979) formulated a series of uniform motions in matrix form between cartesian knot points: one straight line translation with constant linear velocity with respect to the base and two rotations with constant angular velocity. One rotation rotated the approach vector to align the tool, and another rotation rotated the orientation vector of the tool about the tool axis. At the initial and final positions, the linear and angular velocities are zero (at rest). Taylor (1979) developed a more efficient technique by using a quaternion formulation to represent rotation rather than 3×3 rotation matrices to reduce storage requirements and the number of primitive operations. In addition, he reduced the number of interpolation points by employing tracking to ensure that the end effector remained within a 'deadband' region about the desired points. This method is not adopted here since the matrix representation is used for other computations and is generally employed. The inverse transform at each knot point must be performed at a rate sufficiently high enough to drive the joints to track in real-time accurately, yet avoiding the structural resonant frequency of the manipulator to maintain servo stability. Paul (1979) found that the driving frequency should be eight times the resonant frequency. For the Unimate PUMA 560/600, $f_{resonant} \sim 8$ Hz implies a sampling frequency of 60 Hz giving a sample interpolation point every 16 ms. High sampling rate is particularly important for force control to yield desirable dynamic performance. Indeed, Ranger is utilising 200-Hz update rates compared with the Shuttle RMS which employs an update rate of 12.5 Hz limited by the Space Shuttle General Purpose Computer.

There is some evidence for straight line linear path planning in human cognition [Shepard 1984]. Surface optical information about bodies is constrained by the geometric lines of perspective. Similarly, the relative motion of rigid objects are also constrained by kinematic geometry. Although an infinite number of paths exist through which an object may be moved and/or rotated from a position A to a position B, the simplest motion is given through Chasles theorem. A unique axis exists such that an object may be moved from A to B by a rotation about that axis with a simultaneous translation along that axis to generate a minimal 6 DOF helical twist in 3D space (including the limiting cases of circular or rectilinear motion).

9.1 CARTESIAN TRAJECTORY GENERATION

The straight line trajectory is generated to provide intermediate knot points in cartesian space along the trajectory. Generally, it is characterised by a sequence of cartesian task positions/orientations represented by pairs of points j and $j + 1$ expressed in DH matrix form in a time interval $[0,T_j]$:

$$T_{06j} = C_j(t)H_j^j \qquad \text{and} \qquad T_{06j+1} = C_{j+1}(t)H_{j+1}^{j+1}$$

where

$$H = \begin{pmatrix} n & s & a & p \\ 0 & 0 & 0 & 1 \end{pmatrix}$$

$$C(t) = \begin{pmatrix} 1 & -w_z & w_y & v_x \\ w_z & 1 & -w_x & v_y \\ -w_y & w_x & 1 & v_z \\ 0 & 0 & 0 & 1 \end{pmatrix} \delta t$$

$C(t)$ represents the time-dependent transform matrix of the moving working coordinate frame with respect to base coordinates. If H_j is represented with respect to $j + 1$ coordinates:

$$T_{06j} = C_{j+1}(t)H_j^{j+1} \rightarrow H_j^{j+1} = [C_{j+1}(t)]^{-1}T_{06j}$$

The motion between any points j and $j+1$ is from $T_{06j} = C_{j+1}(t)H_j^{j+1}$ to $T_{06j+1} = C_{j+1}(t)H_{j+1}^{j+1}$. Motion from j to $j+1$ is expressed in terms of a drive function $D(h)$ where $h = t/T$ where t = elapsed time and T = total move time:

$$T_{06}(h) = C_{j+1}H_j^{j+1}D(h)$$

At $h = 0$, $D(h = 0) = I$ such that $T_{06j} = H_j^j = H_j^{j+1}D(h = 0)$ and at $h = 1$, $D(h = 1)$ such that $T_{06j+1} = H_{j+1}^{j+1} = H_j^{j+1}D(h = 1)$ which comprise boundary conditions. Hence,

$$D(h = 1) = \left[H_j^{j+1}\right]^{-1} H_{j+1}^{j+1}$$

where

$$H_j^{j+1} = \begin{pmatrix} n^j & s^j & a^j & p^j \\ 0 & 0 & 0 & 1 \end{pmatrix}; \quad H_{j+1}^{j+1} = \begin{pmatrix} n^{j+1} & s^{j+1} & a^{j+1} & p^{j+1} \\ 0 & 0 & 0 & 1 \end{pmatrix}$$

Now,

$$\left[H_j^{j+1}\right]^{-1} = \begin{pmatrix} n_x^j & n_y^j & n_z^j & -(n^T p)^j \\ s_x^j & s_y^j & s_z^j & -(s^T p)^j \\ a_x^j & a_y^j & a_z^j & -(a^T p)^j \\ 0 & 0 & 0 & 1 \end{pmatrix} \equiv \begin{pmatrix} \rho_x^j & \rho_y^j & \rho_z^j & -(\rho^T p)^j \\ 0 & 0 & 0 & 1 \end{pmatrix}$$

$$\rightarrow D(h = 1) = \begin{pmatrix} \rho_x^j & \rho_y^j & \rho_z^j & -(\rho^T p)^j \\ 0 & 0 & 0 & 1 \end{pmatrix}\begin{pmatrix} n^{j+1} & s^{j+1} & a^{j+1} & p^{j+1} \\ 0 & 0 & 0 & 1 \end{pmatrix}$$

$$= \begin{pmatrix} n_j^T n_{j+1} & n_j^T s_{j+1} & n_j^T a_{j+1} & n_j^T(p_{j+1} - p_j) \\ s_j^T n_{j+1} & s_j^T s_{j+1} & s_j^T a_{j+1} & s_j^T(p_{j+1} - p_j) \\ a_j^T n_{j+1} & a_j^T s_{j+1} & a_j^T a_{j+1} & a_j^T(p_{j+1} - p_j) \\ 0 & 0 & 0 & 1 \end{pmatrix}$$

Equivalently, this matrix represents one translation and two rotations: $D(h) = T(h)R^a(h)R^s(h)$, where

$$T(h) = \begin{pmatrix} 1 & 0 & 0 & hx \\ 0 & 1 & 0 & hy \\ 0 & 0 & 1 & hz \\ 0 & 0 & 0 & 1 \end{pmatrix}; \quad R^s(h) = \begin{pmatrix} \cos(h\varphi) & -\sin(h\varphi) & 0 & 0 \\ \sin(h\varphi) & \cos(h\varphi) & 0 & 0 \\ 0 & 0 & 1 & 0 \\ 0 & 0 & 0 & 1 \end{pmatrix}$$

$$R^a(h) = \begin{pmatrix} \sin^2\phi[1-\cos(h\theta)]+\cos(h\theta) & -\sin\phi\cos\phi[1-\cos(h\theta)] & \cos\phi\sin(h\theta) & 0 \\ -\sin\phi\cos\phi[1-\cos(h\theta)] & \cos^2[1-\cos(h\theta)]+\cos(h\theta) & \sin\phi\sin(h\theta) & 0 \\ -\cos\phi\sin(h\theta) & -\sin\phi\sin(h\theta) & \cos(h\theta) & 0 \\ 0 & 0 & 0 & 1 \end{pmatrix}$$

Hence,

$$D(h) = T(h)R^a(h)R^s(h) = \begin{pmatrix} D_{11}(h) & D_{21}(h) & D_{31}(h) & hx \\ D_{12}(h) & D_{22}(h) & D_{32}(h) & hy \\ D_{13}(h) & D_{23}(h) & D_{33}(h) & hz \\ 0 & 0 & 0 & 1 \end{pmatrix} \tag{9.1}$$

where

$$D_{11}(h) = \cos(h\varphi)\left[\sin^2\varphi(1-\cos(h\theta))+\cos(h\theta)\right]-\sin(h\varphi)\left[\sin\phi\cos\phi(1-\cos(h\theta))\right]$$

$$D_{12}(h) = \cos(h\varphi)\left[-\sin\phi\cos\phi(1-\cos(h\theta))\right]+\sin(h\varphi)\left[\cos^2\phi(1-\cos(h\theta))+\cos(h\theta)\right]$$

$$D_{13}(h) = -\cos(h\varphi)\cos\phi\sin(h\theta)-\sin(h\varphi)\sin\phi\sin(h\theta)$$

$$D_{21}(h) = -\sin(h\varphi)\left[\sin^2\varphi(1-\cos(h\theta))+\cos(h\theta)\right]-\cos(h\varphi)\left[\sin\phi\cos\phi(1-\cos(h\theta))\right]$$

$$D_{22}(h) = -\sin(h\varphi)\left[-\sin\phi\cos\phi(1-\cos(h\theta))\right]+\cos(h\varphi)\left[\cos^2\phi(1-\cos(h\theta))+\cos(h\theta)\right]$$

$$D_{23}(h) = \sin(h\varphi)\left[\cos\phi\sin(h\theta)\right]-\cos(h\varphi)\left[\sin\phi\sin(h\theta)\right]$$

$$D_{31}(h) = \cos\phi\sin(h\theta); \quad D_{32}(h) = \sin\phi\sin(h\theta); \quad D_{33}(h) = \cos(h\theta).$$

These expressions for $D(h)$ may be equated for $h = 1$:

$$x = n_j^T(p_{j+1}-p_j); \quad y = s_j^T(p_{j+1}-p_j); \quad z = a_j^T(p_{j+1}-p_j);$$

$$\cos\phi\sin\theta = n_j^T a_{j+1}; \quad \sin\phi\sin\theta = s_j^T a_{j+1}; \quad \cos\theta = a_j^T a_{j+1};$$

$$\rightarrow \tan\phi = \frac{\sin\phi\sin\theta}{\cos\phi\sin\theta} = \frac{s_j^T a_{j+1}}{n_j^T a_{j+1}}$$

$$\rightarrow \tan\theta = \frac{\sin\theta}{\cos\theta} = \frac{\sqrt{(\cos\phi\sin\theta)^2+(\sin\phi\sin\theta)^2}}{\cos\theta} = \frac{\sqrt{\left(n_j^T a_{j+1}\right)^2+\left(s_j^T a_{j+1}\right)^2}}{a_j^T a_{j+1}}$$

Also, $R^s(1) = R^a(1)^{-1}T(1)^{-1}D(1)$:

$$\begin{pmatrix} \cos\varphi & -\sin\varphi & 0 \\ \sin\varphi & \cos\varphi & 0 \\ 0 & 0 & 1 \end{pmatrix} = \begin{pmatrix} s^2\phi(1-c\theta)+c\theta & -s\phi c\phi(1-c\theta) & -c\phi s\theta \\ -s\phi c\phi(1-c\theta) & c^2\phi(1-c\theta)+c\theta & -s\phi s\theta \\ c\phi s\theta & s\phi s\theta & c\theta \end{pmatrix}$$

$$\begin{pmatrix} n_j^T n_{j+1} & n_j^T s_{j+1} & n_j^T a_{j+1} \\ s_j^T n_{j+1} & s_j^T s_{j+1} & s_j^T a_{j+1} \\ a_j^T n_{j+1} & a_j^T s_{j+1} & a_j^T a_{j+1} \end{pmatrix}$$

$$\rightarrow \sin\varphi = \left[-\sin\phi\cos\phi(1-\cos\theta)\right]n_j^T n_{j+1}$$

$$+ s_j^T n_{j+1}\left[\cos^2\phi(1-\cos\theta)+\cos\theta\right] + a_j^T n_{j+1}(-\sin\phi\sin\theta)$$

$$\rightarrow \cos\varphi = \left[-\sin\phi\cos\phi(1-\cos\theta)\right]n_j^T s_{j+1}$$

$$+ s_j^T s_{j+1}\left[\cos^2\phi(1-\cos\theta)+\cos\theta\right] + a_j^T s_{j+1}(-\sin\phi\sin\theta)$$

$$\tan\varphi = \frac{\sin\varphi}{\cos\phi} \rightarrow \varphi = \tan^{-1}\left(\frac{\sin\varphi}{\cos\varphi}\right)$$

This completes the computation of the 4×4 drive function $D(h)$. The drive transform $D(h)$ corresponds to a translation and two rotations to move from position to position.

$$T_{06}(h) = \begin{pmatrix} n_j & s_j & a_j & p_j \\ 0 & 0 & 0 & 1 \end{pmatrix} D(h) = \begin{pmatrix} n_h & s_h & a_h & p_h \\ 0 & 0 & 0 & 1 \end{pmatrix} \qquad (9.2)$$

where

$$n_x^h = n_x^j D_{11}(h) + s_x^j D_{12}(h) + a_x^j D_{13}(h)$$

$$n_y^h = n_y^j D_{11}(h) + s_y^j D_{12}(h) + a_y^j D_{13}(h)$$

$$n_z^h = n_z^j D_{11}(h) + s_z^j D_{12}(h) + a_z^j D_{13}(h)$$

$$s_x^h = n_x^j D_{21}(h) + s_x^j D_{22}(h) + a_x^j D_{23}(h)$$

$$s_y^h = n_y^j D_{21}(h) + s_y^j D_{22}(h) + a_y^j D_{23}(h)$$

$$s_z^h = n_z^j D_{21}(h) + s_z^j D_{22}(h) + a_z^j D_{23}(h)$$

$$a_x^h = n_x^j D_{31}(h) + s_x^j D_{32}(h) + a_x^j D_{33}(h)$$

$$a_y^h = n_y^j D_{31}(h) + s_y^j D_{32}(h) + a_y^j D_{33}(h)$$

$$a_z^h = n_z^j D_{31}(h) + s_z^j D_{32}(h) + a_z^j D_{33}(h)$$

$$p_x^h = n_x^j hx + s_x^j hy + a_x^j hz + p_x^j$$

$$p_y^h = n_y^j hx + s_y^j hy + a_y^j hz + p_y^j$$

$$p_z^h = n_z^j hx + s_z^j hy + a_z^j hz + p_z^j$$

The drive function $D(h)$ may now be used to derive the elements of the 4×4 DH matrix at any desired point along the trajectory between the initial and final points in space of the end effector. This enables knot points to be defined along the trajectory. The minimum number of intermediate trajectory segments is five linking six knot points, each equidistant in cartesian space to provide the initial point (at rest), the departure point, the approach point and the final point (at rest) with acceleration/constant velocity/deceleration phases. For a five-segment trajectory, $h = 0, \frac{1}{5}, \frac{2}{5}, \frac{3}{5}, \frac{4}{5}, 1$.

This formulation will produce constant linear and angular velocities, but at the endpoints, to avoid discontinuities, acceleration and deceleration phases are required. The time to accelerate from one velocity to another requires a time 2τ for both acceleration and deceleration assuming a symmetric constant acceleration profile. Luh & Lin (1981) pointed out that the stop at each line segment extends the execution time of the task. It is possible to initiate velocity change τ before the next point and maintain that acceleration until τ into the new motion segment, i.e. a constant acceleration from $-\tau$ to τ to patch between motion segments. This generates a path of a sequence of straight line segments connected by smooth arcs. There are m segments and $(m - 1)$ intersection points denoted by H_j^j between the $j = 0$ initial point and the $j = m$ final point. During the execution of segment j, the position/orientation of the end effector may be either in transition from segment $j - 1$, in segment j, or in transition to segment $j + 1$.

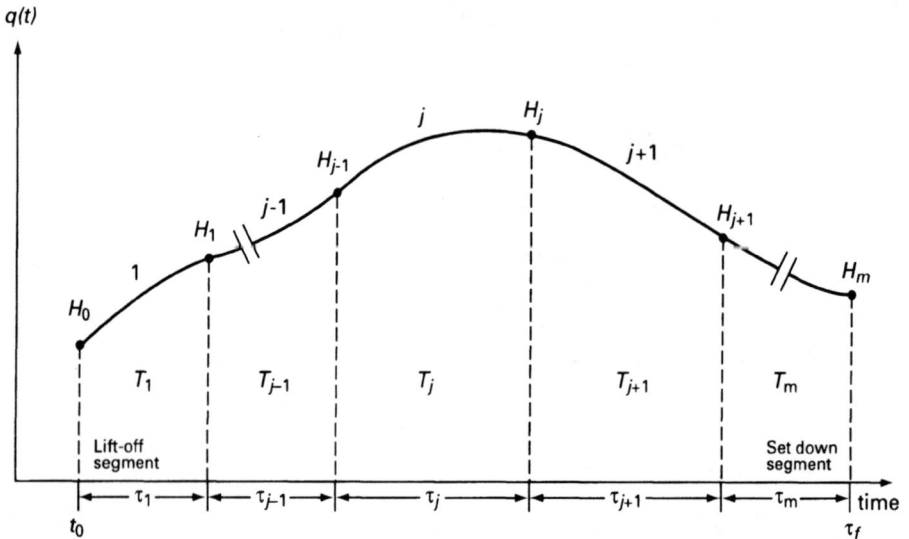

Fig. 9.1. Cartesian trajectory. (Adapted from Fu et al (1987) *Robotics: Control, Sensing, Vision and Intelligence*, with permission from McGraw-Hill.)

Changes of position/orientation of the end effector are related to v_j and w_j with respect to base coordinates. In the midportion of segment j, v_j and w_j are constant; in transition from segment j to $j + 1$, $\dot{v}_j$ and $\dot{w}_j$ are nonzero constants. The hand rests at the initial and final positions/orientations:

$$\left.\begin{aligned} p_k &= p_0, p_m \\ \theta_k &= 0 \end{aligned}\right) \quad \text{and} \quad \left.\begin{aligned} v_k &= 0 \\ w_k &= 0 \end{aligned}\right) \quad \text{for} \quad k < 1 \quad \text{and} \quad k > m \qquad (9.3)$$

For constant velocity segments

$$\left(\text{for} \quad \tau_0 + \sum_{k=1}^{j-1} T_k + \tau_{j-1} < t < \tau_0 + \sum_{k=1}^{j} T_k - \tau_j \right)$$

$$v_j = \frac{p_j - p_{j-1}}{T_j}$$

$$w_j = \left(\frac{\theta_j}{T_j} \right) \left[\frac{1}{2 \sin \theta_j} \begin{pmatrix} a_j^T s_{j-1} - s_j^T a_{j-1} \\ n_j^T a_{j-1} - a_j^T n_{j-1} \\ s_j^T n_{j-1} - n_j^T s_{j-1} \end{pmatrix} \right]$$

where

$$\theta_j = \cos^{-1} \left[\left(n_j^T n_{j-1} + s_j^T s_{j-1} + a_j^T a_{j-1} \right) / 2 \right] \qquad (9.4)$$

For accelerating segments

$$\left(\text{for} \quad \tau_0 + \sum_{k=1}^{j} T_k - \tau_j < t < \tau_0 + \sum_{k=1}^{j} T_k + \tau_j \right)$$

$$\dot{v}_j = \frac{v_{j+1} - v_j}{2\tau_j}$$

$$\dot{w}_j = \frac{w_{j+1} - w_j}{2\tau_j} \qquad (9.5)$$

The total execution time is

$$\tau_0 + \sum_{j=1}^{m} T_j + \tau_m$$

If the the time interval $2\tau_j$ during which velocities change is too long, the path travelled by the end effector may deviate from the nominal path. Hence, a ratio may be fixed between τ_j and T_j with $\tau_{j-1} + \tau_j \leq T_j$.

Example trajectory: Usually, five equal time intervals each satisfying a cubic are selected. The first $2T$ intervals have a nonzero second derivative (acceleration phase). At $2T$ to $3T$ the second derivative is set to zero and a constant first derivative holds for constant maximum velocity. From $3T$ to $5T$, the path again has a nonzero second derivative (deceleration phase). The constant rate portion in this case encompasses one-fifth of the trajectory. In reality, due to rate limitations of the actuators, short acceleration and deceleration phases at the ends of the trajectory with long constant velocity portions are more common. Maximum speed is attained by eliminating the constant rate portion in bang-bang acceleration/deceleration manoeuvres, but these are not usually adopted or indeed obtainable.

For a five-equal-segment trajectory:

$$T = \tau_0 + \sum_{1}^{5} T_j + \tau_5 \rightarrow T_1 = \frac{T}{5} - \tau_0, \quad T_2 = T_3 = T_4 = \frac{T}{5}, \quad T_5 = \frac{T}{5} - \tau_5$$

In such a five-segment trajectory, the first trajectory segment is the guarded departure move, so arbitrarily $\tau_0 = 0.5T_1$; similarly, the last trajectory segment is the guarded approach move, so again arbitrarily $\tau_5 = 0.5T_5$ such that $\tau_0 = \tau_5$. This defines the guarded move at half speed. It is also desirable to accelerate and decelerate the end effector uniformly to the desired speed so $\tau_2 = \tau_3 = 0$ and $|\tau_1| = |\tau_4| = 0.5T$.

For $T = 0$:

$$v_{j=0} = 0$$

$$w_{j=0} = 0$$

For $0 < t < 2\tau_0$:

$$\dot{v}_{j=0} = \frac{v_{j=1}}{2\tau_0}$$

$$\dot{w}_{j-0} = \frac{w_{j=1}}{2\tau_0}$$

For $2\tau_0 < t < \tau_0 + T_1 - \tau_1$:

$$v_{j=1} = \frac{p_{j=1} - p_{j=0}}{T_1}$$

$$w_{j=1} = \frac{\theta_{j=1}}{T_1} \left(\frac{1}{2 \sin \theta_{j=1}} \begin{bmatrix} a_{j=1}^T s_{j=0} - s_j^T a_{j=0} \\ n_{j=1}^T a_{j=0} - a_j^T n_{j=0} \\ s_{j=1}^T n_{j=0} - n_j^T s_{j=0} \end{bmatrix} \right)$$

For $\tau_0 + T_1 - \tau_1 < t < \tau_0 + T_1 + \tau_1$:

$$\dot{v}_{j=1} = \frac{v_{j=2} - v_{j=1}}{2\tau_1}$$

$$\dot{w}_{j=1} = \frac{w_{j=2} - w_{j=1}}{2\tau_2}$$

with

$$\langle v_{j=1} \rangle = \frac{v_{j=2} - v_{j=1}}{2}$$

$$\langle w_{j=1} \rangle = \frac{w_{j=2} - w_{j=1}}{2}$$

at point H_1

For $\tau_0 + T_1 + \tau_1 < t < \tau_0 + T_1 + T_2$:

$$v_{j=2} = \frac{p_{j=2} - p_{j=1}}{T_2}$$

$$w_{j=2} = \frac{\theta_{j=2}}{T_2} \left(\frac{1}{2\sin\theta_{j=2}} \begin{bmatrix} a_{j=2}^T s_1 + s_{j=2}^T a_1 \\ n_{j=2}^T a_{j=1} - a_{j=2}^T n_{j=1} \\ s_{j=2}^T n_{j=1} - n_{j=2}^T s_{j=1} \end{bmatrix} \right)$$

For $\tau_0 + T_1 + T_2 < t < \tau_0 + T_1 + T_2 + T_3$:

$$v_{j=3} = v_{j=2}$$
$$w_{j=3} = w_{j=2}$$

since

$$\dot{v}_{j=2} = 0$$
$$\dot{w}_{j=2} = 0$$

For $\tau_0 + T_1 + T_2 + T_3 < t < \tau_0 + T_1 + T_2 + T_3 + T_4 - \tau_4$:

$$v_{j=4} = v_{j=3}$$
$$w_{j=4} = w_{j=3}$$

since

$$\dot{v}_{j=3} = 0$$
$$\dot{w}_{j=3} = 0$$

For $\tau_0 + T_1 + T_2 + T_3 + T_4 - \tau_4 < t < \tau_0 + T_1 + T_2 + T_3 + T_4 + \tau_4$:

$$\dot{v}_{j=4} = \frac{v_{j=5} - v_{j=4}}{2\tau_4}$$

$$\dot{w}_{j=4} = \frac{w_{j=5} - w_{j=4}}{2\tau_4}$$

with

$$\langle v_{j=4} \rangle = \frac{v_{j=5} - v_{j=4}}{2}$$

$$\langle w_{j=4} \rangle = \frac{w_{j=5} - w_{j=4}}{2}$$

at point H_4

For $\tau_0 + T_1 + T_2 + T_3 + T_4 + \tau_4 < t < \tau_0 + T_1 + T_2 + T_3 + T_4 + T_5 - \tau_4$:

$$v_{j=5} = \frac{p_{j=5} - p_{j=4}}{T_5}$$

$$w_{j=5} = \frac{\theta_{j=5}}{T_5} \left(\frac{1}{2\sin\theta_{j=5}} \begin{bmatrix} a_{j=5}^T s_{j=4} - s_{j=5}^T a_{j=4} \\ n_{j=5}^T a_{j=4} - a_{j=5}^T n_{j=4} \\ s_{j=5}^T n_{j=4} - n_{j=5}^T s_{j=4} \end{bmatrix} \right)$$

For $\tau_0 + T_1 + T_2 + T_3 + T_4 + T_5 - \tau_5 < t < \tau_0 + T_1 + T_2 + T_3 + T_4 + T_5 + \tau_5$:

$$\dot{v}_{j=5} = \frac{-v_{j=5}}{2\tau_5}$$

$$\dot{w}_{j=5} = \frac{-w_{j=5}}{2\tau_5}$$

For $t = T$:

$$v_{j=6} = 0$$

$$w_{j=6} = 0$$

9.2 JOINT TRAJECTORY GENERATION

These methods of cartesian straight line trajectory generation do not allow for actuator limits in their formulation. Actuator torques, accelerations, and velocities at each joint provide constraints to the trajectory yet the path is constained in cartesian coordinates, e.g. the Unimate PUMA 560/600 can exert a maximum of 22 Nm at each of its joints. The manipulator is a highly coupled nonlinear mechanical system, so it is difficult to convert joint torques at the actuators to hand level equivalents. They cannot be converted to cartesian constraints by Jacobian transforms since they are valid only on a point-to-point basis and invalid for continuous paths. Hence, actuator constraints must be applied at joint level. Actuator limits are usually set at reduced torque bands to allow for model error and disturbances. Furthermore, measured variables for feedback are joint rather than cartesian variables, i.e. θ_i and $\dot{\theta}_i$.

For any trajectory, four points are usually specified:

(i)   initial point with zero velocity and acceleration but with specified position;
(ii)   lift-off point out along normal vector to the surface from the initial point with continuous motion;
(iii)   set-down point in along the normal vector to the surface to the final point with continuous motion;
(iv)   final position with zero velocity and acceleration.

At each of these four positions, velocity and acceleration are continuous over the time interval $[0, T_j]$. Joint solutions to m cartesian knot points are computed and it is necessary to interpolate between each joints' knot points. For a 6 DOF robot manipulator, there are six joints and so six approximate interpolation functions between each cartesian knot point. At joint level, a seven degree polynomial function may be used to fit the four knot points with their position, velocity and acceleration boundary conditions. However, such a polynomial is time-consuming to calculate and tends to exhibit 'wandering' from the nominal trajectory and overshoot the trajectory terminal point. Such high-degree polynomials involve high computational burden [Luh & Lin 1984]. The trajectory alternatively may be divided into segments to allow lower degree polynomials to be patched together. The spline curve is a polynomial of degree k with continuity in its derivative of order

$k - 1$ at its interpolation points generated to fit those points. The commonest methods of trajectory segmentation are the 4-3-4 and 5-cubic trajectories. The 4-3-4 three segment trajectory have the first segment modelled as a quartic polynomial from the initial to lift-off points, the second segment by a cubic polynomial from lift-off to set-down point, and the third trajectory element by a cubic polynomial from the set down point to the final point. The five-segment cubic spline trajectory has five segments all modelled by piece-wise smooth cubic spline polynomials linking six knot points, i.e. two additional points must be generated [Lin et al 1983]. The first and second derivative representing joint velocities and accelerations are continuous. In fact, the cubic is the lowest-degree poly-nomial that allows continuity in velocity and acceleration. The cubic polynomial approxi-mates to the cartesian straight line trajectory over a wide range of sampling rates and tracks the straight line path better than the quartic at joint level. Cubic splines may be fitted for the entire trajectory and do not require smoothing at the end points of motion segments. Lin & Chang (1983) suggested that if cartesian knot points are sequentially generated in real-time whereby only a few knot points are known at any one time, the X-spline function which requires only local knot point information for curve fitting, may be more suitable. X-splines are generalisations of the cubic spline and requires only two or three knot points. However, it requires relaxation of the continuity condition of joint acceleration, and discontinuity of the second derivative can cause acceleration jumps at the knot points. This can only be overcome by introducing more knot points, defeating the original purpose of the X-spline technique. It is questionable if there is indeed any advantage at all, since only six set points are required for cubic splines and the target will normally be tracked so that the first and final knot points will be known. The remaining four intermediate points are generated sequentially in turn. Once the first derivative has been calculated, the rest follow rapidly from the cartesian straight line trajector generator. Hence, the five-segment cubic joint trajectory technique appears to have substantial ad-vantages over the others.

The inverse kinematics at each update point are calculated at intervals along the cartesian trajectory to provide a set of joint knot points—interpolation occurs at joint level between the joint knot points using polynomial functions of degree n or less: the closeness to a straight line depends on the resolution of segmentation. The cubic poly-nomial results in minimum jerk to reduce the possibility of structural excitation [Lumia & Wavering 1989]. Lin et al (1983) proposed a generalised cubic polynomial method for n knot points for N joints, i.e. $n - 2$ interpolation points between initial and final knot points joined by $n - 1$ cubic polynomial segments, $Q_j(t)$ to allow jerk, acceleration and velocity constraints of the actuators to be applied at joint level. The cubic polynomials all satisfy required displacement, velocity and acceleration at each knot point and generate continuous displacement, velocity and acceleration over the complete time interval $[t_0, t_n]$. It requires a large number of knot points to be computed offline barring sensor-based real-time control. For real-time operation, trajectory generation must require only a few knot points. Chand & Doty (1985) intoduced a means of generating cubic spline trajectories online by using a limited look-ahead of four points on the trajectory with an error restriction of $<0.5\%$. As noted above, the cubic spline polynomial technique requires six knot points and this is highly compatible with the cartesian methods as well as overcoming some of the more undesirable effects of excessive computation. Lin et al's

(1983) generalised cubic interpolation method is thus adapted to the five segment trajectory:

Normalised time between knot points is given by

$$\tau = \frac{t_{j+1} - t}{T_j} \qquad \text{where} \qquad T_j = t_{j+1} - t_j \qquad \text{and} \qquad t_j \leq t \leq t_{j+1}$$

$$\rightarrow \frac{d\tau}{dt} = \frac{1}{T_j}$$

The cubic polynomial is:

$$Q_j(\tau) = a_{j3}\tau^3 + a_{j2}\tau^2 + a_{j1}\tau + a_{j0} \qquad \text{(cubic polynomial)}$$

$$Q_j'(\tau) = 3a_{j3}\tau^2 + 2a_{j2}\tau + a_{j1} \qquad \text{(parabolic polynomial)}$$

$$Q_j''(\tau) = 6a_{j3}\tau + 2a_{j2} \qquad \text{(linear straight line)}$$

Now,

$$\frac{dQ_j(\tau)}{d\tau} = \left(\frac{dQ_j(\tau)}{dt} \right) \frac{dt}{d\tau} = T_j \left(\frac{dQ_j(\tau)}{dt} \right)$$

$$\frac{d^2Q_j(\tau)}{d\tau^2} = \frac{d}{dt}\left(\frac{dQ_j(\tau)}{d\tau} \right) = \left(\frac{d^2Q_j(\tau)}{dt^2} \right) \frac{d^2t}{d\tau^2} = T_j^2 \frac{d^2Q_j(\tau)}{dt^2}$$

Hence,

$$\dot{\theta}_j = \frac{Q_j'(\tau)}{T_j} = \frac{3a_{j3}\tau^2 + 2a_{j2}\tau + a_{j1}}{T_j}$$

$$\ddot{\theta}_j = \frac{Q_j''(\tau)}{T_j^2} = \frac{6a_{j3}\tau + 2a_{j2}}{T_j^2}$$

At $\tau = 0 \equiv t = t_j$:

$$Q_j(0) = a_{j0} = \theta_j$$

$$\dot{\theta}_j = \frac{Q_j'(0)}{T_j} = \frac{a_{j1}}{T_j} \rightarrow a_{j1} = \dot{\theta}_j T_j$$

$$\ddot{\theta}_j = \frac{Q_j''(0)}{T_j^2} = \frac{2a_{j2}}{T_j^2} \rightarrow a_{j2} = \frac{1}{2}\ddot{\theta}_j T_j^2$$

At $\tau = 1 \equiv t = t_j$:

$$Q_j(1) = \theta_{j+1} = a_{j3} + a_{j2} + a_{j1} + a_{j0} = a_{j3} + \left(\frac{\ddot{\theta}_j T_j^2}{2} \right) + \dot{\theta}_j T_j + \theta_j$$

$$\rightarrow a_{j3} = (\theta_{j+1} - \theta_j) - \left(\frac{1}{2}\ddot{\theta}_j T_j^2\right) - \dot{\theta}_j T_j$$

$$\rightarrow Q_j(\tau) = \left[(\theta_{j+1} - \theta_j) - \left(\frac{1}{2}\ddot{\theta}_j T_j^2\right) - \dot{\theta}_j T_j\right]\tau^3$$

$$+ \left[\frac{1}{2}\ddot{\theta}_j T_j^2\right]\tau^2 + \left[\dot{\theta}_j T_j\right]\tau + \theta_j$$

$$\dot{\theta}_{j+1} = \frac{Q_j'(1)}{T_j} = \frac{3a_{j3} + 2a_{j2} + a_{j1}}{T_j} = \frac{3(\theta_{j+1} - \theta_j)}{T_j} - 2\dot{\theta}_j - \frac{1}{2}\ddot{\theta}_j T_j$$

$$\ddot{\theta}_{j+1} = \frac{Q_j''(1)}{T_j^2} = \frac{6a_{j3} + 2a_{j2}}{T_j^2} = \frac{6(\theta_{j+1} - \theta_j)}{T_j^2} - 2\ddot{\theta}_j - \frac{6\dot{\theta}_j}{T_j}$$

Continuous velocity and acceleration between j and $j + 1$ are obtained from the boundary conditions:

$$\dot{\theta}_{j+1} = \frac{Q_{j+1}'(0)}{T_{j+1}} = \frac{a_{j+1,1}}{T_{j+1}} = \frac{Q_j'(1)}{T_j} \quad \text{and} \quad \ddot{\theta}_{j+1} = \frac{Q_{j+1}''}{T_{j+1}} = \frac{2a_{j+1,2}}{T_{j+1}^2} = \frac{Q_j''(1)}{T_j^2}$$

For a five segment trajectory:

First trajectory segment:

$$Q_1(\tau) = a_{13}\tau^3 + a_{12}\tau^2 + a_{11}\tau + a_{10} \rightarrow Q_1(0) = a_{10} = \theta_0$$

$$\dot{\theta}_0 = \frac{Q_1'(0)}{T_1} = \frac{a_{11}}{T_1} \rightarrow a_{11} = \dot{\theta}_0 T_1$$

$$\ddot{\theta}_0 = \frac{Q_1''(0)}{T_1^2} = \frac{2a_{12}}{T_1^2} \rightarrow a_{12} = \frac{1}{2}\ddot{\theta}_0 T_1^2$$

$$Q_1(1) = a_{13} + \frac{1}{2}\ddot{\theta}_0 T_1^2 + \dot{\theta}_0 T_1 + \theta_0 = \theta_1$$

$$\rightarrow a_{13} = (\theta_1 - \theta_0) - \frac{1}{2}\ddot{\theta}_0 T_1^2 - \dot{\theta}_0 T_1$$

where $\dot{\theta}_0 = \ddot{\theta}_0 = 0$ initially, i.e. the first trajectory segment polynomial:

$$Q_1(\tau) = (\theta_1 - \theta_0)\tau^3 + \theta_0 \quad \text{with} \quad \dot{\theta}_1 = \frac{Q_1'(1)}{T_1} = \frac{3(\theta_1 - \theta_0)}{T_1}$$

$$\text{and} \quad \ddot{\theta}_1 = \frac{Q_1''(0)}{T_1^2} = \frac{6(\theta_1 - \theta_0)}{T_1^2}$$

Second trajectory segment:

$$Q_2(\tau) = a_{23}\tau^3 + a_{22}\tau^2 + a_{21}\tau + a_{20} \rightarrow Q_2(0) = a_{20} = \theta_1$$

$$\dot{\theta}_1 = \frac{Q_2'(0)}{T_2} = \frac{a_{21}}{T_2} \rightarrow a_{21} = \dot{\theta}_1 T_2$$

$$\ddot{\theta}_1 = \frac{Q_2''(0)}{T_2^2} = \frac{2a_{22}}{T_2^2} = \frac{Q_1''(1)}{T_1^2} \rightarrow a_{22} = \frac{1}{2}\ddot{\theta}_1 T_2^2$$

$$Q_2(1) = a_{23} + \frac{1}{2}\ddot{\theta}_1 T_2^2 + \dot{\theta}_1 T_2 + \theta_1 = \theta_2$$

$$\rightarrow a_{23} = (\theta_2 - \theta_1) - \frac{1}{2}\ddot{\theta}_1 T_2^2 - \dot{\theta}_1 T_2$$

i.e. the second trajectory segment polynomial:

$$Q_2(\tau) = \left[(\theta_2 - \theta_1) - \frac{1}{2}\ddot{\theta}_1 T_2^2 - \dot{\theta}_1 T_2\right]\tau^3 + \left[\frac{1}{2}\ddot{\theta}_1 T_2^2\right]\tau^2 + \left[\dot{\theta}_1 T_2\right]\tau + \theta_1$$

with

$$\dot{\theta}_2 = \frac{Q_2'(1)}{T_2} = \frac{3(\theta_2 - \theta_1)}{T_2} - 2\dot{\theta}_1 - \frac{1}{2}\ddot{\theta}_1 T_2 \quad \text{and}$$

$$\ddot{\theta}_2 = \frac{Q_2'(1)}{T_2^{22}} = \frac{6(\theta_2 - \theta_1)}{T_2^2} - 2\ddot{\theta}_1 - \frac{6\dot{\theta}}{T_2}$$

Third trajectory segment:

$$Q_3(\tau) = a_{33}\tau^3 + a_{32}\tau^2 + a_{31}\tau + a_{30} \rightarrow Q_3(0) = a_{30} = \theta_2$$

$$\dot{\theta}_2 = \frac{Q_3'(0)}{T_3} = \frac{a_{31}}{T_3} = \frac{Q_2'(1)}{T_2} \rightarrow a_{31} = \dot{\theta}_2 T_3$$

$$\ddot{\theta}_2 = \frac{Q_3''(0)}{T_3^2} = \frac{2a_{32}}{T_3^2} = \frac{Q_2''(1)}{T_2^2} \rightarrow a_{32} = \frac{1}{2}\ddot{\theta}_2 T_3^2$$

$$Q_3(1) = a_{33} + \frac{1}{2}\ddot{\theta}_2 T_3^2 + \dot{\theta}_2 T_3 = \theta_3 \rightarrow a_{33} = (\theta_3 - \theta_2) - \frac{1}{2}\ddot{\theta}_2 T_3^2 - \dot{\theta}_2 T_3$$

i.e. third trajectory segment polynomial:

$$Q_3(\tau) = \left[(\theta_3 - \theta_2) - \frac{1}{2}\ddot{\theta}_2 T_3^2 - \dot{\theta}_2 T_3\right]\tau^3 + \left[\frac{1}{2}\ddot{\theta}_2 T_3^2\right]\tau^2 + \left[\dot{\theta}_2 T_3\right]\tau + \theta_2$$

with

$$\dot{\theta}_3 = \frac{Q_3'(1)}{T_3} = \frac{3(\theta_3 - \theta_2)}{T_3} - 2\dot{\theta}_2 - \frac{1}{2}\ddot{\theta}_2 T_3 \qquad \text{and}$$

$$\ddot{\theta}_3 = \frac{Q_3''(1)}{T_3^2} = \frac{6(\theta_3 - \theta_2)}{T_3^2} - 2\dot{\theta}_2 - \frac{6\dot{\theta}_2}{2T_3}$$

Fourth trajectory segment:

$$Q_4(\tau) = a_{43}\tau^3 + a_{42}\tau^2 + a_{41}\tau + a_{40} \rightarrow Q_4(0) = a_{40} = \theta_3$$

$$\dot{\theta}_3 = \frac{Q_4'(0)}{T_4} = \frac{a_{41}}{T_4} = \frac{Q_3'(1)}{T_3} \rightarrow a_{41} = \dot{\theta}_3 T_4$$

$$\ddot{\theta}_3 = \frac{Q_4''(0)}{T_4^2} = \frac{2a_{42}}{T_4^2} = \frac{Q_3''(1)}{T_3^2} \rightarrow a_{42} = \frac{1}{2}\ddot{\theta}_3 T_4^2$$

$$Q_4(1) = a_{43} + \frac{1}{2}\ddot{\theta}_3 T_4^2 + \dot{\theta}_3 T_4 + \theta_3 = \theta_4 \rightarrow a_{43} = (\theta_4 - \theta_3) - \frac{1}{2}\ddot{\theta}_3 T_4^2 - \dot{\theta}_3 T_4$$

i.e. fourth trajectory segment polynomial:

$$Q_4(\tau) = \left[(\theta_4 - \theta_3) - \frac{1}{2}\ddot{\theta}_2 T_4^2 - \dot{\theta}_3 T_4\right]\tau^3 + \left[\frac{1}{2}\ddot{\theta}_3 T_4\right]\tau^2 + \left[\dot{\theta}_3 T_4\right]\tau + \theta_3$$

with

$$\dot{\theta}_4 = \frac{Q_4''(1)}{T_4} = \frac{3(\theta_4 - \theta_3)}{T_4} - 2\dot{\theta}_3 - \frac{1}{2}\ddot{\theta}_3 T_4 \qquad \text{and}$$

$$\ddot{\theta}_4 = \frac{Q_4''(1)}{T_4^2} = \frac{6(\theta_4 - \theta_3)}{T_4^2} - 2\dot{\theta}_3 - \frac{6\dot{\theta}_3}{2T_4}$$

Final trajectory segment:

$$Q_5(\tau) = a_{53}\tau^3 + a_{52}\tau^2 + a_{51}\tau + a_{50} \rightarrow Q_5(0) = a_{50} = \theta_4$$

$$\dot{\theta}_4 = \frac{Q_5'(0)}{T_5} = \frac{a_{51}}{T_5} = \frac{Q_4'(1)}{T_4} \rightarrow a_{51} = \dot{\theta}_4 T_5$$

$$\ddot{\theta}_4 = \frac{Q_5''(0)}{T_5^2} = \frac{2a_{52}}{T_5^2} = \frac{Q_4''(1)}{T_4^2} \rightarrow a_{52} = \frac{1}{2}\ddot{\theta}_4 T_5^2$$

$$Q_5(1) = a_{53} + \frac{1}{2}\ddot{\theta}_4 T_5^2 + \dot{\theta}_4 T_5 + \theta_4 = \theta_5 \rightarrow a_{53} = (\theta_5 - \theta_4) - \frac{1}{2}\ddot{\theta}_4 T_5^2 - \dot{\theta}_4 T_5$$

i.e. final trajectory segment polynomial:

$$Q_5(\tau) = \left[(\theta_5 - \theta_4) - \frac{1}{2}\ddot{\theta}_4 T_5^2 - \dot{\theta}_4 T_5\right]\tau^3 + \left[\frac{1}{2}\ddot{\theta}_4 T_5^2\right]\tau^2 + \left[\dot{\theta}_4 T_5\right]\tau + \theta_4$$

with

$$\dot\theta_5 = \frac{Q_5'(1)}{T_5} = \frac{3(\theta_5 - \theta_4)}{T_5} - 2\dot\theta_4 - \frac{1}{2}\ddot\theta_4 T_5 \qquad \text{and}$$

$$\ddot\theta_5 = \frac{Q_5''(1)}{T_5^2} = \frac{6(\theta_5 - \theta_4)}{T_5^2} - 2\ddot\theta_4 - \frac{6\dot\theta_4}{2T_5}$$

where $\dot\theta_5 = \ddot\theta_5 = 0$.

Hence, overall:

$$Q_1(\tau) = \left[(\theta_1 - \theta_0) - \left(\frac{1}{2}\ddot\theta_0 T_1^2\right) - (\dot\theta_0 T_1)\right]\tau^3 + \left[\frac{1}{2}\ddot\theta_0 T_1^2\right]\tau^2 + \left[\dot\theta_0 T_1\right]\tau + \theta_0$$

where

$$\dot\theta_0 = \ddot\theta_0 = 0$$

$$Q_2(\tau) = \left[(\theta_2 - \theta_1) - \left(\frac{1}{2}\ddot\theta_1 T_2^2\right) - (\dot\theta_1 T_2)\right]\tau^3 + \left[\frac{1}{2}\ddot\theta_1 T_2^2\right]\tau^2 + \left[\dot\theta_1 T_2\right]\tau + \theta_1$$

$$Q_3(\tau) = \left[(\theta_3 - \theta_2) - \left(\frac{1}{2}\ddot\theta_2 T_3^2\right) - (\dot\theta_2 T_3)\right]\tau^3 + \left[\frac{1}{2}\ddot\theta_2 T_3^2\right]\tau^2 + \left[\dot\theta_2 T_3\right]\tau + \theta_2$$

$$Q_4(\tau) = \left[(\theta_4 - \theta_3) - \left(\frac{1}{2}\ddot\theta_3 T_4^2\right) - (\dot\theta_3 T_4)\right]\tau^3 + \left[\frac{1}{2}\ddot\theta_3 T_4^2\right]\tau^2 + \left[\dot\theta_3 T_4\right]\tau + \theta_3$$

$$Q_5(\tau) = \left[(\theta_5 - \theta_4) - \left(\frac{1}{2}\ddot\theta_4 T_5^2\right) - (\dot\theta_4 T_5)\right]\tau^3 + \left[\frac{1}{2}\ddot\theta_4 T_5^2\right]\tau^2 + \left[\dot\theta_4 T_5\right]\tau + \theta_4$$

$$(9.6)$$

To maximise speed, the travelling time $t_{j+1} - t_j$ should be minimised within the velocity, acceleration, jerk (rate of change of acceleration) and torque constraints (related to acceleration by $\tau_t = I_{ci}\ddot\theta_i$) for each joint actuator. Kawato et al (1988) stated that minimisation of torque change (i.e. jerk) through the minimisation of an objective function is performed internally in the human brain and it has the form:

$$C_T = \int_{t_0}^{t_f} \sum_{i=1}^{n} \left(\frac{d\tau_i}{dt}\right)^2 dt$$

This is a nonlinear optimisation problem that indirectly minimises energy and produces a bell-shaped speed profile. The simpler cost function to be minimised in this case is:

$$T_j = \sum_{j=1}^{5} t_{j+1} - t_j$$

For each joint $i = 1 \ldots 6$:

$$\left|Q_{ij}'(\tau)\right| \le V_{ij}^{\max}, \qquad \left|Q_{ij}''(\tau)\right| \le A_{ij}^{\max}, \qquad \left|Q_{ij}'''(\tau)\right| \le J_{ij}^{\max}$$

Maximum velocity occurs at the midpoint of the trajectory (arbitrarily if the constant velocity segment is of finite duration), maximum acceleration occurs at the midpoint of the acceleration phases (arbitrarily since the acceleration phases are constant), and maximum jerk occurs at the end points of the trajectory. If the constraints are violated, T_j must be increased by a factor λ such that $T_j^* = \lambda T_j$, i.e.

$$Q_{ij}^{'*}(\tau)\frac{1}{\lambda}Q_{ij}'(\tau); \qquad Q_{ij}^{''*}(\tau) = \frac{1}{\lambda^2}Q_{ij}''(\tau);$$

$$Q_{ij}^{'''*}(\tau)\frac{1}{\lambda^3}Q_{ij}'''(\tau) \qquad \text{where} \qquad \tau = \frac{t_{j+1}-t}{T_j}$$

The velocity, acceleration and jerk are reduced by $1/\lambda$, $1/\lambda^2$, $1/\lambda^3$ respectively to remain within actuator constraints:

$$\lambda = \max\left(1, \lambda_v, \sqrt{\lambda_a}, \sqrt[3]{\lambda_j}\right)$$

where

$$\lambda_v = \frac{\left|Q_{ij}'(\tau)\right|}{V_j^{max}}, \quad \lambda_a = \frac{\left|Q_{ij}''(\tau)\right|}{A_j^{max}}, \quad \lambda_j = \frac{\left|Q_{ij}'''(\tau)\right|}{J_j^{max}}$$

Now, $Q_j(\tau) = a_{j3}\tau^3 + a_{j2}\tau^2 + a_{j1}\tau + a_{j0}$. Hence for each joint i,

$$Q_j'(\tau) = 3a_{j3}\tau^2 + 2a_{j2} + a_{j1}$$

$$= 3\left[(\theta_{j+1}-\theta_j)-\left(\frac{1}{2}\ddot{\theta}_jT_j^2\right)-\dot{\theta}_jT_j\right]\tau^2 + 2\left[\frac{1}{2}\ddot{\theta}_jT_j^2\right]\tau + \left[\dot{\theta}_jT_j\right] \leq V_j^{max}$$

$$Q_j''(\tau) = 6a_{j3}\tau + 2a_{j2}$$

$$= 6\left[(\theta_{j+1}-\theta_j)-\left(\frac{1}{2}\ddot{\theta}_jT_j^2\right)-\dot{\theta}_jT_j\right]\tau + 2\left[\frac{1}{2}\ddot{\theta}_jT_j^2\right] \leq A_j^{max}$$

$$Q_j'''(\tau) = 6a_{j3} = 6\left[(\theta_{j+1}-\theta_j)-\left(\frac{1}{2}\ddot{\theta}_jT_j^2\right)-\dot{\theta}_jT_j\right] \leq J_j^{max} \qquad (9.7)$$

A straightforward choice of λ is 2. This resembles the 'bracketing' procedure which is a version of the root-finding bisection algorithm which is guaranteed to converge on a root if one exists.

9.3 OBSTACLE AVOIDANCE

For any path planning to be effective, it must deal with the problem of obstacles in its work environment and avoid those obstacles to generate collision-free motion and to achieve a safe grasp point. Obstacle avoidance relies heavily on real-time sensory data. The goal is to automatically find a continuous safe path between configurations through a group of obstacles without collision. This is generally known as the piano-mover's problem and in general has a complexity of n^{2^r} where r = number of degrees of freedom of movement and n = number of moving objects [Schwartz & Sharir 1988]. It is desirable to obtain optimal collision-free solutions to the piano-mover's problem. Evidently, such computational complexity is totally infeasible for real-time solution, so any practical approach needs to employ assumptions and simplifications. Henderson & Gruyen (1990) interpreted the piano-mover's problem for a robot agent as essentially an egocentric problem whereby the robot must maintain certain spatial and temporal relations between itself and the world—a task that is routinely undertaken by the animal world.

The global planning methods that have been developed are of two types—the configuration-space (C-space) approach [Lozano-Perez 1981, 1983b] and the free-space approach [Brooks 1983]. The problem may be broken down into two subproblems: Findspace and Findpath. Findspace is concerned with generating a safe robot configuration such that the robot does not impose on any obstacles. Essentially, this boils down to determining whether an object A can be placed inside some specified region R so that it does not collide with any of the objects O_i already in that region. Findpath is concerned with generating a safe path between a set of robot configurations to get from an initial to a final point such that all configurations are safe along the safe path. Essentially this boils down to determining how to move A from one location to another without causing collisions with objects O_j.

In the configuration-space approach, the robot dimensions are reduced to a point while convex polyhedral obstacles are expanded to compensate and stored as lists of vertices. A free-space graph is constructed with arcs connecting the vertices of the expanded obstacles. Safe paths are those which do not intersect any of the obstacles. An A^* algorithm based on a distance-travelled cost function will select the shortest path. A problem with this approach is that it will allow too close a proximity to the obstacles unless an error margin of additional expansion is included. This, however, will cause valid solutions to be lost.

The free-space method considers freeway channels between polygonal obstacles by dividing these free channels into swept-volume overlapping generalised cones. Findpath then reduces to comparing the swept volume of the object with the swept volume of free space. The problem is to find a path from the initial position to the goal position following the spines of the generalised cones and changing from cone to cone at the spine intersection points. Translations along the spines are rotation-free and rotations are restricted to inter-cone intersection points at which all orientation changes are performed. A graph is constructed using distance travelled as a cost function for the A^* algorithm search. The graph nodes represent intersection points with the arcs representing free-space channels. The robot passes through the centre of these channels allowing a safe margin. The drawback with this approach is that it often generates a longer path than necessary and is not well suited to tightly constrained environments. This essentially excludes parts-mating operations which are representative of cluttered environments.

Both approaches have difficulties in three dimensions and in coping with rotational motion. However, both methods have been combined to consolidate some of the benefits of each and resolve some of the difficulties characteristic of each method. The C-space is subdivided into freeways for the hand and freeways for the arm, based on constraints of the freeways for the hand. There still remains the problem that certain arrangements yield no solution restricting its use. All these approaches are hypothesise-and-test methods which lack generality and cannot be used in cluttered environments or environments which are dynamic. Furthermore and more significantly, computation time remains a problem limiting its use to long-horizon time global planning only. For a k degree of freedom robot with m faces and edges whereby the joint range is resolved into r values in a workspace of n faces and edges, the complexity is $O(r^{k-1}(mn)^2)$. As the manipulator configuration changes in body-centred coordinates, the cost of recomputation is prohibitive.

9.3.1 Potential field approach

Obstacle avoidance has traditionally been a computationally expensive capability. However, real-time obstacle avoidance may be achieved by using the artificial potential field method in the low-level control laws [Khatib 1985]. Collision-avoidance is fundamentally a fast-response capability appropriate to real-time control, i.e. it is an algorithmic process (in C or FORTRAN for example) rather than a logical process (in PROLOG or LISP for example). A path-planning algorithm may be used to generate a set of critical points along a global path generated through the techniques described above [Krogh & Thorpe 1986]. These critical points act as subgoals for the potential field algorithm which implements dynamic steering using local feedback information to generate collision-free paths. The two levels of control may be executed in parallel with the path-planner algorithm generating future subgoals.

In the potential field approach, the manipulator is considered to move through a field of forces whereby the goal position provides an attractive force for the end effector and obstacles generate repulsive forces at their surfaces. This will direct the end effector towards the goal but away from the obstacles. The potential field obeys the superposition principle of force summation to generate the resultant potential field. Sensors such as vision sensors provide data concerning the shape and location of obstacles.

The force field is described by:

$$V(q) = V_g(q) + V_0(q)$$

$$F(q) = F_g(q) + F_o(q)$$

where $F = -grad\ V$, $F_g =$ attractive goal force and $F_o =$ repulsive obstacle forces.

These forces may be incorporated directly into the control law to generate joint actuator torques to direct the end effector away from obstacles towards the goal: $\tau = J^T F(q)$ due to the artificial field of forces acting at the end effector.

For the goal point, the potential field is attractive:

$$V_g(q) = \frac{1}{2} K_p^{obj} (q_g - q)^2$$

where $K_p^{obj} =$ proportional gain, $q_g =$ generalised goal position vector with respect to reference coordinates, and $q =$ end effector position vector with respect to reference coordinates.

Goal force, $F_g(q) = -grad\ V_g = K_p^{obj}(q_g - q)$. This attains its minimum value of zero when $q = q_g$. This is a proportional control law so damping is added for stability:

$$F_g(q) = K_p^{obj}(q_g - q) + K_v^{obj}\frac{dq}{dt}$$

such that

$$K_p^{obj}(q_g - q) > K_v^{obj}\dot{q}_{max} \qquad (9.8)$$

For the obstacles, a repulsive potential field generates a barrier at the obstacle surface which falls off rapidly away from the surface of convex polygonal objects to reduce perturbation effects:

$$V_0 = \frac{1}{2}N\left(\frac{1}{p} - \frac{1}{p_0}\right)^2 \quad \text{if } p \leq p_0$$

$$= 0 \qquad\qquad \text{if } p > p_0$$

where $p_0 =$ limiting range of potential field influence $p/(1 - \sqrt{m/N}v_{max}\,p_{max})$, $p =$ shortest distance from end effector to obstacle, $m =$ shortest distance from object centre to object surface, $N =$ potential field strength = constant, $p_{max} =$ clearance width of end effector, and $v_{max} =$ maximum speed of end effector.

Repulsive obstacle force,

$$F_o(x) = -grad\, V_o = N\left[\frac{1}{2}\left(\frac{1}{p} - \frac{l}{p_0}\right)\left(\frac{1}{p}\right)^2\frac{dp}{dx}\right] \quad \text{if } p \leq p_0$$

$$= 0 \qquad\qquad\qquad\qquad\qquad \text{if } p > p_0 \qquad (9.9)$$

F_0 is the force inducing artificial repulsion from the surface of the obstacle. Obstacles may be described by the composition of geometric primitives such as the cylinder or ellipsoid:

$$F_o = \sum_{i=1}^{n} F_{oi}$$

due to the superposition property of the potential field.

Surface n-ellipsoid parallelepiped of dimensions a, b, c:

$$\left(\frac{x}{a}\right)^{2n} + \left(\frac{y}{b}\right)^{2n} + \left(\frac{z}{c}\right)^{2n} = 1$$

Surface n-cylinder of cross section (a, b) and length $2c$:

$$\left(\frac{x}{a}\right)^{2} + \left(\frac{y}{b}\right)^{2} + \left(\frac{z}{c}\right)^{2n} = 1$$

Typically, $n = 4$ gives a good approximation.

The potential field approach as well as being suitable for real time control is also ideally suited for the uncluttered space environment since it forms paths based on known obstacle positions rather than free-space corridor generation. Collision-avoidance of links and joint limit avoidance may also be achieved with this method, but both require a different Jacobian to relate the required reference points to the joint torques. Kweon et al (1992) advocated the artificial potential field as a means of sensor fusion representation. Each sensor generates its own potential field model of any objects and multiple sensor data may be combined as the superposition of repellent forces in the force field. Potential fields provide a generalised environmental medium by virtue of their conceptual proximity to energy measures. Environmental complexity may be quantified using potential energy metrics across both the space and time dimensions.

The potential field method is prone to local minima which occur when the obstacle repulsive forces balance the goal attractive forces. These Lagrange points occur when the end effector is within range of the short-range influence of the obstacle potential field and the long-range attractive goal potential field such that $|F_{goal}| = |F_{obstacle}|$. These points will trap the end effector and may be avoided by using a higher-level global planner (e.g. C-space/free-space approaches) which derives a series of interim points along a path that avoids the local minima. The global planner works by searching for acceptable paths using a cost function based on the shortest distance, e.g. A^* algorithm, and so tends to be slow.

9.3.2 Simulated annealing

Rather than using global planning to overcome the local minima problem, a simulated annealing algorithm may be used [Kirkpatrick et al 1983]. This technique is an optimisation method for finding the minima or maxima for functions with many independent variables such as the NP-hard travelling salesman problem. It is an example of a contribution from pure science to practical engineering. It is based on the Metropolis algorithm which is used in statistical mechanics to numerically simulate the average behaviour of a Gibbs ensemble of many body systems ($\sim 10^{23}$ atoms/cm^3) in thermal equilibrium at a finite temperature. This search algorithm does not suffer from the problem of getting stuck in local minima.

In the gas state, each atom is weighted by its Boltzmann probability $e^{-E_i/k_B T}$ where E_i = energy of atom, k_B = Boltzmann constant, and T = absolute temperature, and this generates a normal Boltzmann distribution of energies. At low temperatures, however, when the ensemble begins to solidify, ground states become dominant with the ground state configurations being weighted by $e^{-N/2}$. In crystal growth, rapid cooling (quenching) at exponential rates such as at $T_n = (T_i/T_0)^n T_0$ causes the system to get out of equilibrium and the solid will form a glass with no crystalline order but with metastable, locally optimal structures far from the globally optimal highly ordered crystalline state. Iterative algorithms and techniques for optimisation problems are equivalent to this process of rapid quenching from high to low temperatures where the cost function plays the role of energy. By allowing only solutions which decrease the cost function, only metastable solutions will be found. High temperatures correspond to gradient descent algorithms, while low temperatures correspond to stochastic searches.

Annealing involves melting the material to high temperature and then lowering the temperature very slowly, allowing a long time near the freezing point to allow the system to achieve thermal equilibrium. This generates the global ground state with a defect-free highly ordered crystalline state. For optimisation problems, simulated annealing can produce solutions using effective temperature as the control variable. It allows uphill steps in the energy cost function. Solutions which yield $\delta E < 0$ are accepted, but also solutions with $\delta E > 0$. The probability of such a solution with $\delta E > 0$ is $P(\delta E) = e^{-\delta E / k_B T}$ and if a random number generator for $0 < \text{Rnd} < 1$ value is less than $P(\delta E)$ the solution is retained, otherwise it is rejected. The system retains thermal equilibrium during slow cooling and will tend towards a Boltzmann distribution without entering non-equilibrium metastable states. This procedure avoids getting stuck in local minima since the stochastic component allows uphill transitions. Simulated annealing is essentially an adaptive divide-and-conquer stategy and provides an interesting model of the way that evolution by natural selection operates. Hence, it is inherently parallel offering the potential for fast computation corresponding to maximum *a posteriori* estimation by maximising entropy across all probabilistic equilibrium states.

9.4 SUMMARY OF MANIPULATOR TRAJECTORY COMPUTATIONS

Trajectory-generation involves defining the trajectory of the manipulator end effector in terms of generalised cartesian position, velocity and acceleration over time between the initial configuration and the final configuration. Once the cartesian trajectory of the end effector has been computed, these variables must be transformed into their joint equivalents in sequence using the methods outlined previously: joint solution to the inverse position kinematics, joint rate solution to the inverse velocity kinematics, joint acceleration solution to the inverse acceleration kinematics, joint torque computation from the Newton–Euler dynamics (with a control law implementation—see Chapter 10). At each cartesian knot point characterised by its inertial cartesian variables $q^*, \dot{q}^*, \ddot{q}^*$, the solutions for finding the joint variable equivalents $\theta, \dot{\theta}, \ddot{\theta}$ must be computed. First, the inverse position kinematics formulation must be employed to find the joint angles corresponding to each cartesian end effector position. The first computation is to transform roll, pitch and yaw coordinates of each cartesian position into their *DH* matrix equivalents of twist (n), turn (s), and tilt (a) rotation vectors. The inverse solution to the *DH* matrix formulation proceeds to find joint angles $\theta = f^{-1}(q^*)$ for any given generalised inertial cartesian position q^* either analytically or numerically (equations 6.6–6.11, 8.8 and 8.9). These equations convert end effector coordinates

$$q = \begin{pmatrix} n & s & a & p^* \\ 0 & 0 & 0 & 1 \end{pmatrix}$$

as given by the 4×4 *DH* matrix into their equivalent respective joint coordinates.

Once the joint angles have been found, they form the basis for further computations. The joint rate inverse kinematic transformation involves inversion of the freeflyer Jacobian matrix $\bar{J}(\theta)$ (which was defined using the cross product method—equation 6.15 and 18.13) such that $\dot{\theta} = \bar{J}(\theta)^{-1} \dot{q}^*$. Matrix inversion is usually numerical utilising Gaussian elimination for example. The freeflyer Jacobian transpose may also be computed simultaneously for use in force control which transform end effector forces f_{ext} into their equivalent joint torques $\tau = \bar{J}(\theta)^T f_{ext}$. The final kinematic transformation is that of acceleration to find joint acceleration equivalents to cartesian accelerations $\ddot{\theta} = \bar{J}^{-1}(\ddot{q} - \dot{\bar{J}}\dot{\theta})$. This involves the use of the freeflyer Jacobian derivative (equations (6.24) and (8.14)) in conjunction with the inverted freeflyer Jacobian matrix. This completes the inverse kinematic solution of finding the joint equivalents $\theta, \dot{\theta}, \ddot{\theta}$ of the specified cartesian end effector kinematic variables $q^*, \dot{q}^*, \ddot{q}^*$ defined by each cartesian trajectory knot point.

Trajectory generation can then proceed with the above computations being performed along the trajectory. The trajectory is defined in terms of the manipulator end effector as a series of generalised inertial position, velocity and acceleration over time between the initial configuration and the final configuration. A straight line cartesian trajectory is generated to provide intermediate knot points in inertial cartesian space along the trajectory. The 4×4 matrix inertial drive function $D^*(h)$ describes the trajectory generation in terms of a single translation and two rotations (equation 9.1). The drive function is used to derive the 4×4 DH matrix at any inertial cartesian point along the trajectory (equation 9.2). The knot points are calculated at a rate limited by the computational capability of the robot control processor with the rate being bounded from below by a factor of the resonant frequency of the manipulator ~60 Hz. The minimum number of intermediate trajectory knot points is six with acceleration/deceleration phases. The departure and approach points are characterised as guarded moves at low speed. This is also the number of points required by the five-segment cubic spline curve-fitting procedure for the corresponding joint trajectory.

The dynamics formulation takes as its input the kinematic joint variables $\theta, \dot{\theta}, \ddot{\theta}$ and outputs the joint torques required to drive the joint actuators to achieve the stipulated cartesian trajectory. The most efficient formulation is the recursive Newton–Euler method. The Newton–Euler dynamics defines the forward kinematic variables (linear and angular velocity and acceleration) of each link and these are propagated from the base of the manipulator to the end effector (equation 7.2). This allows the forces and moments to be calculated for each link about each link centre of mass (equation 7.4). The backward recursion involves propagating the dynamic variables of force and moment for each link from the end effector to the base of the manipulator to compute the joint torques at each joint (equation 7.3).

The computed torque closed-loop control law forms the basis of most types of robotic control system (see Chapter 10, equation 10.7). It also forms part of the hybrid position/force control system. The force control segment involves computation of the required joint torques to apply desired end effector forces and torques on the target object. This is calculated using the freeflyer Jacobian transpose matrix which relates the joint motor torques to the required end effector forces and torques. The computed torque scheme overlies the dynamics formulation to calculate the joint torques which drive the joint motors to realise the specified end effector trajectory. Sensory feedback is required from joint sensors to provide proprioceptive data on the actual joint positions and joint rates. The computed torque technique provides PD control at each joint with linearisation by feedforward compensation of the nonlinear components of the manipulator dynamics. These computations enable the manipulator control system to drive the joint motors to track the desired reference trajectory of the end effector.

9.5 PARALLEL PROCESSING IMPLEMENTATIONS

The need for real-time trajectory generation is critical for robotics applications: robotic algorithms place extreme burdens on computational resources, necessitating the use of enhanced parallel and concurrent computing architectures. It should be noted that the $O(n^4)$ complexity of the Lagrange–Euler dynamic equations imply that parallel processing would be of limited value in reducing the computation time unless large numbers of processors can be used.

The Newton–Euler dynamics with their linear recursive structure may be implemented on p processors for an n link manipulator reducing their complexity to $O(k_1(n/p) + k_2(\log_2 p))$ [Graham 1989]. They are suitable for a parallel pipelined architectural implementation. In pipelined processing, the CPU subdivides the execution of instructions into a sequence of steps which are executed by individual subsidiary hardware unit processors. It operates like an industrial assembly line simultaneously executing the same instructions on different data streams. It is used extensively in high-performance computing. They are particularly suited to floating-point arithmetic computations. The systolic array is a generalisation of pipelining in a multidimensional pipelined array operating under the same control unit [Kung 1982]. It comprises a two dimensional grid (e.g. hexagonal or rectangular) of SIMD processor units with nearest neighbour four-link connections. The partial results of computations are passed onto one or more adjacent processors in the grid. They are essentially optimised for a particular calculation and have a time complexity independent of N because the flow of data through the network is fixed and optimised to a constant value. Input–output is overlapped with computation and each operand is input only once and operated on many times. A hexagonal systolic array staggers the computations so that the correct partial products arrive at the correct processor at the correct time. Such architectures are well suited to matrix and vector computations such as $N \times N$ multiplication. They are extremely fast and efficient but are limited to only a few operations. Speedups can be significant, e.g. $N \times N$ matrix multiplication can be performed in time complexity $O(N)$ rather than $O(N^2)$ as in conventional processors. For a 6 DOF robot manipulator, this would require 36 processor units which for the foreseeable future is not likely to be implemented on space systems. In these approaches, the number of processors required implies a heavy hardware cost and associated mass penalty.

Coprocessor are a SISD (single instruction, single data) architecture comprising specialised slave processors which relieve complex instructions from the main central processor thereby speeding up execution time (often as add-on boards to conventional machines). They do not operate simultaneously with the central processor. Usually, such coprocessors are RISC (reduced instruction set computers) designs. RISC architectures offer increased performance when executing commands as only the most-used instructions are implemented in hardware while others are implemented in software [Silbey 1986, Patterson 1985]. Indeed, less used and more complex instructions may be transferred to the compiler. This reduces operations and addressing modes and allow pipelined execution of instructions generating a five-fold increase in performance. One implementation of the RISC is the INMOS transputer. It is an advanced CMOS

n-substrate microprocessor computer. The transputer is a small (occupying one-tenth the silicon area of standard microprocessors) but complete computer system on a chip. Each transputer can work on a single job using its own local memory to overcome the bus bottleneck inherent in traditional von Neumann sequential architectures of single processor/single memory databus. In addition, transputers may be connected up into network arrays by high-speed serial links to provide good performance with minimal support circuitry with low power requirements [Walker 1985]. The transputer has four bidirectional communications link interfaces for input and output which operate independently and provide memory-to-memory data transfer rates of ~10 Mbps per channel. Its timer has a high clock cycle ~80 MHz. Communication can be almost simultaneous on all links in both directions. Processor speeds average around 10 Mips while dissipating ~1 W. In a network each transputer works on its own process, using its own local memory. Transputers use a concurrent programming language OCCAM to allow parallel processing to reduce waiting time for instructions. The hardware scheduler multiplexes concurrent processes in terms of priority and data availability. OCCAM allows both sequential and concurrent processes to communicate through the pipelined point to point channels and is almost as dense as assembler code. In this way it acts as a RISC processor due to the high-density instruction set. OCCAM's INPUT and OUTPUT primitives allow the sending of messages through interprocess communications channels activated when both transmitting and receiving processes are ready. OCCAM is implemented in a hardware kernel but allows FORTRAN and C programming. The IMS 424 is a 32-bit microprocessor. A 16-bit version, the IMS T222, is also available. Internal local static RAM is provided with 4 kbytes capacity for program and data storage. It has memory interface to allow the use of additional external off-chip RAM up to 4 Gbytes for the T424 and 64 kbytes for the T222 through a 26 Mbyte/s memory interface. If ROM is added it can act as a high-performance microprocessor with execution rates five times higher than standard microprocessors. Three IMS T222s are used on the Cluster spacecraft with 32 kbyte of external RAM (internal RM disabled) and 32 kbyte of ROM.

Thomlinson et al (1987) tested the T424 transputer for SEU (single event upset) threshold and total dose hardness using Californium-252, cobalt-60 and X-ray sources. The SEU threshold was around 3 MeV cm^2/mg and total dose hardness was 40 krad which was considered adequate for LEO where the expected dose rate is around 1 krad/y in LEO. They found that on-chip RAM was the dominant source of soft errors. When data was moved from internal to external memory devices, the error cross-section dropped by a factor of 50. SEU tests only yielded failures in internal memory. The T425 had a high radiation tolerance level at 44 krad. Such tests were extended to the T222 which had a functional failure limit of 10 krad [Cotavelo et al 1990]. The T222 with a disabled on-chip RAM increased its hardness to 40 krad and this could be increased to 70 krad under mission simulation. The T800 transputer is a version of interest to ESA and it comprises a 32-bit microprocessor with 4 kbytes on-chip RAM and an external memory interface. Internal memory was the major source of failure which when disconnected exhibited a 40 krad tolerance compared with 3 krad tolerance with operational RAM. The T800 flew on ENVISAT-1. The T800 has since been superseded by the T805 which exhibits a very low tolerance at 10 krad. The faster the operating speed, the lower the radiation tolerance of transputers. Furthermore, it is necessary to operate transputers with external memory which is highly susceptible to proton-induced SEU. Only when transputer devices are unpowered do their radiation tolerance levels reach 100 krad.

Zalzak & Morris (1991) suggested the use of INMOS transputers for concurrent processing for minimum-time solutions to the trajectory-generation problem. However, transputers are not space-rated and function only in LEO. They are likely to suffer severe degradation in radiation environments such as GEO (geosynchronous equatorial orbit) or HEO (high equatorial orbit). Therefore, they would limit the orbital range of any spacecraft adopting them.

10

Robotic manipulator control systems

As well as mechanism and structure, robot designs are also characterised as systems. Systems are of two types: process plant where material is processed and service networks which process information. Motor control systems in robotics are of the latter type. The control system provides the mapping between sensing of the environment status and the effector actions on the environment. The environment state is successively altered by those actions invoking the need for successive sensing, i.e. a perception–action cycle. In human cognition, sensory–motor cortical maps provide the means for cyclically alternating between attentional (sensory) and orienting (motor) systems during movement to direct the agent towards its target. The cycle implements comparison between a terminal motor map and a proprioceptive motor map to determine the degree of match [Grossberg 1980]. Only when the entire map plan has been executed do the proprioceptive and motor maps match thereby terminating the movement.

It is characteristic of robotics that the sensory processing required by the machine-level controller is concerned with sensory servoing. A robot must have the ability to change configuration and the position, velocity and acceleration of its end effector in real-time. This is a severe constraint on the control system. It is absolutely essential for online control to be realised, that efficient computation of the control algorithms occurs. The situation is hampered by the complex nonlinear dynamics due to the robotic manipulator's articulated structure. The majority of industrial robots are open-loop controlled but this is not sufficient for variable environments where sensory feedback is required. This necessitates closed-loop control whereby information is fed from the system output back to the system input.

The objective of the robot controller is to determine the input torques required to drive the joints from an initial steady state to reach a final steady state of desired output positions with the desired output velocities and accelerations within a prescribed settling time. This requires minimising to zero the difference between the actual performance trajectory and the desired reference trajectory. This discrepancy between the observed and predicted behaviour is the error. The controller applies input torques to reduce this error between actual and desired trajectories to zero. The control law specifies the magnitude and direction of the correction to be applied in response to the error measurements specified as the feedback gains. If exact models of the robot dynamics were available and those models were linearised exactly, then control could be implemented using optimal control techniques, and several workers have used this approach. For example, Luo & Saridis (1985) applied the LQR optimal control technique to robotic manipulator PID control to derive the proportional, integral and derivative feedback gains using a performance index based on quadratic error by casting the Lagrange–Euler formulation into a state space formulation of the dynamics [Athans 1971]. However, modelling errors will occur in such formulations because the robot is a complex, highly nonlinear system for which complete models are not available (e.g. dry coulombic friction at the joints is difficult to model). The use of high gains to render the control insensitive to parameter variations is unsuitable since this can produce instability and excitation as well as possible actuator saturation. It is a characteristic of nonlinear feedback systems that they can exhibit chaotic dynamics under certain conditions [Baillieul et al 1980]. Chaotic behaviour is characteristic of a sensitive dependence on initial conditions, variations of which produce output behaviour that is irregular and disordered. Although random and unpredictable, such behaviours obey mathematical rules. However, chaotic output is not desirable in robotic control systems, although they have been implicated in mammalian neural dynamics.

At each sample period, the control algorithm performs the following functions [Geshke 1983]:

(i)   read sensor outputs of joint position and velocity; compute desired joint angles, velocities and accelerations using the inverse kinematic formulations;

(ii)   compute the model dynamics to calculate the required input torques.

Hand control is normally performed separately from arm/wrist control since hand control can be complex if multijointed universal hands are used instead of parallel jaw grippers. The arm/wrist control problem is to calculate the joint torques to cause the manipulator to track the desired trajectory. This must be computed frequently at $f_{update} \sim k f_{max}$ where f_{max} for most manipulators is 5–10 Hz and $k \sim 5$–8, i.e. $f_{update} \sim 60$ Hz.

10.1 RESOLVED MOTION RATE CONTROL

Resolved motion control specifies all control variables in terms of cartesian end effector coordinates. The chief advantage lies in performing all kinematic control at cartesian hand level which is more suitable for task execution. One difficulty lies in the fact that sensors operate at joint level and forward transformation is required to implement feedback control introducing the possibility of inaccuracy. Resolved motion control is used in teleoperation with the human operator providing closure of the control loop. The major disadvantage is that it takes no account of the dynamics of the system unless these are explicitly included in the control system.

In resolved rate control, $\dot{\theta}$ is specified, θ is measured and $\ddot{\theta}$ is obtained numerically. Present and desired positions are computed by resolved position techniques and corrective cartesian rate $\dot{q}$ is calculated to reduce the position error and joint rates are obtained by inversion of the Jacobian [Whitney 1969, 1972] or as follows:

$$
\begin{pmatrix}
n_x^d(t) & s_x^d(t) & a_x^d(t) & p_x^d(t) \\
n_y^d(t) & s_y^d(t) & a_y^d(t) & p_y^d(t) \\
n_z^d(t) & s_z^d(t) & a_z^d(t) & p_z^d(t) \\
0 & 0 & 0 & 1
\end{pmatrix}^{-1}
\begin{pmatrix}
n_x^d(t+\delta t) & s_x^d(t+\delta t) & a_x^d(t+\delta t) & p_x^d(t+\delta t) \\
n_y^d(t+\delta t) & s_y^d(t+\delta t) & a_y^d(t+\delta t) & p_y^d(t+\delta t) \\
n_z^d(t+\delta t) & s_z^d(t+\delta t) & s_z^d(t+\delta t) & p_z^d(t+\delta t) \\
0 & 0 & 0 & 1
\end{pmatrix}
$$

$$
=
\begin{pmatrix}
1 & -w_z^d & w_y^d & v_x^d \\
w_z^d & 1 & -w_x^d & v_y^d \\
-w_y^d & w_x^d & 1 & v_z^d \\
0 & 0 & 0 & 1
\end{pmatrix}\delta t
$$

such that

$$
\dot{q}^d =
\begin{pmatrix}
v_x^d \\
v_y^d \\
v_z^d \\
w_x^d \\
w_y^d \\
w_z^d
\end{pmatrix}
=
\begin{pmatrix}
n^d(t)[p^d(t+\delta t)-p^d(t)]/\delta t \\
s^d(t)[p^d(t+\delta t)-p^d(t)]/\delta t \\
a^d(t)[p^d(t+\delta t)-p^d(t)]/\delta t \\
\frac{1}{2}[a^d(t).s^d(t+\delta t)-a^d(t+\delta t).s^d(t)]/\delta t \\
\frac{1}{2}[n^d(t).a^d(t+\delta t)-n^d(t+\delta t).a^d(t)]/\delta t \\
\frac{1}{2}[s^d(t).n^d(t+\delta t)-s^d(t+\delta t).n^d(t)]/\delta t
\end{pmatrix}
\tag{10.1}
$$

and $\dot{e} = \dot{q}^d - \dot{q}$ with $\dot{q} = J\dot{\theta}$.

This mode of control is adopted by the Space Shuttle Remote Manipulator System and it is also being suggested for use in future Canadian space robotic systems. It is particularly suited to redundant systems using the Moore–Penrose inverse of the Jacobian. It is the most generally used method of control for teleoperated robots, including human prosthetics, but it is primitive taking no account of the manipulator dynamics.

Resolved motion force control works well only when the load mass is much greater than the manipulator mass [Wu & Paul 1982], i.e. when only negligible joint torques are required to accelerate the links, thereby restricting its validity. Although such a situation may occur for space robotics, this is rarely the case in terrestrial robotics. Furthermore, the method is approximate rather than exact.

10.2 DC MOTOR CONTROL

Actuators at each joint provide the torques to drive the end effector through its required trajectory. The voltage-controlled dc motor is an electromechanical system whereby the mechanical part is coupled to the electrical part and it drives an inertial load. Permanent magnets give a constant magnetic field and the motor is controlled by the current in the armature coil. The system input is the applied voltage to the armature coil and the output is the angular position of the rotor. Internal sensors are used to provide direct information concerning the state of the manipulator joints for the control system. The motor shaft is connected to optical encoder, resolver or potentiometer sensors for position sensing and tachometer sensors for velocity sensing. The actuator dynamics affect robot manipulator performance and so must be used to derive the control gains [Pan & Sharp 1991].

If gear transmission is used, the gear ratio n is given by (for a direct drive motor, $n = 1$):

n = (no. of teeth of input motor shaft)/(no. of teeth of output load shaft)

$$= \frac{N_m}{N_l} = \frac{\theta_l}{\theta_m}$$

where subscript 'l' indicates load and subscript 'm' indicates motor.

Hence, actuator displacement, $\theta_l = n\theta_m$, actuator velocity, $\dot{\theta}_l = n\dot{\theta}_m$, and actuator acceleration, $\ddot{\theta}_l = n\ddot{\theta}_m$.

Consider a single dc motor [Luh 1983a,b; Moya & Seraji 1987]. Total torques comprise torques on the motor and on the load: $\tau = \tau_l + \tau_m$ where $\tau_m = J_m\ddot{\theta}_m + b_m\dot{\theta}_m$ and $\tau_l = J_l\ddot{\theta}_l + b_l\dot{\theta}_l$, where J = moment of inertia and b = viscous friction damping.

This includes the internal torques required to overcome the inertia and the damping effects (assuming that the nonlinear static friction is negligible—typically it is ~0.02 Nm). Conservation of work states that the work done by the load referred to the load shaft $\tau_l\theta_l$ is equal to the work done by the load referred to the motor shaft $\tau_l'\theta_m$. The load inertia is increased by n^2 when referenced to the motor shaft:

$$\tau_l' = \frac{\tau_l\theta_l}{\theta_m} = n\tau_l = n(J_l\ddot{\theta}_l + b_l\dot{\theta}_l) = n^2(J_l\ddot{\theta}_m + b_l\dot{\theta}_m)$$

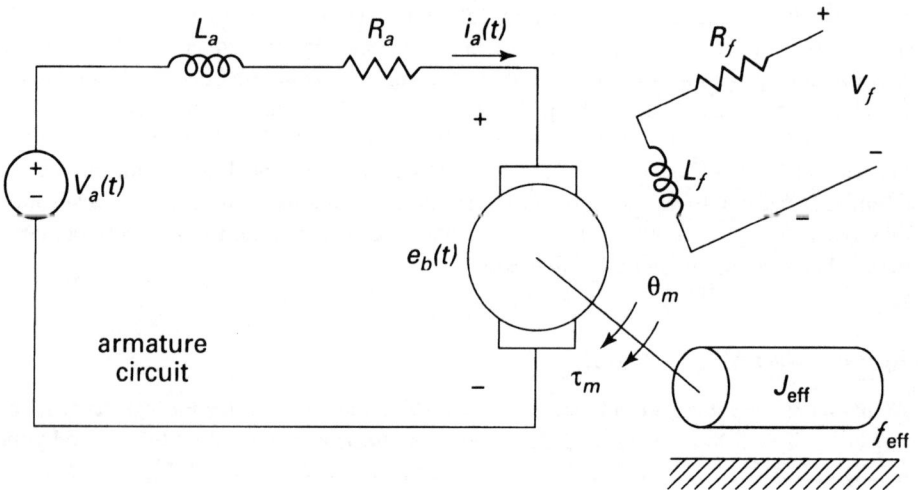

Fig. 10.1. DC motor equivalent circuit (from Fu et al 1987). (Reproduced with permission from Fu et al (1987) *Robotics: Control, Sensing, Vision and Intelligence*, © McGraw-Hill.)

Hence, the torque at the motor shaft,

$$\tau = \tau_m + \tau_l' = J_{eff}\ddot{\theta}_m + b_{eff}\dot{\theta}_m$$

where $J_{eff} = J_m + n^2 J_l =$ effective combined inertia, and $b_{eff} = b_m + n^2 b_l =$ effective viscous friction coefficient.

The torque required to accelerate the inertial load is directly proportional to the armature current: $\tau_m = K_t i$, where $K_t =$ motor torque constant.

Apply Kirchhoff's voltage law to the armature circuit to relate armature current i to the applied control voltage V_i at the terminals of the armature:

$$V_i = R_i i + L_i \frac{di}{dt} + E_b$$

where $R_i =$ armature coil resistance, $L_i =$ armature coil inductance, and $E_b =$ back EMF.

The back EMF varies in proportion to the angular velocity of the motor according to Faraday's law:

$$E_b = K_b \dot{\theta}_m = \frac{K_b}{n}\dot{\theta}_l$$

where $K_b =$ motor electromechanical torque constant.

Armature current,

$$i = \frac{V_i - L_i \dfrac{di}{dt} K_b \dot{\theta}_m}{R_i}$$

For a constant input voltage, the motor will rotate at a constant angular velocity such that the back EMF balances the applied voltage. The angular position output increases at a contant rate. This is characterised by the proportionality gain K_t to generate the electromagnetic torque:

$$\rightarrow \tau = K_t \frac{V_i - L_i \dfrac{di}{dt} K_b \dot{\theta}_m}{R_i}$$

This is the electromechanical transducer equation for the motor. This may be equated to the earlier formulation for the motor torque:

$$J_{eff}\ddot{\theta}_m + b_{eff}\dot{\theta}_m = K_t \frac{V_i - L_i \dfrac{di}{dt} K_b \dot{\theta}_m}{R_i}$$

The electrical time constant $t = L/R$ is much less than the mechanical time constant, so the armature inductance may be considered as negligible.

Apply the Laplace transform:

$$T(s) = s^2 J_{eff}\theta_m(s) + s b_{eff}\theta_m(s) = K_t \left[\frac{V_i(s) - sK_b\theta_m(s)}{R_i} \right] \quad \text{where} \quad s = \sigma + jw$$

Open loop transfer function,

$$G(s) = \frac{\mathscr{L}(\text{output})}{\mathscr{L}(\text{input})} = \frac{\theta_m(s)}{V(s)} = \frac{K_t}{s^2 R_i J_{eff} + s(R_i b_{eff} + K_t K_b)}$$

$$\frac{\theta_l(s)}{V(s)} = \frac{nK_t}{s^2 R_i J_{eff} + (R_i b_{eff} + K_t K_b)} = \frac{nK}{s(T_m s + 1)}$$

where

$$K = \text{motor torque gain constant} = \frac{K_t}{R_i b_{eff} + K_t K_b}$$

$$T_m = \text{motor torque time constant} = \frac{R_i J_{eff}}{R_i b_{eff} + K_t K_b}$$

It is now necessary to close the open-loop control system by introducing feedback. A closed loop may be implemented such that the angular position output is fed back to cancel the input voltage to maintain a constant voltage for a constant angular position. A position (proportional) controller servos the motor in such a way that the actual angular displacements of the joint track the desired angular displacement. The tracking error from the reference (desired) trajectory q^d is given by $|q^d - q| < e$ and for small error, e, the actual (measured) trajectory q must remain in the neighbourhood of q^d for stability. The position (proportional controller output is proportional to the error. The applied voltage of the motor varies linearly with the position error defined as the difference between the desired angular displacement and the angular displacement measured by the position sensor:

$$V_i = \frac{K_p e}{n} = \frac{K_p}{n}(\theta_l^d - \theta_l)$$

where K_p = position feedback gain.

Position feedback provides a deadbeat response to external disturbances which might deflect the manipulator. This is proportional feedback which has the effect of reducing the steady-state error. To minimise the response time and provide stability with no overshoot, damping of flexible modes must be added by including a rate feedback proportional to the velocity error to stabilise oscillatory responses particularly if K_p is large. The actual velocity of the joint is measured by a tachometer sensor:

$$V_i = \frac{K_p}{n} = (\theta_l^d - \theta_l) + \frac{K_v}{n}(\dot\theta_l^d - \dot\theta_l) = \frac{K_p e + K_v \dot e}{n}$$

where K_v = velocity feedback gain.

Velocity feedback based on error-rate to minimise overshoot reduction is particularly important for direct drive motors due to their high inertial loads. Derivative feedback compensates for changing measurements and tracks rapid changes, so reducing transient errors. It adds a phase lead to dampen rapid, oscillatory changes. Such PD control changes the poles of the transfer function to more desirable locations in the left half of the complex frequency domain. An integral term may be introduced into the control law (for PID control) to enable higher velocities to be obtained with lower joint torques [Nicosia & Tomei 1984]. It ensures that the controller output is proportional to the amount of time the error is present, eliminating bias offsets and steady-state errors for high accuracy. It gives high gain at low frequencies by introducing a phase lag. However, an integrator is not usually included in robot motors because the velocities of robot motion are not particularly high in comparison to other applications and integral error compensation can introduce instability. Substitute the feedback into the open-loop transfer function to generate the closed-loop transfer function:

$$\frac{\theta_l(s)}{K_p E + s K_v E} = \frac{nK_t}{s^2 R_i J_{eff} + s(R_i b_{eff} + K_t K_b)}$$

$$\rightarrow G(s) = \frac{\theta_l(s)}{E(s)} = \frac{nK_t(K_p + sK_v)}{s^2 R_i J_{eff} + s(R_i b_{eff} + K_t K_b)}$$

Substitute $E(s) = \theta_l^d - \theta_l$ to obtain the closed-loop transfer function:

$$\frac{\theta_l(s)}{\theta_l^d(s)} = \frac{G(s)}{G(s)+1} = \frac{K_t(K_p + sK_v)}{s^2 R_i J_{eff} + s(R_i b_{eff} + K_t K_b + K_t K_v) + K_t K_p} \quad (10.3)$$

K_p and K_v are chosen so that e and $\dot{e}$ tend to zero exponentially: $\ddot{e} + K_v \dot{e} + K_p e = 0$. High controller gains increase the size of the neighbourhood about the operating point within which tracking errors are exponentially convergent. The transfer functions have the form $G(s) = 1/(s^2 + 2\zeta w_n + w_n^2)$ for a second-order differential equation (damped linear oscillator) $\ddot{y} + 2\zeta w_n \dot{y} + w_n^2 y = 0$ where w_n = characteristic natural frequency (typically $f_n \sim 10$ Hz for a manipulator; ζ = damping ratio. For $\theta_l(s)/\theta_l^d(s)$:

$$w_n = \sqrt{\frac{K_t K_p}{R_i J_{eff}}} \quad \text{and} \quad 2\zeta w_n = \frac{R_i b_{eff} + K_t K_b + K_t K_v}{2\sqrt{J_{eff} R_i K_t K_p}}$$

If $\zeta < 1$, we get a fast response with overshoot, i.e. underdamped response. This increases the possibility of collision. If $\zeta > 1$, we get overdamping with no oscillation to generate zero steady state error. If $\zeta = 1$, we get critical damping with maximum speed, i.e. for stability, $\zeta \geqslant 1$.

$$\text{Position feedback gain, } K_p = \frac{J_{eff} R_i w_n^2}{K_t} > 0 \quad (10.4)$$

$$\text{Velocity feedback gain, } K_v \geqslant \frac{2\sqrt{J_{eff} R_i K_t K_p} - R_i b_{eff} - K_t K_b}{K_t} \quad (10.5)$$

For a direct drive motor, since the motor is mounted in adjacent links directly, the motor inertia is that of the link mounting. Too high a position gain will increase the amplitude of oscillations. At high gain values, an effectively infinite amplitude ratio resonance peak occurs with zero input making the system become unstable. In order not to excitestructural oscillations, Paul (1981) suggested that the natural damping frequency should be less than half the structural oscillation frequency: $w_n \leq 0.5w_s$. The structural resonant frequency depends on the material stiffness and the motor inertia:

$$w_s = \sqrt{\frac{K_s}{J_{eff}}}$$

where K_s = effective stiffness of the joint. Hence,

$$0 \leq K_p \leq \frac{w_s^2 J_{eff} R_i}{4 K_t}$$

Similarly,

$$K_v \geq \frac{R_i w_s \sqrt{J_{eff}} - R_i b_{eff} - K_t K_b}{K_t}$$

For direct drive motors [Asade et al 1983]:

$$\frac{R_i}{K_t} = 15.88 \, \text{V/Nm} \quad \text{for arm motors}$$

$$\frac{R_i}{K_t} = 2.15 \, \text{V/Nm} \quad \text{for wrist motors}$$

$$J_{eff} = 3.3 \, \text{kgm2} \quad \text{and} \quad b_{eff} = 1.4 \, \text{N/m/s}$$

are typical parameters for robot motors.

10.3 COMPUTED TORQUE CONTROL

It is possible to control each robot joint individually using a PD servo system to each joint motor. Indeed, this is the way that the majority of industrial robot manipulators are controlled. However, the dynamic interactions between the joints allow only slow speeds at the joints to keep the nonlinear coupling small. In fact, as many as 30% of 'linear' control loops exhibit oscillations due to nonlineararities from hysteresis, stiction, etc. Nonlinear systems require linearisation about the operating point, in this case the desired trajectory, in the neighbourhood of which the system may be regulated by a linear control system to track the reference input. This may be accomplished by feedforward compensation to linearise the system by decoupling the system into a series of decoupled linear independent subsystems which may be controlled by PD feedback laws [Bejczy & Tarn 1986]. This computed torque technique is a nonlinear control method whereby the dynamic model of the manipulator is used to dynamically decouple the joints through precomputed nonlinear feedforward compensation to cancel the effects of the nonlinear coupling. It is a model-based control method. The nonlinear coupling terms are treated as disturbance torques and fed forward into the controller of each joint. In effect, it is an example of a MIMO system which has been decoupled into a series of SISO systems. As well as feeding forward nonlinear components to compensate for the interaction forces, it simultaneously feeds back deviations from the desired trajectory. It is a PD technique where position and velocity feedback provide corrective torques to decrease the position error and provide damping whilst employing exact multivariable linearisation about the desired trajectory through feedforward compensation to cancel out the nonlinear effects. Each joint acts as an second-order oscillator with natural frequency w and damping ratio ς:

$$\tau_i = \ddot{\theta}_i^d + 2\varsigma w(\dot{\theta}_i^d - \dot{\theta}_i) + w^2(\theta_i^d - \theta_i) \qquad \text{as a single dc motor}$$

The motor torque generated by each dc motor has an additional disturbance torque:

$$\tau_i = J_{eff}\ddot{\theta}_m + b_{eff}\dot{\theta}_m + D \qquad \text{where } D = \text{disturbance torques}$$

Hence,

$$J_{eff}\ddot{\theta}_m + b_{eff}\dot{\theta}_m + D = K_t \frac{V_i - K_b\dot{\theta}_m}{R_i}$$

$$\rightarrow R_i J_{eff}\ddot{\theta}_l + (R_i b_{eff} + K_t K_b)\dot{\theta}_l = nK_t V_i - nR_i D$$

Apply Laplace transforms with $V(s) = (K_p E + sK_v E)/n$:

$$s^2 R_i J_{eff}\theta_l(s) + s[R_i b_{eff} + K_t K_b]\theta_l(s) = K_t\left(K_p E(s) + sK_v E(s)\right) - nR_i D(s)$$

$$\rightarrow \frac{\theta_l(s)}{E(s)} = \frac{K_t(K_p + sK_v) - nR_i D(s)}{s^2 R_i J_{eff} + s(R_i b_{eff} + K_t K_b)}$$

Hence,

$$\theta_l(s) = \frac{K_t(K_p + sK_v)\theta_l^d(s) - nR_i D(s)}{s^2 R_i J_{eff} + s(R_i b_{eff} + K_t K_b + K_t K_v) + K_t K_p}$$

System error derives from the disturbing torques [Fu et al 1987]:

$$E(s) = \theta_l^d(s) - \theta_l(s)$$

$$= \frac{[s^2 J_{eff} R_i + s(R_i b_{eff} + K_t K_b)]\theta_l^d(s) + nR_i D(s)}{s^2 R_i J_{eff} + s(R_i b_{eff} + K_t K_b + K_t K_v) + K_t K_p} \tag{10.6}$$

The disturbance torques are precomputed as centrifugal/coriolis and gravity torques and fed forward into the controller to compensate for their effects such that $D(s) = \tau_c(s) + \tau_g(s) + \tau_e(s) - \tau_{comp}(s)$ where each term respectively refers to coriolis/centrifugal torques, gravity torques, other disturbance torques (e.g. static friction or load-dependent torques), and precomputed torques. The precomputed term depends on the dynamic model of the manipulator which precomputes the gravity and centrifugal/Coriolis torques and possibly the viscous friction torques at the joints. The overall computed torque control law may be represented as in Fig. 10.2 for all the joints.

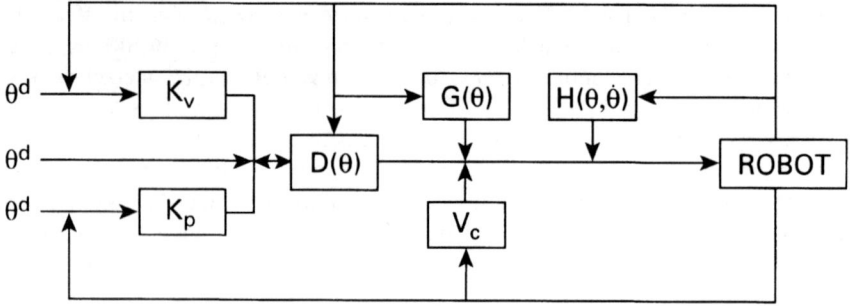

Fig. 10.2. Computed torque control flow. K_p = position gain, K_v = velocity gain, $G(\theta)$ = gravitation torque, $H(\theta,\dot{\theta})$ = coriolis/centrifugal torque, $D(\theta)$ = inertia matrix, and V_c = other disturbance torques.

The steady-state performance of the system is given by the error between the desired output and the actual output after transients have decayed. It is determined by the numerator behaviour of the error as s tends to zero.

A step input of $\theta_l^d(s) = A/s$ commands the manipulator to hold its position constant to illustrate the response to sudden disturbances:

$$e_{ss}(step) = \lim_{s \to 0} \frac{\left[s^2 J_{eff} R_i + s(R_i b_{eff} + K_t K_b)\right] A/s + nR_i D(s)}{s^2 R_i J_{eff} + s(R_i b_{eff} + K_t K_b + K_t K_v) + K_t K_p}$$

$$= \lim_{s \to 0} \frac{nR_i(\tau_c + \tau_g + \tau_e - \tau_{comp})}{s^2 R_i J_{eff} + s(R_i b_{eff} + K_t K_b + K_t K_v) + K_t K_p}$$

If

$$\tau_{comp}(s) = \tau_c(s) + \tau_g(s) \quad \text{then} \quad e_s(step) = \frac{nR_i \tau_e(s)}{K_t K_p} = \frac{4n\tau_e(s)}{w_s^2 J_{eff}}$$

where

$$K_p \leqslant \frac{w_s^2 J_{eff} R_i}{4K_t}$$

A ramp input of $\theta_l^d(s) = A/s^2$ commands the manipulator to move with constant velocity to illustrate its response to time-variable input:

$$e_{ss}(ramp) = \lim_{s \to 0} \frac{\left[s^2 J_{eff} R_i + s(R_i b_{eff} + K_t K_b)\right] \dfrac{A}{s^2} + nR_i D(s)}{s^2 R_i J_{eff} + s(R_i b_{eff} + K_t K_b + K_t K_v) + K_t K_b} = e_{ss}(step)$$

The computed torque control law for all joints has the form in terms of Lagrangian matrix coefficients (the superscript 'd' indicates desired variable values):

$$\tau = D(\theta)\left[\ddot{\theta}^d + K_v(\dot{\theta}^d - \dot{\theta}) + K_p(\theta^d - \theta)\right] + H(\theta,\dot{\theta}) + G(\theta)$$

since

$$u = \ddot{\theta} = \ddot{\theta}^d + K_v(\dot{\theta}^d - \dot{\theta}) + K_p(\theta^d - \theta)$$

where position error, $e = \theta^d - \theta$ and velocity error, $\dot{e} = \dot{\theta}^d - \dot{\theta}$ and K_v and K_p are 6×6 velocity and position feedback gain matrices.

In cartesian coordinates, the commanded joint accelerations are given by such that:

$$u = \ddot{\theta} = J^{-1}\left[\ddot{q}^d + K_v(\dot{q}^d - \dot{q}) + K_p(q^d - q) - \dot{J}\dot{\theta}\right]$$

such that:

$$\tau = D(\theta)J^{-1}\left[\ddot{q}^d + K_v(\dot{q}^d - \dot{q}) + K_p(q^d - q) - \dot{J}\dot{\theta}\right] + H(\theta,\dot{\theta}) + G(\theta)$$

$H(\theta,\dot{\theta}) + G(\theta)$ comprise the feedforward dynamics component—for space systems, $G(\theta) = 0$. For the Newton–Euler equations, an equivalent recursive control law is generated by substitution of $\ddot{\theta}_i$ into the Newton–Euler equations [Fu et al 1987]:

$$\ddot{\theta}_i = \ddot{\theta}_i^d + \sum_{j=1}^{n} K_v^{ij}(\dot{\theta}_j^d - \dot{\theta}_j) + \sum_{j=1}^{n} K_p^{ij}(\theta_j^d - \theta_j) \qquad (10.7)$$

K_v^{ij} and K_p^{ij} are velocity and position feedback gains for joint i. They do not vary from task to task unless the payload varies greatly.

For nonlinear systems, the output response to standard input test signals is no guarantee to its behaviour and it may become sensitive to input variations and exhibit unstable limit cycles (jump resonance) under certain conditions. The commonest non-linearity is physical saturation such as Coulomb friction, hysteresis and, most importantly, stiction. For highly nonlinear systems, a Lyapunov function may be chosen similar to the normalised Hamiltonian $H = KE + PE$. It is a constant of motion (an invariance) and is correlated closely with the concept of the stability condition being a state of minimum energy at equilibrium. The scalar Lyapunov function chosen is based on the KE of the error with velocity replaced with velocity error:

$$V = \frac{1}{2}|\dot{e}|^2 + \frac{1}{2}e^T K_p e \qquad \text{where} \qquad e = \theta - \theta^d$$

$$\dot{V} = \frac{\partial V}{\partial e} = \dot{e}^T\left\{D(\theta)^{-1}\left[-H(\theta,\dot{\theta}) + \tau\right] - \ddot{\theta}^d + K_p e\right\}$$

The computed torque control law:

$$u = \tau = D(\theta)[\ddot{\theta}^d - K_p e - K_v \dot{e}] + H(\theta,\dot{\theta})$$

$$\rightarrow \dot{V} = -\dot{e}^T K_v \dot{e} \leqslant 0$$

Hence, the closed-loop computed torque technique is globally asymptotically stable as the energy derivative is negative (the Lyapunov direct method is similar to the Hurwitz stability criterion). If K_p is large in comparison to the initial kinetic energy error

$\frac{1}{2}\dot{e}^T D(\theta)\dot{e}$, the tracking error converges to zero. The initial KE error may be made zero by choosing the initial desired velocity to equate to the initial actual velocity. The computed torque scheme is stable provided robot parameters are reasonably accurately known and is robust to model errors of <5–10%. For this reason, Abdullah et al (1991) classify the computed torque technique as a robust control scheme. The robustness can be increased by using PID control laws:

$$u = \ddot{\theta}^d + K_p e + K_v \dot{e} + K_i \int e \, dt$$

Moya & Seraji (1987) have shown the computed torque method to provide perfect reference tracking for both transient as well as steady-state modes. The linearisation process depends on virtual perfect modelling of the manipulator dynamics. It will become unstable for large uncertainties >10% by producing amplitude output errors in the dynamics. Inexact cancellation in the feedforward linearisation can result in the loss of stability. In fact, variable payload mass will cause deviations from the model and state measurement sensors are often corrupted with noise. Finally, all else being equal, as long as the available actuator power output is sufficient to generate the required tracking torque, the tracking position and velocity errors should remain small. However, if the available power falls short of the required output, then tracking errors may become large. Astrom (1987) classified the computed torque scheme as an example of the gain scheduling adaptive control technique because the dynamic parameters are determined by the nonlinear feedforward compensation which effectively transforms the system from a nonlinear one to a set of linear subsystems. However, it does require *a priori* information about the system including payload variability.

10.3.1 Resolved acceleration control
This method is in fact a resolved cartesian version of the computed torque technique [Luh et al 1980b]:

$$\ddot{q} = \ddot{q}^d + K_v(\dot{q}^d - \dot{q}) + K_p(q^d - q)$$

where

$$e = \begin{pmatrix} p^d - p \\ \frac{1}{2}(n \times n^d + s \times s^d + a \times a^d) \end{pmatrix} \quad \text{and} \quad \dot{e} = \begin{pmatrix} v^d - v \\ w^d - w \end{pmatrix}$$

If no errors occur in position or velocity, $\ddot{q}$ will track $\ddot{q}^d$ such that the error equation is given by $\ddot{e} + K_v\dot{e} + K_p e = 0$. Now, $\ddot{\theta} = J^{-1}(\ddot{q} - \dot{J}\dot{\theta})$:

$$\rightarrow \ddot{\theta} = J^{-1}\left[\ddot{q}^d + K_v(\dot{q}^d - \dot{q}) + K_p(q^d - q) - \dot{J}\dot{\theta}\right] \tag{10.10}$$

The input torques to each joint are determined from the computed $\ddot{\theta}$ and measured $\dot{\theta}$ and using the Newton–Euler equations or Langrange–Euler equations. For the Lagrange–Euler equations:

$$\tau = D(\theta)\ddot{\theta} + H(\theta,\dot{\theta})$$

$$= D(\theta)\left[J^{-1}\left(\ddot{q}^d + K_v(\dot{q}^d - \dot{q}) + K_p(q^d - q) - \dot{J}\dot{\theta}\right)\right] + H(\dot{\theta},\theta)$$

The resolved acceleration control method is computationally very intensive.

10.4 ADAPTIVE CONTROL METHODS

Adaptive control methods are used on the basis that many dynamic parameters for any physical system to be controlled may not be known *a priori*. Model-based methods such as the computed torque method are sensitive to large uncertainties in physical parameters such as mass, friction, moments of inertia and unmodelled payload parameters. There are other complicating features in freeflying robotic spacecraft such as flexible appendages and fuel sloshing. Tosunoglu & Temar (1987) defined adaptive control as: 'a feedback control system that is adaptive if the gains are selected with online information of the plant outputs and/or plant state variables'. It is a special form of nonlinear feedback control [Astrom 1987]. Adaptive methods of robot control are generally suitable only for Lagrange–Euler formulations of the robot dynamics as they require the explicit structure of those equations. This effectively precludes them from real-time implementation without significant advances in space-rated parallel computational hardware.

Two approaches are generally appropriate for dynamic control of manipulators with variable parameters: robust and adaptive control. Stengel (1991) differentiated between adaptive and robust control schemes: a robust control scheme retains satisfactory performance in the presence of variations to the environmental parameters without changes in the control system parameters; adaptive control schemes involve altering the control system parameters to adjust to off-nominal behaviour in the environment by identifying the actual plant structure online through its parameters during the evolution of the plant and control processes. Robust systems use unique control parameter and gain values which are designed to be stable within a region around the nominal trajectory for certain process parameter ranges. Hence, they merely tolerate disturbances and it is difficult to robustify controllers over a wide parameter range. Although parameters such as friction may vary, the most important variable parameter for a robot manipulator is the payload which will vary from zero up to the maximum payload mass. Indeed, the dynamic parameters of the payload may not be known a priori [Moya & Seraji 1987]. Adaptive behaviour is essential for complex tasks with a high reliance on sensory information to cope with changes in the environment. Indeed, adaptive techniques are considered essential for intelligent control systems as they simplify higher levels of the control heirarchy [Saridis 1979, Sanderson et al 1988]. Walker & Wee (1991a,b) highlighted adaptive control as highly desirable for manipulators in the space environment as their payload variations exceed those characteristic of terrestrial manipulators.

Generally, the adaptive control system monitors the plant input/output to identify its dynamic characteristics and plant parameters through online learning, on the basis of which it adapts its own control parameters such as feedback gains. The existence of globally convergent adaptive control laws have been established [Ortega & Spong 1988]. It is important for stability that modelling errors are small, otherwise nonlinear phenomena such as limit cycles and chaotic dynamics can occur—this illustrates an important point that adaptive control is no substitute for accurate models [Garg 1989]. All adaptive systems are based on the assumption that the parameter estimation process and adaptation rate of the control parameters vary much faster than the plant parameters.

Variable structure methods involve driving the error to a state-space switching surface which is discontinuous. This can generate chattering which can in turn excite unmodelled high frequency dynamics. This is particularly poignant for systems where a degree of flexibility exists. Suction control smooths the discontinuities over time to eliminate the chattering but destroys the asymptotic global stability—asymptotic stability is the tendency to reach equilibrium in the absence of input, regardless of the system's initial state. Hence, such methods are not suitable for robot control or deployable space structure control [Abdullah et al 1991].

There are two major approaches to adaptive robot control (excluding variable structure methods) [Landau 1985, Hsia 1986, Tosunoglu & Temar 1987]—self-tuning adaptive control (STAC) [e.g. Koivo & Guo 1983] and model reference adaptive control (MRAC) [Mufti 1985, Somlo & Cat 1988]. The STAC scheme employs an algorithm similar to the Kalman filter algorithm. Both STAC and MRAC schemes are similar in that the adaptive predictor of STAC may be regarded as an implicit reference model [Landau 1981]. One of the problems with STAC schemes is that the model may possess unstable pole–zero cancellation generating instability [Goodwin et al 1984]. In addition, the arbitrary pole placement methods inherent in STAC schemes are inaccurate, generating tracking errors of up to 14% of the magnitude of link lengths [Tourassis & Neumann 1985].

The model reference adaptive control method is the most popular in robotic manipulator control. Dubowsky & Desforges (1979) derived a MRAC adaptation control algorithm which is suitable for implementation with the Newton-Euler formulation of the robot dynamics to calculate the control gains adaptively. It requires a few additional computations. The MRAC is based on the steepest descent method to minimise a quadratic function of the error between the reference model output and the plant output. By assuming that the coupling between joints to be small, a time-invariant, linear, second-order differential equation may be used to model each joint independently. Thus the formulation may be used with the computed torque control method which assumes linearisation of the manipulator dynamics. The position and velocity feedback gains of the manipulator PD control law for each joint may be adjusted so that the closed-loop performance characteristics of the joints matched those of the reference model joints.

The linearised plant dynamics were given by a spring–damper model:

$$\alpha_i \ddot{\theta}_p + \beta_i \dot{\theta}_p + \theta_p = \theta_i$$

The linear reference model was given by: $a_i \ddot{\theta}_m + b_i \dot{\theta}_m + \theta_m = r_i$ where

$$a_i = \frac{1}{w_n^2} \qquad \text{and} \qquad b_i = \frac{2\varsigma}{w_n}$$

The reference model represents the desired performance and the controller attempts to make the plant track that desired performance. The desired position reference trajectory was given by:

$$r = \frac{1}{K_p} \ddot{\theta}^d + \frac{K_v}{K_p} \dot{\theta} + \theta$$

A gradient method (the MIT rule) was used based on adjusting the parameters to minimise a quadratic function of the output error, e, between the plant and the model. It is a hill-climbing technique such that the gradient with respect to the parameters are calculated to give the desired performance.

MIT rule:

$$\frac{\partial f(e)}{\partial t} = -k \frac{\partial e}{\partial f(s)}$$

where $f(e)$ = control parameter, k = adaptation rate, and $\partial e / \partial f(e)$ = error sensitivity with respect to control parameter.

The MIT rule is effectively a linear filter to compensate the sensitivity derivative from the process input–output relations. A modified least-squares estimate of the error may be given by:

$$f(e) = \frac{1}{2} \int_0^t (w_0 e_i + w_1 \dot{e}_i + w_2 \ddot{e}_i)^2 \, dt$$

where w_i = error weighting factor and e = error.

The steepest descent method minimises the error function:

$$\dot{\alpha}_i = -k \frac{\partial f(e)}{\partial \alpha_i} \approx \frac{\partial f(e)}{\partial a_i} = (w_0 e_i + w_1 \dot{e}_i + w_2 \ddot{e}_i) \left(w_0 \frac{\partial \theta_i}{\partial a_i} + w_1 \frac{\partial \dot{\theta}_i}{\partial a_i} + w_2 \frac{\partial \ddot{\theta}_i}{\partial a_i} \right)$$

$$\beta = -k \frac{\partial f(e)}{\partial \beta_i} \approx \frac{\partial f(e)}{\partial b_i} = (w_0 e_i + w_1 \dot{e}_i + w_2 \ddot{e}_i) \left(w_0 \frac{\partial \theta_i}{\partial b_i} + w_1 \frac{\partial \dot{\theta}_i}{\partial b_i} + w_2 \frac{\partial \ddot{\theta}_i}{\partial b_i} \right)$$

with the gradients calculated from the model by:

$$a_i \frac{\partial \ddot{\theta}_i}{\partial a_i} + b_i \frac{\partial \dot{\theta}_i}{\partial a_i} + \frac{\partial \theta_i}{\partial a_i} = -\ddot{\theta}_i$$

$$a_i \frac{\partial \ddot{\theta}_i}{\partial b_i} + b_i \frac{\partial \dot{\theta}_i}{\partial b_i} + \frac{\partial \theta_i}{\partial b_i} = -\dot{\theta}_i$$

α, β = positive scalar gains.

The PD feedback gains were adjusted as :

$$\dot{K}_p = -\frac{\dot{\alpha}_i K_p}{a_i}$$

$$\dot{K}_v = K_p \dot{\beta}_i - \frac{K_v \dot{\alpha}_i}{a_i}$$

These adaptive gains behave as weighting factors to determine the degree of adaptivity. Zero adaptive gains yield the standard PD control law. As the adaptive gains become larger in magnitude, the controller becomes more rapid in response to disturbances, but also more unstable. By ensuring that $w_1 \ll w_0$, stable behaviour is enhanced. Nicosia & Tomei (1984) showed that the linear decoupled MRAC scheme was asymptotically stable. However, Leahy et al (1987) performed experimental verification of the decoupled MRAC scheme on PUMA robots and found that a lack of velocity reference input degraded the performance of the control system (the above scheme provided only position reference). Similarly, high-speed operation deteriorates the performance of any linear controller since nonlinear terms become appreciable. In fact, the linear MRAC alone generates inferior performance to the computed torque method of control alone, particularly at high speeds [Leahy et al 1987, Leahy 1990]. This illustrates the need for accurate models when using adaptive methods. Hence, the reason why a hybrid linearised adaptive computed torque control technique (HACT) is recommended to overcome the shortcomings of both whereby the computed torque method is implemented to ensure linearisation with adaptation of the control gains on the basis that the linearisation is reasonably effective.

The MRAC scheme has since been extended to incorporate the computed torque method using both the Lyapunov second direct method and Popov hyperstability approaches (similar to the Nyquist stability criterion). These also rely on the cancellation of nonlinearities to generate decoupled linear systems by exact linearisation. They robustify the MRAC to cope with strong joint coupling and both are asymptotically stable [Bayard & Wen 1987, Ortega & Spong 1988]. Craig et al (1986,1987) introduced an adaptive nonlinear MRAC scheme for manipulator control using Lyapunov methods. This CTMRAC uses a nonlinear computed torque PD-based feedback control law in Lagrangian form with a nonlinear compensator to decouple the joints. The algorithm learns via the adaptation rule derived by Lyapunov methods to match the model parameters to those of the actual system. The error equation is equated to the dynamics equation to define the relationship between the servo errors and the dynamic parameter errors. It then uses the computed torque control law to linearise the system so that the joints behave as a set of independent second-order subsystems. Alternatively, the Popov hyperstability approach (similar to the Nyquist criterion) may be used to derive the adaptation algorithm rather than the Lyapunov function and it offers greater robustness [Katbib 1988, Mufti 1985]. Both these methods require the Lagrange–Euler formulation of the manipulator dynamics to be computed, introducing considerable computational burdens.

Asare & Wilson (1987) evaluated 3 MRAC schemes including the CTMRAC and 2 linearised MRAC control methods. The CTMRAC offered high-precision tracking of position and velocity over linear MRAC schemes as well as efficient torque curves of lower magnitude. It was capable of accurate trajectories in the presence of payload variations and modelling inaccuracies but errors do not converge asymptotically to zero but to a bounded region near zero. In fact, in their criticism of linear adaptive control techniques applied to robots, Bayard & Wen (1987) cited the CTMRAC as one of the few approaches which are analytically valid for such a nonlinear system and the performance superiority of the CTMRAC indicates the necessity of employing accurate models. The disadvantage generally of the CTMRAC is that it requires joint acceleration measurement which is not normally performed in robots. The incorporation of angular accelerometers at the joints into the existing sensor suite of position and velocity sensors is not generally adopted. No comparison of the CTMRAC and the HACT methods have been performed but in view of their similar assumptions of linearisation of the manipulator dynamics, they may offer similar performances with HACT overcoming the deficiencies of both the computed torque and model referenced methods.

10.4.1 Neural network methods of robotic control

Artificial neural network methods of adaptive control have been utilised in robotics. The first neural network model for robotic control was CMAC (cerebellar model articulation controller) [Albus 1975, 1979a,b]. It used input–output measurements stored in a look-up memory to provide feedforward control signals. CMAC is based on the architecture of the cerebellum which plays an important role in motor learning by storing associative muscle activation patterns for the control of the limbs and body. Most muscle activity involves the coordination of agonistic and antagonistic muscle stiffness. The cerebellar cortex undergoes plastic changes in synaptic connections during motor learning involving all skeletal muscles of the body. The cerebellum receives a wide variety of sensory inputs including proprioceptive, touch, vestibular, visual and auditory stimuli through its mossy fibres. Narendra & Parthasavathy (1990) suggested that neural networks could be used for parameter estimation in control problems.

Kung & Hwang (1989) considered that artificial neural networks were suitable for robotic path control as it represented a pattern-recognition problem. The path control problem can be defined as the mapping between measured sensor signals into computed actuator commands. It may be implemented as a two-stage process—one neural network produces the desired torque commands from the desired joint trajectory inputs, while the other neural network produces reference torque commands from the actual trajectory inputs. The first stage involves generalised learning. In the first stage, a neural net is trained to output the desired joint motor commands from the desired trajectory. These are used as inputs to drive the manipulator to produce the actual joint trajectory which allows fine-tuning of the system of the form $p(\tau)$ after the generalised learning phase. This second stage involves specialised learning in attempting to minimise the training set required for generalised learning. The actual trajectory provides the input to the second network. The second network derives the reference torque commands in attempting to minimise the error between the reference torques and the desired torques in a least squares sense.

Nguyen & Widrow (1990) used a slightly different technique applicable to robotic manipulators. One multilayer artificial neural network is trained to identify the system dynamics (the emulator) while another multilayer artificial neural network learns to control the actual system dynamics (the controller), both using the back-propagation learning algorithm. Once the emulator has been trained to match the manipulator dynamics, it is used to train the controller network. The controller network learns to drive the emulator from an initial state to a desired state over a number of trials by minimising the error function. The overall objective was to train the controller network to produce the correct control signals to drive the manipulator to the desired trajectory given the current trajectory state. The emulator acts as a feedforward component to the control system while the controller acts as an error feedback component effectively implementing an adaptive computed torque control method. This prompted the use of the two-layer perceptron neural network as equivalent to a linear control strategy (i.e. the PD feedback control law) and the three-layer perceptron neural network as a nonlinear adaptive feedforward strategy [Yabuta & Yamada 1990].

Kuperstein et al (1987) have applied artificial neural networks to robotic control by providing a means for adaptive visual–motor coordination. This involves the process of mapping directly between a sensory input and a motor output through a topological map stored in the connection weights of the network. A neural network was used to adaptively control a visually guided robot manipulator arm using pattern recognition. It essentially learned an adaptive camera-position topographic model of the environment from an ego-centric frame of reference based on sensory–motor reactions over a sequence of trials of different movements. The manipulator was controlled using consistency between the signals used to drive it and the signals received on sensing the results of the movement. The robot learned to correlate visual end effector positions from stereo cameras with joint motor activity signals producing those end effector positions, i.e. joint angle to end effector position mappings. This provided visually guided behaviours.

Although artificial neural networks have been applied to adaptive robot control in various guises, they have yet to be applied extensively in space systems. The lack of verifiability and validation of neural network control behaviour makes their use some-what controversial from a reliability point of view. Generally, space-rated software are validated as algorithms but the world of control systems is inherently heuristic, fuzzy and uncertain. In their favour, neural networks often provide robust control strategies in the face of such uncertainty and may certainly be used if more conventional techniques were used for backup. Their chief disadvantages are the need for lengthy training periods and the necessity for re-training from scratch when system parameters change. Furthermore, neural network configurations such as number of layers, number of nodes per layer and connectivity cannot be selected systematically. It is unlikely that they will be used in the near future for space robotic applications.

10.5 FORCE FEEDBACK CONTROL

The motion of a robotic manipulator is performed in two distinct phases: gross motion control from initial to final position/orientations along a trajectory, and fine control where the end effector dynamically interacts with the target object. Gross motion utilises position/velocity feedback for geometric information while fine motion utilises force feedback for dynamic information [Whitney 1977]. Force control is fundamental to robotics since the fraction of time spent to complete the terminal compliant phase is comparable or greater than that to perform the gross motions. Furthermore, most robotic functions are concerned with the manipulation of objects in the world. Assembly is a basic manufacturing operation that involves the interaction of piece parts. Indeed, robot function may be regarded as a series of desired and undesired collisions with the environ-ment. A trajectory may be divided into sequences of compliant motion joined by guarded moves. Manipulator motion is characterised by two basic states and transitions: (i) the manipulator controls displacement and monitors force in the force space where unpredicted forces indicate error; (ii) the manipulator controls force and monitors displacement in position space whereby any unpredicted displacement indicates error. Switching between control modes occurs on contact or release. On contact, hand forces increase until the desired force is attained and control switches from position to force control. On release, hand forces drop to zero and control switches from force to position control. Paul (1987) introduced into the computer programming language WAVE two commands to differentiate two types of force control methodologies: STOP (terminate current motion when force exceeds a limit—this is the guarded move) and FORCE (pursue the current motions to generate a force of a given value—if zero, this corresponds to free motion). Switching between position/velocity control only and position/force control may be accomplished at each sampling period with activated touch sensors triggering the required formulation. Proximity sensing may be used to provide a smooth transition from vision to touch control. It has been suggested that inclusion of force control reduces task execution time by ~30–50% and that for space operations force control is critical [Varsi 1991].

For manipulation tasks, compliant motion through force control is absolutely essential when the position of a gripper is constrained by the task. The use of position/velocitycontrol only will introduce substantial errors due to the unavoidable sensor and robot parameter inaccuracies. The remote centre compliance wrist assembly is a passive compliant device which is suitable only for a small class of tasks involving misaligments (ie. peg-in-hole tasks). Hence, active force control using force sensors is required when the manipulator makes contact with its environment to precisely control contact forces through closed loop fine motion feedback. To include force control, the general dynamic equation has the form:

$$\tau = D(\theta)\ddot{\theta} + H(\theta,\dot{\theta}) + G(\theta) + J^T F_{ext}$$

There are two main approaches to explicit force feedback control which close the force control loop around the position control loop [Whitney 1987]. The stiffness approach is based on the generalised spring model [Salisbury 1980]. Stiffness is the property that induces restoring forces and torques on a cantilever beam as it is deflected from a nominal position. The difference between the actual and desired effector positions are related to force errors through a stiffness matrix: $\delta f = K_s \delta q$ where δq = displacement from nominal position q_0, $K_s = 6 \times 6$ diagonal stiffness matrix the inverse of which is the compliance matrix (with dimensions of position/force). The generalised damper model treates forces as being in proportion to velocity errors offering advantages of continuity [Whitney 1969, 1977]: $\delta f = K_F \delta \dot{q}$ where $\delta \dot{q}$ = velocity error $(v^d - v)$ with $v = 0$ nominally, $K_F = 6 \times 6$ diagonal damping matrix, the inverse of which is the accommodation matrix (with dimensions of velocity/force). Transformation to joint level is straightforward, suggesting that both methods are virtually equivalent:

$$\dot{q} = \delta q = J\dot{\theta} = J\delta\theta \rightarrow \delta f = K_s J\delta\theta = K_F J\dot{\theta}$$

$$\tau = J^T F = J^T K_s J\delta\theta = J^T K_F J\dot{\theta}$$

In both cases, these explicit methods involve nullifying the contact forces and torques at the end effector by generating position/velocity modification commands. However, in general there is no one-to-one correspondence between contact forces and misalignments in assembly tasks, whereas these methods assume that there is such a linear relation between force variation and position/velocity error. Both approaches exhibit sluggish behaviour in stiff environments [Whitney 1987].

Whitney (1987) investigated the stability of force control. The manipulator interacts with its environment which behaves as a spring producing a reaction force/torque on the end effector. Both the environment and force sensors have natural frequency responses. The environment stiffness K_e is often high ~10^6 N/m which produces little damping but give high natural frequency responses generating poor stability. Low stiffness environments still give K_e ~ 10^5 N/m and such enivironments are often determined by both the object stiffness K_0 and the manipulator compliance K_m:

$$\frac{1}{K_e} = \frac{1}{K_o} + \frac{1}{K_m} \rightarrow K_e = \frac{K_m K_o}{K_m + K_o}$$

For aluminium objects, $K_0 = EA/l$ where E = Young's modulus = 6.7×10^4 Nmm^2, A = cross sectional area and l = length. Generally $K_o \gg K_m$, hence, K_e ~ K_m. A system is stable if $K_e K_F \delta t < 1$ or $K_m K_F \delta t < 1$ where δt = sensor sample rate, $K_F = \text{diag}(K_f K_\tau)$ such that $K_f = R^3 K_\tau$ where R is the distance between the contact end point and the coordinate centre of freedom (usually the wrist) [Whitney 1977]. Stability requires $K_F K_e$ to be small, so, to deal with stiff environments with high K_e, K_F must be small. Generally, a fixed force gain is not ideal: the value of K_F will differ depending on the contact dynamics between the manipulator and the environment. As stiffness of contact increases, so K_F should be reduced. Without *a priori* knowledge of the environmental stiffness, K_F should be set as low as possible, but in softer environments this will result in slower response times and so extended task completion times. Typically, $K_F = \text{diag}(0.005, 0.0001)$. At the transition point between position and force control at contact, collision energy is absorbed by the natural compliance of the system and dissipated with the possibility of destructive consequences. Force sensors are typically the least stiff parts of the manipulator and the most fragile: the Scheinmann force sensing wrist required a force overload mechanism to protect it from damage. A force sensor has a typical stiffness $K_s = 5 \times 10^4$ N/m. The time constant of interaction is much less than that of the feedback controller response. The frequency of force measurement must be as high as possible for stability of the closed loop. There is the limitation imposed by the computation to intervals ~20 ms. Hence, only resonant frequencies ~60 Hz are capable of being tracked by the controller. However, such a control sample will generate a time delay in the control loop between force measurement and feedback. The manipulator hand is supported by spring-like force sensors and the compliance of the reaction surfaces, so the hand behaves as an oscillator causing small amplitude oscillations on the force signal. This oscillation will be persistent and will arise particularly at the contact-make and contact-break phases of manipulation. Hand natural resonant frequency,

$$f = \frac{1}{2\pi} \sqrt{\frac{K_e + K_s}{m_{wrist}}} \approx 300 \text{ Hz}$$

with m_{wrist} = wrist mass = 0.6 kg. These oscillations with durations of microseconds cannot be responded to by software control loops due to their high frequency relative to the control cycle. A low bandpass filter may overcome this if the amplitudes are low. However, for large input forces, a low bandpass filter is not capable of compensation. Wen & Murphy (1991) suggested that although derivative force feedback cannot be used since this requires gross arm motions, integral force feedback may be used as it can represent the virtual motions of compliance. An integral force feedback control law acts as a filter to counter instability. It removes bias forces by introducing a saturation nonlinearity and reduces the effective gain for large force errors making the controller stable. For good transient response and disturbance rejection in an infinitely stiff environment, the integral gain should be large, but if the environment exhibits any flexibility (i.e. any real environment), the integral gain should be small. One way to reduce $K_F K_e$ is to introduce passive compliance at the wrist to reduce the effective environmental stiffness, i.e. a remote centre compliance (RCC). If force/torque is applied at the compliance centre, translation/ rotation of the device reduces the effective stiffness of the environment. However, the RCC is susceptible to damage if large displacements are forced upon it and it is limited to misaligments ~1 mm or ~1°. The instrumented RCC device is an RCC effectively mounted to a force/torque sensor for monitoring displacements to prevent damage from excessive displacements. The RCC has low inertia such that contact energies are low and can be absorbed by the passive compliance. Such devices can be locked for position control to provide the required stiffness.

10.5.1 Hybrid position/force control

Hybrid position/force control is reckoned to be superior to other forms of force control enabling tasks of considerable complexity to be performed [Raibert & Craig 1981; Paul 1987]. Mason (1981) introduced the concept of C-(constraint) space to break down manipulator tasks into basic components defined by contact surfaces. Assembly operations are constrained due to contact between the workpieces and the end effector such that the number of degrees of freedom is reduced so that arbitrary motion cannot be specified. The constraint exerts a reaction force on the end effector and can generate large joint forces which need to be controlled. For each task configuration, there exists a compliance frame from which the task is best described. This compliance frame {C} may be fixed to the end effector cartesian coordinate frame or reside in the object reference cartesian coordinate frame centred at its centre of mass. The C-space is the space of n-parameters defined by the task configuration. In this C-space, there are N natural constraints which result from the geometry of the task and N orthogonal artificial constraints imposed by the desired trajectory of manipulation. Two subspaces s_f and s_c exist in the C-space. In the free subspace s_f, the configuration of the manipulator is such that no contact between the manipulator and the target object occurs. In the constrained subspace s_c, the manipulator is in contact with the object. Both subspaces are separated by the C-surface which defines the task geometry. Hence, the C-space concept identifies three separate states: free space motion, contact and force exertion. The manipulator cannot cross the C-surface into s_c—it can only lie in s_f or on the C-surface since s_c is defined by the object space. If on the C-surface, the manipulator can move over the C-surface tangentially (position-controlled motion). Displacement along the normal of the C-surface changes the magnitude of the contact forces since the object generates reaction forces on the end effector (force-controlled motion). Hence, natural constraints partition the DOF through position constraints normal to the C-surface and through force constraints tangential to the C-surface. Artificial constraints specify the desired position and force trajectories with force constraints along the C-surface normals and position constraints along the C-surface tangents. The direction in which the manipulator moves defines whether the degree of freedom is controlled by force feedback or position feedback. Constraint of any degrees of freedom implies force control in the constrained directions while position control is used for those directions which produce unconstrained displacement motion.

This is the concept behind the hybrid position/force control methodology where position control is applied along the cartesian C-surface tangent and force control along the cartesian C-surface normal [Raibert & Craig 1981]. The position vector is constrained in the directions complementary to the constraint direction of the force vector. The task is then characterised by motion in certain directions and the exertion of force in the remaining directions. The hybrid mode allows force feedback to produce corrective motion to compensate for position inaccuracies. It satisfies position and force trajectory constraints simultaneously. During hybrid position/force control each degree of freedom is controlled by only one loop such that both position and force feedback loops act cooperatively to control each joint. Each joint thus contributes to the control of position and force at the end effector, i.e. the end effector space is partitioned into two orthogonal domain components which are complementary. Central to the hybrid concept is the use of a diagonal compliance selection matrix S which selects which degrees of freedom are force-controlled and which are position-controlled. This is also the disadvantage of the scheme, in that the selection matrix which defines the task geometry cannot be algorithmically generated for a particular task. However, for many well-defined tasks the S matrix is well known.

$$S = \mathrm{diag}(s_1 \ldots s_n)$$

where $s_j = 1$ for force control; $s_j = 0$ for position control.

The selection matrix is implemented thus:

$$\tau_i = \sum_{j=1}^{n} s_j \varphi_{ij} \delta f_j + (1 - s_j) \pi_{ij} \delta q_j$$

where

$\varphi =$ force transfer function,

$\pi =$ position transfer function, and

$\delta f, \delta q =$ force/position errors.   (10.9)

In general, additional coordinate transforms are required to define the constraint frame, but for the majority of manipulation tasks, the constraint frame may be chosen to coincide at the end effector, e.g. peg-in-hole tasks are specified in end effector tip coordinates. The scheme is implemented at cartesian level, so cartesian position errors are calculated from measured joint coordinates and subtracted from the desired trajectory while cartesian force errors are obtained directly. Furthermore, Uchiyama et al (1987) stressed the need for a velocity feedback component in the hybrid scheme.

$$\delta q = q^d - q = p_n^d - p_n = \text{position error}$$

$$= \text{difference between desired and actual cartesian positions}$$

$$\delta \dot{q} = \dot{q}^d - \dot{q} = v_n^d - J\dot{\theta} = \text{velocity error}$$

$$= \text{difference between desired and actual cartesian velocities}$$

$$\delta f = F_{ext}^d - F_{ext} = \text{force error}$$

$$= \text{difference between desired and actual cartesian forces}$$

The position/velocity errors are set to zero in the force-controlled directions and force errors are set to zero in the position controlled directions by the selection matrix and the errors are mapped into position- and force-controlled subspaces.

$$e_q = (I - S)\delta q$$

$$\dot{e}_q = (I - S)\delta \dot{q}$$

$$e_f = S \, \delta f$$

These errors are then transformed back into joint coordinates using the inverse and transposed Jacobian for use in the joint referenced control law.

$$e_\theta = J^{-1} e_q$$

$$\dot{e}_\theta = J^{-1} \dot{e}_q$$

$$e_\tau = J^T e_f$$

Positioned-controlled torque,

$$\tau_p = D(\theta)\left[\ddot{\theta}^d + K_p e_\theta + K_v \dot{e}_\theta\right] + H(\theta, \dot{\theta}) \tag{10.10}$$

Force-controlled torque,

$$\tau_f = J^T F_{ext}^d + K_F e_\tau + K_{Fi} \int e_\tau \, dt \tag{10.11}$$

where K_F and K_{Fi} are 6×6 diagonal force feedback matrices and they depend on the stiffness of the environment. $J^T F_{ext}^d$ is the feedforward force component for reference tracking.

$$\text{Complete hybrid position/force control law:} \quad \tau = \tau_p + \tau_f \tag{10.12}$$

The use of a PD type position-control law in conjunction with a PI type forcecontrol law makes the hybrid control law of the PID type. The hybrid scheme exhibits a stable and accurate response to step inputs with very little overshoot in position and force trajectories [Raibert & Craig 1981]. Zheng & Paul (1985) noted that the hybrid scheme required the J^{-1} transformation to joint coordinates from cartesian coordinates which imposes a heavy computational burden. To reduce the computational load, they adapted the hybrid scheme to operate directly in joint coordinates by introducing a joint compliance matrix $C = J^{-1} S J$ to cast the hybrid control law in cartesian form. This was far more efficient and yields a control law of the form:

$$\tau = D(\theta)\left\{\ddot{\theta} + (I - C)\left[K_v(\dot{\theta}^d - \dot{\theta}) + K_p(\theta^d - \theta)\right]\right\} + H(\theta, \dot{\theta})$$

$$+ J^T F_{ext}^d + (C)\left[K_F J^T (F_{ext}^d - F_{ext}) + K_{Fi} J^T \int (F_{ext}^d - F_{ext})dt\right] \qquad (10.13)$$

West & Asade (1985) characterised hybrid position/force control as being based on constraints imposed by contact with the environment reducing the number of degrees of freedom similar to a closed kinematic chain. They established a $6 \times m$ contact Jacobian for such a closed kinematic chain such that $\dot{\theta} = (I - J_c^+ J_c)\dot{q}$ defines the velocity control loop and $\tau = (J_c^+ J_c)^T F_{ext}$ defines the force control loop. A diagonal selection matrix is required to select between the loops. The method is considerably more complex than the usual approach to hybrid position/force control with little apparent advantage.

Impedance control is a methodology for robotic control based on a biological paradigm [Hogan et al 1985]. The cerebellum stores muscle activation patterns learned from the association of proprioceptive stimuli indicating co-contraction of agonist–antagonist muscle lengths. Modulation of contraction in the timing and amplitude of muscle activity regulates the viscoelastic properties of the joints to move between postures. Muscle behave as tunable springs by virtue of their arrangement around joints in agonistic/antagonistic pairs. The elasticity of the muscle is determined by its activation level which determines their length-extension α which in turn defines the equilibrium position and stiffness of the joint. This may be modelled as a potential function of the joint angle, the negative derivative of which is the generalised force. The postural force field is characterised by the product of a displacement vector and a stiffness matrix. Activation of agonistic/antagonistic muscles determines the potential minimum (determined by the α ratio) and the potential minimum curvature (determined by the α sum). Displacement from the potential minimum results in a restoring torque independent of feedback. Thus all control acts through a single time-dependent potential function with the muscles themselves computing the joint torques. The function of the nervous system is to transform the desired trajectory into a sequence of joint equilibrium positions and stiffnesses determined by a global time-dependent potential function. This method of hybrid position/force (impedance) control is still under development though offers great promise. It provides a unified framework for all kinds of manipulator motion involving dynamic interactions while preserving stability during interactions with the dynamic environment. However, it is important to note that biologically, such control of joint viscoelastic properties is achieved by feedforward open-loop mechanisms from learned cerebellar inverse dynamics models due to delays and instabilities in online feedback mechanisms.

10.5.2 Adaptive force control
Seraji (1987a,b) applied the Lyapunov-based model reference approach to adjust the feedback gains of the force control loop. Adaptive techniques are particularly suited to force control as the end effector forces and torques may vary widely and require online control gain adaptation. The dynamic model was that of a mass-spring damper to model the force applied to the end effector:

$$M_0\ddot{q} + D_0\dot{q} + K_e q = F_q$$

Now the force exerted by the end effector to the environment:

$$F = K_e q \rightarrow \frac{M_0}{K_e} \ddot{F} + \frac{D_0}{K_e} \dot{F} + F = F_q$$

Equate this to a PID force control law:

$$\frac{M_0}{K_e} \ddot{F} + \frac{D_0}{K_e} \dot{F} + F = F^d + K_{fp} e_f + K_{fi} \int_0^t e_f dt + K_{fv} \dot{e}_f$$

$$\ddot{F} + \frac{D_0}{M_0} \dot{F} + \frac{K_e}{M_0 F} F = F^d + K_{fp} e_f + K_{fi} \int_0^t e_f \, dt + K_{fv} \dot{e}_f$$

Now, $e_f = F^d - F$ with $\dot{e}_f = -\dot{F}$ and $\ddot{e}_f = -\ddot{F}$:

$$\ddot{e}_f + M_0^{-1}(D_0 + K_{fv})\dot{e}_f + M_0^{-1}(I + K_{fp})e + M_0^{-1} K_{fi} \int_0^t e_f \, dt = 0$$

Hence,

$$\dot{e}_f = \begin{pmatrix} 0 & I & 0 \\ 0 & 0 & I \\ -M_0^{-1} K_{fi} & -M_0^{-1}(I + K_{fp}) & -M_0^{-1}(D_0 + K_{fv}) \end{pmatrix} e_f = A e_f$$

where

$$e_f = \begin{pmatrix} \int e_f \, dt \\ e_f \\ \dot{e}_f \end{pmatrix}$$

Lyapunov analysis may be used to derive the adaptation law through the energy-like Lyapunov function. The Lyapunov function is given by $V = e_f^T Q e_f$. There exists a positive definite matrix M which satisfies the Lyapunov equation:

$$MA + A^T M = -Q$$

Q is diagonal so $M > 0$ may be found analytically.
We have:

$$\begin{pmatrix} M_1 & M_2 & M_3 \\ M_2 & M_4 & M_5 \\ M_3 & M_5 & M_6 \end{pmatrix} \begin{pmatrix} 0 & I & 0 \\ 0 & 0 & I \\ -A_1 & -A_2 & -A_3 \end{pmatrix} + \begin{pmatrix} 0 & 0 & -A_1 \\ I & 0 & -A_2 \\ 0 & I & -A_3 \end{pmatrix} \begin{pmatrix} M_1 & M_2 & M_3 \\ M_2 & M_4 & M_5 \\ M_3 & M_5 & M_6 \end{pmatrix} =$$

$$= \begin{pmatrix} -2Q_1 & 0 & 0 \\ 0 & -2Q_2 & 0 \\ 0 & 0 & -2Q_3 \end{pmatrix}$$

$$-2M_3A_1 = -2Q_1$$

$$M_1 - M_3A_2 - M_5A_1 = 0$$

$$M_2 - M_3A_3 - M_6A_1 = 0 \qquad\qquad M_3 = A_1^{-1}Q_1$$

$$2M_2 - 2M_5A_2 = -2Q_2 \qquad \rightarrow M_5 = (A_2A_3 - A_1)^{-1}\left[(A_1^{-1}A_3^2 - A_1)Q_1 + A_3Q_2 + A_1Q_3\right]$$

$$M_3 - M_6A_2 + M_4 - A_3M_5 = 0 \qquad M_6 = (A_2A_3 - A_1)^{-1}[A_1^{-1}A_3Q_1 + Q_2 + A_2Q_3]$$

$$2M_5 - 2M_6A_3 = -2Q_3$$

The adaptation laws are:

$$\dot{K}_{fi} = \alpha_1 q(t)\int_0^t e_f\, dt + \frac{d}{dt}\left[q(t)\int_0^t e_f\, dt\right]$$

$$\dot{K}_{fp} = \beta_1 q(t)e_f + \beta_2 \frac{d}{dt}\left[q(t)e_f\right]$$

$$\dot{K}_{fv} = \gamma_1 q(t)\dot{e}_f + \gamma_2 \frac{d}{dt}\left[q(t)\dot{e}_f\right]$$

where

$$q(t) = M_3\int_0^t e_f\, dt + M_5 e_f + M_6\dot{e}_f = \text{weighted force error vector}$$

$$\alpha, \beta, \gamma = \text{positive scalar gains}$$

Hence,

$$K_{fi} = K_{fi}(0) + \alpha_1\int_0^t q(t)\int_0^t e_f\, dt + \alpha_2 q(t)\int_0^t e_f\, dt$$

$$K_{fp} = K_{fp}(0) + \beta_1\int_0^t q(t)e_f\, dt + \beta_2 q(t)e_f$$

$$K_{fv} = K_{fv}(0) + \gamma_1\int_0^t q(t)\dot{e}_f\, dt + \gamma_2 q(t)\dot{e}_f \qquad\qquad (10.14)$$

Adaptive force control will be particularly critical for space-based in-orbit servicing missions.

10.5.3 Grasp planning

Choosing a grasp configuration on a part is similar to the Findspace problem for a safe configuration amongst a set of obstacles but involves additional constraints including [Latombe 1984, Volz et al 1984]:

(i)   the end effector fingers must be in contact with the grasp points on gripping;
(ii)   the configuration must be reachable at both initial and destination points without interference between them;
(iii)   the grasp points must be stable such that the object is immovable in the grasp.

For a good grip both contact dimensions and finger separation should be as large as is feasible. If world coordinates do not coincide with the manipulator base coordinates and a different reference frame is defined as the world reference, a relational transform is required to relate the reference and base frame coordinates.

Stability may require the centre of mass of the object (if it is small) to be located between the jaws of the end effector. Stability however is ultimately determined in the general case by a lack of slip between the fingers and the object. Parallel planar surfaces whose distance from each other is less than the maximum parallel jaw finger opening are suitable as grasp points and the internal faces of the fingers must overlap the grasp surfaces with sufficient contact area. Other possible grasp pairs include face and parallel edge, face and vertex, and parallel edges for cylinders. However, two parallel faces are the preferred legal grasp configuration with opposing digits orthogonal to the target surface (similar to finger and thumb). For small objects, almost all grasps are stable. Tactile sensors mounted onto the inner surfaces of the fingers can provide feedback on the direction, magnitude and location of incident forces. The grasp plane is defined as the plane parallel to the faces being grappled and midway between them. When approaching a grasp point, the fingers remain parallel to the grasp plane and centred about it, but otherwise free to rotate and translate in the plane. This minimises the risk of collision with the object to be grasped. The grasp plane is bounded by a rectangle whose size determines the range of motions allowed by the hand during grasping. The rectangular volume that the hand can sweep out while constrained to move in the grasp plane is the grasp volume. The motion of the hand is a translation along the free motion vector connecting the finger and object grasp points in the grasp plane without rotation. Grasping is heavily dependent on features. Features may be defined as specific geometric configurations found on the surfaces, edges or corners of a workpiece which modify or aid in achieving a given function [Vijaykumar & Arbib 1987]. Specific features of objects dictate how these objects may be assembled together. Global features include shape, length, width, height, volume, surface area, location and local features such as holes. If the origin of the coordinate object frame lies it the object's centre of mass, then local features are specified in terms of these coordinates. This allows a representation of features and the relationships between them. This is the primary problem—of identifying positions and orientations of features on objects relative to each other.

For assembly operations, certain heuristic requirements must be met:

(i)   the features of the female connector must be hollow;
(ii) the dimensions of the female connector must be at least as large as the male connector.

Other heuristics may include: closeness of fit such that the difference between each pair of mating features should be small and the features should be symmetric. For example, insertion operations: insert feature k of object i into feature l of object j with conditions:

(i)   normal axes of feature k and feature l must be colinear and opposed;
(ii)   bottom surface of insertion feature k will be against the bottom surface of containing feature l;
(iii) if insert and container are cylindrical, their cylindrical surfaces will be in contact;

(iv)   centre of insertion feature k will be a distance equal to the depth of feature l away from the centre of feature l along the direction of the normal axis of feature j with respect to object coordinate frame origin.

A CAD database may be used to store the grip points in the target object. This grip list would be generated offline from geometric and topological data within the CAD database. Selection of the grip point is generated online. Certain grip positions may be eliminated offline: grips on sets of vertices (subject to rotation) and grips on edge/vertex combinations (subject to twisting). Grip positions are ordered in terms of the quality of the grip position (i.e. resilience to slippage and twisting). The motion primitive for grasping comprises of:

(i)   move the end effector to the approach position;
(ii)   open the end effector jaws;
(iii)   guarded move at half-speed to the grasp position;
(iv)   centre the end effector on the grasp position;
(v)   grasp the object with a suitable pinching force;
(vi)   check slip sensors.

The peg-in-hole/screw-in-hole task is one of the commonest assembly tasks and is used as the standard for force control validation for assembly (indeed, it is fundamental to the ORU exchange operation). The compliance frame is at the tip of the tool with the z-direction defined as pointing to the hole from the tool tip coinciding with the hole axis.

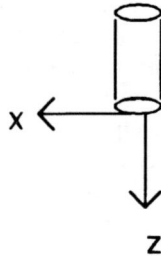

$$v_x = 0, f_y = 0 \qquad\qquad v_y = 0, f_x = 0$$

Natural constraints : $w_x = 0, f_z = 0$ Artificial constraints : $v_z = p\alpha, n_x = 0$

$$w_y = 0, n_z = 0 \qquad\qquad w_z = \alpha, n_y = 0$$

where α = rate of turn constant and p = pitch.

Two degrees of freedom are position-controlled: rotation about the z-axis and translation along the z-axis, so $S = (1,1,0,1,1,0)$. This may be implemented in a robot control language such as WAVE as:

```
COMPLY FORCE X 0
COMPLY FORCE Y 0
COMPLY TORQUE X 0
COMPLY TORQUE Y 0
```

MOVE FORCE Z UNTIL STOP FORCE Z FMAX
MOVE TORQUE Z UNTIL STOP TORQUE Z NMAX

Lateral and angular errors are corrected in response to chamfer contact forces. Wedging occurs when the peg becomes locked at the beginning of insertion such that further force will deform the peg or the hole. The insertion depth l is constrained by $l \leqslant 2\mu r$ where μ = friction coefficient and r = peg radius. Jamming occurs when the contact forces do not correct misalignments when

$$\frac{nr/f_z}{\lambda \sin \theta} + \frac{f_x/f_z}{(\lambda/\lambda + 1)\sin \theta} \leqslant 1$$

where $\lambda = l/2\mu r$. Forces at the peg tip correspond to lateral errors which are corrected by translation and torques at the peg tip correspond to angular errors which are corrected by rotation.

10.5.4 Closed-chain configuration force control

Once a closed kinematic chain is formed by the load, the load couples the dynamics of the two arms. The dynamic equations then have an additional term which represent the interaction forces. Several workers have recommended that when the closed-chain configuration is adopted the individual dynamic equations of the two coordinating robots be combined to form a unified dynamic equation [Zheng 1989]. This is not desirable as this introduces computational burdens and the philosophy behind the formulation considered here is to decompose the dual-manipulator problem into two single-manipulator problems. The dynamics of each manipulator may then be considered separately by monitoring the interaction forces and moments at the end effector applied to the object. The control of each manipulator may then be computed in parallel by dedicated microprocessors for each arm. The dynamics of the two manipulators may be represented as (using the Lagrange representation for illustration—the same arguments apply to the Newton–Euler method) [Hayati 1986, 1988]:

$$D^l(\theta^l)\ddot{\theta}^l + H^l(\theta^l, \dot{\theta}^l) = (J^l)^T F^l_{ext} = \tau^l$$

Force control is fundamental to cooperative tasks involving more than one arm. Hayati (1986) and Alberts & Soloway (1988) employed wrist force sensors on both arms to measure interaction forces to exert desired forces of compression, tension or torsion on a rigidly held workpiece. The force controller regulated the forces and torques at each arm and the desired interaction force bias was maintained. Two approaches for closed-chain control are possible: (i) the master arm is position controlled with the slave arm force controlled to provide compliance; (ii) symmetric hybrid position/force control for both arms. Master/slave architectures lose the major benefit of dual arm control in that the arms do not share the load. Tao et al (1987) proposed a master/slave configuration whereby the master arm is position-controlled whilst the slave arm adopted the hybrid position/force-control method with a force sensor. Seraji (1987c) considered three approaches to dual arm control: position–position where both arms are position-controlled only, position–hybrid where one arm is position-controlled and the other is hybrid-controlled, and hybrid–hybrid where both arms were hybrid-controlled. He found that position–position control required passive compliance in order not to generate excessive forces on the load and that the position–hybrid system created fluctuations in the force profile due to vibrations in the position-controlled arm. The hybrid–hybrid condition generated a smooth force profile giving the best performance. Hybrid control of both arms has been pursued as the most flexible approach to closed-chain control since arm roles may be interchanged readily. Furthermore, control can be applied to each arm independently even in the closed-chain configuration provided the kinematic constraints are observed.

Uchiyama et al (1987) and Garg (1988) used hybrid velocity/force control for both arms equipped with wrist force sensors to give superior performance over position/force control. There are several ways to implement the hybrid scheme and velocity provides a natural mapping between position and force: $\dot{\theta} = J^{-1}[SK_p(q^d - q) + (I - S)(F^d - F)]$. This may be extended for PID control. Uchiyama & Dauchez (1988) and Meier & Graf (1991) extended this hybrid scheme using as central controllers to symmetric non-master slave scheme to include the coupled kinematics of the closed chain.

Carignan & Akin (1988) cited load distribution between manipulators as being particularly important for dual-manipulator freeflyer robots in space. Load distribution is a notion pertaining to both manipulators and if implemented should be implemented at ahigher level than the individual controllers, e.g. the Activity Controller. Orin & Oh (1981) considered the force distribution between two robot arms and a single payload. An infinite number of solutions exist due to the redundant nature of the problem (12 actuators to control 6 DOF of the payload). They applied constraints including maximum torques, maximum reaction forces and sufficient friction at the end effectors. They used a linear programming technique to minimise a cost function based on energy expenditure to minimise the power requirement within these constraints. Two criticisms have been levelled against the linear programming technique: firstly, the computation time ~10–20 s for the solution precludes its use for real-time operation [Alberts & Soloway 1988, Soloway & Alberts 1989]; secondly, some of the joints are deactivated at each solution— as the robot configuration changes, the joint torques are repeatedly switched on/off creating excitation of the structure [Zheng & Luh 1988].The desired forces and moments required to move the payload may be expressed as:

$$F_{ext}^d = m_{n+1}\dot{v}_{n+1}$$

$$N_{ext}^d = I_{n+1}\dot{w}_{n+1} + w_{n+1} \times I_{n+1}w_{n+1}$$

These are the sum of the external forces/moments at the end effector [Uchiyama & Dauchez 1988; Orin & Oh 1981]:

$$F_{ext}^d = \sum_{l=1}^{2} f_{ext}^l$$

$$N_{ext}^d = \sum_{l=1}^{2} n_{ext}^l + r_{n+1}^l \times f_{ext}^l \qquad (10.15)$$

In more compact form: $F^d = H \prod$ where

$$F^d = \begin{pmatrix} F^d_{ext} \\ N^d_{ext} \end{pmatrix}$$

$$H = \begin{pmatrix} I & 0 & I & 0 \\ r^{l=1}_{n+1} \times & I & r^{l=2}_{n+1} \times & I \end{pmatrix}$$

$$\Pi = \begin{pmatrix} f^{l=1}_{ext} \\ n^{l=1}_{ext} \\ f^{l=2}_{ext} \\ n^{l=2}_{ext} \end{pmatrix} \tag{10.16}$$

[Alberts & Soloway 1988, Soloway & Alberts 1989].

The H-matrix comprises the contact force map and the values of the required forces to move the payload provide the threshold limits to the applied forces on the object.

Hayati (1986) and Alberts & Soloway (1988) defined a quadratic function of a weighting matrix W to be minimised to minimise the cartesian forces applied at the end effector to reduce the joint torques required: $Q = \Pi^T W \Pi$ where $W = \mathrm{diag}(\alpha_1 \alpha_2 \alpha_3 \alpha_4)$, $\Pi = H^{-1} F^d$, α = weight matrix for translational and rotational components of each arm. The coefficients α_1 and α_3 relate to applied forces to each arm respectively while α_2 and α_4 relate to the applied moments to each arm respectively. The joint torques vary in proportion to motor current which in turn varies as the root of power consumption. However, although a better quadratic might be $Q = \tau^T W \tau$, representing joint space load distribution with minimum power consumption, the considerable computational overhead of such a function precludes its use [Soloway & Alberts 1989]. In any case, the cartesian forces will lead to an indirect energy minimisation through the Jacobian transpose with reduced stresses on the payload. Using Lagrangian multiplier λ:

$$\tilde{Q} = \Pi^T W \Pi - \lambda^T \left(H \Pi - F^d \right)$$

For mimimum $\tilde{Q}$:

$$\frac{\partial \tilde{Q}}{\partial \Pi} = 0 \rightarrow 2 \Pi^T W - \lambda^T H = 0$$

$$\rightarrow \Pi^T = \frac{1}{2} W^{-1} \lambda^T H = \frac{1}{2} W^{-1} \lambda H^T$$

Now

$$F^d = H \Pi = \frac{1}{2} H W^{-1} H^T \lambda \rightarrow \lambda = 2(H W^{-1} H^T)^{-1} F^d$$

Hence,

$$\Pi = \frac{1}{2} W^{-1} H^T 2 (H W^{-1} H^T)^{-1} F^d = W^{-1} H^T (H W^{-1} H^T) F^d$$

This will generate the desired end effector forces and torques to be applied to the payload. Carignan & Akin (1988) considered the torque distribution between the two arms in a closed kinematic chain when both manipulators adopt the hybrid scheme. They minimised an energy cost function similar to the one above to give an optimal torque distribution. By transmitting torques through the payload, the joint torques required for motion were reduced while minimising the internal forces on the load. Zheng & Luh (1988) suggested that the load $M_{n+1}\ddot{q}$ should be distributed evenly between each manipulator to maximise the load-carrying capacity (assuming identical arms), where

$$M_{n+1} = \begin{pmatrix} m_{n+1}I_3 & 0 \\ 0 & R_n^0 I_{cn+1} R_0^n \end{pmatrix}$$

and $\ddot{q} = J\ddot{\theta} + \dot{J}\dot{\theta}$, i.e. $F_{ext}^{l=1} = F_{ext}^{l=2}$, or equivalently

$$\begin{pmatrix} f_{ext}^{l=1} \\ n_{ext}^{l=1} \end{pmatrix} = \begin{pmatrix} f_{ext}^{l=2} \\ n_{ext}^{l=2} \end{pmatrix} = 0.5 \begin{pmatrix} F_{ext}^d \\ N_{ext}^d \end{pmatrix}$$

with $\alpha = 0.5$ for equal load-sharing.

Alberts & Soloway (1988) suggested a more general distribution of load sharing with a portion α to one manipulator and $(1 - \alpha)$ to the other manipulator, i.e.

$$F_{ext}^{l=1} = \begin{pmatrix} f_{ext}^{l=1} \\ n_{ext}^{l=1} \end{pmatrix} = \alpha \begin{pmatrix} F_{ext}^d \\ N_{ext}^d \end{pmatrix} \qquad \text{and} \qquad F_{ext}^{l=2} = \begin{pmatrix} f_{ext}^{l=2} \\ n_{ext}^{l=2} \end{pmatrix} = (1 - \alpha) \begin{pmatrix} F_{ext}^d \\ N_{ext}^d \end{pmatrix}$$

so that internal forces,

$$F_{int} = \begin{pmatrix} f_{int} \\ n_{int} \end{pmatrix} = \begin{pmatrix} f_{ext}^{l=2} \\ n_{ext}^{l=2} \end{pmatrix} - \begin{pmatrix} f_{ext}^{l=1} \\ n_{ext}^{l=1} \end{pmatrix}$$

The weight α is selected according to the load-carrying capacity of each manipulator. The generation of applied moments n_{ext}^l may then be penalised to have moments being generated preferably through the $r_{n+1}^l \times f_{ext}^l$ term. The weights α_2 and α_4 may be set to be K times as large as α_1 and α_3 such that $K > 1$.

Thus far, consideration has only been given to the motion-causing components of force—internal forces and torques may build up if not controlled. These forces are generated when manipulators apply forces and torques against each other through the payload. Their sum is zero, not producing any motion but they do induce stresses in the payload. In general:

$$F_{int} = H^+ \prod + (I - H^+ H) \prod$$

where $(I - H^+ H$ = null space matrix and H^+ = Moore–Penrose inverse of the contact force map.

Using a PI force control law:

$$F_{\text{int}} = \begin{pmatrix} f_{\text{int}} \\ n_{\text{int}} \end{pmatrix} = F_{\text{int}}^d + K_f(F_{\text{int}}^d - F_{\text{int}}) + K_{fi}\int(F_{\text{int}}^d - F_{\text{int}})dt$$

i.e.

$$F_T = F^d + F_{\text{int}}. \tag{10.17}$$

Several possibilities exist for internal force control [Kopf 1989]:

(i)   Apply zero internal force criterion where $F_{\text{int}}^d = 0$—this is computationally the fastest approach as the remaining two approaches are computationally intensive;
(ii) Apply a minimum strain energy criterion—if the stiffness in each direction are equal, this method defaults to the zero internal force case;
(iii) Apply a minimum power criterion with weighting factor $R/(K_t)^2$.

Kopf (1989) found that the minimum strain method gave $1/200$ of the strain of the minimum power method and $1/250$ of the strain of the zero internal force method. However, power is required to implement the zero internal force and the mimimum power method yielded $1/9$ of the torque of the minimum strain approach. This corroborates Zheng & Luh's (1988) observation that energy minimisation does not yield minimum forces exerted at the end effector. Since power is likely to be a limiting factor for space systems, good power usage is required. The power minimisation method yields as much as $9/10$ of the torques of the zero internal force method, a saving of only 10% at a high computational cost. In view of this and the requirement for real-time operation, the zero internal force method appears be the approach of choice.

Pittelkau (1988) stressed the need for adaptive control using an adaptive identifier for load-sharing of an unknown payload and used minimisation of internal forces as a performance index. An adaptive identifier adjusts the load sharing at each control interval according to the force feedback. Force at each arm:

$$F_{ext}^{l=1} + F_{ext}^{l=2} = M_{n+1}\ddot{q}$$

The load $M_{n+1}\ddot{q}$ is apportioned to each manipulator so that the forces and torques required for each manipulator can be determined. The interactive forces and torques are given by:

$$F_{ext}^{l=1} = \alpha M_{n+1}\ddot{q} = \alpha F_{ext}^{l=2} - (I-\alpha)F_{ext}^{l=1}$$

$$F_{ext}^{l=2} = (I-\alpha)M_{n+1}q$$

$$= -\alpha F_{ext}^{l=2} + (I-\alpha)F_{ext}^{l=1} \quad \text{such that} \quad F_{ext}^{l=2} = F_{ext}^{l=1} = 0$$

If all the diagonal components of $W = \text{diag } \alpha_i$ are equal, then the respective end effector frames can be used as the task oriented frame of reference. The performance index to be minimised with respect to α is the sum of the weighted-squared joint torques: $Q = \frac{1}{2}\tau^T W\tau$ where

$$\tau = \begin{pmatrix} \tau_{l=1} \\ \tau_{l=2} \end{pmatrix} = \text{matrix of joint torques for each arm}$$

Now,

$$\frac{\partial Q}{\partial \alpha} = \left(\frac{\partial \tau}{\partial \alpha}\right)^T W\tau = 0 \quad \text{and} \quad \frac{\partial^2 Q}{\partial \alpha^2} = \left(\frac{\partial \tau}{\partial \alpha}\right)^T W\left(\frac{\partial \tau}{\partial \alpha}\right) > 0$$

Now,

$$\frac{\partial \tau_{l=1}}{\partial \alpha} = J_{l=1}^T M_{n+1}\ddot{q} = J_{l=1}^T\left(F_{ext}^{l=1} + F_{ext}^{l=2}\right)$$

$$\frac{\partial \tau_{l=2}}{\partial \alpha} = -J_{l=2}^T M_{n+1}\ddot{q} = -J_{l=2}^T\left(F_{ext}^{l=1} + F_{ext}^{l=2}\right)$$

Hence,

$$\frac{\partial \tau}{\partial \alpha} = \begin{pmatrix} J_{l=1}^T\left(F_{ext}^{l=1} + F_{ext}^{l=2}\right) \\ -J_{l=2}^T\left(F_{ext}^{l=2} + F_{ext}^{l=1}\right) \end{pmatrix} \tag{10.18}$$

Newton's method may now be used to find the root α to minimise Q with seed $\alpha_0 = 0.5$:

$$\alpha_{k+1} = \alpha_k - \mu\left(\frac{\partial Q/\partial \alpha}{\partial^2 Q/\partial \alpha^2}\right) = \alpha_k - \mu\left(\frac{\left(\frac{\partial \tau}{\partial \alpha}\right)^T W\tau}{\left(\frac{\partial \tau}{\partial \alpha}\right)^T W\left(\frac{\partial \tau}{\partial \alpha}\right)}\right)$$

where $\partial \tau/\partial \alpha$ is given above and $0 < \mu < 1$ controls the convergence rate. This method is computationally intensive, however.

11

Robotic sensorimotor subsystem

Having considered the analytic aspects of robotic control, we turn in this chapter to consider the sensor/actuator subsystems of payload support, particularly with regard to the proposed ATLAS freeflyer example introduced earlier. Sensing provides feedback information about the health of the agent through internal sensors and about the state of the environment through external sensors. Sensors and actuators comprise the input/output of the control system as a whole, and provide the interface for the digital electronic control system to the analogue non-electronic world. They provide the means for coupling to the environment. The speed of sensory processing is critical as it essentially limits the rate at which the control system can operate in response to sensory data and actuate to alter that environment.

Robot sensors may be divided into two major types: internal sensors which provide proprioceptive information about the current state of the manipulator, and external sensors which provide exteroceptive information about the current state of the environment in which the robot is embedded. External sensors may be further subdivided into contact sensors which sense adjacent space directly and ambient sensors which sense a larger spatial range. Contact sensors are used for avoidance behaviour only while ambient sensors can be used also for approach behaviours. However, ambient sensors provide ambiguous signals as they integrate environmental information from a region of space which is non-invertible.

11.1 ROBOTIC SENSOR TECHNOLOGIES

Robot sensing of the environment for perceptual information is central to effective control and actuation. The first generation of robots were 'senseless' but this is not sufficient for flexible behaviour. Sensors provide monitoring of task execution and the state of the external world in which the task takes place. The real world in which a robot performs is variable and sensory information is required to monitor how it changes. The sensor system provides a degree of adaptability to enable compensation for deviations of the real world from the control model due to kinematic and dynamic uncertainties. The system must predict the expected outcome of its actions, and the difference between expected and actual outcomes can be determined. Once an error state has been determined the appropriate recovery action can be taken. Sensor properties are determined by their linearity, sensitivity, accuracy, resolution and frequency response. As sensors vary in terms of range, directionality, sensitivity, resolution, specificity and accuracy, a suite of sensors is usually required. An array of sensors may be used in complementary fashion to extract information reliably.

The sensing sequence comprises two stages: transduction where the environment properties' effect on the sensor (its state parameter) is converted to an electrical signal (measurement), and signal processing. Sensor bias is a constant offset which can be as high as 25% of the signal range, but the bias is constant so calibration can be used to eliminate it. Although the most basic electrical measurements are current and voltage, transduction usually involves the measurement process generating a variable voltage rather than current to avoid capacitive charging effects. If the output to the detector is current, current-to-voltage conversion may be accomplished by dropping the current across a load resistor and measuring the voltage.

INPUT   OUTPUT

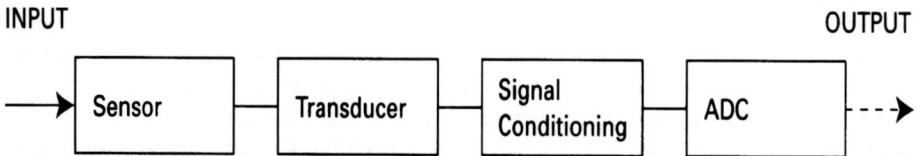

Fig. 11.1. Sensory processing sequence.

The three most important measurements are displacement, temperature, and radiation. Displacement measurement is of fundamental importance as it provides the basis for velocity, acceleration, strain, force and pressure measurements sometimes in conjunction with time measurements by piezoelectric quartz oscillators and/or Schmitt triggers. Measuring instruments are generally based on moving coil meters which measure a deflecting torque $F = BilN \times r$ which is dampened to prevent oscillations. There is no basic difference between ammeters and voltmeters—both are current devices with low and high resistance respectively. The movement of a conducting coil in a uniform radial magnetic field of a permanent magnet which is opposed by a spring giving a displacement proportional to the coil current is extensively exploited in moving coil instruments and meters. When the magnetic flux linking the coil changes, it produces an EMF given by $\varepsilon = N(d\phi/dt)$ which on integration yields

$$V_0 = -\frac{1}{CR}\int V_i \, dt$$

11.1.1 Proprioceptive sensors
Variable resistance potentiometers are cheap and rugged. Potentiometers can be used to measure voltage precisely without taking up current. A potentiometer is basically a thin resistance slide wire with a sliding tapping key connected to a galvanometer. Angular

position in robot joints may be measured by variable resistance rotary potentiometers by measuring the variation in resistance between the track end and a cursor marking the extension of the joint. They have a high degree of linearity and good resolution. Potentiometers are speed limited to ~10 rpm rotations and 1 m/s translation speeds (for prismatic joints) and suffer from frictional wear. The commonest terrestrial displacement sensor is the linear and rotary variable transformer in which the coupling between the primary and secondary coils depends on the position of a movable ferromagnetic core. Synchroresolvers are based on the variable rotary transformer with a fixed outer coil and rotating inner coil and measure angular position by electromagnetic principles thereby resembling motors. An oscillator provides two AC voltages with fixed frequency and a 90° phase separation as input to the windings on the stator core in quadrature. The rotor is also supplied with a sinusoidal AC voltage via the inductance coil around it. The revolving transformer transmits energy to the rotor without the need for brushes. The coupling between the stator and rotor depends on the relative coil positions and the rotor provides a measurable sinusoidal voltage output. Synchroresolvers are absolute encoders and give resolutions ~5 arcsec. Synchroresolvers which are robust in hostile environments are used by the nuclear industry for teleoperative robotics and are well suited to high temperature and high radiation environments up to 150°C. Optical encoders are the most common shaft displacement sensors for space robots—their digital nature ensures that no ADC (analogue-to-digital conversion) is required. Absolute optical encoders are based on binary codes to represent position. They are high-precision rotary devices mounted onto the shaft to generate a digital word (usually based on a binary 4-bit Gray code) identifying the position of the shaft. Such Gray codes are based on single bit changes from one state to the next offering an accuracy of 0.5 bits. The change in the least significant bit brings the device to the next state. The device is constructed from a slit source composed of a rotating disc of multiple concentric rings which projects light patterns onto a photocell array. The patterns depend on the arrangement of light contrasts through the disc and this depends on the code. These devices are accurate and fast, and the lack of contact keeps drive torques low. However, they are subject to noise corruption and are fragile and inaccurate under vibration. Incremental encoders are more common than absolute encoders. Incremental encoders require only three photocells and only two or three patterned rings with slots. Movement either side of a reference position produces a stream of light pulses onto the photocells as the light is patterned by the two inner rings as they pass through them. Position is determined by the number of pulses. They require re-zeroing after loss of power. Optical encoders offer the highest accuracies and are tolerant of high electromagnetic interference environments.

Velocity sensing involves the generation of a dc voltage, V, due to the relative motion of an electrical conductor in a cylindrical magnetic field of flux ϕ (the principle of the electric generator): $V = Blv$. The tachometer is the rotational equivalent with a rotating coil between the magnetic poles. The rotation speed is measured by coupling a dc generator to the wheel shaft of the motor which acts as the armature. The EMF generated by the dc generator is proportional to the rotation speed if the field flux is maintained constant—generally, the magnetic field is generated by a small permanent magnet. Thus the dc generator coupled to the moving joint produces an induced voltage (measured by a high-resistance voltmeter) in proportion to the speed of the rotor. This is the dc velocity sensor.

An ac tachometric generator coupled to the moving joint produces a voltage with an oscillation frequency in proportion to the speed of the rotor. Generally, a dc tachometer is chosen for robotic joint motors, and in spacecraft this is the obvious choice given the dc power available. Time differentiation of optical encoder output using RC resistance differentiating circuits to obtain velocity is another common method of generating velocity feedback.

Acceleration can be measured from a tachogenerator by using an RC resistance differentiating circuit so that the voltage across the resistance is proportional to acceleration. Dedicated accelerometers use the piezoelectric effect similar to force sensors—a known test mass is mounted on a disc of piezoelectric material such as lead titanate or lead zirconate. Generally, accelerometers are not used in robotic manipulator sensing.

11.1.2 Force–torque sensors

Force-torque sensors provide large increases in manipulator position accuracy as small variations in relative position between the end effector and the environment generate large contact forces. Hence active force sensing provides the means for high-resolution manipulation. Together with the proprioceptive sensors they provide the means for servo-level control feedback information. Force–torque sensors are used for fine tuning after gross motions have been executed. They are particularly essential in assembly tasks which requires force feedback for accurate fine motions. They can sense end effector gripping forces on a manipulated object which may be fed back to the servocontroller.

Strain gauges are the basis of force sensors and they exhibit electrical resistance which varies with length and in turn with applied force. The strain gauge measures the extension of the material when the material is subjected to stress. It comprises a length of nichrome or cupronickel wire is arranged in a meandering 'W' pattern which increases the length of the sensor. It is cemented to the substrate link to be measured for force with epoxy adhesive. The wire electrical resistance changes as the wire is subjected to stress by mechanical deformation which is measured by a Wheatstone bridge. To alleviate temperature differences which cause resistance changes, a second identical strain gauge is placed nearby in the unstressed state. Calibration is through the guage factor, G:

$$G = \frac{\Delta R/R}{\varepsilon} \approx 2.2 \text{ typically}$$

where $R = \rho l/A$ = resistance and $\varepsilon = dl/l$ = axial strain.

For Ca/Ni/Mn alloy, $G = 2.1$. Semiconductor gauges of doped silicon or germanium offer superior sensitivities but are highly temperature-dependent. Force sensing at the wrist eliminates friction and damping uncertainties inherent in joint torque sensing. One such wrist sensor is the Scheinmann wrist. The Scheinmann wrist is a 6 DOF assembly comprising 16 semiconductor strain gauges mounted onto four cross webbings arranged in a Maltese cross configuration with four gauges per webbing [Shimano & Roth 1975]. These are wired into eight voltage divider pairs to give differences in strain between opposite sides of the web and so measure small elastic deflections in the wrist. The wrist force sensors transform forces and moments at the hand into displacements at the compliant wrist. The sensors are capable of resolving forces at hand level of ~0.05 N and have a

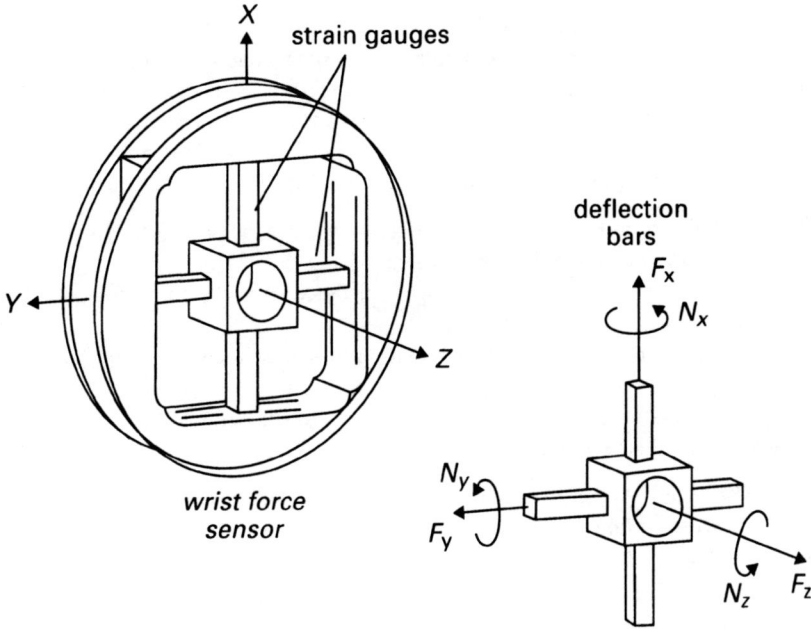

Fig. 11.2. Wrist force–torque sensor (from Fu et al 1987). Reproduced from Fu et al (1987), *Robotics: Control, Sensing, Vision and Intelligence* © McGraw-Hill.

maximum force/torque capability of 150 N/300 Ncm. The device has a disc geometry with a diameter of 5.3 cm and a thickness of 2.5 cm.

Coupling effects may be eliminated by introducing a calibration matrix to convert sensor readings to force vectors at the hand. A calibration matrix and offset matrix is necessary to transform sensor readings into a force vector. It usually assumes that there is no coupling between force components (as justified by basic theories of elasticity). The calibration matrix gives the force relative to the wrist and is determined experimentally but the offset matrix is required to resolve the forces at the end effector tool endpoint which lies a distance (dx, dy, dz) from the sensors:

$$\begin{pmatrix} F_x^{hand} \\ F_y^{hand} \\ F_z^{hand} \\ N_x^{hand} \\ N_y^{hand} \\ N_z^{hand} \end{pmatrix} = \begin{pmatrix} 1 & 0 & 0 & 0 & 0 & 0 \\ 0 & 1 & 0 & 0 & 0 & 0 \\ 0 & 0 & 1 & 0 & 0 & 0 \\ 0 & -d_z & d_y & 1 & 0 & 0 \\ d_z & 0 & -d_x & 0 & 1 & 0 \\ -d_y & d_x & 0 & 0 & 0 & 1 \end{pmatrix} \begin{pmatrix} F_x \\ F_y \\ F_z \\ N_x \\ N_y \\ N_z \end{pmatrix} = \begin{pmatrix} F_x \\ F_y \\ F_z \\ -d_z F_y + d_y F_z + N_x \\ d_z F_x - d_x F_z + N_y \\ -d_y F_x + d_x F_y + N_z \end{pmatrix}$$

$$(11.1)$$

This is then converted to joint torque requirements through the Jacobian transpose as implemented in the robot controller. Force sensor data generally requires filtering and estimation to remove unwanted noise and generate usable feedback information. The problem of manipulator weight which tends to corrupt force data does not occur in zero-gravity space environments making force control a particularly promising capability.

The wrist force–torque sensor configuration exhibits most of the desirable requirements for force sensors:

(i)   high stiffness to reduce compliant deflection so that the accuracy of hand positioning is not reduced due to disturbing forces;

(ii)   compact design to minimise sensor wrist distance from the end effector and reduce lever arm to reduce position errors;

(iii) good linearity to ensure ease of calculation of sensor readings using simple matrices;

(iv)   very low hysteresis and internal friction to maximise the sensitivity and repeatability of sensor readings.

JPL (NASA's Jet Propulsion Laboratory) has also designed a 6-dimensional force–torque sensor with a Maltese cross configuration similar to the Scheinmann wrist with a dynamic range of 0.5–300 N.

11.1.3 Tactile sensors

Touch sensing is an essential complement to vision which is less useful at the 0.01–1 mm ranges though human visual acuity is capable of perceiving ~3 arcsec [Holliday 1993]. Touch is of primary importance possibly more important than vision for fine manipulation and may be regarded as a form of refined force sensing [Paul 1987]. Exploratory groping of surfaces with a tactile sensor is an efficient way of determining 3D structure and vision can be employed to provide guidance to the touch sensor. Indeed blind people use touch extensively and can survive adequately in the world—it is debatable whether a person devoid of touch could do so well. Touch is an active, exploratory process in which sensing and actuation are intimately linked such that the actuation process through movement of the hand and fingers define the sensory date directly. Taction does not suffer from inoperability in darkness, or from problems due to shadows, lighting constraints, colour or obscuration, though it does suffer from restricted views of the object. The simplest form of touch sensing is the binary microswitch which may be used to initiate grasping on detecting the presence of an object. It is not considered adequate for high performance. Slip sensing is also required. Slippage can occur due to gravitational loading (absent in space) or from manipulating objects with high inertias. Slip can be prevented by increasing normal gripping force before the tangential force reaches the static friction (stiction) between the object and the end effector so that the normal force exceeds stiction. Slip sensors must not only detect the fact of slip but its direction and rate also. JPL have constructed a slip sensor which uses a circular conductive plate attached to a needle [Bejczy 1980]. There are 16 electrical contact points distributed evenly in a circle under the plate. The needle contacts a sphere with irregular dimples over its surface which is in turn supported by a bearing. Any object pressed against the sphere will turn the sphere as slip occurs. The rolling sphere oscillates the needle causing the circular plate to touch one

object slip

dimpled ball

contacts
(16 places)

conductive disk

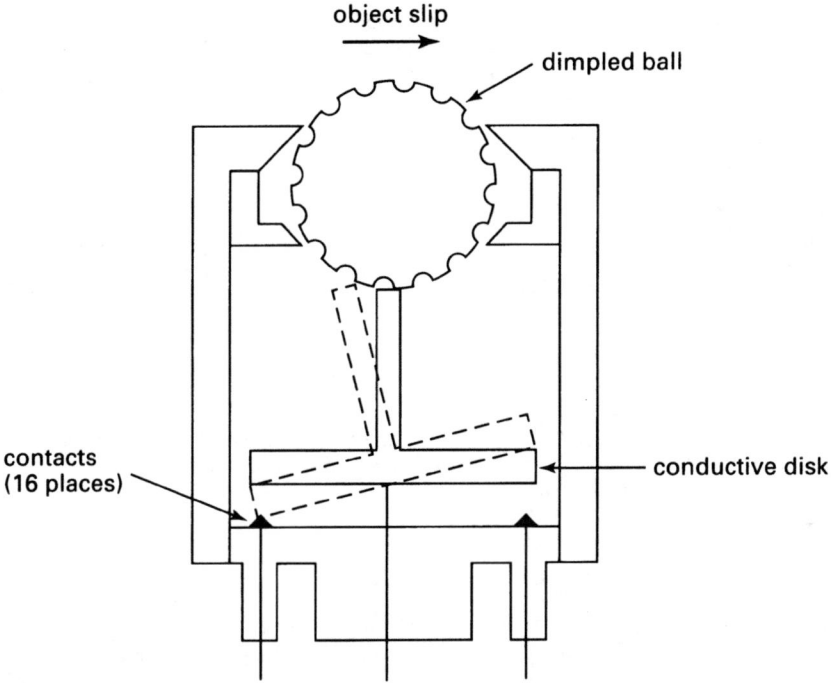

Fig. 11.3. JPL slip sensor. Reproduced, with permission, from Bejczy (1980) 'Sensors and man–
machine interface for advanced teleoperation', *Sci.* **208**, 1327–1335. © 1980 American
Associaton for the Advancement of Science.

of the 16 electrical contact points. Which of the contact points is closed depends on the
direction of the roll of the sphere, i.e. 16 directions can be discerned.

Tactile sensors may take the form of an 'artificial skin' distributed as arrays of
pressure-sensitive pads over the robot links on a thin, flexible compliant substrate for
continuous tactile sensing in all directions. The transducers must be durable and wear-
resistant particularly to slip friction. Harmon (1982) discussed the most desirable
properties for touch sensors. Fast transduction and response times are required ~1–10 ms
compared with human reaction times of ~200 ms. Resolution requires 1–2 mm spacing
between sensors corresponding to 5×10 to 10×20 elementary points for a fingertip—
this brackets the human fingertip characteristic of 10×15 points. The dynamic range
required is from 1 g to 1 kg with a sensitivity of 1 g. Hysteresis should be negligible.
Presently available arrays have good spatial resolution but do not provide all of the other
requirements. Resistive and conductive elastomer materials are promising if the dynamic
range can be increased and hysteresis reduced. They are cheap, heat-resistant and easily
fabricated. One example is the use of carbon fibre inserted into orthogonal silicone rubber
sheets. The carbon fibre responds to pressure by decreased electrical resistance while the
silicone rubber conducts current in response to pressure according to decreased resistance
at the strip interfaces. Semiconductor materials permit small devices and preprocessing

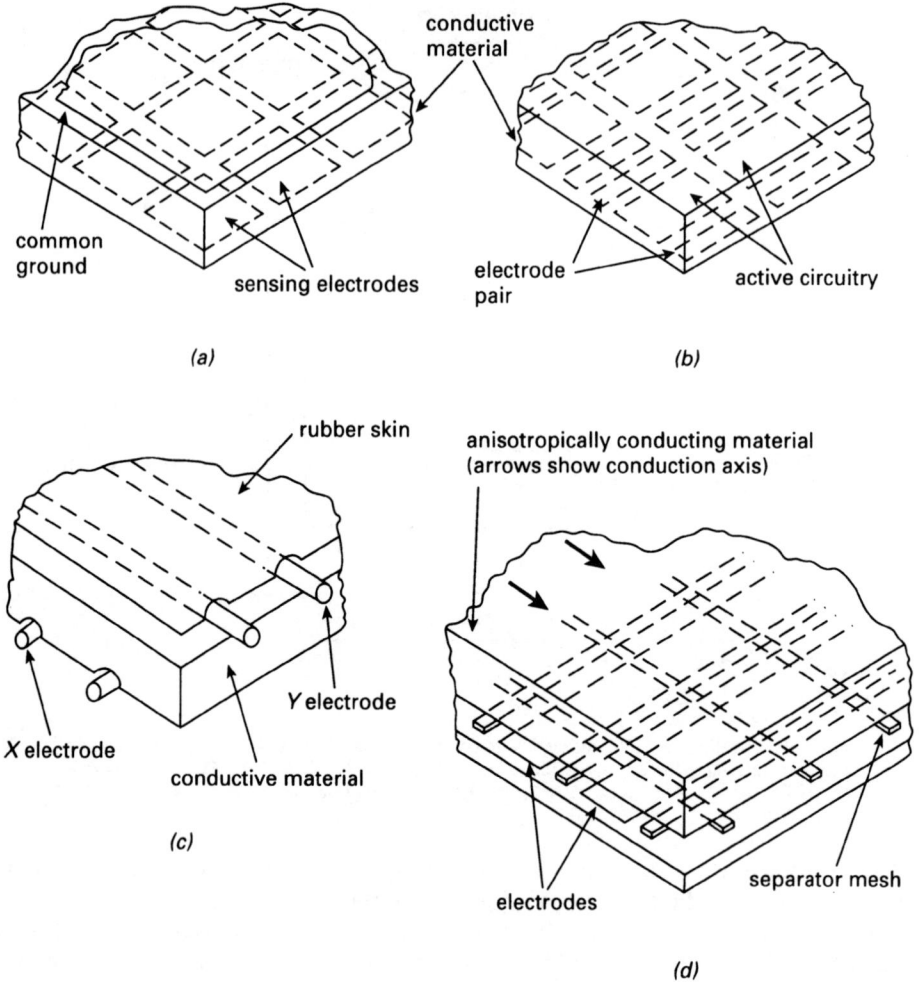

Fig. 11.4a. Different approaches to artificial skin and their implementation as tactile sensors. Reproduced from Fu et al (1987), *Robotics, Control, Sensing, Vision and Intelligence,* © McGraw-Hill.

but they suffer from fragility. However, they are fast developing a into new trend in sensor technology, that of microsensors.

Piezoelectric materials are insulators which produce charges on their surface when mechanically deformed, a property determined by the piezoelectric coefficient $(\Delta l/l)(\Delta E/E)$, e.g. quartz has a piezoelectric coefficient of 2×10^{-12} m/V. Lead titanate and lead zirconate have much higher values. The crystal deformation produces an EMF which can be measured in conjunction with a high-gain dc amplifier with high input impedance. Piezoelectric polymers are subject to electric charge generation in proportion to the displacement resulting from an applied force such that integration of the current

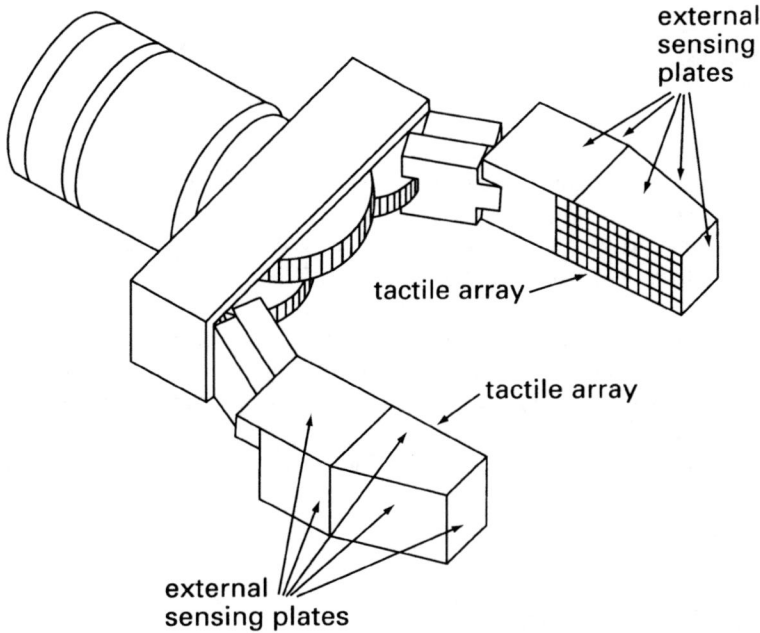

Fig. 11.4b. Artificial skin implementation as tactile sensors. Reproduced with permission from
Fu et al (1987), *Robotics: Control, Sensing, Vision and Intelligence*, © McGraw-Hill.

value gives the force value. Polymeric piezoelectric film transduction has great potential
with high signal-to-noise ratios and good flexibility but is again fragile. An example of
this technology is the use of polyvinylidene fluoride (PVF_2). When stretched to four
times its natural length it achieves a piezoelectric state. Piezoresistive transduction which
also undergoes decreased electrical resistance under pressure also has great potential but
the sensors tend to be rigid, stiff and slippery. However, they are robust to high temperatures ~1100°F and offer low hysteresis. Magnetoresistive materials (e.g. Ni-based metals
such as Rumalloy 81/19 Ni–Fe) respond to external forces by altered resistance due to
altered magnetic fields. Tactile sensors have a tendency to suffer from wear and tear but
for space applications wear and tear, is not the constraint that it is in industrial applications.

A sheep-shearing robot has been used successfully which uses compliance sensors to
take account of inaccuracies of a geometric model of sheep due to breathing. It gives a
very low injury rate to the sheep from the shearing cutters due to these compliance
sensors. Additional future capabilities should include temperature and humidity sensing
for distinguishing materials and for incipient slip estimation. Processing of touch image
data is similar to visual image processing but less complex, but it also produces inferior
quality images. Pressure gradient sensing may provide edge and hole detection. The
processing of tactile data really requires a somatotopic map of the skin surfaces for object
form identification. Such a representation could be overlaid on a visual representation in

a one-to-one equivalence manner as a combined tactile–visual map fusing the two data representations.

11.1.4 Proximity sensors

Proximity sensing is necessary for the detection of the near presence of approaching objects or obstacles. Vision is unsatisfactory at very close quarters due to problems of blockage in 3D scenes. The working range for such sensors varies up to ~1 m. Simple LED sensors may be used to detect the presence or absence of an object within the end effector jaws. LEDs are active emitters based on II-IV semiconductor materials providing red, amber, green and blue point sources of light. Photodetectors to detect the light source are either photoresistive or photovoltaic. Photoresistive sensors rely on the generation of conduction electrons due to the incidence of electromagnetic radiation through the photo-electric effect. They are bulky, fragile and power-hungry. Photovoltaic sensors are p-n junction semiconductors in which radiation causes splitting of electron–hole pairs in the p-n junction. They are dark current limited. The presence of an object will interrupt the

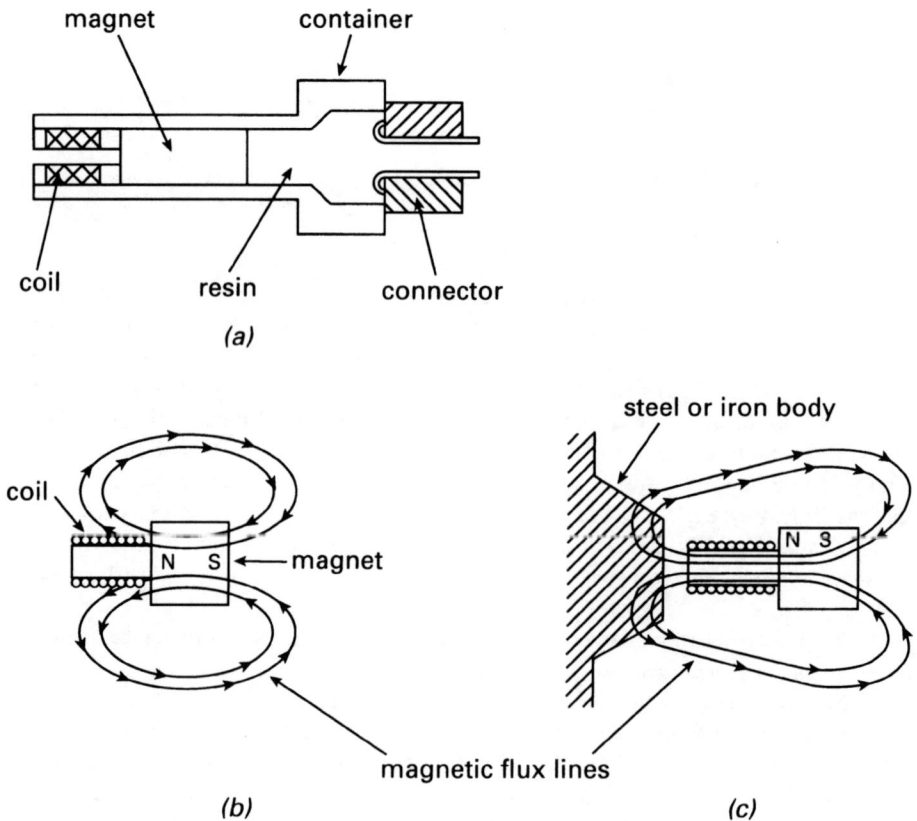

Fig. 11.5. Inductive sensor and its operation. Reproduced from Fu et al (1987) with permission from Società Italiana di Fisica © 1981.

light, constituting a binary Yes/No sensory system. Ultrasonic and acoustic proximity detectors cannot be used in the vacuum of space so are not considered further, though they may be suitable for operation in planetary atmospheres given adequate modelling and calibration. Inductive proximity sensors can only sense ferromagnetic metallic objects since they operate using a low-frequency oscillator circuit which generates an oscillating magnetic field in the coil to induce eddy currents in the conducting target. They have such short range that they effectively comprise near contact switches. This is the principle of the metal detector.

Hence, capacitative proximity detectors which can detect most materials are preferred—although spacecraft material is primarily metallic, new composite materials are being increasingly employed. Capacitative sensors detect the beginning or end of oscillations due to changes in the capacitance of a resonant circuit caused by change in the permittivity of a dielectric material in the sensing capacitor. They can usually detect over distances of about a millimetre and are usually employed between end effector fingers. Such proximity sensors can provide a smoother transition between tactile and vision sensing.

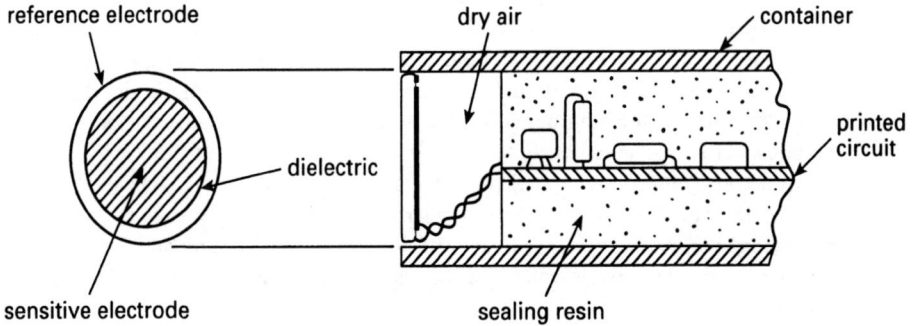

Fig. 11.6. Capacitative sensor. Reproduced from Fu et al (1987) with permission from Società Italiana di Fisica © 1981.

11.1.5 Range sensors

Range sensing is employed for sensor range location. Ultrasonic radar is unsuitable for space application, though suitable for planets with atmospheres. To date, range sensing in space has been dominated by microwave radar systems. There are two modes of microwave radar operation: synthetic aperture mode and sounder mode. The sounder transmits a pulse straight to the object and detects the return signal. Radar rangefinding involves determining the time between the emission of a pulse and its return. It requires the object to be detected to have a high dielectric constant for radar reflection. Most airborne radars are pulse Doppler radars—they do not transmit and receive at the same time so the signal transmission and receiver noise are uncorrelated [Wandell 1989]. The time between succeeding pulses (pulse repetition interval), $\tau = 1/f_r$ where f_r = pulse repetition frequency = number of emitted radar pulses/second ~100 Hz – 100 kHz. The peak power, P, and pulse width, τ, determine the energy of a single pulse. Detection range, R, is increased by increasing the pulse repetition frequency, pulse interval (~5000 μs – 100 ms typically) and peak power: $R = ct/2$. The Doppler effect allows the measurement of range rates by Doppler frequency shift $\Delta f = -2(dR/dv)/\lambda$ where λ = carrier wavelength. The Doppler effect may be employed to make the radar blind to stationary chatter by responding only to moving objects. Pulse radars are well established in RVD (rendezvous and docking) operations for spacecraft. The US manned Gemini mission series used an active L-band noncoherent pulse radar transponder for range and range-rate, angle and angle rate data to enable manned docking enhanced with lights for visual observation. The Shuttle uses a small K-band RVD pulse Doppler radar to measure range and angle and range- and angle-rates to find spacecraft to within ~20 m/0.3 m/s, 8 mrad/0.14 mrad/s accuracies. It can skin track to 12 nmi if the target is uncooperative or to 300 nmi if the target is cooperative with a transponder. However, the Shuttle radars have high peak powers of 1–5.5 kW. Most planetary probe radars have much more limited powers available, e.g. Magellan's imaging radar had a peak power of 325 W. A radar comprises four main sections: transmitter and receiver magnetrons, transmit–receive duplexer switch, and slotted waveguide antenna, plus data processor. A ~100 ns duration pulse of megahertz frequency emission is delivered by the transmitter to the antenna through the duplexer with the receiver disconnected. After a short delay, the duplexer disconnects the transmitter according to a timing oscillator unit, and connects the receiver so any reflected pulse may be detected. The spiked pulse is shaped into a constant amplitude, flat-topped square wave by a pulse-forming network so that the return pulses are recognisable. The duplexer allows a single antenna to serve as both receiver and transmitter by isolating transmitter/receiver signals. A typical intermediate frequency is ~30 MHz with a bandwidth of 5 MHz. The NASA radar transponder comprised a single transmitter antenna and three receiver antennae to measure the differences in time of arrival of the received pulses: $\Delta t = 2R/c$. The pulse radar performance is dictated by the radar equation [Wandell 1989]:

$$P_{radar} = \frac{P_t G_t}{L_t} \frac{1}{4\pi r^2 L_m} \sigma \frac{G_r \lambda^2}{4\pi r^2 L_r} \frac{1}{L_p}$$

where P_t = transmitter power, G_t = transmitter gain, L_t = transmitter losses, L_r = receiver losses, r = range, σ = target cross-section ~3 m^2 for a spacecraft typically, L_m = medium losses, G_r = receiver gain, λ = radar wavelength, and L_p=polarisation losses.

Free space losses are due to the inverse square distance of transmission and receipt. High performance requires high transmitter power or large aperture area. Travelling wave tubes offer high powers and represents the usual technology adopted for space radar systems rather than the solid state approach used in onboard communication systems. For a round trip, the flux received is thus in inverse proportion to the fourth power of the distance to the target. Radar detection capability depends on the S/N (signal-to-noise) ratio

$$\frac{S}{N} = \frac{P_t G_t A_r \sigma}{(4\pi)^2 R^4 k T_e B}$$

such that

$$P_r = kT_eB\frac{S}{N}$$

and +10 dB is usually the minimum value of S/N required. S/N ratio can be enhanced by pulse compression. Compression produces a pulse of high bandwidth compression ratios ~100 (i.e. an S/N improvement of 20 dB). Frequency modulation with chirp is common but binary digital PCM is more appropriate with digital computer systems. A received power P_r of 4 dB would provide a range of 40 km. Their resolution is frequency-dependent limited by the beamwidth $\theta = 1.22\lambda/D$. Microwave radars can determine the shapes of metallic surfaces which optical sensors and laser rangefinders have difficulty with due to their specular reflection properties. NASA's microwave radars operate at 100 GHz in order to detect and track objects moving at velocities ~0.03–1.5 m/s (typical of astronauts in an MMU). They offer great potential for docking systems due to their small size. They can operate at several carrier frequencies allowing them to measure backscattering for various polarisations to determine target object structures. They can also penetrate multilayer thermal blankets. Electronically steered phased arrays of many dipoles eliminates the inertia problems of slewing the radar assembly. Each dipole is fed with different phased signals providing the means for altering the direction of the main radiation lobe. Although their space pedigree in RVD operations is well established, the increasing usage of GPS (Global Positioning System) combined with the high powers required for microwave radar make them a suboptimal choice in terms of their inherent mass penalties.

Optical (laser) radars have considerably greater accuracy than microwave (maser) radars. LIDAR (light detection and ranging) is a time of flight optical range detection scheme between an optical tansmitter and receiver telescope [Jarvis 1983]. Detector response times of ~50 ns are required for accuracies in rangefinding of the order of metres. Two possibilities are available. A pulsed laser light time of flight between emitted and reflected beams is measured when detected by a photomultiplier. Alternatively, a continuous wave light beam undergoes a phase shift between transmitted and received signals which is measured. LIDAR systems employ a narrow beam with little beam spread and the wavelength of the light is less than the object surface undulation dimensions. Hence, beam bounce from false inclines and specular reflection will not occur. Some of the reflected rays strike the receiver and the time of flight measurement gives the distance to the target. The laser scan provides information on angles and positions of objects within its field of view, generating depth and height profiles of objects in the scene [Krishen 1989]. Doppler signal processing can also generate motion data. Laser radar has a range of 20 km for uncooperative targets (without transmitting beacons) and 200 km for cooperative targets [Korf 1982]. Laser radar is simple to implement and they require low computational overheads. Optical heads (lenses) may be mounted inside the end effector jaws of the robot. Fibre optic cables with low attenuation may connect the light source and the detector to the optic heads. This technique degrades in performance when weak reflections from the object occur. It is often used with other sensors such as stereovision as a complementary sensor for cross-checking depth measurement. Volpe & Ivler (1994) suggested that LIDAR offers the best accuracy/range combination of all proximity sensors. It is likely to become used extensively in space robotics in the near future.

11.1.6 Rover scientific instruments

Scientific instruments comprise the principal payload for robotic planetary rovers in the form of an integrated suite of scientific sensors pertinent to the task of surface exploration. For detailed geochemical analysis of rock-forming minerals and atmospheric constituents, planetary orbiters are limited to ~3 km resolutions for mineral deposit identification.

The sensors that provide spacecraft telemetry information are often similar to primitive scientific sensors, e.g. temperature/pressure measurement. Space instrumentation must be small in volume, mass and power requirement [Surkov 1997]. They must function over a wide range of temperatures and be robust to continuous and shock loading. They must operate autonomously, usually by preprogrammed sequences with automated error recovery, and transmit their data to a remote receiver. Passive sensors include piezoelectric crystals (for pressure measurement) and thermocouples (for temperature measurement). Powered sensors include variable resistors, inductors and capacitors. Temperature sensors are common on robotic spacecraft and rovers, and indeed are used for telemetry data on all spacecraft for thermal control. There are three main types of temperature sensor. Resistive coils are wires of platinum, nickel or copper with highly temperature-dependent resistivity. Thermistors are semiconductors with temperature-dependent resistivity which is usually negative. They are more sensitive ~4%/°C than resistive coils but limited in temperature range. Most temperature-measurement devices are based on the thermocouple which is a thermoelectric sensor. A heated junction between a pair of wires of dissimilar metals such as Pt and Rh/Pt alloy produces a flow of electrons across the junction. If a temperature difference exists between cold and hot junctions, then an EMF is generated, causing a current to flow through the circuit dependent on the temperature. For temperatures up to 750°C, iron and nickel–copper alloy is used, providing an EMF with resolution 0.054 mV/°C.

We shall examine some specialised scientific sensor technology typically deployed on planetary landers and rovers here. The Martian lander Viking launched in 1976 included an onboard biological laboratory with a volume of only 0.3 m^3. There were three biological experiments to measure metabolic activity of the Martian soil as well as the first gas chromatograph–mass spectrometer to be utilised onboard a spacecraft. The need for small, compact and lightweight sensors and instruments is paramount for all spacecraft, and the gas chromatograph–mass spectrometer is a unique and versatile instrument. For terrestrial bodies like Mars, mineralogical and chemical analysis of rock forms the basis of planetary studies. In particular, hydrates, carbonates, cherts, evaporites and phosphates can be diagnostic of aqueous processes. Biologically mediated processes would be indicated by phosphates and certain carbonates. Isotopic ratios for the biochemical elements C, H, O, N, and S would also be required to ascertain biological processes.

Differential thermal analysis provides mineralogical information as it can distinguish among clays, silicates, feldspars, zeolites, glasses and evaporites. The temperature of the sample is measured with reference to an inert material while the two samples are heated. Chemical reactions cause a change in the sample's temperature which indicates the reagents present.

X-ray spectrometry can detect the major rock-forming elements and their concentrations. The distribution of emitted X-rays are characteristic of the element. The Viking

X-ray spectrometer was an active instrument which used Fe-55 and Cd-109 radioisotope X-ray sources on the constituent elements of the rock in comparison with a calibration library of pre-stored spectra. The alpha-proton-X-ray spectrometer (APXS) can provide data on the chemical composition of rock particularly low mass number elements. The instrument comprises of a sensor head which is deployed against the rock or soil sample using a robotic mechanism due to the limited transmission distances of alpha particles. The sensor head contains a radioactive alpha-particle source (typically Cm-244 with a half-life of 18.1 years) and three detectors: an alpha particle Si detector, a proton Si detector, and an X-ray Si-PIN detector. The APXS measures the three inter-actions of alpha particles bombardment with the sample—Rutherford elastic back-scattering of alpha particles (for light elements C, N and O detection), the production of protons due to the alpha-proton nuclear reactions (for rock-forming F, Na, Mg, Al, Ca, Si and S detection), and natural X-ray emission due to the excitation of atomic structures (for Na and heavier elements). The front silicon detector is the alpha particle detector and has a thickness of 35 mm to stop alpha particles of 6.5 MeV or less (the maximum energy of Cm-244 backscatter alpha particles is 5.8 MeV). However, it is transparent to protons of energy greater than 1.6 MeV. The second silicon detector is 320 mm thick and will stop all protons with an energy up to 6 MeV. Threshold logic can discriminate between proton and alpha particle events. The X-ray detector comprises a Si PIN photodiode mounted onto a Beryllia substrate with a thin Be front window. Typical integration time is 600 minutes for a single sample. These three multiple and simultaneous measurements allow the determination of the composition of all elements except H. An APXS instrument was included in the Mars Pathfinder rover which was a copy of the ones flown on the Russian Vega and Phobos missions.

A gamma-ray spectrometer measures the distribution of C, O, N, Cl, Fe, Ti, Si, Ni, Mg, Al, Ca, K, Th, U and O on the planetary surface—the detection of natural gamma rays due to radionuclides U, Th and K determine the radioative inventory of the planet (gamma ray secondary emissions also result from cosmic ray bombardment of C, O, N, Mg, Si, Cl, Ca, Mn, Fe, Ni). Used in conjunction with a neutron spectrometer, it can detect H by measuring the ratio of epithermal (intermediate) to thermal (fast) secondary neutron flux emission due to bombardment of hydrogen atoms by fast neutrons of 10^6–10^7 eV from cosmic rays. The neutron spectrometer often indicates the presence of water as the H source (as used on Lunar Prospector) and is also sensitive to Fe and Ti deposits. Alternatively, neutron activation can artificially bombard rock to emit gamma rays.

The Mossbauer spectrometer which utilises nuclear resonance to give mineral identification. The Mossbauer effect is based on resonant absorption of emitted gamma rays by Fe-57 according to the electronic structure of the sample mineral. The main parts of the instrument are the gamma-ray and X-ray detectors that measure the backscattered radiation. The radiation source is typically based on Co/Rh. Nuclear magnetic resonance detects nuclear spin state splitting of Fe-57 due to the influence of magnetic materials, the resonant frequencies being dependent on the mineral species. The Mossbauer spectrometer requires long integration times over around 12 hours.

Gas chromatography (GC) is used for the analysis of gaseous species, either in natural gaseous form or evolved gases from pyrolysis. Of interest for exobiology are organics, water, CO_2, NO_x, NH_3, H_2S, and SO_x. Gas chromatography is based on separating chemical mixtures by passing them through a stationary column of material to which different chemical species bind with varying degrees of strength. As the binding is reversible, a chemical equilibrium is set up where those that bind most effectively spend more time bound to the material and take the longest time to pass through. The time to pass through is characteristic of the chemical species. Gas chromatography is often coupled with mass spectrometry in a combined gas chromatograph/mass spectrometer (GCMS) instrument as used on the Viking Mars lander. Furthermore, the gas chromatograph/mass spectrometer of Huygens is the heaviest of its 40-kg, six-instrument suite, at 19.5 kg, as well as the most important with regard to organic chemistry. It is the second GCMS instrument to be flown and is designed to give the atmospheric chemical composition profile of the Titan atmosphere.

Mass spectrometry provides isotopic abundances including C^{13}/C^{12} ratios. This ratio is depressed by 2–4% in organically derived materials due to organic synthesis. The chief disadvantage of chromatography is that it cannot separate enantiomers due to their similarity to each other. The sample can be heated in a pyrolysis chamber to evolve the gaseous species from solid/liquid samples. Combined with pyrolysis, the GCMS can measure evolved chemical species from solid, liquid or aerosol particles. Stepped pyrolysis (e.g. the Viking GCMS pyrolyser step-heated up to 773 K) generates molecular gases without fragmentation and allows the detection of organics such as amino acids, purines, pyramidines, fatty acids, etc. The Huygens Aerosol Collector/Pyrolyser will collect filtered aerosols for chemical composition analysis by the GCMS subsequent to a three-step pyrolysis at different temperatures up to 600°C. Different compounds of different molecular weights (and so compositions) pass through the GC column at a unique rate thereby separating them. Each compound emerging from the column is directed into the mass spectrometer for identification. The mass spectrometer ionises the gas, accelerates the ions through a magnetic field which separates out the different mass numbers of the gas atoms. Mass spectrometry requires high resolution ~8000× to resolve neutron–proton mass differences implying large instruments with high power requirements. Spacecraft payload mass spectrometers are usually much less accurate and are designed to differentiate between chemical species rather than isotopes. Capillary electrophoresis may be used with samples with ionisable functional groups. The gas sample is injected into a stream of a carrier gas, typically He which is passed into a column of absorbent material. The transit time through the column characterises the compound.

11.1.7 Signal processing

Signal processing is the transformation of the electrical signal generated by transduction into useful information by amplification and filtering to allow interpretation, i.e. the extraction of the required data. Analogue signals have time-variable voltages. Amplification and filtering is usually followed by conversion from analogue to digital form by sampling and quantisation for further higher-level processing. Transmission of high-level sensor attributes rather than raw sensor data is preferable due to the much lower bandwidth of transmission required. Preprocessing of sensor information in the spacecraft reduces the transmission bandwidth of the data link and reduces the burden of computation on the ground. Fast sensor data processing is required for real-time control.

A fundamental process in signal conditioning is filtering the signal to remove noise and improve signal-to-noise ratios (e.g. RC filter, Chebyshev/Butterworth filters, impulse response filters). Digital electronics is now superseding analogue electronics in many applications. Comparators and ADCs (analogue to digital converters) are common input devices by which analogue signals can drive digital circuits. These days, rather than electronically filtering a time-varying voltage signal, the digitised signal voltage $y(t)$ may be passed through a convolving digital filter algorithm implemented in software. The FIR (finite impulse response) and IIR (infinite impulse response) filters are the basis of many digital signal processing functions. Whereas analogue filters can suffer from drift and instability, the digital approach allows fine-tuned filtering by adjusting the convolving coefficients. Furthermore, digital filters can track low-frequency signals accurately. Interfacing to digital circuits involve extensive use of buffers to convert data into serial pulses. Data and address buses allow the transfer of data. Coaxial cables are normally used for digital signal transmission between devices. ASCII is transmitted as 8-bit serial strings. The RS-232 standard for transmission of data between computer and terminals sends data asynchronously with start and stop bits flanking the 8-bit ASCII code as a 10-bit group and transmitted at a fixed serial rate such as 1200, 2400 and 4800 baud. The RS-422 and IEEE-488 standard allows for faster serial data rates from 9600 baud to 38 400 baud. Spectrum analysers are instruments used for analysing the frequency domain behaviour of signals. The majority are digital and are based on the discrete FFT (fast Fourier transform) which is a numerical algorithm for implementing the Fourier transform as a truncated summation across a finite number of digitised data points over a finite sample time T. Many digital signal processors are based on dedicated digital processing chips with real-time performance. Most have an ADC front end and DAC back end for use with analogue electronics. They rapidly compute the many sums of products involved these computations. The Harvard bus separates instructions and data allowing the CPU to fetch both from memory simultaneously. The core of these digital signal processors is the multiply/accumulate unit which solves the many sum of products computations. A decentralised processing unit for each sensor would perform rapid local data processing including signal processing and data compression while acting as buffers to the main spacecraft computer [Gough & Wooliscroft 1989]. High-level sensory processing effectively comprises the input processor which transforms the sensory data into a symbolic format suitable for an inference engine to effect knowledge manipulation (which may be located on the ground) [Taylor et al 1987]. The output processor transforms output commands to the control system as control signals.

Artificial neural networks (ANN) have been used for space instrumentation—a software implementation to perform rapid autocorrelation function computations was flown on the Space Shuttle in 1992 [Gough 1993]. A 32-bit neuron input network was implemented on RAM memory units acting as neurons for pattern recognition. The sparsely connected network was implemented with N bit addresses for each RAM unit. The 32-point input layer was randomly connected to 16 associative memory units which provided 8-bit outputs. Ninety-six training patterns were used to train the network. The ANN took up 4 kbytes of PROM and used an 8-MHz 80C86 processor to provide rapid computations in 4 ms.

Signal processing is a huge field which cannot be covered here. Suffice it to say that electronic hardware and computer software offer a range of techniques and methodologies in the selection of sensory signal processing.

11.1.8 Sensor fusion

Sensors are required to deliver data about various properties of the environment. A distributed and diverse sensor suite enables a system to cope with failures gracefully due to functional redundancy. They can also provide multiple information sources which operate in parallel. Each sensor output is degraded by noise so if multiple sensors are used to determine the same property, uncertainty is reduced. Furthermore one modality may suffer degradation under certain conditions but by using additional modalities such problems can be eliminated. Multisensor fusion is a natural extension of using multiple sensors for verification. For complex environments, it is desirable to simultaneously integrate such sensory data from diverse multiple sources. Multisensor data fusion improves the overall performance of interpretation and classification by providing a more reliable and robust estimation of the environment than each sensor considered alone. Sensor suites offer redundant information with different fidelities or offer complementary information of different modalities. It decreases feature ambiguity and uncertainty due to noise, and reduces search spaces during matching at later processing stages. In robotics, the commonest fusion proposals have been the fusion of stereoscopic vision sensors, the fusion of vision information with rangefinder information, and the fusion of visual and tactile data. Although multisensor fusion is believed to occur in the human brain, ventriloquism seems to indicate that visual information dominates over auditory information in the fusion process [Luo & Kay 1989]. Furthermore, space adaptation syndrome (space sickness) also suggests that sensory fusion processes are not complete with vision dominating over vestibular data, the mismatch being the cause of the nausea characteristic of the condition. Such hierarchical dominance between sensors is a natural one in robotics which avoids the many problems of sensor fusion. Vision may have high priority over gross motions with proximity sensors taking over dominance during near-contact guarded motions until tactile sensors become dominant during touch manipulation. Several problems are immediately apparent in sensor fusion [Pau 1987]:

(i)   different sensors provide different physical transduction mechanisms, different locations, with different sensitivities to the environments to which they are exposed, have different bandwidths, different spectral detectivity profiles and differing data processing rates;

(ii)   registration is required of data points of one sensor corresponding to data points on another sensor;

(iii) merging data from widely differing forms of representation into a unified representation medium.

Sensor fusion can occur at different levels from the integration of raw data to the integration of object properties of a world model. Fusion essentially involves the integration of different sensory modalities into a single representational format such as a world model. Learning takes place when a world model is updated with new sensory information. For orthogonal or sequential data, such symbolic fusion is sufficient: orthogonality implies no correlation between data while sequential data implies using one sensor to guide another sensor in sequence, e.g. for different resolutions. However, if the measurement data is intersecting, data merging techniques are required. Indeed, redundant information may be fused at a lower level of representation than complementary information which requires fusion at symbolic level.

High-level integration implies a two-stage process involving signal processing of the diverse raw data and then combining it at a 'fusion' centre [Thomopoulos 1990]. An example of such high-level symbolic fusion may be production rules in expert systems. Production rules may be used to symbolically represent relations between object features and fusion may occur when two or more production rules referring to the same object are combined during logical inferencing to form one rule. Another possible unified mode of representation is to use structural feature graphs to represent relations between attributes and properties corresponding to a pictorial description such that topological information of spatial relationships is retained of the objects encountered. Merging of structural graphs according to their match based on neighbourhood relations may be performed in a relatively straightforward manner. Indeed, these two methods are in fact equivalent as production rules may create schemas to represent objects within graphical network hierarchies formed by the schemas. A common central environment for the integration of sensor data may be provided by a distributed blackboard architecture to provide high-level sensor fusion with each sensor module passing messages to the blackboard [Harmon et al 1986]. This architecture provides a means for dealing with uncertainty in visual scenes regarding important objects within the scene. The blackboard provides centralised control of multiple specialist knowledge sources. It provides the mechanism for multisensor fusion at different hierarchical levels [Clemint et al 1993]. Each specialist sensor module is independent but they all share information by communication through the global blackboard database. Objects may be described in terms of features and relations between these features. The blackboard may be organised as a semantic network data structure to hierarchically structure objects and their properties. The sensors update the blackboard with new information to cause blackboard state changes. The detection of specific objects requires *a priori* knowledge of the types of objects encountered. Radiometric properties of an object can be determined from its composition. Geometric properties of objects can also aid identification, particularly with respect to spatial relationships between parts. Shafer et al (1986) called their blackboard model of sensor fusion a whiteboard since it represented a topological map of the scene using potential fields and the modules were separate expert systems dedicated to mobile robotic behaviour for their NAVLAB mobile robot: CAPTAIN, NAVIGATOR, PILOT and PERCEPTION expert subsystems. Sensor fusion was implemented by integrating through the whiteboard map with other processes.

For similar modalities, redundant information may be fused at a lower level of representation. If redundant data is in conflict, a mechanism for assigning confidence to each data value is needed. Confidence may be assigned to provide weighted averages with weightings to each sensor value. Heuristic techniques involve a supervisory approach according to sensor reliability—this is an ad hoc approach. Hence, a probability based approach is normally used such as Bayes theorem to find likelihood ratios with the weights determined by confidence values. The simplest fusion strategy involves raw sensor measurements of the same property obtained from multiple sensors being combined (direct fusion) [Hackett & Shah 1990]. Before fusion can occur, consistency between diverse data sources must be determined to check that they represent the same physical entity. The Mahalanobis distance which should be small for the same representative object can determine this:

$$d = \frac{1}{2}(x_i - x_j)^T C_k^{-1}(x_i - x_j)$$ (11.2)

where x_k = sensor output and C_k = variance–covariance matrix.

For two sensors with scalar outputs this reduces to:

$$d = \frac{(x_1 - x_2)^2}{\sqrt{\sigma_1^2 + \sigma_2^2}}$$ (11.3)

where σ_i = standard deviation of sensor measurement.

The simplest way to fuse data is to use Bayes's maximum likelihood ratio to model the sensor uncertainty. Bayes's theorem assigns probability on the basis of evidence. Bayes's rule is given by:

$$p(\theta|X) = \frac{p(X|\theta)p(\theta)}{p(X)}$$ (11.4)

where X = sensor output, θ = object property, $p(X|\theta)$ = probability of output being X given property θ, and $p(\theta|X)$ = probability of property being θ given output X.

$p(X|\theta)$ is determined from the sensor model. The best estimate of object property θ using k sensor readings is given by the likelihood estimate with θ chosen so that it is maximised:

$$p\left(\sum_{i=1}^{k} X_i|\theta\right) = \sum_{i=1}^{k} p(X_i|\theta)$$

As logarithms are easier to manipulate:

$$L(\theta) = \sum_{i=1}^{k} \log[p(X_i|\theta)] \quad \text{where} \quad p(X_i|\theta) = \frac{1}{\sqrt{(2\pi)^n C_i}} e^{-\frac{1}{2}(X_i - \theta)^T C_i^{-1}(X_i - \theta)}$$

Hence,

$$L(\theta) = \sum_{i=1}^{k} \left(-\frac{1}{2}\log[(2\pi)^n C_i] - \frac{1}{2}(X_i - \theta)^T C_i^{-1}(X_i - \theta)\right)$$ (11.5)

The best estimate $\hat{\theta}$ is found by differentiating $L(\theta)$ with respect to θ and equating to zero:

$$\hat{\theta} = \frac{\sum_{i=1}^{k} C_i^{-1} X_i}{\sum_{i=1}^{k} C_i^{-1}}$$

For two sensors with scalar outputs:

$$\hat{\theta} = \frac{(\sigma_2^2)x_1 + (\sigma_1^2)x_2}{\sigma_1^2 + \sigma_2^2} \tag{11.6}$$

More complex fusion strategies involve such techniques as the weighted least-squares fit and the recursive Kalman filter. Modelling of multisensor fusion is very important. Measurement is usually modelled as a Gaussian probability distribution requiring an estimate of the mean and the variance–covariance matrix. The recursive Kalman filter estimation algorithm can be used to merge data from multiple images into a composite by assuming Gaussian errors [Porill 1988]. For a normal distribution the probability distribution function is described by its mean and covariance. For a measurement, the expected error $\langle x \rangle = \bar{x}$ and the covariant measurement matrix is given by:

$$C = \text{cov}(x) = \left\langle (x - \bar{x})(x - \bar{x})^T \right\rangle = \left\langle xx^T \right\rangle - \bar{x}\bar{x}^T$$

This assumes a normal distribution function of the form:

$$p(x) = \frac{1}{(2\pi)^2 \det S} e^{-\frac{1}{2}(x-\bar{x})^T C^{-1}(x-\bar{x})}$$

A composite object measurement is characterised by $\bar{X} = \langle x_1 \ldots x_n \rangle^T$ and $C = \text{diag}(c_1 \ldots c_n)$ corresponding to each primitive descriptor. It is necessary to find the optimal update of $\hat{X}$ and $\hat{C}$ given a measurement of Z related to X by $Z = HX + u$ where H = plant matrix and u = measurement error with zero mean and known covariance. If no previous measurements have been made $\hat{X} = 0$ and $\hat{C} = I$. The optimal update rules are given by:

$$\hat{C}' = \hat{C} - \frac{(\hat{C}h)(\hat{C}h)^T}{h^T \hat{C}h} \quad \text{and} \quad \hat{X}' = \hat{X} - K(Z + h^T \hat{X})$$

where $K = \hat{C}h / (h^T \hat{C}h)$ and h = sensor dynamics matrix.

11.1.9 Summary

We have examined the sensory payload technologies available to support robotic in-orbit servicers or planetary rovers. Such sensory methods will determine the capabilities of the mission, so are central to its design. Indeed, it might be stated that a robot is only as good (in terms of its performance) as its senses. However, sensors are incumbent with overheads such as computational support requirements, so the sensor suite selection must be judicious.

11.2 ROBOT ACTUATOR TECHNOLOGIES

An actuator may be defined as a device that transforms electrical or mechanical energy into controllable motion. Actuating motors act as the output effectors of the manipulator to alter its position with respect to the environment reference frame by providing movement.

11.2.1 Manipulator joint motors

Pneumatic and hydraulic actuators utilise the potential energy stored in a fluid when under pressure to produce low-speed fluid flow. Mineral oils tend to be employed as hydraulic fluid and compressed air as the pneumatic fluid. Even though hydraulic systems offer the highest power/weight ratios, they suffer from leakage tendencies between connecting pipes and have a strict requirement for the filtration of particles larger than 5 μm in size to avoid blocking pipes and jamming spool valves. Hydraulic systems also have a tendency to be large and heavy and they generate torques as a dynamic function of control input: $\tau = f(u)$. Pneumatic systems suffer from dry coulomb friction, and gas compressibility can give rise to oscillatory behaviour and inaccurate actuation. Electric motors convert electrical energy into mechanical energy (specifically electrical current into rotational torque)—they have several advantages: they are straightforward to control, they are accurate, they have fast response and their wiring is more easily accommodated than pipes. Electric motors produce a rotational output in response to an electrical input and have high efficiency with low losses, particularly in variable-speed applications. Electric motors come in three classical types and several non-classical types. Classical machines include dc motors, ac synchronous motors and ac induction motors. Non-classical motors include stepper motors, brushless dc motors and switched reluctance motors, and these types are characterised by relying on power electronics to control movement of the active coils.

Most industrial motive power is provided by ac induction motors due to the efficiency of ac power transmission at high voltages and the ease of stepping up or stepping down voltages using transformers. They are simple in design with three windings on the stator and a simple rotor. The ac motor generally uses slip rings instead of commutators and adopts stationary active coils with rotating field poles to avoid large power transfer through the slip rings—slip rings can generate friction torque, slowing the motor, so are best avoided through the use of rotary transformers which do not require mechanical contact. The alternating current in the windings produces a rotating magnetic field which causes the rotor to rotate. The synchronous ac motor comprises of a magnetised rotor which revolves in synchronisation with the rotating magnetic field produced in the windings of the stator. Its speed is therefore in proportion to the ac supply frequency to the stator. DC supply to the rotor windings is supplied through shaft-mounted rectifier supplied by the ac source. The ac induction motor uses a rotor that rotates more slowly than the rotating magnetic field produced in the stator to create a relative velocity in the stator field with respect to the rotor. This moving field induces current in the rotor conductor which interacts with the rotating field to produce torque. The highest speed achievable by an ac induction motor using a 50-Hz supply is 3000 rpm. Small convertors to dc power supplies which convert dc to ac allow the use of ac motors as these are more reliable, easier to maintain and more efficient than dc motors. Some spacecraft solar array bearing and power transfer assemblies (BAPTA) for manoeuvring solar arrays have been powered by rotary transformers without mechanical contact (such as slip rings) requiring ac power. This requires dc/ac converters to be installed. However, as present-day spacecraft adopt dc power generation, dc electric motors are generally used as they are the easiest to run from batteries and ac motors generally require at least 300 W of power.

DC motors offer accurate closed-loop control with torque independent of position and rotor speed and generate torque in proportion to the control input: $\tau = ku$. A shunt motor has its field in parallel with the armature giving constant speed independent of the voltage under a range of loads. A series motor has its field in series with the armature so that its speed is inverse to the torque load. DC motors have lower power/weight ratios than hydraulic actuators since the frame forms part of the magnetic circuit but their benefits make them the only feasible choice for space robotic applications. In standard dc motors, the armature coils are wound on a magnetic Fe cylinder. The magnetic field is produced either by coils on the iron poles or by permanent magnets. The armature coil windings and commutator segments are interconnected so that the conductors carry current continuously through the brush terminals. Most fundamentally, standard motors have a radial magnetic field with axial current flow which interact and exert torques on the armature. The iron core is laminated to reduce eddy current losses. In permanent magnet motors, the magnetic field flux is constant, but in a wound-field motors, the magnetic flux is variable. The use of permanent magnets rather than electromagnetic field coils allows the generation of a stationary field flux in the gap between the poles of the stator rather than a variable field flux. This enables a more flexible design and a torque output that is proportional to the armature current. DC motors suffer from load affecting the speed and current of the motor. Motor speed may be controlled by varying the armature voltage. DC motors have good mechanical properties with no deformation problems and long thermal time constants. Most small mobile dc motors are rated at 12 V, 24 V or 48 V. Carbon brushes in sliding contact with the revolving commutator are subject to friction and wear so standard dc motors are rarely used.

The choice of magnetic materials include ALNICO (alloy of Al, Fe, Co, Ni, and Cu) which offer ~1 T flux densities, ferrite magnets (compounds of Fe, Ba or Sr oxides) which offer ~0.5 T fields (the most common) or rare earth (e.g. samarium) cobalt permanent magnets which offer in excess of 2-T fields. Neodymium–iron–boron magnets offer even superior performance to SmCo magnets but suffer from large temperature-variable properties, limiting them to maximum temperatures of 150–200°C. Such high fields cause temperature increases due to significant I^2R Joule effect losses at the armature which impose the limitation. Although cooling with oil can alleviate this somewhat (for terrestrial manipulators), the importance of a thermal conductive path for heat dissipation requires that the motor is mounted on a metal platform, imposing mass penalties. For high rigidity, actuators with high inertia are required and revolute joints offer better rigidity than prismatic joints. The armature comprises a current-carrying coil and the commutator provides a fixed distribution of current, i.e. the coil current changes direction when the motor passes through the neutral line such that all torque contributions produce a unidirectional torque. The commutation electronics typically include a Hall-effect sensor, signal conditioning and amplifier circuitry. Commutation limits the velocity of the motor beyond which sparking occurs—performance levels are bounded by ~0.1–10 000 rpm.

Brushless dc motors are related to stepper motors in that the electronic switching of the stator current is controlled by rotor position rather than commutators. Semiconductor transistor switches eliminate brushes for high reliability and the absence of sparking. The rotor carries the permanent magnetic field and the armature coils act as the stator to keep the switches stationary. The switches reverse the current in the armature coils as the field poles rotate. This requires sensing of the rotor position for control of the switches. As the ac conveying stator coils create a rotating magnetic field synchronous with the rotor, the frequency is determined by the rotor speed.

Stepper motors utilise a rotor which moves through precise, fixed angular steps. The rotor motion is determined by computer-generated dc pulses through the armature winding with the number of pulses defining the incremental angular position. They are similar to synchronous ac motors with the exception that the supply is switched dc rather than sinusoidal ac. They are therefore digital devices and have typical resolutions ~1.5°. Typically in all motors, resolution and speed are inversely related. Most stepper motors are variable reluctance devices which use the alignment torque principle such that the rotor undergoes torque to rotate it in alignment with the stator poles where reluctance is minimised. A single pulse applied to the motor causes it to switch to the next step position. If the stepping rate is higher than the rotor natural frequency, the rotor motion is continuous. They are usually associated with open-loop control but closed-loop control with incrementable encoders offer even better performance under variable loads. Their advantage is their ability to stay in one position rigidly—the holding torque. They suffer from a lack of linearity, long time constants, a tendency to oscillate at 50–150 steps/s and tend to suffer from considerable Joule losses. Their torque outputs are position-dependent and their discontinuous acceleration and deceleration are undesirable. They are however, generally adopted for antenna pointing through one- or two-axis alignment.

Four types of dc motor are available: standard (described above), disc, bell and toroidal motors. Disc-shaped armatures have coils arrayed radially rotating in an axial magnetic field in opposing pairs, so the coils are perpendicular to the axis of rotation producing torque on the armature. The field and current directions are interchanged in comparison with the standard dc motor. They are compact and are favourable for space operation (e.g. large antenna pointing on EDRSS). They have low armature inductance and low inertia but low thermal time constants, necessitating mounting on a metal platform. Both disc and standard motors offer high torque/weight ratios and are comparable in performance except that disc motors offer better transient qualities. Bell armature motors have magnets mounted onto the rotor with the windings mounted within the stator to reduce the mechanical time constant but, although they have low power requirements, they are torque-restricted due to the small surface area of the magnets. Toroidal motors mount the motor directly on the shaft with magnets on both rotor and stator. They are capable of large torques with low rotation speeds and low torques at high speeds.

De Peuter (1994) described a brushless 'reluctance' motor capable of producing ~30 Nm torques which has been developed for ESA which generates its own magnetic reduction gearbox. Reluctance motors are hybrids of synchronous and stepper motors. They do not contain any permanent magnets. While the stator is similar to that of a dc motor, it acts as an electromagnet. The rotor consists of iron laminates which is attached to the electromagnetic stator. The reluctance of the magnetic circuit decreases as the rotor aligns with the stator pole to a minimum when the rotor is in line with the stator. An ac applied to the motor creates a rotating magnetic field stored in the air gap along the magnetic flux path. The ESA reluctance motor comprises a coaxial assembly of discs of magnetic material aligned with the rotational axis of the motor. Alternate discs are fixed to the motor housing to form the stator while the remaining discs are attached to the rotor and drive shaft of the motor. Each disc has a number of equally spaced radial slots filled with material of higher magnetic reluctance. These structures determine the path of magnetic flux through the discs. Magnetic flux follows the path between the rotor and stator whereby the total average reluctance is minimum. Applying ac to the stator windings creates a rotating magnetic field and a rotating magnetic flux in the gap between the rotor and stator. The ratio of rotor slots to stator slots determines the gear reduction. The movement of the rotor required to maintain the magnetic circuit which tracks the rotating magnetic field is much slower than that of the field itself, thereby effecting gear reduction.

The motor of choice for spacecraft application is the electronically commutated brushless dc motor (ac is not considered because it lacks a demonstrable space pedigree). The absence of brushes eliminates coulomb friction and power is transmitted through Be–Cu roller contacts rather than slip rings. Brushless dc motors require high stiffness (minimum natural frequency ~10 Hz) and the choice of standard, disc, bell and toroidal motor depends on the application. Disc motors have the space pedigree. Beryllium may be used for the mechanical parts by virtue of its low thermal expansion coefficient (similar to steel), low density, good thermal conductivity and very high modulus of elasticity. Any lubrication that is employed must be hermetically sealed.

11.2.2 Gear transmission systems

The transmission of rotational motion from one shaft of a machine to another is a universal requirement in motors. Just as transformers are required to 'gear' down current and increase voltage in inverse ratio at constant power so motors as mechanical systems require similar matching. Due to high motor shaft speeds and low torques, step-down gear transmission (typically with ratios of ~50:1–100:1) is usually required for larger torque outputs and lower speeds (as power $P = \tau \upsilon$ = constant). If the shafts to be coupled are not colinear then belts, chains or gears are required. The most common and versatile choice is the toothed gear for rotational motion transmission. Stresses in gear trains can be high, culminating in failure. Failures may result in shearing at or near the root of the tooth due to repeated bending stresses when the force shared by two teeth is transferred to a single tooth during a portion of the meshing cycle. Alternatively and more frequently, damage to the involute profile may occur due to tooth shear stresses accompanying compression stresses which in turn reduce the effective length of the path of contact. Backlash errors occur when gears do not mesh perfectly but have spaces between them. Severe torsional vibration can arise in transmission systems. If there is backlash, it is desirable that the amplitude of the vibration torque does not exceed the mean torque so the gear is not subject to torque reversal and impact loads on the gear tooth. Robot joint rigidity and positioning accuracy are determined by the quality of the transmission system. Reversible systems require electromechanical brakes to maintain joint orientation when the actuator supply is cut off (fail-safe). Epicyclic gear trains are more versatile than standard spur gears but higher gear ratios may be obtained with worm gears.

The harmonic drive is used for electric motors and offer large gear ratios with no backlash. Around the high-speed central shaft are mounted ball-bearings and a satellite flexspline. The flexible flexspline has fewer teeth than the planetary fixed spline in which it is contained so that around 10% of the gear teeth are meshed. They offer good mechanical resistance and rigidity as well as reduction coefficients as high as 320 for a single stage. Power losses in lubricated gearing is usually very low since tooth friction is less than 1% of the full load transmission power. It is friction that causes power losses which are dissipated as heat. Lubrication can cause the friction coefficient between metal at ~0.3 typically to drop to 0.01% of the dry surface level. KG80 is a hydrocarbon oil that is commonly used for lubrication in space motors. Lubeco 905 dry is used in the Shuttle RMS. In order to reduce the bulkiness at joints which would restrict visibility all manipulator joint gearing is usually mounted into the shoulder of the manipulator.

Direct drive motors do not utilise mechanical transmission between the actuator and the link, offering the advantage of compactness and low mass [Asade et al 1983]. 'Quasi-direct drives' are self-contained units of motor, transmission gearing and measurement devices. The direct drive dc motor is mounted in a pressurised oil-filled housing. It simplifies servo-control by eliminating the gear transmission. This significantly reduces nonlinear transmission errors such as friction, backlash, compressibility, meshing, wear and gyroscopic reaction effects. Furthermore, transmission systems are sensitive to over-loading and are intolerant of the collisions required in force control. The direct drive motor offers high mechanical stiffness, enhancing accuracy, repeatability and linearity. The direct drive motor is directly coupled at the joint with the rotor attached to one link and its stator attached to the adjacent link connected through a brush ring. It is backdriveable and may be torque-controlled rather than position-controlled. By using samarium-cobalt magnets which give 3–10 times as much magnetic energy as ferrite magnets, high torques may be generated. The problem of variable loads usually associated with direct drive motors can be easily compensated for within the controller (supposedly—not such a trivial problem as has been shown). Direct drive motors are particularly suitable for wrist actuation. The direct drive robot manipulator tested by Asade et al (1983) had a 56-cm diameter motor in the first joint with 204-Nm peak torque, a 30-cm diameter motor in the second and third joints with 136-Nm peak torques, while the wrist motors were 23-cm diameter motors with 54-Nm peak torques.

New materials are being used for high magnetic energy density permanent magnets. These include neodymium–iron–boron ($Nd_2Fe_{14}B$) alloys which can generate magnetic fields ~10 T. By reducing the grain alloy diameters to around 20 nm even greater fields could be produced. Hence, there is great potential for direct drive motors for use in robotic joints eliminating the need for transmission systems.

11.2.3 End effectors/tool sets

End effectors tend to be customised and can comprise up to 20–30% of the total manipulator cost. The manipulator arm is usually fitted with a tool holder to simplify the attachment of the end effector in the form of a plate mounting with holes for bayonet pins. This enables interchangeability of tools and end effectors within a standard wrist mount with standard latch/unlatch fixtures. End effector types should include general-purpose hands, power tools, etc. The human hand is a single tool designed through natural selection to perform multiple functions, of which prehensile, manipulation and sensory functions are the most important [Salisbury 1988]. It can form a hook grip, a scissor grip, multi-fingered chuck, a squeeze grip and multiple geometric grips through its capability to mould to many geometries. Prehensile capability allows the holding of an object in a controlled state relative to the hand. It requires the application of sufficient force on the object to hold it stationary. It imposes constraints by contact and imposes structural constraints and frictional constraints. It is best to minimise the reliance on friction. Manipulation implies holding an object securely relative to the wrist for the translation of heavy objects and the rotation of small objects with respect to the wrist. Sensing includes the detection of contact surfaces on the object. There must be sufficient contacting surfaces and sufficient applied force to avoid slippage. At least two fingers are required to grasp an object. In almost 50% of cases of assembly, this two-point contact is sufficient to maintain an object in position. All two-fingered hands are actuated by a single actuator to ensure symmetric movement. Such parallel jaw grippers are single-degree-of-freedom mechanisms and they come in a variety of types and are usually designed for both internal gripping (for hollow objects) and external gripping. The screw-type parallel jaw gripper has a central helical screw and nut which, when moved away from the fingers, pull the jaws together.

Universal grippers with three or more fingers have multiple joints and can apply uniform pressure on surfaces which are arbitrarily shaped. Torque is applied by a grip wire which is transmitted to the end effector tip by a pulley arrangement. If pressure-sensitive material is used over the inner surface of the hand, generic geometric shapes can be discriminated. The incorporation of compliant force sensors into each finger allows fine-scale response to applied forces. Mechanical hands attempt to emulate the human hand but none can yield the full twenty-two degrees of freedom of the human hand except perhaps the Robonaut hand. For robotic mechanical hands, three fingers each with three joints are generally sufficient for most tasks [Salisbury & Craig 1982]. Nine degrees of freedom is the minimum number required for secure grasping and arbitrary movement of a grasped object. Hence, each finger represents a mini-manipulator of three degrees of freedom. Each finger joint has prelocked mechanical spring end stops and the fingers are operated by a single wire attached to each phalange link. Mechanical hands may be able to grasp and use conventional hand tools designed for human use. Anthropomorphic design has advantages for ease of control and operation, e.g. Robonaut hand. Other such hands include the Utah/MIT hand and the Stanford/JPL hand, both of which have three fingers of three joints each. Further developments are providing smart end effectors with integrated sensing and processing. In addition to general-purpose manipulation, any space robot will be required to perform a multitude of tasks and this requires a tool complement of specialised devices which may be interchanged.

The remote centre compliance (RCC) linkage was developed at the US Draper Laboratories to provide a multi-axis 'float' to allow misalignments of around 1 mm lateral error and 1o angular error for peg-in-hole tasks. Three parallel rods allow lateral translation movement while three concurrent rods allow angular movement at the tip to correct misalignments. Assembly comprises mostly of parts mating. In view of this, the RCC is designed particularly for peg-in-hole insertions with chamfered corners which are the most common of assembly tasks. Indeed, around 35% of all assembly tasks are simple peg-in-hole operations and 25% are screw insertions which is variant of the peg insertion

[Nevins & Whitney 1980]. The instrumented RCC provides active sensor feedback to the RCC passive mechanical structure. Transducers measure the distances between the centre shaft and the support case of the RCC by projecting a laser beam onto three two-dimensional photodetectors inside the device. The RCC has a mass of 0.45 kg, a focal length from the base of 20 cm, a lateral stiffness of 100 N/cm and a torsional stiffness of 0.1 Nm/mrad.

An important task for a space robot is the exchange of ORUs or other equipment and this involves negotiating latches. The Standard Dextrous Grasp Fixture (SDGF) include the H-fixture and the Micro-fixture. NASA SSP 30550 Robotic System Integration Standards for the ISS establishes robotic compatibility standards for ORUs, etc. [Parrish 1999]. Typical tools include the Robot Micro-Conical Tool and the Socket Extension Tool. The Micro-Fixture is a standard mechanical interface between the ORU tool changeout mechanism and ORUs. The latching of removable equipment onto a mounting base must be achieved by pure mechanical action—sensor-assisted control has the disadvantage of increasing the complexity through added hardware and electronics. Design simplicity is the most important factor for latches to minimise the number of components and thereby the number of potential failure points to contribute to overall reliability. In addition, this also helps to minimise mass. Translational and rotational motion should be along a single axis rather than over multiple axes. Positive latching is also desirable such that any force which acts to detach an ORU in fact tends to tighten the latch mechanism. Automated locking should also be be incorporated so that drive power may be shut down when the latch is closed to remove loads from the drive mechanism. Latching must also be quick and be able to be activated and reversed automatically. Contact must be rigid with relative motions minimised. Guiding systems such as tapered rails should be used to compensate for misalignments due to manufacturing tolerances, differential expansion due to thermal gradients and positioning errors. Alignment devices such as pins or cones will not separate or mate cleanly in unison under such conditions. To prevent this, inherent compliance is required to increase the operational reliability. Dedicated retendors effectively decouple latch activation from other operations. Finally, maximum load dispersion to large structural components being joined should be incorporated while minimising load concentration on the latch mechanism itself to minimise load distortion. Typical grapple fixtures have a conical shape with a radial tooth which reacts to the applied torque. The gripper possesses a conical protrusion which mates with the grapple fixture. The rocking jaws close around the grapple fixture. Latching occurs in a number of steps ñ guiding to the target position, capturing with misalignment correction, and closing the interface with a pre-defined preload.

The RMS end effector closes three snare cables around the grapple pin, drawing it in until close contact is made. A grapple load of up to 499 kg is applied to the grapple pin. The SSRMS is equipped with a specialised end effector—the Latching End Effector (LEE)—that can latch onto specialised fixtures—Power Data Grapple Fixtures (PDGF)—located around the ISS (Fig. 11.7a). Similarly to the Shuttle RMS, the SSRMS uses a three-snare wire mechanism to close the end effector and lock it. These, however, will be replaced by higher load-bearing mechanical latches similar to the ESA latch described next. A pair of curvic couplings ensure precise alignment. The PDGF is a standardised interface mounted at various locations on the ISS comprising a grapple pin and two pairs

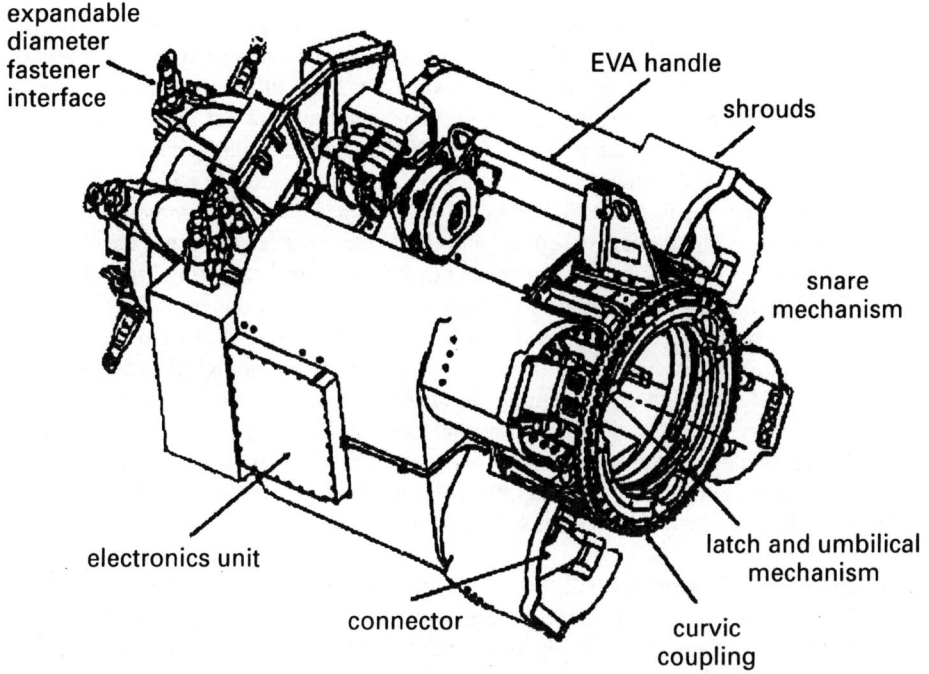

(a)

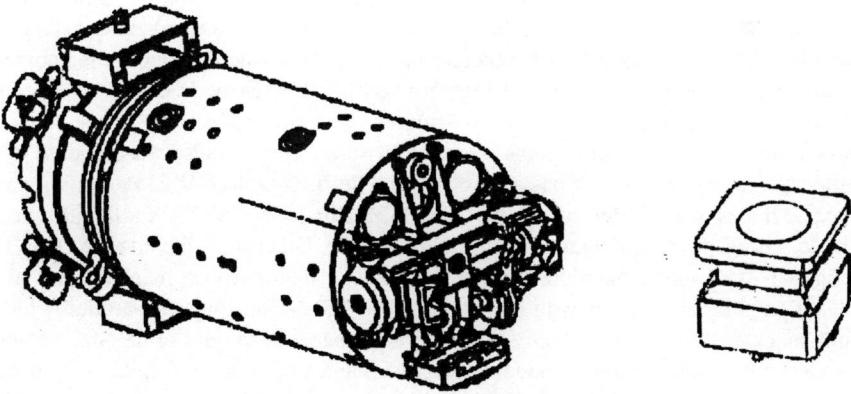

(b)

Fig. 11.7. (a) SSRMS latching end effector; (b) ORU tool changeout mechanism. Reproduced with permission of MacDonald Dettweiler Robotics Ltd: http://www.mdrobotics.ca

of umbilical connectors for power and data transfer. The SPDM end effector comprises an ORU/Tool Changeout Mechanism (OTCM) which can operate specialised robotic tools such as dedicated wrenches and socket extensions (Fig. 11.7b). It comprises a set of parallel jaw grippers for grasping micro-fixtures, a retractable nut drive, a motorised socket wrench for retracting the micro-fixture retaining bolt, and a light/camera assembly.

A new ESA latch has been designed which operates automatically with simple attach/ detach with a high loading capacity [Panin 1992]. It includes self-alignment capability, positive latching and automatic resetting, and has a relatively large capture envelope and short insertion stroke. The latched member comprises a pin protracting from the ORU base. The pin has a circumferential groove for positive engagement with two hook-shaped levers pinned at a common revolute joint. Bales at the end of each lever fit into the ORU pin groove. A central pin from the latch palm slides through the revolute joint. When the ORU pin contacts the central pin at the palm of the latch a preloaded spring acts between the levers to close them around the ORU pin. Unlatching is initiated by powering an actuator to move the control pin to disengage and open the latch levers. The whole assembly is capable of rotating about the revolute joint providing the mechanism with compliance. Hence the ESA latch represents another possible member of ATLAS's tool suite.

Although robotic servicers will tend to use dedicated tooling, the end effector gripper should be compatible with the use of the existing set of EVA tooling (as in Robonaut). Several versions of the EVA power tool exist: the standard EVA power tool, the HST EVA power tool, the mini EVA power tool and the smart EVA power tool (under development). The baseline taken here is the HST EVA power tool as this is based on the standard EVA power tool but has improvements which are to be included in the smart EVA power tool. The EVA HST power tool is now the standard cylindrical power tool design. It has a mass of 3.4 lb, of which 0.28 lb is battery, and has a length of 12.2 in and a diameter of 2.75 in. It has two operating speeds at 20 rpm and 60 rpm for tightening and loosening with stall torques ranging from 30 to 300 in-lb at a variety of settings. Its power source is a 7.2-V rechargeable NiCd battery pack. It is lubricated by fluorinated polyester vacuumised grease. It is constructed from a glass-filled lexan body chassis with copper and steel gearing in a polyimide gear housing covered with reflective Al tape. This tool would form one of the central tools of any tool suite. There is also the standard MMS Module Servicing Tool which has a mass of 70.5 lb. It has a 16.5 V EMU battery power source and has a diameter of $14.75 \times 14.3 \times 35.2$ in (width $\times$ height $\times$ length). It has variable settings for torque ranges from 25 to 160 ft-lb. Other tools include the ORU Tool Changeout Mechanism mentioned earlier. This is a multipurpose mechanism that includes a parallel jaw gripper with a grip force of 4450 N and finger separation width of 14 cm, an extensible 7/16″ socket wrench device capable of torques of 68 Nm, a connector socket for power, data and video connections, and two overhead lights and an overview camera. This tool is specifically designed for the robotic segment of the ISS and is designed for heavy loads rather than astronaut EVA-type operations. To enable walking between the nodes of a truss structure specialised end effectors would be required. As each node is essentially a sphere with threaded holes spaced at 45°, Xu et al (1992) proposed a motor-driven screw to engage the threaded holes on such nodes to provide a large contact force for stability (about 1800 N).

Planetary missions will require their own dedicated set of tools for surface exploration [Boyd & Clark 1992]. A coring drill is essential to sample columns of the subsurface soil—such a tool would be the central implement in core sampling for sample return missions. The Apollo Lunar Surface Drill used a rotary percussive coring drill with tungsten carbide cutters and titanium drill stem. A typical rotary drill requires 20–30 W to produce a 10 mm core in hard rock. Modern approaches use low-power, compact ultrasonic vibration ~20 kHz impact devices based on stacked piezoelectric motors with few moving parts—they require 25–30% of the power and generate 20–30 times the push force of rotary drills. They operate like jackhammers so do not rotate, 'drill walk', or exhibit low-frequency chatter, and the drill bit does not require sharpening. A 0.4 kg drillhead mounted on a 4 kg platform requiring 3-ion axial load and 3 W of power has demonstrated 25 mm cores [Bar-Cohen et al 1999, Das et al 1999].

11.3 MICRO-ELECTROMECHANICAL SYSTEMS (MEMS)

Micro-electromechanical systems (MEMS) offer a dramatic development in space robotics as an enabling technology. Space exploration has traditionally provided a driver towards the miniaturisation of technology to reduce the mass of launching payloads into space (and so cost) and such techniques have extensive applications in the medical, environmental and automotive industries. Miniaturisation of sensors, actuators and microelectronics to micrometre size offer reduced mass, volume (and so launch costs) and power requirements, and the potential for batch processing and mass production of common off-the-shelf integrated microsystems with increased reliability [de Aragon 1994]. The world sensor market has been estimated at £25B/y in the twenty-first century. Advances in silicon integrated circuit technology and fabrication have led to the development of high-performance, low-cost and mass-produced sensors. There has been a tendency to employ dedicated electronics integrated with the sensor to make them 'smart', e.g. in the internal combustion engine [Wolber & Wise 1979]. Such dedicated processors may be integrated by common interfaces to ensure the millisecond processing time required for real-time performance within a dynamic environment. Such smart sensors are a prerequisite requirement for operation in an unstructured and variable 'real-world' environment to provide feedback information on the state of both the robot and the world. They offer utilisation of nonlinear transduction to provide greater dynamic range and/or sensitivity.

MEMS are devices with physical dimensions of less than 1 mm which operate on mechanical principles [Angell et al 1983]. Microsensing is a new advanced sensor technology that is rapidly becoming available based on electronic integrated circuit batch-fabrication technology. The next generation of sensors will be based on silicon which is abundant and cheap. Silicon is an effective material for transducing many physical phenomena such as light intensity, pressure, force and temperature. Furthermore, pure silicon has the hardness, Young's modulus of elasticity of 1.9×10^{12} dyne/cm^2 and tensile strength of 6.9×10^{10} dyne/cm^2 similar to stainless steel, i.e. several times that of iron or glass. Corrosion-resistant, tough, wear-resistant thin films such as SiC or Si$_3$N$_4$ can be used to provide a casing for such silicon devices [Petersen 1982]. Si$_3$N$_4$ has a hardness second only to diamond.

Pressure sensing is the best developed of such sensors, most of which are based on the piezoelectric effect. They provide 60 times the sensitivity of geometrical deformation sensors. A thin piezoresistive membrane ~10–50 μm is etched on a silicon wafer with four silicon strain-gauge resistors, ion-implanted in a Wheatstone bridge configuration with aluminium interconnects. The electrical resistance of the diaphragm changes when the diaphragm flexes. MOS-based amplifying electronics can be implemented on chip in a similar manner. Accelerometers may also be constructed with piezoresistive materials by replacing the membrane with a silicon oxide cantilever attached to a gold layer proof mass at the end of the cantilever sealed in a cavity. Vibration in the direction of the accelerometer's axis of symmetry generating a shear stress in the piezoelectric crystal. This deforms the crystal, altering its electric polarizability and generating an electric current due to the crystal's asymmetric strain tensor. Alternatively, a metal layer may be deposited on top of the oxide cantilever so that the metal layer and the Si layer at the bottom of the well act as two plates of a capacitor. The voltage output is proportional to its acceleration. Such capacitative pressure sensors are more sensitive than piezoresistive sensors. The resonant frequency of the cantilever beam is given by: $f_R = 0.162(t/l^2)\sqrt{E/\rho}$. Once the resonant frequency is measured, and the dimensions of the beam known, Young's modulus of the beam material can be computed. A MEMS-based pressure sensor based on a Si capacitative transducer was included in the Deep Space 2 mission with a pressure range of 0–12 mbar as a meteorological pressure sensor. It drew a power of only 20 mW and had an operational temperature range of –80–50°C. New manufacturing techniques allow the construction of suspended or movable 3D structures. Thin diaphragms, cantilever beams and other 3D mechanical microstructures can thus be etched in silicon or other materials. Mechanical sensors with cantilevers and diaphragms form the largest type of microsensor for displacement and pressure measurements.

Such silicon-based microsensors are able to utilise piezoresistive and piezoelectric effects in relation to applied stress, capacitance changes from pressure and thermal resistance which varies as a function of temperature, all of which may be used to measure pressure and pressure-derived parameters such as force, acceleration and fluid flow. Other capabilities include magnetoresistance and Hall effect voltage generation by applied magnetic fields. Ion-selective FET Si chemical/biological sensors offer the capability of measuring the concentration of ionic and gaseous species though these are less well developed and are presently limited to gas detection, humidity and pH measurements. Thin film processes have been developed to measure thermal, radiation, mechanical, magnetic and chemical signals.

Microsensors are small and of high accuracy with reduced noise and leakage. Such solid-state sensors produce low-level analogue outputs, so signal conditioning electronics such as amplifiers are required. In addition, they offer the capability of secondary parameter (e.g. temperature) compensation through the use of standard electronic components such as resistors, diodes and transistors—normally resistive bridge circuits are used to compensate for temperature variations. They are constructed similarly to VLSI microprocessor chips through conventional photolithography and so allow the close integration of electronic and mechanical components. Even complex signal processing functions such as A/D conversion may be achieved through dedicated processors integrated and manufactured with the sensor. This gives them the name 'smart' sensors in that the transducer that converts the physical quantity into the electrical voltage signal is integrated onto the same chip and the signal conditioning electronics. MEMS technology exploits the conversion of electrical signals and mechanical forces.

The development of Si microactuators, ~μm^2–mm^2, are a logical extension of the more established microsensor technology. Such microactuators are mechanical devices that can allow motion control over very small distances ~0.01–10 μm. There are ten actuation methods of transforming energy into motion [Gilbertson & Busch 1996]. Electromagnetic devices include electric motors, solenoids and relays. They exhibit fast operational speeds with high efficiency but require perpendicularity between a current conductor and the moving element, making it difficult to manufacture in planar silicon. An example is the micro-valve which uses a small electromagnetic coil wrapped around a silicon micromechanical valve structure. Electrostatic devices use charge buildup due to free electron flow and include silicon micro-stepper motors with a rotor ~100 μm in diameter operating at 25–36 V with speeds of 2500–15 000 rpm to generate torques of up to 13 pNm. Electrostatic devices exert great forces over very short distances with low current consumption. These devices offer ~10^3-Hz cycle rates. Thermomechanical devices utilise the expansion/contraction of materials during temperature changes such as bimetallic thermostats. A micro Stirling cycle gas heat engine with high efficiency has been constructed which can act in reverse as a cooling system. Bimetallic cantilever microactuators of gold on silicon with length ~500 μm and deflection ~100 μm using 200 mW of power. A 200 μm long cantilever microactuator of silicon, silicon dioxide and p-doped silicon can produce deflections of 4 μm. Temperature changes may be induced by resistive heating, ultrasonic heating or radioactive heating. Examples include paraffin heat pipes. Phase change devices use dimension changes of materials between different states of matter such as a micro-steam engine which has been developed which uses small ~2–60 μm resistive heaters to heat a non-conductive F-based cooling fluid to create small gas bubble of ~34–76 μm diameter to lift an etched Si plate up by 25 μm using only ~20–40 mW of power. It can be fully cycled at 10 Hz. Phase change devices offer very high forces but do require cooling for reverse transformation. A multi-element pin-jointed crank has been constructed with the crank arm being 50 μm long with the pin joint being made of a hub around which another element is free to rotate. Such micromechanical devices are based on the construction of micrometre-sized structures including valves, turbopumps, gear-wheels, rotor-based motors and micromanipulators. Integration of microactuators with microsensors offer complete systems and a powerful miniaturisation technology. However, Si requires protection in hostile chemical environments. Gear trains, motors and actuators at the micrometre scale have been constructed using Ni and Cr metal alloys, ceramics and plastics for different environments. Copper vapour lasers may be used for micromachining of diverse materials. They operate at visible wavelengths (either 578 nm or 510 nm at yellow and green respectively) with high pulse rates ~40 kHz offering ~10 μm deep cutting with no heat damage due to glassification to peripheral areas. These lasers have high gain and efficiencies and visible wavelength operation implies that material reflectivity is low at ~50%. A 12:18 electrostatic micromotor of thickness 2.5 μm with a rotor ~100 μm diameter and a stator-rotor gap of 2 μm can provide torques of ~0.25 μN [Lawes 1998]. IBM inkjet nozzles are also manufactured via MEMS fabrication methods. Superconductor materials exhibit diamagnetism (Meissner effect)—this is the ability to reflect external magnetic fields. In fact, some non-conducting materials such as Bi, Si and graphite possess this ability at low temperatures. The Meissner effect could be exploited in MEMS devices to levitate magnets against gravity such as frictionless, self-levitating bearings to prevent stiction and friction without lubrication. A microscale high temperature superconducting linear actuator of YBCO has been demonstrated in the form of a $1 \times 1 \times 10$ mm slider levitating 1 mm and being driven horizontally with 30 mg of force. The most developed technologies are electromagnetic, electrostatic and piezoelectric devices typically requiring ~100 V to generate proportional forces with rapid actuation and low power consumption.

It is the integration of electronic and mechanical systems, including signal processing electronics on a single silicon wafer, that gives these devices the name microelectromechanical systems (MEMS). Function often imposes a limit to physical size reduction, e.g. an antenna requires a certain aperture area in order to function effectively. MEMS technologies often involve redesign of macro-system technologies—often rotating masses can be replaced with vibrating combs and tuning forks in microgyroscopes and micro-accelerometers. Friction is the biggest problem for MEMS which can destroy their functioning. Stiction is a similar problem where small electric charges can pull components together and weak chemical bonds can form between them. Small electromagnets must usually be incorporated to levitate objects and components to keep them operational. Another approach is to coat the components with thin films of Teflon. Distributed arrays of MEMS offer a integrated systems approach for spacecraft payloads, although they may not offer sufficient forces and torques for some applications.

The chief problem with space-rated MEMS in space is their high susceptibility to radiation. However, the US Air Force has developed a technology demonstration programme Miniature Technology Sensor Integration (TSI) to develop microsensor technologies in missile detection and tracking for the BMDO programme. These include visible imaging spectrometers and short and medium infrared imagers, much of the experience coming from the Clementine programme. The Space Bioreactor of dimensions $84 \times 60 \times 60$ mm with a 3-V electric power requirement is one example of suchtechnologies which includes a micropump for fluid circulation and microsensors to monitor parameter changes in the reactor chamber (STS-65). A MEMs-based capillary gas chromatograph device has been developed for *in-situ* chemical analysis of amino acids [Spiering et al 1998, Angell et al 1983, Harrison et al 1993]. Such technologies offer great promise for planetary landers and rovers. The gas chromatograph separates, identifies and measures the quantity of a gas mixture sample. A 5-nl gas sample is injected by an electrokinetic pump from a sample vial through a valve into a long capillary column which is lined with silicone oil in which different gases have different degrees of solubility. The time a gas remains adsorbed depends on its solubility. The gas chromatograph has a long spiral capillary column of 1.5 cm length and $10 \times 30~\mu$m cross-section formed by adding a groove ~200 μm wide and 40 μm deep etched in the Si wafer, followed by chemical vapour deposition of silicon nitride onto the wafer. Then the wafer was anodically bonded to the glass capillary. The glass column and Si wafer are heated to 400°C with a negative voltage ~1200 V applied to the glass to anodically bond the two surfaces. The negative charge of the glass fuses the glass and silicon together to form a hermetic seal. The detector is a thin film metal resistor on a thermally isolating membrane of Pyrex glass. A constant electric current is passed through the resistor and as the gas sample has a lower thermal conductivity than the He carrier gas, the resistance increases when the sample gas is output through the column, thereby providing a measure of the thermal conductivity of the gas. The total integrated chemical laboratory device is fabricated on a 5-cm^2 wafer and involves no moving parts.

MEMS devices offer great promise for miniature on-chip sensors, instruments and laboratories for exploratory spacecraft generally and robotic rovers in particular.

12

Vision and image processing

We devote a complete chapter to vision sensing as it represents a primary modality in robotics—it is used for remote sensing for obstacle avoidance and navigation. It will dominate much of the processing requirements onboard the in-orbit servicing spacecraft, and will dominate the telemetry traffic to the ground. Initially, we consider vision sensor hardware, then vision processing which will, in conjunction with the control system dominate the computational processing requirements.

12.1 VISION SENSORS

Vision is the most information-rich form of sensory data. Indeed, it is one of the primary modalities for dealing with a variable environment at a distance, so all robotic systems for space application will implement some form of visual sensing. Stars like the Sun emit principally in the visible spectrum due to its surface temperature of 6000 K—hence, the utility of vision for extracting distant information within solar system environments.

Vision provides the mechanism to obtain distance information to enable strategic rather than tactical behaviour. Vision may be defined as the process that produces a representation of the external world from images that is not cluttered with irrelevant information. Robotic 3D vision is a fundamental requirement for the recognition of objects, the determination of object position and orientation in space and the extraction of relevant features of the object for visual servoing. It is essential for motion planning in unstructured and dynamic environments and tracking of moving objects requires ~10 frames/second. Vision sensors can provide closed loop servo-control feedback through the potential field method. There is a distinction between passive vision such as the camera and active vision such as LIDAR. The space environment exhibits variable intense light and dark periods and shadowing. At 300 km altitude the light intensity varies from one hour of extreme brightness to half-an-hour of complete darkness with up to 10^8 intensity ratios [Krishen 1989]. The absence of any atmosphere prevents scattering of light, generating large contrasts of intensity which are characteristic of this environment. Most spacecraft are white providing high reflectivity which can cause flaring on camera images. Docking requirements impose the need for visual perception within a 30° cone for 50 m distance [Krishen 1989]. Beyond this, range data may be used.

Vidicon cameras are vacuum-tube-based instruments which suffer from drift, noise, limited resolution and spatial distortion so vision systems tend to be based on Si solid state CCD (charge-coupled devices) or CID (charge-injection devices) arrays of photosensitive cells with a range of spatial resolutions. The CCD camera effectively emulates the receptive fields of the human retina in that it comprises independent spatially distributed channels sensitive to intensity variations. One-dimensional arrays are used in line scanning mode on planetary landers. A CCD 512×512 pixel array is typical. Such silicon arrays are the basis of the CCD camera which offer sensitivity to 400–1100 nm light with efficiencies of around 90%. Light is input to the CCD analogue shift register where electronic charge accumulates. Light incident onto the silicon forms electron–hole charge pairs within the semiconductor cells at the depletion layer. Electronic charge is trapped in the potential well produced by the electrodes. Electrodes are insulated from each other by very highly doped p-type material. Each pixel is a potential well acting as a capacitor. The electrode is insulated from the semiconductor by a thin silicon dioxide layer ~100 nm, similar to a MOS transistor. A small positive voltage ~10 V drives the positive holes in p-type silicon substrate away from the upper depletion layer while attracting electron minority carriers into the thin upper depletion layer ~10 nm below the upper electrode. Electrons accumulate and a photocurrent is generated which is linearly related to the incident flux of photons in a photomultiplier tube. The total number of electrons is read out by charge-coupling the detecting electrodes to a single read-out electrode. The electron charge proportionality to the light intensity in each depletion layer provides the means for a spatially digitised electronic image. To recover the electron image, charge coupling is adopted. Charge coupling transfer loss can be minimised using buried channel techniques rather than surface channel techniques. When the voltages either side of the electrodes are reduced to, for example, 2 V, the hole depletion layer disappears under those electrodes, and electron charge below the adjacent higher voltage electrodes diffuses under the potential difference. So by sequentially altering the charging voltages along the electrodes, the stored charge can be moved through the device to the output electrode for measurement. Three separate voltage cycles are required to move charges along the array. The order of appearance of electron charge at the output electrode is related to the spatial position of the initial storage electrodes. An array of sensor electrodes lie above the array of charge-coupling elements, e.g. photosensor with gate switching electrode sandwiched between two array elements. During operation, the sensors are held at 0 V with the gate at 2 V for electrons to accumulate in the photosensor. After exposure, the gate electrode is switched to 10 V while the charge coupling electrodes are held at 2 V. For the charge to be driven to the charge coupling electrode, the sensor electrode is dropped to 2V. A photoemitter such as KBr or NaCl is coated onto the cathode primary at −1000 V potential. The photocurrent is amplified by the photomultiplier with gains ~10^6–10^7. The photomultiplier accelerates the photoelectrons from the cathode photoemitter (usually coated with cesium antimonide paint) by an electric field to a secondary electron emitter coated with electron emitting KCl or BeO. Dynodes are at successively more positive potential than the cathode by 100 V at each stage until the anode is reached.

Bailly et al's (1985) CCD camera is composed of 288 lines of 384 pixels per line generating an image area of 6.6×8.8 mm. The sampling rate of the vision system for

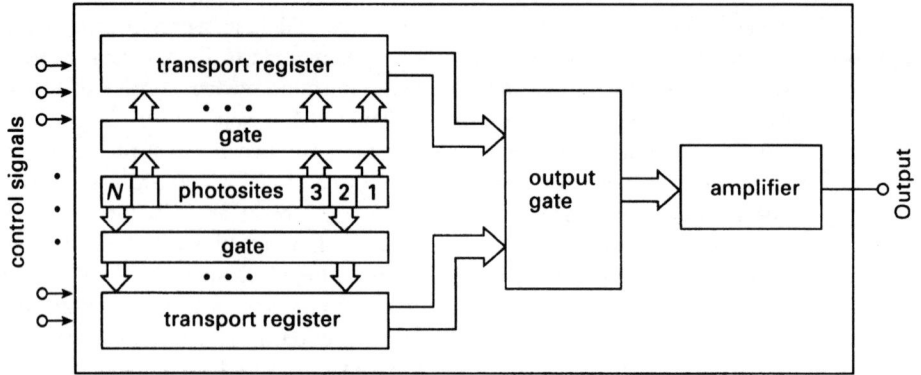

(a)

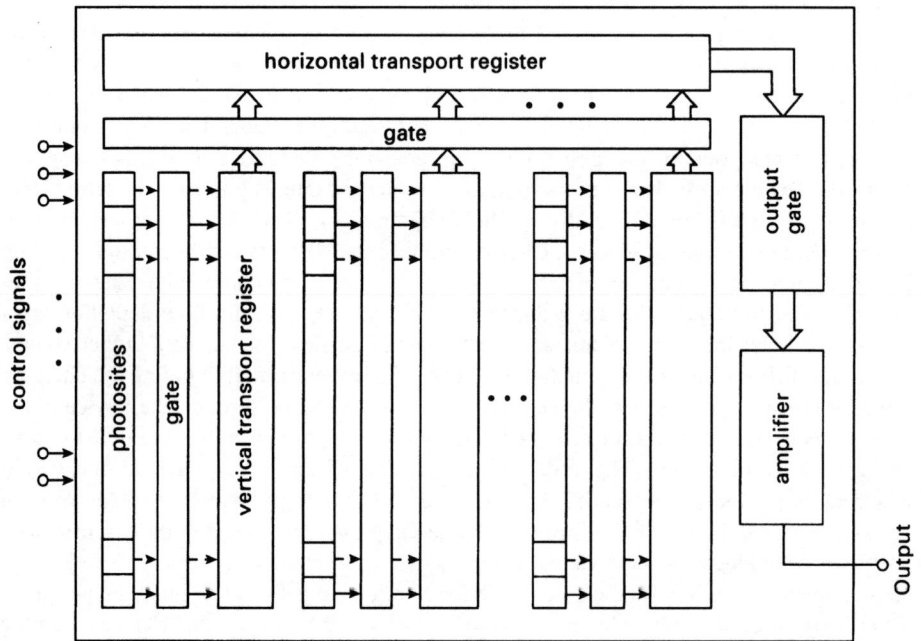

(b)

Fig. 12.1. CCD imaging area sensor (from Fu et al 1987). Reproduced with permission from Fu
et al 1998), *Robotics: Control, Sensing, Vision and Intelligence* © McGraw-Hill.

effective real-time operations should be 20 Hz minimum [Lin et al 1989]. Standard
cameras can produce video images of either 525 lines with 30 frames/s (American EIA
standard) or 625 lines at 25 frames/s (European CCIR standard). Each TV frame comprises around 250 kbytes. For computer processing, the camera must be digitised at a

sample rate of 14.8 MHz for a CCIR system. A robot controller may be connected to each TV camera by an RS232 link standard which uses a 25-wire connector. CCD cameras should have pan, tilt, zoom and focus control as fundamental capabilities. CCD photo-sensitive arrays are fast and robust, have high geometric fidelity and good resolution and do not suffer appreciably from distortion, blooming or electronic drift. The solid state CCD 2-dimensional array has low mass, small volume and low power consumption, offering high signal-to-noise ratios with efficiencies close to 100% (50 times that of photographic emulsion). The image matrix and the memory matrix are separate. After the integration time of less than 0.5 s, the charge packets from the photosensitive image matrix are transferred into the memory matrix with line-to-line shifting within the image pixels. In the memory matrix, the pixel contents are read out sequentially. Thus the splitting of the optical image unit and the electronic memory unit allows a low mass optical head and easy implementation [Bailly et al 1985]. The CCD provides the basis for feature detection through extracting edges and lines from the image. Colour imaging may be provided as colour is often a good approximation to reflectance. Sharp changes in reflectance on a surface usually indicate object boundaries or edges. While a grey scale image comprises 8-bit pixels (giving $2^8 = 256$ brightness levels), colour images typically require 24 bits—8 bits each for red, green and blue components. Quantisation introduces quantisation noise indicated by the $S/N = 2\log_{10}(2^n\sqrt{12})$ where n = degree of quantisation—8-bit quantisation has a S/N ratio limit of 56 dB. The standard photovisual filter is usually the UBV filter which ranges across the visible and near UV regimes offering tripeak (red, green and blue) maximum response at 3650, 4400, and 5500 A. For ultraviolet wavelengths (900–3000 A), detection is usually as for optical but with LiF coatings.

Automatic zoom allows the detection of 2 mm objects at 1 m distance. Automatic iris, variable scan rates, automated gain and brightness control offer the capability of handling high intensity sunlight up to ~13 000 ft-cdl. Mounting should give pan and tilt capabilities. An LCD polarising filter of identical resolution to the CCD may be placed over the CCD array to reduce specular reflection and glare and provide bloom protection. Any excess light would be blocked by the LCD shade.

Illumination is an important aspect of visual sensing and should be such that the complexity of the image is minimised and information extraction is enhanced [Gonzalez & Safabakhsh 1986]. Indeed, illumination may be considered critical as the Solar Maximum Repair Mission was hampered by restrictions of the normal camera light levels. An intensity of 10 kft-cdl at the workspace is the minimum necessary for robotic and EVA operations. Diffuse lighting from several directions are suitable for objects with smooth regular surfaces—this may be obtained temporally or spatially. Backlighting is ideal in applications where silhouettes are sufficient for recognition but not suitable for space application. A spatially modulated light source can project stripes or grids on an object. This structured light usually comprises a laser strip emitted onto the object. This generates a distinct contour of the object and the object's curvature distorts the light pattern providing geometric information of 3D shape and depth profile. Changes in distance along the strip are represented by curvatures and edges. Directional lighting is suited for the inspection of surfaces to detect flaws. Lasers are preferred to filamentary light due to their brightness, coherence, directionality and monochromaticity. No ambient light is required with structured light. Stroboscopic lighting eliminates blurring and infra-red/ultraviolet glare. The general call is for highly flexible lighting arrangements which may be adapted to suit variable situations.

Sensitive low-resolution infrared vision offers night capability for identification in complete darkness such as during eclipse periods. There are two types of infrared detector: thermal detectors employ the pyroelectric effect to detect temperature changes while quantum detectors use semiconductor materials [Philips Electronics Ltd 1989]. Thermal imaging involves scanning at 25 times per second and most commercial images are of 70 lines with 100 colours for CCIRTV compatibility. Although thermal detectors operate at ambient temperatures without the need for cooling, high-quality thermal imaging requires the use of quantum infrared detection. Quantum detection utilises semiconductors with bandgap energies of 0.25 eV for 3–5 μm detection and 0.1 eV for 9–13 μm detection. The discontinuity is caused by atmospheric absorption of intervening wavebands—a condition which will not affect any in-orbit servicer. Such infrared sensors require cryogenic tem-peratures to reduce the dark current due to thermal excitation. A popular material is cadmium mercury telluride (CMT) which is tailorable across the whole waveband with the source coupled to a CCD multiplexer. CMT detectors require cooling to 193 K for 3—5 μm detection and 80 K for 8–13 μm detection. Infrared response can be acquired in CCD arrays. Silicon materials with bandgaps of 1.1 eV may also be tailored through impurity doping with iridium for 3-5 μm sensitivity and Ga for 8–13 μm sensitivity with cooling to 25 K. Photoconductive modes of infrared detection are generally employed over photovoltaic modes of detection. For wavelengths of 1 μm – 1 mm, photoconductive devices use PbS for 1–4 μm (with liquid nitrogen cooling at T ~ 77 K) and doped Ge for >4 μm (with liquid He at T ~ 2 K). InSb has sensitivity to 1–5.5 μm. Hence, doped silicon requires much lower cryogenic temperatures. The figure of merit for quantum infrared detectors is given by:

$$D = \frac{V_s}{V_n} \frac{\sqrt{A\Delta f}}{P} \tag{12.1}$$

where V_s, V_n = rms signal voltage, rms noise voltage, A = detector element area, P = rms incident power, and δf = frequency range.

Infrared detectors are also used for the detection of space-based carbon dioxide laser communications links operating at 10.6 μm. Infrared detectors require cryogenic cooling and this is usually accomplished with Stirling cycle cryocoolers using liquid nitrogen, imposing mass penalties to their employment.

The human eye is characterised by a large concentration of ~10^5 optical fibres in ±0.5–1° foveal region of the retina along the optic axis with high resolution. Away from the fovea the distribution of optical fibres becomes more diffuse and so of lower resolu-tion through the majority of the visual field. Active control of attention and the non-uniform resolution provides orders of magnitude improvement in image-processing requirements, hence the element density of pixels of the CCD may be patterned after the human eye with high resolution FOV of 1–2° and low peripheral resolution view of ~100°. Gaze shifting and gaze holding movements of the eye provide a means to rapidly aim this narrow FOV. Fast saccadic eye movements move the fovea between different targets in the visual field while active gaze control keeps the fovea on a specific target. The retina and the regions of the brain that control the six muscles of the eyeball in pairs corresponding to the three axes of rotation are connected in the optokinetic response neurons in a negative feedback loop. There is also a feedforward vestibulo-ocular reflex which uses the vestibular apparatus of the inner ear to detect movement of the head which responds much faster than the optokinetic feedback loop. A similar kind of control system architecture would be required for artificial gazing. Ballard (1991) advocated an animate vision paradigm based on this function. It would require an invariant object centred frame of reference (e.g. the local inertial frame). Two cameras with foveated vision which are able to pan and tilt in synchronicity, each with independent saccading ability at 3 Hz and convergence control in the yaw axis could be used. A vestibulor-ocular reflex with smooth pursuit of targets would provide robust static and moving target tracking. At any instant only a small number of features in the environment that are relevant to the particular task are registered and new features are incrementally registered only when required. This approach to vision would dramatically decrease the requirement for data transmission at any instant to the ground, but it is not clear if this approach would be suitable for teleoperation or even virtual reality methods as such gaze shifting and gaze holding would require tight slaving of the remote cameras to the teleoperator eye movements which average saccade movements 4–5 times per second—the remote camera inertia and articulation motors could not match those of the human eyeball.

CCD cameras should be mounted in the following configurations [Elfving 1990a,b]:

(i)   a mechanical panoramic mobile camera with pan, tilt and zoom capability for differing viewing angles of the overall scene;

(i)   a minimum of two stereoscopic camera mounted on the robot spacecraft between the two manipulators orientable about two axes (pan and tilt) to give general over-head and frontal view and illumination;

(ii)   each manipulator requires one camera/light assembly mounted onto the wrists to illuminate the workspace for close proximity vision. The second manipulator camera/light assembly can provide overhead or sideways lighting or even backlighting for small objects. In addition, a small imager/ pencil beam light may be incorporated within the palm of the general-purpose gripper (eye-in-hand configuration).

The feature-extraction algorithms used in sensory processing for vision may be implemented in a high-level programming language for running on a general-purpose micro-processor. However, specialised hardware can be used to perform edge detection and thresholding at TV video rates of ~30 frames/s for real-time operation. Such dedicated vision processing electronics is essential for real-time performance.

12.2 IMAGE PROCESSING

Vision is a sensory modality based on the description of scenes representative of the functional properties of the physical world. Vision is an active process that imposes interpretation of the data in terms of models. Such model-based interpretation is complex and a balance is required between data-directed and goal-directed processing. Bottom-up processing is guided by top-down processing—local data is used to generate initial interpretation which is used to eliminate inconsistencies, e.g. the Kaniszu triangle. Subjective contour illusions in human vision appear to be a product of this process whereby an edge is perceived due to expectation despite the absence of local evidence. Humans often infer on the basis of the known characteristics of familiar objects, implying some form of model-based approach. A hierarchy of representations corresponding to the processing steps transforms a grey-level input image into a symbolic scene description. Such a succession of representation levels allows the appropriate representations to use the information available for the required purpose, e.g. from reactive alert response to complex event understanding.

Vision processing is an information processing problem that transforms a sensor input comprising a 2D array of brightness levels representing the 2D projection of a 3D scene into an output that comprises a description of the 3D scene, the objects in it and their relationships, i.e. vision is based on the measurement of light intensity against a background noise level. All 2D lines in an image correspond to a possible projection of an infinite number of 3D objects—this is the inverse optics problem. Three-dimensional information is lost in the projection to 2D and the single brightness level is determined by the incident illumination, reflecting characteristics and orientation of the viewed surface in relation to the viewer and the light source. The transformation of a 3D world into 2D intensity distributions by an optical system involves a transfer function which is only vaguely determined due to the changing interaction between optics, illumination and surface characteristics of the imaged object. Human visual processing is highly robust to highly variable viewing conditions, significant levels of noise, highly variable degrees of occlusion of objects, and highly variable orientations. Humans cope by using extensive *a priori* knowledge and inferencing illustrating the necessity for model-based object recognition. Hence, models with *a priori* knowledge are required to provide consistent expectations in terms of goals to decode the scene [Barrow & Tenenbaum 1981]. The overall goal of picture recognition and scene analysis is to extract useful information from images. Vision processing is a hierarchical process which may be characterised by four main levels as defined by Marr [Aleksander & Morton 1987]:

(i)   two-dimensional intensity pixel image of local features;
(ii)   extraction of boundaries between different intensity levels (primal sketch);
(iii) viewer-centred determination of global features and distances through stereoscopic imaging with respect to illumination conditions (2.5D sketch);
(iv)   object-centred 3D representation describing objects delineated by their surfaces and their relations to each other.

12.2.1 Low-level vision processing

The first stage in vision processing is the digitisation of the input field into arrays of pixel values of image intensity $I(x, y)$ according to its grey level. A 2D array camera generates a 2D grey-level visual image array of light intensity values $I(x, y)$—typically resolved to 8 bits to give 256 possible values. The light intensity at a point in the image is given by: $I = ER \cos i$ where E = incident illumination, R = surface reflectance (albedo), i = angle of viewing incidence from the local surface normal. The vision processing initially involves preprocessing to filter out noise, segmentation of regions into parts representing objects, property measurement of invariants, shape analysis of features and boundaries and structural analysis for model matching [Rosenfeld & Weszka 1976, Rosenfeld 1981, Albus 1984, Rosenfeld 1986].

The basic philosophy behind image processing is to delineate objects in the world from the background based on finding their edges, which have high contrast from the background [Charniak & McDermott 1985]. This is the first step to defining object boundaries prior to their categorisation and recognition. Preprocessing involves transforming the image to correct distortions due to blurring and noise by performing grey-level adjustments. One method is through the use of convolution masks as convolution represents a good model of linear physical processes. This is a $m \times m$ matrix kernal which is convolved over the $n \times n$ image matrix I to generate a filtered $n \times n$ image matrix I':

$$I'_{ij} = \sum_{p,q} k_{p,q} I_{i-1,j-q} \qquad \text{where} \qquad l \leqslant i, j \leqslant n, l \leqslant p, q \leqslant m \qquad (12.2)$$

The top left-hand corner of the mask is superimposed over each pixel in the image in turn. The gradients between pixels are calculated by using mask coefficients in a weighted sum of the pixel and its neighbours:

$$\Delta_x \text{ gradient}: \quad \begin{array}{cc} -1 & 1 \\ 0 & 0 \end{array} \qquad \Delta_y \text{ gradient}: \quad \begin{array}{cc} -1 & 0 \\ 1 & 0 \end{array}$$

To calculate diagonal directions at 45° and 135° for diagonal neighbours only, the Roberts cross operator gradient operator can be used. It is sensitive to rates of change to detect discontinuities and may be used to generate a gradient edge map of the form:

$$\sqrt{\left(\frac{\partial f}{\partial x}\right)^2 + \left(\frac{\partial f}{\partial y}\right)^2} \qquad \text{in the direction} \qquad \tan^{-1}\left(\frac{\partial f / \partial y}{\partial f / \partial x}\right)$$

This may be applied through the convolution masks:

$$\begin{array}{cc} 0 & 1 \\ -1 & 0 \end{array} \qquad \text{and} \qquad \begin{array}{cc} 1 & 0 \\ 0 & -1 \end{array}$$

These edge detectors are 2×2 masks but larger masks reduce noise through local averaging. A 3×3 spatial convolution mask (Sobel edge operator) is centred on each pixel in turn to compute the gradient of intensity change to neighbouring pixels which replace the intensity value of the central pixel of the form:

$$\sqrt{(\Delta_x f)^2 + (\Delta_y f)^2} \qquad (12.3)$$

where

$$\Delta_x f = [f(x-1, y-1) + 2f(x, y-1) + f(x+1, y-1) - f(x-1, y+1)$$

$$+ 2f(x, y+1) + f(x+1, y+1)]$$

$$\Delta_y f = \left[f(x+1,y+1) + 2f(x+1,y) + f(x+1,y-1) - f(x-1,y+1) \right.$$
$$\left. + 2f(x-1,y) + f(x-1,y-1) \right]$$

The Sobel convolution mask invokes a centre-surround operation to make it sensitive to all edges:

−1	0	1		1	2	1	
−2	0	2	and	0	0	0	
−1	0	1		−1	−2	1	

This procedure is repeated at each pixel—constant illumination will generate small values. Such edge detection masks reduce the effects of blurring. For a 512×512 pixel image, a 3×3 kernel can be convolved over all the pixels in 1.6 ms using specialised hardware.

Image deblurring techniques assume that blurring is a diffusion process satisfying

$$\frac{\partial f}{\partial t} = k\left(\frac{\partial^2 f}{\partial x^2} + \frac{\partial^2 f}{\partial y^2} \right) = k\nabla^2 f$$

where $\nabla^2 f = L$ is the Laplacian operator. The blurred picture $\hat{f}$ is related to the sharp image by: $f = \hat{f} - K\nabla^2 \hat{f}$ where $0 < K < 1$. In the discrete case, gradient measurements are given by:

$$\nabla^2 f(x,y) = \Delta_x f(x,y) - \Delta_x f(x-1,y) + \Delta_y f(x,y) - \Delta_y f(x,y-1)$$
$$= f(x+1,y) + f(x-1,y) - 2f(x,y) + f(x,y+1)$$
$$+ f(x,y-1) - 2f(x,y)$$
$$= f(x+1,y) + f(x-1,y) + f(x,y+1) + f(x,y-1) - 4f(x,y) \quad (12.4)$$

A more complex Laplacian would include diagonal neighbours of $f(x+1, y+1)$, $f(x-1, y-1)$, $f(x-1, y+1)$, and $f(x+1, y-1)$. Filtering noise involves replacing points that differ from neighbouring points more than a threshold amount by the average of its neighbours. The average of a pixel and its four neighbours $f(x \pm 1, y)$ and $f(x, y \pm 1)$ is given by:

$$\nabla^2 f(x,y) = 5\left[\tfrac{1}{5}\left(f(x,y) + f(x-1,y) + f(x+1,y) + f(x,y-1) \right.\right.$$
$$\left.\left. + f(x,y+1) - f(x,y) \right) \right]$$
$$= 5\left[\hat{f}_{av} - f \right] \quad (12.5)$$

If $K = 1/5$ then $f = 2\hat{f} - \hat{f}_{av}$. This strengthens the high-frequency content of the signal relative to the low-frequency content, generating a crisper image. The convolution mask for the Laplacian operator emphasises the central pixel at the expense of outer pixels which has a strong biological analogue:

$$
\begin{array}{ccc}
-\frac{1}{8} & -\frac{1}{8} & -\frac{1}{8} \\[6pt]
-\frac{1}{8} & 1 & -\frac{1}{8} \\[6pt]
-\frac{1}{8} & -\frac{1}{8} & -\frac{1}{8}
\end{array}
$$

It is isotropic as it utilises all its nearest neighbours weighted equally in filtering noise. A smoothing mask replaces each grey-level image with an average of itself and its neighbours:

$$
\begin{array}{ccc}
\frac{1}{9} & \frac{1}{9} & \frac{1}{9} \\[6pt]
\frac{1}{9} & \frac{1}{9} & \frac{1}{9} \\[6pt]
\frac{1}{9} & \frac{1}{9} & \frac{1}{9}
\end{array}
$$

For space application, it is desirable to reduce the effects of illumination variation. The grey level is proportional to the product of reflectivity r and illumination i which are dominated by high-frequency and low-frequency components respectively: $\log I(x, y) = \log E(x, y) + \log R(x, y)$. A Gaussian convolution mask is an isotropic mask with a cross-section that yields a weight profile with a Gaussian normal distribution.

$$
\text{Gaussian weight,} \quad w = \frac{e^{-r^2}}{2\sigma^2}
$$

where r = pixel distance from the centre of the mask, and σ = mask width.

The weighting drops to 60% of maximum when $r = \sigma$. The weights are normalised so that all weights sum to unity. A two-dimensional mask is generated by multiplying elements of the one-dimensional mask together. The Gaussian convolution mask removes noise optimally. As small-scale texture contains a lot of grey-level variations at high spatial frequencies, smoothing removes high spatial frequencies without distorting lower spatial frequencies. The width of the mask sets the scale below which variation is removed, but above which variation is retained. This process is equivalent to that which occurs in retinal processing.

Low-level feature extraction is obtained through segmentation through the separation of an image into different regions of brightness homogeneity which comprises the next level of processing on the basis that features are bounded by large charges in grey-level values between pixels [Gonzalez & Safabaksch 1986]. This usually involves thresholding, edge detection and template matching. Thresholding generates the binary edge map of light/dark binary values. Global thresholding is used for scenes where objects vary significantly from the background by using a single constant intensity threshold value (e.g. in backlighting conditions). Local thresholding is used in situation involving grey-scale images where contrasts may not be high and the threshold value depends on the local average intensity (e.g. in diffuse lighting conditions). Dynamic thresholding depends on the position of the pixel so that proximate point intensities determine the threshold value.

A centre-surround mask can locate grey-level boundaries to generate the edge map. The difference-of-Gaussian mask is an approximation of convolution with a Gaussian mask with a centre-surround mask (Laplacian of the Gaussian). The output of the centre-surround mask becomes positive and negative either side of the zero-crossings (light intensity boundaries). Thresholding into a binary image makes the boundaries evident. Canny's edge detector combines Gaussian smoothing with horizontal and vertical difference operators to provide a smoothed gradient direction at each pixel. Local maximum is then found and thresholding is applied to generate a binary image.

If the illumination to the scene can be controlled and the object occupies a large fraction of the scene, the threshold may be set by examining a histogram of the image which indicates how often each grey level occurs in the image. The histogram should have two peaks, one for the background grey level and one for the object grey level due to their relative high frequencies. The grey level at which to set the threshold is the intermediate rare grey level forming the valley between the two peaks. Alternatively, the image may be divided into blocks with a threshold picked for each block. Small objects are not easy to extract from their backgrounds by thresholding because the histogram may be comparatively flat and many objects may have different intensities. Hence, edge detection is required when high contrasts in grey levels occur at object boundaries whereby two dimensional derivative operators perform edge detection for outlining by computing the brightness gradient between image points. The gradient values are significant only in regions of intensity discontinuity such as edges. The complete outline of the object is obtained by detection of the edges between light and dark regions of the grey-level image, i.e. the boundary of the object against its background.

The Hough transform can be used to detect the alignment of points of high gradient corresponding to line elements which may be grouped into lines. Artificial objects (including spacecraft) often have straight or circular edges which project as straight and elliptical boundaries. The Hough transform is a global method for linking such edges. It maps data into a parameter space and a search is made in this parameter space for data clusters. It finds the infinite straight or curved lines on which the image edges lie, and the portion of the line between two ends is a line segment. If we take a point (x, y) in the image, all straight lines that pass through the point have the form: $y = mx + c$, or equivalently, $c = -mx + y$. Each line with a different gradient and intercept corresponds to one of the points on the line in (m, c) parameter space. If two pixels p an q in (x, y) space lie on the same single line, there is an intersection of the two lines p and q in (m, c) parameter space, i.e. all pixels of a single line lie at a single point in (m, c) parameter space. For curved lines, the line may be represented as: $r = x\cos\theta + y\sin\theta$. Circles may be represented as $r^2 = (x - a)^2 + (y - b)^2$. A point in (x, y) can be represented as a curve in (r, θ) parameter space. Hence, (m, c) space may be represented as a two-dimensional accumulator array, and (a, b, r) space as a three-dimensional array. A general ellipse requires a five-dimensional parameter space. A search for the elements of the accumulator array produces large values where they correspond to lines in the original edge map. To cope with more general, non-elliptical closed contour shapes, active contour models may be used. A series of control points are connected by straight lines. The active contour is specified by the number and coordinates (x, y) of each of those control points. An energy function is used to reduce the energy of the active contour by moving the control points:

$$E_{contour} = E_{int} + E_{ext}$$

Internal energy is dependent on intrinsic properties such as curvature and length. External energy depends on externally defined constraints. The motion of the active contour is dependent on simulated forces. The elastic energy is proportional to the total length via the sum of squares of distances between adjacent control points:

$$E_{el} = K_1 \sum_{i=1}^{n} (x_i - x_{i-1})^2 + (y_i - y_{i-1})^2$$

where K_1 = constant dictating degree of elasticity, and $x, y_{i,i-1}$ = control point coordinates i and $i - 1$. This generates an elastic force on the ith control point:

$$F_{el} = -2K_1\big[(x_i - x_{i-1}) + (x_i - x_{i+1}) + (y_i - y_{i-1}) + (y_i - y_{i+1})\big] \qquad (12.6)$$

Control points are pulled towards their nearest neighbours so that the force is towards the average position, i.e. moved inwards while smoothing. Outlying points are pulled in fastest, and the highest curvatures are smoothed fastest. The active contour surrounds the objects of the image. The external energy function is defined as the negative of the sum of grey levels of the pixels. Minimising the function moves the contour towards the brightest parts of the image:

$$E_{ext} = -K_2 \sum_{i=1}^{n} I(x_i, y_i)$$

$$\rightarrow F_{ext} = \frac{K_2}{2}\big(I(x_{i+1}, y_i) - I(x_{i-1}, y_i) + I(x_i, y_{i+1}) - I(x_i, y_{i-1})\big) \qquad (12.7)$$

The control points are pulled in the direction of the grey level gradient. After the internal force is applied, the external force alone is applied to close the active contour. Active contours can be implemented which average over more pixels, or even more sophisticated variants. This method is compatible with the potential field method of obstacle avoidance.

An object is completely determined by its boundary curve. Protrusions and indentations may be detected by characteristic sequences of intensity values, e.g. corners, spurs, notches, etc. Clustering involves generating smooth continuous curves from the edge segments. To distinguish between objects it is necessary to label object pixels to the same object. Labels are assigned to sets of objects that are mutually connected in a curved S-border following process. Recognition is basically a labelling process. A chain code can completely specify object borders. It specifies the boundary as a sequence of horizontal, vertical and diagonal curves of length 1 or $\sqrt{2}$. It involves numbering line segments between points so that the whole boundary of the curve is represented by coordinates specifying the start point and the sequence of moves to follow the border. Each move requires only 3 bits while the start point requires only $2 \log_2 n$ bits. This is much more compact a representation than the binary image which requires n_2 bits. Relative poses between 3D edges are characterised by four parameters:

(i)   α—the angle between lines supporting edge segments;
(ii)   d—the length between lines measured along the area perpendicular to the lines;
(iii) a_1 and a_2—the distance from the base of the common normal to the nearest end of
        the edge.

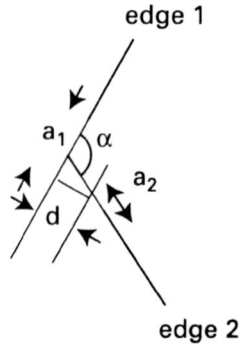

Fig. 12.2. Line segment poses.

The Clowes–Huffman vertex identification process does not require explicit object
model prototypes to interpret line drawings of polyhedral scenes [Nau 1983, Charniak &
McDermott 1985]. Polyhedral objects have flat faces bounded by straight edges. Every
three-dimensional edge is associated with two faces either side of each edge. If only one
face is visible, the edge is occluding. Catalogues of vertices with their spatial coordinates
(x, y, z) are used and have been expanded to include cracks, occlusions and shadow
edges. It provides a means to label vertices of 2D line drawings of 3D objects comprised
of flat surfaces. Vertices are three-faced corners formed by three meeting edges at which
each line represents a boundary and may be convex (+) or concave (−). This provides
local evidence that adjacent regions belong to the same body. Natural constraints limit the
number of vertices that are physically realisable on a polyhedral object since they are the
intersection of two or three faces and these can be used to determine the object shape.
There are 208 mathematical combinations of vertex representations. However, there are
four broad types of vertex representing physically possible junctions: the arrow vertex,
the T vertex, the fork Y, and the L-peak vertex. The L-junction represents only two edges
rather than the three edges in all other junctions. T junctions occur only when one object
occludes another. The real world essentially constrains the number of possibilities. For a
given type of vertex in line drawings of polyhedral solids only certain combinations of
interpretations of edges are physically possible. Each line may be either an occluding
edge, a convex edge or a concave edge. Every vertex and line must have a single label
that is globally consistent for the line drawing to represent a real 3D object. Obviously,
Mobius strip type constructions are eliminated. The search for consistent labelling of
objects is restricted by the constraint that every line must be assigned the same label by
the junction interpretations at its two ends. To avoid a combinatorial explosion in the
search for consistent interpretations, a parallel local filter can be applied. Much of vision
processing can be cast into parallel processes and most of the perception stage in human

cognition is massively parallel in nature. Waltz's constraint propagation scheme enabled interpretation of a 2D line drawing as a 3D scene property by consistently labelling vertices of objects in parallel. The local filter assigned a set of possible interpretations to each vertex and considered pairs connected by a common line to eliminate locally inconsistent vertex interpretations to converge rapidly to a unique solution. For instance, a silhouette of a brick comprises of six vertices which alternate between L- and arrow-vertices and an internal Y vertex (see Fig. 12.3). The Y-vertex for instance differentiates the cuboid from the pyramid. A hole may be regarded as a connected curve that does not touch the border of the image.

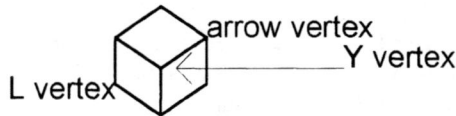

Fig. 12.3. Vertices of a brick.

For objects which are not uniform in brightness, neither thresholding nor edge detection are suitable because within the object grey levels may have high contrast due to holes and internal edges defining the 'coarseness' of an object. The object is distinguishable from the background by its characteristic pattern of grey levels representing its visual texture. Certain local patterns of grey levels tend to occur in the neighbourhood of each pixel. The image can be segmented into differently textured regions to separate the clusters. Texture is important in defining the 3D surface structure of an object. Roughness can be quantified through the Hausdorff dimension 0 (rough) $< h < 1$ (smooth). The dimensions of texture scale inversely with distance and the direction of tilt is foreshortened by the cosine of the angle between the surface normal to the line of sight due to perspective when a surface is viewed obliquely. The orientation of elements comprising textures tend to be isotropically distributed and so orientation can be extracted from the most isotropically elliptical texture elements.

Vision is more complex than pure brightness intensity processing: humans use physical characteristics such as reflectance and colour of 3D surfaces to determine the composition and structure of objects for general-purpose visual identification. Surface reflectance gives the albedo defined as the ratio of total reflected to total incident illumination. Because range data is uncorrupted by reflectance variations, scanning laser range finders may use this to generate accurate range information unaffected by the ambient illumination. Intensity values of the image array is underconstrained, so additional constraints must be applied [Aloimonos & Rosenfeld 1991]. Recovery of E, R and i from an intensity array I requires additional constraints as well as distance and orientation. Variation in shading may be used to determine the 3D shape of a uniformly diffusing (Lambertian) surface with constant albedo illuminated by a point source to generate constant illumination:

$$\frac{dE}{E} = \frac{dI}{I} + \frac{dR}{R} + \frac{d(\cos(i))}{\cos(i)} \rightarrow \frac{dE}{E} = \frac{d(\cos(i))}{\cos(i)}$$

since $dI = dR = 0$. This enables recovery of shape given surface orientation given that most surfaces are piecewise smooth (from texture analysis). The Lambertian surface reflectance is only approximately true for real surfaces. Often illumination varies slowly and continuously over the scene while albedo tends to be piecewise constant in being uniform over patches but discontinuous at patch boundaries: $dE/E = dI/I$ within regions and $dE/E = dR/R$ at boundaries. Indeed, albedo changes are often indicative of occluding edges representing step discontinuities from one surface material to another. This is based on the retinex theory of colour vision and illustrates the way humans estimate colour under a wide variety of intensities by using surface reflectance at different wavelength distributions. Red, green and blue digital colour images are arrays of triplet values across three spectral bands. Indeed, the UBV filter used in astronomical telescopes emulates strongly the sensitivity of the human eye to colour. Segmentation into regions of different colours partitions the 'colour' space. This is the process used in multispectral scanners in remote sensing. Surface colour is estimated by recovering albedo independently in the three primary colour spectral bands. Colour changes occur when albedo becomes discontinuous. Hence, a multitude of techniques are required for robust low-level visual processing.

Segmentation involves grouping pixels together into regions of similar intensity values. The final segmentation divides regions of similar properties from other regions by high rates of intensity change. The scene is partitioned into regions of approximately uniform brightness corresponding to surfaces or texture regions which are locally consistent by iteratively subdividing it. For real world scenes, region growing is preferred over segmentation since it does not require straight boundaries such as curves and is well suited to texture segmentation and analysis. The initial set of small regions is based on seed pixels which are merged by comparison with neighbouring pixels. This merging process continues with segments growing until the uniformity condition holds for the region by adding in similar neighbouring pixels. The region is grown until it stops. This process is performed simultaneously across all regions as similar regions gradually coalesce. Adjacent regions are then successively merged when their boundary contrasts or texture constraints are low until only boundaries of strong contrast remain. Regions are then grouped them into objects which as a whole will not be uniform. Once regions have been extracted from the image, it is possible to extract properties such as global characteristics as connectedness, compactness, etc. The region is delineated by its closed boundary which may be specified by parametric curves, its centre and radii, or a vector pair (n, d) where n = directional vector, d = vector to closest point on the line to the origin. It is possible to specify a point on the line as $p = d + tn$.

Global features such as area, perimeter, centroid-perimeter radii, first and second moments and number and area of holes may be extracted. The area of the object image is the number of pixels in the object image. The perimeter of the object is the number of steps required to follow its border. The extent of an object is the length of a line in that direction—the greatest extent is its diameter. The extraction of such global features requires normalisation of the image using autocorrelation functions. Correlation is a time domain measure of similarity between waveforms—if the two waveforms are time-shifted versions of each other, the procedure is autocorrelation:

$$R_{xx} = \frac{1}{T} \int_0^T x(t_1) x(t_2)\, dt \qquad (12.8)$$

When $t_1 = t_2$, this reduces to the mean square value of the signal which represents the maximum autocorrelation. The cross-correlation function applies correlation to different waveforms:

$$R_{xy} = \frac{1}{T} \int_0^T x(t_1) y(t_2)\, dt \qquad (12.9)$$

It is commonly used in radar where transmitted and received signals are cross-correlated to extract the 'echo'. Correlation provides a best fit matching criterion for comparison with prototype templates.

Normalisation involves extracting invariant properties under common equivalence transformations representing permanent properties of the environment. The set of characteristic measurements (features) are extracted from f and stored as $P(f)$ where

$$P(f) = \iint h(x, y) f(x, y)\, dx\, dy,$$

e.g.

$$C_{ff}(u, v) = \int \int_{-\infty}^{\infty} f(x, y) f(x + u, y + v)\, dx\, dy$$

is independent of translation and

$$P_{ff}(s, \varphi) = \int_0^{2\pi} \int_{-\infty}^{\infty} f(r, \theta) f(sr, \theta + \varphi) r\, dr\, d\theta$$

is independent of rotation. For digital images, these are difficult to use so moments are used to extract the desired global features and define translation and rotation invariant properties. Moments represent a class of Fourier power spectrum techniques. Such frequency domain filtering is more efficient than correlation computations, and auto- or cross-correlation functions are obtainable from the Fourier transform frequency-domain analysis. Ignoring negligible Fourier coefficients particularly high-frequency components can reduce image storage density with up to 10:1 compression ratios because their effects are distributed over the entire image. Fourier analysis of images involves the extraction of the power spectrum which is defined as the square of the modulus of the Fourier transform of intensity. Moment generation may be defined for a grey level intensity array:

$$M_{ij} = \sum f(x, y) x^i y^j \qquad (12.10)$$

where $i + j = $ order of moment, and $f(x, y) = $ grey-level weighting. Zero to second-order moments enable the calculation of perimeter, outline area, centroid (centre of mass if even mass distribution), angle to major axis, elliptical cross-section, area/(perimeter)^2 (this measures the compactness of an object or its inverse complexity), area/(thickness)^2 (elongatedness). These quantities are required for manipulation such as for finding grapple points.

$$\text{Area,} \quad A = \iint f(x, y)\, dx\, dy$$

Centre of mass,

$$\bar{x} = \frac{M_x}{A} \quad \text{where} \quad M_x = \iint xf(x,y)\,dx\,dy$$

$$\bar{y} = \frac{M_y}{A} \quad \text{where} \quad M_y = \iint yf(x,y)\,dx\,dy \tag{12.11}$$

Moment of inertia,

$$I = \iint r^2(x,y)f(x,y)\,dx\,dy = \iint (x\sin\theta - y\cos\theta + \rho)^2 f(x,y)\,dx\,dy \tag{12.12}$$

Orientation angle,

$$\theta = \frac{1}{2}\sin^{-1}\left(\frac{2M_{xy}}{\sqrt{4M_{xy}^2 M_{xx} - M_{yy}^2}}\right)$$

where

$$M_{xx} = \iint (x - \bar{x})^2 f(x,y)\,dx\,dy$$

$$M_{xy} = \iint (x - \bar{x})(y - \bar{y})f(x,y)\,dx\,dy$$

$$M_{yy} = \iint (y - \bar{y})^2 f(x,y)\,dx\,dy \tag{12.13}$$

The principal axis is the line through the centroid about which the moment of inertia is least. Elevation and azimuth angles of the centroid can also be generated. The moments are shown in continuous form but equivalent versions exist where the integral is replaced by the summation over all pixels. Higher-order moments may be combined to form Gibson invariants which correspond to permanent properties of the environment despite changes in illumination to characterise objects for discrimination purposes. If objects are partly obscured, then local features such as holes and corners may be used for recognition.

12.2.2 High-level vision processing

Higher levels of visual processing involve object representations at a symbolic level with global invariant properties. This level of processing is highly goal and expectation driven. Low-level vision processing is highly under-constrained so *a priori* knowledge is necessary to impose constraints to enable solutions to be found [Poggio et al 1985]. There are three main methods for such pattern recognition: template matching, decision theoretic and syntactic approaches (we shall not be covering decision-theoretic or syntactic approaches here) [Fu 1982, Fu & Rosenfeld 1976]. Template matching is model-based [Binford 1982].

Recognition between the scene and model is a classic search problem which is computationally intensive. Ultimately, all image classification is based on fixed sets of model prototypes which comprise templates against which the image must be normalised with respect to size and position and then matched. This requires the set of objects to be encountered to be known in advance. Interpretation is interleaved with the signal processing steps by using partial matches with models which are used to verify object interpretation hypotheses by finding statistical evidence of features that may have been missed in the signal analysis. The outline is then matched with a computer model prototype using statistical similarity measures such as least-squares estimation. Recognition occurs by matching the input representation to parts of the model template prototypes using statistical methods. Symbolic abstractions include polygonal surfaces that correspond to regions that have the same constant properties and continuously varying properties. Surface fragments may be merged together into larger regions until they enclose a volume with a given shape and size in an object centred coordinate frame.

Contrasts in images representative of objects are characterised by curvature, colinearity, symmetry, parallelism and cotermination [Biederman 1987]. Objects are segmented by regions of sharp concavity, especially at cusps where there are discontinuities in curvature. When gaps exist, humans assume continuity of curvature, colinearity, symmetry and cotermination, i.e. humans fill in the gaps in contours during low-level perception. These volumes may be based on generalised cylinders defined by a 2D curve that specifies a central axis and a cross-sectional area or radius. There are 24 generalised conical 'geons' from which all volumetric shapes can be built using 81 attachment relations defining their spatial relationships—two geons can construct over 10 000 objects and three geons can construct 306 billion objects (recognition-by-components). Generalised cylinders are based on translational invariance through a hierarchy of congruence relations. These primitive parts may be classified to form clusters which represent entire objects of possibly complex structure linking the primitive structures with set-theoretic operators such as union, intersection and difference. It has been estimated that there are some 30 000 commonly experienced objects or categories of object in everyday hunter-gatherer, agricultural and urban life. This would require learning an average of 4–5 objects per day for 18 years. This is the peak rate of word acquisition by children at their peak between the ages of 2 to 6. There are certain natural objects which are not well modelled by geons but require fractal description, but as the in-orbit servicing environment is likely to be populated by man-made objects only, and the planetary environment is likely to be populated by well-defined geological structures, this should be sufficient. A 3D object may be represented as a solid polyhedral object via a set of edges formed by surfaces bounding its volume, e.g. geometric CAD database model. CAD/CAM models of objects to be manipulated could provide the data needed to perform manipulation. Once object recognition is accomplished this is sufficient for navigation purposes. Next, objects must be related to scenes to determine object locations and grouped into higher-level configurations to represent events and scenes. Indeed, objects and assemblies are defined in terms of the shape of their parts and their relations to each other. Objects often intersect with each other to produce concave edges in the image which can be extracted. The construction of objects in the scene allows the use of a blocks world which represents simplified scenes as comprised of simple poyhedral objects of approximately uniform brightness with brightness discontinuities at their boundaries (edges). The scene is a description of a hierarchy of objects and object groups. In reality, shadows, texture and light scattering will complicate the scene.

An alternative description particularly for higher levels of representation is to characterise objects as symbolic expressions in a relational graph of attributes whose nodes use primitive object properties and whose arcs are relationships. The relational graph is invariant to rotation and size. Such attribute relational graphs allow identification of the object by graph matching if the CAD model is converted into a graphical representation. Flynn & Jain (1991) tackled the problem of generating relational graph representations from the constructive geometry models of IGES standard CAD databases. They used a geometrical inference engine to automatically build the relational graphs which contained a hierarchy of geometric primitives connected by binary relations (e.g. proximity, orientation, relations between surfaces) from the CAD models. This simplified the matching process dramatically in a depth-first search while making the interfacing to expert system-based planning systems easier to implement. Every object must be registered in the scene description into aggregate objects and this requires complex reasoning processes and knowledge. At this level, vision blends into cognition and artificial intelligence. Since a knowledge base may be represented as a hierarchical network of schema models, so a hierarchical description of a scene may be represented. Each model represents a class of objects and each schema represents a particular resolution of the scene forming a compositional hierarachy of schemata.

Template matching is the simplest form of pattern recognition but complex patterns such as those encountered in planetary environments may be analysed using decision-theoretic and/or syntactic approaches [Fu 1982, Fu & Rosenfeld 1976]. Template matching assumes invariance in rotation and translation and image congruence. Different orientations and illumination patterns tend to corrupt the matching process. However, model-based approaches to robotic vision are generally adopted particularly when the objects to be manipulated are known in advance or when object data can be uploaded to the planetary rover.

12.2.3 Stereovision and range estimation

The use of stereoscopic binocular vision from two different positions through two cameras allows three-dimensional range estimation at close quarters. Such visual range sensors should be mounted such that their sensing axes are parallel to camera optical axes. There is a loss of information in the projection from a three-dimensional object into a two-dimensional image making it difficult to deduce 3D information from the 2D image since the image-world mapping is not one-to-one. In general, a 2D pinhole camera image has image coordinates (x, y) which are related to world coordinates (X, Y, Z) of point P_i by [Trivedi et al 1990, Abidi et al 1991]:

$$\begin{pmatrix} c_1 \\ c_2 \\ c_3 \\ c_4 \end{pmatrix} = \begin{pmatrix} a_1 & a_2 & a_3 & a_{10} \\ a_4 & a_5 & a_6 & a_{11} \\ a_7 & a_8 & a_9 & a_{12} \\ a_{13} & a_{14} & a_{15} & a_{16} \end{pmatrix} \begin{pmatrix} X \\ Y \\ Z \\ 1 \end{pmatrix} \qquad (12.14)$$

where $x = c_1/c_4$ and $y = c_2/c_4$, and $A = 4 \times 4$ world to camera coordinates transform.

The camera model describes the relation between the camera coordinates and the image coordinates:

$$x = X_c \frac{f}{-Z_c} \quad \text{and} \quad y = Y_c \frac{f}{-Z_c} \tag{12.15}$$

where $(X_c Y_c Z_c)^T$ = object position with respect to camera coordinates, and f = camera focal length.

These are the perspective projection equations derived from similar triangles geometry (law of sines). For a directed camera where the coordinate system of the camera centres on the lens centre with its z axis pointing towards the object of interest:

$$\begin{pmatrix} X \\ Y \\ Z \\ 1 \end{pmatrix} = \begin{pmatrix} a_1 & a_2 & a_3 & a_{10} \\ a_4 & a_5 & a_6 & a_{11} \\ a_7 & a_8 & a_9 & a_{12} \\ 0 & 0 & 0 & 1 \end{pmatrix} \begin{pmatrix} X_c \\ Y_c \\ Z_c \\ 1 \end{pmatrix} \tag{12.16}$$

The A-matrix represents the transform from world coordinates (X, Y, Z) to camera coordinates (X_c, Y_c, Z_c). This is not soluble from a single image. Stereovision overcomes this by using two images separated by binocular parallax. Since each image is a projection, the intersection of each projection dictates the location of the surface point. Each point on a surface generates corresponding points in each of two images, each determining a unique light ray. Intersection of the rays determines the location of the surface according to geometric constraints. Geometric triangulation may yield continuous surfaces so that the distances to neighbouring points are minimised.

Three dimensional information is derived from the disparity of corresponding regions in the two images from the two cameras. This angular disparity between the two images together with a known separation distance between the cameras enables the range computation. The procedure is common in nature and is the mode of distance estimation used by primates. Three steps are required for the measurement of stereo disparity [Marr & Poggio 1976]. First, a location on the scene must be selected from one image. Second, the same location must be identified in the other image. Third, the disparity between the two image points must be measured. This requires the identification of corresponding points in the images in the two cameras so that the images may be matched. This is the major problem in stereovision (the correspondence problem). The feature must have certain properties: it must be unique or, for multiple features, the disparity must vary smoothly across the object. It requires a given feature in one image to be registered with a corresponding feature in the other image such that they both correspond to the same feature in the 3D object (hence the name, the correspondence problem) [Marr & Poggio 1976, Ballard et al 1983]. Several constraints may be applicable. Corresponding points must share the same elevation y (equipolar constraint). A normalised measure of the mismatch between two digital images f and g may be given by a cross-correlation or FFT (fast Fourier transform) between f and g search and target arrays. Two-dimensional FFT is important in image processing to filter high spectral components to reduce noise and low

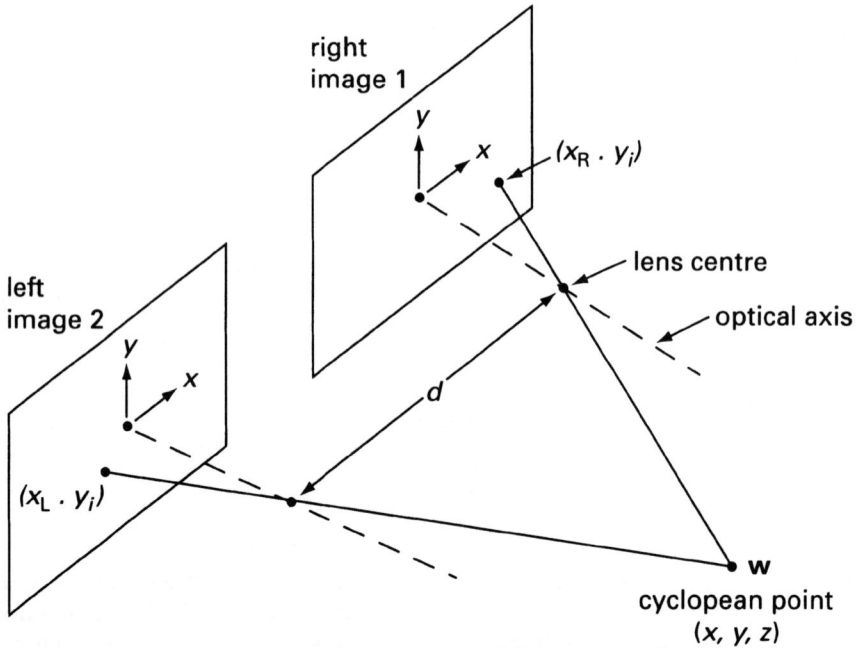

Fig. 12.4. Geometry of stereo imaging.

spectral components from the image or to convolve the image with a point spread function (convolution mask). By applying the FFT many of the Fourier coefficients are negligible thereby reducing storage and processing requirements and providing 10:1 compression rates for images. The Fourier transform matrix multiplication has $O(n^2)$ complexity but the FFT reduces this to $O(N \log_2 N)$—it utilises the Cooley–Tukey algorithm which is a divide-and-conquer strategy based on shuffling array computations. A cooperative algorithm comprising a set of iterative difference equations can also solve the correspondence problem. Such an algorithm may be implemented as a multilayer artificial neural network perhaps as a special-purpose board. The use of an artificial neural network mapping model is an effective way to provide the means of implementing parallel mutual constraint satisfaction. A neuron is assigned to each possible conjunction of an image element of one camera and an image element of the other camera. Inhibitory links can apply the uniqueness constraint so that line-of-sight units are suppressed while the continuity constraint of adjacent image points at similar depths may be implemented using excitatory links. Mutually consistent nodes would excite each other while incompatible nodes would inhibit each other.

The stereo configuration is usually a convergent parallel camera system with focal length f separated by an inter-camera distance d with their optical axes parallel to each other. From the midpoint (X, Y, Z) in cyclopean coordinates:

$$X = \frac{D(x_L + x_R)}{2(x_L - x_R)}, \quad Y = \frac{dy}{x_L - x_R}, \quad Z = \frac{df}{x_L - x_R} \qquad (12.17)$$

where $x_L - x_R$ = stereo disparity. From the perspective projection equations:

$$x_L = \frac{(x + d/2)}{z}, \qquad x_R = \frac{(x - d/2)}{z} \tag{12.18}$$

where z = depth of object relative to the camera. A point in the world coordinate system (x, y, z) projects onto a point (x_L, y_i) in the left image and (x_R, y_i) in the right image where

$$x_L = \frac{f(x + d/2)}{z}, \qquad y_i = \frac{fy}{z}, \qquad x_R = \frac{f(x - d/2)}{z}$$

The coordinates (x_R, y_i, x_R) refer to coordinates in disparity space defined as $x_L - x_R$. Now, in fact, optical axes of the cameras converge to a fixation point at a depth of fixation z_0, i.e. zero disparity. We compute the disparity of the object at the fixation point if the camera optical axes were parallel:

$$X = \frac{d(x_L + x_R)}{2(x_L - x_R + fd/z_0)}, \qquad Y = \frac{dy}{(x_L - x_R + fd/z_0)},$$

$$Z = \frac{df}{(x_L - x_R + fd/z_0)} \tag{12.19}$$

Using well defined markings makes the process simpler. Spinning objects with characteristic dimensions of 1 m can be tracked at ~10 rpm unaided by markers, but if visual markings are available tracking capability is increased to 15 rpm. The disparity is then computed to extract the position of the feature in 3D space. Motion provides a means for range estimation using a single camera—indeed, humans use purposive eye movements to observe objects at multiple vantage points. Six landmarker points allow solution to the matrix A that characterises the problem. Each point provides a pair of linear equations in the matrix elements. Certain properties of A, however, allow the number of marker points to be reduced to four P_0, P_1, P_2, P_3 with their respective images p_0, p_1, p_2, p_3. The A-matrix may be regarded as a translation and three rotations. Once A is known, c_i is known for each target point and so therefore the 3D coordinates of the target is derived. The first step in the derivation of A is the transformation of world coordinates of point P_i to intermediate standard position coordinates by matrix S such that P_0 lies at the origin, P_1 lies on the positive X axis and P_2 lies on the positive Y axis.

$$S = (R_\alpha R_\beta R_\gamma P)^T$$

where R_α = rotation of α about the X axis, R_β = rotation of β about the Y axis, and R_γ = rotation of γ about the Z axis.

The second step in the derivation of A is the transformation of the real camera image to an ideal image coordinate I with its origin at the lens centre but with its x axis coinciding with the X axis of the standard intermediate coordinates, i.e. p_0 is at the image origin and p_1 is on the positive x axis:

$$I^{-1} = R_c R_a R_b$$

where R_a = tilt of a about the new X axis, R_b = pan of b about the new Y axis, and R_c = roll of c about the new Z axis.

The third step is to relate standard intermediate coordinates to the ideal coordinates, i.e. the transform of the target points at standard coordinates $(X_i^\alpha Y_i^\alpha Z_i^\alpha)$ viewed by a camera at the ideal coordinates $(x_i^\beta y_i^\beta)$:

$$\left(X_i^\beta Y_i^\beta Z_i^\beta\right)^T = C\left(X_i^\alpha Y_i^\alpha Z_i^\alpha\right)^T$$

$$\left(x_i^\beta y_i^\beta\right)^T = \left(-F\frac{X_i^\beta}{Z_i^\beta}, \; -F\frac{Y_i^\beta}{Z_i^\beta}\right)^T \tag{12.20}$$

Once these three matrices are known then $A = \Gamma^{-1}CS$ [Abidi & Gonzalez 1990, Abidi et al 1991]. This method may be used as a backup method of visual depth derivation to the stereovision system advocated if one of the stereo cameras fails. Stereovision has an operating range of 5–50 cm and increases teleoperation speeds by ~10% [Varsi 1990].

12.3 POTENTIAL FOR REMOTE SENSING APPLICATIONS

Many of these onboard distance sensory capabilities (particularly vision) may allow the use of these sensors for low resolution (1:500 000) wide area Earth-oriented remote sensing purposes during the dormant phases of the ATLAS mission (which comprise the vast majority of its mission life) [Simonett 1982]. The environment is now an issue of major political importance and continuous monitoring of the environment is necessary to track climatic changes particularly resulting from the greenhouse effect. Sun-synchronous polar orbit at ~800–1000 km allows a satellite to scan the entire planet's surface passing over the same point at the same local time for constant illumination over a period of weeks, months or years (e.g. Landsat, SPOT). The fact that polar orbit is seen as the optimal orbit for global scanning and the perceived increase in its use for Earth observation satellites suggests that this orbit would be a natural one for ATLAS operations. Overview images allow effective targeting of regions for more detailed high-resolution images through the detection of relevant phenomena prior to identification and analysis. ATLAS may offer low-resolution remote sensing data to complement the data from dedicated high-resolution satellites, or even in tandem with such satellites.

A particularly important capability offered by the ATLAS sensor suite that is not yet available to Earth Observation satellites is the potential use of lasers for active sensing of the Earth's atmosphere using short light pulses to obtain altitude resolved information with high vertical resolution. Such sensors are essentially LIDAR techniques which utilise a laser as the light source to actively illuminate the atmosphere in conjunction with a receiving telescope which allows its use during night phases of the orbit as well as the sunlit portions to probe the upper atmosphere without absorption from the optically thick lower layers [Lutz & Armandillo 1992]. A fraction of the emitted radiation is backscattered by the atmosphere which is detected by the receiver. The distance of the LIDAR to the target is determined from the time of flight, so the return signal contains samples from different regions of the atmosphere which can be resolved. They offer greater vertical resolution than microwave atmospheric sounders.

13

Autonomous robot task planning

Autonomous control systems should be capable of performing well under significant uncertainties in both the system and the environment for extended periods of time without external interaction. Space is a relatively unstructured environment characterised by incomplete *a priori* data. The generic feature of the intelligent system is that it has the capacity to deal with uncertainties in its environment. The control problem is fundamental to all cognitive processes and intelligent behaviour (i.e. real-time problem solving and planning) [Schmidt et al 1978]. For space systems, autonomy is highly desirable as it reduces ground system requirements and costs which are a significant fraction of the operating costs of space missions. There are many different degrees of autonomy but space vehicle autonomy is highly demanding. The best way to achieve autonomy is to utilise high-level decision making techniques, i.e. intelligence to provide a predictive capability—intelligent control is the means to achieve autonomy in adaptive systems [Antsaklis 1991]. This allows the anticipation of the outcome of behaviour and allows goal-oriented actions to be implemented to achieve some purpose. This involves the process of internalisation of external reality to facilitate its simulation. Planning may be regarded as internalised motion which constructs the potential consequences of actions. The incorporation of intelligence at higher levels of the control system implies the need for flexible decision-making and reasoning. This in turn implies that artificial intelligence (AI) techniques should be implemented at these higher levels of control, particularly for problem solving. Other means of installing autonomy are through reactive behaviour control. Intelligence is particularly suited to anthropomorphic robots designed to perform human-type tasks in environments which may be hazardous to human operators [Aleksander & Morton 1987]. It is also suited to space operations where the large distances involved can invoke intolerable time delays to remote human teleoperators. Indeed, intelligence is absolutely necessary for any tasks beyond simple, highly repetitive tasks in a totally structured and predictable environment.

At a minimum, intelligence requires the ability to sense the environment, make decisions and produce actions to adjust to new situations and function reliably. In fact, this is precisely the definition of a control system. The ability to plan which implies the ability to predict future events forms part of this control system. A robotic system comprises three subsystems: a perception subsystem and an action subsystem linked by a decision/

control subsystem. Hence, there is a central role to be played by the decision subsystem in an intelligent robotic system. At a more developed level, intelligence includes the ability to integrate sensory data, recognise and classify objects and their relationships, represent knowledge in an integrated world model, reason and make decisions, and plan future events. A world model M has the structure $\langle \theta, P, E, A, T, B, G, \varphi \rangle$ where θ = a set of actions, P = set of other agents, E = set of primitive events, $A \in E \rightarrow P$ = agent of events, $T \subseteq E$ = set of possible events/worlds, $B \subseteq T \times P$ = set of beliefs, $G \subseteq T \times P$ = set of goals, φ = predicate interpreter. Indeed, McCarthy & Hayes (1969) considered that intelligence absolutely requires an adequate model of the world to enable the agent to perform tasks in the external world according to its goals and abilities and if necessary add information from the external world observed through the senses to the internal world model. The world model representation provides the means for overcoming the limited information transmission rates from the sensors by imposing structure and order on sensory inputs. Perception limits the information extracted from sensory signals due to bounded rationality limitations. The information extracted must so reflect important properties and object interactions such as object function and context rather than mere physical properties. Although incomplete, the world model uses past information to predict the future and constrains subsequent sensory processing based on expectations. Intelligence integrates knowledge and sensory feedback information in a goal-directed control system that plans purposeful actions. Furthermore, purposiveness requires anticipation or expectancy concerning the choices in achieving goals. Ultimately, the degree of intelligence will be determined by the computational power available, the algorithmic sophistication of the programs and the amount of memory.

The robotic system provides the most demanding test of AI techniques since it requires the intelligent control system to interact with the real world via a physical robot body. Sensory input and motor output are not analogues of the standard peripheral read and write commands in a computer programming language. The input system processes raw sensory input signals and transforms them into symbols necessary for input to a symbol processor; the output converts the processor output of symbols into actuator control signals. The physical properties of sensors and actuators cannot be divorced from the general problem of artificial intelligence [Brady 1985]. There is a strong coupling between the robot agent and the environment. The physical embodiment of the information processor (the body) provides a buffer to the external environment [Fritz et al 1989]. Dreyfus (1967) suggested that bodies are a necessary prerequisite for intelligence and that disembodied computing systems will never exhibit truly intelligent behaviour. The body provides a fundamental system of reference for cognition.

A major difficulty lies in the presently limited capabilities of space-rated hardware. However, notwithstanding this, autonomous control software may be implemented to run from the ground station. Due to the long characteristic time scales of higher levels of control, the time delay to remote sites should not present any difficulties. Direct teleoperative solutions are not optimal for planetary rovers for Mars and beyond where the roundtrip delay time can be up to 40 minutes. Ideally, a supervisory control system should be implemented and as capabilities are enhanced so the control system should become more autonomous over time. The first complete and integrated robot system capable of perceiving its environment, acting on its environment and planning actions to achieve its goals using a polyhedral world model was HANDEY [Lozano Perez 1987]. Although ground station automation has its costs in software development, flight operations involves a substantial cost in personnel. The high complexity of automated planning software is suited to ground control and these usually involve expert systems to manage space operations through planning and scheduling of remote routine and reflexive space robotic operations on the ground. High computing power is necessary for automation of ground systems and for detailed simulation of spacecraft operations. This reduces the reliance on personnel for long-duration missions.

The ultimate goal of artificial intelligence (AI) research is to impart human level intelligence to computing machines. As pointed out by Sutherland (1986), AI has been characterised by different paradigmic approaches. Although this is an expression of the natural tendency towards specialisation and diversity, it is more a matter of choice of preference rather than imperatives of methodology. Although the main concern here is with AI applications to engineering, AI theory of emulating human cognition is of particular relevance to robotics which itself seeks to emulate human manipulation capabilities. Most AI engineering systems tend to assume the man-in-the-loop to attend to difficult decision making functions rather than truly autonomous stand-alone operation. For space systems, it is desirable to eliminate the man in the loop and impart the machine with the full spectrum of human capabilities due to the expense of maintaining ground crew and the remote distances often encountered.

Any intelligent agent has access to two kinds of information [Kelly & Bonner 1985]:

(i)   A *posteriori* information—this kind of information is available before task execution and includes knowledge of the contents of the environment, the relationships between them and the strategies required to perform the commanded task. It is static and is usually comprised of a preprogrammed data/knowledge base (e.g. CAD/CAM).

(ii) A *priori* information—this kind of information is available only after task execution has been initiated and includes additional knowledge about the environment gathered during task execution. It can verify expectations and inform of deviations from expectations. This form of knowledge is dynamic and is gained from sensors.

Both types of knowledge are required by an intelligent system and need to be integrated into a unified representation.

13.1 LOGIC PROGRAMMING

Early expectations in AI of 'mechanised thought' spurred by advances in automated chess playing and mathematical reasoning were unrealised but certain key developments were instrumental [McCarthy 1961]. The central concept linking all these developments was that intelligent behaviour could be represented by a computer program as a model of human reasoning. The importance of symbolic processing is based on the Physical Symbol Hypothesis which states that symbol manipulation is a necessary and sufficient condition for general intelligent behaviour [Newell et al 1958, Newell & Simon 1976], i.e. intelligence is based on rule-based symbol processing. A symbolic representation of a situation in the environment involves a transformation to map the outside world and the internal representation of it. Formal logic holds a central place in AI, the simplest form of which is propositional logic. Logic essentially constructs formal representational models with the structure of the real world in terms of arbitrary conceptual objects (symbols) which have certain relationships between them. Inference engines are a major component of knowledge processing architectures such as expert systems and they employ logic-based deduction techniques. Knowledge representation almost exclusively uses a logical language.

There is a standard form of first-order predicate logic (Horn clause logic) that limits the conclusion of an implication to at most a single atom. Horn clause logic is computationally equivalent to the universal Turing machine. Atomic formulas which specify hypotheses, however, may be combined conjunctively: H_1& ... &H_n where H_i is the condition of the hypothesis. This has the advantage that logical deductions are carried out automatically. PROLOG is a logic programming language based on Horn clause logic and PROLOG rules are similar to context-free grammar rules in that they have one element in the consequent, and several elements in the antecedent [Genesareth & Ginsberg 1985, Davis 1985]. PROLOG allows facts about objects and relations to be represented as sentences in its formal logic syntax. The basic unit of the Horn clause is the atomic formula which states a specific predicate relation for a group of objects: Rel (x, y) where x and y are constants, variables or expressions such that x is related to y. Any formula of predicate logic can be transformed into Horn clause form by:

(i)   replacing $A \leftrightarrow B$ with $A \rightarrow B$ and $B \rightarrow A$;
(ii)   replacing $A \rightarrow B$ with $\sim A \vee B$;
(iii) replacing $\sim(A \wedge B)$ with $\sim A \wedge \sim B$ and $\sim(A \vee B)$ with $\sim A \wedge \sim B$;
(iv)   replacing $\sim \exists x P$ with $\forall x \sim P$, and $\sim \forall x P$ with $\exists x \sim P$;
(v)   use the conjunctive normal form with conjunction of disjunctives by replacing $(A \wedge B) \wedge C$ with $(A \wedge C) \wedge (B \wedge C)$, and $(A \wedge B) \wedge C$ with $(A \wedge C) \wedge (B \wedge C)$.

More complex facts can be expressed with the reverse implication between predications: $q:- (p_1, ..., p_n)$ represents that the truth of p_i implies the truth of q where :- represent 'if' relating conclusion:- (condition), e.g. rel1(x, y) :- rel(x, z)&rel3(z, y). Hence q is true if all of $p_1, ..., p_n$ are true. Truth value solutions follow via *modus ponens* directly. Both declarative facts and production rules can be encoded in Prolog [Dahl 1983]. A declarative representation may be encoded as—holds (assertion,state); a production rule may have the form—holds (assertion1, state1) if holds (assertion2, state2).

The PROLOG logic program consists of a set of such Horn clausal expressions. If there are no conditions then the relationship is asserted to hold unconditionally. Every variable is assumed to be universally quantified over the clause in which it appears. A list can also be indicated explicitly by enclosing the elements of the list in square brackets separated by commas. PROLOG encodes a set of beliefs or facts about the world and allows the computation to derive conclusions that are logically implied by those beliefs using the rules of inference.

A PROLOG logic program implements a path-finding approach to problem solving where the problem domain is represented as the initial state of the world, the desired goal as the final state of the world and the problem solver finds a path as a set of actions (subgoals) that transform the initial state to the final state. The PROLOG interpreter usually uses a depth first backward chaining search strategy to explore the search space, but PROLOG rule clauses can be run in forward chaining mode as production rules. Queries of the form: '$p_1 \dots p_n$?' will extract the logical reasoning sequences of the conclusion: 'p_i if $q_1 \dots q_m$'. It implements automated resolution theorem proving similar to natural language parsing as the inference search process is implicit in the language rather than requiring an explicit search program. Hence, inferencing and control are automatic and there is no need for the separation of the deductive component from the factual component as PROLOG can represent both inference rules and factual knowledge. Multiple inference paths may be possible and the theorem prover will try alternatives and automatically backtrack on failure. PROLOG can implement AND/OR graphs for decision-making. Different paths will often result in different binding values to the variables of the original goal. It is desirable to limit the search space and PROLOG uses clause ordering to achieve this. The cut mark! allows the user to control backtracking by pruning the search space for a procedure under certain conditions. The side effect is to freeze all choices made after the parent goal. Alternatively, a general metalevel control language such as MRS may be used. PROLOG has generally superseded LISP which was based on Church's λ-calculus. The λ-calculus was developed as an alternative to set theory for the foundation of mathematics. Unlike the Turing machine which is sequential, λ-calculus is an abstract model of computation that retains implicit parallelism. LISP is a functional language that is procedural in nature while PROLOG is a rule-based language that is declarative in nature. PROLOG allows tasks to be represented with [ ] and perform list-processing based on recursive functions similar to LISP. Several attempts have been made to combine functional LISP with logical PROLOG programming languages through semantic unification: POPLOG, LOGLISP but PROLOG can represent lists of objects recursively [Wah et al 1989]. PARLOG and Concurrent PROLOG are PROLOG version which allow parallel computation by adding special constructs for parallelisation [Davis 1985].

13.2 ROBOTIC PLAN GENERATION

Higher levels of control imply the need for decision making capability in terms of tasks and objects to be manipulated and this implies the need for plan generation. Plan generation is basically the process of specifying the ordering of actions to achieve desired goals. Plan generation is very much part of the control problem in that it generates the input to the control system as a sequence of commands to achieve the desired goals and monitors the plan execution for feedback. The artificial intelligence paradigm is based on storing knowledge as a set of symbols which are manipulated according to a set of formal algorithmic rules—it is based on the assumption that all knowledge can be so formalised, i.e. automated logical reasoning [Simon 1978, Post & Sage 1990]. Plan generation has been traditionally viewed as the generalisation of the programming language process in that the plan represents a possible solution to a problem through the execution of an effective well-defined procedure, i.e. planning is an algorithmic problem-solving process. The purpose of planning is to generate a sequence of actions to achieve given objectives and goals. Ultimately, plan generation may be regarded as a form of self-programming and this is a major goal in robotics to develop autonomous capabilities to machines. As we shall see, AI in its broadest context as cognition may be regarded as a sub-discipline of robotics as robotic implementation of AI imposes its most stringent test.

13.2.1 Plan generation as search

Plan generation as generated problem-solving may be characterised as a state-space search for actions which when applied to the outside world change it into a new goal state. Indeed, AI methodology has been characterised *in toto* as search, since much of it is concerned with the exploration of a range of possible actions or inferences in the pursuit of well-defined goals or solutions [Charniak & McDermott 1985]. A typical application is in chess-playing systems based on search and evaluation. Search involves producing a tree of possibilities of moves. Whereas humans use tightly focussed searches based on geometrical pattern recognition based on accumulated knowledge (a parallel process) in playing chess, the machine utilises brute-force search strategies (a serial process) [Berliner 1978]. Logical assertions may be used as symbolic structures to describe the state of the universe of discourse (the world model) with respect to the effects of all actions expressed as logical implications. First-order predicate calculus is usually used as the means of describing the world's state.

The problem space comprises a set of symbolic structures or states and a set of operators to change states. The initial world state represents the global representation of the problem to be solved. Each action generates new possible actions creating a branching search tree of possible states. The complete plan is a path through the search space from the initial to the final desired goal state representing a sequence of actions tracing a path from the root of the tree to the desired end node (the solution). The actions act as operators which provide the means for conditional operations to change the state of the world. The conditional operation 'if p then q else r' allows program branching into alternatives to form the search tree—the condition is state-dependent and specifies how the condition changes its state. Hence, operators by representing the effects of actions on the state of the world are expressed as logical implications based on rules of inference such as Robinson's resolution principle. Such operators comprise of three parts: the precondition list, the effect list and the body of the operator which defines the logical relations between the condition and the effect. In this way, preconditions decide the applicability of operators, effects determine the effects of their execution, and their bodies describe the means to achieve the effects in relation to the world model. An operator is applicable if the operator's preconditions hold in the world model. Planning is a search through the space of inferences to find a temporal sequence of operators such that the first operator in the sequence is applicable to the initial world model and the final operator produces a world model in which the goal is true, i.e. the world model is transformed through a series of successive states linked by the operators (actions or inferences). Preconditions must be satisfied before the goal can be attained. Each precondition is essentially a subgoal which is further reduced by further branching of the goal tree until the terminal goal state is reached. This process is a process of problem reduction or subgoaling using state-space search to find the sequence of preconditions that must be met before the goal can be achieved. It is a divide and conquer strategy.

This hierarchical approach to planning is characterised by simplified tasks which decompose into finer-grained detailed subplans. Subgoaling involves transforming task level specifications into robot manipulator level specifications forming a goal tree. The problem is recursively reduced to a series of totally ordered subproblems using the problem-reduction operators until a series of elementary motion primitives suitable for input to a robot control system or programming language compiler is attained. The plan is then a representation of actions for achieving the goal or task.

State spaces can be searched either in a forward direction from the initial state and applying operators to find a path to the goal state (data-driven strategy) or in a backward direction by starting from the goal state and applying inverse operators to find a path to the initial state (goal-driven strategy). Several methods exist to find a path from the root of the tree to the desired goal. The depth-first exhaustive search involves moving down each path from the root down the tree until the goals at the bottom of the tree are reached in turn until the desired goal is attained. If the required goal is not reached, the tree is backtracked until such a path is found. The breadth-first exhaustive search involves examining every node/state at one level before progressing down the the next level until the desired goal is reached. The depth-first and breadth-first searches are blind searches which are complete and exhaustive but the computation times explode exponentially with the dimensions of the goal tree [Aleksander & Morton 1987]. It was this problem of combinatorial explosion that led Sir James Lighthill to declare that AI systems would be totally ineffective. Chess-playing programs restrict the depths to which they explore and evaluate moves by using a scoring system [Michie 1973]. This cost functional can determine the optimal course of action:

$$C_f = C_g + C_h$$

where C_g = cost from initial node to candidate node, and C_h = cost from candidate node to goal node.

The best first search combines the two approaches of depth-first/breadth-first search with minimising cost functions which measure the distance between nodes of the goal tree to find the shortest path. The minimum distance between the candidate node and the goal node should be found as the optimal solution. The best first search allows switching between paths between depth-first and breadth-first approaches. At each step, the most promising node is selected. If one of the nodes chosen generates nodes that are less promising, another node at the same level is selected thereby changing the depth-first search into a breadth-first search. If the nodes are better than earlier choices, the search backtracks as necessary. This graph-traversing best first A^* algorithm can be used to find the minimum cost solution, i.e. a minimax principle [Fu et al 1987, Charniak & McDermott 1985]. It is a branch-and-bound graph search algorithm which uses the evaluation function to order nodes, combining lower bound remaining distance estimation with dynamic programming principles, and is effective at dealing with average case situations though it fares less well under worst-case scenarios. Minimaxing assigns one player a positive value and the other a negative value. Each node represents the state of the game. Values assigned to each node benefit either the maximiser or the minimiser. These search programs are similar to those used in games playing programs such as chess. The planner passes the nominal task plan as a sequence of executable primitives to a buffer for execution by the control system. The AI planner generally performs this planning process offline and monitors the task execution online with error diagnosis.

Some problems in a world model may be reduced to more than one problem reduction operator resulting in alternative subproblems. This generates AND/OR tree inference nets where some goals are satisfied only when all the immediate subgoals are satisfied (conjunctive AND nodes) and other goals which require only one of several subgoals to be satisfied (disjunctive OR nodes). The branching factors should be as small as possible to keep the AND/OR tree manageable. Each node represents a set of possible world states and each action is a transition from a set of possible world states to a set of possible resultant world states. The solution to the planning problem should again lead to the goal in a finite time and minimise the maximum path length. The AND/OR approach was used by Sanderson et al (1988) to generate sequential plans for assembly/disassembly sequence tasks in satellite servicing such as those exemplified by the Solar Maximum Repair Mission. The AND/OR graph represents the assembly sequences with cost functions based on entropy measures. The assembly product was taken to comprise several parts joined together in a stable configuration. This provides a natural representation of the hierarchical assembly process of pairs of parts and modular subassemblies [Nevins et al 1987]. A CAD database provided a geometric description of the assembly and its parts, their dynamic properties, geometric constraints, assembly relationships and parameters, and sensor-specific information. Such are needed to identify potential jigging and gripping surfaces, clearances and tolerances. Relational graphs were used to derive the relative positions and orientations of parts and subassemblies. The assembly planner used the geometric and physical constraints to derive feasible operational sequences of motion primitives using precedence requirements. The feasibility of sequences were based on contact and attachment relations according to a hierarchy of operational preconditions:

(i)   release of attachments;
(ii)   stability of subassemblies;
(iii)   separability of subassemblies.

The AND/OR graph provides a natural representation of feasible assembly sequences. Each node of the AND/OR graph corresponded to a configuration subassembly which is connected by operations represented as a relational structure. Arcs corresponded to disassembly operations which if reversed provided the re-assembly sequence. The assembly/disassembly of the product was regarded as a backward path search using the AO* algorithm in the state space of all possible configurations of parts [Fu et al 1987]. The AO^* algorithm is similar to the A^* algorithm modified to handle the AND transition. As the branching from the initial to the goal state (forward chaining) exceeded that from the goal to the initial state (backward chaining), a backward search was employed. Weights were assigned to arcs using criteria of operational complexity and subassembly degree of freedom to reduce the search space. Any task may be represented by:

Replace ⟨part⟩ *in* ⟨product⟩

(i)   *disassemble* ⟨product⟩ *until* ⟨part⟩ *is an independent substrate*
(ii)   *replace* ⟨old part⟩ *with* ⟨new part⟩
(iii)   *reassemble* ⟨product⟩

The assembly plan is input to the task planner which extends the assembly plan and produces a detailed task plan that describes detailed actions. The task plan comprises three phases: grasping which requires a list of grasp sites, collision-free trajectory generation, and fine-motion strategy planning for insertion. An assembly world model provides a runtime description of the current state of the parts being assembled, i.e. parts status and assembly completion status. The Solar Maximum Repair Mission involved the replacement of the main electronics box (see Fig. 13.1).

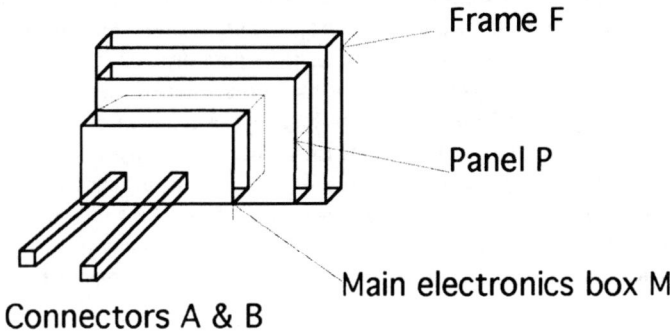

Fig. 13.1. Solar Maximum Repair Mission main electronics box assembly.

The solution tree to disassemble the satellite is found by the path search:

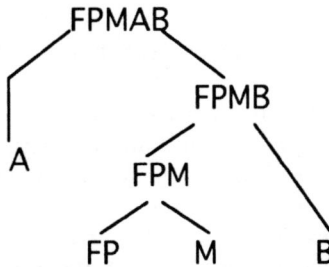

In general, automated process/assembly planning reduces the lead time, increases consistency and more efficient production of the final product.

The domain-independent GPS (General Problem Solver) system represented the states of the world as logical assertions of first-order predicate calculus. GPS used means–end analysis as a domain-independent heuristic to compare the current initial world state (the problem) with the desired final world state (the solution) and then search for procedural operators that reduced that difference [Charniak & McDermott 1985]. Means–end analysis was used recursive to incrementally bring the current state closer to the goal state. GPS used depth first search and forward chaining from the current state to the goal state using its 'find-and-reduce-difference heuristic'. GPS was the basis of STRIPS (STanford Research Institute Problem Solver) planner [Fikes et al 1972]. STRIPS has provided the basis for many subsequent planners and it has a central role in planning systems [Simon 1991]. The state of the world was modelled as a set of first-order predicate logical clauses. Primitive actions were coded as sets of routines in the form of predicate calculus. The values of logical arguments depended on the truth or falsity of the assertions. These formulas represented theorems about the state of the world. Each action was an operator schemata which acted to change the state of the world towards the final goal state by describing the actions in terms of preconditions and effects. The sequence of actions selected thus comprised a search for the correct sequence of actions to achieve the goal.

The operator was characterised by an ADD-list which listed the consequent changes to add to the world model by the application of the operator, a DELETE-list which listed antecedent deletions to the world model by the application of the operator, and a PRE-CONDITION that provided the context of logical formulae to be fulfilled so that the operator may be applied. The ADD and DELETE list effectively described the assertions that would be true after execution, i.e. the effects. These properties of each operator localised the effects of each operator. Plans were built from these operator schemata in an ordered sequence through the search space of intermediate states from the initial to the final goal state by transforming the world model from one intermediate state to another. The operators were essentially sets of condition–action propositions like production rules. The conditions defined the circumstances that must exist before the operator may be fired. The actions define the effects on the world and the body of the rule describes the decomposition of actions into subactions. The plan was a tree of nodes representing a hierarchy of actions and their temporal order down to the level of primitives. Propositions represented the initial state and goal state of the world and a library of possible actions.

Backward chain inferencing is generally the method of plan generation from the goal to the initial state due to the excessive fan-out of forward chaining. It ensures that the preconditions of each action in a plan are satisfied, or if not, it finds an action that has this precondition as its effect and places it in the plan. This backward chaining regression from the goal to the initial state is recursive. Backtracking may be chronological or dependency-directed. Chronological backtracking involves making a choice at a node until a failure is encountered. On failure, the most recent choice is backed up to and an alternative is selected until a failure is encountered. Dependency-directed backtracking backs up to the choice responsible for the failure rather than the most recent one. This process is valid for search spaces which are nearly decomposable into independent subgoals. A sequence of actions transforms the initial state of the world through a series of intermediate states into the goal state. Resolution theorem proving generates subgoals from the preconditions of the operators. The specific plan as a sequence may be generalised by replacing the parameter constants by operator variables and 'chunked' and stored for future use as a single macrooperator. Thus a repertoire of possible actions could be built up as fixed sequences of simpler actions.

STRIPS was integrated with an execution supervisor PLANEX to monitor plan execution at run time to allow for modification of the plan. Positive surprises occurred when some part of the plan was no longer required and so could be omitted, while negative surprises involved the failure of executed parts of the plan and so required either re-execution of the failed part of the plan repeatedly or replanning for corrective action. Some tasks may be undertaken on failure by repetition of a sequence of actions until it succeeds. This iterative type of planning revolves around a 'while-do' program statement which repeats an iterative sequence until a certain condition is verified. Replanning involves the substitution of new solutions to the failed plan. Davis & Comacho (1984) implemented a version of STRIPS in PROLOG. The chief problem of STRIPS and all its variants is that they are capable of dealing with only simple planning problems. Problems such as the Towers of Hanoi problem which require overcoming obstacles to goal reduction are not soluble by these methods based on reducing searches. The Towers of Hanoi problem comprises three pegs and a set of different-sized discs stacked on one of the pegs in decreasing order of size. The problem is to transfer all the discs from one peg to another with the constraints that:

(i)   one disc is moved at a time;
(ii)   the larger disc cannot be placed onto a smaller disc.

The environment is constantly changing and reasoning in the real world must reflect this. STRIPS and similar planning programs suffer from the 'frame' problem which arises out of the dynamic nature of the world [Hayes 1973]. In the process of planning, each action changes certain specific features of the situation. The frame problem arises from the need to represent those aspects of the world which remain invariant during state changes. The difficulty is in keeping track of the consequences of performing actions and the required alteration to the world model representation in real time. In addition, as the world model becomes more complex so the computation time required to update it explodes exponentially. It is necessary to avoid re-inferring assertions for each new instant of time. Since successive states of the world are logically independent, the difficulty is in deciding which statements continue to be true and persist after the performance of a given action. Indeed, generally almost all the assertions will continue to be true after the action has been applied as most of the world is unaffected by events as they tend to make only localised changes to the world. It is necessary to state that actions change only features of the situation to which it directly refers. STRIPS localises the effect of actions so that almost all the statements continue to be true to the next state unchanged unless explicitly eliminated. One possibility is to include 'always true' frame axioms to explicitly represent the invariants but this limits flexibility and nullifies the inherent logical correctness of the planning process. Furthermore, the number of such frame axioms becomes very large if the world model is complex. The performance of actions must qualify what does not change when actions are performed through frame axioms. This also introduces the possibility that frame axioms may become invalid as they will vary from the type of state change. This approach was used in STRIPS and many subsequent planners. If the components of the world interact, an action can have a global effect on the world, e.g. if a block is withdrawn from a stack, the whole stack falls. Hence, it is important to note interactions and relationships between the world's components.

SHRDLU was a planning program similar to STRIPS designed to demonstrate natural language communication which operated in a limited 'blocks' microworld of simple 3D objects such as cubes and pyramids on a flat surface in an attempt to tackle the problem of context [Winograd 1980]. In moving the differently coloured and shaped blocks around with a simulated robot arm, SHRDLU was forced to devise plans to generate the required configuration of blocks. The Microplanner program directly encoded facts about the world as procedures that operated on the world model. Its MOVER procedure was based around commands of the form: PUT (blockname_1) ON (blockname_2). SHRDLU planned a sequence of actions to move block-by-block to obey the command. It had at its disposal 11 subroutines such as GRASP, PUT, CLEAR, MOVE, etc. Each subaction in turn was made up of a sequence of more detailed subactions until the level of servocontrol was reached. Futhermore, before any command could be executed, certain preconditions had to be satisfied so the program compared goals and subgoals with a set of postconditions associated with each action which specified the configuration of the blocks world that the action should produce. If the configuration satisfied the condition of the goal or subgoal, the action became part of the plan. The MOVER goal tree was represented as an AND/OR graph such that some goals were satisfied only when all the immediate subgoals were satisfied (AND nodes) and other goals were satisfied if only one immediate subgoal was satisfied (OR nodes). The entire process of constructing a sequence of actions from the goal state back to the state where all the precondition actions necessary to initiate this chain were satisfied was a backward chaining process as adopted in STRIPS. The plan to achieve the goal was the reverse of the sequence of actions formed during the search. This illustrates the hierarchical nature of planning motor actions. The simulated microworld was limited, however. The blocks world was characterised only by geometric shapes and dimensions. The central problem of the perceptual processing interface between the real-world objects and symbolic manipulation was not addressed. The real world is far more complex, uncertain and dynamic than the blocks microworld.

Another important problem is that of the conjunction of interacting subgoals whereby, if any one of the subproblems is solved independently, the solutions to others may be destroyed [Chapman 1987]. Some goals are conjunctive goals which satisfy several constraints simultaneously. The traditional strategy for achieving goals P and Q is to first derive a solution to achieve P first then modify it to achieve Q also. But this will not necessarily work as actions can have side-effects (variant on the 'frame problem'). Subgoals may also have different priorities which may also in turn change according to the situation. Motor actions may have long-term consequences which may affect future motor actions. It is often necessary to achieve several goals simultaneously particularly in complex tasks. Such goal interactions are ubiquitous in planning. One strategy for dealing with this is to reach subgoals independently and then treat interactions as they arise. Re-ordering of subobjectives will eliminate the interaction problem in many cases, e.g. Sussman's HACKER which was based on SHRDLU. HACKER achieved this through a form of trial-and-error learning. It broke down goals into subgoals while critics monitored the planning and debugged the programs using rules to avoid known errors. The debuggers detected subgoals that had been performed and then been undone. These subgoals were then avoided by reordering. This method is inefficient when many subobjectives occur.

WARPLAN and RSTRIPS used objective regression to generate successive partial solutions which did not interfere with other goals all ready achieved but allowed further refinement during execution, i.e. previous goals were protected [Camarinha Matos 1987]. If conflict arose, a new operator was inserted into a previous plan. This did not optimise plans, however. Adaptability in the control of problem-solving behaviour is characteristic of intelligent agents such that control is used to adapt planning behaviour to solve the problem situation. A dynamic planning strategy generates partial plans for real-time problem solving which are successively refined or replanned. The system must adopt the plans relevant to the changing situations. While problem solving is implemented, plans must evolve during the problem-solving process. By adopting the notion of least commitment choices, plans are deferred to lower levels of the hierarchy until a decision is required to be made so that successful planning and execution of the task may continue. This least commitment strategy of partial plans is the best approach for dealing with conjunctive goals. Decisions of ordering are taken only when interference with other plans will definitely not occur. General goals are addressed first with details being filled in later [Hendler et al 1990]. Hence, planning may be linear whereby the sequence is totally ordered or nonlinear whereby the sequence is partially ordered. This method of partially ordered nonlinear planning reduces the necessity for backtracking and allows concurrent actions to be performed in parallel, e.g. NOAH which was the first nonlinear planner that used partial ordering on subgoals until execution time at which point the complete plan with total ordering must be decided—this approach was characterised by several later planners. The NOAH system saved the state of the solution at each point where there were alternative actions and kept a record of alternative choices. NOAH was a planner that used a procedural network to describe relationships between actions so that the search space is not a set of world states but a space of partial plans. Declarative knowledge represented the effects of planned actions and procedural knowledge of the domain represented the actions of the operators in generating subgoals. The procedural network was a hierarchical system to cope with the complexity of such planning with different successive levels of representation such that goals were successively expanded into more detailed subgoals down to robot-level control operator primitives. It was similar to STRIPS but included a table of multiple effects to detect interactions between goals.

Camarinha-Matos (1987) stressed the need for reasoning about resources such as time and abilities since a major source of conflict is competition for the use of scarce resources. This implies the need for priority considerations. ABSTRIPS assigned a level of criticality to each precondition so that plans were generated in descending order of criticality with emphasis on those critical for the mission. By assigning levels of criticality to each precondition, those of maximum importance to the plan's success were reached before less critical subgoals.

Many planners work entirely in block worlds, e.g. HACKER, NOAH, and even reach a high degree of sophistication, e.g. BUILD. Fahlman (1974) used the block world as an internal world model for a planner system BUILD for sophisticated construction of blocks into various erections by a one-handed robot. The BUILD program was based on heuristic search procedures and operated on the blocks world to produce a properly sequenced plan for converting the present state of the world to a goal state. It was capable of constructing sets of subassemblies of blocks and by grasping the supporting block could transport the entire subassembly to the desired point. It was also capable of using extra blocks as temporary supports, simple scaffolding and counterweights. The basic command was MOVE (blockname) AT (current_location) AT (new_location). The top-level goal was BUILD which assembled a plan from the current situation to the final desired situation as a properly ordered series of PLACE situations through backward chaining similar to SHRDLU. Block worlds are poor representations of the real-world. ABSTRIPS is a real world planning system and can modify precomputed plans to account for deviations in the real world from the model. Unfortunately, blocks worlds are insufficient to describe natural environments such as planetary surfaces encountered by rovers, but may be sufficient for modelling in-orbit servicing scenarios.

The advantage of an hierarchical representation is that it allows a distinction between goal priorities and strategic level planning. Allen & Litman (1986) advocated the use of metaplans, i.e. sets of plans about the planning process itself independently of the domain itself. The metaplan defines constraints such as resource limitations and the contexts of applicability of plans. Metaplans include the plan as a knowledge domain in itself, information concerning the introduction of plans, the execution of plans, the abandoning of plans, etc., independent of the plan domain. Such metaplanning allows the planner to reason not only about the goal but also about the various techniques available for plan generation. Goals (metaplans) are given priorities with the highest priority goals being considered first. Metaknowledge (knowledge about knowledge) may be implemented independently of object-level knowledge. Metaknowledge in the form of metarules to guide and constrain searches of knowledge bases of object-level statements make effective use of heuristics. Chunks of metaknowledge may be embedded into the object-level knowledge as objects, using the object-level rules to communicate between the metalevel objects (object-level programming). The metalevel representation can devise proof strategies, control inferencing and increase the expressive power of knowledge and implement control knowledge (heuristics). Metaknowledge increases the power of inferencing and reduces the length of object-level proofs. This allows the procedural use of declarative knowledge. Metaknowledge provides the means for chunking object-level sequences into metalevel objects which do not need to be recomputed as a limited form of learning. Weybrauch (1980) described FOL (first-order logic) as an expert system whose expertise was reasoning, i.e. a meta-reasoning system.

As the world changes over time, it is necessary for the planning system to represent time. Time is a major problem for planning systems that regard the world's evolution as a sequence of states with actions causing state changes in the world. A plan is essentially a representation of some aspects of the future. Most actions take time and several can occur at once. Time specifies when conditions should be achieved and for how long they should be maintained. However, many temporal constraints are imprecise due to their relative nature, a lack of a clear relationship between temporal markers and different temporal scales. Planning of activities occurs in a temporal universe with deadlines and external events. Each activity is defined by a temporal window with a start time and a duration. Since much temporal knowledge is relative rather than involving absolute dates and times represented by temporal coordinates, Allen & Litman (1986) introduced a temporal logic based on the temporal interval to accommodate inexact relationships between intervals through a disjunction of primitive ordering relations such as $X < Y$ (X before Y), $Y > X$ (Y after X), $X = Y$ (X concurrent with Y), and X o Y (X overlaps Y). This allows the formation of a constraint network comprising directed acyclic graphs of nodes representing interval connected by arcs which relate the intervals. A hierarchical structure offered a means of dealing with varying time scales. The time order respects the subgoaling hierarchy.

DEVISER is a 4000-line INTERLISP planner designed for Voyager spacecraft mission sequencing (though adaptable to different missions) which adopted a sophisticated notion of time to generate partially ordered plans and it was based strongly on NOAH [Vere 1983]. It generates parallel plans to achieve goals with time constraints imposed. Similar to STRIPS, it implemented productions with preconditions, a delete-list and an add-list. It generated plans by backward chaining from unordered subgoals. Conflicts are resolved by ordering formerly unordered plans. For instance, subgoals which were parallel may be resolved sequentially. The syntax may be given by (believe A S I) which means that agent A believes sentence S to be true during the time interval I. Actions occur during intervals and these intervals are ordered. Every perception of the environment, every internal inference and every command to the actuators is given a time stamp. Typically, windows provide an upper and lower time bound for activities. They may be explicitly specified or computed dynamically during plan generation for scheduling. DEVISER simplifies the process of conflict resolution. The default window is the 'anytime' window with an eternal duration. A variant on DEVISER is Spaceworld which provides an environment for which an autonomous spacecraft such as Voyager can schedule photographing objects in deep space and transmitting the information to Earth. MARVEL (multimission automation for real-time verification of spacecraft engineering link) has been used on Voyager since 1989 to monitore spacecraft telemetry to detect departures from expected behaviour to reduce operator workload.

13.2.2 Plan generation using expert systems

Planning viewed as a search problem suffers from certain problems. The search space can become large and grow exponentially and searching for a path can involve exponential computation. Furthermore, finding an optimal plan in even a simple blocks world is an NP hard problem and so fundamentally intractable due to the conditional dependency of effects of actions on the input situations and the frame problem of discovering which propositions are unchanged by an action and derived side-effects. When search trees become large, the computational cost of solution may become impracticably large. Chess playing is a simple, closed-world planning problem that is a formal logic system amenable to 'mechanisation'. All possible moves from a given starting position are searched to a given depth (number of moves ahead). The evaluation of possible paths through the search tree is done through a scoring system based on the value of each piece and the threat to each piece that can be captured. The search program attempts to maximise its score. To calculate the score for every legal move and the opponent's possible responses even at low depth becomes impossible: the average number of alternative possible moves is 35 and an exhaustive search that looks only three moves ahead requires a search through $3^{35} = 2 \times 10^9$ possibilities and for a whole game $\sim 10^{120}$ choices. This is similar to the 'travelling salesman problem' (TSP) who must find the optimal route between n cities, visiting each only once—indeed, TSP is algorithmically fundamental to the whole class of NP problems. As there are $n!$ possible combinations of routes to search, this combinatorial explosion becomes prohibitive even for small n.

One way to direct the search pattern in promising directions is by using informal heuristics to constrain the search and increase the search efficiency. These heuristics are vague truisms and are used to generate weak forms of implication 'if p then q'. This conditional statement expresses a causal relationship between the antecedent and the consequent but the consequent does not necessarily follow from the truth of the antecedent, leading to conclusions of uncertain validity. However, heuristics can provide enough structure to allow the construction of tractable domain-specific planners for certain domains, i.e. in well-behaved worlds. For the travelling salesman problem, a city is selected and the nearest city to it is selected, and this process is repeated until all the cities have been visited. This heuristic reduces the search time from $n!$ to n. Although an optimal solution will not be found, this method yields a satisfactory solution (satisficing). Such heuristics or task-dependent 'rules of thumb' effectively prune the search tree by excluding unlikely routes and reduce the computational overhead by concentrating on more plausible paths. This is essentially a 'beam' searching approach enabled by constraint propagation by the heuristics. The $\alpha\beta$ pruning process discards whole chunks of the search tree. Heuristic knowledge can direct search patterns towards the most promising routes to improve the problem solving or planning process and cope with its NP hard intractability. Modern chess-playing programs use heuristics to select a few of the most promising routes to a greater depth. In this way, domain knowledge may structure the planning process.

The search process may be considered to be a 'hill-climbing' process such that the program with inputs λ_i and outputs $E(\lambda_i)$ attempts to find the direction of maximum gradient ascent $\partial E/\partial \lambda_i$. But such methods are suited only to solution spaces with smooth topologies. Heuristics help to isolate possible islands ('mesas') without exhaustive calculation. However, heuristics tend to find only locally optimal solutions rather than global optima. This is adequate if such local optima are deep but if the solution space is fairly flat (the 'mesa phenomenon' [Minsky 1961]) such hill-climbing is inadequate. Domain-independent planning must be relaxed and is inadequate for real-world planning due to the exponential nature of path finding through a solution space. The real world involves actions having several possible outcomes depending on factors beyond the agent's control or knowledge. All changes made by actions must be represented as postconditions—many actions cannot be formalised in this way since they require logically deductive generation of propositions from the postconditions. The use of knowledge in domain-specific planning becomes necessary. There is a space–time tradeoff in using heuristics since they require memory space. An extension to this notion is the expansion of heuristics through the implementation of the expert system using a domain-specific knowledge base.

AI has undergone a paradigm shift from power-based inferential search techniques to knowledge-rich strategies which offer a means to reduce the search requirements. This shift represents the Knowledge Principle whereby search and reasoning alone is regarded as insufficient for intelligent behaviour—a great deal of knowledge of the world in which the agent operates is required [Lenat & Feigenbaum 1991]. Knowledge is one of the prerequisites for intelligent behaviour and in effect represents the agent's learning history. Much of this knowledge should be sufficiently general to enable its use for multiple but specific domains (the Breadth Principle). Such knowledge imposes descriptions of regularities in the world including declarative maps, planning constraints, etc. Human thought is a pragmatic device based on factual content rather than logical syllogism, i.e. it is knowledge-driven. Mental models are context-dependent with conceptual knowledge forming the central utility in high-level cognition. This is model-based reasoning which draws heavily on world knowledge rather than logical inferencing and assumes that the mental models correspond to the real world. Systems composed of a few inference rules but with a lot of declarative knowledge present a very powerful process due to the large combinatorial number of ways facts can be combined using those rules. This knowledge-based approach to problem solving encodes programs that describe real-world relationships through constraints and rules to provide solutions by outlining sequences of actions while utilising a reasonable expenditure of resources.

Expert systems are programs that solve problems normally requiring the knowledge and skills of human experts [Hayes-Roth 1984, 1985a, b, Hayes-Roth & Jacobstein 1994, Klisker 1986]. Essentially, they model human decision-making processes through the use of knowledge bases. Knowledge bases differ from databases in that databases contain numerical information whereas knowledge bases store knowledge in a compact higher level of representation (typically symbolic) of multiple data sets. The knowledge base acts as the agent's model of the world, albeit incomplete. Problems are described using objects, attributes and relations. The objects are data structures which represent knowledge about conceptual entities. These objects can be organised into hierarchical classes and each class has attributes associated with it defining its characteristics. The objects are organised to show the relationships between them. The knowledge base loosely represents stable long-term memory (LTM) and they are limited to specific problem domains. Real-world data in a dynamic global working memory database (loosely correlating to short-term memory (STM)) must be matched against knowledge stored in the knowledge base. Events comprise the inputs to the working memory which is acted upon by an inference engine. This requires the expert system to search the knowledge base for knowledge that matches the data using heuristic control procedures encoded as rules of 'good judgement' to avoid the combinatorial explosion by narrowing the search. The inference engine is independent of the knowledge base and emulates deductive behaviour through logical procedures for searching and deriving conclusions from the sets of facts in working memory. The inference engine interprets the current state of working memory and applies the relevant logic procedures which specify the changes to be made. It uses predicate logic to construct deductive proofs to make inferences. Hence, working memory functions as the world model containing all the task relevant data representing the state of the world containing operating knowledge (from the knowledge base) and environmental knowledge (from the sensors) (see Fig. 13.2). There are several types of knowledge that are used: knowledge about objects, knowledge about actions and events, knowledge about performance, and knowledge about internal procedures (metaknowledge).

Production rules for state representation are particularly suitable for use in robot planning and are the commonest form of logic representation [Charniak & McDermott 1985]. A production system consists of a knowledge base containing all the task-relevant information in the form of a set of production rules. Production rules represent relatively independent chunks of knowledge and are used by the inference engine to infer solutions to problems. Production rules define relationships between the rule's terms through

Production rules	KNOWLEDGE
input → **Working memory**	**Declarative facts**	BASE

data ↑ ↓ update

| **Rule interpreter** | **Rule/data selection** | INFERENCE ENGINE |

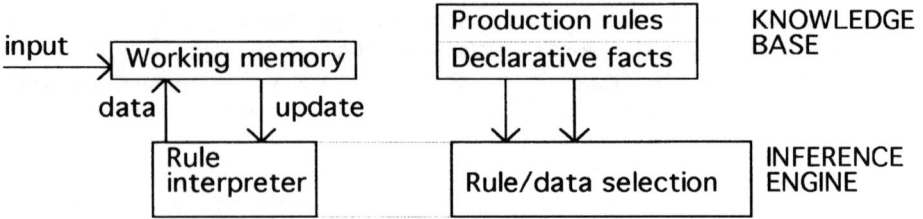

Fig. 13.2. Knowledge-based expert system model.

implication. These rules consist of one or more antecedent precondition statements linked to one or more consequent action statements (condition–action rules). They essentially model plausible hypothesis–conclusion inferencing logic for problem solving through deduction. Production rules are a subset of predicate logic with the addition of prescriptive components to indicate how the rules are used during reasoning. They can be used to model routines with particular input–output characteristics compactly as stimulus–response behaviours. Clauses may be combined using Boolean logic operators. The rules may be represented as: if $\langle$condition $P_1 \dots P_n\rangle$ then $\langle$action $Q_1 \dots Q_m\rangle$. The condition is usually the conjunction of a number of predicates while the action is usually single predicate. They provide a means for making hypothetical statements about the world. They incorporate practical conditional if–then knowledge and can solve reasonably complex problems by combining the rules into sequences and indicate explicit lines of reasoning. The precondition statements determine the applicability of the rules in terms of the conditions that must be true for implementation of the rule. The action statements are the operators which determine the effects of the action if the rule is applied to the world model. Production rules make the assumption that intelligent behaviour may be formalised in this way by viewing the world as logical elements independent of their context. Inferencing may be accomplished by forward or backward chain searches. Forward chaining utilises existing data to generate further declarations by triggering all the relevant 'if' clauses that are true and continuing until a goal is reached. Backward chaining begins at the required goal and identifies the consequents that are required to satisfy the goal, generating subgoals. Expert systems may be both event-driven and goal-driven or equivalently they may be used in forward chaining or backward chaining mode. Event-driven behaviour occurs when data or events in the working memory produce a matching antecedent to invoke the rules (forward chaining). Perceptual productions comprise of sensory stimuli conditions and actions which operate in forward chaining mode to invoke activity from stimuli to action. Goal-driven behaviour occurs when expectations or goals are used to match the consequences to control rule execution (backward chaining).

The production rules effectively embody the inference process in a specialised programming language such as OPS5. They may operate similarly to deductive logic to derive new facts as well as actions. Given the truth of the antecedent, the conclusion may be inferred which in turn may match the antecedents of other rules generating a chain of inference. Hence, the system can effectively form an inference net of multirooted AND/OR nodes. If the antecedent is satisfied, then the consequent which may define an action or a conclusion is applied. Hence they may model deductive reasoning and perception–action behaviour. Thereby production systems can model inferences from specific observations, generalisations and categorisations of data, necessary and sufficient conditions for achieving goals, and the consequences of hypothetical situations. If the antecedent can be satisfied then the consequent which defines an action or a conclusion is applied. If the consequent is an action, the action is executed while if it is a conclusion, the conclusion is inferred.

In a production rule-based expert system, a rule interpreter (inference engine) decides which rules are satisfied by the data in the working memory database. Domain-independence is achieved by separating the domain-specific knowledge from the mechanisms of applying the knowledge, i.e. the domain-specific knowledge lies in the knowledge base while the inference engine uses general techniques. When all conditions in the rule are satisfied, the rule is triggered to perform the action specified. If more than one rule is satisfied and contradiction occurs, conflict resolution priority strategies are employed: specificity—use the most specific rule before the most general, i.e. the rule with the most constraining conditions; strength—use the rule with the greatest certainty factor; recency—use the most recently fired rule; goal-orientedness—use the rule that satisfies the highest priority goal. The global dynamic working memory stores temporary assertions from earlier rule-based inferences and these are changed by data entering in from the environment and by the execution of the conclusions from the firing of matched rules. The working memory contains the state of the problem solving process, i.e. it is a common world model data structure capable of simulating the planned actions. An iterative cycle is performed by identifying the rules that are relevant to the problem followed by applying one or more of those rules to solve the problem. On every cycle the conditions of each production rule are matched against the current state of the working memory. One or more of the matching rules are selected for application and fired. Its actions change the state of the working memory thereby enabling new production rules to become applicable on the next cycle.

Production rules are essentially active demons that are invoked by incoming data in the working memory. Hence, the process is data-directed such that satisfaction of the condition part ensures that the action part will follow which updates the working memory. Since the production rules are self-contained modules, the rule base can be incrementally added to in order to accommodate new knowledge up to a few thousand rules. Production rules often have numerical weights assigned to them to quantify the level of belief in the consequent given that the antecedent is true. These weights (certainty factors) comprise numerical confidence ratings which quantify plausibility and are the basis of heuristic control to guide the program search for plausible inferences. The certainty factor is a measure of the association between premises and actions to model uncertainty. The basis of determining uncertainty is probability theory either through Bayesian statistics (usually) or Dempster–Shafer theory of plausible inference (rarely as it is complex to compute) to give a truth value as a probability on the real unit interval [0, 1]. Alternatively, fuzzy logic may be used based on possibility theory [Tong 1977, Maiers & Sherif 1985]. Most often, degree of certainty is quantified in terms of probability via Bayes theorem which determines the likelihood of an occurrence, e.g. PROSPECTOR, a mineral exploration expert system [Duda, Hart & Nilsson 1976].

The rule interpreter invokes pattern matching to copy the rules into the working memory. For more sophisticated production-based expert systems the working memory may accommodate different lines of abstraction. The use of cooperating specialist rule bases is an extension of the distributed paradigm to the expert system and this allows multiple conclusions to be conflicting. Metarules can provide heuristics adaptive control of resource-limited systems by expressing preferences and priorities in rules.

Models can be constructed in the working memory to represent candidate solutions and rules are executed when their hypotheses correspond to expectations. This allows relative focussing of attention. The diagnostic mode uses backward chaining whereby they work backwards from an hypothesis to test the antecedent cause through a plausible chain of inference, e.g. for medical diagnosis of bacterial infection (MYCIN), mineral prospecting (PROSPECTOR), electronic component troubleshooting (ACE) or organic molecular analysis (DENDRAL). Such expert systems are designed to capture world-class expert knowledge which takes the human expert some ten or more years to achieve. Other successful implementations of rule-based expert systems include the DARPA (Defense Advanced Research Projects Agency) ALV (Autonomous Land Vehicle) which is an autonomous mobile vehicle for battlefield operations. NAVEX is an expert system for the interpretation of space navigation data. The general characteristic is that such expert systems have limited specialised domains of application.

The plan generation process involves the use of a search tree from the initial world state and the production rule represents the operators. Each production rule whose antecedent is matched by the current state the generates a branch or possible course of action with the new state determined by the consequent of the production rule. This process continues and expands until a state is reached that matches the desired goal conditions. Hence, the search generates a plan of action from the start node to the goal node. Moore et al (1987) discussed the use of expert systems for real-time process plant control while Drabble (1991) advocated AI techniques for spacecraft control and automation of mission operations using knowledge-based planning by expert systems. The spacecraft itself and the environment in which it operates are comparatively well structured (no unpredictable free agents around, i.e. people) so it is well suited to automation to reduce the human workload of spacecraft monitoring. Indeed, AI techniques have been used for the automation of fault isolation and diagnosis on the Space Shuttle thruster reaction control system [Georgeff & Lansky 1986].

Pattern matching that is used by the rule interpreter is a basic operation in symbolic processing and around 90% of execution time in production-based expert systems is spent in this phase. It proceeds through unification which is a form of bidirectional pattern matching used in logic programming and it determines whether two terms can be made identical by finding a set of substitutions for variables in terms. Hence, it is highly desirable to perform such pattern recognition through hardware. Widely varying field length and data uncertainty precludes exact pattern matching so best matching must be used. Neural networks are the best performers at best matching through association. A major bottleneck also occurs in knowledge base retrieval due to input/output limitations. There is thus a trend towards hardware inferencing machines and software knowledge bases.

The chief difficulty of production rules is that they are inflexible to iteration and recursion. They are incapable learning as they lack the means to perform induction to abstract properties from examples. They are difficult to modify or extend while maintaining consistency (truth maintenance). As with all domain-dependent expert systems, they are brittle and limited to their specialised universe of discourse and tend to fail dramatically in problems outside their area of expertise and in unforeseen circumstances.

There are two main forms of knowledge representation commonly used in expert systems, frames and production rules, of which the latter are the most popular [Nau 1983]. Frames are schematic, modular data structures for representing generic concepts concerning objects, classes of objects and situations. Frames provide a natural representation for declarative knowledge (knowledge about something) while production rules are usually used for procedural knowledge (knowledge about how to do something) [Fikes & Kehlar 1985]. Frames and production rules may be integrated into a hybrid representation scheme so that frames can represent procedural action-oriented knowledge. Production rules can be attached to frame slots as values to be invoked as demons when the frame is invoked. This ability to attach rules or rule classes as demons to frame slots provides great flexibility in controlling reasoning, particularly for reactive functions. Conceptual dependency structures can be built with groups of production rules appearing in different concepts. Indeed, this type of representation has been under development as an intelligent aid to human satellite operators in the expert diagnosis and correction of satellite malfunctions (STAR-PLAN).

The task planning system must have a description of the objects to be manipulated, the task environment, the robot capabilities, the initial and desired states of the world (world model). The world model is altered by both internal problem solving processes (reasoning) and observations of the external world. Observation involves adding new data and axioms representing events to the set of logical statements representing knowledge from the knowledge base. A geometric description such as CAD can provide object information and information about the robot capabilities while other information must be provided through implemented knowledge bases about the environment in which it acts. A CAD database is an object library prepared offline containing data about objects represented graphically as constructive solid models formed by combinations of primitive objects [Rembold & Dillman 1985]. It can generate information concerning geometric dimensions of a metrological and topological nature (size, shape, adjacency relations) and physical characteristics such as inertia, mass, friction coefficients, tolerances, etc. This information can be used to infer relationships between objects such as attachment points, stable positions for grip and approach/depart vectors. The dynamic world model may be represented as a hierarchical semantic network structure which may be updated by correct insertion into the structure of sensory data. The semantic network provides functional relations between semantic entities and objects and task primitives. Goals may be defined by adding nodes representing the robot with the additional relations shown in Fig. 13.3. The structures represent the state of both the environment and the robot. Expert systems and CAD databases require a common language to transform computer-readable CADinformation into expert system compatible data structures and this requires a high-level interface which can transform CAD solid geometry models into relational graphs [Flynn & Jain 1991].

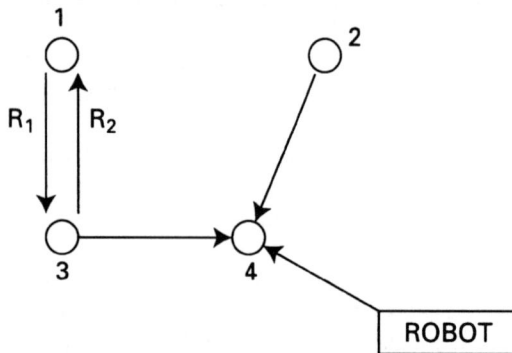

Fig. 13.3. Goal determination.

13.3 BEHAVIOUR-BASED REFLEXIVE CONTROL

Theorem proving and logical inferencing have in general been insufficient in providing human level capabilities in machines particularly concerning low-level skills such as pattern recognition, learning and robust behaviours so ubiquitously exhibited by the animal kingdom—typically functions best suited to connectionist approaches. Animal behaviour is highly adaptive to the changing environment. Even with 1–100 Mips hardware developments, such has not been achieved. Traditional planning systems based on logical reasoning with world models exhibit brittleness and slow performance. A typical mobile rover implementing AI planning takes up to 10 minutes to act on an image frame. They cannot perform adequately in new situations and their learning capabilities are limited. The problem is that the outside world is only partially observable and undergoes continuous change. Particularly important is the ability to cope with unexpected events which cannot be anticipated but which must be dealt with by concurrent contingency routines. Planning is inadequate in situations where events occur unexpectedly. Recovery usually requires replanning and in a dynamic environment this will be required almost continuously. Autonomy requires a variety of different behaviours. This is the Animat approach to robotics which simulates biological approaches by focussing on complete systems [Wilson 1991]. This is opposed to the Artificial Intelligence approach which is based on the Physical Symbol Hypothesis. Such symbolic techniques are inflexible by themselves in coping with unforeseen circumstances and are brittle in the face of deviations in the real world [Beer et al 1990]—much adaptable behaviour in the real world does not involve logical reasoning.

Robot behaviour may be categorised as moving between points in an environment whilst avoiding collisions with objects. Animal behaviour rather than human reasoning may be more appropriate as the basis for developing machine autonomy. Agre & Chapman (1990) outlined their plan-as-communication view in which plans are merely one resource amongst many used for improvisation. This views the agent as performing goal-directed activity through participating in the world rather than controlling it through planning. They outlined their theory of activity that routine activity is independent of plans. Their system PENGO used contingencies to improvise moment to moment in constant interaction with its environment. These contingencies were patterns of directed activity routines without building world models. It did, however, use visualisation of a series of scenarios. It regularly invoked and interleaved its repertoire of activity routines. The complexity of its actions reflected the complexity of the environment which essentially combinatorially increased the action possibilities. As a dynamic system, the mobile agent indulges in motor interaction with the environment and its sensory feedback is a product of both the environment and its previous actions. This is very similar to reactive behaviour-based methods. Learning took place through the agent's interactions with the environment through the incremental evolution of a dependency network of rules which converted perception to action similar to memory-based reasoning.

Brooks (1990) proposed a 'nouvelle' AI paradigm called the Physical Grounding Hypothesis such that autonomy requires representation to be grounded in the physical world since the world is its own best model—the incompleteness of simulated world models can miss important parameters. The upshot of this is to couple perception and action directly without an intervening world model and to minimise internal symbol processing. This denies the linear sense–think–act cycle philosophy of subdividing intelligent architectures into perceptual-central-motor subsystems. A single object can generate large variations in sensory signals. This perceptual aliasing problem is inherent in attempting to map highly variable sensory stimuli patterns onto the same internal symbolic representation. Perceptual classification is viewed as part of the sensory–motor coordination process. The robot agent can move around so sensory patterns will vary according to distance from the object, the viewing direction and illumination. Hence, only a small segment of the world is filtered by the perception process for analysis, and this perception process determines behaviour. Only a small part of the environment is relevant for behaviour. What is perceived is determined by the sense organs, the transduction process, and the process of learning to discriminate important data. Autonomous robots interact with the external environment and their behaviour is determined by the stimuli they receive as input. Physical embodiment is a necessary component of such interaction and it is through such coupling with the world that effective behaviour emerges. There is no central global control. Global communication arises through local interactions with the environment. Situated robotics provides existence proofs for the emergence of structured behaviours (symbols) through the interaction of primitive behaviours in the real world.

Anderson & Donath (1990) used multiple components of reflexive behaviours which operated concurrently and independently. They found that high-level global behavioural patterns emerge which are not ascribable to individual simpler components at lower levels. Control structures emerged from the distributed loosely coupled behavioural modules through parallel local interactions of a nonlinear nature. This organisation gave rise to more complex and irreducible behaviours. They used avoidance behaviours to cause motion away from a stimulus and attractive behaviours to cause motion towards a stimulus. Each behaviour was mapped onto a potential field where repulsive forces protected objects and attractive forces guided the robot towards specific locations.

Generalised wandering behaviour was an exploratory mode of behaviour. Feather (1967) suggested that intelligent behaviour is information seeking and that this provides the basis for curiosity as modelled by such wandering behaviour. It is through exploration that cognitive maps of the environment are constructed and stored as the basis for navigation in complex environments. There is little doubt that as well as propositional forms of representation, humans as animals utilise pictorial forms of representation through cognitive maps of spatial relations. Such spatio-temporal maps provide data on the availability of food, their location and temporal variations. On the basis of this information, predictive monitoring of the environment is possible and so is the derivation of behaviours that are likely to achieve success. Passive avoidance behaviour halted movement similar to the 'freeze' behaviour in many animals. Active avoidance behaviour avoided approaching objects similar to flight behaviour in birds. These two avoidance behaviours effectively model predator–prey behaviour. Location attractive behaviour was regarded as a component of migratory habits of some animals. Object attraction in the direction of an object simulates food-seeking or mate-seeking behaviour. Open space attraction for unobstructed travel is also characteristic of animals which use range data for this purpose. Mataric (1995) suggested that a basic set of behaviours comprises: wandering, avoidance, following, aggregation, dispersion and homing for individual and groups of mobile robots. The behaviours obstacle-avoidance and target-following imply discrimination ability. Each behaviour is selected according to environmental cues. The behaviours may be combined to generate higher-level behaviours. Wandering and avoidance generate safe-wandering. Wandering, avoidance and aggregation combined produces flocking. The addition of homing produces flocking towards a specific location. General foraging can be accomplished by combining different behaviours triggered by different conditions, e.g. temporal switching between avoidance, dispersion, following, homing and wandering. Hence, group behaviours can be generated by local environmental conditions. Additional basic behaviours for robotic arms could include grasping and dropping. Emergent behaviours such as wandering and navigation arise through coupling between each behaviour component. This coupling may be temporal or spatial. Spatial ordering involves concurrent activation of the component behaviours. Anderson & Donath (1990) held each of the primitive reflexive behaviours to be spatially ordered. The difficulty with spatially ordered behaviours is that they can cause conflict resulting in erratic behaviour. The total number of unique behaviours emergent from different primitive behaviours may be assumed to be a binomial function in that i possible combinations of n items taken k at a time is given by:

$$i = \sum_{k=0}^{n} \frac{n!}{k!(n-k)!}$$

Alternative orderings include hierarchical coupling of primitive behaviours by adding temporal ordering. Alternatively, behaviours may be weighted according to a learning algorithm. If temporal ordering is included through hierarchical coupling, the number of potential emergent behaviours will be increased. The behavioural modules are usually considered to be hierarchically organised such that each instinct is decomposed into finer-grained behaviours. Instincts at the same hierarchical level compete with each other.

Thresholding behaviours eliminates the possibility of chaotic behavioural responses. Spatially distributed ordering introduces the potential for cyclic behaviour due to the deterministic nature of mobile response and the inability to respond to events over several time periods. This is a consequence of a lack of memory which limits response flexibility when a reaction to an input may be dependent on previous inputs. Memory provides the capability of reacting to events over a number of time intervals and so to altering behaviour on the basis of previous behaviour.

Brooks (1986, 1989) introduced a subsumption architecture of hierarchical behaviour-based units to provide robust and reflexive robot performance with a priority organisation to avoid multiple activations of mutually exclusive behaviours (see Fig. 13.4). The problem is decomposed vertically into a number of levels of task competence based on behavioural units rather than functional modules. The behaviours as functional units comprise well-defined control laws which are goal-oriented. Each task module (typically implemented as finite state machines) is connected directly to the outside world via sensors and actuators and operate in parallel. The control system is built bottom-up incrementally and each level of competence includes the lower level as a subset. Simple tasks are solved first with new layers of competence added for greater complexity. This is an incremental approach. Individual layers can work on individual goals concurrently. Each level is implemented and debugged from the bottom up analogous to the evolutionary process of adding layers of greater behavioural complexity to animal brains to provide greater capabilities. Each level specifies a behavioural pattern which directly links perception to action. Level 0 comprises a Collide module which senses repellent forces around objects and a Runaway module which invokes avoidance behaviour when the repulsive forces exceed a threshold. With level 0 control only, a mobile robot remains stationary until a moving obstacle approaches it. Level 1 includes a Wander module to generate new headings periodically and an Avoid module which accepts a force vector input from level 0 and suppresses output from the Runaway module. With level 1 and level 0 control operating, a mobile robot randomly moves off in some direction while avoiding obstacles. Level 2 attempts to find corridors of free space to traverse and the

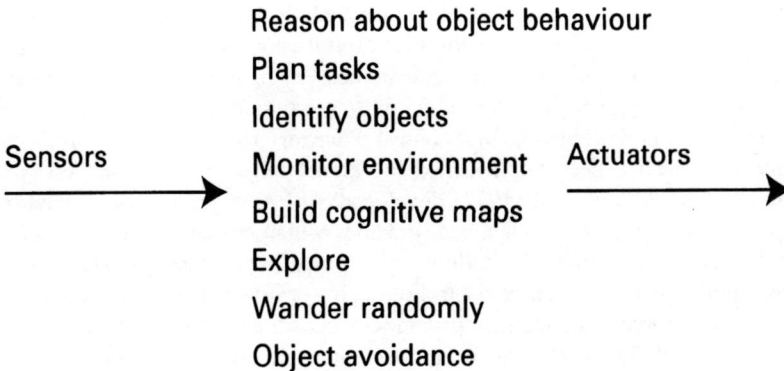

Reason about object behaviour

Plan tasks

Identify objects

Sensors   **Monitor environment**   **Actuators**
———————▶   ———————▶
                               Build cognitive maps

Explore

Wander randomly

Object avoidance

Fig. 13.4. Subsumption architecture task decomposition (adapted from Brooks 1986). Reproduced with permissionm of IEEE © 1986 IEEE.

Pathplan module attempts to reach specified goals. With level 0, 1 and 2 control, the robot finds a corridor to a specified goal at a distance and moves towards it. This level of behaviour may be likened to exploratory behaviour for the provision of information. Levels 3 and above are more plan-based 'intelligent' behaviours. Level 3 builds a cognitive map of the environment to plan routes between landmarks. Level 4 monitors dynamic changes in the environment. Level 5 identifies objects in the environment and reasons about tasks to be performed on them. Level 6 formulates plans to change the state of the world as required. Level 7 reasons about object behaviour in the world and modifies plans accordingly. Each level runs on its own processor and each processor represent a finite state machine running completely asynchronously with no shared global memory or central control. Each module represents a layer directly connecting perception to action operating asynchronously in a data-driven strategy. Each module abstracts only those aspects of the environment which are relevant to its operation (egocentric invariants). Each module has input and output lines for message passing. The output line from one module is connected to input lines of one or more other modules. Each level examines data from the lower level. It is permitted to inject data (advice) into the lower level to suppress the normal data flow output from the lower level by actively interfering and issuing replacement data. This generates conflict resolution. There is no overall representation of the world. Complexity of behaviour is a reflection of the environment rather than of the agent and the state of the world completely determines the actions of the autonomous agent [Brooks 1990]. As primitive behaviours operate concurrently, complex higher-level but unplanned behavioural patterns emerge, attributable to none of the individual simple behaviours alone, imparting reactive behaviour to dynamic changes in the environment.

Brooks (1989) implemented the behaviour based control system on a 1-kg six-legged robot with a length of 35 cm (Genghis) as a network of 57 finite state machine processors. A typical processor is the MIT Media Lab miniboard 2.0 with an 8-bit Motorola 6811 neurochip CPU, 256 bytes of internal RAM and 12 kbytes of programmable ROM. Each leg was controlled separately and rough terrain was compensated by force monitoring in each leg. Whiskers anticipated obstacles and infrared sensors detected moving objects to which it was attracted. Robust walking behaviours were produced by the distributed system with little central coordination and it was capable of robust steering and target following. Collisionless navigation is thus achieved using simple computational processes. Hence, complex behaviours emerged from the network of simple behaviours with little central control. Planetary rover prototypes implementing behaviour control have been implemented in various forms, such as the Rocky series. Miller (1992) and Miller et al (1992) developed ALFA, a programming environment comprising of networks of computational modules which communicate with each other, sensors and actuators through dataflow communications channels. This allows the implementation of behaviour-based control through a programming environment.

The potential force-field method provides a possible unified representation for a cognitive map and for combining behaviours at each level in a sensor fusion-type process independent of sensory modality. Arkin (1987) used a form of behaviour-based control whereby the primitive behaviours were implemented as multiple concurrent and independent motor schemas capable of communicating and coping with conflicting data.

Reactive navigation of a mobile robot emerged through the instantiation of these motor schemas. The schemas were generic specifications of appropriate patterns of behavioural actions codified in an organised data structure. The primitive schema behaviours may be combined to yield more complex behaviours. Schemas are invoked by sensor data and they provide expectations for the appropiate motor actions. Each schema outputs a potential force field vector:

Avoid obstacle schema:

$$O = \begin{cases} 0 & d > S \\ \dfrac{S-d}{S-r} \times G & \text{for} \quad r < d \leq S \\ \infty & d \leq r \end{cases}$$

where S = sphere of influence of obstacle, r = radius of obstacle, G = gain, and d = distance of robot from obstacle centre.

Path-following schema:

$$V = \begin{cases} P & d > w/2 \\ \dfrac{d}{w/2} \times G & \text{for} \quad d \leq w/2 \end{cases}$$

where w = path width, P = offpath gain, G = onpath gain, and d = distance of robot from path centre.

Each schema had an activation level which acted as a threshold for instantiation and this activation level may be controlled by altering the gains. Each schema is independent, and they are activated in parallel without layering and their potential field outputs are summed. Summing requires communication to resolve conflicting forces and this may be accomplished through a blackboard. This essentially provided a world model/cognitive map. Local minima may be overcome using a background noise component to produce a low-magnitude random force vector to remove the robot from undesirable equilibrium points. The noise schema in conjunction with the Avoid-obstacles schema generates exploratory behaviour. These motor schemas drive the robot to interact with its environment to satisfy the goals generated from a planning system.

13.4 HYBRID PLANNING METHODS

Behaviour-based control methodologies are ideal for low-level online reactive behaviours to be invoked during plan execution to provide robustness to planning. However, behaviour-based techniques are insufficient by themselves for in-orbit servicing, though of great potential for autonomous planetary rovers. Rovers in particular are exposed to unstructured, variable environments and require extensive reactive capabilities. In-orbit servicing, however, requires task-oriented knowledge, planning, reasoning and problem-solving capabilities. Environmental complexity imposes the same level of complexity on the robot agent. Some environments are pure stimulus–response environments with complete determinism such that they provide immediate indications of the optimal action to be pursued by the robot agent. More complex environments exhibit delayed response and even a degree of indeterminacy. Thornton (1997) pointed out that as obstacle motion increases in mobile robot environments, so the environmental cues to the sensory system increase in complexity requiring the ability to anticipate. As the system complexity increases, the number of potential interactions between subsystems increases exponentially requiring the hardwiring of an exponentially increasing number of 'look-up table' behaviours. The ability to cope with dynamic obstacles using static behaviours is adequate while the speed of an obstacle is less than that of the robot, i.e. the static assumption is valid as an approximation. This assumption breaks down when the obstacle motion exceeds that of the robot. This introduces the need for anticipation. Anticipation implies the need for knowledge of the possible consequences of its actions to avoid collision, i.e. it requires memory of past experiences and the ability to plan and reason ahead. Behaviour-based robots lack explicit goals and goal manipulation. In-orbit servicing would require symbolic concept representation. A situation-driven set of behaviours is only effective if enough environmental cues are available to determine what actions are applicable. If a task requires knowledge about the world and requires reasoning and memory rather than just perception and survival, then reactive behaviour is insufficient.

Similarly, robots must interact with the real physical world which is imprecisely modelled and so requires the use of sensors during plan execution to gather previously unknown data from the environment to ensure successful task completion. Unforeseen circumstances caused by asynchronous spontaneously triggered events in the real universe may invalidate the present plan. Not all events in the world are governed by predictable cause-and-effect relationships. Agents outside the control of the robot can cause changes in the environment. The real world is thus too complex to model in detail. The real world is highly dynamic so significant differences between the planned and real states are to be expected. Indeed, planning may be regarded as a game between the robot and Nature. The robot chooses motor control signals while nature chooses the sensor signals. For complex tasks, too little information may be available to pre-compile solutions and this may result in computationally explosive numbers of potential situations and solutions. Planning involving conditional effects is undecidable such that a plan cannot be guaranteed to succeed without execution time additions to the plan. Hence, the planner must work in close cooperation with the execution supervisor which monitors the results of the application of the plan in real time. Rather than replanning from scratch which is time-consuming and wasteful, a more efficient approach is to empower the executor with the necessary decision capability to repair plans dynamically about recovery procedures and error correction. This must be online and event-driven. Overrides can behave as interrupts to halt the execution of preplanned sequences. This involves tighter coupling between sensing and action through reactive behaviours to improve responsiveness in dynamic environments. Sensors then require local processing capabilities to allow reactive low-level decision capabilities to trigger interrupts when unusual alarm situations occur. Partial hierarchical planning where plan formation and plan execution is interleaved is necessary for effective planning [Georgeff & Lansky 1986]. This allows a reactive component to be implemented to cope with changes in a dynamic environment and modify plans to achieve goals when necessary. Such reactive procedures are invoked by events and change the focus of attention. Hence, both goal-directed and event-directed behaviours are possible due to the interleaving process. They modelled the action sequences of a plan as a sequence of world states produced by the behaviour of the agent. However, a dynamic unstructured environment may require a vast number of contingencies. An autonomous system requires both goal-oriented planning and robust reaction. In short, the need is to hybridise or integrate the 'sybaritic' approach which emphasises embodiment in the real world environment through sensors and actuators and GOFAI (Good Old-Fashioned AI) which emphasises symbol manipulation as the basis of rational problem-solving. Indeed, robotic implementation of autonomous capability is inherently a 'sybaritic' approach to AI.

Payton et al (1990) have considered the problem of integrating high-level planning with lower-level reactive behaviours. Reactive behaviours require tight coupling between sensing and action so they viewed the lowest level of the control hierarchy as a subsumption-type architecture. Plans were viewed as resources for advice rather than as constraints on predetermined courses of action. This allowed opportunistic advantage to be taken in unexpected situations. At each instant of time, the best choice of goals was selected without discarding alternative plans. The internalised plans were represented as gradient fields computed from graph searches to minimise the abstraction requirement. The low-level behaviours implemented were obstacle avoidance and navigation (wandering) but without subsumption-blocking of behaviours allowing each behaviour to communicate with each other. Malcolm & Smithers (1990) were also concerned with interfacing a PROLOG planning system with a plan execution system of behavioural modules. The planner was grounded through a single hierarchy of behavioural modules with tight sensing and action coupling rather than with a twin hierarchy of sensing and action (e.g. NASREM). Their SOMASS robotic assembly system comprised a cognitive planner implemented in Prolog in hierarchical format which utilised dependency directed backtracking and a subcognitive plan execution system. The plan execution system was essentially charged with the function of handling the uncertainties of the real world. The more competent the plan execution system the less complex the plan needed to be. The subcognitive plan execution system was built from behavioural modules which encapsulated useful behaviours with certain functional capabilities. By handling uncertainty at a low level within the plan execution system, the planner was able to work in an ideal and certain world.

Maes (1990) suggested that explicit goal handling with reactive capabilities eliminates the need for re-planning or re-programming with altered goals. Goals effectively bias choices of action (action selection). One approach to such action selection is through emergence via the local interactions between action-oriented modules. Each module represents an operator of the classical planner and may be described by the quadruple $(c_i, a_i, d_i, \alpha_i)$ where c_i = pre-condition list, a_i = action effect list, d_i = action effect delete list, α_i = activation level. The module is executed when all the preconditions are true and the activation level exceeds the threshold. Modules are linked into a network with three types of link between modules: successor links, predecessor links and conflictor links. Successor links exist when every proposition p that is a member of x's add list is also a member of the precondition list of y. Predecessor links exist for the inverse relation. Conflictor links exist when every proposition p that a member of the delete list of y is

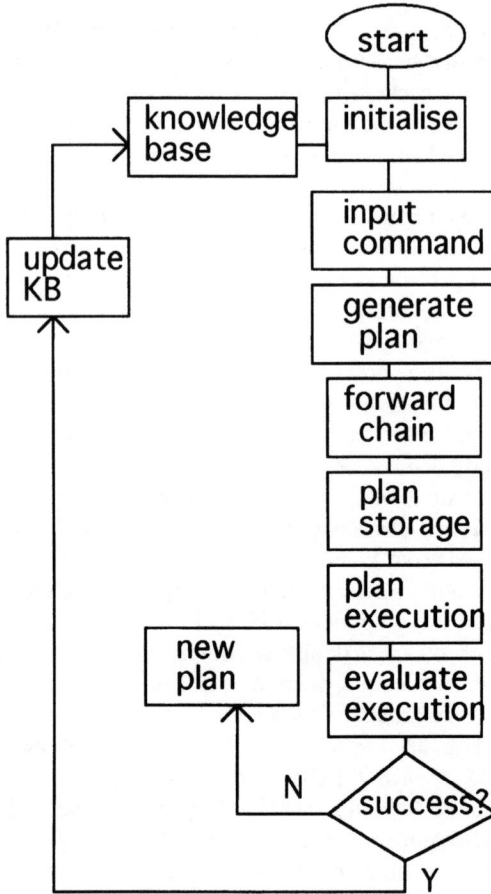

Fig. 13.5. Robotic planning flowchart.

also a member of the precondition list of x. Modules use these links to activate and inhibit each other with inhibition providing the means to deal with interacting subgoals. Activation energy is determined by the input from the current situation according to its match to modules and by goals which determine the activation threshold of goals. Protected goals remove activation from modules that would undo them. Spreading activation across the network of modules equilibrates to the best actions determined by the current situation and the current goals of the agent. Successor links propagate action forward to successor modules and predecessor links propagate activation backwards to predecessor modules by a fraction of their respective originator module activation energies. Conflictor links decrease the activation level by a fraction of the originator module's activation energy when a module undoes a true precondition of another module. All inputs of activation from a module is weighted by $1/n$ where n = number of propositions in the precondition/ add/delete list is appropriate. As the state of the environment changes, so too do all the activation patterns. Global parameters may be used to alter the activation dynamics through threshold adjustment, input activation adjustment, goal activation adjustment, and conflict activation adjustment. The preconditions act as subgoals. A sequence of modules are highly activated which transform the current situation into the goal situation through forward spreading from the current state and backward spreading from the goal state. As the links are only local among modules, different paths may be evaluated in parallel. The system is a marker-passing system without variable use in that objects in the environment are represented by only those features relevant to the agent. Hence, new operator modules are not required when encountering new objects. Goals are specified in terms of functional constraints on the objects in the environment. The control structure is emergent when particular actions are activated. The interaction dynamics between action-oriented modules establish the sequence of actions in a distributed manner in response to environmental conditions and global goals. The local interactions between components forms an emergent global structure.

Mataric (1992) suggested that rather than utilising hybrid approaches with separate reactive and planning modules, it is possible to maintain a map representation based on discovered landmarks distributed over the subsumption architecture of behaviours. This represents a fully integrated approach. The robot's control system comprised of behaviours specific to basic navigation, landmark detection and map computation. Basic navigation and obstacle avoidance involved strolling forward unless an object was in front of the robot, turning away from objects, and aligning with boundaries to follow them. Landmark detection involved matching sensory data to stored landmark signatures. The mapping algorithm mapped the structure of the environment based on the spatial relationships of the landmarks. The map was coarse and encoded as a topological graph with each node representing a landmark and the links representing adjacency. Each land-mark was a behavioural set of rules and the links operated as message-passing connec-tions. Each landmark behaviour received inputs from the landmark detector, while sending outputs to other landmark behaviours. Localisation involved comparing stored landmark signatures with the current sensory data. Expectation was generated by spread-ing activation through the graph, so adjacent landmarks were primed forming a path to the goal landmark, i.e. path planning.

Handelman et al (1990) introduced a robot control methodology modelled on human motor skills learning. There are three phases of skills learning of sensorimotor reflexes:

(i)   the early cognitive phase where the task is approached consciously through inferencing by the beginner;
(ii)   the intermediate associative phase where response patterns begin to emerge;
(iii)   the final autonomous phase when the execution becomes automatic and reflexive for the expert.

The skills learning process includes a shift from conscious declarative processing to automated procedural processing, i.e. the declarative representation becomes 'chunked' into a more condensed procedural representation through a compilation mechanism. An alternative representation of motor skills acquisition is through the repeated associative learning of input/output pairs which gradually determines the connection weights of an artificial neural network which becomes fixed on completion of the learning process.

Such a method of implementing motor skills learning to robots should offer high levels of dexterity and adaptability. A three-level hierarchy was defined for providing reflexive behaviour. A rule-based execution monitor teaches a CMAC (cerebellar model architecure controller) network how to accomplish a task by observing the rule-based task execution. The knowledge-base represents declarative knowledge with inferencing abilities while the neural network represents associative procedural knowledge with pattern-matching abilities. The neural network effectively captures the causal relationships between the robot and the environment. The knowledge base determines how to accomplish the control objectives using the rules. These rules include the learning strategies, task execution processes, network training data and performance monitoring. It teaches the neural network to accomplish the same tasks by allowing them to observe the task execution. They learn through generalisation at which point they assume control responsibility. The neural net then fine-tunes its performance during task execution through reinforcement learning. The knowledge base continuously evaluates the network's performance. If errors occur due to the changing dynamic environment, the knowledge-base is re-engaged and the network re-trained. The basic control system implements low-level servo-reflex on the innermost loop, reflex modulators in the middle loop to provide gain adjustment, and execution monitoring to supervise reflex modulator learning.

The servo-level implements a standard computed torque control law. The rules in the knowledge-base are independent of the dynamic model of the manipulator. A goal-directed backward search is implemented by an inference engine through the knowledge-base of parameters, production rules and procedures. Parameters are values relevant to the control objective while the rules express relationships and dependencies between parameters. Within the conditions and actions are procedures that invoke time-critical control tasks such as low-level joint reflexes and control laws. The rules provide the means to build higher-level control actions out of the procedure building blocks to generate coordinated motion. All manoeuvres are implemented as a series of steps. The neural network implementation is a CMAC which performs a perceptron-like table lookup

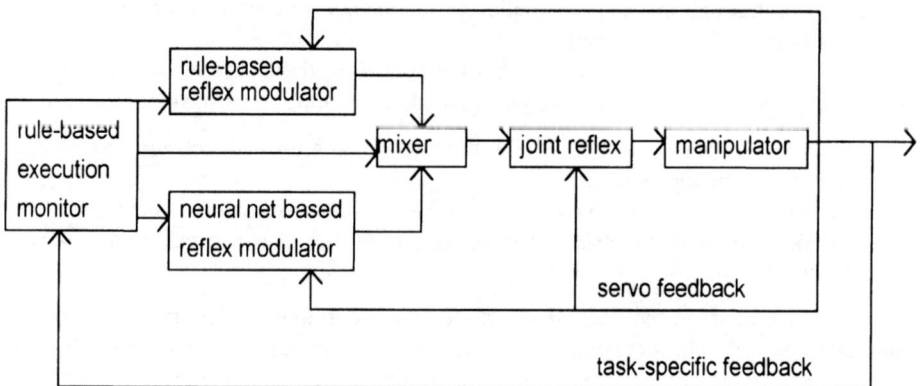

Fig. 13.6. Motor skills learning architecture (from Handelman et al 1990). Reproduced with permission from IEEE © 1990 IEEE.

for generating nonlinear mapping functions between multiple input–output variables. The knowledge base control executor provides the means for it to learn by example. The necessary generalisations of sensorimotor behaviour are accomplished through connection weight adjustments via a steepest descent updating algorithm. The CMAC learns to duplicate the rule-based reflex commands. When the CMAC's outputs match those of the rule-based outputs, learning ceases and the CMAC outputs directly to the joint reflexes during manipulation. Once trained, the input values cause the CMAC to output control gains and commands to the servocontrol system of the manipulator. Thus the CMAC reflexively executes the required movements. If the CMAC output deteriorates subsequently, control switches back to the knowledge-base while CMAC undergoes further learning. Hence, knowledge bases can be used to supervise and train artificial neural networks and indeed, the neural network offers a more robust performance by fine-tuning its performance through reinforcement learning. It is a highly promising technique for use in space-based systems as it offers an integrated approach utilising the neural net for its robustness and the knowledge-base as a reliable backup.

13.5 ROBOT CONTROL ARCHITECTURE

We shall now examine the robot control architecture which comprises the organisation of multiple elements of the control system. Diverse sources of knowledge need to be integrated into a useful world model of the environment so that intelligent decisions based on the available information can be made to enable efficient task completion. Intelligent control is the mathematical problem of finding the right sequence of decisions and controls to achieve the required task with minimum supervision from a human operator. Some form of intelligent control architecture is required to provide an effective model for the generation of intelligent behaviour. As problems become more complex, modularisation becomes necessary to localise the problem into portions. Composition is also required to re-integrate the module outputs.

13.5.1 Distributed artificial intelligence
Distributed computational architectures are used on spacecraft when most of the computational power loads reside in the subsystems with the central computer acting as a coordinator. It is characteristic of single bulk architectures that as they become more complex, bureacracy develops such that increasing proportions of the increased capacity are required for self-administration. The modularity of distributed computing offers greater adaptability and extensibility. In conventional distributed computing each node processes information almost autonomously without interaction. However, distributed expert systems attempt to solve problems that cannot be decomposed into independent, isolated subproblems by cooperative exchange of information through network communication protocols. It provides a means of speeding up the generation of the solution to combinatorial problems by up to an order of magnitude enabling the potential for real-time response following the principle of bounded rationality. DIA (distributed artificial intelligence) offers system reliability with graceful degradation if one or more agents should fail, eliminating single points of failure. Task sharing involves nodes sharing the computational load of problem-solving while results-sharing involves nodes assisting

each other by sharing partial results [Decker 1987, Smith 1980]. When subproblems cannot be solved by independent agents working separately without communication, results sharing must be adopted. This provides the means for simultaneous mutual constraint satisfaction on the different knowledge sources to operate in parallel and influence each other. The organisation of distributed AI ranges from data-driven heterarchical teams to goal-driven hierarchical structures.

13.5.2 Hierarchical architectures

The hierarchical approach to intelligent control architectures involves task decomposition to allow control of information without overloading any subsystems. This involves breaking down commands at each level into subcommands until at the lowest level they represent joint servocontrol. Inputs to each level are taken from higher-level outputs. Higher up the hierarchy, control problems become increasingly difficult to characterise, involving increasing amounts of symbolic rather than numeric data. At these higher levels, optimality becomes either difficult or impossible to implement and these levels replace optimality with the concept of adequacy or 'satisficing'. Low-level control functions require fast response times for real-time control while high-level cognitive functions require slower response times. Hence, at the lowest level are the autonomous motor control functions and at the highest level are selectable algorithms which require decisions to be made. Hierarchies avoid the exponential growth of increasingly detailed information in a system and the consequent increase in computational requirements.

Saridis (1983) proposed a hierarchical intelligent control scheme which utilises high-level decision-making capabilities with a low-level control system. It was a three layer hierarchically structured control architecture based on the principle of 'decreasing precision with increasing intelligence' (though each layer may be subdivided into more than one sublayer): the organisation level, the coordination level, and the execution level (see Fig. 13.7). The precision at each level defines the control hierarchy as a 'resolutional' hierarchy, i.e. from lower to higher levels there is a decrease in the use of numeric–algorithmic techniques and an increase in symbolic decision-making methods and an increase in time constants. Similarly there is a decrease in bandwidth and control rate (granularity) which avoids an exponential growth of increasingly detailed information further up the hierarchy.

The organisation level at the top of the hierarchy defines the tasks to be executed and decomposes them into subtasks in order of execution and supervises the performance of the overall system. This level uses techniques typical of AI such as knowledge-based expert systems. It may either be autonomous or interpret input commands by a human operator. This is the level of purposeful decision making, reasoning and planning, data storage and retrieval from static knowledge based long term memory (LTM), i.e. goal-directed structured information processing and management by generating expectations. This level involves clear task formulation with little or no precision and adopts learning by updating knowledge bases by storing new information. Machine reasoning associates compiled input commands derived from user commands to the required activities; machine planning formulates complete and compatible ordered activities to execute the required tasks; machine decision making selects the ordered activities with the highest probabilities of success; machine learning updates the knowledge bases; machine memory

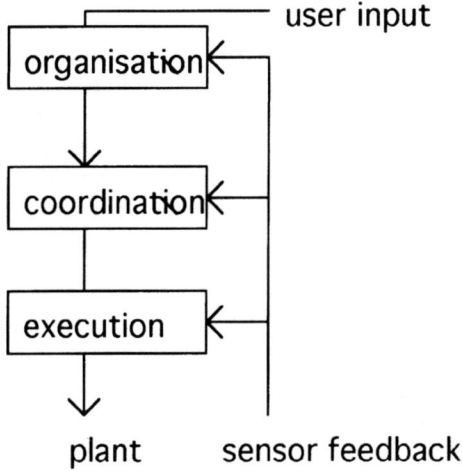

Fig. 13.7. Hierarchical approach to robotic control.

exchange retrieves and stores information from LTM. Little *a priori* information is used at this level but it provides expectations for lower levels, i.e. goal-driven. These functions are performed sequentially.

The coordination level receives instructions from the organisation level and coordinates the execution by imposing performance criteria and penalty functions which may be layered to improve the system's performance (e.g. selection of control gains, obstacle avoidance). It acts as the intermediate interface between the organisation level and the execution level by dispatching organisational information to the execution level. It acts as a buffer between the two levels similar to short-term memory (STM) and coordinates subtasks and communicates selective feedback to the organisation level. Both the organisation level and the coordination level should have learning capabilities. The coordination level uses techniques typical of operations research such as dynamic programming and involves less knowledge and more organisation. It maximises the efficiency of task execution using penalty functions and involves reactive behaviours to compensate for unexpected events. It synchronises operations and coordinates tasks, i.e. an Activity Coordinator. This level utilises a mixture of *a priori* and *a posteriori* information.

The execution level is the run-time control of the motion execution of multiple parallel local tasks with high precision. This level uses techniques typical of control theory such as optimal control which minimises an energy functional. This level requires the numeric mathematical model. This level utilises primarily *a priori* information, i.e. data-driven. Feedback from the coordination to the organisation level is offline while feedback from the execution level to the coordination level is online. Indeed the environment must be monitored continuously by sensors for feedback. Feedback between each level permits replanning or re-execution. Sensor feedback at different levels of the hierarchy is critical such that the sensor feedback forms part of the control loop *in toto*. Valavanis & Yuan

(1988) and Kim (1989) suggested that the organisation level should be characterised by a knowledge-based expert system operated on by an inference engine, the coordination level should be a distributed processing system of loosely coupled dedicated parallel processors, and the execution level be a series of specific hardware components. The coordination level also requires a parallel processing task scheduler to synchronise parallel tasks for minimum time performance, i.e. individual actuators are assigned to individual controllers. The system as a whole should be capable of intelligent functions to include reasoning, decision making and knowledge utilisation in order to perform:

(i)   task selection and planning from user requirements;
(ii)   formulation of complete task plans;
(iii)   supervised/unsupervised learning;
(iv)   automated updating of the knowledge base;
(v)   simulation runs of plans for verification.

Valavannis & Saridis (1985, 1988) and Saridis (1988) discussed the nature of intelligent control as reducing the uncertainty that results from incomplete *a priori* knowledge in the environment of operation and the state of the machine itself some of which may be managed or reduced by certain control strategies. They introduced the concept of entropy as a common analytic measure to characterise the functioning of all levels of the hierarchy from the intelligent activity of the higher levels to the precision of the lower levels. Uncertainty may be treated as a probabilistic distribution (entropy) function. For higher levels, information processing involves the production of entropy through the amount of structured information (knowledge) processed, i.e. facts and heuristics at these levels reduce uncertainty.

Meystel (1988) proposed a nested multiloop hierarchy with top-down resolution refinement using successive approximations. Upper levels with their low resolution of state have slower time scales for planning activities. The lower levels with their high resolutions of state have faster time scales for dynamic activities. There is a gradual change from offline planning to online real-time control. Although at all levels there are independent control loops, vertical links connect the levels to provide consistency of overall operation. The three-level hierarchy was referred to as the Planner–Navigator–Pilot hierarchy with perception–plan–actuation control legs. The Planner–Navigator–Pilot levels were also linked horizontally at each level (see Fig. 13.8). Each lower level is constrained by solutions from the upper levels. At any level if constraints are violated on the basis of new environmental conditions and new information this new information should be submitted to the next upper level for updating to generate a new set of constraints. Planning is connected to control by intermediate decision making levels. The sensory hierarchy from the preprocessing level involves a sequence of zooming operations, i.e. the focussing of attention to within the constraints of the limited computing power at each level. This is a form of recursive generalisation and such generalisations are required to efficiently use computing resources through attention. The knowledge-based system is also hierarchically determined by resolution of representation. Sensory information flows up the hierarchy in bottom-up data-driven fashion while commands flow down the hierarchy in top-down goal-driven fashion.

PERCEPTION	CONTROL	GOALS
priority selection	cost parameter	planner
object	avoidance	navigator
entities	reflexive	pilot
preprocessing	servocontrol	execution
sensors	environment	actuators

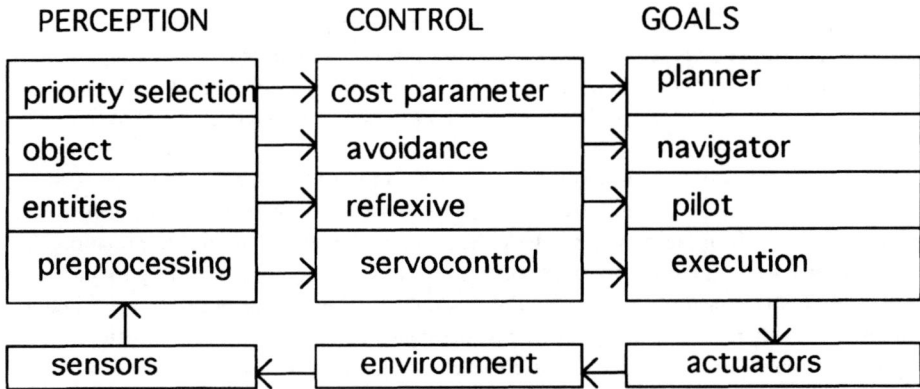

Fig. 13.8. Planner–Navigator–Pilot hierarchy.

13.5.3 Blackboard architectures

The blackboard model is inspired by cognitive psychology in that it essentially models aspects of human cognition: knowledge sources represent permanent long-term memory while the blackboard represents short-term working memory. Blackboard systems use multiple knowledge sources to analyse different aspects of a complex problem. The blackboard control architecture provides a uniform system which integrates a number of diverse, specialised and independent knowledge sources which communicate through a common global database (blackboard). The essence of the blackboard system which uses a common memory between multiple knowledge sources is that it provides the means for results sharing when subproblems cannot be solved by independent non-communicating agents. Different specialists contribute to different aspects of the plan which are posted to the blackboard and incorporated into the plan. All domain-specific data (including partial solutions) are posted onto the blackboard by each knowledge source for accessibility allowing the broadcast of events to all the knowledge sources via the blackboard. A master control program examines the blackboard and schedules the component sub-solutions into composite solutions. The blackboard is a high-level operating system that integrates the control of the distributed component knowledge sources according to its 'master' plan. An executive program controls the activation of each knowledge source in turn when an event occurs. The system goals determine the plan which specifies the tasks to perform and in which order according to priorities and time constraints. No individual agent can solve the problem alone so the blackboard acts as a shared global database to allow communication and cooperation. The agents are knowledge sources that can react to and modify the blackboard data structures. Each knowledge source comprises procedural condition–action production rules for event-driven processing. The knowledge sources are independent and autonomous yet allow simultaneous indirect cooperation amongst the knowledge sources through information in the blackboard. Each knowledge source is guided by strategic plans to provide metalevel control. The blackboard is a centralised database which allows hypothesise-and-test processes. The system maintains a set of hypotheses based on its world model and uses these to focus processing effort on expected events. Each knowledge source either applies hypotheses or tests hypotheses on the blackboard. Knowledge sources can modify posted plans based on subsequent contributions from other sources. Executive knowledge sources apply inference techniques to generate solution elements on the blackboard and knowledge sources respond to, generate, and modify solution elements on the blackboard. The blackboard system is then both data- and goal-driven. A set of solution elements on the blackboard constitutes a partial plan. Complementary or alternative plans can coexist and one or more partial plans can be merged to form more complete plans. Each agent develops partial interpretations and hypotheses based on their incomplete data. Solutions can be constructed by the aggregation of mutually constraining partial solutions on a blackboard using hypothesise and test strategies. Such results sharing based on different perspectives is data-directed. Partial hypotheses are proposed and tested for plausibility at each stage of the processing. Results sharing facilitates solutions to problems that cannot be subdivided into subtasks. Results from one agent influences and constrains results from other nodes.

Silverman et al (1989) used a blackboard system as a NASA testbed for autonomous ground control centre operations for spacecraft support. Typically the system would replace the ground facility manager, the spacecraft job scheduler, the facility operator and the repairman. Hearsay II is a continuous speech understanding blackboard system comprising a set of functionally distinct cooperating knowledge source experts which operate in parallel, a blackboard hierarchically divided into sound, word and sentence levels, a priority-based task scheduler, and a focussing mechanism for metalevel control. The global blackboard records solution elements to the control problem as data structures with attributes and values. The blackboard status triggers knowledge source production rules thereby contributing solution elements to the shared problem. Each knowledge source is knowledgeable about some aspect of language processing: syntax, semantics and lexical.

Problem-solving may be characterised as a search through a large problem space which exhibits exponential explosion in computational complexity. Partitioning the search space and using multiple agents to independently search each partition is one approach. However, it is possible to generate a superlinear speedup and computational increase with polynomial performance. Cooperation through a central blackboard accessible by all agents alters the search problem by introducing massive speedup over using non-cooperative agents [Clearwater et al 1991]. Hence, the blackboard is an optimal solution. One such blackboard model we have seen is NASREM. The NASREM architecture is so general that virtually any conceivable architecture could be made to fit it. Fig. 13.9 shows a compatible control architecture which would be suitable for ATLAS. The planner would be implemented at the ground station and all levels below and including the coordinator (with the reactive component) would be implemented on the spacecraft.

The blackboard system lends itself to a modular arrangement of knowledge sources to manipulate diverse types of information for the guidance and control of the ATLAS spacecraft from strategic mission and task planning to sensory perception and the communication of sensory information. Each knowledge source module may be restricted to correspond to the vehicle's major subsystems and their interaction with the space environment. The top-level control executive is the mission planner which performs several functions. It must store command plans uploaded from the ground station through the communications link. It must monitor the performance of plan execution in the

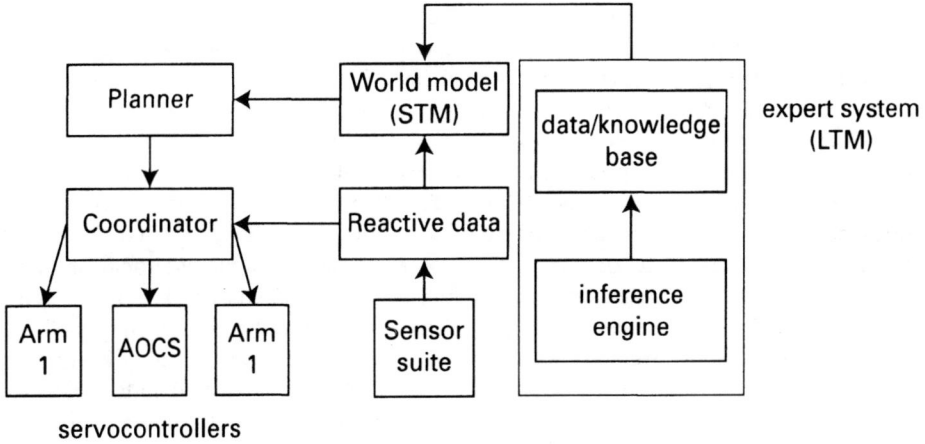

Fig. 13.9. Proposed ATLAS control hierarchy.

environment and all internal subsystems to cope with unscheduled manoeuvres. The reaction component requires rapid computation and reaction time for near-instant response (characteristic of algorithmic programs using languages such as C or ADA) while the planning component requires a logical decision-making structure typical of expert systems (characteristic of logic programs using such languages as PROLOG). The planning capability must be interleaved with the lower control level functions to adapt the mission plan to anomalies in system performance and environmental exigencies. At the lowest level of the control hierarchy is the real-time servocontrol system providing the sensor/actuator interface. This hierarchical control implementation would be suited to distribution through the communications link between the ground station and the space-craft.

13.6 FUTURE DEVELOPMENTS IN ROBOTIC AUTONOMY

It has been pointed out that the space sector imposes a strong requirement for unmanned, near-autonomous robotics in which robotic spacecraft, be they planetary rovers or in-orbit servicing spacecraft, must operate with high levels of reliability and robustness within complex environments, despite their high complexity, in order to achieve their mission objectives [Hobbs et al 1996]. As the system complexity increases, the number of potential interactions between the subsystems and the uncertainty of the operating environment increases exponentially. Any potential interaction that has not been antici-pated can lead to catastrophic failure. To this end, they suggest that co-evolutionary methods may be profitably utilised to incrementally evolve increasingly complex robot control systems within dynamic environments. Here, we present a few examples of such approaches to robust robotic control.

Dorigo (1993) implemented behaviour-based control methods using genetic algorithms (GA). The GAs provided a mechanism for reinforcement learning based on the performance of robotic behaviour and generated feedback concerning the success or failure of actions. The genetic algorithms were implemented as classifier systems in the form of condition–action rules constructed from 'chromosomes' based on symbol strings of the form: if ⟨sensor data⟩ and ⟨current goal⟩ then ⟨action⟩ formed from a three-valued alphabet $A = (0, 1, *)$. The classifier rule is composed of a number of chromosomes which represent the condition–action information as a string of three chromosomes. The first two chromosomes comprise the condition part while the third is the action part. The classifier may send messages to the effectors directly or to other classifiers. A set of such rules comprises the knowledge base and new rules may be generated from past experience. A message list collects all the messages to and from the environment via an input–output interface of detectors and effectors. Environmental messages in the message list are matched against the condition parts of the classifier rules and those that match are activated in parallel. A feedback mechanism provides reinforcement learning via reward/ punishment. An apportionment of credit algorithm distributes the payoff to the rules that performed the actions. The GA provides a means for rule discovery through modification of the population of solutions. A fitness function relates rule solutions to performance. The value of this fitness function is determined by the rules' usefulness as calculated by the apportionment of credit algorithm. The worst classifiers are replaced by *m* new classifiers. In this way, an internal world model is built up corresponding to factors in the external environment. The apportionment of credit algorithm classifies rules according to their usefulness (determined by the fitness function). A variable real valued strength is associated to each classifier. When a classifier invokes an action on the environment a payoff is generated the value of which depends on the effectiveness of the action. For multiple and sequential rules, the bucket brigade algorithm passes reward back through the decision chains so that the payoff flows backwards. The system may be implemented as several classifier subsystems acting in parallel where each classifier subsystem learns a simple behaviour (such as avoidance of obstacles, grasping of objects, etc.) through interaction with the environment. Sensors monitor the environment while rules determine which actions are to be taken. The system as a whole learns to coordinate its activities at a higher hierarchical level. As hierarchical structures learn faster than single-level distrib-uted architectures, an hierarchical structure of behaviours is produced. The hierarchical structure has the lowest-level classifier system, learning basic behaviours from sensory data, while those at a higher level learn to coordinate those behaviours with no direct access to the environment. Such coordination learning leads to more complex behavioural sequences. Each classifier system receives excitatory and inhibitory signals from the classifier systems connected to it. Classifiers at the same level compete. This enables the system to distribute behavioural knowledge across the architecture. The experimental unit was successful at following lights and avoiding heat sources. The coordination level learned a switching procedure between the two low-level behaviours. Essentially it may be regarded as a coarsely structured neural network. Indeed, it may be that the higher-level coordinating behaviours could have subsequently evolved symbolic reasoning capable of manipulating the classifier rules as a superstructure built onto the instinctive behavioural base. This approach was extended and applied to a small robot AutonoMouse which had two eyes. Each classifier system of 100–500 rules was implemented on a distributed set of transputers as its brain, each with a specific behaviour arranged hierarchically (Alecsys) [Dorigo 1995]. It successfully exhibited simple light following, evasion of sounds, homing, and coordination of switching between these simple behaviours.

Koza (1993) applied the technique of genetic programming for the control of autonomous mobile robots. Genetic programs operate on the level of symbolic expressions. A computer program comprises of a set of functions (e.g. logical operations). The computer program represents the solution to a problem. A population of computer programs comprises a generation. The initial population of computer programs is random. Each program is executed and assigned a fitness value according to how well it solves the problem. New computer programs are created by swapping and recombining random segments of two programs. This process will produce increasingly fitter populations of programs. This process continues until an adequate computer program solution has been found or until a fixed number of generations have been run. Autonomous robots with five functional behaviours with inputs from sonar sensors were tested. The initial program population comprised four primitive functional functional behaviours representing STROLL, AVOID, ALIGN, CORRECT comprising some 150 program functions such as 'move back a fixed distance', 'turn right/left a fixed angle', 'move forward a fixed distance', etc. The purpose was to evolve the subsumption architecture. The subsumption architecture may be represented by the IF conditional to implement a decreasing priority of behaviours:

If S_1 then CORRECT
If S_2 then ALIGN
If S_3 then AVOID
If S_4 then STROLL

Fitness was defined as the maximum coverage of wall perimeter in the shortest time. The best 10% of parents surviving into the next generation intact and the cross-over was applied to the other 90% of the programs to create new programs. A computer program was succesfully evolved which exhibited the subsumption architecture via the composition of the conditional IF statement in performing wall following of an irregularly shaped wall.

De Garis (1991) used GAs to evolve neural network modules (GenNets) to construct hierarchical control systems. Each artificially evolved neural net is regarded as a functional black box which performs a specific behaviour, i.e. a specific control system. Fitness is a measure of the quality of the behaviour. Each neural net module input is connected to each sensor and each output unit is connected to each motor. Synaptic weights and thresholds can be coded and evolved as genetic algorithm chromosome values. The GA has better convergence properties than gradient-based techniques such as backpropagation in the face of noise, reducing training times by two orders of magnitude. The GA searches the solution space globally while gradient techniques perform local searches in the vicinity of the current solution. It finds the best combination of connection weights rather than just each weight independently. A population of genetic algorithms has a proportion (typically 20%) which survive through cross-over into the next generation undergoing 10% random mutation. A typical fitness criterion for a wheeled rover is given by:

$$F = V + \left(1 - \sqrt{DV}\right) + (1 - I)$$

where V = average rotation speed of opposing wheels, DV = difference between signed speed values of opposing wheels, and I = activation value of the sensor with the highest input. This maximises speed, movement in a straight line and avoidance of obstacles. The fitness function should compute information that is available to robot through its sensors only. The N neurons of the neural net are labelled 1 to N ($N \sim$ 10–15 being typical) and N^2 interconnections if the net is fully connected. A number of binary places P are allocated to specify the weight values and one bit to specify its sign. For N^2 connections, the binary string requires $N^2(P + 1)$ bits to specify all the connection weights. It is possible therefore to specify the GenNet from its chromosome representation. GenNet modules may be constructed into hierarchical structures to exhibit arbitrarily complex behaviour. Each GenNet with its specific behaviour may be evolved separately and then organised appropriately. Control GenNets are used to direct and command functional GenNets in that the outputs of the control nets serve as inputs to the functional nets. Control GenNets are evolved once the functional GenNets have been frozen. Such hierarchical control may be multiply layered. This methodology has been applied to an artificial lizard, LIZZY which reacts to three types of creatures each of which emits a characteristic frequency: mate, predator and prey. Its behaviour varied according to the signal it received and its strength (indicating its distance) via its a suite of detector, logic and effector GenNets. GenNet weights were stored in ROM and the weights were used to connect up and control physical detectors and effectors.

Mataric & Cliff (1997) described Khepera, a wheeled rover with eight infrared proximity sensors. A model simulation Khepsim serves as the environment to evolve GA-based neural network controllers which can be downloaded to the real robot. If evolved in simulation, there will be a mismatch between simulated and real sensory data. Simulation does not describe the physical laws of interaction in the real world and physical sensors and actuators are subject to uncertainty. However, after initial evolution in the simulated environment, only a few additional generations are usually required to achieve successful behaviour in the real world—adaptive fine-tuning. It is required that the level of noise in the simulation matches that of the real world. Most of the evolutionary process is performed on the simulator with a further few generations of evolution on the robot itself. The use of simulation significantly speeds up the learning process. Typical tasks included obstacle avoidance and goal-seeking. Grasping behaviour also has been achieved using a 5-input/4-output neural network. The inputs were from two frontal sensors, left and right sensors, and a gripper sensor. The outputs included two wheel velocities, pick-up procedure and release procedure. The fitness function included parameters concerned with robot–target distance, target position relative to the robot, presence of an object in the gripper and its release, etc. Learning gripping behaviour was more complex than most other rover-type behaviours as it involved a multitude of sensors and tasks including moving the robot close enough to the target to be grasped. The fitness function became complex and subgoals were embedded in them to produce the required behaviours. The species adaptation algorithm (SAGA) may overcome this problem by utilising variable length genomes [Husbands et al 1995a,b]. Initially, short, simple genotypes may be used to generate elementary behaviours, and the length increased until sufficiently complex behaviours are evolved. A specialised cross-over procedure is used with these variable-length genomes. Further work has involved the evolution of visually guided behaviours on GA-based recurrent neural network control systems incorporating excitatory and inhibitory mapping between vision and actuators [Husbands et al 1995b]. A number of circular receptive fields of given radius and angular position $r\phi$ are represented on the genome which provides the basis for interpreting visual inputs through the construction of vector flowfields representing robot motion. Incremental evolution whereby increasingly challenging tasks are solved sequentially rather than always starting from random populations offers better performance.

Handley (1993) introduced Genetic Planner which used GA-based artificial selection to breed computer programs that generate plans to purposively control a mobile robot. The genetic planner used cross-over recombination and artificial selection through a fitness measure to breed computer programs. A world model comprised a set of proce-dural operators which operated on the set of predicates that defined the world model. The world model predicates were of the form: at(object, x, y), with-respect-to(object1,object2). The procedural operators included: 'turn left/right 90', 'move forward a fixed distance', etc. The fitness function was defined as a distance-between-objects measure. The agent used the procedural operators to generate robot action plans utilising a goal-oriented fitness measure and a world simulation. The genetic planner used the operators to execute candidate computer program plans and the fitness function evaluated the state of the world after each simulation. As the genetic planner did not reason about the world, it randomly generated plans and ran them until it came up with effective plans. This type of capability has use in trajectory planning.

These embryonic approaches to robotic learning may have use in robotic rovers of the future, but their scaleability to more complex autonomous behaviours such as that involved in in-orbit servicing manipulation is not clear.

13.7 SUMMARY

It is desirable to make planning more robust through survivability, i.e. reactivity. How-ever, reactive systems cannot perform complex tasks. Hence, a mixed mode of planning and reactivity is necessary. A well-balanced combination of centralised and distributed control is required to generate efficient and reliable behaviour with distributed control exhibiting rapid responses and centralised control exhibiting slower responses typical of higher hierarchical levels [Courtois 1985]. The executive and planning process are inter-leaved so the executive can post changes to the world to a blackboard world model which may alter the truth values of the logical assertions already stored in the blackboard working memory, thereby affecting the current context, e.g. SEPIA (Situated Execution of Planning & Improvisation in Alternatives) [Segre 1991]. A conflict invokes replanning to eliminate the conflict. Reactivity should be implemented on the spacecraft while planning should be implemented on the ground. Such a system is compatible with the NASREM computing architecture.

14

Spacecraft orbit control system

Having considered the robotic payload and robotic support systems, we turn now to the spacecraft bus segment (with particular emphasis on ATLAS, our design example) and consider the impact of the robotic payload segment on the spacecraft bus design. The first onboard subsystem we consider is the orbit control system comprising the systems that are concerned with orbital manoeuvres. Initially, we look at orbital mechanics in general and the rocket propulsion system that acts as the orbital actuators. Finally, we examine the in-orbit servicing mission profile which will define selection of both the orbit maneouvring requirements and the onboard propulsion system. The mission profile in particular characterises the type of mission that a robotic in-orbit servicer such as ATLAS must achieve, indicating the challenges of orbit manoeuvres, phasing orbits and RVD (rendezvous and docking) operations. These factors define the fuel budget as one of the critical design parameters for the spacecraft.

14.1 ORBITAL MECHANICS

The orbit for most spacecraft is well-defined as they are usually chosen for their ground track, but for a servicer spacecraft such as ATLAS the orbit will be variable. Below geostationary orbit (GEO) satellite ground tracks shift westward due to the Earth's rotation. GEO is used to maintain the same position relative to the Earth for communication, navigation and meteorological satellites. It is an equatorial orbit as an inclination other than zero will generate a figure-of-eight ground track. The Molniya orbit exhibits a slow apogee passage over high latitude ground stations. Low Earth orbit (LEO), which may be equatorial, inclined or polar, gives complete coverage of the Earth's surface and is used by Earth observation satellites and some meteorological satellites (such satellites are typically polar). This orbit is usually Sun-synchronous to ensure similar illumination conditions at the same time every day. Ground coverage is not an issue for in-orbit servicers such as ATLAS since its application is space-oriented rather than Earth-oriented. The issue of concern is ATLAS's target market. However, since the ISS will be in the 400 km altitude region, this may be taken as a nominal reference orbit. For an equivalent GEO mission profile, any robotic servicer would require the addition of off-the-shelf solid rocket boosters for LEO–GEO transfer, e.g. the Inertial Upper Stage (IUS) or the Payload Assist Module (PAM). The variable orbit robotic servicer resembles in many respects the OMV (Orbital Manoeuvring Vehicle) type spacecraft application. For an LEO to HEO (high Earth orbit, typically 1200–1400 km) transfer the Δv requirement varies from ~0.6 to 1.5 km/s depending on the precise orbit transfer. For most spacecraft, the perigee burn in orbit raising is generally considered as part of the launch phase and the apogee boost is generally considered as part of the station acquisition phase both of which precede the operational orbit phase. Inclination changes cannot be effected by launchers economically so this is part of the station acquisition phase. Although station-keeping requires only ~200–400 m/s Δv per year, evasive manoeuvres will require typically ~150–4600 m/s Δv depending on the severity of the manoeuvre. For ATLAS, the perigee/apogee/inclination burns form part of the in-orbit operational phase. For the ATLAS spacecraft, the operational orbits are considered to vary from ~200–1200 km altitude, generating a Δv requirement of 0.53 km/s for a one-way manoeuvre. A two-way manoeuvre requires a Δv of 1.06 km/s. ATLAS would thus require refuelling periodically with small launched payload fuel tanks, particularly if large Δv manoeuvres are required—this highlights some of the logistic difficulties that arise in re-tasking spacecraft and the importance of providing infrastructure services in space.

First of all, a brief introduction to orbit mechanics is presented which define spacecraft trajectories [Larson & Wertz 1992, Agrawal 1980]. Orbit mechanics is defined by Newton's law of universal gravitation $\ddot{r} = -\mu/r^2$ which describes a conservative force with gravitational energy $U = -\mu/r$ which is constant at a particular orbital altitude where $\mu = GM_{Earth}$.

The orbit describes a conic section with constant angular momentum $h = r^2\dot{\theta}$ such that:

$$r = \frac{\left(h^2/\mu\right)}{1 + e\cos v} = a(1 - e^2)/(1 + \cos v) \qquad (14.1)$$

where h = angular momentum vector, e = eccentricity of orbit, and $v = \theta_0$ = true anomaly (position of the satellite in orbit).

This is a statement of Kepler's first law that if two objects in space interact gravitationally then each will describe an orbit that is a conic section with the centre of mass at one focus. The energy of a satellite is constant along its conic section orbit as given by the *vis-à-viva* equation:

$$E = \frac{1}{2}U - \frac{\mu}{r} = -\frac{\mu}{2a}$$

The exchange of kinetic and potential energy will cause the satellite to slow down as it gains altitude. Conservation of energy and conservation of angular momentum can yield the orbital velocity as given by:

$$v = \sqrt{\mu\left(\frac{2}{r} - \frac{1}{a}\right)}$$

The orbital velocity formulation is a statement of Kepler's second law that if two objects are orbiting each other due to gravitational interaction their orbits will sweep out equal areas in equal intervals of time. The formulation reduces to $v_{circ} = \sqrt{\mu/r}$ for circular orbits and to $v_{parab} = \sqrt{2\mu/r}$ for parabolic orbits. The Keplerian orbital elements are six in number and together completely and uniquely describe the orbit with respect to an Earth-centred inertial reference frame. They are thus:

(i)   a=semimajor axis which determines the size of the orbit:

$$a = -\frac{\mu}{2}\left[\frac{v^2}{2} - \frac{\mu}{r}\right] = \frac{r_a + r_p}{2} \qquad \text{where} \qquad r = \left[\frac{a(1 - e^2)}{1 + e\cos\theta}\right]$$

(ii) e = eccentricity which determines the shape of the orbit:

$$e = 1 - \frac{r_p}{a} = \frac{r_a}{a} - 1 = \frac{r_a - r_p}{r_a + r_p}$$

where r_p = radius of perigee = $a(1 - e)$, r_a = radius of apogee = $a(1 + e)$, $e = 0$ for circular orbit, $0 < e < 1$ for elliptical orbit, $e = 1$ for parabolic orbit, $e > 1$ for hyperbolic orbit;

(iii) i = inclination which determines the plane of the orbit with respect to the reference plane (Earth's equatorial plane for Earth orbit missions)

$$i = \cos^{-1}\left(\frac{h_z}{h}\right)$$

where $h = r \times v$ = specific angular momentum;

(iv) Ω = right ascension or longitude of the ascending node which locates the plane of the orbit with respect to the background stars usually taken from the first point of Aries (this defines the direction in which the Sun ascends through the plane of the ecliptic at the vernal equinox);

(v) w = argument of perigee which locates the perigee point with respect to the sky (line of apsides);

(vi) θ_0 = true anomaly or argument of perigee which locates the orbiting body in its orbit.

The time-dependent position of the orbiting object is given by:

$$\dot{\theta} = \frac{\mu^2}{h^3}(1 + e\cos\theta)^2 \tag{14.2}$$

Position and velocity data allows the computation of all six Keplerian elements and so characterise the orbit completely. The velocity along the orbit will vary between

$$v_p = \sqrt{\frac{2\mu r_a}{(r_a + r_p)r_p}}$$

at perigee and

$$v_a = v_p \frac{r_p}{r_a}$$

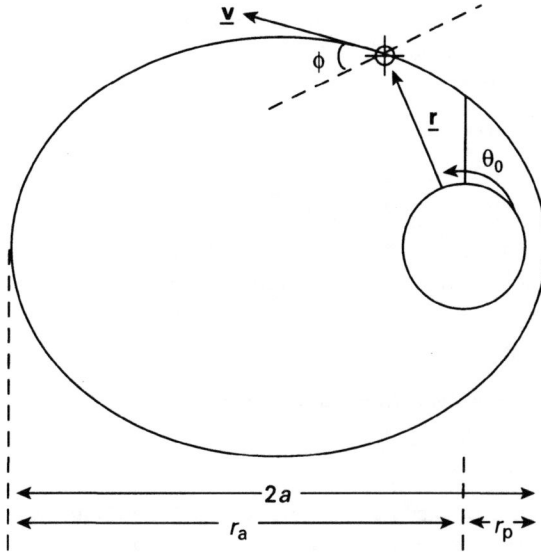

Fig. 14.1. Elliptic geometry and orbit parameters.

at apogee. Kepler's third law is defined in terms of the Keplerian orbit parameters: if two objects in space revolve around each other due to gravitational attraction, the squares of their orbit periods are proportional to the cubes of the semimajor axes of their elliptical orbits:

$$P = 2\pi \sqrt{\left(\frac{a^3}{\mu}\right)} \tag{14.3}$$

From an Earth station on the Earth's surface, the reference coordinates have their fundamental plane in the local horizon and use the observables:

(i)   Azimuth, $Az = \tan^{-1}(r_x/r_y)$ which defines the angle of the orbiting object in the horizontal plane;

(ii)   Elevation,

$$El = \tan^{-1}\left(\frac{r_z}{\sqrt{r_x^2 + r_y^2}}\right)$$

which defines the angle of the orbiting object in the vertical plane;

(iii)   Range, $R = \sqrt{r_x^2 + r_y^2 + r_z^2}$ which describes the vectorial distance from the site to the spacecraft.

These observables may be related to:

(i)   the difference between the longitudes of the spacecraft and the ground station $\delta\lambda$;
(ii)   the ground station latitude φ;
(iii) the spacecraft latitude δ;

by spherical trigonometry

$$Az = \tan^{-1}\left[\frac{\sin\delta\lambda}{\cos\varphi\tan\delta - \sin\varphi\cos\delta\lambda}\right]$$

$$El = \tan^{-1}\left[\frac{\sin h - R_E/r}{\cosh}\right] \qquad \text{where} \qquad h = \sin^{-1}(s\delta s\varphi + c\delta c\varphi c\delta\lambda)$$

R_E = Earth radius

r = satellite distance from Earth centre

$$R - \frac{r - R\sinh}{\cos(h - El)}$$

For planetary missions, patched conics are often used in orbital trajectory computations whereby the trajectory is divided into legs for each planetary body encountered [Bate et al 1986]. Every planet has a sphere of influence within which its gravitational field dominates the orbit mechanics of the spacecraft. The Earth's sphere of influence is around 1×10^6 km in diameter. Outside of this region, the Sun dominates in gravitational attraction. The final leg comprises the sphere of influence of the target body. Gravity assists are common in planetary missions to minimise the fuel consumption in deflecting the spacecraft into new trajectories. Their geometries are given by:

$$\sin(\alpha/2) = \left(1 + r_p v_\infty^2/\mu\right)^{-1} \tag{4.14}$$

where α = deflection angle between inbound and outbound asymptotes, r_p = pericentre radius, v_∞ = hyperbolic excess velocity.

Given that fuel is a critical commodity, it is desirable to reduce fuel requirements during orbit transfers, and one way to achieve this for an Earth-orbiting spacecraft like an in-orbit servicer such as ATLAS which is required to undergo multiple orbit transfers is through aero-assisted orbit transfer. Aero-assisted orbit transfer involves the use of aerodynamic forces to produce an orbital altitude or orbit plane change for OTV (orbital transfer vehicle) applications particularly during deorbiting [Walberg 1985]. Orbital plane changes are particularly Δv demanding.

Multipass aerobraking occurs at high altitude ~70–90 km in a near-free molecular flow environment. This method of orbit circularisation by aerobraking has been demonstrated by the US Atmospheric Explorer. Each atmospheric pass lowers the orbit apogee until the orbit is circularised through progressive velocity decrements. Low drag implies the requirement for multiple passes. Hence, multipass aerobraking spacecraft need have no aerodynamic surfaces apart from a deployable drag brake to provide the frontal area and shield the rest of the spacecraft from the rarefied molecular flow. Since aerodynamic heating is low, aerobrakes can be constructed from metallic or ceramic refractory materials for reusability rather than requiring ablation. The other advantage of multiple pass aerobraking is that deployable drag brakes are required instead of complete aeroshell enclosures. A deployable drag brake of similar structure to a ribbed antenna may be deployed from an annular housing for aerobraking at high altitudes and retracted for re-use. The lifting brake provides a large drag area, a large effective nose radius, low mass, and a low ballistic coefficient for shallow penetration into the atmosphere. Trajectory shaping is produced through roll modulation. The lifting brake is evidently the method of choice for an OTV though it is limited to absolute plane changes ~2° with a one- or two-pass capability. The only disadvantage of multiple passes is that it takes much longer than the single-pass manoeuvre. However, aerobraking, doubles the roundtrip LEO–GEO pay-load capacity of an all-propulsive vehicle.

Aero-assisted orbit transfer is a young technology, and given the additional constraints on the systems design of the robotic freeflyer spacecraft, it is unlikely to be adopted in the foreseeable future despite the gains of reducing fuel expenditure.

14.2 ROCKET PROPULSION SYSTEM

Satellites which require large changes in orbital parameters to rendezvous and dock with other spacecraft need fixed or gimballed propulsion thrusters which may be throttled as required for changes in translational velocity. Autonomous burns for trajectory manoeuvres and corrections require an onboard GNC (guidance navigation and control) computer. A rocket motor comprises a thrust chamber, a propellant feed system and propellant storage tanks (typically of Ti). Liquid fuel and oxidiser is pumped into the combustion chamber where chemical reactions cause the resultant gas to expand. The rocket performance is governed by the Tsiolkovsky rocket equation which is derivable from Newton's second law [Cornelisse et al 1979]:

$$\Delta v = v_{ex} \ln R \tag{14.5}$$

where R = mass ratio = m_i / m_f, m_i = initial mass = spacecraft loaded mass = $m_{pl} + m_{sb} + m_p$, m_f = final mass = spacecraft dry mass = $m_{pl} + m_{sb}$, v_{ex} = propellant exhaust velocity ~2–4 km/s for chemical systems and ~30 km/s for electric systems, m_{sb} = mass of spacecraft bus, m_{pl} = mass of payload, m_p = mass of pro-pellant.

The structural ratio may be defined as:

$$\pi = \frac{(m_{sb} + m_{pl})}{m_p} \sim 5\text{–}20\% \text{ typcally} \tag{14.6}$$

The payload ratio may be defined as:

$$p = \frac{m_{pl}}{m_i} \tag{14.7}$$

The specific impulse defines the fuel performance to produce maximum thrust for a given mass-flow rate of fuel and is defined from Newton's second law as:

$$I_{sp} = \frac{v_{eq}}{g} = \frac{F}{\dot{m}g} \tag{14.8}$$

where v_{eq} = equivalent exhaust velocity, $\dot{m} = (m_f - m_p)/t_{burn}$ = propellant flow rate, and g = acceleration due to gravity, F = thrust.

It is desirable to have a low mass ratio as close to 1 as possible, i.e. a large exhaust velocity or specific impulse. If the spacecraft accelerates at constantly at **a** the spacecraft mass will vary as $m(t) = m_i e^{-at/v_{ex}} = m_f e^{at/v_{ex}}$.

Another useful performance parameter is jet power,

$$P = \frac{1}{2}\dot{m}(gI_{sp})^2 \tag{14.9}$$

Thrust,

$$F = m\frac{dv}{dt} = \dot{m}v_{ex} + A_{ex}(p_{ex} - p_\infty) \tag{14.10}$$

where A_{ex} = exit nozzle area, p_{ex} = nozzle exit pressure, p_∞ = ambient pressure, and $v_{ex} = \sqrt{2C_p(T_c - T_{ex})}$.

For maximum thrust,

$$\frac{dF}{dA_{ex}} = p_{ex} - p_\infty = 0 \rightarrow p_{ex} = p_\infty$$

If the exit pressure of the nozzle exceeds the ambient pressure it is underexpanded, while if is less than the ambient pressure it is overexpanded. If the optimal condition holds, then $F = \dot{m}v_{ex} \approx \dot{m}v_{eq}$. Hence,

$$\Delta v = gI_{sp} \ln\left(\frac{m_i}{m_i - m_p}\right) \qquad \text{where} \qquad m_p = m_i - m_f$$

$$\rightarrow m_f = m_i e^{-\Delta v/gI_{sp}}$$

$$\rightarrow m_p = m_i\left(1 - e^{-\Delta v/gI_{sp}}\right) \tag{14.11}$$

An alternative representation which translates directly to the orbital manoeuvre and so is valid for electric thruster spiral orbit manoeuvres is given by:

$$\frac{m_f}{m_i} = \exp\left[\frac{\sqrt{\mu}}{gI_{sp}}\left(\frac{1}{\sqrt{a_f}} - \frac{1}{\sqrt{a_i}}\right)\right] \tag{14.12}$$

where μ = gravitational constant = $3.99 \times 10^{14}\,\mathrm{m}^3/\mathrm{s}^2$, a_i = initial semimajor axis, and a_f = final semimajor axis.

The Δv budget provides an estimation of the cost of the mission. Ideally, the propellant mass should be less than the spacecraft dry mass and refuelling may be necessary. Attitude control requires ~3–10% of the total propellant mass if thrusters are used for attitude control.

The requirement for $p_{ex} = p_\infty$ determines the design of the rocket nozzle [Cornelisse et al 1979]. For a launch vehicle during ascent, the optimum exit pressure varies with the ambient pressure. In addition, they suffer from gravity effects and drag ~ 1.5–2.0 km/s. In a vacuum $p_\infty \approx 0$, so for in-orbit operation A_e must be large to give high thrust which introduces mass penalties. For good performance, a high combustion temperature T_c and high combustion pressure P_c are required with low molecular weight (M_r) fuels since $I_{sp} \propto \sqrt{T_c/M_r}$ and characteristic velocity (which measures the energy) $v_{char} = (P_c A_{th})/\dot{m}$. The characteristic velocity of monopropellant hydrazine N_2H_4 is 1.33 km/s; for bipropellant N_2O_4/MMH (monomethyl hydrazine), it is 1.62 km/s while for cryogenic LO_2/LH_2 it is 2.36 km/s. The convergent–divergent de Laval nozzle accelerates the hot combustion gases from the combustion chamber to give supersonic flow. The axial thrust is given by: $F = \lambda \dot{m} v_{ex}$ where $\lambda = \frac{1}{2}(1 + \cos\alpha)$ and α = divergence angle. The bell nozzle has a short design and expands the flow adiabatically at high angle and allows the gases to exit in a near-axial direction reducing the divergence losses. Extendible exit cone nozzles can provide a solution for motors operating in both air and vacuum without being so large as to introduce packaging difficulties during launch. The extendible nozzle is in two sections where the second section is translated from the stored position to the operational position by springs or plungers. They are not compatible with regenerative cooling—regeneratively cooled systems are restricted to large liquid-fuelled engines (i.e. launchers). For smaller engines as would be employed for orbit control, radiation cooling is adopted by using low mass refractory metals such as niobium or carbon-carbon materials which allow high operating temperatures within the chamber/nozzle which act as heat sinks.

The thrust coefficient measures the energy conversion coefficient to exhaust velocity:

$$c_f = \frac{F}{P_c A_{th}} = 1.6 \text{ for } \varepsilon = 30:$$

$$= 1.86 \text{ for } \varepsilon = 200:1$$

where

$$\varepsilon = \frac{A_{ex}}{A_{th}} = \frac{\rho_{th} v_{th}}{\rho_{ex} v_{ex}} = \text{ nozzle exit area to throat expansion ratio } \sim 20$$

$$v_{th} = \sqrt{\frac{\mathcal{P}_{th}}{\rho_{th}}} = \text{ critical throat velocity}$$

$$\gamma = \frac{c_p}{c_v} \sim 1.21\text{–}1.26 \text{ for most fuels and oxidisers}$$

Solid propellants (eg. Al powder with ammonium perchlorate, NH_4ClO_3 dispersed in a plastic binder such as polybutadiene) are inflexible since they are limited to using all the propellant in a single burn but they are simple and reliable and burn smoothly at constant rate on the exposed surface of the charge. They offer high mass rates but low $I_{sp} \sim 250$ s and are often used for LEO/GEO perigee/apogee kick motors (e.g. PAM-D apogee kick motor). Multiple trajectory requirements imply multiple burns which in turn imply the need for using liquid fuel with throttling and restart capability. Liquid-fuelled systems tend to be more complex than solid fuel systems with moving parts but offer greater control and flexibility. They offer high combustion temperatures and low molecular weights. Cold gas systems which utilise a relatively inert gas such as nitrogen, argon or propane yield low thrust levels: orbital maneouvering requires $\sim 10^3$ N for ~ 60 s and the LEO–GEO transfer requires a Δv of 4.3 km/s. Although cold gas systems are suitable for stationkeeping $\sim$mN, they are not suitable for large-scale orbit transfers. For orbit–orbit transfers orbit and attitude control, non-cryogenic liquid fuels tend to be adopted. Pressurised He keeps the fuel under pressure at the outlet valve. Monopropellants are sufficient for orbit and attitude control. An example is the decomposition of anhydrous hydrazine N_2H_4 by catalytic action (see Fig. 14.2(a)). The Shell 405 catalyst comprises an Al substrate with finely divided iridium on the surface of the catalytic bed:

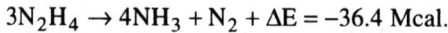

$$3N_2H_4 \rightarrow 4NH_3 + N_2 + \Delta E = -36.4 \text{ Mcal.}$$

A fraction ~ 0.4 of the ammonia decomposes further:

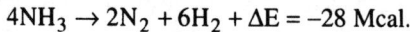

$$4NH_3 \rightarrow 2N_2 + 6H_2 + \Delta E = -28 \text{ Mcal.}$$

Electric resistive heaters are required to maintain moderate bed temperatures for long life applications to avoid catalytic degradation. This fuel will give an I_{sp} of 200–250 s (nominally 235 s).

For larger Δv for orbit transfers, bipropellants which employ fuel and oxidiser should be employed, eg. fuel monomethyl hydrazine (MMH)/oxidiser nitrogen tetroxide (N_2O_4) mixture gives $I_{sp} = 310$ s (i.e. $\sim 50\%$ greater than for monopropellants) (see Fig. 14.2(b)). The oxidiser/fuel stoichimetric ratio is 1.64 for the N_2O_4/MMH mixture and the combustion temperature T_c is 3460 K. It eliminates the need for an ignition system. This is a popular choice for deep space missions to the outer solar system which often adopt this type of propellant with a 400 N main engine thruster. However, these ammonia-based fuels are highly corrosive and plume impingement on the spacecraft must be avoided. Hydrazine has the advantage that it can be used as both monopropellant or bipropellant fuels. Multifunctional monopropellant propulsion systems may give 4–5 orders of magnitude thrust variability without the need for throttling [Sabroff 1968]. An example is monopropellant hydrazine and hydrogen peroxide which offers three operating modes:

(i)   low thrust operation—the decomposition products are stored in the plenum and used as cold gas jets;

(ii)   intermediate thrust operation—the monopropellant fuel is used for thrusting;

(iii) high thrust operation—the monopropellant may be used in bipropellant mode with the peroxide and mixed in a rocket thrust chamber.

Furthermore, their performance may be improved by using electrothermal resistojets. Electrothermal resistojet thrusters use resistance heater elements to convert electricity to heat and increase the exhaust velocity of the gases thereby increasing the specific impulse of the propellant. Electrical energy is used to increase the exit velocity of the gas usually H_2, N_2, NH_3 or any combination of these gases or hydrazine giving up to $\sim 100\%$ increases in I_{sp} to ~ 400 s for catalytic hydrazine systems (with potential increases up to 600 s). For hydrazine systems, the fuel is heated electrically to trigger exothermic decomposion at ~ 550 K into its products which are then superheated ohmically to ~ 2000 K before ejection. Alternatively, electric arc discharge heating to 20 000 K can

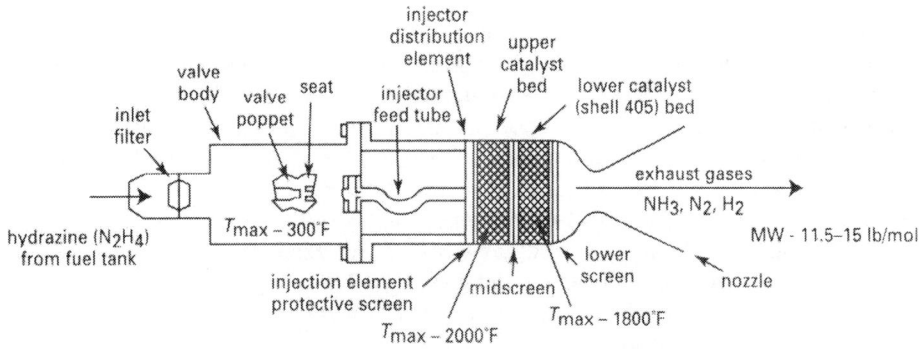

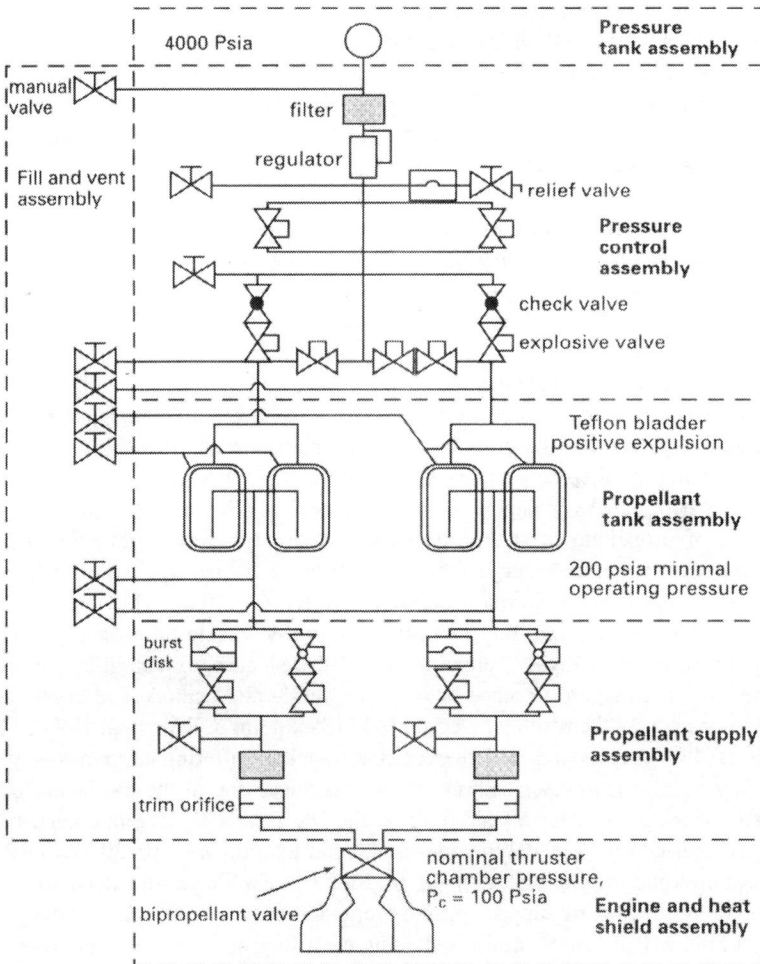

Fig. 14.2. (a) Hydrazine thruster (from Werte & Larson 1999); (b) MMH/N_2O_4 bipropellant propulsion system (from Barter 1999). Reproduced with permission from TRW Space & Electronics Group.

create a plasma which is accelerated through the supersonic nozzle. This doubles the I_{sp} rating from a conventional resistojet to ~800 s. This is three times the I_{sp} available to a conventional hydrazine thruster. The hydrazine resistojet also avoids the problem of deterioration of the catalyst bed in conventional thrusters. Resistojet thrusters have been used on Intelsat V and several commercial comsats for stationkeeping. Resistojets are capable of throttle ratios of 10:1 and have very high expansion ratios ~5000:1 which eliminates corruption of sensors by jet plumes. They also have the advantage that if the heater fails they can be used as cold gas thrusters. An example is the LeRC hydrogen resistojet which has the following properties: $I_{sp} = 710$ s, $F_{thrust} = 5.25$ N, $P = 30$ kW, $\eta = 0.61$. They also have the advantage that their power consumption is low compared with fully electric propulsion. The catalytic thrusting mode may be adopted as backup. They have been advocated for use with ion thrusters to optimise I_{sp} for power-limited scenarios such as for a LEO–GEO space tug.

Fuel tanks should be arranged symmetrically about the vehicle thrust axis to avoid shifts in propellant distribution as fuel is consumed. The spacecraft may employ from a few to ~30 thrust units to provide a thrust ~500 N sufficient for orbit transfer. One such thrust unit is the 0.4 kg NASA 4.5 N hydrazine thruster with a power requirement of ~5 W/thruster. These thrusters may be gimballed typically to ±7° to align the thrust vector through the spacecraft during burns. Gimballing of thrusters allows account to be taken for changes in spacecraft mass distribution due to fuel expenditure.

14.2.1 Electric propulsion systems

Chemical fuels are energy-limited since the energy is stored in the molecular structure of the fuel and released through combustion. The maximum I_{sp} attainable with chemical systems is limited by the energy released in chemical reactions and LO_2/LH_2 offers one of the best performances available with an $I_{sp} = 475$ s. This energy limitation imposes large ratios of propellant to spacecraft mass. Electric propulsion thrusters are power-limited and overcome the energy limitation offering higher I_{sp}, e.g. the T5 thruster operated at 25 mN thrust required a nominal power of ~640 W. This imposes a consequent decrease in the required fuel-mass fraction by increasing burn-times. Present electric propulsion methods involve creating energy by electrically charging the propellant gas, generating a voltage difference between an anode and cathode and accelerating the propellant ions across the resultant electric field [Sackheim & Rosenthal 1994]. Electrical energy is used for accelerating and ejecting the propellant offering large mass savings due to their high I_{sp}. A neutraliser cathode introduces electrons to the ion beam to prevent spacecraft charging. They have been used on the US Telstar IV among others. If electric thrusters were used on 12 y lifetime comsats in conjunction with 10–20 kW solar arrays, the comsat lifetime would be extended to 15–17 y. Similarly, the need to reposition satellites for re-tasking to support military operations can limit spacecraft operational lifetime but this effect can be minimised using electric propulsion. As well as repositioning capability electric propulsion provides a loiter capability for critical observations and the maintenance of relative position for constellations.

Ion thrusters generally offer up to an order of magnitude increase in I_{sp} over chemical systems ~3000 s. We shall consider the electrostatic and radio frequency ion thrusters which accelerate charged propellant ions electrostatically in an electric field [Clark 1975]. There are variants of ion thrusters including colloidal thrusters, magneto-electrostatic containment (Hall) thrusters, contact ionisation thrusters, and magnetoplasma dynamic thrusters, but are in lesser degrees of development. Electrostatic devices employ a high current pulsed discharge from a capacitor to ionise and accelerate the propellant. The Kaufman ion thruster is the most highly developed electric propulsion system (barring the resistojet) and its performance increases as its size is scaled up though the use of multiple small thrusters is preferred over a single large thruster. Propellant gas atoms are injected into the discharge chamber through the rear walls at low chamber pressure $\sim 10^{-3}$ torr. In the Kaufmann ion thruster, the injected propellant atoms are ionised into an almost fully ionised neutral plasma by a dc electric arc discharge between a central cylindrical hollow cathode which emits electrons and a cylindrical concentric anode (see Fig. 14.3). An axially divergent symmetric solenoid magnetic field nozzle ~10 G is applied by permanent magnet pole pieces at the downstream end of the anode and around the cathode to increase the efficiency of ion generation to ~80% by increasing their spiral path length in the discharge chamber. It also forms a magnetic nozzle to limit ion and electron migration to the outer chamber walls. An electric field is imposed between a pair of aligned electrode grids at the end of the discharge chamber due to the potential difference between them. A screen electrode is maintained at the same positive potential as the cathode. A negatively biased accelerator grid downstream of the screen grid accelerates the positive ions of the plasma across the ~kV potential difference. The resultant ion beam produces thrust and a second smaller external cathode downstream of the accelerator grid supplies electrons for neutralisation of the ion beam. Exhaust velocity is determined by the available power and the grid breakdown voltage. They are 80% efficient. The propellant may be any material that is gaseous at 400°C and is usually Hg but Ar and Xe are also suitable as alternatives to minimise environmental effects as well as simplifying thruster design, particularly for geocentric missions. Note, however, that Hg comprises a very effective gamma-ray shield if used with nuclear fission power systems. Unfortunately, Xe has a high cost and limited availability, but Kr/Xe combinations retain high thruster performance over the use of Kr alone while utilising the low cost of Kr. These ion engines operate at low thrusts to maintain a low power supply requirement and so operate over long continuous thrust burn periods. Kaufman ion thrusters have been flight tested as secondary propulsion systems and are suitable for sizes up to ~1–2 m diameter appropriate for primary propulsion. Ion thrusters using Xe have been adopted for Intelsat VIII, Telstar IV, ETS-VI and Eureca. Kaufman ion thrusters have been flight tested as secondary propulsion systems and are suitable for sizes up to ~1–2 m diameter appropriate for primary propulsion.

Electric power input for electric thrusters is given by $P = (FI_{sp}g)/2\eta$ for a beam power of $P = \frac{1}{2}\dot{m}v_{ex}^2 = \frac{1}{2}Fv_{ex}$ so operation at low thrust minimises the power requirement [Mickelson 1967]. Heavy propellants maximise the mass/charge and minimise the required thruster size. Lower mass propellants, however. reduce the voltages required for the accelerator grid. Performance can be increased by increasing the beam accelerating discharge potential, beam current density and the flow rate. For an ion thruster, exhaust velocity:

$$v_{ex} = \sqrt{\frac{2eV_T}{m_i}} \tag{14.13}$$

where V_T = threshold voltage ~5 kV, m_i = ion mass, $F = \eta v_{ex} \dot{m}$ = thrust, η = efficiency ~0.9, and $\dot{m}$ = mass flow rate. Hence

$$F = \dot{m}v_{ex} = \frac{\dot{m}e}{m_i}\sqrt{\frac{2m_iV_T}{e}} \quad \text{where} \quad I = \text{total current} = \frac{\dot{m}e}{m_i} \tag{14.14}$$

Rf ion thrusters (RIT) operate similarly to the Kaufmann thruster but use an rf ionising discharge plasma generated by an rf-generator induction coil rather than dc discharge (see Fig. 14.4). The excited free electrons injected by the neutraliser ionise the propellant gas. However, rf coupling is inefficient and lifetime limitations can occur due to shorting of the rf field by metallic coatings being sputtered onto the interior of the discharge chamber. The 10 cm diameter German RIT-10 has a thrust level of 15–25 mN for an input power of 700 W giving an I_{sp} of 3000–3100 s .

The 10 cm diameter UK-10 electrostatic ion thruster of mass 2 kg is an example offering an I_{sp} of 30 km/s—it uses an Hg propellant to provide 10 mN of thrust, but there is a variant which uses Xe to provide 20–25 mN of thrust with a power of 700 W giving an I_{sp} of 3400 s. A scaled-up 25-cm version of mass 7 kg (UK-25) has a thrust of 230 mN, giving a specific impulse of 4900 s but it requires a power input of 6.25 kW (power-to-thrust ratio of 30 W/mN). The UK-25 was designed as a primary propulsion unit for orbit maneouvres. Secondary propulsion required thrust levels ~10–30 mN while primary propulsion for orbit changes requires thrust levels ~200–300 mN, e.g. the UK-25. These thrust levels can be made higher by varying propellant flow rates and power supply voltages or currents.

The UK DERA (Defence & Evaluation Research Agency, formerly RAE) T5 Kaufmann ion thruster (10 cm) has the following characteristics [Fearn 1982]:

Thrust:	10–25 mN (typically, but 70 mN is achievable)
Exhaust velocity:	30–35 km/s
Ion beam current:	80 mA
Anode potential:	42 V
Discharge current:	1.0 A
Efficiency:	0.74
Power:	230 W

In general, electric thrusters tend to suffer from sputtering erosion of the cathode and anode element materials which increases as currents are increased. This is caused by ions falling from the discharge region back onto the cathode surface and so sputtering away the electrode material and this tends to limit durability. Thruster cathode lifetimes scale as $t(kh) = 53.7P_0(kW)^{-0.56}$. Sputtering can be reduced by using Hg propellant as Hg vapour is absorbed onto the accelerating grid which acts as a protective layer. Other less noxious fuels such as Xe are becoming commoner but they do not have this desirable property. Ar and Xe give better performance with efficiencies ~70% and simplified design and are not a harmful contaminants like Hg is. However, long lifetimes can be

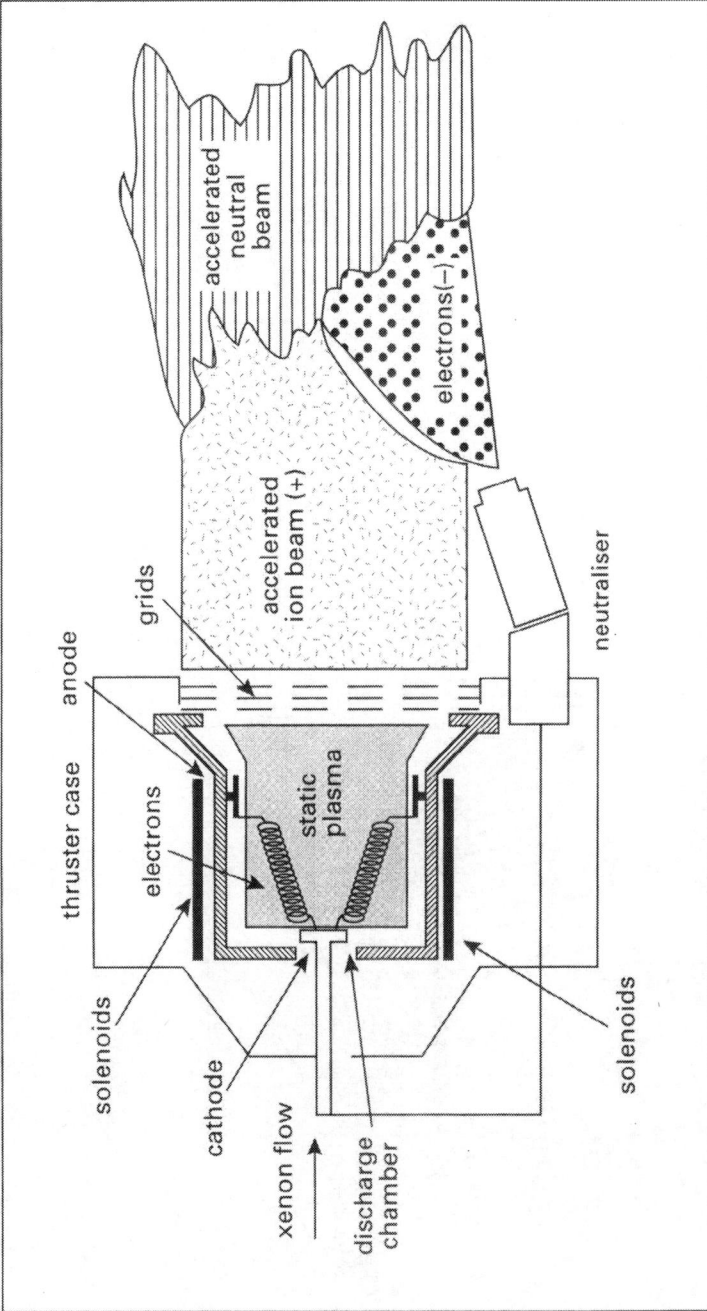

Fig. 14.3. Electrostatic ion thruster (reprinted from *ESA Bulletin*, August 1988).

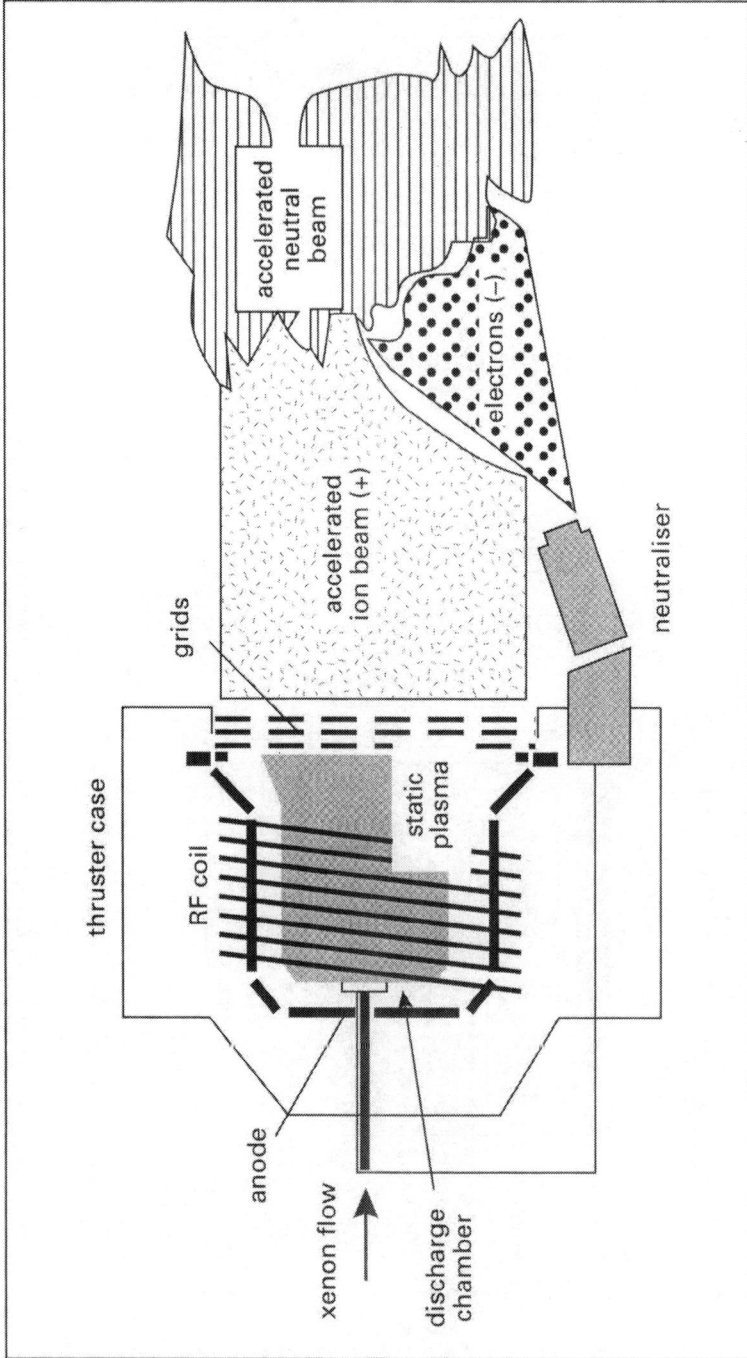

Fig. 14.4. Radio-frequency ion thruster (reprinted from *ESA Bulletin*, August 1998).

Thruster	Kaufman	Rf	MESC	MPD	Colloid	FE	CI
Thrust (mN)	200	160	17	140×10^3	0.5	1.5	1.5
Power (kW)	5.4	3.6	0.34	6000	0.01	0.16	0.12
I_{sp}(s)	5000	3360	3270	2400	2000	9000	6700
Efficiency	0.92	0.83	0.88	0.31	0.7	0.8	0.7

Fig. 14.5. Properties for a 10-cm thruster (averaged from Fearn 1982).

maintained by using spring mechanisms to continuously replenish the surfaces and by ensuring that the anode voltages <40 V. Furthermore the electromagnetic field and propellant fluxes can induce environmental impacts such as sputter erosion of exposed surfaces, interactions between the ambient space plasma and the ejected plasma causing spacecraft charging and arcing, and electromagnetic interference due to the electromagnetic fields generated as a possible hazard to radio signal propagation (this final effect does not appear to be significant particularly at higher frequencies). High voltages can also cause electrical breakdown in dual grid systems—this, however, may be alleviated by adopting a four-grid system whereby the extraction of ions and their acceleration are separated. The K5 thruster has an operational lifetime of 10 000 h.

The SERT-C spacecraft was designed (but never flown) with ion thruster technology for LEO to GEO orbit raising [Fearn 1982]. With a spacecraft mass of 818 kg, it was to use 3×30 cm ion thrusters for orbit raising and 4×8 cm ion thrusters for attitude control and station-keeping powered by two solar arrays of 2.74×16.2 m for 9 kW BOL and 4.5 kW EOL power raising. The 30-cm ion thruster was the 10 kg NASA/Hughes thruster with $V_T = 5$ kV, $I = 4$ A, $P = 20$ kW and $\eta = 0.9$. The use of Hg gives an $I_{sp} = 630$ s and $F = 0.58$ N while Ar gives an $I_{sp} = 1430$ s and $F = 0.26$ N. Each thruster module had a mass of 55 kg including tankage and waste heat radiator. ESA's Artemis (Advanced Relay & Technology Mission) is a test satellite for data relay and land mobile communications applications. It uses ion propulsion units for north–south station-keeping. The British UK-10 (based on the T5 thruster) and the German RIT-10 were adopted, comprising thruster, neutraliser, power supply and control unit, propellant tank and flow control unit. Two sets of each thruster were mounted onto Artemis.

For an LEO–GEO standard orbit-raising manoeuvre, continuous low thrust electric propulsion tangential to the trajectory would require ~30–100 days to complete the manoeuvre in a spiral trajectory with a thrust of ~10^{-4} g. This is one reason why they have been advocated for interplanetary missions. However, continuous electric thrusters may offer great fuel savings for multiple orbit spacecraft due to their high specific impulses, particularly for shorter orbit repositioning transfers ~1000 km. Continuous thrust results in an increased Δv requirement of ~20% over standard manoeuvres. All these systems except the electrothermal resistojet system impose large power requirements.

A second design approach for Atlas may utilise more advanced technologies. A Kaufman Ar ion propulsion system may be adopted with regenerative fuel cells. The flexible solar array size is enlarged to cope with ion thruster power requirements. They are deployed while operating the thrusters as the thrusts will be low during orbit transfer but will be retracted during manipulation. Fuel cells power the spacecraft during the manipulation phase. Peak power requirements are now given by the thrusters.

Peak power $P = FI_{sp}/2\eta$ where $\eta = 0.9$, $I_{sp} = 6000$ s, and $F = 0.5$ N. Hence $P = 1.7$ kW per thruster. Assuming three 30-cm thrusters implies the peak power load is 5 kW for the propulsion system alone. The trend towards the use of higher performance propulsion increases the time between refuelling assuming the same propellant mass is loaded:

$$\Delta v = -gI_{sp} \ln\left(1 - \frac{m_p}{m_i}\right) = 22.45 \text{ km/s}$$

This is sufficient for over two LEO–GEO two-way transfers (one-way Δv of 4.2 km/s). Hence, this may provide a prototypical spacecraft bus for an off-the-shelf OMV Space Tug. As pointed out by Wertz et al (1988) if such a system used low thrusts as characterised by electric propulsion the payload could deploy all its arrays, antennae, etc. before operational orbit insertion to high orbit, offering greater reliability for system checkout as LEO is accessible to in-orbit maintenance by astronauts. However, no such limitation is imposed by ATLAS. Due to the limitation on solar array size for a robotic freeflyer this approach appears to rule out electric propulsion systems as an option for an EVA-equivalent servicer (though for a larger robotic servicer this may not apply as nuclear power sources may become attractive). This design is an alternative to our baseline but reliability considerations suggest that established technologies should be adopted for the baseline design of the spacecraft bus of ATLAS.

14.3 MISSION PROFILE

A typical mission profile for the robotic freeflyer ATLAS consists of several phases with rendezvous and docking (RVD) required [Claudinon et al 1985]. MacInnes (1992) suggested that both target and interceptor perform a cooperative rendezvous to minimise the interceptor fuel consumption. It is conceivable, however, that the thrusters or attitude control system of the target is inoperative (e.g. Solar Max had no attitude control available). Furthermore, it is likely that no gain is in fact achieved as the target's operational capability is reduced unless refuelled. Fallin (1975) considered the problem of minimising Δv requirements for several intersatellite transfers between different orbits. It is highly unlikely that this situation of multiple missions will be required, since intersatellite transfers will be determined according to the operational priority of missions rather than according to Δv minimisation. Tangential thrusting changes the semimajor axis, eccentricity and argument of perigee; radial thrust changes the eccentricity and argument of perigee; thrust to the poles changes the inclination, longitude of ascending node and argument of perigee. In general continuous impulse efficiency is not constant along the orbit so optimal techniques are used to calculate the thrust sequence to change the required orbit using the minimum amount of fuel.

Rendezvous may be defined as a series of manoeuvres designed to bring two vehicles into close proximity to ultimately enable docking such that the two vehicles become mechanically joined. The mission profile phases are given by:

(i) Spacecraft manoeuvres to rendezvous with another spacecraft demand critical timing. Orbit rendezvous and interception in space and time require phasing orbits with coelliptic flight patterns. Orbit rendezvous requires interception of the target spacecraft by the chaser spacecraft and this involves phasing orbits. It was during the Gemini missions that the coelliptic flight pattern was developed. It was used during Apollo for the lunar module rendezvous with the circumlunar CSM (command/service module) and is still adopted by the Space Shuttle. The interceptor coasts to catch up with the target and when the line of sight is ~25° above the interceptor's local level, the final intercept manoeuvre is executed using either onboard radar or optical guidance. As the two spacecraft approach, the chaser adjusts its orbit to match its position more closely to that of the target. The phasing manoeuvres of the interceptor vehicle from below and behind the target bring the interceptor to within ~100–500 m of the target to avoid the possibility of a collision. The below and behind manoeuvre will also be the relevant manoeuvre if aero-assisted deorbiting is adopted. The chaser spacecraft is in a slightly lower orbit than the target and since it has a slightly shorter period, it can intercept or overtake the target. During the early days of the space effort, orbital rendezvous and docking were performed within ~1 h of orbit insertion. The wait time in the initial orbit was given by:

$$t = \frac{\varphi_i - \varphi_f + 2k\pi}{w_c - w_t} \tag{14.15}$$

where $\varphi_i - \varphi_f = 180 - \alpha =$ phase angle between target and interceptor for rendezvous, $k =$ number of rendezvous required, $w_c =$ angular velocity of interceptor/chaser, $w_t =$ angular velocity of target, and $\alpha =$ lead angle from interceptor to target.

However, the single-orbit chase pattern is prone to failure so 24-h or 48-h chase patterns have tended to become common. If the chase orbit has a smaller semimajor axis than the target orbit then it will catch up so that the target is directly above the chaser spacecraft at which point the chaser can execute its first manoeuvre. The standard orbital mechanics orbit transfer between two non-intersecting orbits characterises the homing phase of the manoeuvre [Chande & Newcomb 1985, Kaplan 1976]. To inject into a desired orbit from a parking orbit at least two burns are required. The two-impulse Hohmann transfer is the minimum energy manoeuvre to patch between two circular concentric orbits and involves the minimum fuel expenditure in traversing a central angle of 180°. From a circular parking orbit, a perigee burn increases the initial orbit circular velocity by Δv_p to facilitate the transfer by opening the transfer ellipse tangential to and intersecting both orbits:

$$\Delta v_p = \sqrt{\frac{\mu}{r_p}}\left(\sqrt{\frac{2r_a}{r_a + r_p}} - 1\right) \tag{14.16}$$

from $v_p = \sqrt{\mu/r_p} =$ perigee circular velocity with period, t_p, where $t_p = 2\pi\sqrt{a^3/\mu}$, $\mu = GM_{Earth}$, $r_p =$ perigee radius from Earth centre (Earth radius = 6383.2 km), $r_a =$ apogee radius from Earth centre.

The transfer orbit has a period,

$$t_{trans} = 2\pi\sqrt{\frac{a^3}{\mu}} = 2\pi\sqrt{\frac{\left[(r_a + r_p)/2\right]^3}{\mu}} \tag{14.17}$$

At the apogee, another thrust is applied to circularise the elliptical transfer orbit by incrementing the velocity by Δv_a :

$$\Delta v_a = \sqrt{\frac{\mu}{r_a}}\left(1 - \sqrt{\frac{2r_p}{r_a + r_p}}\right) \quad \text{to} \quad v_a = \sqrt{\frac{\mu}{r_a}} = \text{apogee circular velocity}$$

$$\tag{14.18}$$

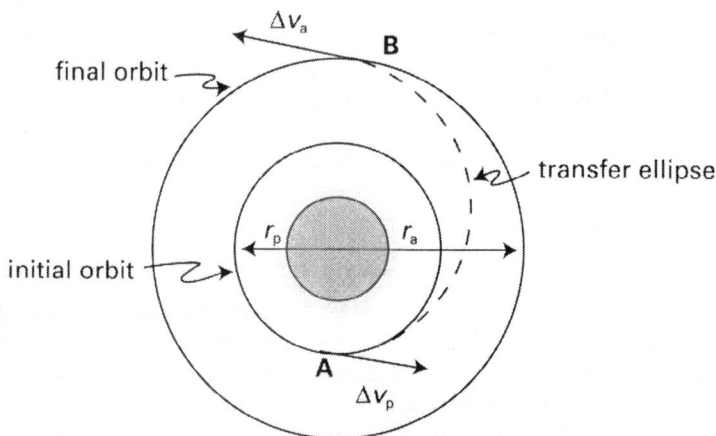

Fig. 14.6. Hohmann transfer ellipse.

The total velocity change is given by: $\Delta v = \Delta v_p + \Delta v_a$. The time of flight for the Hohmann transfer given by: $\Delta t = \pi\sqrt{a^3/\mu}$ where a = orbit semimajor axis.

For an orbital plane change, the angular momentum vector must be altered by applying thrust parallel to that vector and perpendicular to the orbit plane at equatorial crossing such that the orbit precesses by angle $\Delta i : \Delta v_i = [2v_i \sin(\Delta i/2)]$. However, such plane rotations (e.g. equatorial to polar manoeuvre) involve the use of much fuel (an inclination change of 1° requires ~208 m/s Δv at LEO) so it would be desirable to use aerodynamic manoeuvres for plane changes such that the lift vector is rotated out of the plane of the trajectory without or at least minimising propulsive burns. The most efficient method for plane changes while minimising fuel expenditure is to change inclination at the apogee burn circularisation manoeuvre since the spacecraft velocity is lower at the apogee than at perigee:

$$\Delta v_{a/i} = \left(2v_a^2 + v_a^2 - 2\Delta v_a v_a \cos \Delta i\right)^{1/2} \tag{14.19}$$

where Δv_a = initial transfer orbit circular velocity, and v_a = final orbit circular velocity.

Time for the complete transfer, $\Delta t = \pi \sqrt{a^3/\mu}$.

If low thrust electric impulse manoeuvres were used at ~0.001 g the LEO–GEO transfer would take several months—such manoeuvres are not impulsive and generate spiral trajectories.

This phase of the mission profile is obstacle-free and would take the freeflyer to within ~10–30 km below the target and coplanar with its orbit. The double coelliptic manoeuvre accomplishes this in two burns: first to within 30 km and second to within 15 km of the target, e.g. Space Shuttle. This preliminary phasing manoeuvre is analysed in geocentric inertial reference coordinates. It has been suggested that nodal regression may be utilised to reduce the Δv requirements [Robertson et al 1988]. Nodal regression $\delta\Omega$ varies with altitude and inclination and the resultant differential nodal regression rates between orbits may be exploited. Coplanar situations occur when the ATLAS orbit plane rotates above or below the target orbit plane. On alignment, a Hohmann transfer may be executed. Similarly, offsetting inclinations may be exploited particularly at near-polar orbits. The initial interception is to an orbit some 20 km below the target coplanar with its orbit.

The two-impulse rendezvous motion of the chaser in the target orbital frame are given by the Clohessy–Wiltshire equations if both orbits are circular with similar semimajor axes and inclinations. They are essentially Hill's equations equated to zero, ie. no external forces are applied to the chaser spacecraft [Claudinon et al 1985]. Eccentricity may be accommodated in the equations. These and subsequent manoeuvres involve the closure of the two vehicles separated by small distances relative to the orbit dimensions with respect to geocentric inertial coordinates. Hence, a non-inertial frame of reference is used fixed in the target vehicle to enable the chase vehicle manoeuvres to be determined with respect to the target vehicle. A chaser spacecraft has coordinates r, θ, z with respect to a target spacecraft and the target-referenced coordinate axes rotate with orbital velocity $w_0 = \sqrt{\mu/r_0^3}$. This provides a set of linear constant coefficient differential equations (Clohessy-Wiltshire equations):

$$\left. \begin{array}{l} \ddot{r} - 2w_0 r_0 \dot{\theta} - 3w_0^2 r = 0 \\[2mm] r_0 \ddot{\theta} + 2w_0 \dot{r} = 0 \\[2mm] \ddot{z} + w_0^2 z = 0 \end{array} \right\} \quad \text{where} \quad \left\{ \begin{array}{l} r = x \\[2mm] y = r_0\theta \end{array} \right\} \quad \text{for cartesian representation}$$

$$(14.20)$$

The solution to these equations give the drift rates for the chaser to enable rendezvous with the target. Initial conditions provide constraints for their solution such that z_0, r_0, θ_0 are the initial coordinates of the chaser spacecraft. The solutions are given by the time rate of change of the coordinates and relative velocities:

$$z(t) = z_0 \cos w_0 t + \frac{\dot{z}_0}{w_0} \sin w_0 t$$

$$r(t) = -\left(\frac{2}{w_0}r_0\dot{\theta}_0 + 3r_0\right)\cos w_0 t + \frac{\dot{r}_0}{w_0}\sin w_0 t + 4r_0 + \frac{2}{w_0}r_0\dot{\theta}_0 \qquad (14.21)$$

$$\theta(t) = \theta_0 - \left(3\dot{\theta}_0 + \frac{6w_0 r_0}{r_0}\right)t + \left(\frac{4\dot{\theta}_0}{w_0} + \frac{6r_0}{r_0}\right)\sin w_0 t + \frac{2\dot{r}_0}{w_0 r_0}\cos w_0 t - \frac{2\dot{r}_0}{w_0 r_0}$$

$$\dot{r}(t) = \left(2r_0\dot{\theta}_0 + 3w_0 r_0\right)\sin w_0 t + \dot{r}_0 \cos w_0 t$$

$$\dot{z}(t) = -z_0 w_0 \sin w_0 t + z_0 \cos w_0 t \qquad (14.22)$$

$$\dot{\theta}(t) = \left(-3\dot{\theta}_0 - \frac{6w_0 r_0}{r_0}\right) + \left(\frac{6w_0 r_0}{r_0} + 4\dot{\theta}_0\right)\cos w_0 t - \frac{2\dot{r}_0}{r_0}\sin w_0 t$$

The out-of-plane component δz of the chaser oscillates with the orbital period and is removed by waiting until the drift reaches $z = 0$ and thrusting with acceleration $\ddot{z}$ such that $\dot{z} = 0$. If the chaser is below the target as is normally the case, a perigee-raising manoeuvre adjusts the chaser orbit. A coelliptic manoeuvre at the chaser orbit apogee will allow the chaser to match the target. The terminal phase involves a two-impulse final trajectory. A closed-loop terminal proportional controller reduces the range and range rate to zero to allow the transfer trajectory to be maintained. As the final closure occurs braking manoeuvres reduce the residual velocity to zero near the target.

Rendezvous is complete when the chaser and target are seperated by <100 m with zero relative velocity. Chaser manoeuvre acceleration is dominated by orbit mechanics for a rectilinear manoeuvre $\ddot{r}$. The only passive station-keeping positions are behind or in front of the target in its orbit as radial or out-of-plane differences will cause oscillations of the chaser with respect to the target. The Clohessy–Wiltshire equations describe how to reduce to zero the spacecraft relative motion assuming the impulse durations to be instantaneous. The result is a series of coasting trajectories [Simmons et al 1990]. The spacecraft is now at point P_1 which is a stable standoff point on the target orbit. Station-keeping at P_1 enables the interceptor to perform further trajectory planning prior to execution on the basis of sensor information and allows waiting for advantageous manoeuvre conditions (e.g. ground station visibility). Onboard trajectory planning is required to preserve spacecraft autonomy without violating the mission constraints such as maximum time of flight, fuel efficiency, etc. The trajectory planner creates a series of intermediate states between the current state and the target state defining a planned orbit trajectory with an impulse profile of all the Δv thrusts that need to be performed. The number of impulses should be restricted to a maximum of four and this will be sufficient to generate closing velocities down to 0.01 ft/s. A closed-loop autopilot may execute additional corrective firings if the current state strays too far outside a specified deadband around the reference trajectory. Guidance for intercept and rendezvous requires guidance radar and/or optical lidar, gyroscopic reference devices and accelerometer sensors for range, range rate, elevation and azimuth angles along the line of sight to the target.

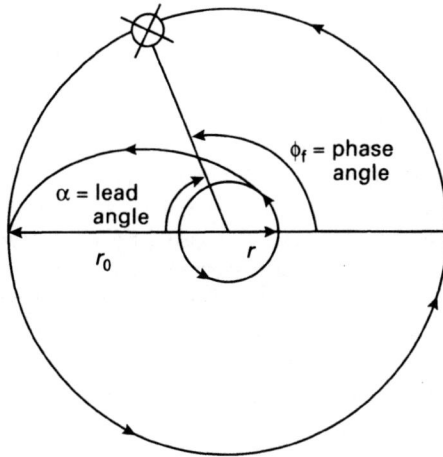

Fig. 14.7. RVD geometry between two co-elliptic orbits.

(ii)   This is the final approach phase which will take the interceptor to within ~50 m of the target spacecraft. It is not unusual for up to six correctional burns to be required to complete docking. A flyaround manoeuvre from point P_1 may be required to acquire an approach corridor along the spin axis (usually the major axis of inertia) if the target is spin-stabilised (see Fig. 14.8). If there are no unbalanced forces on the target an arbitrary rotation of the x, y and z axes of the target may be resolved into a constant rotation about a single resultant spin axis. This spin axis will comprise the approach corridor vector. If the spin axis lies along the orbit plane then the flyaround is not necessary (unlikely). In general, the spin axis lies out of the orbit plane along the orbit normal so flyaround is most likely to be necessary when dealing with spin-stabilised spacecraft. Flyaround will

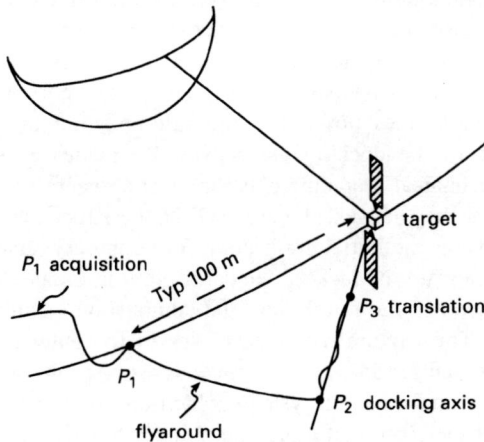

Fig. 14.8. Rendezvous trajectory (from Claudinon et al 1985).

require complex computational facilities and position/attitude sensors to enable a trajec-tory profile to be calculated. After the flyaround phase, the interceptor lies at point P_2 along the selected docking axis. The motion of both bodies are virtually identical and assumed to be rectilinear with no net relative translational velocity. Active stationkeeping is required until the next maneouvre. If the target is a spinner the next manoeuvre must be to spin up the interceptor to a constant angular rate with the flyaround manoeuvre placing the interceptor perhaps 100 m below the target aligned with its spin axis. The rotation of a spin-stabilised target may be determined by visual tracking. Human reactions are too slow to track and grapple spin-stabilised spacecraft which have rotation speeds of up to 60 rpm. Astronauts are generally limited to ~1 rpm [Schenker 1988]. Feature matching of images and internal models gives the orientation of the target spacecraft with its grip windows. The attitude control system of the interceptor spins it up to match the rotation of the target such that their relative rotational spins are zero while aligned along the target's spin axis.

(iii) This is the final translation phase along the docking axis to point P_3. Visual inspection of the docking points by TV cameras is maintained. Trajectory control is closed-loop from the range sensors and translation is performed at a predefined range/range rate profile. The last few metres is executed at constant velocity with accuracies ~2 cm and 0.2° to the point P_3. Up to this point the manipulator arms have been in a stowed configuration and once P_3 is reached the target fixtures lie within the workspace of the manipulators at less than 2 m distance. The chaser spacecraft must then align its heading vector with the object's grapple vectors so that the manipulators are positioned such that they may grapple the grapple points by straight line trajectories. The arms move to the set position and then to their projected entry points to the grapple points on the target. Obstacle avoidance may be necessary if satellite appendages impose possible collision hazards. Wang (1987) proposed a grappling technique (the acquisition phase) such that the joints of the manipulators are locked and the vehicle translates in a straight line so that the manipulators intercept the object at predefined points and then grasp those points. This procedure is highly inflexible in that it comprises a standard spacecraft docking manoeuvre which is expensive in fuel due to the tight constraints on docking accuracy. Furthermore, it does not utilise the full potential of the manipulators in eliminating this phase of the docking manoeuvre. De Peuter et al (1993) noted that limited options for mechanical grappling exist on most spacecraft due to thermal insula-tion covering. However, they suggested that for GEO operations the nozzle of the apogee boost motor could serve as the initial grasp point for temporary attachment. A dedicated capture tool to accommodate the lack of standardisation in apogee nozzles comprises a stinger which is inserted via the nozzle into the combustion chamber and expanded to provide a firm grasp. The capture tool then clasps to the outer ring of the nozzle to achieve greater stiffness. Unfortunately, this manoeuvre grasps the spacecraft at the oppo-site side to the payload and other subsystems which are the most likely candidates for repair. They suggested therefore that a grapple point is found subsequently (since most spacecraft are now being designed to be serviced by astronauts) or if necessary build a small lightweight truss structure of its own to use as 'scaffolding', but this imposes further mass constraints. This technique, however, is well-suited to GEO satellites with apogee kick motors.

(iv)   After the rendezvous phase when the chase vehicle approaches the target the capture phase ensues comprising of a soft dock for initial alignment within the capture envelope followed by a hard dock for the final alignment and rigidising of the composite structure. Berthing may be defined as the grasping of a payload with a manipulator and manoeuvring it to an active spacecraft at the berthing point for locking. Docking involves the active spacecraft flying into a docking port on a passive spacecraft such that the two are joined rigidly [AIAA 1991]. Rendezvous may be cooperative (as in Apollo–Soyuz mission) or uncooperative (as in Solar Maximum mission). Cooperative spacecraft carry transponders and lights to aid rendezvous.

Spacecraft rendezvous and docking (RVD) are central capabilities for any complex space activity. The techniques outlined here were developed during the manned Gemini missions preceding the Apollo missions to the Moon. Gemini's precursor Mercury (1961– 63) placed six spacecraft into space (two suborbital lobs and four low orbital flights) and was designed merely to place men into space and recover them. Gemini was orders of magnitude more complex and was designed to establish crew flights of up to 14 days, permit an astronaut to leave the spacecraft in a spacesuit and rendezvous with another spacecraft by manoeuvring in orbit. All these activities had to be perfected in preparation for a lunar expedition. To this end, Gemini spacecraft were the first manned spacecraft to use onboard computers for onboard inertial guidance during ascent, rendezvous, orbital flight and re-entry. All these objectives were achieved in ten two-man Gemini flights in 1965 and 1966. The first RVD was performed by Gemini VI in 1965 with a Gemini VII target that had been launched in advance. This was followed by Gemini VIII, IX, X and XI RVD missions with unmanned Agena upper stage targets. The Agena D target vehicle had a spinal docking collar comprising of a docking cone. Without the lessons learned during the Gemini programme in RVD, men would never have landed on the Moon nor inhabited space stations. The Soviets have considerable experience in RVD. They have had a space station in orbit continuously since Salyut-1 was launched in 1971. Six more were launched through to Salyut-7 in 1983. They were fully refuellable and resuppliable, unlike Skylab. The Mir space station was launched 1987 and is a modified Salyut with an airlock and docking ports at one end and a module on the other. The Soyuz series of spacecraft first launched in 1967 similar to the Gemini act as ferries for their space stations having completed more than 50 flights. The Soyuz have demonstrated the capability for automated docking using radar closed-loop control since 1967 (though not always successfully). The Shuttle is not really designed for close soft docking but utilises the RMS for this purpose, though it has more recently undergone docking with the Mir Space Station and will undergo docking with the ISS.

Mechanical interfaces for docking should be simple, highly reliable and durable. They should have emergency disconnect provisions by manual or autonomous means. Such interfaces should be in line with the centres of masses of the two objects being joined. The load capacity of the interface should be at least 1.5 times the maximum load that the docked elements are capable of exerting. Separation should occur with minimum tip-off dynamics. The interface should be capable of coping with lateral and angular misalignments. Several types of docking mechanism are used in spacecraft docking: (i) three-point berthing ring; (ii) cone and ring docking adaptor; (iii) probe and drogue docking system. Relevant to manipulators are grapple fixtures which comprise of a pin marked to give range and orientation feedback to a vision system:

Below are listed NASA's standard tolerance levels:

Grasping tolerances:
  Angular capture misalignment (pitch, yaw): ±5°
  Angular capture misalignment (roll): ±5°
  Linear misalignment (*x* axis, *yz* plane): ±10 mm
  Maximum linear contact velocity (along *x* axis): 0.3 m/s

Berthing contact tolerances:
  Lateral misalignment (*x, y, z* axes): ±76 mm
  Angular misalignment (roll, pitch, yaw): ±1.5°
  Linear velocity(*x, y, z* axes): ±0.05 m/s
  Angular velocity (roll): ±0.2°/s
  Angular velocity (pitch, yaw): ±0.52°/s

Some of these tolerances are quite stringent and illustrate how manipulators can save fuel which would otherwise be expended in attaining such tight constraints. Manipulators are inherently capable of accommodating fine tolerances during manipulation as the rendezvous phase is required only to place the target within the manipulator workspace. This reduces the fuel penalty in achieving precise docking. It also allows the accommodation of the soft dock by incorporating most of the thrusting through orbital mechanics rather than braking thrusts and off-axis thrusts.

(v) Manipulator grasping effectively performs the docking manoeuvre culminating in the mechanical joining of the two spacecraft. As the grapple fixtures are within the work volume of the manipulators they accelerate to the speed of the predefined grapple points and make contact ensuring zero relative velocity and acceleration between them. At the actual grasp operation both position, velocity and acceleration of the manipulators must match those of the grapple points to ensure soft docking with minimised contact forces. This implies the need for a tight sensor feedback loop. If velocities and accelerations do not match within a narrow envelope slippage may occur. If the relative velocity is within a given limit perturbing forces at the end effector tip can be minimised at specific stopping distances [Krishen 1989] with the braking rule given by:

$$v(t) = \frac{v_p}{2}\left(1 + \cos\left[\frac{\pi t}{T_b}\right]\right)$$

For very short stopping distances and minimised end effector forces the relative velocity must be <0.1 ft/s (0.03 m/s). At contact, the wrist force/torque sensors give the contact forces. Control switches from visual/proximity guidance to the force sensors. The open-loop topology of the system instantaneously and discontinuously changes to the closed-loop topology [Marcyk & Bellazzi 1989]. The force controller reduces the speed of the grapple points to zero within a specified time frame and within maximum force constraints. This capture manoeuvre is absolutely basic to any servicing operations. The discontinuous change from open loop to closed loop highlights the necessity of using a consistent unified control methodology throughout the operation. If alternating control strategies are used the switching between them may cause instability problems.

(vi)   Once contact has been made there exists a single rigid body of the combined target and the robot interceptor. Thrusters and attitude control actuators on ATLAS (and/or target satellite if it is of large mass and cooperative) provide torques to passivate the complete system and suppress tumbling within a finite time sufficiently long not to exceed actuator capabilities. Ideally, latching arrangements from a docking mount should be available so that after grasping the target a hard dock may be initiated between the mating parts through a standardised ball and socket (ATLAS stabiliser/target docking port). If this is not the case then one arm must retain the target spacecraft in a locked mode while the other arm performs the required manipulation tasks. The manipulators should be positioned directly opposite the work place [Adams et al 1987]. After docking, the system may be translated to any desired location if necessary. From this point on the payload acquisition/docking phase proceeds to the payload transfer/servicing phase [Marcyk & Bellazzi 1989].

(vi)   Once the manipulation and servicing tasks have been performed ATLAS must decouple from the target satellite with zero net kinetic energy. Retraction of the manipulators may accomplish this through the reverse of the soft dock/grasp. Either ATLAS or the target may reposition and reorient the target spacecraft correctly for its continued operation.

15

Spacecraft attitude control system

The ACS (Attitude Control System) is a major component of the spacecraft subsystems and comprises sensors, processors and actuators for implementing spacecraft attitude control. It is common to integrate the onboard data handling (OBDH) subsystem and the ACS into the integrated control and data system (ICDS) by utilising a centralised processing architecture for the onboard computer unit with dedicated interface equipment (e.g. Artemis). The ACS is the subsystem with which the robot control system interacts most strongly and so is considered in detail here. Its design is highly dependent on the mission profile and it is usually the most complex subsystem of the spacecraft. Spacecraft have traditionally been designed with attention focussed initially on structural loading with the control system being designed separately. However, the spacecraft structure and its control system often interact: Explorer 1 which discovered the van Allen radiation belts became unstable when its whip antennae dissipated the spacecraft's spin energy (1958); Mariner 10 experienced roll axis instabilities due to interaction between the controller and the flexible solar array torque tubes (1973); HST encountered thermally induced vibration at the day/night terminator, requiring modifications in the control system (1992).

The introduction of manipulators to a spacecraft will introduce further interactions between the spacecraft-payload structure and the control system. Any free kinematic chain without internal gyroscopic motions in space will end up spinning about its greatest moment of inertia if there is relative motion between the rigidly or elastically attached links [Pringle 1966]. Hence, any robotic manipulator will require some form of attitude control to prevent this from occurring to any spacecraft to which the manipulator is attached (the mounting spacecraft). The control system is then the central component of the robotic spacecraft design. Furthermore, the autonomy constraint on the spacecraft requires the controller to handle a large range of dynamics, and the reliability constraint on the spacecraft requires the controller to be robust to large parameter variations. For a complex robotic spacecraft then, control is the central issue.

The attitude control system is the angular momentum management system and is central to any spacecraft to maintain mission requirements [Larson & Wertz 1992, Wertz 1978]. It is especially critical in the design of a robotic servicer spacecraft. It drives the overall spacecraft design since all subsystems interface to the attitude and orbital control system and a freeflying robot interceptor is no exception. Control–structural interaction problems also occur due to appendage flexure and thermal distortion both of which it is desirable to minimise.

15.1 IN-ORBIT ATTITUDE PERTURBATIONS

All spacecraft are subject to disturbing torques such as aerodynamic, gravity-gradient, solar radiation pressure, magnetic torques and fuel sloshing $\sim 10^{-4}$–10^{-5} Nm which, although small, will induce attitude errors. Disturbance torques on the spacecraft are either internal (fuel slosh, crew motion, machinery and actuator operation, thermal loads, etc.) or external (atmospheric, magnetic, solar radiation pressure, gravity gradient, docking impacts). Internal torques only affect momentum distribution but external torques affect the total angular momentum. Fluid sloshing mode frequencies in microgravity are generally very low in the range 0.1–0.01 Hz. Many of the effects of internal torques are due to uncertainty in centre of mass location and thrust misalignments. Typical values of uncertainty in centre of mass location ~ 1–3 cm and typical thrust misaligments are ~ 0.1–0.5°. Furthermore, since typically half the initial spacecraft mass is liquid fuel, fuel slosh has a significant effect on energy dissipation and inertia properties of the spacecraft. Sloshing, however, can be controlled by baffles. External torques may be cyclic or secular for a particular orbit but, for a robotic freeflyer which undergoes orbit and attitude changes, they will be secular during operational phases and cyclic during dormant phases. Solar array flexure will be a significant external torque. In comparison to docking impacts and grappling manoeuvres which will cause reaction moments on the spacecraft mount, these external moment sources will be negligible. Hence, usually these disturbances are unmodelled (with the exception of solar array flexibility which will be significant).

Aerodynamic forces are dominant at altitudes less than 500 km where the atmospheric density is sufficient to cause drag and orbit degradation and circularisation of highly eccentric orbits. Atmospheric drag limits practical orbits to greater than 120 km altitude while orbits over 600 km altitude allow lifetimes in excess of 10 y without reboost. If the orbit is elliptical most drag occurs at perigee and tends to reduce the apogee towards circularisation at the perigee. Aerodynamic forces exert torques when there exists a moment arm between the centre of pressure and the centre of mass of the spacecraft.

$$\tau = r_{cp} \times F_{aero} \qquad \text{where} \qquad F_{aero} = \frac{1}{2}\rho v^2 A c_D \tag{15.1}$$

The moment arm between the centre of pressure and centre of mass r_{cp} varies with the shape of the body. For a cylinder $r_{cp} = 1.3$, and for a cone $r_{cp} = 2 \sin^2 \alpha$ with a cone half-angle α. The aerodynamic coefficient c_D varies with altitude between 2.0–2.5. At 400 km altitude, $c_D = 2.2$ typically. Below 200 km altitude, c_D rises rapidly due to a rapid increase in air density. A typical satellite cross-section area is 2 m^2.

The change in orbit period due to orbit decay:

$$\frac{\delta\tau}{\tau} = \frac{-3\pi\rho r}{\left(M/Ac_D\right)} \tag{15.2}$$

where $(M/Ac_D) =$ vehicle ballistic parameter, 10–100 kg/m^2 being typical for a spacecraft.

The Earth's gravitational field approximately obeys the Newtonian gravitation law:

$$F = G \frac{M_{Earth} m}{r^2} = \frac{\mu_{Earth} m}{r^2}$$

However, gravity gradient torques occur at all orbits and are due to the object experiencing different forces of attraction on different parts of the spacecraft due to firstly to the spherical nature of the Earth's gravitational field, and secondly, due to variations in the gravity field component arising from the Earth's oblateness. However, the sphericity of the Earth's gravitational field will dominate gravity gradient torques creating a variable gravitational field of forces on different parts of the spacecraft. Torques will tend to rotate the spacecraft to align its axis of minimum inertia with the local vertical and thereby stabilise it in low orbits:

$$\tau = \left[\left(\frac{3\mu}{R^3} \right) \times I \right] \theta \tag{15.3}$$

where $\mu = GM = 3.99 \times 10^{14}$ Nm^2/kg, $M = 5.97 \times 10^{24}$ kg $=$ mass of the Earth, $G =$ universal gravitation constant $= 6.6 \times 10^{-11}$ Nm^2/kg^2, $\theta =$ maximum deviation of z-axis from local vertical to Earth centre, and $R =$ orbit radius in Earth radii.

Solar radiation torques dominate above $\sim$1000 km altitude particularly at GEO at an altitude of $\sim$36 000 km. They are caused by solar radiation pressure as the rate of change of momentum per unit area. Each proton carries momentum $p = h\nu/c$. Solar arrays contribute a major component of solar pressure torques and should be mounted symmetrically to locate the centre of area as close to the centre of mass as possible. Solar radiation pressure is the basis for solar sail propulsion without fuel expenditure using very large reflective sails of aluminium-coated Mylar for instance (area–mass ratio is typically around 20). The low thrust entails the use of continuous spiral trajectories using the radiation pressure component in the direction of desired thrust. They are generally limited to above 1000 km altitudes due to atmospheric drag on the sail.

$$\tau = r_{cp} \times F_{solar} \qquad \text{where} \qquad F_{solar} = (1 + K) p_s A_\perp \tag{15.4}$$

The surface reflectivity K varies between 0 for a black body and 1 for a mirror and for a spacecraft is taken to be $\sim$0.5 typically. The solar pressure constant $p_s = I_{solar}/c = 4.5 \times 10^{-6}$ N/m^2 at 1 AU. The solar constant I_s has a value of 1358 W/m^2 at 1 AU. The projected illuminated area perpendicular to the sun vector $A_\perp = A \cos i$ where $i =$ angle of incidence. Solar sailing has been used for three-axis attitude control through the rotation of solar panels for the Eurostar spacecraft bus.

Any system in which masses are coupled to each other by nonrigid members which move in relation to each other is capable of vibrating due to periodic forces on the structure of the system. If the relative motion between these masses give rise to restoring forces as functions of relative displacement and/or amplitude (e.g. between elastic coupling members) then vibrations may occur at more than one frequency. The vibration mode shapes comprise the most fundamental aspect of the structural dynamics.

Particularly in multi-mass systems the total motion will be a combination of several modes including fundamental and overtone modes. Solar arrays, antennas and antenna support structures are often deployable and being large are subject to structural flexure. In addition, forced vibrations are endemic in rotating machines such as electric motors. Essentially deployables may be represented as fixed-free beams. The dynamic behaviour is given by the eigenvalue frequencies. The lowest natural frequency of flexible components should be at least an order of magnitude greater than the rigid body frequencies to allow neglect of flexibility. If the forcing frequency is close to the system natural frequency, resonance will occur generating responses of potentially infinite amplitude limited by the system's damping characteristics. The simplest model of a structure modelled as a fixed-free beam is that of a uniform cantilevered beam of mass M length l and flexural stiffness EI (= 170 Nm^2 for stiff booms on typical spinning spacecraft) and it oscillates with simple harmonic motion $\ddot{\theta} = -w^2\theta$ with $w_n = \sqrt{k/M}$.

$$\text{Modal frequency } \quad f \approx \frac{1}{2\pi}\sqrt{\frac{k}{M}} = \frac{1}{2\pi}\sqrt{\frac{3EI}{(M + 0.243\rho l)l^3}} \qquad (15.5)$$

where I = area moment of inertia of cross-section, K = stiffness constant, E = Young's modulus, and ρ = density.

It is necessary to have high structural damping to minimise structural vibration and a wide separation between the control frequencies and structural frequencies to avoid structural excitation through resonance which occurs at $w_r = \sqrt{1 - 2\beta^2}$ with amplitude $A_r = F/2k\beta$, damping constant C, and damping ratio, $\beta = C/(2\sqrt{Mk}) \leqslant 0.02$ typically. Hence the modal frequency should be an order of magnitude greater than the rigid body control frequency (~60 Hz). The transfer function of a flexible body has a rigid body component with flexibility effects being introduced as a feedforward component. Hence flexible mode angular momenta may be added to the general rigid body dynamics to model flexible appendages.

For flexible spacecraft, finite element analysis may be used to discretise the equations of motion. Real structures have almost an infinite number of degrees of freedom due to the infinite number of points on them. Geometric shapes of space vehicles are usually highly irregular, complex and multi-bodied and so they defy analytic investigation of the natural frequencies and mode shapes of the spacecraft and its components as a whole. Modelling methods assume a finite number of degrees of freedom by using shape functions for deformation between a finite number of points on the structure. For each degree of freedom, there is an associated natural frequency but only the first few natural frequencies are of interest as the rest are usually well above the experienced forcing frequencies. The finite element method is extensively used for structural analysis with the structure represented as an assembly of many discrete elements where each element locally models a part of the structure. The total assembly of continuous elements constitutes the finite element model. Such methods are well-suited to computer simulation, e.g. NASTRAN.

One promising approach to active structural vibration suppression in the presence of disturbance torques is through the use of multilayer perceptron neural networks using backpropagation [Game 1993]. The artificial neural network may be used to model the inverse dynamics of the system from input-output pairs of a training set. However, as stated earlier, neural networks are not much favoured in space applications.

15.2 ATTITUDE ACTUATION SYSTEMS

The primary driver to the ACS is pointing accuracy and stability which defines the need for control of attitude in the event of perturbations which act to change the orbital and attitude parameters of the spacecraft. The spacecraft is required to make station-keeping corrections to maintain these parameters within allowable bounds. The use of onboard autonomy allows for ground station outage, but ground-based control reduces the requirement for onboard electronics. The cost of the ACS comprises ~10–20% of the total satellite cost and in some cases ground control may be more cost-effective. The requirement for onboard autonomy may impose an additional ~3–8% to the total satellite cost. However, in this case for the robotic freeflyer, ground control is not a cost-effective solution. For long mission lifetimes, onboard autonomy is preferred for cost-effectiveness as the cost of ground control becomes excess to that of increased onboard hardware and complexity [Sabroff 1968]. High pointing accuracy dictates the need for inertial orientation star sensors or star mappers and signal processing electronics for conversion to Earth-referenced coordinates. For instance, high pointing accuracy is required for directional communications antennas since the half-power attenuation near the edge of the bandwidth may increase dramatically with small pointing errors. If the receiver centre is not located at the centre of a transmitter beam of narrow beamwidth, small pointing errors will reduce the gain by: $G\,(\mathrm{dB}) = -12(e/\theta)^2$ where e = pointing error and θ = antenna half-power beamwidth.

A robotic freeflying spacecraft will have to move between multiple reference attitudes of operation so the spacecraft will be required to reorient itself. Such a spacecraft will comprise multiple payload segments including sensors each requiring different variable orientations. Such instruments will require gimballing with respect to the spacecraft. Multiple attitude requirements impose heavy constraints for independent pointing, e.g. solar panels must be Sun-oriented, directional antennae must Earth-pointing, attitude sensors require star, Earth or Sun orientations, and thermal radiators and sensitive instruments must be deep space pointed away from the Sun. Such directional components require two-axis degrees of freedom gimballing to provide the pointing capability. Gimballing is thus the only flexible way to cope with multiple orientation requirements.

Passive systems which employ environmental torque sources for momentum control such as gravity gradient, aerodynamic and magnetic systems do not provide sufficient accuracy (limited to ~1–10°) for the precision pointing required of a robotic freeflyer satellite. Gravity gradient techniques are popular for stabilisation of very large structures since they exploit the tendency of a body to align its axis of minimum inertia to the local vertical. An alternative technique for attitude stabilisation is spin stabilisation whereby the spacecraft body rotates about its principal axis of maximum inertia. The rate of precession ϖ of the angular momentum vector $\boldsymbol{H}$ obeys $\tau = \varpi \times H$ which imparts gyroscopic rigidity to the momentum bias of $\boldsymbol{H}$ giving it resistance to changes in attitude. This technique necessitates the use of omnidirectional antennas for Earth pointing communications which are characterised by zero antenna gain. Dual spin stabilisation, however, allows the use of an Earth-pointed antenna on a despun platform. The spacecraft rotates at a high spin rate typically ~60 rpm to provide the gyroscopic stiffness while the despun platform coupled through a bearing assembly rotates at 1 rev/day in the spacecraft orbit to keep the antenna Earth-pointed. All momentum bias techniques introduce an oscillating nutation mode which must be damped to reduce wobbling. For many missions, the normal to the orbit plane (pitch axis) is the direction chosen for momentum bias. For such a fixed reference trajectory, the local vertical rotates 90° every quarter orbit such that yaw errors are changed to roll errors every 90° and back again every 180°. Both spin and dual-spin stabilisation are suitable only for spacecraft with single attitude reference trajectories during the spacecraft operational lifetime. They are totally unsuitable for a robotic freeflyer application which requires multiple attitudes and slewing during operation. The gyroscopic stiffness imparted by spinning is obviously not desirable since changes in attitude will be required during a typical robotic servicing mission. Whereas most operational spacecraft to date have adopted nominal pointing direction with fixed attitude, the robotic freeflyer will need to attain multiple attitude orientations. Hence, the freeflying robot interceptor will require active three-axis stabilisation to provide the flexibility for multiple arbitrary attitude reference trajectories involving frequent large-angle slewing manoeuvres about the spacecraft centre of mass. Indeed, three-axis stabilisation has become the method of choice for large comsats. Three-axis stabilised spacecraft tend to be rectangular or polygonal rather than cylindrical and this offers greater flexibility for packaging components.

However, spin stabilisation capability ~10 rpm may be desirable during Δv thrusting parallel to the spin axis during orbit insertions to maintain a stable orientation to reduce the effect of any misalignments of the thrust vector from the spacecraft centre of mass which would otherwise cause the vehicle to veer off course or tumble. In fact, rather than spinning up the whole vehicle, spinning up the lower section of the spacecraft and maintaining the upper section of the spacecraft despun through a bearing assembly reduces the torquing requirement. This is the dual-spin configuration which provides gyroscopic stiffness from the spinning section while the despun platform is maintained stationary by counter-rotation. This configuration requires nutation damping by an energy dissipation mechanism in the despun section. After the circularisation of the apogee burn the spacecraft then undergoes a typical sequence of acquisition manoeuvres [Kaplan 1976]. There is an initial despin to ~1°/s. The normal means of satellite despinning following orbit injection thrusting is through the use of yo-yos or gas thrusters. Yo-yos are expendable masses which are attached by cords initially wrapped around the spacecraft body axis. Centrifugal forces pull them away until a split hinge enables the masses to escape carrying away the rotational kinetic energy. Yo-yos can be used only once, and so are not suitable for multiple orbit thrusting. The initial despin is followed by dewobbling followed again by a final despin to zero. In a typical spacecraft solar panel deployment and Sun acquisition is then achieved. Earth acquisition of antennas and star acquisition is then accomplished. If liquid fuel is used for thrusting, gimballing the thruster allows the use of three-axis stabilisation during thrusting. Alternatively, spin stabilisation may be employed during dormant modes between operations to reduce power consumption by three-axis control actuators.

One mode of three-axis stabilisation is through gas jet thrusters which exert torques by expending fuel. They are usually mounted in clusters for attitude control. Four thrusters are required to produce torque about one axis, i.e. 12 thrusters are usually required for three-axis stabilisation: $\tau = F_{thrust} \times l$ where l = moment arm, but four thrusters can be adopted if fuel consumption is not an issue (e.g. XMM). Two pulses are required for start and stop. Cold gas thrusters can produce <1 N thrust and have a low specific impulse <100 s. For high impulses, hot gas systems which operate by chemical reactions use mono- or bi-propellants (e.g. hydrazine derivatives) and are capable of producing 1×10^4 N thrusts. The propellant mass required for attitude change is determined by the pulse length t of the thrust and the thrust F_{thrust}: $m_p = (F_{thrust} \times t)/I_{sp}$. Limit cycling ~0.001–0.1°/s occurs due to the discrete on––off pulse mode of thrusting via the Schmitt trigger. This is a major deficiency in the accuracy of control for precise attitude maintenance. The standard NASA thruster pair has a 4.5 N operating thrust with ~20 000–50 0000 total pulse rating with pulse length t ~ 0.02–0.1 s.

Thrusters should be avoided due to their excessive fuel overhead in bang-bang thrusting, particularly for multiple slewing and rapid re-orientations. The only feasible choice of three-axis control is through internal momentum transfer devices which generate internal torques and redistribute momentum within the spacecraft such as reaction wheels, momentum wheels or control moment gyros. Built-in torque motors provide momentum storage as a means of transferring momentum from one part of the spacecraft to another without changing the momentum of the vehicle as a whole. They do not suffer from pulse effects and do not expend scarce fuel. They are a fundamental requisite for a robotic system which undergoes rapid mass and moment of inertia changes. Momentum wheel dynamics are similar to those of a dual-spin system (whereby the spinning component is viewed as a large momentum wheel along the spacecraft spin axis) except that the angular momentum of the dual-spin spacecraft is significantly higher than that of a three-axis stabilised spacecraft with fixed momentum wheels. Momentum wheels are nominally driven at high constant speed (~10^3 rpm) to provide momentum bias and also tend to impart a degree of gyroscopic stiffness to attitude. The speed can be increased or lowered to within 10% of nominal speed to compensate for external torques ~0.4–40 Nm. Momentum bias in comparison with zero momentum bias lowers the total control torque capability before saturation occurs and introduces undesirable gyroscopic nutation. Furthermore, they have the disadvantage of greater wear due to operation at constant motion. The Japanese ETS-VII used momentum wheels, however, which have a maximum torque capability of ~1–5 Nm with an accuracy of ~0.1°. Reaction wheels are nominally static (zero-bias momentum) but free to rotate and they generate torques by being driven in either direction with electric motors at the required angular velocities and accelerations in response to attitude disturbances or commands. The wheels generate equal and opposite internal torques to the spacecraft in which they are rigidly mounted to counteract the external torques on the spacecraft thereby keeping the attitude fixed in inertial space. They provide a reasonably fast response ~0.1–1°/s and good accuracy ~0.001°. They do not suffer from the dead zone error problem inherent in thruster pulse control. However, they can suffer from jerking through low-speed regimes where the wheel response tends to be nonlinear. Reaction wheels are large in volume and in mass, as well as being limited in torquing capability which is set by its inertia, i.e. size and speed. However, Dougherty et al (1971) calculated that the mass of a three-axis reaction wheel system may be less than that of a comparable momentum wheel system for high angular momentum capacities. Their big advantage is that they are simple to control since each wheel controls each vehicle reference axis degree of freedom independently. Three reaction wheels mounted orthogonally aligned along each spacecraft principal axis suffices to control spacecraft pitch, yaw and roll attitudes, though usually a fourth skewed wheel is included for redundancy, e.g. NASA's Standard Reaction Wheel configuration. The fourth wheel is aligned equally inclined from the three principal axes, forming a tetrahedral configuration, and the fourth wheel substantially increases the momentum storage capacity of the attitude control system. The NASA Standard Reaction Wheel system has a momentum capacity of 20 Nm s, a torque capability of 0.5 Nm, a mass of 25 kg, and a power consumption of 7.5 W at peak. In general, they are limited to ~300–1000 Nms momentum capacity and to maximum torques of ~1–5 Nm with the power requirement varying up to ~100 W/Nm when torquing.

Longman Lindberg & Zedd (1987), Marcyk & Bellazzi (1989) and Ellery (1996) suggested that reaction wheels offer insufficient torque capabilities to compensate for spacecraft base reaction moments exerted by manipulator movements unless the robot manipulators are operated prohibitively slowly. Furthermore, they are insufficient for high-rate slewing manoeuvres at $>1°/s$. The control moment gyro (CMG) is an advanced form of momentum wheel based on gyroscopic action. The wheel is mounted on a shaft free to rotate mounted onto gimbal bearings. It is spun at high speed, mounted in either one or two gimbals fixed perpendicular to the spin axis of the wheel and the rotor is free to rotate about the gimbal axes. Torque motors fitted to the gimbals apply control torques to displace the gimbal angles in such a way that the inertial orientation of the momentum wheel spin axis is altered to redirect the rotor angular momentum. The torques applied at the gimbals cause a change in the angular momentum vector, the rate of change of which is proportional to the applied couple. The wheel spins at a constant rate $~10^3$ rpm so the direction of the angular momentum vector undergoes change with a precession of $\bar{\omega}$ about the spin axis and a constant nutation of θ to generate an equal and opposite reaction torque to the spacecraft thereby redistributing its angular momentum. The torque gain is such that small torques exerted about the gimbal axes produce a much larger torque on the vehicle because the gimbal torques have only to overcome the gimbal and rotor inertias which are small whereas the output torques are dependent on the rotor momentum and gimbal rate. The gimbal motors are brushless dc motors with permanent magnet rotors. Potentiometers provide gimbal feedback for torque gain compensation in the servo system. Only two double-gimballed CMGs are required for three-axis stabilisation. For each CMG, the constant speed flywheel is held in the inner gimbal which is coupled through a pivot which in turn is held fixed to the spacecraft [O'Connor & Morine 1969, Liska 1986]. The incorporation of fluid supports improves the accuracy and the life of the CMG by eliminating bearing friction. In fact, one double-gimballed CMG allows control torques to be applied about all three axes through two-degree-of-freedom gyrotorquing and one-degree-of-freedom wheel speed control (such as a reaction wheel or momentum wheel) [Salutin & Bainum 1983]. It is usual, however, to operate the CMG at constant flywheel speed for two-degree-of-freedom control only. However, they can provide graceful degradation by being operated in two modes:

(i)   CMG mode—wheel speed is constant and the spin axis is rotated by the gimbal motors for two-degree-of-freedom gyrotorquing;

(ii)   reaction wheel mode—the gimbals are locked and wheel speed may be varied for single-degree-of-freedom wheel acceleration.

The use of CMGs in Skylab (with a nominal fixed momentum of 3000 Nm s) for precise pointing of the ATM (a solar telescope) mounting platform demonstrated their effectiveness in controlling the station attitude to within ~1 arcsec in spite of crew movements [Kaplan 1976, Chubb et al 1975]. The Skylab attitude control system comprised three double-gimballed CMGs, each of mass 200 kg, oriented to each of the three vehicle axes providing 100% redundancy. Kennel (1982) recommended parallel mounting such that all outer gimbal axes are parallel. This allows identical mounting interfaces such as brackets, harnesses, etc. CMGs have excellent performance bounds generating torque capabilities ~1000 times that of reaction wheels (up to ~5–10 kNm). They offer rapid response for high slew rates, high-accuracy attitude control and superior dynamic ranges to reaction wheels. They typically require ~15–30 W standby power and ~0.2 W/Nm when torquing. Essentially, they are rate gyroscopes which operate in reverse [Yarber et al 1966]. However, double-gimballed CMGs do suffer from limited gimbal angles and so limiting the attitude change (which is proportional to the gimbal angle deflection). Furthermore, they tend to drift as part of the angular momentum of the CMG unit resides in the gimbals due to their finite mass which causes them to move with respect to inertial space. Paradiso (1991) proposed the use of three single-gimballed CMGs for three-axis stabilisation. Three is the minimum number required with one controllable degree of freedom per axis. These offer larger torque capabilities and high momentum capacities for lower cost, power, mass and reliability. However, for a cluster of n single-gimballed CMGs, there exist 2^n gimbal angle singularities in the momentum space which are difficult to predict and avoid [Vadali et al 1989]. Due to this, they require far more complex control laws than the double-gimballed type. The double-gimballed CMG only exhibits singularity in the antiparallel condition when the gimbal angle approaches $\pm 90°$. At this singular condition, the double-gimballed CMG can no longer produce torque, but this situation is simple to monitor predict and avoid.

CMGs are much more complex to control than reaction wheels. The choice of actuator will ultimately depend on the size of the robot and the payload masses to be manipulated, though the CMG does provide a greater dynamic range of capabilities and force control will necessitate their employment over other actuators [Ellery 1996].

Example:

	Mass (kg)	RPM (10^3)	Ang. mom. capacity (Nm s)	Max output torque(Nm)	Max gimbal rate (°/s)	Size
Bendix DGCMG	253	4–12	1400–4000	237	5–30	1.1m diam.
HST RW	48	3	<100	0.82	—	0.6m diam.

All momentum transfer devices behave as momentum stores effectively transferring external momentum to the internal wheels. This will lead to eventual saturation such that the rotor spins at the maximum motor drive rate and the stored momentum has to be offloaded by momentum dumping to desaturate the internal torquers. The maximum angular momentum stored in the wheels is proportional to the area under the torque curve which is used to size the wheel. Momentum dumping is accomplished by imposing external torques on the spacecraft. In LEO, this may be accomplished by magnetorquers apart from in the plane of the magnetic equator by exerting an external torque on the spacecraft without using consumables [Kamm 1961, Burrow 1961]. Since the geographical equator and the magnetic equator do not coincide this does not preclude equatorial orbits—in such orbits the magnetic field direction varies by ±120°. The magnetorquer generates a magnetic field derived from three orthogonal current-carrying electromagnetic coils aligned along each spacecraft principal axis. The generated dipole field interacts with the Earth's local magnetic field generating a couple on the spacecraft. Since a couple is produced, magnetorquers are not sensitive to movements of the spacecraft centre of mass. They require flux gate magnetometers onboard to determine the local magnetic field magnitude and direction (at 200km, the Earth's magnetic field strength is ~0.3 G) and at 200 km altitude are limited to 5° accuracy but this is sufficient for dumping (but not for attitude sensing). Magnetic flux detecting SQUIDS make highly sensitive magnetometers and magnetic gradiometers [Clarke 1989].

Torque on a current carrying coil: $\tau = M \times B$, where B = Earth's magnetic flux, and M = controllable spacecraft magnetic dipole moment. The values of B and M need to be determined. The Earth's magnetic field is a complex structure which comprises a steady-state main field and a 10% secular variable component. Periodic variations arise due to solar flux in particular. The field is confined by solar wind leading to night/day variations, seasonal variations and solar cycle variations. Similarly, solar radiation ionises the upper atmosphere creating current flow which produces a solar variable field. In spite of this, Earth's magnetic field may be approximately modelled as a non-rotating dipole located at the Earth's centre. The most common magnetic field model is the tilted dipole model with the magnetic equator at 13° to the geographical equator, North magnetic pole located at 79° north latitude and 70° west longitude and South magnetic pole at 79° south latitude and 110° east longitude:

$$\begin{pmatrix} B_{north} \\ B_{east} \\ B_{vertical} \end{pmatrix} = -\left(\frac{6378 \text{ km}}{R_{Earth} + h} \right)^3 \begin{pmatrix} -c\varphi & s\varphi c\lambda & s\varphi s\lambda \\ 0 & s\lambda & -c\lambda \\ -2s\varphi & -2c\varphi c\lambda & -2c\varphi c\lambda \end{pmatrix} \begin{pmatrix} 29900 \\ 1900nT \\ -5530 \end{pmatrix} \quad (15.6)$$

where φ = latitude, λ = longitude, R_{earth} = Earth radius, and h = altitude. These coordinates must be transformed into spacecraft coordinates to determine the magnetic torques on the spacecraft.

The controllable magnetic dipole moment, $M = p|NIA|$, where $p = (-1, 0, 1)^T$ = polarity, N = number of coil turns, I = current, and A = coil area.

Resistance, $R = (Nrl)/A$, where r = resistivity, and l = perimeter length of coil per turn.

Mass of the coils, $m = NlA\rho$, where ρ = density.
Since power $P = I^2R$, then

$$M = \sqrt{\frac{Pm}{rl}}\,\frac{A}{l}$$

For a circular geometry which provides the maximum magnetic moment, $A = \pi l^2/4$ and $l = \pi D$, so

$$M = \frac{D}{4}\sqrt{\frac{Pm}{rl}} \tag{15.7}$$

Such devices allow continuous momentum dumping without fuel expenditure so that the velocity control loop can operate around a nominal constant value. The maximum momentum storage required for continuous torque input $H_m = H_0(2/\sin^2 \delta\theta - 1)$ where $\delta\theta$ = maximum change in field direction over a quarter orbit (e.g. 90° for polar) and the momentum input $\dot{H} = 4H_0/T$ where T = orbital period. For GEO, momentum dumping must be achieved through mass expulsion thrusting. Ion propulsion may reduce the mass requirement and can be used in near continuous low thrust mode.

15.3 SPACECRAFT ATTITUDE DYNAMICS

Prior to discussing attitude control algorithms, a brief introduction to spacecraft attitude dynamics is given here. All spacecraft attitude computations require an inertial reference. An inertial frame in the strictest sense is that which is fixed and non-accelerating with respect to the stars, i.e. with fixed right ascension and declination with respect to the stellar background. The celestial sphere is then an infinitely distant sphere onto which the stars are projected. In practice, however, for orbital motion about the Earth a sufficient choice of inertial origin is the centre of the Earth (geocentric inertial system) with one axis x fixed along the first point of Aries (direction of the vernal equinox where the Sun is placed against the stellar background on March 21—the vernal equinox marks the intersection of the Earth's equatorial plane and the Earth's ecliptic plane of orbit around the Sun and moves westward along the ecliptic), a second axis in the equatorial plane y and a third axis z to complete the right-hand orthogonal system (in the direction of the winter solstice—the Sun's position on the first day of winter) marking the celestial sphere from the North pole.

In fact, the frame of reference fixed to the Earth's centre is not really inertial as it orbits the Sun with period one year. The Earth revolves around the Sun in 365 d, 6 h and 9 min with respect to the stars. The Earth rotates on its axis every 24 h (solar day) while its rotation period relative to the stars (sidereal day) is almost 4 min shorter as the Earth must rotate an extra $1/365$ of a full revolution to complete the solar day. There is no inertial frame (as shown by Einstein) since the Sun rotates around the Galactic centre and the Galaxy is in relative motion with respect to other galaxies on a hierarchy of distance scales. However, the geocentric system is used and spherical coordinates may be used to define the position of a satellite (see Fig. 15.1(a)): r = radial distance of the satellite from

Northern hemisphere

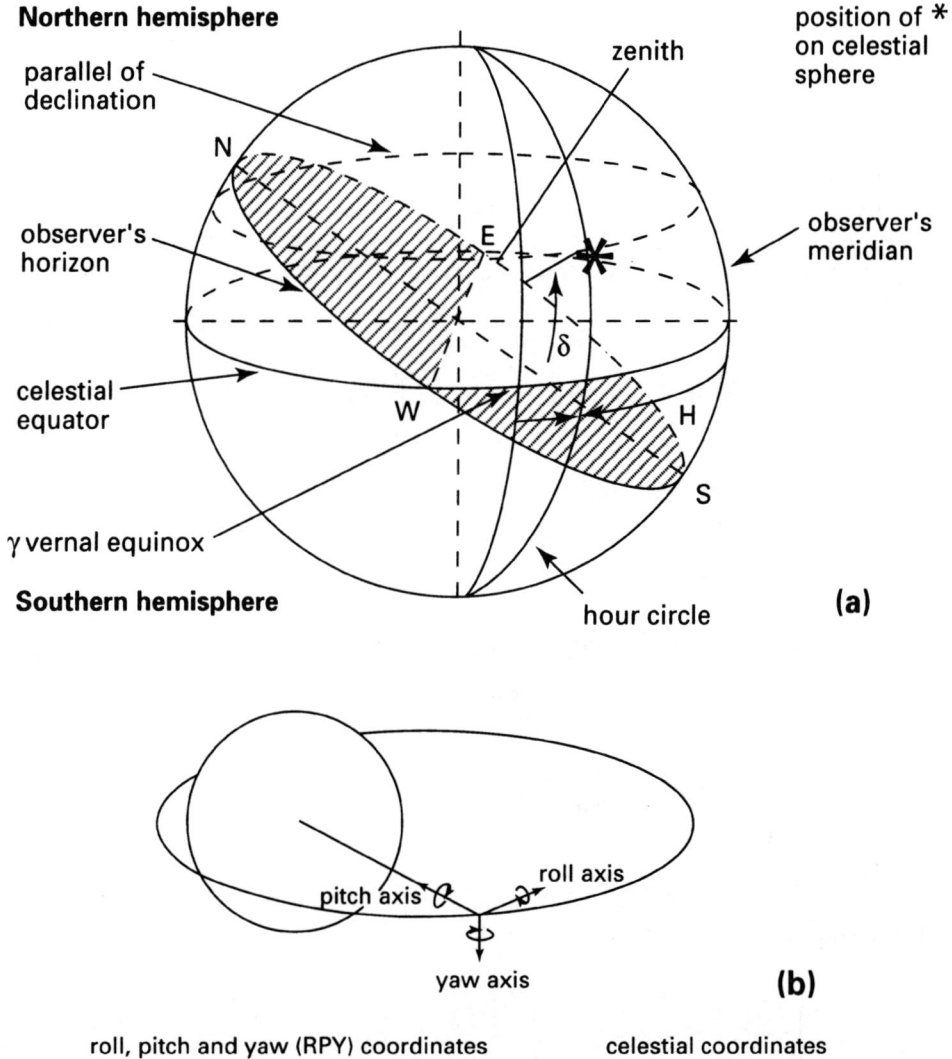

Fig. 15.1. (a) Coordinate systems (reproduced from Barter (1999) with permission from TRW Space & Electronics Group) (b) roll–pitch–yaw axes.

the geocentre; α = right ascension (angle in the equatorial plane from the vernal equinox eastwards); δ = declination (angle from the equatorial plane of the radius vector).

The orientation of a body with respect to an inertial reference frame may be uniquely described by the Euler angle set which denotes a Z–X–Z rotation sequence (ψ, θ, φ) where φ = longitude/right ascension of the ascending node; δ = inclination; ψ = argument of perigee which describe the orientation of Earth-orbiting space objects relative to the Earth's equatorial plane:

$$\begin{pmatrix} x_{inertial} \\ y_{inertial} \\ z_{inertial} \end{pmatrix} = R \begin{pmatrix} x_{body} \\ y_{body} \\ z_{body} \end{pmatrix}$$

where $R = 3 \times 3$ orthonormal rotation matrix with nine direction cosines completely specifying the orientation of the body relative to inertial coordinates subject to six orthogonality constraints. Hence three is the minimum number of parameters ψ, δ, φ required to characterise attitude. For the Euler angle description [Weisel 1989, Thomson 1986]:

$$R_{\varphi\theta\varphi} = \begin{pmatrix} c\psi & s\psi & 0 \\ -s\psi & c\psi & 0 \\ 0 & 0 & 1 \end{pmatrix} \begin{pmatrix} 1 & 0 & 0 \\ 0 & c\theta & s\theta \\ 0 & -s\theta & c\theta \end{pmatrix} \begin{pmatrix} c\varphi & s\varphi & 0 \\ -s\varphi & c\varphi & 0 \\ 0 & 0 & 1 \end{pmatrix}$$

$$= \begin{pmatrix} c\varphi c\psi - s\varphi c\theta s\psi & c\varphi s\psi - s\varphi c\theta c\psi & s\varphi s\theta \\ -s\varphi c\psi - c\varphi c\theta s\psi & -s\varphi s\psi + c\varphi c\theta c\psi & s\varphi s\theta \\ s\theta s\psi & -s\theta c\psi & c\theta \end{pmatrix} \qquad (15.8)$$

The inertial attitude of the spacecraft may be calculated by:

$$\psi = \tan^{-1}\left(\frac{R_{13}}{R_{33}}\right) = \tan^{-1}\left(\frac{s\theta s\psi}{s\theta c\psi}\right)$$

$$\varphi = \tan^{-1}\left(\frac{R_{31}}{R_{32}}\right) = \tan^{-1}\left(\frac{s\varphi s\theta}{c\varphi s\theta}\right)$$

$$\theta = \cos^{-1} R_{33} = \cos^{-1} c\theta$$

For rotation rates, the transforms are given by:

$$\begin{pmatrix} w_x \\ w_y \\ w_z \end{pmatrix} = \begin{pmatrix} s\theta s\varphi & s\varphi & 0 \\ s\theta c\varphi & -s\varphi & 0 \\ c\theta & 0 & 1 \end{pmatrix} \begin{pmatrix} \dot\psi \\ \dot\theta \\ \dot\varphi \end{pmatrix}$$

and

$$\begin{pmatrix} \dot\psi \\ \dot\theta \\ \dot\varphi \end{pmatrix} = \frac{1}{s\theta} \begin{pmatrix} s\varphi & c\varphi & 0 \\ c\varphi s\theta & -s\varphi s\theta & 0 \\ -s\varphi c\theta & -c\varphi c\theta & s\theta \end{pmatrix} \begin{pmatrix} w_x \\ w_y \\ w_z \end{pmatrix}$$

Euler rates are not orthogonal so transposing the transformations do not yield the correct formulations for the inverse transformations. The classical Euler set ψ, θ, φ becomes singular at $\theta = 0$, $\pm \pi$, i.e. for level flight. An alternative Euler set which does not suffer from this is the aircraft Z–Y–X sequence of yaw, pitch and roll from a local vertical which only becomes singular at $\beta = \pm\pi/2$ corresponding to vertical flight (see Fig. 15.1(b)). Since vertical flight is not usually encountered during in-orbit operations this is usually the set adopted. The roll axis X is the direction of flight, the pitch axis Y is normal to the orbit plane and the yaw axis Z is directed to the Earth's centre. This coordinate frame rotates about the pitch axis with respect to the inertially fixed coordinate frame at an angular rate of 1 rev/day.

$$R_{\gamma\beta\alpha} = \begin{pmatrix} c\gamma & -s\gamma & 0 \\ s\gamma & c\gamma & 0 \\ 0 & 0 & 1 \end{pmatrix} \begin{pmatrix} c\beta & 0 & -s\beta \\ 0 & 1 & 0 \\ -s\beta & 0 & c\beta \end{pmatrix} \begin{pmatrix} 1 & 0 & 0 \\ 0 & c\alpha & -s\alpha \\ 0 & s\alpha & c\alpha \end{pmatrix}$$

$$= \begin{pmatrix} c\gamma c\beta & -s\gamma c\alpha + c\gamma s\beta s\alpha & s\gamma s\alpha + c\gamma s\beta c\alpha \\ s\gamma c\beta & c\gamma c\alpha + s\gamma s\beta s\alpha & -c\gamma s\alpha + s\gamma s\beta c\alpha \\ -s\beta & c\beta s\alpha & c\beta c\alpha \end{pmatrix} \qquad (15.9)$$

The inertial attitude of the spacecraft is computed simply:

$$\gamma = \tan^{-1}\left(\frac{R_{12}}{R_{11}}\right) = \tan^{-1}\left(\frac{s\gamma c\beta}{c\gamma c\beta}\right)$$

$$\beta = \sin^{-1}(-R_{13}) = \sin^{-1}(s\beta)$$

$$\alpha = \tan^{-1}\left(\frac{R_{23}}{R_{33}}\right) = \tan^{-1}\left(\frac{c\beta s\alpha}{c\beta c\alpha}\right)$$

For rotation rates with respect to body referenced coordinates, the transforms are given by:

$$\begin{pmatrix} w_x \\ w_y \\ w_z \end{pmatrix} = \begin{pmatrix} c\gamma c\beta & -s\gamma & 0 \\ s\gamma c\beta & c\gamma & 0 \\ -s\beta & 0 & 1 \end{pmatrix} \begin{pmatrix} \dot{\alpha} \\ \dot{\beta} \\ \dot{\gamma} \end{pmatrix}$$

and

$$\begin{pmatrix} \dot{\alpha} \\ \dot{\beta} \\ \dot{\gamma} \end{pmatrix} = \frac{1}{c\beta} \begin{pmatrix} c\gamma & s\gamma & 0 \\ -s\gamma c\beta & c\gamma c\beta & 0 \\ c\gamma s\beta & s\gamma s\beta & c\beta \end{pmatrix} \begin{pmatrix} w_x \\ w_y \\ w_z \end{pmatrix} \qquad (15.10)$$

Differentiation gives the rotational acceleration transforms with respect to body referenced coordinates:

$$\dot{w}_x = \ddot{\alpha}c\gamma c\beta - \dot{\alpha}s\gamma c\beta - \dot{\alpha}c\gamma s\beta - \ddot{\beta}s\gamma - \dot{\beta}c\gamma \approx \ddot{\alpha} - \ddot{\beta}\gamma - \dot{\beta}\dot{\gamma}$$

$$\dot{w}_y = \ddot{\alpha}s\gamma c\beta + \dot{\alpha}c\gamma c\beta - \dot{\alpha}s\gamma s\beta + \ddot{\beta}c\gamma - \dot{\beta}s\gamma \approx \ddot{\alpha}\gamma + \dot{\alpha}\dot{\gamma} + \ddot{\beta}$$

$$\dot{w}_z = \ddot{\alpha}s\beta - \dot{\alpha}c\beta + \ddot{\gamma} \approx -\ddot{\alpha}\beta - \dot{\alpha}\dot{\beta} + \ddot{\gamma}$$

$$\begin{pmatrix} \dot{w}_x \\ \dot{w}_y \\ \dot{w}_z \end{pmatrix} = \begin{pmatrix} c\beta c\gamma & -s\gamma & 0 \\ s\gamma c\beta & c\gamma & 0 \\ -s\beta & 0 & 1 \end{pmatrix} \begin{pmatrix} \ddot{\alpha} \\ \ddot{\beta} \\ \ddot{\gamma} \end{pmatrix} + \begin{pmatrix} -s\gamma c\beta - c\gamma s\beta & -c\gamma & 0 \\ c\gamma c\beta - s\gamma s\beta & -s\gamma & 0 \\ -c\beta & 0 & 0 \end{pmatrix} \begin{pmatrix} \dot{\alpha} \\ \dot{\beta} \\ \dot{\gamma} \end{pmatrix} \qquad (15.11)$$

Inversely, inertial rotational accelerations are given by:

$$\ddot{\alpha} = \frac{1}{c\beta}\left[c\gamma \dot{w}_x + s\gamma \dot{w}_y - s\gamma w_x + c\gamma w_y + s\beta\dot{\alpha} \right] \approx \dot{w}_x + \dot{\gamma}\dot{w}_y + \gamma\dot{w}_y$$

$$\ddot{\beta} = \frac{1}{c\beta}\left[-s\gamma c\beta\dot{w}_x + c\gamma c\beta\dot{w}_y + (s\gamma s\beta - c\gamma c\beta)w_x - (s\gamma c\beta + c\gamma s\beta)w_y + s\beta\dot{\beta} \right]$$

$$\approx -\dot{\gamma}w_x - \gamma\dot{w}_x + \dot{w}_y$$

$$\ddot{\gamma} = \frac{1}{c\beta}\left[c\gamma s\beta\dot{w}_x + s\gamma s\beta\dot{w}_y + c\beta\dot{w}_z + (c\gamma c\beta - s\gamma s\beta)\dot{w}_x \right.$$

$$\left. + (c\gamma s\beta + s\gamma c\beta)w_y - s\beta w_z + s\beta\dot{\gamma} \right]$$

$$\approx \dot{\beta}w_x + \beta\dot{w}_x + \dot{\gamma}w_y + \gamma\dot{w}_y + \dot{w}_z$$

$$\begin{pmatrix} \ddot{\alpha} \\ \ddot{\beta} \\ \ddot{\gamma} \end{pmatrix} = \frac{1}{c\beta}\left[\begin{pmatrix} c\gamma & s\gamma & 0 \\ -s\gamma c\beta & c\gamma c\beta & 0 \\ c\gamma s\beta & s\gamma s\beta & c\beta \end{pmatrix} \begin{pmatrix} \dot{w}_x \\ \dot{w}_y \\ \dot{w}_z \end{pmatrix} \right.$$

$$\left. + \begin{pmatrix} -s\gamma & c\gamma & 0 \\ -c_{\beta\gamma} & -s_{\beta\gamma} & 0 \\ c_{\beta\gamma} & s_{\beta\gamma} & -s\beta \end{pmatrix} \begin{pmatrix} w_x \\ w_y \\ w_z \end{pmatrix} + s\beta \begin{pmatrix} \dot{\alpha} \\ \dot{\beta} \\ \dot{\gamma} \end{pmatrix} \right] \qquad (15.12)$$

It is possible to avoid singularities altogether by using the Euler four-parameter set (similar to quaternions [Klumpp 1976]) which is based on Euler's principal rotation theorem that the most general displacement of a rigid body can be obtained by first translating a body parallel to an arbitrary axis then rotating the body about the axis in a single rotation. No three-parameter set is free from singularities but the Euler set may be chosen such that orientational singularities will not be entered into or by switching between representations. Traditionally, spacecraft have adopted the roll–pitch–yaw Euler convention. The vertical flight configuration is an unlikely scenario but if it did occur the representation could switch to the classical Euler set.

The rigid body dynamics are formulated in vehicle body coordinates and angular momentum of such a body may be defined as:

$$
\begin{pmatrix} H_x \\ H_y \\ H_z \end{pmatrix} = \begin{pmatrix} I_x & -I_{xy} & -I_{xz} \\ -I_{xy} & I_y & -I_{yz} \\ -I_{xz} & -I_{yz} & I_z \end{pmatrix} \begin{pmatrix} w_x \\ w_y \\ w_z \end{pmatrix}
$$

where

$$
I_x = \int (y^2 + z^2)dm \quad I_{xy} = I_{yx} = \int xy.dm
$$

$$
I_y = \int (x^2 + z^2)dm \quad I_{xz} = I_{zx} = \int xz.dm
$$

$$
I_z = \int (x^2 + y^2)dm \quad I_{yz} = I_{zy} = \int yz.dm
$$

The axis of a body of revolution is a principal axis and any transverse axis passing through the centre of mass is a principal axis. For principal axis body coordinates centred on the body centre of mass, the principle axes include the maximum and minimum inertias. With respect to spacecraft principal axis coordinates, the inertia products become zero, i.e. $I = \mathrm{diag}(I_{ii})$—this is a simplification as non-zero off-diagonal inertia products will occur in real engineering systems:

$$
H_x = I_x w_x
$$

$$
H_y = I_y w_y
$$

$$
H_z = I_z w_z
$$

For a general rigid body, $I_x > I_y > I_z$; for a disc, $I_z > I_x$; for a rod, $I_z < I_x$.

The moments associated with angular momentum changes are defined by the differential of angular momentum with respect to time:

$$
N = \dot{H} = \left(\frac{dH}{dt} \right)_{body} + w \times H = I\dot{w} + w \times Iw
$$

Explicitly, this gives Euler's moment equations describing the external moments acting on the spacecraft in spacecraft coordinates and its response.

$$
\begin{pmatrix} N_x \\ N_y \\ N_z \end{pmatrix} = \begin{pmatrix} I_x \dot{w}_x \\ I_y \dot{w}_y \\ I_z \dot{w}_z \end{pmatrix} + \begin{pmatrix} (I_z - I_y)w_y w_z \\ (I_x - I_z)w_x w_z \\ (I_y - I_x)w_x w_y \end{pmatrix} - \begin{pmatrix} I_{xy}(\dot{w}_y - w_x w_z) + I_{xz}(\dot{w}_z + w_x w_y) \\ I_{yz}(\dot{w}_z - w_x w_y) + I_{xy}(\dot{w}_x + w_y w_z) \\ I_{xz}(\dot{w}_x - w_y w_z) + I_{yz}(\dot{w}_z + w_x w_z) \end{pmatrix}
$$

$$
(15.13)
$$

With respect to principal axis coordinates (i.e. zero off-diagonal inertia elements):

$$
\begin{pmatrix} N_x \\ N_y \\ N_z \end{pmatrix} = \begin{pmatrix} I_x \dot{w}_x \\ I_y \dot{w}_y \\ I_z \dot{w}_z \end{pmatrix} + \begin{pmatrix} (I_z - I_y)w_y w_z \\ (I_x - I_z)w_x w_z \\ (I_y - I_x)w_x w_y \end{pmatrix}
$$

$$
(15.14)
$$

If the external moments N acting on the spacecraft are zero (assuming external moments are modelled separately):

$$I_x \dot{w}_x = (I_y - I_z) w_y w_z$$

$$I_y \dot{w}_y = (I_z - I_x) w_x w_z$$

$$I_z \dot{w}_z = (I_x - I_y) w_x w_y$$

Analytic solutions are not readily available but it is possible to gain some kind of qualitative understanding by a graphical representation [Weisel 1989].

Conservation of momentum is given by: $H^2 = I_x^2 w_x^2 + I_y^2 w_y^2 + I_z^2 w_z^2$

Conservation of energy is given by: $2K = I_x w_x^2 + I_y w_y^2 + I_z w_z^2$

This enables a series of coupled equations in w to be derived:

$$w_y^2 = \left(\frac{2I_z K - H^2}{I_y I_z - I_y^2} \right) - \left(\frac{I_x I_z - I_y^2}{I_y I_z - I_y^2} \right) w_x^2 \qquad w_z^2 = \left(\frac{2I_y K - H^2}{I_y I_z - I_z^2} \right) - \left(\frac{I_x I_y - I_x^2}{I_y I_z - I_z^2} \right) w_x^2$$

$$w_x^2 = \left(\frac{2I_z K - H^2}{I_x I_z - I_x^2} \right) - \left(\frac{I_y I_z - I_x^2}{I_x I_z - I_x^2} \right) w_y^2 \qquad w_z^2 = \left(\frac{H^2 - 2I_x K}{I_z^2 - I_x I_z} \right) - \left(\frac{I_y I_z - I_z^2}{I_z^2 - I_x I_z} \right) w_y^2$$

$$w_x^2 = \left(\frac{2I_y K - H^2}{I_x I_y - I_x^2} \right) - \left(\frac{I_y I_z - I_z^2}{I_x I_y - I_x^2} \right) w_z^2 \qquad w_y^2 = \left(\frac{H^2 - 2I_x K}{I_y^2 - I_x I_y} \right) - \left(\frac{I_z^2 - I_x I_y}{I_y^2 - I_x I_y} \right) w_z^2$$

The conservation equations describe three-dimensional ellipsoids in the body frame of reference:

$$\frac{w_x^2}{(H/I_x)^2} + \frac{w_y^2}{(H/I_y)^2} + \frac{w_z^2}{(H/I_z)^2} = 1$$

$$\frac{w_x^2}{(2K/I_x)} + \frac{w_y^2}{(2K/I_y)} + \frac{w_z^2}{(2K/I_z)} = 1$$

The angular momentum ellipsoid and the kinetic energy ellipsoid have different axis lengths. The angular momentum ellipsoid axis length is given by

$$\left(\frac{H}{I_x}, \frac{H}{I_y}, \frac{H}{I_z} \right)$$

while the kinetic energy ellipsoid axis length is given by

$$\left(\sqrt{\frac{2K}{I_x}}, \sqrt{\frac{2K}{I_y}}, \sqrt{\frac{2K}{I_z}} \right)$$

The angular velocity vector is given by the intersection between the two ellipsoids (polhodes).

15.4 ATTITUDE CONTROL ALGORITHMS

The attitude control system utilises the attitude dynamics formulation to maintain a stable spacecraft orientation through the adoption of attitude actuator devices (Fig. 15.2). For the purpose of analysis, reaction wheel control is discussed here for simplicity with CMGs given a brief look over later. Although CMGs would be the actuation method of choice for ATLAS their control is much more complex than that for reaction wheels. Spacecraft dynamics with reaction wheels has the same form as for a general rigid body but with two additional coupling terms to account for nonzero wheel speeds and accelerations. The angular momentum of a rigid body with spinning wheels is the sum of the angular momentum of the rigid body and the angular momentum due to the spinning wheels. The wheels stabilise the spacecraft about any body-fixed principal axis. The total angular momentum of the system with angular momentum conservation is given by [Umetani & Yoshida 1989, Junkins & Turner 1986]:

$$H = H_s + H_w + H_m = 0 \tag{15.15}$$

with respect to vehicle body coordinates where $H_s = I_s w_s$ = spacecraft angular momentum, $H_w = I_w(w_s + \Omega)$ = reaction wheel angular momentum = $H_{w1} + H_{w2} + H_{w3}$ for three wheels, $H_m = I_m(w_s + w_m)$ = manipulator angular momentum, I_s = inertia matrix of asymmetric spacecraft, and I_m = inertia matrix of robot manipulator, i.e. $H_s + H_w = -H_m$.

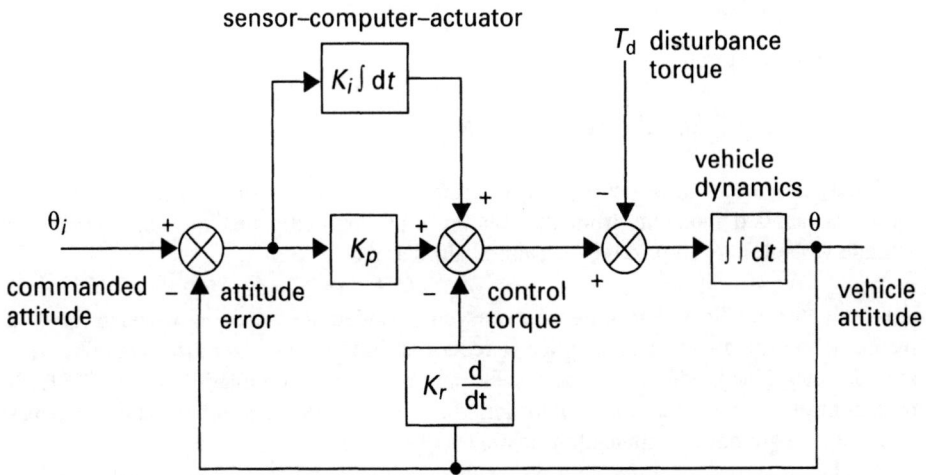

Position, integral and rate (derivative) control–"PID"

Fig. 15.2. Basic attitude control system (from Barter 1999). Reproduced with permission from TRW Space Electronics Group.

Assuming each of three reaction wheels to be identical and each coinciding with each vehicle principal axis:

$$
\begin{pmatrix} I_x^s w_x^s \\ I_y^s w_y^s \\ I_z^s w_z^s \end{pmatrix} + \begin{pmatrix} I_a^w (w_x^s + \Omega_x) \\ I_t^w w_y^s \\ I_t^w w_z^s \end{pmatrix} + \begin{pmatrix} I_t^w w_x^s \\ I_a^w (w_y^s + \Omega_y) \\ I_t^w w_z^s \end{pmatrix} + \begin{pmatrix} I_t^w w_x^s \\ I_t^w w_y^s \\ I_a^w (w_z^s + \Omega_z) \end{pmatrix} = - \begin{pmatrix} H_x^m \\ H_y^m \\ H_z^m \end{pmatrix}
$$

$$(15.16)$$

where Ω = relative wheel speed, I_a^w = axial inertia of axisymmetric wheel = I_z^w, and I_t^w = transverse inertia of axisymmetric wheel = $I_x^w = I_y^w$. Equivalently:

$$
\begin{pmatrix} I_x^* w_x^s \\ I_y^* w_y^s \\ I_z^* w_z^s \end{pmatrix} + \begin{pmatrix} I_a^w \Omega_x \\ I_a^w \Omega_y \\ I_a^w \Omega_z \end{pmatrix} = - \begin{pmatrix} H_x^m \\ H_y^m \\ H_z^m \end{pmatrix}
$$

$$(15.17)$$

where $I_i^* = (I_i^s + I_a^w + 2I_t^w) =$ composite spacecraft/wheel inertia.

(i) For an attitude-controlled platform, $w_s = 0$ nominally. Differentiating with respect to time and noting that $\dot{H}_m = -N_r$:

$$
\begin{pmatrix} I_a^w \dot{\Omega}_x \\ I_a^w \dot{\Omega}_y \\ I_a^w \dot{\Omega}_z \end{pmatrix} = - \begin{pmatrix} N_{rx} \\ N_{ry} \\ N_{rz} \end{pmatrix}
$$

$$(15.18)$$

The wheel motor torque,

$$
\tau_m^w = \frac{d}{dt}\left(I_a^w \Omega\right) \quad \text{so} \quad \tau_m^w = -N_r
$$

The body-mounted wheels drive the spacecraft in a large angle slew manoeuvre to counteract the applied moments from the robotic arms (assuming that the nominal desired attitude variables are zero so the measured values are the errors).

(ii) Once the feedforward component has commanded the reaction wheel torques, a feedback control law must be supplied to reduce attitude errors to zero from nominal zero attitude rates. The feedforward component may be used in conjunction with a PD feedback control law for the attitude controller, i.e. attitude control and robot control schemes outlined here are both computed torque modes of control.

Similar equations are used with the angular velocity and angular acceleration values given by the attitude sensors to generate errors from nominal.

$$
H = H_s + H_w = 0
$$

where $w_s \neq 0$.

$$H = \begin{pmatrix} I_x^* w_x^s \\ I_y^* w_y^s \\ I_z^* w_z^s \end{pmatrix} + \begin{pmatrix} h_x \\ h_y \\ h_z \end{pmatrix} = 0$$

where $h_i = I_a^w \Omega$ = relative wheel momentum with respect to the spacecraft.

Differentiate with respect to time according to Euler's equation of rotational motion to derive the moments associated with angular momentum changes defined by the differential of angular momentum over time:

$$\dot{H} = \dot{H}_s + \dot{H}_w = I^* \dot{w}_s + \dot{h} + w_s \times (I^* w_s + h)$$

$$= I^* \dot{w}_s + w_s \times I^* w_s + I_a^w \dot{\Omega} + w_s \times I_a^w \Omega = 0$$

where

$$\dot{H}_s = \begin{pmatrix} I_x^* \dot{w}_x^s + (I_z^* - I_y^*) w_y^s w_z^s \\ I_y^* \dot{w}_y^s + (I_x^* - I_z^*) w_x^s w_z^s \\ I_z^* \dot{w}_z^s + (I_y^* - I_x^*) w_x^s w_y^s \end{pmatrix} \quad \text{and} \quad \dot{H}_w = \begin{pmatrix} \dot{h}_x + h_z w_y^s - h_y w_z^s \\ \dot{h}_y + h_x w_z^s - h_z w_x^s \\ \dot{h}_z + h_y w_x^s - h_x w_y^s \end{pmatrix} \quad (15.19)$$

These are first-order Euler differential equations of the general form

$$N = \dot{H} = \left(\frac{dH}{dt} \right)_{body} + w \times H = I\dot{w} + w \times Iw$$

which describe the external moments acting on the spacecraft in spacecraft principal axis coordinates [Agrawal 1980]. They are coupled and nonlinear and provide the body-referenced angular velocities and accelerations with which to drive reaction wheels. A strapdown optical inertial navigation system with rate gyroscopes and sensors rigidly attached to the vehicle can give angular rates with respect to the vehicle body frame, w, $\dot{w}$ from which inertial rates may be computed. Integration provides angular position values. The variation of parameters approach to attitude dynamics allows the use of the torque-free solution even when when small torques are acting on the spacecraft [Wertz 1978].

Now, wheel motor torque,

$$\tau_m^w = \dot{h}_w = \frac{d}{dt} \left(I_a^w (w + \Omega) \right) = I_a^w \dot{w}_s + \dot{h} = I_a^w (\dot{w}_s + \dot{\Omega})$$

$$\begin{pmatrix} \tau_x^m \\ \tau_y^m \\ \tau_z^m \end{pmatrix} = - \begin{pmatrix} (I_x^* - I_a^w) \dot{w}_x^s \\ (I_y^* - I_a^w) \dot{w}_y^s \\ (I_z^* - I_a^w) \dot{w}_z^s \end{pmatrix} + \begin{pmatrix} (I_y^* - I_z^*) w_y^s w_z^s \\ (I_z^* - I_x^*) w_z^s w_x^s \\ (I_x^* - I_y^*) w_x^s w_y^s \end{pmatrix} + \begin{pmatrix} h_y w_z^s - h_z w_y^s \\ h_z w_x^s - h_x w_z^s \\ h_x w_y^s - h_y w_x^s \end{pmatrix} \quad (15.20)$$

Now,

$$
\begin{pmatrix} \dot{w}_x^s \\ \dot{w}_y^s \\ \dot{w}_z^s \end{pmatrix} = \begin{pmatrix} (I_x^* - I_a^w)^{-1}\big[(I_y^* - I_z^*)w_y^s w_z^s - h_z w_y^s + h_y w_z^s - u_x\big] \\ (I_y^* - I_a^w)^{-1}\big[(I_z^* - I_x^*)w_x^s w_z^s - h_x w_z^s + h_z w_x^s - u_y\big] \\ (I_z^* - I_a^w)^{-1}\big[(I_x^* - I_y^*)w_x^s w_y^s - h_y w_x^s + h_x w_y^s - u_z\big] \end{pmatrix}
$$

and

$$
\begin{pmatrix} \dot{h}_x \\ \dot{h}_y \\ \dot{h}_z \end{pmatrix} = \begin{pmatrix} u_x \\ u_y \\ u_z \end{pmatrix} - I_a^w \begin{pmatrix} \dot{w}_x^s \\ \dot{w}_y^s \\ \dot{w}_z^s \end{pmatrix}
$$

The wheel dynamics may now be given:

$$
\begin{pmatrix} I_a^w \dot{\Omega}_x \\ I_a^w \dot{\Omega}_y \\ I_a^w \dot{\Omega}_z \end{pmatrix} = \begin{pmatrix} u_x - I_a^w \dot{w}_x^s \\ u_y - I_a^w \dot{w}_y^s \\ u_z - I_a^w \dot{w}_z^s \end{pmatrix}
$$

$$
= -I_a^w \begin{pmatrix} (I_x^* - I_a^w)^{-1}\big[(I_y^* - I_z^*)w_y^s w_z^s - h_z w_y^s + h_y w_z^s\big] \\ (I_y^* - I_a^w)^{-1}\big[(I_z^* - I_x^*)w_x^s w_y^s - h_x w_z^s + h_z w_x^s\big] \\ (I_z^* - I_a^w)^{-1}\big[(I_x^* - I_y^*)w_x^s w_y^s - h_y w_x^s + h_x w_y^s\big] \end{pmatrix}
$$

$$
+ \begin{pmatrix} I_x^*(I_x^* - I_a^w)^{-1} u_x \\ I_y^*(I_y^* - I_a^w)^{-1} u_y \\ I_z^*(I_z^* - I_a^w)^{-1} u_z \end{pmatrix} \tag{15.21}
$$

The simplest way to transfer angular momentum from the spacecraft to the rotor for slewing is to let the wheel relative momentum increase linearly such that $h =$ constant. Euler's equations may yield analytic solutions using Jacobian elliptical functions but the usual procedure for solving Euler's equations in spacecraft attitude control systems is by applying the Runge–Kutta numerical integration method which approximates a Taylor series extrapolation of the differential equation function by evaluating the first derivative at points within the required interval. A fourth-order algorithm with a fixed time step can yield solutions ~msec comparable to robotic algorithm solutions.

For the differential equation $dy/dt = f(y,t)$:

$$
y_{n+1} = y_n + \frac{h}{6}(k_1 + 2k_2 + 2k_3 + k_4)
$$

where

$$k_1 = f(y_n, t_n),$$

$$k_2 = f(t_n + \tfrac{1}{2}h, y_n + \tfrac{1}{2}hk_1),$$

$$k_3 = f(t_n + \tfrac{1}{2}h, y_n + \tfrac{1}{2}hk_2),$$

$$k_4 = f(t_n + \tfrac{1}{2}h, y_n + \tfrac{1}{2}hk_3), \qquad \text{and} \quad h = \text{step size.}$$

The method is equivalent to Simpson's rule, is stable and does not require a starting procedure. However, to minimise round-off errors, the Runge–Kutta–Fehlberg algorithm may be used [Hardacre 1999].

For a complete description of the rotational dynamics, a set of kinematic Euler angle representations are required to provide the body attitude with respect to inertial co-ordinates. The commonest Euler set for Earth orbit operations are the roll–pitch–yaw formulations from a local vertical which relate inertial roll-pitch-yaw rates $(\dot{\alpha}, \dot{\beta}, \dot{\gamma})$ to body-referenced rates. The roll axis x lies in the direction of flight, pitch axis y is normal to the orbit plane, and the yaw axis z points towards the Earth centre:

$$w_x = \dot{\alpha}\cos\gamma\cos\beta - \dot{\beta}\sin\gamma$$

$$w_y = \dot{\alpha}\sin\gamma\cos\beta + \dot{\beta}\cos\gamma \qquad (15.22)$$

$$w_z = -\dot{\alpha}\sin\beta + \dot{\gamma}$$

After the slewing acquisition stage is complete and the feedforward component has been realised, attitude errors will be small, so linearisation is permitted for the feedback part of the controller eliminating problems of singularity such that $\cos a \sim 1$ and $\sin a \sim a$:

$$w_{0x} = \dot{\alpha} - \gamma\dot{\beta}$$

$$w_{0y} = \dot{\alpha}\gamma + \dot{\beta} \qquad (15.23)$$

$$w_{0z} = \dot{\gamma} - \beta\dot{\alpha}$$

This may be differentiated to yield:

$$\dot{w}_{0x} = \ddot{\alpha} - \dot{\beta}\dot{\gamma}$$

$$\dot{w}_{0y} = \ddot{\alpha}\gamma + \dot{\alpha}\dot{\gamma} \qquad (15.24)$$

$$\dot{w}_{0z} = \ddot{\alpha}\beta - \dot{\alpha}\dot{\beta} + \ddot{\gamma}$$

These may be substituted into (15.21):

$$
\begin{pmatrix} I_x^*(\ddot{\alpha} - w\dot{\beta}\dot{\gamma}) \\ I_y^*(\dot{\alpha}\gamma + \dot{\alpha}\dot{\gamma}) \\ I_z^*(\ddot{\gamma} - \ddot{\alpha}\beta - \dot{\alpha}\dot{\beta}) \end{pmatrix} + \begin{pmatrix} (I_z^* - I_y^*)(\dot{\alpha}\gamma + \dot{\beta})(\dot{\gamma} - \dot{\beta}\alpha) \\ (I_x^* - I_z^*)(\dot{\alpha} - \dot{\beta}\gamma)(\dot{\gamma} - \dot{\beta}\alpha) \\ (I_y^* - I_z^*)(\dot{\alpha} - \dot{\beta}\gamma)(\dot{\alpha}\gamma + \dot{\beta}) \end{pmatrix}
$$

$$
+ \begin{pmatrix} \dot{h}_x + h_z(\dot{\alpha}\gamma + \dot{\beta}) - h_y(\dot{\gamma} - \dot{\beta}\alpha) \\ \dot{h}_y + h_x(\gamma - \dot{\beta}\alpha) - h_z(\dot{\alpha} - \dot{\beta}\gamma) \\ \dot{h}_z + h_y(\dot{\alpha} - \dot{\beta}\gamma) - h_x(\dot{\alpha}\gamma + \dot{\beta}) \end{pmatrix} = \begin{pmatrix} 0 \\ 0 \\ 0 \end{pmatrix} \qquad (15.25)
$$

This is the general case and is very complex requiring control to be implemented in vehicle body coordinates. In general, a nominal reference trajectory may be selected such that the spacecraft follows a normal orbit trajectory similar to traditional spacecraft. This will be valid in the majority of cases. By assuming that the coupling terms are negligible (in the small angle limit), this allows a simplified form to be used for control in Earth-referenced inertial coordinates. Orbital motion about Earth is a rate of change of pitch causing the body-referenced axes to rotate at an orbital rate of

$$
\dot{\beta}_0 = w_{orb} = \sqrt{GM_{earth} / R_{earth+altitude}^3}
$$

$$
\begin{pmatrix} w_{0x} \\ w_{0y} \\ w_{0z} \end{pmatrix} = \begin{pmatrix} \dot{\alpha} \\ \dot{\beta} \\ \dot{\gamma} \end{pmatrix} + \begin{pmatrix} 1 & \gamma & -\beta \\ -\gamma & 1 & \alpha \\ \beta & -\alpha & 1 \end{pmatrix} \begin{pmatrix} 0 \\ -w_0 \\ 0 \end{pmatrix} = \begin{pmatrix} \dot{\alpha} - \gamma w_0 \\ \dot{\beta} - w_0 \\ \dot{\gamma} + \alpha w_0 \end{pmatrix}
$$

This may be substituted into the Euler equations:

$$
\begin{pmatrix} I_x^*\ddot{\alpha} + (I_z^* - I_y^*)(\dot{\beta} - w_0)(\dot{\gamma} + \alpha w_0) \\ I_y^*\ddot{\beta} + (I_x^* - I_z^*)(\dot{\alpha} - \gamma w_0)(\dot{\gamma} + \alpha w_0) \\ I_z^*\ddot{\gamma} + (I_y^* - I_x^*)(\dot{\alpha} - \gamma w_0)(\dot{\beta} - w_0) \end{pmatrix} + \begin{pmatrix} \dot{h}_x + h_z(\dot{\beta} - w_0) - h_y(\dot{\gamma} + \alpha w_0) \\ \dot{h}_y + h_x(\dot{\gamma} + \alpha w_0) - h_z(\dot{\alpha} - \gamma w_0) \\ \dot{h}_z + h_y(\dot{\alpha} - \gamma w_0) - h_x(\dot{\beta} - w_0) \end{pmatrix} = \begin{pmatrix} 0 \\ 0 \\ 0 \end{pmatrix}
$$

$$
(15.26)
$$

Rearranging the terms into the relevant inertial angular rates allows the use of an inertially derived PD control law of the form:

$$
\begin{pmatrix} I_x^*\ddot{\alpha} + K_p\alpha + K_v\dot{\alpha} \\ I_y^*\ddot{\beta} + K_p\beta + K_v\dot{\beta} \\ I_z^*\ddot{\gamma} + K_p\gamma + K_v\dot{\gamma} \end{pmatrix} = \begin{pmatrix} 0 \\ 0 \\ 0 \end{pmatrix}
$$

Even when incorporating the simplicity of reaction wheels, the computations are fairly complex.

15.4.1 Control moment gyroscope control

The complexity of the control moment gyroscope is introduced here for which all the control laws can be adapted. Powell et al (1971) derived a CMG control system including the gimbal steering law, the momentum distribution law and momentum desaturation laws. The gimbal steering law provides the gimbal rate commands in response to error inputs. The gimbal rates in turn produce the torque on the space vehicle to reduce those errors to zero. The momentum distribution law involves utilising the redundant degrees of freedom to provide optimal relative momentum by changing the gimbal angles without changing the stored angular momentum. The antiparallel condition occurs when all CMG angular momentum vectors are colinear. As gimbal stops inhibit inner and outer gimbal angles resulting in degraded performance, the momentum distribution law minimises the gimbal angles (particularly the inner gimbal) with singularity avoidance while keeping the total angular momentum constant. The momentum desaturation law kicks in when actuator saturation is approached. CMG desaturation is then imposed to dump the angular momentum when the inner gimbal angle approaches $\delta \to \pm 90°$. It is generally assumed that the gyroscopic moments of the gimbals are negligible with respect to the gyroscopic moments of the rotor. For a single double-gimballed CMG with roll and yaw gimbal angles:

$$h_x = c\delta s\zeta h_w$$

$$h_y = -c\delta c\zeta h_w$$

$$h_z = -s\delta h_w$$

where δ, ζ = pivot angles of inner and outer gimbals about x and z coordinates respectively, and h_w = flywheel momentum bias.

The angular momentum of a spacecraft and a single double-gimballed CMG in vehicle coordinates is given by:

$$H = \begin{pmatrix} I_x w_x + c\delta s\zeta h_w \\ I_y w_y - c\delta c\zeta h_w \\ I_z w_z - s\delta h_w \end{pmatrix} \tag{15.27}$$

Differentiate the angular momentum equation:

$$\begin{pmatrix} N_x \\ N_y \\ N_z \end{pmatrix} = \begin{pmatrix} I_x \dot{w}_x + (I_z - I_y)w_y w_z + \dot{h}_x \\ I_y \dot{w}_y + (I_x - I_z)w_x w_z + \dot{h}_y \\ I_z \dot{w}_z + (I_y - I_z)w_x w_y + \dot{h}_z \end{pmatrix} \tag{15.28}$$

The CMG is rotated at the commanded rate to produce reaction torques $N = -w \times H$ by gyroscopic action with respect to inertial coordinates.

For the inner gimbal: $I_x \ddot{\alpha}_x + K(Ts+1)[\dot{\alpha}^d - \dot{\alpha}] + h_w w_z c\delta - h_w w_y s\delta$

For the outer gimbal: $I_z \ddot{\gamma} + K(Ts+1)[\dot{\gamma}^d - \dot{\gamma}] - h_w w_x c\delta$ (15.29)

The dynamics are calculated with respect to vehicle body-fixed CMG mounting coordinates assuming that the relative gimbal rates are much greater than inertial rates and accelerations for simplification. The outer gimbal motion (subscript o) is specified with respect to moving CMG mounting coordinates (subscript b) by a relative gimbal rotation about the CMG mounting z-coordinate:

$$\begin{pmatrix} x_o \\ y_o \\ z_o \end{pmatrix} = \begin{pmatrix} c\zeta & s\zeta & 0 \\ -s\zeta & c\zeta & 0 \\ 0 & 0 & 1 \end{pmatrix} \begin{pmatrix} x_b \\ y_b \\ z_b \end{pmatrix} \tag{15.30}$$

The inner gimbal motion (subscript i) is specified as a rotation about the x coordinate of the outer gimbal:

$$\begin{pmatrix} x_i \\ y_i \\ z_i \end{pmatrix} = \begin{pmatrix} 1 & 0 & 0 \\ 0 & c\delta & s\delta \\ 0 & -s\delta & c\delta \end{pmatrix} \begin{pmatrix} x_o \\ y_o \\ z_o \end{pmatrix} \rightarrow \begin{pmatrix} x_i \\ y_i \\ z_i \end{pmatrix} = \begin{pmatrix} c\zeta & s\zeta & 0 \\ -c\delta s\zeta & c\zeta c\delta & s\delta \\ s\delta s\zeta & -s\delta c\zeta & c\delta \end{pmatrix} \begin{pmatrix} x_b \\ y_b \\ z_b \end{pmatrix} \tag{15.31}$$

The torque produced by the CMG is the rate of change of angular momentum in CMG housing coordinates. If the CMG has a constant spin rate with spin axis about the inner gimbal y coordinates:

$$\begin{pmatrix} N_x^i \\ N_y^i \\ N_z^i \end{pmatrix} = \begin{pmatrix} w_x^i \\ w_y^i \\ w_z^i \end{pmatrix} \times \begin{pmatrix} 0 \\ H_y \\ 0 \end{pmatrix} \tag{15.32}$$

where
$$\begin{pmatrix} w_x^i \\ w_y^i \\ w_z^i \end{pmatrix} = \begin{pmatrix} \dot{\delta} \\ 0 \\ 0 \end{pmatrix} + \begin{pmatrix} 1 & 0 & 0 \\ 0 & c\delta & s\delta \\ 0 & -s\delta & c\delta \end{pmatrix} \begin{pmatrix} w_x^o \\ w_y^o \\ w_z^o \end{pmatrix}$$ and

$$\begin{pmatrix} w_x^o \\ w_y^o \\ w_z^o \end{pmatrix} = \begin{pmatrix} 0 \\ 0 \\ \dot{\zeta} \end{pmatrix} + \begin{pmatrix} c\zeta & s\zeta & 0 \\ -s\zeta & c\zeta & 0 \\ 0 & 0 & 1 \end{pmatrix} \begin{pmatrix} w_x^b \\ w_y^b \\ w_z^b \end{pmatrix}$$

The inner gimbal motions are dependent on the outer gimbal motions producing strong control torque cross-coupling between the CMG axes. The inner gimbal applies an opposite reaction torque to the outer gimbal. This gives the inner gimbal torques in inner gimbal coordinates. To convert to CMG base coordinates:

$$\begin{pmatrix} N_x^b \\ N_y^b \\ N_z^b \end{pmatrix} = \begin{pmatrix} c\zeta & -c\delta s\zeta & s\delta s\zeta \\ s\zeta & c\zeta c\delta & -s\delta c\zeta \\ 0 & s\delta & c\delta \end{pmatrix} \begin{pmatrix} N_x^i \\ N_y^i \\ N_z^i \end{pmatrix} \tag{15.33}$$

This gives the torques in CMG coordinates. The transformation into spacecraft vehicle coordinates depends on the mounting orientation of the CMG with respect to the vehicle C_j which comprises a 3×3 rotation matrix of zero and unity elements if $90°$ and $180°$ orientations from the CMG to the vehicle are used:

$$N^v = \sum_{j=1}^{n} C_j^{-1} N_j^b$$

for n CMG units. The vector difference between torques from individual CMGs and the total desired torque is minimised to provide near-optimal distribution of relative momentum. As the CMG flywheel speed is constant, the torque of each CMG is perpendicular to the spin angular momentum H. The cross product law is a simple steering law and has the form: Gimbal rate for $CMG\ j$, $w_{ij} = KH_{CMGj} \times e_\tau$, where e_τ = torque difference error, w_i = angular rate of inner gimbal, and H_{CMGj} = angular momentum of $CMG\ j$.

All vectors are transformed from vehicle coordinates into inner gimbal space for error control. The gimbal rate commands are driven by the CMG torque error differences. The CMG control laws are much more complex than those for reaction wheels but CMGs offer greater actuation capabilities, and as we have seen are required for reaction compensation while using manipulator force control [Ellery 1996].

15.4.2 Optimal attitude slewing

It is without doubt that any spacecraft, a robotic servicer being no exception, that large angle slewing manoeuvres will be required to re-orient the spacecraft for a given purpose. Large attitude manoeuvres with negligible final error can be generated but often with an unacceptably long maneuvre time. Control motor torques are used to drive the spacecraft motion to zero such that all the system angular momentum and energy are absorbed into the rotor spin about the inertially fixed angular momentum vector. Constant motor torques are required to provide constant rotor acceleration for the whole trajectory $\dot{h} = I_a^w \dot{\Omega} = $ constant and this will require counter-rotation manoeuvres of the spacecraft unless switching is employed to maintain a second phase of control when $H_s = h + \sqrt{2I^s E^s}$ where $E^s = \frac{1}{2} I w_s^2$ to provide a constant rotor angular momentum relative to the spacecraft, i.e. $h = I_a^w \Omega = $ constant. In fact no sequential combination of constant and zero torque control signals will achieve a slewing manoeuvre exactly in finite time with $w(t_f) = 0$. This suggests the use of optimal control using a performance index minimising $\int_{t_0}^{t_f} u^2\, dt$ [Junkins & Turner 1986]. Such torque-shaping allows the wheel rates to be constrained smoothly reducing excitations of flexural degrees of freedom and suppressing vibration and structural deformation during large-angle manoeuvres. Such optimal control torques result in ~50% reduction in slewing angles. The NOVA-1 spacecraft was launched in 1981 and executed several large-angle slew manoeuvres in minimum time using optimal control techniques. This method of optimal slewing is suited to the feedforward component of reaction torque compensation from the robot control system to the spacecraft bus attitude control system in response the robotic manipulator movements.

Body-referenced coordinates may be mapped to inertial coordinates by:

$$w = \Phi\dot{\theta}$$

where $\theta = (\alpha, \beta, \gamma)^T$, and $\dot{w} = I^{-1}w \times Iw + I^{-1}u$. Differentiating:

$$\dot{w} = \dot{\Phi}\dot{\theta} + \Phi\ddot{\theta}$$

$$\rightarrow \ddot{\theta} = \Phi^{-1}\left[-I^{-1}(w \times Iw) - \dot{\Phi}\dot{\theta}\right] = \Phi^{-1}\left[-I^{-1}(\Phi\dot{\theta} \times I\Phi\dot{\theta}) - \dot{\Phi}\dot{\theta}\right] \qquad (15.34)$$

Hence the attitude control problem may be represented as:

$$\dot{w} = A(w) + Bu$$

$$y = Cw$$

where

$$A(w) = \begin{pmatrix} -I_x^{-1}(I_z - I_y)w_y w_z & 0 & 0 \\ 0 & -I_y^{-1}(I_x - I_z)w_x w_z & 0 \\ 0 & 0 & -I_z^{-1}(I_y - I_x)w_x w_y \end{pmatrix}$$

$$B = \begin{pmatrix} I_x^{-1} & 0 & 0 \\ 0 & I_y^{-1} & 0 \\ 0 & 0 & I_z^{-1} \end{pmatrix} \qquad (15.35)$$

The performance index to be minimised is given by: $J = \frac{1}{2}\int_0^t (u^T Ru + w^T Qw)dt$

The Hamiltonian function is given by: $H = \frac{1}{2}(u^T u + w^T Qw) + \lambda^T[A(w) + Bu]$ (15.36)

From Pontryagin's principle:

$$\dot{\lambda} = -\left(\frac{\partial H}{\partial w}\right)^T = -\left(\frac{\partial[A(w) + Bu]}{\partial w}\right)^T \lambda$$

$$\dot{w} = \left(\frac{\partial H}{\partial \lambda}\right) = A(w) + Bu$$

$$\frac{\partial H}{\partial u} = 0 = Ru + B^T\lambda$$

$\rightarrow$

$$u = I\dot{w} + \frac{1}{2}R^{-1}B^T\lambda$$

$$\dot{w} = Aw - BR^{-1}B^T\lambda \qquad (15.37)$$

$$\dot{\lambda} = -Qw - A^T\lambda$$

Applying this to a three-axis stabilised reaction wheel system for simplicity with no external moments acting:

$$\begin{pmatrix} I_x \dot{w}_x + \dot{h}_x \\ I_y \dot{w}_y + \dot{h}_x \\ I_z \dot{w}_z + \dot{h}_y \end{pmatrix} = \begin{pmatrix} (I_y - I_z)w_y w_z - h_z w_y + h_y w_z \\ (I_z - I_x)w_x w_z - h_x w_z + h_z w_x \\ (I_x - I_y)w_x w_y - h_y w_x + h_x w_y \end{pmatrix}$$

where

$$I = I^* + I_a^w 2 I_t^w$$

$$h = I_a^w \Omega \tag{15.38}$$

Motor torque

$$u = \frac{d(I_a^w w)}{dt} = I_a^w \dot{w} + \dot{h}$$

yields:

$$\begin{pmatrix} (I_x - I_a^w)\dot{w}_x \\ (I_y - I_a^w)\dot{w}_y \\ (I_z - I_a^w)\dot{w}_z \end{pmatrix} = \begin{pmatrix} (I_y - I_z)w_y w_z - h_z w_y + h_y w_z - u_x \\ (I_z - I_x)w_z w_x - h_x w_z + h_z w_x - u_y \\ (I_x - I_y)w_x w_y - h_y w_x + h_x w_y - u_z \end{pmatrix} \tag{15.38}$$

where $\dot{h} = -I_a^w \dot{w} + u$.

The ideal performance index to be minimised is: $J = \frac{1}{2}\int_0^T (q_1 u^2 + q_2 \dot{u}^2 + q_3 \ddot{u}^2)\,dt$ to provide smooth manoeuvres without jerks at the start and stop points but generates greater peak torques than $J = \frac{1}{2}\int_0^T q u^2\,dt$ where q = diagonal weight matrix (= diag(I) usually). This gives a measure of electrical energy expenditure: $E = \int_0^T qu\,dt$. Boundary conditions are given by

$$w_x(T) = w_y(T) = w_z(T) = 0$$

$$w_x(0) = w_y(0) = w_z(0) = 0$$

The Hamiltonian is given by: $H = \frac{1}{2}u^2 + \lambda \dot{w}$ where λ = Lagrange multiplier costate

$$H = \frac{1}{2}\begin{pmatrix} u_x^2 \\ u_y^2 \\ u_z^2 \end{pmatrix} \begin{pmatrix} (I_x - I_a) \\ (I_y - I_a) \\ (I_z - I_a) \end{pmatrix} + \frac{1}{2}\begin{pmatrix} u_x^2 \\ u_y^2 \\ u_z^2 \end{pmatrix} + \begin{pmatrix} \lambda_x[-H_{bz}w_y + H_{by}w_z - u_x] \\ \lambda_y[-H_{bx}w_z + H_{bz}w_x - u_y] \\ \lambda_z[-H_{by}w_x + H_{bx}w_y - u_z] \end{pmatrix}$$

where $H_b = Iw + h$. Now,

$$\frac{\partial H}{\partial u} = 0 = u = \left[\frac{\lambda}{(I - I_a)}\right] \quad \text{and} \quad \frac{\partial^2 H}{\partial u^2} = 1 > 0 :$$

$$\begin{pmatrix} u_x \\ u_y \\ u_z \end{pmatrix} = \begin{pmatrix} \lambda_x/(I_x - I_a) \\ \lambda_y/(I_y - I_a) \\ \lambda_z/(I_z - I_a) \end{pmatrix}$$

$$\rightarrow \begin{pmatrix} \dot{w}_x \\ \dot{w}_y \\ \dot{w}_z \end{pmatrix} = \begin{pmatrix} (-H_{bz}w_y + H_{by}w_z)/(I_x - I_a) - \left(\lambda_x/(I_x - I_a)^2\right) \\ (-H_{bx}w_z + H_{bz}w_x)/(I_y - I_a) - \left(\lambda_y/(I_y - I_a)^2\right) \\ (-H_{by}w_z + H_{bx}w_y)/(I_z - I_a) - \left(\lambda_z/(I_z - I_a)^2\right) \end{pmatrix} \qquad (15.40)$$

Pontryagin's principle states that $\dot{\lambda} = -(\partial H/\partial w)^T$:

$$\begin{pmatrix} \dot{\lambda}_x \\ \dot{\lambda}_y \\ \dot{\lambda}_z \end{pmatrix} = \begin{pmatrix} -\lambda_y H_{bz}/(I_y - I_a) + \lambda_z H_{by}/(I_z - I_a) \\ -\lambda_z H_{bx}/(I_z - I_a) + \lambda_x H_{bz}/(I_x - I_a) \\ -\lambda_y H_{bx}/(I_y - I_a) + \lambda_x H_{by}/(I_x - I_a) \end{pmatrix} \qquad (15.41)$$

Wheel speed is given by: $\Omega = (H_b - I_w)/I_a$.

The reaction moment compensation slewing manoeuvres of the ATLAS in-orbit servicer should be optimal for torque-shaping to minimise disturbances to the spacecraft mount and the positioning of the end-effectors.

15.5 INERTIAL NAVIGATION

Navigation is the process of determining vehicle linear and angular position and velocity of motion of the space vehicle by sensors so that desired objectives are achieved. It involves the measurement of a six-element state vector of position and velocity relative to a reference frame (usually inertial frame of reference fixed with respect to the stars) obtained through double integration of acceleration measurements. Guidance is defined as the process of measuring and steering the propulsive vector of the spacecraft such that its resultant trajectory will intercept its target. Navigation data allows guidance by using the current state vector estimate to derive a vector to the required location which is converted to propulsion system thrust magnitude and direction. Indeed, this is the purpose of the control system which essentially maps the data from the sensors to the desired actuation manoeuvres. Five functions for guidance, navigation and control (GNC) are required. Firstly, the desired results are determined (of orbit and attitude). Secondly, there is the determination of the actual vehicle situation (by sensors). Thirdly, comparison of the actual situation with the desired situation enables the determination of the required deviations. Fourthly, the deviation determines the corrections needed. Application of these corrections into a control system as inputs to generate changes in the vehicle motion as the output defines the final function. The first four functions comprise the navigation process. The final process is the guidance segment to alter linear and angular positions and rates of the spacecraft using actuators (thrusters for position and internal torquers for attitude). Overall the complete sequence represents the control system. Orbit and attitude determination is therefore a fundamental part of the GNC subsystem and this involves the determination of the values to the parameters which completely specify the orbiting body's motion relative to known points. This enables the steering requirements to be computed to attain the required destination points. Attitude determination by sensors precedes attitude control and all attitude control is fundamentally limited by the sensor accuracy. Ground tracking using radar techniques to determine the Doppler shifts to calculate range and range rate parameters of azimuth, elevation and slant range requires an rf beacon/transponder on the spacecraft. Ground radar tracking stations can provide ~3 m and ~1 mm/s accuracies. These are smoothed using Kalman filtering techniques and are converted to the required orbital parameters using numerical techniques to solve the nonlinear equations that relate them. For autonomy of navigation, however, onboard measurements should be made by a vehicle-mounted sensor system to provide the two vector directions required for orbit and attitude determination.

15.5.1 Guidance navigation and control (GNC) sensors

For orbital guidance the primary sensor is the accelerometer which produces an output in proportion to the rectilinear acceleration applied along the sensor input axis according to Newton's second law of motion. The accelerometer comprises of a proof mass whose motion along a principal axis is restrained by springs. The output signal is in the form of pulses, the interval between which is proportional to the magnitude of the applied acceleration. The output is amplified and fed back to provide a restoring force to counter-act the acceleration, so that the inertial mass remains in its equilibrium position. Clocks are used onboard to measure time using atomic oscillations to enable integration of acceleration measurements. Three such single-axis accelerometers may be mounted such that their input axes form an orthogonal triad. Typically, the accelerometer comprises a proof mass supported by a pendulum or electrostatically supported. Modern accelero-meters use quartz crystal flexure-mode resonators in a beam configuration to restrain the proof mass. This is the vibrating beam accelerometer the behaviour of which depends on its tension. Increased tension causes the resonant frequency to increase. They require no bias tension which is a source of instability and also respond to compression [Kuritsky & Goldstein 1983].

The most accurate reference is the fixed star background of known catalogue posi-tions. Star trackers are used as inertial azimuth and elevation devices to represent the star line of sight at the observer's position/orientation to give good accuracies ~0.01°. A pair of skewed star trackers can provide such attitude information by using different stars. It is best to select two stars ~90° apart in azimuth to obtain optimal position/orientation fix performance. If the star sensor is gimballed with known attitude it can act as a space sextant system to give position and attitude information. For large slewing manoeuvres, a wide FOV or gimballed tracking of the reference star is required. Typically, the star sensor has a small FOV ~6° to minimise the effect of sky background so two degree-of-freedom gimballing in azimuth and elevation is usually employed to maintain the star in the centred position and achieve an effective optical access FOV of ~95°. Vehicle attitude body coordinates are derivable from the gimbal angles and inertial information from the sensor readings. The star sensor is a solid-state device comprising a CCD array of pixels to locate the star with respect to the sensor axis. It is essentially an optical telescope of the folded cassegrain design. Image resolution is the main performance parameter and is a function of aperture size-to-wavelength ratio. A star map catalogue of stellar properties and locations is stored onboard to identify stars in the sensor FOV. The star tracker can also be used as a Sun sensor if special optical filters are used. They tend to have high power and processing requirements and are usually of high mass. They are also sensitive to sunlight (without filters) and to Earth-reflected light. They can only use bright stars normal to the Sun such as Polaris or Canopus (3rd/4th magnitude) to minimise tracking errors due to noise and possible confusion between stars. Canopus is used because it lies at a very large angle from the Sun 15° south of the ecliptic while Polaris lies at the north pole. However, such high-magnitude star trackers suffer from long time gaps between observations (up to 15–30 min). Dimmer stars ~7th/8th magnitude can maintain stars constantly in their FOV but identification can be problematic. A star catalogue is necessary containing accurate data on stellar positions and magnitudes—the GSFC (Goodard Spaceflight Centre) SKYMAP records 25 000 stars with magnitudes greater than 9.0. Star trackers form a fundamental component of three-axis stabilisation and provide the basis for periodic calibration of onboard gyroscopes from external stellar observations.

Inertial navigation development by gyroscopes is synonymous with rocketry development since World War II highlighting their central importance in missiles, rockets and spacecraft alike. Gyroscopes may be employed to provide 'memory' for star trackers with insufficient capability to maintain tracking of the reference stars. This star outage is caused by the local vertical changing direction as the satellite moves about its orbit. Gyroscopes require periodic recalibration of angular position from the external reference sensors due to drift ~0.01°/h to fix the drift. Relatively frequent star observations virtually eliminate the effects of gyro drift due to friction in the gimbal bearings. A gyroscope is a high spin rate rotor whose spin axis is free to precess about a perpendicular axis of symmetry. The rate of rotation of the outer gimbal assembly fixed to the spacecraft (the input axis) is proportional to the rate of rotation of the gyro inner gimbal about the output axis as the outer gimbal is prevented from rotating by mounting it in a chassis. The angular momentum vector of the rotor spin is constant with respect to the inertial frame unless an external torque is applied. The applied torque is related to the precessional angular velocity of the spin axis in the plane of the spin and input axes which tends to align the spin axis with the input axis. The rate gyroscope measures the vehicle rotation rate by the displacement of the gimbal about the output axis: $\dot{\theta} = -K\dot{\psi}$. This may be integrated by an onboard computer for inertial orientation information. The rate integrating gyroscope integrates the vehicle angular rate such that the gyroscope output is in proportion to the inertial rotation: $\theta = -K(\psi - \psi_0)$. However saturation (gimbal lock) at $\theta = \pm 90°$ can occur during large slew manoeuvres. Although the rate integrating gyroscope is more accurate and rugged, the rate gyroscope is preferred since the maximum vehicle rates for saturation can be made very large. The gyroscope rotor may be mounted in gimbals on an inertial supporting platform attached to the spacecraft so that the spacecraft rotation causes a change in gimbal angles. Three gyroscopes are mounted on the supporting inertial platform. Three single-gimballed gyroscopes or two double-gimballed gyroscopes are required for three-axis sensing. The gyroscopes detect the changes in attitude relative to the gyroscope rotors and the inertial platform is torqued by electric motors to compensate for the applied torques to the spacecraft and nullify the gimbal

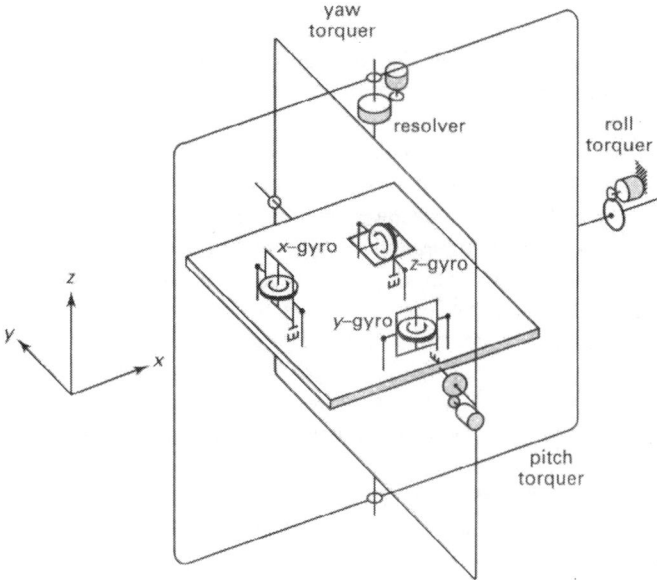

(a)

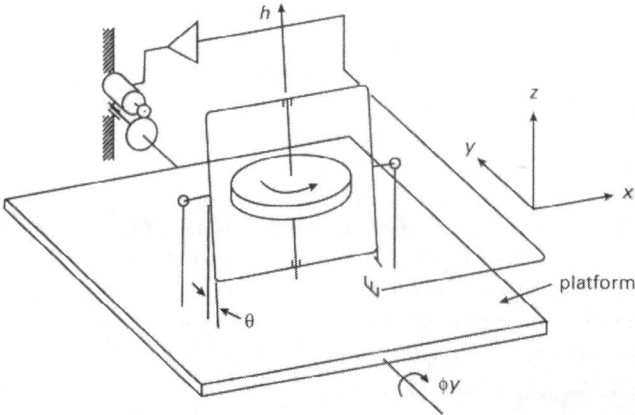

(b)

Fig. 15.3. Three-axis gyroscopic platform (from Thomson 1986). Reproduced with permission
from Dover Publications Inc.

angles or gimbal rates such that the platform remains stationary in inertial space. The applied torques required to maintain equilibrium are used as a measure of the attitude motion. If the platform is mounted in a fluid bearing spherical cavity under neutral buoyancy conditions the counteracting torques are applied by small fluid jets.

The strapped-down inertial platform has three single-degree-of-freedom gyroscopes with their precession axes perpendicular to each rotational degree of freedom and three accelerometers kept in the same position relative to the assembly unit which in turn is attached directly to the spacecraft frame without the use of gimbals [Baldassani 1968, Kuritsky & Goldstein 1983]. Inertial rates are given in terms of the vehicle body frame of roll, pitch and yaw axes. The inertial system is the initial position of the roll, pitch and yaw axes. The body relative attitude rates may be numerically integrated to yield the vehicle attitude with respect to inertial coordinates directly using the Runge–Kutta algorithm to solve Euler's equations, i.e. compute a virtual inertially stabilised platform. Strapdown systems offer less massive and mechanically simpler systems than inertially gimballed gyroscopes and higher reliability and lower cost results. This is the result of trading off physical simplicity of the device with greater usage of computing power. Typically the strapdown gyro comprises a spinning rotor and a hydrodynamic float bearing, hermetically sealed in the main housing without gimbals. The typical construction material is boron carbide for the rotor and motor shaft and beryllium for the main structure. They represent the 'brickwall' technique such that the accelerometer and gyroscope form independent instrument packages with physical and electrical isolation from other instrument packages.

The strapdown laser gyro is a non-mechanical inertial sensor which exhibits a simple construction and high reliability due to the lack of moving parts and high robustness [Matthews & Bates 1978, Kuritsky & Goldstein 1983]. They have high dynamic range up to ~600°/sec, low power consumption and low thermal conductivity, eliminating the need for temperature control. The laser gyro is based on the Sagnac effect. A closed-loop path such as a triangular coil arrangement is provided along which a light beam is generated from a source and sent in opposite directions (usually laser light). Special relativity states that the light will always propagate at c with respect to inertial space independent of the motion of the source and the receiver. The ring laser gyro is based on the fact that a Fabry–Perot-type cavity resonator forming a closed optical path will generate counter-directed laser beams at the same frequencies between the reflectors. For example an He–Ne gas plasma may be optically pumped into stimulated emission at optical frequencies. If the closed optical path is rotated inertially perpendicular to the plane of the travelling wave the oppositely directed laser beam will oscillate with different frequencies within the closed optical cavity:

Beat frequency $\delta f = 4A\Omega/(L\lambda)$

where A = area of the laser cavity, L = perimeter of the cavity, λ = emitted radiation wavelength, and Ω = inertial rotation rate.

The frequency difference may be measured via an optical lightwave interference pattern. Lock-in occurs when the two frequencies are close, i.e. at low rotation rates ~100°/h such that the two frequencies approach each other and the system behaves as a coupled oscillator. The commonest way to overcome this problem is the 'dither' technique. A symmetric mechanical angular rocking of the gyro is maintained between two bias points +B to –B about the gyro input axis by a stiff flexure suspension acting as a coiled spring. This produces a rate about the gyro input axis causing the gyro to oscillate about the lock-in zone with the bias averaging to zero and eliminating the lock-in. However, a random drift error is generated over a time δt which integrates to ~0.001°/h to give an attitude error of $\delta\theta \approx \int_t^{t+\delta t} \Omega dt + 0.001$. This random error sets a limit on the performance of the ring laser gyro which increases with the square root of time unless recalibrated.

Fibre-optic gyros are simpler than ring laser gyros, are more reliable, and do not suffer from lock-in problems. A closed-path optical fibre is fed with two opposing laser beams in the 1.3–1.6 μm wavelength range for low losses from a laser diode. If the closed path is rotated in inertial space a counter-rotating beam of light experiences a different path length from the reference beam. The time of flight for the light to travel the length of the fibre-optic cable changes if the coil rotates about its axis. This time difference is exhibited as a difference in phase between the counter-travelling laser beams forming an optical interference pattern with the fringe shift sensed by a photocell. At the end of the path the two beams are out of phase and this phase difference depends on the angular rate of change of attitude and the perpendicular plane of the two beams applied to the closed loop.

$$\text{Phase shift} \quad \delta\varphi = \frac{2\pi L D \Omega}{\lambda_0 c_0}$$

where L = optical fibre path length, D = coil diameter, Ω = rotation rate, λ_0 = wavelength of light *in vacuo*, and c_0 = speed of light *in vacuo*.

Three such fibre-optic laser gyros may be used in strapdown mode along the roll, pitch and yaw axes for three-axis attitude information.

Finally, the NAVSTAR GPS (NAVigation System using Timing And Ranging Global Positioning System) radio link through an onboard GPS receiver may be used for accurate determination of orbital elements prior to orbit transfer and to bound errors that may occur in the inertial navigation system. A similar system is the Russian GLONASS. The first onboard GPS receiver was flown in Landsat. GPS is a US DoD (Department of Defense) system of 24 three-axis stabilised satellites in constellation (21 active units plus three in-orbit spares in six different 12-h period orbital planes at 20 000 km altitude inclined at 55°, i.e. nominally four satellites in each plane equally spaced at 60° apart) costing ~$10b. GPS effectively replaced the US Navy's Transit series which was one of the most successful military satellite systems ever designed, having given over 20 y service. The satellite orbits repeat almost the same ground track once per day. Its original development was to allow the precise delivery of weapons with the satellites broadcasting time coded signals to derive accurate position and velocity estimates. Any fixed or mobile user can view a minimum of four satellites at all times to provide 24-h navigation information for global coverage. LEO satellites can track six or more GPS satellites. Four GPS satellite signals can compute satellite position and velocity in 3D and time. Each satellite emits a coherent L band signal and carries a highly stable atomic clock for deriving ranging signals. It uses geometric triangulation and radio ranging with onboard atomic clocks (with a frequency stability of one part in 1013 per day) to generate accurate position and velocity information to a mobile ground user receiver (with an unsynchronised quartz crystal clock with a stability of one part in 10^8 per day). The time shift T between the satellite signal and the receiver clock represents the difference between the time of travel of light (which proportional to the range from the satellite) and the lag of the user clock. Each satellite radiates two carrier frequencies: L_1 at 1575.42 MHz and L_2 at 1227.60 MHz. Each carrier is modulated by a fast 10.32 MHz bandwidth spread precision code (P code) which is securely encrypted for military applications for very high accuracies ~18 m. The L_1 carrier is also modulated with a 1.023 MHz bandwidth spread 1023 chirp sequence code (Clear/Acquisition code) which is available to general users offering 100 m, 5 cm/s and ~300 ns accuracies. However, accuracies of ~5m and 1cm/s are achievable by using multiple differential techniques, through a reference station equipped with a GPS receiver with precisely known coordinates—this allows systematic tracking of relative positioning of the remote position with respect to the reference station. Both codes are pseudorandom binary sequences and are examples of spread spectrum communications. The receiver needs to have information about the code sequences for each satellite for triangulation purposes. The receiver performs cross-correlation to acquire the satellite signal and demodulates the navigation data from the satellite orbit parameters. Three satellites give X and Y coordinates anywhere on the Earth's surface and a fourth satellite can give elevation in addition. GPS offers full coverage of navigation data without the need for worldwide distributions of Doppler ground tracking stations. To be usable by spacecraft for accurate range, range rate and tracking information, the spacecraft must carry a GPS receiver. Permanent ground-based receivers are installed into a network in Spain, French Guiana, Sweden, Australia and Kenya.

15.5.2 Kalman filtering

Errors occur due to computational time delays T between measurement and correction. A periodic low-frequency cyclic departure and restoration to and from the average attitude will occur in consequence. There is a limit to the frequency response of the spacecraft due to the limited bandwidth $\sim 1/T$ Hz. Errors will also occur due to measurement, calibration and alignment inaccuracies and noise from the sensors causing a high-frequency jitter. These effects cannot be sensed by the control loop and require filtering to remove the jitter. The jitter is treated as random with zero mean Gaussian distribution (white noise). Every sample period, the sensor outputs are filtered to estimate the current values. An onboard floating-point microcomputer is required to process the information from the respective sensors using an efficient filtering algorithm to provide a least-squares method of filtering. The best fit to the collected data generates the current estimate of the integrated orbital/attitude parameters. The integrated inertial system thereby eliminates the effects of gyro drift bias and noise in star sensors. The extended Kalman filter is the commonest approach to data fusion from separate data sources. Indeed, the first Kalman filter ever flown was aboard the Apollo Lunar Module primary computer to process rendezvous data for rendezvous with the Command Service Module in lunar orbit. New measurements are made periodically which are statistically combined with the present estimates and its error by optimal recursive Kalman estimation [Gewal et al 1991].

The Kalman filter uses a dynamic model for the time evolution of the system and a model of the sensor measurement system to find the most accurate estimation of the system's state, using a linear estimator based on the present and past measurements [Raol & Sinha 1985, Lefferts et al 1982]. It is a recursive predictive updating method that

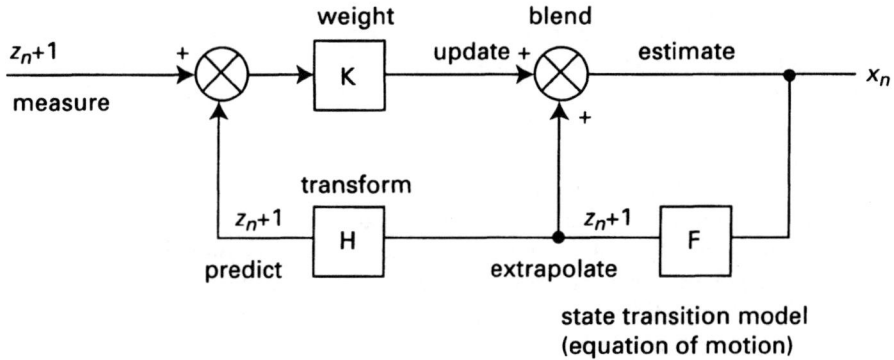

Fig. 15.4. Kalman filter (from Barter 1999). Reproduced with permission from TRW Space & Electronics Group.

estimates the parameters of a model with new measurements while incorporating the effects of noise (assumed to be Gaussian white noise). The error between parameter estimates and measurements of those parameters is used to update the parameter estimates through a gain matrix. Its chief characteristic is its use of the statistics of the sensor measurements (measurement error covariance). It can be used to determine the optimum mixing of gyroscopic and star tracker data. The estimation problem may be represented by the state vector variation $\dot{x}(t)$ which depends on the nonlinear process $f[x(t)]$ with Gaussian process (white) noise $n(t)$ and the observed measurement $y(t)$ which depends on the nonlinear measurement function $h[x(t)]$ and the Gaussian measurement (white) noise $v(t)$. The state and measurement equations are:

$$\dot{x}(t) = f[x(t)] + n(t)$$
$$y(t) = h[x(t)] + v(t)$$

(15.42)

The extended Kalman filter is a recursive technique that provides an estimate update of the state vector $\hat{x}$ and so is suitable for real-time implementation, particularly if a multi-processor architecture is used [Gary 1987]. The filtered estimate of x_i is given by $\hat{x}_i = x(t_i|t_i)$:

$$\hat{x}_i = \bar{x}_i + K_i(y_i - \hat{y}_i)$$

(15.43)

where $K_i = P_i H_i^T (H_i P_i H_i^T + R_i)^{-1}$ = Kalman filter gain, $P_{i+1} = \Phi_i P_i \Phi_i^T + \Phi_i Q_i \Phi_i^T$ = covariance matrix, $\bar{x}_{i+1} = x(t_{i+1}|t_i)$ = predicted estimate of

$\hat{y}_i = h(x_i)$ = predicted measurement of y_i, P_i = estimate error covariance matrix satisfying the Ricatti equation, $H_i = \partial h/\partial x^T$ = measurement matrix of sensitivity, Q_i = process noise covariance matrix, R_i = measurement noise covariance matrix, and $\Phi_i = e^{F(t_i - t_{i-1})}$ = state transition matrix, where $F = \partial f/\partial x^T$.

The conventional Kalman filter suffers from numerical stability problems so other techniques are required as the filter can diverge under certain conditions [Roal & Sinha

1985]. Numerical errors can be propagated and become divergent—this is particularly the case if the linear model and the nonlinear process differ by more than 10%. This can be overcome by factoring the error covariance matrix to improve the numerical stability. The UD factorisation method factorises P thus: $P = UDU^T$, where U = unit upper triangular matrix, and D = diagonal matrix. This comprises the extended Kalman filter. It is algebraically equivalent to the standard Kalman filter and when used with single-precision arithmetic matches the performance of the conventional filter using double-precision arithmetic. Other divergences occur due to non-Gaussian noise and when non-linearities are present. This can arise because the attitude problem is nonlinear but the Kalman filter assumes linearity. The largest computational burden is imposed by computation of the transition matrix and the contribution of process noise to the state covariance matrix. These problems may be alleviated by using a robust adaptive Kalman filter which has been found to be computationally flexible and efficient [Roal & Sinha 1985].

15.6 SUMMARY

This completes the consideration of the spacecraft attitude control system. Evidently, through the robotic feedforward compensation scheme, the robot control system interacts directly with the spacecraft attitude control system. The attitude control system scheme resembles the computed torque technique in that error feedback is implemented by the dedicated attitude control system while the external effects of the robot are fed forward for compensation. Standard attitude control techniques may be employed with the proviso that reaction moment feedforward compensation is required. The compexity of the attitude control system will not have gone unnoticed. To this end, ESA has its General Purpose Satellite Simulation Package (GPSSP) for real-time simulation of satellites particularly attitude dynamics and control, including modelling of telemetry/command links to and from the ground station.

16

Spacecraft avionics (C&DH) subsystems

Spacecraft avionics is essentially synonymous with the command and data handling (C&DH) component of the spacecraft subsystems, namely tracking, telemetry and command (TT&C) processing and the flight computers. Nominally, power conditioning and distribution also make up part of the avionics complement, but this has been relegated to the next chapter with power generation and storage. The spacecraft avionics implements onboard control functions for the spacecraft, which in the case of robotic spacecraft, be it an in-orbit servicer such as ATLAS or a planetary rover, are particularly demanding by virtue of the complexity of the control system and the requirement for real-time performance.

16.1 ONBOARD FLIGHT COMPUTER TECHNOLOGIES

Both military and space activities were instrumental in the miniaturisation of microelectronics for reduced mass and power requirements [Noyce 1977]. Microprocessors are the self-contained ALUs (arithmetic and logic unit) of the computer mounted onto a single Si chip. The DEC PDP/8 was the first minicomputer (1965). The first generation microprocessor was heralded by the 4-bit Intel 4004 CPU of 2300 transistors per chip in 1971, followed by the 8-bit second-generation chip in 1972, the 16-bit third-generation Intel 8048 chip in 1974 and the 32-bit fourth-generation 8086 chip in 1981 [Gupta & Toong 1983]. Moore's law states that the complexity and performance of integrated circuits doubles every 18 months. Presently, we have reached a capability of the 32-bit microcomputer with 10^7 gates per chip. The chip–component integration level has accelerated: one component per chip (1959), 10 components per chip (1964), 1000 components per chip (1969), 32 000 components per chip (1975), and 250 000 components per chip (1980), etc. LSI technology generated a 10^6 increase in computing power in only 20 y, providing whole systems on a chip, and this has increased by a further 10^6 with VLSI. Microprocessors incorporated the CPU on a chip and mini-computers have provided in addition memory and input–output functions on a single chip. Feature sizes today are ~1.0 μm decreasing at 30% per year. From 1970 to the mid-1980s, there was an approximately log linear increase in ips (instructions per second) rates and a linear increases in processor speeds, memory sizes and communication rates.

Since then, however, the increases have been approximately exponential. Processor speeds are currently at several hundred Mips, memory sizes at several Gbytes, and communication rates at ~50 kbps due to the development of data compression techniques. Clock cycles have been increasing rapidly—at present up to 500–1000 MHz. Virtual memory using disc storage to augment the main memory, read/write RAM and ROMs and small fast cache memories have also enhanced microprocessor performance. The inclusion of memory (RAM, ROM, EPROM and EEPROM), input/output functions and other auxiliary functions onto the chip have marked the advent of the microcomputer.

The Intel 8086 microprocessor was the workhorse of space systems for a long time with its 6200 transisors per chip capable of executing 770 Kips. Spacecraft payloads usually involve some form of sensor-based instrumentation which requires processing [Gough & Wooliscroft 1989]. A sensor head accepts data on a specific phenomenon and a dedicated processor processes the data. Most physical quantities in the real world are analogue and so vary continuously. Human senses are for instance analogue. Typical analogue devices include sensors, interfaces, op-amps and voltage regulators. To be integrated with digital devices analogue-to-digital and digital-to-analogue converters (ADC/DAC) are required and are extensively used in communications systems. This processed data is input to the spacecraft telemetry system and downloaded to the ground station which may perform various algorithmic processes on the processed data. Most telemetry systems use 8-bit data words and so are limited to 256 values, whereas most instrument output ranges exceed this. Logarithmic data compression is required to compact the data into 1 byte. Hence there is a need for onboard computer systems. The onboard computer must generally perform real-time control and navigation, health monitoring and housekeeping, command processing and subsystem management, communications and signal processing. Rapid response capability generally entails space-based onboard processing. Autonomy requires the spacecraft to function with little interaction with the ground. This requires high reliability with low failure rates $<10^{-6}$ bps plus fault tolerance. The hardware usually comprises the databuses and processors with algorithms implemented in software. Heat dissipation in convectionless space can be problematic even for low-power CMOS processes which can generate up to 100 mW. Thermal conductive heat sinks should be implemented in the main spacecraft structure.

The instruction set is the machine code used by the processor to interface between the processor and the software. The operating system executive manages and schedules mission-specific applications software and other functions, and provides interrupt handling, resource management and fault detection. The operating system runtime kernel supports higher-order languages. The operating system input/output handlers control data flow. BITE (Built-In TEst) diagnostics identify faults, failures and recovery. The operating system mathematics utilities are implemented as a library of routines used by several functional elements in the applications software. Assembler languages are more efficient than higher-level language software which require ~25% more coding but are more difficult to maintain and develop. Minimal memory and the lack of off-the-shelf compilers used to cause almost exclusive use of assembly coding in spacecraft software. Compilers essentially map high-level instructions to groups of electronic component hardware. However, with applications programs becoming increasingly complex, high-level languages are being increasingly used. The choice of high-level language is wide:

FORTRAN for scientific applications; C is almost the commercial standard; ADA is the DoD MIL-STD-1815A standard. ADA in particular is becoming popular and modern flight computers for attitude control are almost exclusively standardised in ADA. Application software is difficult to test completely as not all situations that may occur in all possible combinations can be anticipated in advance. Galileo's complete C&DH software comprised 35 000 lines of code (including 7000 lines for fault protection), plus 37 000 lines of code for attitude control and articulation (including 5 500 lines for fault protection). Software quality is an issue that is represents a well-known but difficult problem in terms of reliability of function. It is generally accepted that one in every 300 lines of code even after testing possesses a bug. The most international set of software engineering standards is the EN ISO 9000 series. Yourdon process methodology is the generally adopted systems software analysis and design tool. There are standardised tools available for software design such as computer-assisted software engineering (CASE) tools. The object-oriented approach to programming development attempts to reduce bugs by breaking down the programs into component parts down to the smallest level of detail. Each component is an object characterised by a name, its attributes, and its operation. The objects are related to each other through well-defined links into a hierarchical model. The software system comprises of a collection of algorithms defining processes each of which may be executed synchronously or asynchronously. These processes can be modelled as Petri nets for analysis. Object-oriented methodologies are not yet widespread.

Application software	Size (kword)		Throughput (kips)
	Code	Data	
Communications/C&DH	2.0	6.5	10.0
Attitude sensing	11.5	16.7	93.5
Attitude control	26.9	9.5	103.2
Autonomy	15.0	10.0	20.0
Fault handling	6.0	11.0	20.0
Power/thermal management	2.0	2.0	8.0
	63.4	55.7	254.2

Operating system	Size (kword)		Throughput (kips)
	Code	Data	
Executive	3.5	2	$0.3n$ with $n = 200$ tasks/s (60)
Runtime kernel	8.0	4.0	
I/O device handlers	2.0	0.7	$0.05m$ with $m =$ no. of commands/s (80)
BITE	0.7	0.4	0.5
Math utilities	1.2	0.2	
	15.4	7.3	140.5

Fig. 16.1. Typical computational requirements (adapted from Larson & Wertz 1992).

Up until recently, a typical onboard space-flight computer required as a minimum a 150-kword size (+25% margin) and a 400 kips processor speed (+30% margin) [Larson & Wertz 1992].

Several computer memory types are available. In volatile RAM (random access memory), data is lost if power is removed. Dynamic RAMs allow charge to be stored on a capacitor for a few milliseconds for high-density MB information storage. Almost 33% of all devices made are DRAMs. They are read and refreshed every few milliseconds before the charge leaks off. The 64 Mb DRAM can store a volume of the *Encyclopaedia Britannica* enabled by the development of trench and stacked capacitor structures. Static RAMs are based on the latch and do not require refreshing. They are used as on-chip caches close to the CPU to minimise data transport acrosss the databus by virtue of their greater access speeds over DRAMs. Non-volatile ROM (read only memory) contains permanent data patterns stored during manufacture through metallisation for bipolar devices or through gate oxidation for MOS devices. PROMs (programmable ROM) allow data patterns to be stored after manufacture, but they remain permanent by blowing a fusible link in the emitter connection. Software for initialisation and contingency operations to control hardware which are not subject to modification are usually permanently held as microprograms in non-volatile ROM (firmware) to free RAM for functions requiring more flexibility. Onboard programs may need modification or reloading due to upgrades or corruption. Programs may be stored in PROM and copied into RAM once operational. There has been a general trend to increase the spacecraft capabilities by placing all sophisticated algorithms in software rather than hardware or as fixed sequences of ROM firmware to give the flexibility offered by reprogrammability of RAM. Ground commands are written directly to RAM. Magnetic cores and plated wire memories have been used for spaceborne non-volatile RAM but suffer from a lack of support maintainability. EPROMs (electrically PROM) allow charge to be stored in the gate insulator for long periods. Memory is erased using ultraviolet radiation. Ultraviolet EPROMs are unsuited to space application due to the ionising radiation environment of space. EEPROMs (electrically erasable PROM) are non-volatile memories which may be selectively erased and rewritten electrically. They store charge on a floating conductive region in the middle of the MOS gate oxide. EEPROMs are ideal due to their non-volatility and reprogrammability [Rodgers & Stroll 1986]. They give a bit density of 64k/chip and software may be modified a limited number of times due to the repeated stress of high voltage during erase/write cycles. A bit is erased by applying a positive signal to the gate relative to the substrate inducing a positive trapped charge. Most EEPROMs have around 10-y lifetimes in ground-based applications with a limit of 10 000 erase/write cycles. Like other MOS devices, EEPROMs are susceptible to radiation and a spacecraft can accumulate up to 10^6 rads (Si) during its operational lifetime. N-channel EEPROMs are more radiation resistant than p-channel devices and are tolerant up to 10^4 rads (Si). The Westinghouse BORAM 6164 EEPROM is a radiation-hardened version.

Program storage possibilities include tape recorders, magnetic bubble memory and solid-state memory [Jones et al 1982]. Tape recorders are slow to read and solid-state memories have moderate capacities and low power requirements. Magnetic bubble memories are good for data storage in difficult environments as they are robust but require relatively high power. They are non-volatile and store data through magnetic domain state. As an applied B-field rotates, the bubble follows it. They offer high densities of storage, rapid access rates and fairly radiation tolerant. Optical storage offers high density for non-volatile storage but cannot be reprogrammed.

	Magnetic tape	Bubble memory	Solid-state memory
Storage density	10^7 b/m^3	10^6 b/m3	10^6 b/m^3
Storage capacity	10^{10} bits	10^7 bits	10^8 bits
Operating temperature	0–50°C	–55 to 85°C	–55 to 85°C
Power requirement (W)	15	10	3
Package mass (kg)	4	12	6
Package volume (cm^3)	2250	4500	7000
Size (cm)	$15 \times 20 \times 8$	$20 \times 14 \times 17$	$20 \times 28 \times 12$
BER	10^{-6}	10^{-8}	10^{-10}

Fig. 16.2. Computer memory devices (adapted from Larson & Wertz 1992).

16.2 RADIATION-HARDENED ELECTRONICS

A spacecraft structure constitutes an irregularly distributed shield to radiation. This radiation originates from galactic cosmic rays, trapped particles of the Van Allen radiation belt and solar flux. Galactic cosmic rays consist of highly energetic heavy nuclei with energies of ~GeV from supernovae explosions. Solar flares are also a source of mostly 100 MeV protons which last for several days at a time. From sunspot minimum to sunspot maximum during the 11-y solar cycle, exposure varies from 5 to 20 rem. Solar flares generate up to 1000 rem. Such radiations have a deterimental effect on spacecraft avionics—the energy of computer logic is ~10 MeV so such radiations can disrupt the operation of semiconductor circuits. CMOS FET is generally the choice for space system electronics. It is based on complementary pairs (n and p channel) of transistors on a common Si substrate and it offers low power dissipation as it dissipates power during switching only. Their operational characteristics are dependent on the electrical properties of the insulator oxide layers which when subjected to ionising radiations can trap charges.

Deposited energy density describes the effects related to localisation as measured by radiation absorbed dose in units of either $Gy = 1 \text{ J/kg}$ or $rad = 100 \text{ erg/s} = 0.01 \text{ Gy}$ ($1 \text{ rad} = 6.25 \times 10^7 \text{ MeV}$). Radiation absorbed dose (rad) is the product of particle fluence and stopping power of the material. One rad is equivalent to 3×10^7 electrons/cm^2 at 2 MeV or 4×10^6 protons/cm^2 at 300 MeV. Unless precautions are taken, both transient and accumulated damage to electronic components will occur generating performance degradation. Mechanisms of radiation interaction with the spacecraft components vary. Two types are particles impinging on electronic devices: photons and particles [Srour & McGarrity 1988, Stassinopoulos & Raymond 1988]. Accumulated damage from proton and electron exposure to electronics will limit the spacecraft lifetime while transients from protons or cosmic rays can disrupt operation possibly catastrophically. Shielding is limited for single-event upsets (SEU). Displacement damage (crystal lattice vacancy creation), ionisation (electron–hole pair creation and trapping), and direct nuclear reactions can be caused by high- energy particles and high-energy photons.

As far as photon flux is concerned, for Si with $Z = 14$, the photoelectric effect dominates at $E < 50$ keV, Compton scattering at 50 keV $< E < 20$ MeV, and pair production at $E > 20$ MeV. Accumulated effects are due to ionisation and atomic displacement from photon and electron exposure. Transient effects are due to ionisation from single high-energy cosmic rays and protons. The Earth's surface has a background of 0.1 rem/y. Electron exposure mostly causes ionisation damage while protons mostly cause both ionisation and displacement damage. High-energy protons of 10 MeV+ can also cause SEE (single-event effects, a form of localised ionisation damage) and the incidence of SEU (single event upsets) increases sharply above 30 MeV. Single heavy particles from cosmic rays of energies ranging between $\sim 10^5$–10^{16} eV will produce a dense ionisation track of electron–hole pairs if incident on a semiconductor device. This can cause transient upsets or even permanent damage such as latchup.

Protons are two orders of magnitude more damaging than neutrons and neutrons are in turn two orders of magnitude more damaging than electrons and electrons are an order of magnitude more damaging than gamma-rays. Rutherford scattering dominates at low energy recoils $\leqslant 5$ keV and nuclear elastic scattering dominate with recoils $\geqslant 10$ keV. Electrons require >150 keV to produce displacement in Si while protons require only >100 eV. Ionisation is the dominant effect over displacement. Displacement faults occur when particles are absorbed or scattered by nuclei in the target material. Defects resulting from this create recombination centres and reduce minority carrier mobility. Hence, bipolar transistors are most affected by reduced current gain. Majority carrier MOS are not significantly affected. Ionisation involves the creation of electron–hole carrier pairs in the target material by photons and charged particles. Si generates one electron–hole pair per 3.6 eV absorbed. Alpha particles of $\sim$MeV energies are absorbed in 25 μm of Si creating $\sim 10^6$ electron–hole pairs. Mesons in cosmic rays can generate $\sim 10^5$ electron-hole pairs. These carriers then diffuse into capacitors and lattice defects.

Trapped charges in defects cause shifts in device potentials and degrade carrier mobility. Surface devices such as MOS are the most affected. Ionisation causes photocurrents which can either cause transient SEU (single-event upset) which can be reset, or permanent latchup which causes loss of device function. High-energy particles alter the crystalline structure of the solid-state device so that they no longer function correctly. An SEU rate of 5×10^{-10}/d is typical due to charge trapping in lattice defects. Typically, shielding (such as spot shielding using high density materials such as copper or tantalum) is the solution but shielding is insufficient against heavier galactic cosmic rays. The primary effect on CMOS is to produce electron–hole pairs in the SiO_2 insulator between the gate electrode where the threshold voltage is applied and the Si substrate. This reduces the required threshold voltage to enable current to flow between the source and drain. For low-energy particles, recombination of pairs is high, but for high-energy particles the charge yields approach 90%. Charges which do not recombine are transported through the SiO_2 insulator by the applied electric field. Electrons are transported rapidly ($\sim$ns) but holes remain near the point of generation and are transported only slowly through the oxide (in $\sim$s). This causes a slow recovery in the threshold voltage from the initial negative shift provided that the fraction of holes trapped at the Si/SiO_2

interface is less than 0.5. For most unhardened devices, this fraction is usually around ~0.2–0.4 and for hardened devices it is normally less than 0.1. Holes trapped at the interface can be annealed by injecting electrons from the Si valence bands by tunnelling through reduction of the SiO_2 thickness to ~100 A, generating a slow long-term recovery of the threshold voltage. Alternatively, they can be heated to 100–125°C for several hundred hours to anneal out the trapped charge, but this is impracticable. At LSI/VLSI integration levels radiation-induced leakage currents become the dominant failure mechanism. Radiation causes a positive charge buildup in the isolated oxide region and large leakage currents will occur in commercial CMOS at 10^4 rad but hardening enables tolerance to 10^5 rad. Slow recovery occurs in the threshold voltage shift and the leakage current in the insulator region. In low-dose-rate environments, the slow recovery of the threshold voltage will limit the leakage currents. Other passive devices may be affected by radiation dose. Capacitors suffer from voltage buildup and dielectric conductivity changes. Back-to-back double capacitors reduce this effect. Resistors suffer very few radiation effects.

Up to 50% of SEU soft errors could be catastrophic. Commercial microprocessors typically suffer SEU rates of 10^{-6} SEU/bit/day in LEO. Low-energy particles $\leqslant 10$ MeV lose energy much faster than high-energy particles $\geqslant 10$ MeV. The 140 MeV Fe ion is considered the worst culprit for SEU. The energy loss rate is given by the linear energy transfer (LET) function dE/dx defined as the energy deposition per unit length of the semiconductor active region. The dose is proportional to the fluence of particles and the LET. LET peaks at around 1 MeV and then decreases. Charge collection from ionised tracks occurs through drift in the equilibrium depletion region due to carrier charge current. This is caused by ion diffusion beyond the depletion region from the excess charge carrier concentration gradient. Charge funnelling also occurs from the spread of field lines in the substrate. This creates a ion track plasma shunt which acts as a conduit between drain–body junctions for charge transfer and exchange. All these contribute to soft errors by reducing the drain potential difference. Permanent damage is induced through latchup which can lead to burnout through thermal runaway especially if the device is smaller than 0.5 micrometres unless the power supply is switched off. This results from heavy ions creating a high current mode that arises from the existence of parasitic bipolar n-p-n and p-n-p structures that form a four-layer silicon-controlled p-n-p-n or n-p-n-p rectifier structure. It is activated into a low-impedance, high-current state by a triggering ionising pulse that forward biasses the control junction. Typically, 1×10^5 ions/cm^2 can generate latchup in non-radiation-hard electronics.

Radiation-hardened electronics are typically resistant to a LET of 100 MeV cm^2/mg. Radiation damage may be alleviated by increasing threshold voltage gains, heavier doping, reducing gate oxide thickness, using epitaxial growth during manufacture and charge injection annealing. The total dose effects on threshold voltage and channel mobility are governed by the gate oxide where electron–hole pair generation causes holes to get trapped in the oxide layer reducing the n and p channel threshold voltages. Hardening is a process-oriented effect such that oxide and oxide/Si interface is controlled to minimise trappped holes in the oxide. The oxide thickness is reduced to 100 A to allow electron tunnelling to eliminate the trapped holes in the oxide. NAND gates have greater tolerance to failure by minimising the impact of n-device leakage by eliminating parallel n-devices.

A thin epitaxial layer of CMOS helps to prevent latchup due to the close proximity of a heavily doped substrate which acts as a collector. Coating integrated circuits with polyimides provide protection against radioactive decay products such as alpha-particles with energies below 10 MeV but not heavier ions. Shielding is often ineffective against heavy particle-induced SEU. In a storage element, SEU will cause a bit-flip soft error and loss of information. Error correction codes can provide a degree of error rejection but circuit hardening is generally required. The only way to effectively protect electronics is to use structured substrates formed by epitaxial insulation. A thin, lightly doped epitaxial active region over a heavily doped substrate effectively truncates ion tracks. Retrograde well techniques involve doping the deep edge of the well more highly than the surface, reducing the impact of ion shunts to the substrate and its charge transfer efficiency. Both thin epitaxial and retrograde well methods may be combined to limit the maximum amount of radiation-induced photocurrent collection and latchup.

CMOS/SOS (silicon-on-sapphire) technology is based on fabricating single-crystal Si on an insulating dielectric substrate of sapphire [Kerns & Shafer 1988]. It is a subclass of silicon-on-insulator (SOI) technology (such as silicon dioxide) which are used extensively in high-frequency applications, e.g. the 32-bit Dassault FLAME processor which hardened to more than 1 Mrad. In SOS technology, inter-device Si is etched away, providing complete electrical isolation between active devices which form islands so only the parts of an ionised track penetrating Si can be collected by the device, thereby preventing latchup. It offers the advantages of CMOS technology with the robustness to total dose ionisation and latchup. SOS does reduce SEU sensitivity by 10^3 to 10^4 but does not totally eliminate it. VLSI offer microprocessors on a chip using CMOS/SOS technology. The problem with this level of integration is the power dissipation problem and conducting thermal energy away from the electronics. Unhardened electronics are subject failure at 1 krad(Si) of protons and 10^4 rad of electrons, while hardened electronics will fail at 10 Mrad(Si) of electrons and 100 krad(Si) of protons.

ESA's radiation database provides a list of CMOS components that are qualified for low and moderate dose rate exposures to 10 krad (Si). Unhardened CMOS are survivable up to 5 krad (Si) total dose. TTL bipolar devices which are inherently radiation-hard to 10^6 rad (Si) may be substituted for CMOS parts for certain mission-critical hardware or firmware. For testing purposes a 1 MeV electron flux is used as a standard radiation environment. Damage coefficients relate the damage caused by electrons at different energies to those at 1 MeV. Furthermore statistical radiation average of 10 MeV protons produce the same damage as 1 MeV electrons at 3000 times the fluence of 10 MeV protons. As electrons provide the majority of the dose deposited in components Al spot shielding may be employed: 250 mm of Al will provide an effective shield to electrons with energies exceeding 3 MeV. Further shielding yields bremsstrahlung due to interaction of the ionising radiation with the shield.

CMOS (soft)	10^3–10^4 rad
CMOS (hard)	10^4–10^6 rad
CMOS/SOS (soft)	10^3–10^4 rad
CMOS/SOS (hard)	$>10^5$ rad

For testing of electronic components, Co-60 gamma ray emitters are used as they provide a good simulation of the space environment. Hardening electronic circuits can constitute up to 2–5% of the total satellite cost.

16.3 SPACE COMPUTATIONAL ARCHITECTURES

The 6-month duration Mariner mission of 1967 which performed a Venus flyby was the last deep space probe to fly without an onboard computer processor. It executed its command sequences through magnetic sequencers initialised at launch. The first digital computer to fly in space was on the Gemini manned missions and was the most powerful airborne computer of the time. The first computer processor to be flown on an exploratory space probe was on Mariner in 1969 which performed a flyby of Mars. However, it was flown as backup to the primary magnetic sequencer system. Onboard space computers evolved rapidly from the Saturn 1 booster computer (1959–62) to the Space Shuttle computer development programme (1972–75) [Cooper & Chow 1976]. During this 3-year period computing speed increased by 160 times and memory capacity increased by 13 times, while the mass of computers dropped by a factor of one-half and the volume by a factor of one-quarter. Circuit density increased from 0.2 gates/device to 500 gates/device and memory density increased from 12 to 146 Mb/m^3 while access times dropped from 375 ms to 5 ms and processing speeds increased from 3 kflops to 480 kflops. Those trends have been increasing at a near-geometric rate ever since.

The Apollo mission avionics led directly to the computer architecture of the Space Shuttle based on five General-Purpose Computers (GPC). Each has a package mass of 30 kg, utilising 350 W of power. They are 32-bit architectures with 64 kword core memory. Their performance provides 1.2 μs add times and 5.7 μs multiply times together with floating-point arithmetic capability. Indeed, Space Shuttle avionics and computers have served as a model for a generation of spacecraft computers. The NASA Standard Spacecraft Computer-1 (NSSC-1) is a standard ASIC-based computer of the Multi-Modular Spacecraft developed from its predecessors, the Onboard Processor (OBP) which was flown on Copernicus in 1972, and the Advanced Onboard Processor (AOP) which was flown on the IUE in 1975 [Kayton 1989]. The NSSC-1 has been used on the Solar Maximum Mission, Hubble Space Telescope and Landsat. It is an 18-bit architecture with plated wire memory of up to 64 kwords in modules of 8 k. It has a package mass of 10 kg and consumes 30 W of power. Its performance gives 5 μs add time and 38 μs multiply time but does not support floating-point arithmetic. The DF-224 was used on the Hubble Space Telescope in a triple cpu configuration to enhance the NSSC-1 capability during the first servicing mission. The DF-224 is a fixed point, 24-bit computer architecture designed for attitude control and health monitoring. Its clock speed is 1.25 MHz and has a 64 kwords of plated wire memory preprogrammed in assembler code. It has a package mass of 50 kg drawing 80 W of power. The 16-bit space-rated processors, the Intel 16 MHz C&DH 80386-16 and the ACS 80386-16 processors, provides the basis for the Space Station Data Management System using high-level language programs such as C and ADA rather than assembler code. The current generation of UoSATs such as TMSAT use the Intel 80C386 microprocessor backed up by the Intel 80C186 processor and two T800/T803 transputers for image processing [Underwood 2000]. The third HST servicing mission will replace HST's present processors with the 50-MHz 80486 processor which is currently the most powerful NASA radiation-hardened CPU available. The ERC32 chipset is a space-qualified version of the SPARC V7 processor developed by ESA [Taylor et al 1998]. It has 8 Mbyte of SRAM, 4 Mbyte of EEPROM and six RS-422 serial interfaces. A mass memory of 50 Mbyte of EEPROM may be included in addition. The total package mass is 3.4 kg with 28-V dc power supply. The Mars Pathfinder mission used the commercial radiation-hardened 32-bit RISC RS6000 processor with an order of magnitude increase in performance over 16-bit processors. One potential approach to avoid using radiation-hardened versions of commercial processors is to use error-correction codes in conjunction with redundancy. Cross-strapping between the prime unit and the cold spare ensures that there is no single-point failure mode. The goal for onboard computer systems is to provide radiation-hardened 100 Mips/W at 10–100 W power scales.

There are seven major characteristics of spacecraft computers [Theis 1983]: throughput (kips), memory size, input/output capabilities, electric power requirements, reliability and availability/qualification (radiation hardness). Centralised architectures have point to point interfaces with central coordination and suffer from large wiring harnesses. Instructions are executed in sequence but the real world is comprised of many processes and activities occurring concurrently. For high processing powers, groups of microprocessors each with their own memory connected by an isolated data bus are required with a command response protocol or token-passing protocol. Such distributed techniques provide a degree of redundancy and fault tolerance. This is particularly crucial for real-time control operations [Rennel 1980]. Different subsystems perform different complex functions and the use of dedicated distributed computer architectures offer superior performance. Onboard data handling architectures may be a star-connection network (e.g. ECS) or databus-based (e.g. Olympus) (see Fig. 16.3). The star connection employs multiple wires, unlike the bus architecture which is limited to one or two wires, offering a simplified wiring harness and greater power and size efficiency for large systems. The two wire bus separates interrogation and response functions. A central terminal issues instructions on the interrogation bus. If necessary communication protocols can allow configuration into multilevel hierarchical architectures.

The USAF MIL-STD-1750A 16-bit Standard Instruction Set Architecture has become the support tool standard for spacecraft. It provides 1 Mips processing speed, a non-volatile memory capacity of 1 Mb and 70 ns access time from RAM. It employs 1.5 μm VLSI feature sizes based on radiation-hardened CMOS/SOS (silicon-on-sapphire) manufacturing. CMOS/SOS is inherently radiation-hard since neutrons tend to degrade any Si lattice by generating recombination sites. This reduces minority carrier lifetimes, but CMOS is a majority carrier technology and so is relatively unaffected by neutron flux. CMOS power consumption is given by $P = kCV^2$, where k = proportionality constant dependent on the frequency of switching and number of switching transistors, C = device capacitance, and V = operating voltage. For example, a reduction of operating voltage from 5 to 3.3 V results in a 50% reduction in power dissipation. ESA has introduced the MA31750 microprocessor system comprising a CPU and memory management system to be interfaced to an error detection and correction unit, direct memory access controller, 16-bit processor support circuit, and a maskable 64-kword ROM. Its CPU comprises

Centralised architecture

Distributed architectures

RING architecture　　　　　　　　　　　　**BUS architecture**

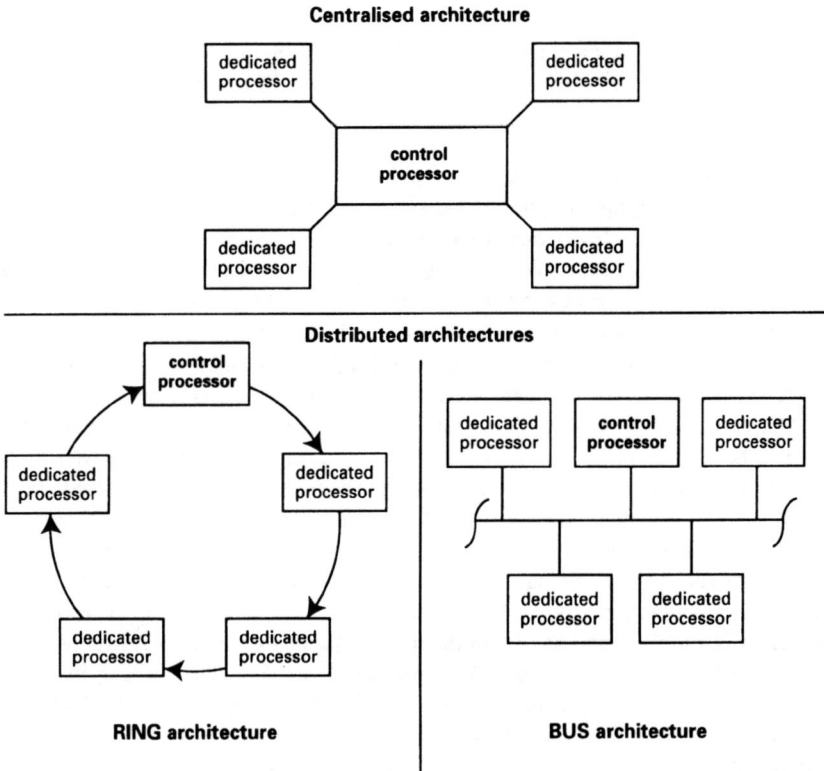

Fig. 16.3. Onboard computational architectures.

Computer	GPC	ATAC-16 ms	TDY750	GVSC
Application	Shuttle	Galileo	MILSTAR	Planned
Word length (bits)	36	16	16	16
Package size (cm^3)	24 000	13 000	40 000	4400
Package mass (kg)	26.8	11.3	43.5	11.3
Power (W)	350	50	200	30
Throughput (kips)	200	500	450	4500
Memory (kwords)	32	64	512	2000

Fig. 16.4. Comparison of space-rated computers (adapted from Larson & Wertz 1992).

2×10^5 transistors in a lead housing and it is capable of 3 Mips while dissipating only 300 mW (cf. RISC processor offers ~20 Mips dissipating 10 W, i.e. 2 Mips/W). It is a CMOS/SOS system and has 500-krad hardness with a packaged processor mass of 20 kg and a power requirement of ~50 W. To meet the radiation requirements, hardness levels of 100 krads to 1 Mrad (Si) total dose are required. To this end, GEC Plessey

Semiconductors manufacture CMOS/SOS radiation hardened circuits with 2.5 μm and 1.5 μm feature sizes which can tolerate $>10^6$ rad (Si) total dose over a 10-y operational life and have an SEU rate of 4×10^{-11} bit errors/day. They are capable of sustaining 10^{15} n/cm^2 neutron flux. Their MAS281 MIL-STD 1750A microprocessor is based on the McDonnell Douglas MDC281 offering <1 Mips and 64k RAM. Transistors are isolated from each other by the insulating sapphire substrate, making it impossible to suffer latchup. The offline storage requirements are of $\sim 10^9$ bits. Off-the-shelf computers reduce the hard-ware cost over customised hardware but customised radiation-hardened hardware is more common. Generally, computational power is severely restricted in space in having only 10% of the processing capabilities of a typical terrestrial workstation.

High-performance electronic computers are particularly susceptible to transient error due to tight timing margins and to permanent faults due to their complexity [Siewiorek 1984, 1991]. There has been a general trend towards using more electronic components over electromechanical ones and computer systems are mission-critical—their malfunction could be catastrophic. Furthermore, the most stringent requirements occur in real-time control systems where recovery time from faults must be minimised. In digital communications, an error rate of 10^{-5} is regarded as the standard acceptable. Fault-avoidance techniques such as the adoption of high-reliability components, limiting gate fan-out, increased burn-in, etc. are insufficient for complex computer systems.

Computer systems must be fault-tolerant to preserve the continued correct execution of programs in the presence of certain operational faults due to physical failures generating output errors. The error is the manifestation of a fault. The fault is the erroneous hardware or software state while the failure is the physical source of the fault. Furthermore, built-in test (BITE) error-detection and re-try techniques must be employed as well as redundant power supplies. A typical overall procedure on fault occurrence is as follows: fault detection by consistency and parity checks on memory, data, address and control information will find the fault and invoke a pause re-try to attempt to correct the problem if it is transient; if this fails software processes are rolled back to a point before the error and then restarted using backup files.

Error-detection techniques usually involve duplication with two identical copies of hardware running the copies of the same computation and the results compared for consistency. This provides both hardware and software redundancy. Duplication may take one of two forms: a loosely coupled multicomputer system involves replication of memory, data and processor each system with its own operating system communicating across a LAN; a tightly coupled multicomputer system involves the replication of data and processor with a single memory, a single operating system and programs with minimal overhead.

Alternatively, parity codes which use information redundancy may be employed involving the addition of an extra odd or even bit to represent parity in a group of bits to detect errors. Parity is effective for detecting transmission errors but it is more difficult to use when data is transformed by arithmetic operations. EDAC (error detection and correction) codes are employed extensively in space processors. A typical implementation is a hardware-based hamming block coder/decoder which can correct single bit errors due to SEU events. Restarts are repeatedly executed on a program that has failed and so are suitable only for transient faults. Roll-back requires backup systems and files to record process and state information to allow the detection of the fault and recovery to be implemented. This backup system is a kind of short-term memory to provide recency recording for error detection. Checkpointing may be used such that during program execution the process state is saved at specific checkpoints. Roll-back then occurs to the last checkpoint. The program is then restarted to redo the computations. Such restart recovery programs are generally installed in the operating system for automated invocation in the event of error. Such software-controlled recovery requires recovery software to remain operational in the presence of faults. Generally, this is not sufficient for fault tolerance.

The JPL STAR (self-test and repair) space computer is a primary example of a multimission space computer designed for long period operation without human intervention. It was designed though never flown to enable Voyager 2 to complete the 10-y 'Grand Tour' of the outer solar system. A test and repair processor module (TARP) controlled recovery and employed memory-copying backup software. Error-correcting codes for error control in communication and storage and checkpointing for roll-back were also implemented. Duplication was used to detect errors in logic processing through triple redundancy and majority voting with two standby spares. Although STAR was never flown, it provided the model for the Galileo Jupiter flyby probe.

Spacecraft components and spacecraft assembly and testing are usually conducted in clean rooms where a unidirectional flow of air is maintained through particle filters. Quality of air is determined by the number of particles of (0.5 μm per litre of air, e.g. class 100 has 3.5 particles/litre. This class requires full suits to be worn with only the eyes exposed. Class 1000 allows facial exposure while class 100 000 allows just clean coats and over-shoes. For electronic components which are highly sensitive to static voltages on human skin, workers must be grounded to a conducting floor with wrist and ankle straps. Humidity levels of 40–50% are maintained as a compromise between the tendency for charge accumulation in dry climates and the tendency for corrosion in humid environments.

16.4 COMMAND AND DATA HANDLING (C&DH)

The TT&C (tracking, telemetry and command) subsystem is subdivided into an rf part utilising the communications subsystem and a C&DH (command and data handling) subsystem. TT&C provides the means for monitoring and controlling the satellite by receiving, decoding and distributing spacecraft commands and formatting the spacecraft telemetry. The TT&C transponder is part of the communications payload. The number of functions and the accuracy at which they must be monitored determines the telemetry data rate. Telemetry makes measurements inside the satellite and transmits them to the ground station via the lower downlink frequency band. This requires the spacecraft to be instrumented with sensors to measure their environments to ascertain events and subsystem states. Such data may include: equipment temperature, propellant pressure, power voltages and currents, GNC data, all operating states, deployment states and redundancy states. Most spacecraft involve hundreds of parameters, some with low sample rates ~1/30s (housekeeping) but some with high sample rates ~1–4/s (e.g. payload data and attitude pointing). Each telemetry sensor is sampled in sequence with a multiplexer combining all the telemetry data into a single bit stream. Multiplexing is the process of combining a number of information-bearing signals with different characteristics into a single transmission. The sampling rate is on average low ~1 Hz and formatted words are typically 8 bits long. To carry separate messages in a single voltage variation it is necessary to share either the available time of transmission or the available frequency bandwidth (time or frequency division multiplexing TDM/FDM) which effectively combines bandpass and bandstop filtering. TDM transmits information as a time series of pulses. Sensor outputs are fed to an electronic commutator which sequentially samples each sensor output. Electronic commutators can multiplex ~60 input channels at a rate of ~5–40 channel samples/s. This method is limited in frequency response and is usually used for low frequency data such as scientific data. Pulses can be sent directly to the modulator or used with FDM. FDM subdivides the frequency bandwidth for parallel transmission. It uses voltage-controlled subcarrier oscillators separated by frequency-selective filters. The stream frequency modulates the voltage-controlled oscillators for transmission. Signal voltages of variable amplitude varies the output of the oscillator frequency representing the signal voltage. One subcarrier band oscillator is used for each signal. Higher data rate streams are allocated higher bandwidths, e.g. video imaging. The outputs of the oscillators are fed to a single amplifier to be combined across all frequencies. Hence, multiple streams can be multiplexed as a single bit stream onto a single carrier. FDM can suffer from intermodulation interference (cross-talk) but the introduction of guard bands overcomes this. The most common terrestrial multiplexing method is frequency division multiplexing though time division multi-plexing is becoming popular.

Command instructions are the means by which control is established. An Earth-based antenna uplinks commands to the satellite onboard receiver. Telecommands are received, filtered, amplified, demodulated and synchronised. Demodulation measures variations to extract the information. Synchronisation aligns the time scales of spatially separated periodic processes. A signal is operated on during a given time interval and this requires time and frequency synchronisation. Generally, an Earth terminal clock is accessed by the satellite clock so that transmissions are initiated at the correct time to arrive at the satellite receiver in synchronisation with the satellite clock. After decoding the commands are distributed to their addresses. The downlink rate ~40 bps–10 kbps (average ~2.4 kbps) is generally much higher than the uplink rate ~8 bps–2 kbps (average 1 kbps) and is the power-limited link. Furthermore, the BER for uplink is typically ~10^{-5} but for the downlink is typically lower at 10^{-4}. The AOCS and payload typically generate the most telemetry and invoke the most need for commands. ERS-1 had two telemetry systems. A standard TT&C system provided housekeeping at 2 kbps on the S-band. The science image data from the payload however required a higher data rate which was provided through a dedicated data telemetry link of 105 Mbps at the X-band. Telemetry data is collected, multiplexed and formatted. The telemetry subcarrier is modulated and then transmitted. Commands are transmitted as digital data blocks which include a spacecraft identification code, synchronising bits, device address word and the command itself with parity bits. All remote units are characterised by a unique address with the central unit distributing commands to the designated units. The command may be time-tagged with the time specified within a command word. They have a frame structure of 16 bits with

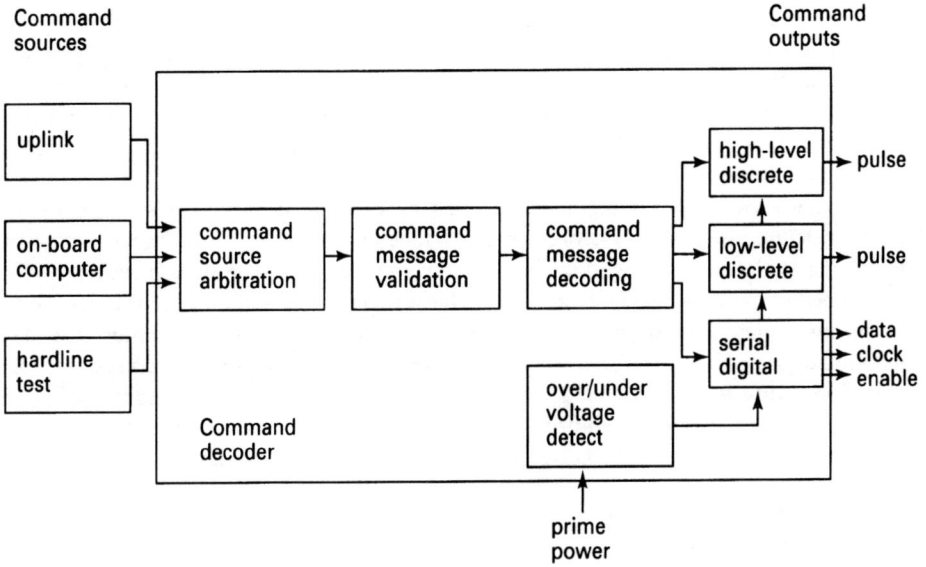

(a)

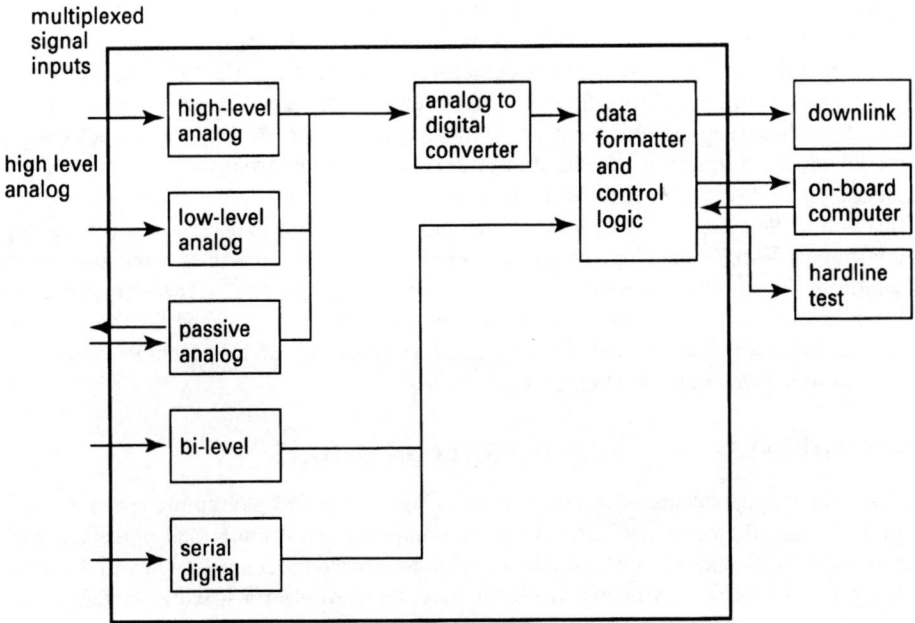

(b)

Fig 16.5: (a) Command Decoder; (b) Data Handling Unit (from Wertz & Larson 1992)

specific fields. A typical command message is ~48–64 bits long. They are of two types: on–off pulse relay commands and memory load word commands. All commands should conform to the Inter-Range Instrumentation Group Standard 106. Telemetry is usually represented by 8-bit words (analogue sensor signals are digitised by ADC). It is formatted whereby each position is assigned a particular parameter. The subcarrier is modulated via PSK.

Tracking involves observing and collecting data to plot the spacecraft trajectory. To track a satellite, the ground station measures range and range rate data for navigation by measuring the return propagation time of the uplinked signal. Range measurement is determined by round-trip time of a signal offering accuracies ~1 m. For non-GEO orbits, the ground antenna must track the satellite across the horizon and requires a widely distributed ground network to cover the complete orbit. Optical measurements may be used if visibility is good but radar trackers can work in any weather conditions. Range rate is based on the Doppler shift in frequency of the received signal due to the relative motion of the transmitter and receiver such that the observed frequency:

$$f = f_0 \left(\frac{\sqrt{1 - v^2/c^2}}{(1 - v/\cos\theta)} \right)$$

where f_0 = reference frequency, v = relative velocity of transmitter/receiver, c = speed of light, and θ = angle between source and line-of-sight vector.

The satellite transmitter downlink carrier phase is coherently phase-locked to the received uplink carrier allowing the Doppler frequency of the carrier to be measured from the ratio of the downlink-to-uplink frequency. The ground phase modulates a pseudorandom ranging code onto the command uplink which the receiver detects, demodulates and retransmits on the transmitter carrier to the ground. The code serves to identify the pulse and determine the time of flight. This determines the range rate as a function of time from the line of sight velocity changes and requires a receive–transmit transponder. Angle tracking measures azimuth and elevation angles from the ground station to the satellite and requires an rf beacon on the spacecraft. The slant range and Doppler shift enable the spacecraft trajectory to be determined for the GNC subsystem. The accuracy of Doppler shift is 0.5 mm/s. More details of the rf component of the spacecraft is given in the next chapter.

16.5 ADVANCES IN ELECTRONICS TECHNOLOGIES

The technological advances alluded to here as having a bearing on future space missions are GaAs-based electronics, specialised computational architectures and optoelectronics. N-type GaAs devices are a recent development to replace Si technology due to their high electron mobility and switching speeds of ~10 ps and so offer a speed advantage. However, their p-type carrier mobility is similar to Si so if complementary GaAs devices are used some of the speed advantage is lost. GaAs has far more radiation tolerance than Si. The surface Fermi level is confined close to the midgap energy where charge trapping occurs due to the lack of dielectric interfaces. This makes them inherently radition-hard to total dose exposure. GaAs technology is based on high mobility metal-semiconductor

FETs (heterostructure FET, HFET) [Kerns & Shafer 1988]. They have rapid gate propagation times ~200 ps, but their fabrication is limited at present to LSI though VLSI is starting to become available. They have a wider operating temperature range than Si from −200°C to 200°C. Several elements such as Al, In and P may be substituted for some of the Ga and/or As for different bandgaps, e.g. AlGaAs has a larger bandgap than GaAs. Due to the bandgap differences between AlGaAs and GaAs layers, a potential well is formed on the GaAs side of the AlGaAs–GaAs boundary. Mobile electrons from the AlGaAs layer form a two-dimensional high-density electron gas in the GaAs layer. Such a high electron mobility and high transductance device exhibits rapid channel turn on with a low supply voltage ~1 V and low power dissipation through the substrate. The HFET technology is insensitive to total ionising dose but is still susceptible to SEU. They are more radiation-tolerant than Si components such that 1 rad (GaAs) = 1.06 (Si). Displacement damage only becomes significant at ~10 rad (GaAs) because of GaAs's shorter carrier lifetime. Furthermore, GaAs FETs are tolerant to total dose due to the lack of gate dielectric insulator to suffer charge buildup. This limits complementary FET GaAs to pn junction gate technology with lower power and lower speeds. However, SEUs still affect GaAs technologies and GaAs have higher power dissipation than CMOS devices. A compromise is to use complementary FET p-n junction technology. The total circuit volume sensitive to SEU can be reduced by minimising the total source, drain and gate overlap areas by making electrical contacts with a deposited metal insulated by a dielectric layer from the substrate. Hence, HFETs offer great promise for radiation-hardened electronics.

Computational speedup by parallelisation is often achievable by avoiding the fetch-and-load bottleneck of the von Neumann serial architecture. Programmable logic arrays (PLA) are specialised layouts for implementing switching logic with an area-efficient structure. They can implement sums or products efficiently. ASICs (application-specific integrated circuits) are based on gate arrays providing systems-on-a-chip technology. Multiple functions can be integrated into a single ASIC. They are custom-built for specific tasks. They are smaller, cheaper, faster and use less power than programmable general-purpose microprocessors offering some ten times increased performance over high end workstations. They can be used to implement forward error-correction such as Reed–Solomon coding, FIR filtering, phase shift key modulation, and image processing functions such as Sobel edge detection, FFT, and Hough transforms. A core operation in DSP is multiplication through array processing. Typically, the main CMOS processor is based on 10^6 gate integration per chip with on-chip DRAM/SRAM operating at 80-MHz clock rates. They are, however, inflexible. Field Gate Programmable Arrays (FGPA) can overcome some of these limitations by virtue of their reconfigurability [Hutchings & Wirthlin 1996, Villasenov & Mangione-Smith 1997].

Both ASICs and FGPAs have been utilised for TT&C (tracking telemetry and command) core functions for ESA's TEAMSAT housed in its OBDH (onboard data handling) box [Habinc et al 1998]. The OBDH box had a mass of 3.2 kg and a power requirement of 7 W and contained a TC board and TM board. ESA's radiation-hard ASICs are silicon-on-sapphire chips with a mass of 15 g for PSK modem functions and RS (Reed–Solomon) convolution encoding/decoding functions. Obvious applications for FGPAs include rapid pattern matching by comparing input patterns to large numbers of stored templates by reconfiguring itself for each data set [Hauck 1997]. FPGAs have also been used to implement Data Encryption Standard (DES) codes utilising 56-bit long encryption keys to encrypt 64-bit data blocks. A set of permutations translates the 56-bit key into 16 subkeys. The FPGA can reconfigure itself to compute each subkey. The chief disadvantage of FPGAs is their lack of on-chip memory to store intermediate results of computations. Data transfers between the FPGA and external memory slows computations down significantly but FPGAs with large memory holding multiple configurations are under development. Mirchandani & Figueiredo (1998) have suggested that FPGA parallel processing architectures would be suitable for TT&C gateways to translate between different retrieval protocols such as science data and housekeeping data at high data rates ⩾ 150 Mbps. Whereas general-purpose microprocessors are well-suited to irregular tasks with a large number of distinct operations, FPGAs are suited to regular tasks which involve a limited number of operations repeatedly. Hybrid architectures involve utilising FPGAs as coprocessors coupled to conventional microprocessors—90% of computation time is spent on 10% of code in most programs.

C&DH in more complex spacecraft will become more highly distributed with interconnection from a number of devices. Distribution is required which impacts on the wiring harness mass and volume. Both point-to-point and multipoint distribution is likely to be required. Usage of optical fibres for high data rates in communications networks such as ground station LANs is also attractive for onboard C&DH functions for large numbers of connections and high data rates >1 Mbps. NASA has developed a 500-Mbps fibre-optic transmitter/receiver pair which reduces the size of comparable commercial systems by two-thirds for use in point-to-point communications aboard spacecraft. Both NASA and the US DoD have flown optical experiments with long duration exposures and neither optical fibres nor laser sources exhibit significant degradation over 15 y, but photodetectors are susceptible to increased dark current due to radiation exposure and so would require hardening. Internal data distribution is critical when extensive onboard processing is adopted in a distributed manner—bandwidth constraints and electrical crosstalk can limit the data rate and multiple distribution capability. The number of pixels or pin-outs that can be supported by any system is given by the space-bandwidth product and satellite imaging requires $\sim 10^8$ pixel levels [Hinton 1988]. Optics have no pin-out limitations and are ideally suited to parallel image processing. Fibre-optics are suitable for internal spacecraft point-to-point communications and have advantages over copper cables in mass savings, immunity to noise, immunity to aliassing, modularity, immunity to EMI (electromagnetic interference) and EMP (electromagnetic pulse) and high data rates up to 10^{12} bps typically. Fibre-optic gyroscopes are used on spacecraft and represents the start of the introduction coherent fibre technology in spacecraft. The sensor introduces a phase shift delay in the signal in response to disturbances and the phase shift is sensed by homodyne mixing of the output path with a reference path in a photodiode detector. As well as large ~Gbps bandwidths with innate parallelism of connectivity, fibre-optics do not suffer from capacitance penalties and can communicate at light speed without crosstalk interference. However, light does not necessarily require physical connection. Such free space communications offers simplicity by eliminating the need for a wiring harness. Optical processing offers gate switching speeds of 5–10 ps but electro-optical systems are likely to be the first mode of supplementation.

17

Spacecraft communications (TT&C) subsystem

The spacecraft bus radio frequency (rf) communication subsystem provides the basis for TT&C (tracking, telemetry and command) functions in communicating with the spacecraft. Radio frequency communications systems use electromagnetic energy to transfer information across large distances at the speed of light without the need for a medium. Satellite communications differs from Earth-based communications because of the long range and large relative velocity between the transmitter and receiver. In fact, the communications link budget is a major design parameter for the spacecraft [Agrawal 1980, Larson & Wertz 1992].

The onboard communication subsystem receives and demodulates uplink commands (e.g. motor ignition, attitude manoeuvres, equipment activation, mechanism deployment) and modulates and transmits downlink housekeeping telemetry (e.g. fuel pressure, temperature, operating status, voltage and current levels) and payload data. The command and data handling (C&DH) subsystem receives and distributes commands and collects and transmits telemetry to the ground via the communications subsystem. The transmitter used for telemetry can also act as a known source for ground tracking. Hence, the onboard rf (radio frequency) system provides TT&C (tracking, telemetry and command) functions. During launch and early orbit phases of the spacecraft mission, ground control sends mission commands for the deployment of instrumentation at precise times. During the rest of the mission lifetime, commands are regularly uploaded to reconfigure mission parameters according to requirements. These mission parameters that are uploaded are often not defined until after launch. Similarly, telemetry is downloaded to the ground. During the launch and early orbit phases, telemetry provides feedback on the effectiveness of the commands that have been uploaded. During the rest of the mission, telemetry provides payload data as well as spacecraft health monitoring. For a robotic in-orbit servicer such as ATLAS (and indeed for a planetary rover), the TT&C communications system provides the control link from the ground station to the remote spacecraft—it provides a fundamental constraint on the nature of the man—machine interface and, so, on the performance of the remote robot. In particular for such robotic space missions, telemetry to the ground station will be dominated by bandwidth-hungry image data.

17.1 SPACE COMMUNICATION FREQUENCY BANDS

The first consideration is that of frequency selection. The Earth's atmosphere has transparent windows in the visible waveband from 300 nm, bounded by ozone absorption, to 1.4 μm, bounded by water and carbon dioxide absorption, and in the radio microwave waveband from 8 mm, bounded by oxygen absorption, to 15 m, bounded by ionospheric reflection. The ITU (International Telecommunications Union), a specialised agency of the UN, allocates radio frequency wavebands for satellite communications. The ITU operates on a 'first-come, first-served' basis. The rf carrier frequency determines the transmitter power, antenna size and beamwidth, which in turn affects the satellite's mass and size. Satellite communications are effectively limited to the 1–50 GHz waveband. Atmospheric absorption increases with frequency due to O_2 and H_2O which generates losses of ~4 dB at 2–6 GHz. Above 10 GHz, rain and cloud attenuation becomes important, particularly at 8–20 GHz generating losses ~10 dB to ~40 dB at higher frequencies near 30 GHz as water vapour has an absorption line at 22.3 GHz. Above 50 GHz, attenuation due to O_2 absorption bands, particularly at 60 GHz, becomes too great for practical use. The ionosphere from 80–1000 km causes depolarising Faraday rotation of signals below 1 GHz. As Faraday rotation of linearly polarised electric fields is significant below 1 GHz, helical antennas are used to generate circularly polarised signals. However, even then below 1 GHz, ionospheric activity causes fluctuations in amplitude and phase, particularly below 200 MHz, so these frequencies are not used generally.

Terrestrial radioband:

VLF:	3 kHz–30 kHz
LF:	30 kHz–300 kHz
MF:	300 kHz–3 MHz
VHF:	30 MHz–300 MHz
UHF:	300 MHz–1.5 GHz

Space waveband:

UHF:	L band	1.5 GHz–2 GHz
	S band	2 GHz–4 GHz
SHF:	C band	4 GHz–8 GHz
	X band	8 GHz–11 GHz
	Ku band	11 GHz–18 GHz
	K band	18 GHz–27 GHz
	Ka band	27 GHz–36 GHz
EHF:	Q band	36 GHz–46 GHz
	V band	46 GHz–60 GHz

In general, higher frequencies are preferred due to their greater bandwidth and higher throughput capacity. Military communications use the UHF band between 0.2–0.45 GHz, which is the minimum frequency usable, and the EHF band. The C band and Ku band are allocated to comsat services—both are near saturation, forcing a trend for comsats to use the Ka band. The Ka band offers negligible S/N degradation due to signal scintillation in solar plasmas—a useful property for near-solar missions. The L-band is the general choice for LEO mobile communication services. Satellite crosslinks may use the 60 GHz V band which would otherwise be severely attenuated by the Earth's atmosphere. This would allow antenna apertures in the $\leqslant 1$ m range with $\sim 1/10°$ beamwidths. Low-noise receivers are being developed for this frequency for GEO intersatellite links to receive 2 W of power. GaAs amplifiers are being developed for this frequency to achieve 10% power efficiency at 5–10 W output levels. Non-comsat spacecraft often use the S band for TT&C [Wilkins 1989]. Typically, the S band signal comprises a 2-GHz carrier, phase-modulated by an 8-kHz or a 16-kHz subcarrier, which itself is phase-shift key modulated by telemetry/command data. The S-band 2-GHz signal provides minimal propagation losses through the Earth's atmosphere of <1 dB. The X band is used by deep space science probes for high data rates, though it is being considered for near-Earth TT&C as the S band becomes increasingly saturated. Absorption still allows satisfactory perform-ance in the X band, provided that there is little rain, but above the Ku band, absorption becomes high. The TT&C (communications) subsystem is limited by fractional band-width which is defined as $\delta f / f_0$ where f_0 = band centrepoint and δf = bandwidth required by signal. At 10 GHz, a typical $\delta f / f_0$ of 1% gives 100 MHz bandwidth. For ATLAS, we adopt the S-band frequency band for compatibility with other systems.

17.2 ONBOARD RADIO-FREQUENCY (RF) DEVICES

A typical TT&C communications system comprises a single antenna, a diplexer, two crosslinked transponders with transmit/receive functions and two command decoders in hot redundancy configuration [Winton et al 1995]. The interface is via the onboard data handling (OBDH) system. The basic communications payload unit common to all space-craft is the transponder which is a combined receiver/transmitter. It converts a received signal from its uplink frequency to its downlink frequency and amplifies it.

Antenna size and transmitter power impact heavily on spacecraft mass and are major cost drivers. Three-axis stabilised spacecraft often mount omnidirectional antennas on top and bottom sides of the spacecraft for early orbit and emergency operations, but a single directional antenna for receive and transmit is more common as well as desirable. Typically, on an interplanetary robotic spacecraft, a high-gain parabolic antenna is adopted in the Cassegrain configuration. Ominidirectional low-gain antennas are often mounted above the high-gain antenna subreflector for backup (e.g. Voyager and Galileo).

The rf antenna radiates electromagnetic radiation into freespace, powered by the feed which comprises a transmission line or waveguide with a given impedance. A waveguide is a pipe with a rectangular or circular cross-section. At microwave frequencies, the transmission line generates conduction losses, so waveguides are used for microwave power transmission. A voltage wave is transported along the transmission line by reflec-tion rather than conduction to transfer power with minimal loss. The transmission line is typically a coaxial cable of two conductors, arranged concentrically, separated by an insulation layer. Energy is transmitted as a transverse electromagnetic wave. The voltage maximum is maintained at the centre of the waveguide and the wave travels at the group velocity which is lower than the free space velocity due to the 'zigzag' path of the wavefront. The characteristic impedance of the transmission line is constant

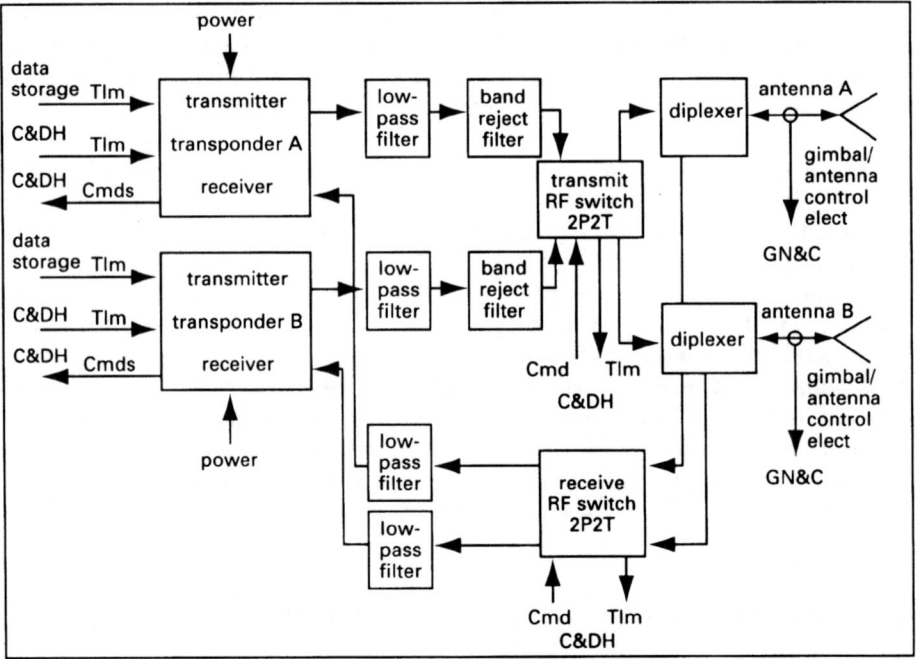

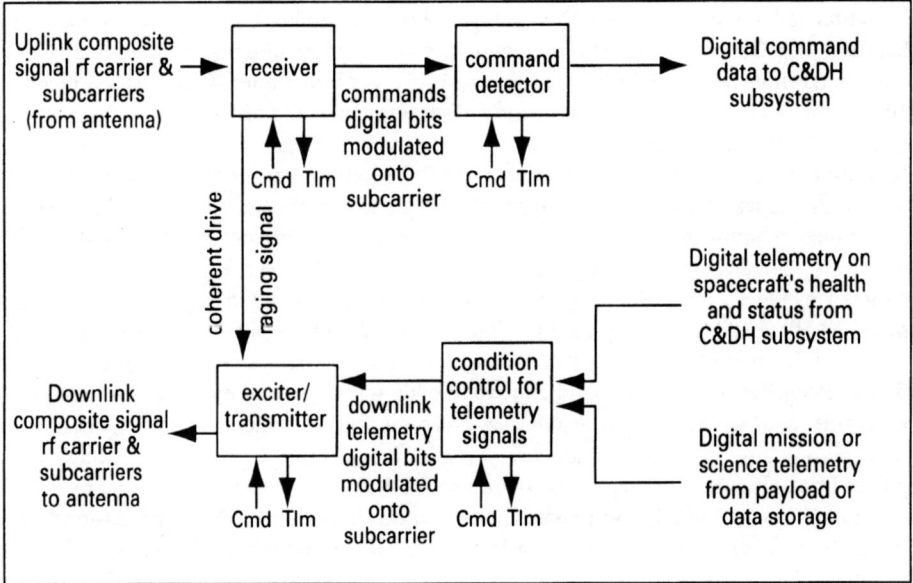

Fig. 17.1 (a) Generic communications subsystem for TT&C; (b) typical transponder function
(from Wertz & Larson 1999).

$Z = E/H = \sqrt{\mu/\varepsilon} = \sqrt{L/C}$ independent of ac frequency. The antenna must match the impedance of the feed to the impedance of freespace Z_0. If an impedance mismatch occurs between the source and load, a reflected wave superposes over the travelling wave creating a standing wave and loss of power from the load. There are two modes of propagation: the TE (transverse electric) mode when there is no E-field component of the wave in the direction of propagation, and the TM (transverse magnetic) mode when there is no H-field component of the wave in the direction of propagation. Fibre-optic cables are the light equivalent to the microwave waveguide.

The most basic form of rf transmitter is the half-wave dipole which provides a near-isotropic radiation pattern with a maximum gain factor of ~1.6. A variant on the half-wave dipole is the half-wave slot used in antennas for vhf and higher frequencies. The E and H fields of the slot are identical to those of a dipole of the same size as the slot but with interchanged E and H fields. While the E field of the slot antenna is symmetric, it is the H field which is symmetric in the dipole. Both are typical of terrestrial systems. Another variant on the half-wave dipole is the rectangular flared waveguide (the pyramidal horn). The waveguide launches an electromagnetic wave as a pure wave with transverse electric E and H field components. The flaring increases the aperture size for controlled radiation patterns. When the wave enters the flared section, its phase front curves so that the phase at the edges are retarded from the phase at the centre, thereby spreading the radiation pattern. The characteristic impedance of TE10 mode longitudinal B-field components approach that of freespace at 377 Ω.

Several other types of antenna are possible, parabolic reflector, helical antenna and horn antenna being the most common. Helical antenna suffer from wide beams and low gains <14 dB. High-gain antennas allow high data rates >1 kbps with low transmitter power and narrow antenna beamwidths ~0.1°. Reflectors are preferred over lenses and phased arrays are preferred for low mass, low complexity and low cost, especially for large diameters >0.5 m. The circular horn feed uses a circular waveguide. It is commonly used as the primary source to uniformly illuminate a parabolic reflector to give higher gain with a narrower beam in either the Gregorian or the Cassegrain configuration.

The power pattern from a parabolic reflector is a pencil beam. This increases the power gain of the antenna. To reduce spillage and noise pickup, the Cassegrain configuration employs a secondary hyperboloid reflector at the virtual focus of the paraboloid to illuminate the main dish uniformly. For a parabolic reflector, the maximum gain is 15–65 dB with efficiency $\eta = 0.6$ and half-power beamwidth (3 dB) of 21 f/D. Hence, reflector antenna performance is governed by $f/D \sim 0.8$–1.6. They confine much of the radiated power to a 3-dB beamwidth with high boresight gain and low sidelobes. The parabolic reflector typically comprises of wire mesh with spacing of less than $\lambda/20$ in order to achieve high resolution of less than $\lambda/8$. Several possible parabolic configurations are possible. The centre feed mounts have high sidelobe levels and long transmission paths from the feed to the reflector. The parabolic reflector is usually a Cassegrain design as it offers short transmission paths but still suffers from high sidelobe levels which degrade directivity. To eliminate beam blockage by the antenna feed which causes loss of gain due to scattering, it can be offset from the boresight direction of the reflector to direct the main beam away from the feed. This configuration offers reduced sidelobe levels. The feed may then be mounted in the spacecraft. A variant of this with an offset

Antenna type	Parabolic reflector	Helix	Horn
	D	0.8λ $\overset{\longleftarrow L \longrightarrow}{\text{(helix coil)}}$ D $C = \pi D$	D $h = \dfrac{D^2}{3\lambda}$ $C = \pi D$

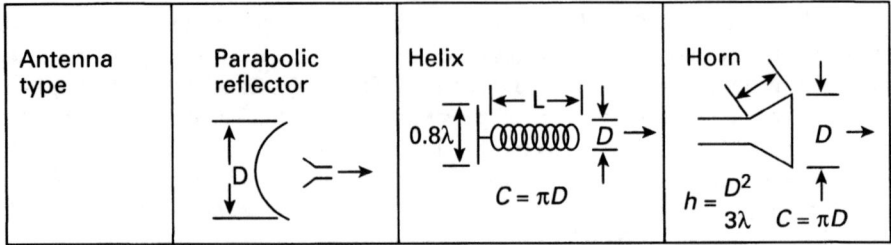

Fig. 17.2. Spacecraft antenna types (from Wertz & Larson 1999).

feed array involves the use of a shaped subreflector to allow scanning of up to 10°. However, for wide area coverage by an electrically steerable antenna a centre-mounted parabolic reflector such as the centre feed Cassegrain is required. The Cassegrain configuration suffers from spherical aberration and coma. The Richey–Chretien has a hyperboloid primary and steeper hyperboloid secondary reflector. It does not suffer coma or spherical aberration. Large antennas require folding and deployment and are often flexible, requiring figure control—this is not desirable outside astronomical applications. The surface shape must be very accurate for parabolic and phased arrays to within $\lambda/20$ and for hf this is very demanding. Current antennae are capable of transmitting data at rates of $\sim 10^9$ bps.

The receiver is usually a superheterodyne receiver but may sometimes be a tuned rf amplifier. The tuned rf system is an rf amplifier tuned to a narrow bandwidth around the transmitting frequency $f = 1/RC$ followed by demodulation and amplification stages. The receiver uses a phase-locked loop system. The receiver provides a frequency reference keyed to the received signal. The received uplink is detected and locked onto. The signal is mixed with a constant frequency oscillator with an oscillation frequency near the main signal frequency. The crystal oscillator which is resonant at one frequency is used as the reference frequency. This modulates the oscillator frequency with the data signal. Mixing is a frequency translation process. The signal centred on f_0 is mixed with a lower frequency f_1 from the local oscillator. After mixing, the signal is centred on an intermediate beat frequency $(f_0 - f_1)$. This can be separated out by a low-pass filter. Amplification of the signal is easier at lower frequencies. A tuner amplifier amplifies the narrowband of frequencies centred on the resonant frequency of the tuned circuit. The signal is then rectified, integrated and outputted. A variable frequency oscillator provides the output. The two frequencies are presented to a phase comparator which determines which is the highest frequency. It generates an error voltage which is fed back to the input to the variable frequency oscillator through a low-pass filter to ensure that the frequencies are the same. The superheterodyne shifts the received signal to a lower frequency than the transmitted frequency. The same antenna may be used for both transmit and receive. This requires the antenna to be optimised over two separate spacecraft frequency bands and an external microwave device must be used to duplex the transmit and receive signals to interface with the transponder processing unit and isolate the transmitter from the receiver.

The transponder transposes the uplink frequency to the downlink frequency using a frequency mixer. A stable onboard local oscillator provides the frequency difference to down-convert the uplink to the downlink frequency. A bandpass filter precedes the receiver stage so that only uplink signals are received and out-of-band signals are rejected. The receiver amplifier then boosts the signal before the rf channels are demultiplexed into into individual channels. The simplest type of transponder is the satellite repeater which retransmits received messages. A regenerative repeater demodulates and reconstitutes the signal while a nonregenerative repeater only amplifies the signal without transformation. Nonregenerative repeaters can be used with many different modulation formats but regenerative repeaters operate in only one modulation format. It is expected that the present commercial use of nonregenerative repeaters will give way to regenerative repeaters as they decouple the uplink and the downlink so that uplink noise is not retransmitted on the downlink, thereby offering improved performance with more flexible traffic routing capabilities.

The transmitter generates a carrier typically in the S band or X band which can be modulated with data encoded in a subcarrier. Transmitters may be either travelling-wave tube amplifiers (TWTA) or solid-state power amplifiers. High-power amplifiers' vacuum tubes are either travelling-wave tubes or klystrons. Much microwave circuitry is still based on valve technology. Thermionic devices employ cathodes which release electrons thermally. The simplest is the diode, comprising an encapsulated filament cathode and anode collector. A triode amplifier is constructed by adding a control grid between the anode and cathode to alter the electron beam current density. For UHF wavebands, the triode suffers from inter-electrode reactances, so klystrons are usually used which vary the electron beam velocity. The output is excited to oscillation at the input signal frequency giving large power gains and continuous operation. A magnetron is a microwave oscillator used for pulsed operation (for example, as required in radar). For higher powers and frequencies, travelling-wave tubes are adopted. Travelling-wave tubes are still common on spacecraft in offering high-gain microwave amplification with low noise for continuous output. The travelling-wave tube comprises two transmission lines connected by transistors along their length. The input signal travels along one transmission line to provide a control signal to the transistors. The control signal causes the transistors to regulate the current supplied to the second transmission line. Constructive current superposition causes an amplified signal to travel down the output line. The TWT utilises the interaction between the waveguide standing wave and the electrons composing the standing wave. Electrons travel faster than the wave and are slowed by the wave, thereby amplifying the wave. Semiconductors offer faster response times but are power-limited, so are used at lower powers and frequencies.

Many transmitters are now GaAs FETs due to their good linearity, high reliability, low mass and high power outputs at lower voltages. Output signals are combined by the output multiplexer before transmission to the ground. Multiplexing involves adding multiple signals into a composite signal without mutual interference similar to Fourier superposition. GaAs technology with its high switching speed, greater radiation resistance and low power consumption is being adapted for satellite communications hardware for multiplexing, modulation and ADC functions. MICs (microwave integrated circuits) are based on microstrip transmission lines and a number of devices. Surface acoustic wave (SAW) devices are based on the piezoelectric effect. A voltage pulse causes the crystal to oscillate at its resonant frequency generating electrical oscillation at that frequency (bandpass filter). MESFETs (mesa-FETs) are JFET amplifiers which employ Schottky barrier gates between the source and drain to overcome charge storage problems and the modulation of semiconductor channel con-ductivities at microwave frequencies. IMPATT (impact avalanche and transit time) diodes introduce a time delay to negative resistance avalanche current in generating microwave amplifier gain. The avalanche is thus time-dependent. Monolithic MICs (MMIC) are devices already fabricated onto the substrate, offering greater miniaturisation. Microstrip antennas are small. A frontal area of conductor is placed over a dielectric substrate with a backplane conductor. The front surface conductor is excited by the microstrip trans-mission line. Such patches can be arrayed and generate the phase profile of a parabolic reflector despite being on a flat surface. They can thus form phased arrays of individual radiating elements. The simultaneously formed multiple, non-interfering spot beams ~1° are steered electronically. The antenna elements form colinear arrays spaced $\lambda/2$ apart. The Mills cross detector uses two colinear arrays perpendicular to each other to generate narrow fan beams orthogonal to each other. Their intersection is a narrow vertical pencil beam ~1° across. The pencil beam is isolated by subtracting the beam in-phase and out-of-phase patterns. Outside the beams, the signals are mutually incoherent. The array elements are fed with signals in phase with each other, so that signals from each element arrive in phase with each other at the receiver. By altering the phase shift between the array elements, the pencil beam boresight direction is displaced by an angle θ so the radio signal can be steered electronically without physical motion. This technique simplifies antenna packaging and eliminates heavy mechanical steering while adding programmability to beam shaping. They do, however, introduce computational complexities. Such phased-array systems dominate in Ku band and C band satellites.

17.3 SIGNAL CODING AND MODULATION

Signal-processing functions on TT&C data are essential. To accommodate onboard signal-processing techniques, MICs and MMICs have been developed to achieve high reliability, small size and enhanced performance. Early satellites transmitted raw data to the ground station but there has been a trend for satellite onboard processing to reduce the required data rate. Onboard decision-making also increases response times and reduces ground support equipment and operations costs at the price of making the spacecraft itself more costly and complex. There are fundamental limitations on communications efficiency, these being the Nyquist minimum bandwidth theorem and the Shannon capacity limit theorem. The Nyquist theorem states that the minimum theoretical bandwidth required to transmit x symbols/s without intersymbol interference is $x/2$ Hz. In practice, this limit is in fact x Hz, i.e. 1 symbol/s/Hz throughput limit. Data transmission rate R through a noisy channel is fundamentally limited by Shannon's information capacity theorem: $R \leqslant C = B \log_2(1 + S/N)$ where $N = N_0 B$, B = bandwidth and R = bit rate. Hence

$$R \leqslant B \log_2 \left[1 + \frac{E_b}{N_0} \left(\frac{C}{B} \right) \right] \tag{17.1}$$

Data rate is a measure of information transfer rate between the satellite and the ground. There are three main categories of signals: data, audio and video. Robotic systems will impose high data rate requirements for telemetry data, audio and video transmission [Krishen 1987]. Video requires Mbps data rates. The inclusion of colour and multiple video cameras can increase this dramatically, necessitating the use of extensive data compression which can reduce the video rate to ~400 kbps from a camera. The Shuttle RMS has a data link rate of 600 kbps while ESA is offering video services at 384 kbps. The primary constraint on data rate for an in-orbit servicer such as ATLAS will be video data, which may be given by [Larson & Wertz 1992]:

$$\text{Image sensor data rate,} \quad R = \frac{B}{\eta} \left(\frac{C}{N} \right) = \frac{\theta_x \theta_y r^2 s b}{f d^2 t}$$

where $\theta_x \theta_y$ = 2D scene dimensions, r = distance to target, s = number of samples/pixel (typically ~1.4–1.8), b = number of bits/sample (2^b amplitude levels), f = frame efficiency = time fraction for data transmission ~0.9–0.95 typically, d = minimum diameter of pixel image projected to target, and t = raster scan time.

The bandwidth efficiency is determined by R/B. It is possible to transmit over a channel of capacity C at a rate $R \leqslant C$. The asymptotic approach to C depends on the complexity of the coding scheme for arbitrarily small error rates. If the channel noise is Gaussian, then $C = \log_2 \sqrt{1 + 2E/N}$ where E = message energy/bit and N = noise energy/bit. Now, since $E \ll N$, then $C = (1/\ln 2)(E/N)$. The required E_b/N_0 approaches the theoretical Shannon limit of -1.6 dB as B increases to infinity. For $E_b/N_0 > -1.6$ dB, coding schemes allow communication with zero error.

An uncoded PSK channel requires ~9.6 dB to achieve a BER of ~10^{-5}. Hence, there is a potential coding gain of 11.2 dB that is theoretically possible (at the expense of increasing bandwidth). Efficient coding offers improvements to approach the Shannon limit to within ~4 dB within channel bandwidth and capacity limits. In fact these limits are beyond practical realisation because they assume strict band-limited channels but some signal power is lost outside the bandwidth main spectral lobe since the power density for a single pulse is

$$S(f) = T \left[\frac{\sin \pi (f - f_c) T}{\pi (f - f_c)} \right]^2 \tag{17.2}$$

where f_c = carrier frequency, and T = symbol duration.

A digital communications system transmits a series of m symbols by transmitting a digital voltage or current waveform $s_i(t)$ for $i = 1 \ldots m$ for a period T for each symbol with a low probability of error p_e. The probability of symbol error is related to the probability of bit error p_b by

$$\frac{p_b}{p_e} = \frac{2^{i-1}}{2^i - 1}$$

The data rate is defined as $R = (1/T)\log_2 m = (k/T)$ bps where $m = 2^k$. For a binary signal $k = 1$, so $p_e = p_b$ and $m = 2$. The ratio E_b/N_0 where $E_b =$ signal energy per bit and $N_0 =$ noise density at the receiver, characterises the digital communications system performance. It it related to the signal average power to noise average power ratio by:

$$\frac{E_b}{N_0} = \frac{ST}{N_0} = \frac{S}{N_0 R} = \left(\frac{S}{N}\right)\left(\frac{B}{R}\right) \tag{17.3}$$

where $S =$ average signal power, $T =$ bit duration, $R =$ bit rate, $N = N_0 B$, and $B =$ signal bandwidth. There is a tradeoff between maximising R and minimising P_b, minimising B and minimising E_b/N_0 or equivalently the required power.

For transmission, the signal is transformed from the source to the transmitter. The receiver reverses this transformation for the destination. Extensive onboard signal processing is required to allow complex functions to be carried out without undue dependency on the ground. A communications system comprises two major components:

(i)   a transmitter comprising a frequency up conversion stage, a power amplifier and an antenna;

(ii)   a receiver comprising an antenna, a low-noise amplifier, and a frequency down conversion stage.

Between these two components signal-processing functions are required which means formatting and modulation/demodulation (MODEM) essentially, plus several other optional supporting functions [Sklar 1983a,b]. This may be modelled as a discrete memoryless channel. The transmit process proceeds from formatting through modulation to transmission while the receive process proceeds inversely from receipt to demodulation to format. The additional signal-processing functions that may and usually are included are: source encoding, channel coding, encryption, multiplexing, and frequency spreading, some of which we shall consider briefly here.

Modulation refers to the process of varying a characteristic of the carrier waveform to include characteristics of the information that is to be transmitted i.e. multiplexing the information with the carrier to match the signal to the channel. This multiplying the time domain signal with the carrier sinusoid shifts the signal spectrum by the carrier frequency. The modulating signal bandwidth is typically less than 10% of the carrier frequency to suppress noise, i.e. $0.01 < Bf_c < 0.1$ so large bandwidths require high frequencies. Analogue modulation is the process whereby some characteristic of the rf carrier waveform is varied in accordance with another baseband information signal waveform. The rf carrier wave of an (ac) sinusoidal signal $y = A(t)\sin[w(t)t + \varphi(t)]$ is usually represented as a voltage variation: $V(t) = V_0 \cos(w_c t + \varphi_c)$. The sinusoidal carrier signal may be modulated to convey information through three features of the carrier. The input signal varies the characteristics of the carrier (amplitude, phase, frequency or, less commonly, polarisation) according to the message symbol. Orthogonally polarised signals of linear horizontal/vertical or left/right circular polarisation components may be generated by polarising the feeds to allow the available satellite frequencies to be used twice. Hence, the effective bandwidth of transmission is doubled. AM (amplitude modulation) is orthogonal to PM (phase modulation) and FM (frequency modulation) as it modulates the phasor in the radial direction while PM and FM modulate the phasor in the angular direction. Hence, PM and FM are inter-related. The simplest form of analogue modulation is AM where the high-frequency carrier amplitude is varied by the transmitted data.

$$\text{AM: } V = V_0(1 + m\cos w_m t)\cos w_c t \qquad (17.4)$$

where $m = V_m/V_c$ = modulation index ≤ 1.

Three frequencies are contained in the modulated signal: $(w_m - w_c)$ = lower sideband frequency, w_c = carrier frequency, and $(w_c + w_m)$ = upper sideband frequency.

$$\text{FM: } V = V_0 \cos[w_c t + m_f \cos w_m t] \qquad (17.5)$$

where $m_f = \Delta f/f_m$ = modulation index $\gg 1$, and $f_m = w_m/2\pi$ = maximum modulating baseband source frequency.

$$\text{PM: } V = V_0 \cos(w_c t + m\cos w_m t) \qquad (17.6)$$

where m = modulation index.

AM is wasteful of power and requires large transmitters as the carrier conveys no information and the message base bandwidth is B and the bandwidth of the AM wave is $2B$ (50% of the transmitted power resides in the carrier). In AM, noise has a wide bandwidth and may be erroneously interpreted as part of the signal. AM may be used to transmit relatively low data rate signals, but high data rate signals require FM or PM within which the effects of noise is negligible and can be removed by passing the signal through a low-pass filter. FM and PM have improved performance due to the greater bandwidth, but they tend to suffer from threshold effects. In FM, the higher the modulation index the higher the bandwidth and the energy of the signal is spread over a wider range of frequencies. As the S/N ratio decreases, the performance degrades slowly until a threshold S/N ratio is reached at which point it drops off rapidly. Although AM systems require more S/N for a given performance its performance degrades gracefully as S/N ratio is reduced. FM gives double side bandwidth $B = 2(m_f + 1)f_m$. FM offers a tradeoff between power and bandwidth:

$$B = 2(f_k + f_m) \qquad \text{(Carson's rule—17.7)}.$$

where f_k = peak frequency deviation, and f_m = highest modulating source frequency.

For radio broadcast, for example, f_k = 75 kHz and f_m = 15 kHz, so B = 180 kHz. Transmission and detection of variations in phase is difficult, so it is not used for analogue signals, but is reserved for digital data in the form of phase changes in steps of 90° and 180°. Quality of service is defined by the S/N ratio after demodulation and is typically 30 dB for telephones. It is defined as: SNR (dB) = $10\log_{10}(P_{signal}/P_{noise})$. Before demodulation, this translates as the C/N ratio: $C/N = P_r/(kT_s B) \sim 18$ dB for a typical C band TV link for example. Signal to noise ratio for FM is related to C/N by:

$$S/N = 3(m_f + 1)m_f^2\left(\frac{C}{N}\right) \qquad (17.8)$$

The efficiency of a communications system is given by the ratio E/N_0 where $N = N_0 B$ representing the energy per bit of information to the noise power density. When B approaches infinity, E/N_0 approaches $\ln 2 = 0.693$. In fact, a higher value of E/N_0 will be the minimum realisable. E/N_0 characterises the efficiency of any modulation–demodulation process. The TV standard for a reasonable picture normally requires $S/N \geqslant 45$ dB and $C/N \geqslant 14.5$ dB.

17.3.1 Formatting and source encoding

Digital signal processing is concerned with the representation of signals as number sequences and their processing to extract the parameters of a signal in a usable form. Formatting and source encoding alters the information source to ensure efficient transmission over the channel with minimum source bit rate. It removes information redundancy and performs information compression. Pulse duration is short in comparison with the duration between pulses, so it is power-efficient (and allows time division multiplexing by interleaving of different signals). Formatting performs ADC (analogue-to-digital conversion) for compatibility for digital processing. The analogue signal is 'modulated' into a digital binary waveform as digital signals are less susceptible to distortion, interference and noise giving lower error rates:

$$S/N = \frac{3(2^{n-1})}{1 + 4BER(2^{2^n} - 1)}$$

Hence, the source analogue signal undergoes two stages of modulation: the demodulator design determines the S/N required to meet a given received power against the antenna gain and noise temperature. In biological communications systems, evolution has favoured, virtually exclusively, frequency modulation for analogue signals and pulse code modulation for digital signals, e.g. sinusoidal-wave-emitting electric fish encode through FM while pulse-emitting electric fish encode through PCM (pulse code modulation) [Richards 1985].

PCM involves periodically sampling (quantising) the analogue signal into 8-bit binary words at a rate according to the Nyquist criterion for perfect waveform reconstruction: $f_s \geqslant 2f_{max}$, where f_{max} is the highest frequency component in the signal waveform spectrum. If sampling is done at less than the Nyquist criterion then aliassing errors within the message band can occur. For voice circuits, $f_m \approx 3.4$ kHz so $f_s \approx 8$ kHz. This generates a digitised word composed of a series of bits the length of which depends on the precision required. The amplitude of each sample is quantised into one of a finite number of predetermined levels determined by the transmission bandwidth, each level representing a digital symbol. ADC divides the total amplitude range into M quantisation levels where $M = 2^n$ with $n = $ number of bits/word. The signal-to-noise ratio is given by $S/N = 10 \log_{10}(s/D)^2$ where $s^2 = $ variance in the input sample and $D^2 = $ mean squared distortion. The quantisation value is proportional to the average value in the sample period. If $n = 2$ then there are four distinct levels possible: (00), (01), (10), (11). Increasing the number of quantisation levels increases the transmission bandwidth and improves the reconstruction fidelity. The overall information rate is given by $2f_m n$. PCM is the most widely used digital format that converts quantised samples into code groups of two amplitude level pulses using fixed amplitudes. Each pulse represents a quantised amplitude value in binary. The maximum quantisation error is $\pm 0.5 \Delta V$ where $\Delta V = V/M = V/2^n$ = the uniform quantisation step and V = full input signal voltage. The signal to quantisation noise power ratio is $\sim M^2 \sim 2^{2^n}$. In uniform quantisation all internal ranges are equal in size with $D = \Delta^2/12$ where Δ = stepsize of quantisation. PCM offers an SNR (dB) of $6n - 7.2$.

Adaptive PCM (APCM) improves performance such that 32 kbps APCM performs as well as 64 kbps PCM, thereby doubling the baseband capacity. It is primarily used for high data rates, >2.4 kbps, to reduce inter-symbol interference which occurs in all PCM modes. Most high-speed modems use robust least-squares equalisation where the signal is weighted by equaliser coefficients. Flash ADC operates at rates ~10–100 MHz but is noisy (though it is well suited to online operation). Integrated ADC operates at only ~10–100 Hz but generates clean data (more suited to scientific data). JPL uses 256 shades of colour for imaging (i.e. number of bits/sample, $n = 8$) and 32 shades of grey for low-definition B&W TV giving maximum quantisation errors of <0.4% and <1.5% respectively and signal-to-noise power ratios of 48 dB and 30 dB respectively. An 8-bit word encoding for each voltage sample is sufficient for good reconstruction of voice. The data rate is given by the product of the number of samples/s and the number samples/bit. PCM therefore requires 64 kbps for each voice circuit.

	Max. input freq. f_m	Sample freq. f	No. bits/sample n	Data rate R (bps)
Voice	3.6 kHz	8 kHz	7	64k
Colour TV	4.6 MHz	8.8 MHz	5	44M

ADC organises all the data into a single bit stream of 8-bit or 16-bit words depending on the resolution required. PCM combined with FM is a common combination for satellite digital modulation. Source encoding performs data compression by removing redundancy in the message to reduce the data set. A compression ratio of 2:1 is achievable through non-linear quantisation. Quantisation levels are distributed non-unifirmly across the quantisation range. They are denser near the zero level and sparser near the maximum levels so that quantisation noise is reduced for lower-intensity signals. The effect of quantisation noise is reduced by varying the size of quantisation steps for weaker signals to give coarser quantisation for near-peak signals. The μ-law or A-law logarithmic compression are two examples of nonlinear quantisation which may be achieved by dedicated hardware circuits. This involves spreading low amplitude sample signals which have higher probability over a larger amplitude range of quantisation while shrinking higher amplitudes of lower probability into a smaller range such that

$$\Delta_i = \frac{2V}{Ng(x)}$$

where $g(x) = V/b$. Now $D = (bs)^2/3N^2$ where b = constant so that $S/N = 3(N/b)^2$ = constant which provides robustness to quantisation by making it insensitive to the probability distribution function of the input signals which are not known *a priori* [Gersho 1977].

The integral of $g(x)$ is $F(x)$ which is a function of input x which varies from $-V$ to $+V$. This is not physically realisable and so may be approximated by:

$$F(x) = \frac{V \log(1 + \mu x/V)}{\log(1 + \mu)}$$

$$\text{where } \mu = \text{empirical constant} \gg 1 \qquad (\mu\text{-law})$$

$$F(x) = \frac{Ax}{1 + \log A} \qquad \text{for } 0 < x < V/A \qquad (A\text{-law})$$

$$= \frac{V + V \log(Ax/V)}{1 + \log A} \qquad \text{for } V/A < x < V \qquad (17.9)$$

Both the A-law and the μ-law have similar robustness.

DPCM (differential PCM) performs source encoding to provide data compression. It involves comparing the current sample with the previous sample and only the difference between the two samples is encoded rather than the actual amplitudes: $D = X_n - X_{n-1}$. The amplitude variation between samples is much less than the total amplitude variation so fewer bits are required—5 bits rather than 7 bits in standard PCM, so the bandwidth requirement is reduced by $2/7 = 28\%$. In fact, DPCM may encode the difference between the current amplitude and a linear predicted amplitude which has been estimated from m quantised past values: $D = X_n - \hat{X}$. The predictor lowers the variance of the signal to be quantised by removing the correlation between successive inputs by subtracting estimation values from each sample. Backwards estimation is preferred over forward estimation because estimation is calculated from quantised values rather than input signals and estimation does not have to be transmitted to the receiver. The transmitter transmits the difference between the input signal and the predicted value. Optimisation may be achieved through a Kalman filter algorithm. It uses the weighted sum of m past values to predict each present sample and transmits the difference from the actual amplitude as an error signal. The waveform is modelled as the output of a linear process of significant features (such as pitch in speech). The weights minimise the average energy in the error signal. For speech, linear predictive coding can produce acceptable quality at 2.4 kbps and high quality at 7.2 kbps, compared with 56 kbps adopted with standard commercial telephone PCM. The weights of the predictor may be adaptively altered to match the changing spectral properties and signal variance. Linear predictive coding can achieve ~10 data compression. Garrett (1983) compared PCM with PPM (pulse positive modulation). PPM divides each unit time interval T into n states of equal duration only one of which contains a signal pulse, i.e. it is PCM with redundancy. PPM has no channel capacity limit and is most suited for deep space communications.

17.3.2 Channel coding

NASA's Pioneer 9 was the first deep space probe mission that utilised error-correcting codes. The simplest way to ensure error-detection and correction is to repeat messages a number of times and utilise majority voting to decide the correct message sequence. However, this is inefficient as the rate of information transfer is reduced by the number of repetitions. Channel encoding involves improving p_e performance and reliability by inserting systematic redundancy through coding at the expense of bandwidth and/or power. Typical redundancy level of such codes is 0.25–0.3. Forward error-correction coding reduces E_b/N_0, thereby reducing the transmitter power and antenna size. Most spacecraft use some form of forward error-correction coding.

Two types of channel coding are used: waveform coding and structured sequences. Waveform coding improves transmission performance while structured sequences encode data in structures with redundancy to detect errors. Waveform encoding uses orthogonal signals to improve the p_b at the expense of power. Structured sequences involve embedding data with structured redundancy and may be either block codes or convolution codes [Bhargava 1983]. Reed–Solomon coding is a block code that is suited to bursty channels characterised by mobile systems exhibiting significant Doppler shifts. Reed–Solomon (RS) coding can be used for concatenated coding by forming an outer code with an inner convolution code to increase coding gain, yet reducing coding complexity, e.g. a 1/6 inner convolution code and an (255, 223) outer RS code was adopted on Cassini. The first mission to use concatenated codes was NASA's Voyager. Galileo used a long constraint length (14, $\frac{1}{4}$) convolution code to provide an extra 2 dB towards compensation for its limited data rate resultant from the failure of its primary antenna to open. For moderate to high data rates ~10 kbps–10 Mbps, convolution codes with Viterbi decoding offer adequate performance. At very high data rates >20 Mbps, concatenated Reed–Solomon coding with an inner convolution code is necessary. There are dedicated RS encoder/decoder chips available. Parallel concatenation recursive convolution codes (turbo codes) achieve close (within 0.35 dB) to the theoretical Shannon limit performance for a BER of 10^{-5}. These have been suggested for ESA's Rosetta probe to give maximum coding gains [Calzolari et al 1998]. Turbo codes are formed by two convolutional encoders and one interleaver. The input data sequence enters the first encoder to generate an encoded sequence. The input sequence is also input in parallel to the interleaver and then to the second encoder to generate a further coded sequence. The two encoded sequences are then concatenated.

17.3.3 Digital modulation

The digital data bit stream with PCM format varies or keys the analogue subcarrier signal. If each bit is encoded onto the carrier independently of other bits in the bit stream then the digital modulation is binary. The digital equivalents of AM, FM and PM are the respective shift-keying modulation techniques which are used in PCM digital signals (ASK, FSK, PSK). There is also an ASK/PSK hybrid called QAM (quadrature amplitude modulation). Different modulation techniques have different spectrum widths and so different bandwidths. To demodulate reliably, the energy required per bit E_b must exceed N_0 by a finite amount and this usually stipulates a BER of ~10^{-5}. Any arbitrary set of waveforms $s_i(t)$ of duration $0 \leqslant t \leqslant T$ is modulated thus:

$$\text{PSK}: s_i(t) = \sqrt{\frac{2E}{T}} \cos\left(w_0 t + \frac{2ni}{m}\right)$$

$$\text{FSK}: s_i(t) = \sqrt{\frac{2E}{T}} \cos w_i t$$

$$\text{ASK}: \ s_i(t) = \sqrt{\frac{2E_i}{T}} \cos w_0 t$$

$$\text{ASK/PSK (QAM)}: \ s_i(t) = \sqrt{\frac{2E_i}{T}} \cos(w_0 t + \varphi_i) \qquad (17.10)$$

where wi = integral multiple of $2\pi/T$, and E = energy.

For k source bits, the modulator selects one from $m = 2^k$ waveforms for each k-bit sequence, where m is the symbol set size and k is the number of bits that each symbol represents. For $k = 1$ and $m = 2$ the modulation schemes are binary. If the receiver uses knowledge of the carrier wave's phase to detect the signal, the detection is coherent and the receiver is phase-locked to the transmitter, otherwise it is noncoherent. Noncoherent detection does not require carrier phase estimation for reference and offers reduced complexity of receiver implementation at the expense of increased p_e. Noncoherent systems are 3 dB poorer in performance than coherent systems.

In binary amplitude shift keying (BASK) the amplitude is varied between two values to represent 1 or 0. On–off keying is the simplest example of ASK but ASK is rarely used except in low S/N conditions due to carrier power wastage as in AM and its vulnerability to fluctuations in signal strength. Binary frequency shift keying (BFSK) uses two close carrier frequencies F_1 and F_2 separated by the minimum data rate frequency to represent 0 or 1 and transmits the sinusoid at one of those two frequencies according to the bit value—information is represented by discrete frequency shifts. Multiple FSK sets the carrier to a number of frequencies to offer efficiencies of 3 bps/Hz. FSK can be coherent or noncoherent. Noncoherent FSK offers higher E_b/N_0 ratios than coherent FSK for a given power. FSK is demodulated by measuring the received power at each possible frequency and selecting the frequency with the highest power. Variations in carrier phase introduced by the transmission channel do not degrade the link performance. Phase shift keying (PSK) is more common than FSK or ASK.

In PSK, information is represented as discrete shifts in phase. Binary phase shift keying (BPSK) uses the two allowable opposite phases of the carrier 0 and 180° to represent binary 1 or 0. BPSK is ASK with the carrier amplitude multiplied by ±1. To demodulate BPSK a carrier reference phase signal is required so it is a coherent technique. BPSK is demodulated by measuring the received carrier phase. Two carrier signals in quadrature can be modulated/demodulated separately. This is QPSK which modulates two BPSK signals independently in phase and in quadrature (orthogonal):

$$s(t) = \frac{1}{\sqrt{2}} A_I(t) \cos\left(2\pi f_c t + \frac{\pi}{4}\right) + \frac{1}{\sqrt{2}} A_Q(t) \sin\left(2\pi f_c t + \frac{\pi}{4}\right) \qquad (17.11)$$

The only difference is in the alignment of the two bit streams of even and odd bits transmitted at a rate of $1/2T$. They may be synchronously aligned. Hence, QPSK takes two bits to define four symbols corresponding to 0°, 90°, 180° and 270° and offering double the bandwidth efficiency of BPSK (of 1 bps/Hz) to 2 bps/Hz. Alternatively, the alignment of the two in-phase and quadrature bit streams may be offset by an amount T. This is offset QPSK (OQPSK) which introduces a delay of one bit period to the Q channel modulator, avoiding the large 180° output phase change. It avoids out-of-band interference and performs better than QPSK in the presence of noise. It retains low spectral sidelobes and uses rectangular pulse shapes. PSK suffers from phase distortion which degrades the signal which may be reduced by differential PSK: this is a 'pseudo-noncoherent' form of PSK represented by:

$$s_i(t) = \sqrt{\frac{2E}{T}} \cos(w_0 t + \varphi_i)$$

No change in the carrier phase is made for binary 0 but a 180° phase reversal represents binary 1. At the receiver the carrier phase is compared with the phase of the previous signal which represents the reference phase for demodulation. The difference in phase between two successive waveforms carries the information. As DPSK compares the present noisy signal with an earlier noisy signal, there is a 3-dB degradation of DPSK from PSK, but it offers reduced receiver complexity.

MSK (minimum shift keying) is a spectrally efficient modulation technique [Pasupathy 1979]. As space communications links are typically power-limited, it is necessary to maximise the channel bandwidth efficiency. MSK is a special case of continuous phase coherent FSK with modulation index $h = 0.5$ for orthogonal signalling, i.e. it requires only $1/2T$ Hz frequency separation and has a performance equivalent to that of coherent PSK. It is an extension of OQPSK with a sinusoidal pulse weighting replacing the rectangular waveforms usually employed:

$$s(t) = A_I(t)\cos\left(\frac{\pi t}{2T}\right)\cos 2\pi f_c t + A_Q(t)\sin\left(\frac{\pi t}{2T}\right)\sin 2\pi f_c t \qquad (17.12)$$

This signal represents FSK with signal frequencies $F1 = f_c + (1/4T)$ and $F2 = f_c - (1/4T)$ such that the frequency separation is $\Delta f = F1 - F2 = 1/2T$, i.e. half the bit rate. This allows the two FSK signals to be coherently orthogonal. MSK has low sidelobes, with the bandwidth containing 99% of the power, with $B = 1.2/T$ compared with QPSK and OQPSK with $B = 8/T$. Hence, MSK is spectrally more efficient. It is coherent in that detection of the MSK is based on observation over $2T$ nullifying the 3-dB disadvantage of noncoherent FSK over coherent PSK, i.e. the receiver has memory so that the decoding process occurs over two successive bit intervals similarly to DPSK.

Optimal digital detection for a required S/N ratio implies matched filter conditions when the signal bandwidth equates to the noise bandwidth. A matched filter transfer function $H(f)$ is the complex conjugate function of the signal spectrum to be processed. The smaller E_b/N_o, the more efficient the modulation detection process. Linear digital filtering has two main forms: infinite impulse response (IIR) and finite impulse response (FIR). FIR filters may be designed with perfectly linear phase while the IIR filter cannot. However, IIR techniques are more efficient. Filters may be low-pass, high-pass, bandpass, multiple bandpass or bandstop according to the frequency requirements.

Coherent PSK and MSK produces the smallest error rate of all modulation techniques and noncoherent FSK produces the largest error rate of all the modulation techniques. Space systems tend to be power limited and utilise bandwidth in favour of power. This favours BFSK and QPSK. Noncoherent BFSK is most often used in military satellites and older US satellites. Coherent BPSK and QPSK are particularly common in civilian spacecraft and are the standards adopted by ESA for the modulation of TT&C data. MSK is used for military tactical radio and extremely low frequency underwater communication, and has been used on some comsats (e.g. AT&T domestic satellites) but is not in common usage on spacecraft generally. Well-established techniques enable a satellite to transmit voice at 16 kbps rather than 64 kbps.

17.3.4 Spread spectrum techniques

In 1986, a satellite TV broadcast in the USA was hacked into by an unauthorised hacker called 'Captain Midnight', illustrating the need for security from interference. Code division multiplexing (CDM) uses spread spectrum techniques where the transmission is combined with a pseudorandom code of fixed length to cause the transmission to spread across a wide frequency waveband exceeding the bandwidth required to transmit the information by as much as 20–100 times [Pritchard 1977, Cook & Marsh 1988]. The carrier phase of the transmitted signal is abruptly changed according to the code sequence. The speed of the code sequence is known as the chipping rate. The amount of spreading is determined by the ratio of chips per bit of information. The signal is spread out over the bandwidth through the channel such that bandwidth $B = 2R_c/R$ where R_c = chip rate, R = information rate and power gain drops as $G = B/R$. Digital data is logically modulo-2 added (EX-Ored) with the faster pseudorandom code. This allows multiple signals to occupy the same bandwidth simultaneously without interference. Spread spectrum signals transmit at much lower spectral power density than narrowband transmitters and represents inefficient use of bandwidth. The technique represents orthogonality in both time and frequency domains and offers access flexibility at the cost of more complex signal processing. It is essentially a frequency spreading technique that allows multiple signals occupying the same bandwidth to be sent simultaneously over the same bandwidth by encoding the signals such that individual streams may be separated at the receiver. The modulated carrier $s_i(t) = A_i(t)\cos(w_0 t + \varphi_i(t))$ is multiplied by a pseudorandom code function $g_i(t)$ and the resultant convolution signal $g_i(t)s_i(t)$ is transmitted over the channel. The signal at the receiver is a combination

$$\sum_{i=1}^{n} g_i(t)s_i(t)$$

If the signal $s_i(t)$ has a narrow bandwidth compared with the wideband signal spread code $g_i(t)$, then the resultant signal has a bandwidth close to that of $g_i(t)$. Each signal is frequency spread by its own code. Using the same pseudorandom code, cross-correlation techniques enable extraction of the data from the artificially imposed 'noise'. The receiver multiplies the incoming signal by $g_i(t)$ to despread the signal and extract the desired signal $g_i^2(t)s_i(t)$ since the unwanted signals $g_i(t)g_j(t)s_j(t)$ with $i \neq j$ will yield zero due to their orthogonality. If any interference or jamming occurs at the receiver, the spread spectrum technique spreads the jamming signal to the bandwidth of the spreading signal. DSPs (digital signal processors) and ASICs (application-specific integrated circuits) provide high-speed synchronisation and decorrelation at the receiver. Spread spectrum techniques may be implemented directly by pseudonoise techniques where spreading is achieved through multiplication of the signal by a binary pseudorandom sequence. Frequency hopping involves the carrier frequency of the transmitter hopping abruptly to different frequencies within the spreading bandwidth at certain intervals according to the pseudorandom code sequence (this requires noncoherent demodulation). Such frequency hopping is used extensively in biological communications to minimise the information transfer to predators [Richards 1985]. CDM is used in military systems due to the hardening effect against jamming and the security afforded by the pseudorandom code. Its use in commercial systems is becoming necessary as protection against outside unintentional or malicious interference. It also forms the basis of the GPS coding scheme. It is therefore considered almost essential in modern spacecraft.

TDMA (time division multiple access) offers the possibility of short compressed bursts of information transfer to the ground and may be utilised by ATLAS during non-operational periods (which will dominate the mission lifetime). Otherwise during operational periods FDMA (frequency division multiple access) is the only viable alternative as the downlink data from various sensors must be obtained in parallel. The adoption of spread-spectrum methods during operations ensures no interference from outside to secure the channel. This provides a degree of security.

17.4 SATELLITE LINK BUDGET

The satellite link budget is a critical design parameter for a spacecraft. Higher data rate implies higher transmitter power and antenna size requirements. High data rates can be reduced by using data-compression techniques. All such techniques can significantly reduce transmitter power and antenna size. Transmitter power generally imposes the highest load on the power system (in comsats) and video data rate provides the maximum data rate requirement. The data rate is constrained by the bandwidth since there is a tradeoff between bandwidth and power which are both scarce resources.

Satellite transmitter power is independent of the satellite altitude for constant coverage. At low altitudes, the required power is reduced because the area coverage in view is smaller. Furthermore, the power relative to the carrier frequency is independent between 200 MHz and 20 GHz in the absence of rain. The communications link budget is used to determine the received useful power and determines the communications system performance. The radio-frequency energy received at the receiver is a function of distance from the transmitter according to the inverse square law plus losses and noise. The performance is measured by the desired power to noise ratio, S/N. Losses will occur if some of the signal is scattered, reflected or absorbed along its transmission path. Noise occurs when unwanted energy is injected into the link, e.g. thermal noise radiated by the antenna due to random electron motion in the conducting medium (assumed to be white Gaussian additive noise). Transmission noise causes 0.4 dB degradation of the downlink and the contribution of uplink to noise is ~10 dB less than that of the downlink. A maximum power flux density limit exists at the Earth's surface ~-148 dBW/m^2/4 kHz which is imposed on satellite transmitters so that satellites do not interfere with existing terrestrial microwave links.

The radiation pattern of an antenna defines its directional capability—for a pencil beam, most of the radiated energy is contrained in the main lobe. The flux density of a microwave wavefront over 4π steradians at a distance r from an isotropically radiating omnidirectional antenna (zero gain) is given by: $\varphi = P_t/(4\pi r^2)$ [Bargellini 1978, Pritchard 1977]. A typical omnidirectional antenna is the vertical dipole. Whereas an omnidirectional antenna radiates power in all directions, a directional antenna concentrates its power in the direction (θ, ϕ) defined by its boresight direction. A narrow beamwidth increases the power flux delivered by transmitting the signal with antenna gain G_t which is dependent on the solid angle over which it transmits. The radiated power forms a well-defined set of lobes of high power flux separated by low-intensity wells. The main lobe is defined as the direction of transmission while the sidelobes represent wasted power. The radian angle of the first null from the boresight is $1.22\lambda/D$. The gain of a directional antenna such as a parabolic antenna is defined with reference to the 0 dB gain of an isotropic omnidirectional radiator over 4π steradians. The power gain may be defined as the ratio of the power flux density to the centre of spot coverage to the power flux density from the isotropic radiator. Hence, a smaller coverage area requires less power for the same received power by varying G_t. At higher frequencies, the gain increases rapidly, requiring better shaping of the antenna surface.

$$\text{Power flux density from transmitter } \varphi = \frac{P_t G_t}{4\pi r^2} \tag{17.13}$$

where EIRP = effective isotropic radiated power, $= P_t G_t$, and $G_t = 4\pi A_t/\lambda^2$ = transmitter gain

$$\text{Total power received from transmitter, } P_r = \varphi A_e = \frac{P_t G_t A_e}{4\pi r^2} \tag{17.14}$$

where $A_e = \eta A_r$ = effective area of receiver = total power absorbed/incident power flux density, e.g. the half-wave dipole has an effective area of $0.13\lambda^2$, $A_r = (\pi/4)D^2$ = physical area of receiver for a circular aperture, and η = antenna efficiency = gain/directivity ~0.4–0.8 typically (average for paraboloid ~0.6).

The effective antenna area accounts for power lost through radiation, scattering, spillover, etc. Now the receiver gain is given by

$$G_r = \frac{A_e}{\lambda^2/4\pi} = \frac{4\pi A_e}{\lambda^2}$$

and the received power is reduced by the transmission path loss L_a and the free space loss L_s the latter being the dominant loss.

$$\text{Power received from transmitter } P_r = \frac{P_t G_t G_r}{L_a}\left(\frac{\lambda}{4\pi r}\right)^2 \tag{17.15}$$

where L_s = free space loss $= (\lambda/4\pi r)^2$, and L_a = transmission path loss.

Usually, power ratings are expressed in dB: S(dBW) = 10 $\log_{10}P$(W). For GEO, free space losses vary around ~195–215 dB between 4–30 GHz according to the frequency. The additional transmission path losses for a GEO link are given by Boltzmann's constant, $k = 1/L_a = 1.38 \times 10^{-23}$ J/K = +228.6 dBW/K/Hz. The beamwidth of a directional antenna is the angle between the half-power (3 dB) points relative to the peak boresight power. The edge coverage for a circular antenna is usually ~4.2 dB less than at the beam centre. For a circular antenna, the beamwidth of a parabolic dish is given by $\delta = 100\lambda/d$ which may be used to estimate the antenna diameter d, e.g.

$$\text{Diameter of receiver antenna} \quad d = \frac{100\lambda}{\delta} \tag{17.16}$$

where $\delta =$ beamwidth $\sim 2°$.

For 1 GHz, $\lambda = c/f = 3$ cm $\rightarrow d = 15$ m for a 2° beamwidth. There is a tradeoff between antenna beamwidth and antenna gain.

$$\text{For a parabolic reflector} \quad G \sim \frac{7A}{\lambda^2} \quad \text{and} \quad A_e \sim 0.6A_r.$$

Resulting receiver gain,

$$G_r = \frac{7\left(\pi l^2/4\right)}{\lambda^2} = 1.4 \times 10^4 = 41 \text{ dB} \tag{17.17}$$

The signal power is corrupted by noise particularly thermal (Johnson) noise which degrades the ability of the receiver to estimate the transmitted signal. This thermal noise is caused by electrons in the receiver circuits being thermally excited above absolute zero by dissipation of power causing random voltages and currents with a Gaussian distribution. For a resistor equivalent, rms voltage $V_{rms} = 2\sqrt{kTR}$.

Thermal noise power is given by

$$N = \frac{hf/2}{e^{hf/kT} - 1} \text{ W/Hz}$$

but at high temperatures, $h_f \ll T$ as $T_r \sim 200$ K and $T_t \sim 1000$ K, so noise power in bandwidth B:

$$N = kT_s B \tag{17.18}$$

where $B =$ frequency bandwidth ~ 27 MHz for video signals, $k =$ Boltzmann's constant, $T_s =$ device effective noise temperature ~ 290 K typically, and $N_0 = kT_s =$ uniform spectral noise density.

Typically, the system noise for the downlink in the range ~ 2–20 GHz is 28 dBK, in the uplink range ~ 2–20 GHz is 31 dBK, and in the crosslink range of 60 GHz is 33 dBK. Thermal noise is only of secondary importance in optical devices where shot noise is dominant. Thermal noise may be limited by minimising the receiver bandwidth to ~ 3 dB. GaAs FET devices have low noise temperatures ~ 70 K at 4 GHz without cooling. The carrier (received power) to noise ratio gives a measure of S/N at the receiver input:

$$C/N_0 = P_t G_t \left(\frac{\lambda}{4\pi r}\right)^2 \left(\frac{1}{L_a}\right) \left(\frac{G_r}{T_s}\right) \left(\frac{1}{kB}\right) \quad \text{and} \quad C/N = \left(\frac{C}{N_0}\right) \left(\frac{1}{B}\right) \tag{17.19}$$

This determines the ability to detect the signal in the presence of noise (assumed Gaussian). The factor G_r/T_s determines the radio receiver sensitivity. A high carrier frequency allows a large bandwidth for high data rates. If all the received power derives

from the modulating signal, the received power is related to the data rate R by:
$P_r = C = RE_b$ such that [Bargellini 1978]:

$$\frac{E_b}{N_0} = \frac{BP_tG_tA_e}{4\pi r^2 NR} \quad \text{where} \quad \frac{E_b}{N_0} = \frac{B}{R}\left(\frac{S}{N}\right)$$

A high signal to noise ratio increases the reliability of signal detection and the fidelity of the information received. Increased transmitter power and directionality of the transmitter antenna increase the range of the signal. The link margin parameter quantifies the difference between the required E_b/N_0 to yield a specific p_b and received E_b/N_0:

$$\frac{C}{N_0} = \left(\frac{E_b}{N_0}\right)_{rec} R = M\left(\frac{E_b}{N_0}\right)_{req} R \rightarrow M$$

$$= EIRP + G_r - \left(\frac{E_b}{N_0}\right)_{req} - R - kT - L_p - L_a \quad \text{(in dB)} \quad (17.20)$$

This is the link budget—an increase in link margin can be achieved by increases in spacecraft and/or ground antenna diameter. Intelsat systems operate with a 4–5 dB margin at the C band and higher margins are usually adopted at higher frequencies due to greater atmospheric losses. The typical transmitter power for a 64 kbps audio band is ~1 W (comsats support many such carriers so their power requirements are usually in excess of 10 W). ERS-1 transmitted data at a rate ~100 Mbps from 10 m diameter antenna with a transmitter power of ~3 kW at a frequency of 5 GHz. The demodulator design determines the S/N required to meet a given received power against the antenna gain and noise temperature. G/T and C/N may be related by:

$$\frac{C}{N} = \frac{\lambda^2 \varphi}{4\pi kB}\left(\frac{G}{T}\right)$$

where $C = (\lambda^2/4\pi)G_t\,\varphi$, and $N = k(T_r + T_t)B$.

This enables computation of the required transmitter power to deliver the required performance.

17.5 DEDICATED TT&C SERVICES

TT&C is a critical spacecraft capability—if it fails, so does the mission. NASA's Space Tracking & Data Network was designed as an S band TT&C system for LEO unmanned civil spacecraft. It comprised interferometer trackers each of which comprised several ground antennas ~100 m apart as the baselines. Intelsat operates its own TT&C (telemetry, telecommand and command) system, eliminating the need to pay for services, but it has a low capacity ~100–250 bps for the command uplink channel and a ~1000–4800 bps capacity for the telemetry downlink. Intelsat V and VI have respectively 12 000 and 30 000 telephone circuits at a cost of ~$500 each—expensive compared with other options.

The TDRSS (tracking and data relay satellite system) is a constellation of three GEO relay satellites plus three in-orbit spares to provide continuous tracking and data relay to spacecraft in LEO. They reside over 41°W, 171°W and 275°W centred on the single White Sands Complex ground station in New Mexico, providing near-continuous 85–100% real-time worldwide coverage to LEO satellites with TT&C services with retransmission to other sites in real-time by the NASCOM network (there is a zone of exclusion over the Indian Ocean below 1200 km altitude). The first TDRS was launched in 1984 and the system became operational with the launch of TDRS-4 at 41°W after a delay imposed by the Space Shuttle *Challenger* disaster in 1986 which destroyed TDRS-2. TDRS-3 was already at 171°W with the in-orbit spare TDRS-1 at 79°W. The TDRSS will eventually be replaced by the US Advanced TDRSS. Each spacecraft has an axial rotation limit of ±265°, an elevation rotation limit of –5° to +95°, a maximum slew rate of 7°/s, an antenna gain of 46 dB, a beamwidth of 0.7°, and an autotracking capability of ±30′. They perform data relay by retransmitting data received through the onboard receive–transmit transponder payload effectively providing GEO communications capability to LEO satellites. TDRSS offers two access bands: two single-access S band antennas of 4.9 m diameter, one multiple access S band antenna, and a single-access K band antenna. The use of TDRSS requires the adoption of TDRSS compatible transmit–receive S band or K band transponders on the spacecraft (in fact two should be employed for redundancy). The S band link can communicate with up to 24 satellites simultaneously and the K band link relays telemetry from any of those spacecraft to the White Sands terminal. The S band offers 300 kbps for the forward path command link at 2.025–2.120 GHz band and 1.2 Mbps for the return path telemetry downlink at 2.2–2.3 GHz. The S band also offers Doppler tracking capability. The K band offers 25 Mbps for the command uplink at 13.775 GHz and 300 Mbps for the telemetry downlink at 15.35 GHz. The uplink signal is modulated with a pseudorandom noise spectrum spreading code and PSK. Hence, the K band offers the greatest data rate capabilities and it requires transmission with a 36 MHz or 72 MHz bandwidth. In fact, the Space Shuttle communications system utilises both the S band and the K band capabilities of TDRSS for its operations. The maximum data rate capability is in fact limited to 48 Mbps by the ground routes leading to the Goddard Spaceflight Centre and Johnson Space Centre from which land links may be provided to the user remote control centre. This eliminates the need for multiple long-distance transmission and introduces the desirability of using an extended LAN. The user schedule time is typically short due to Earth orbit view times. User requests are scheduled electronically in a volatile scheduling system. Comsats can provide high-capacity excess links to remote ground stations by leasing telephone lines. However, major costs can be incurred by leasing telephone lines over long distances so such methods are not popular. A new TDRSS is planned to operate in a cluster architecture offering the same service capability as the current TDRSS due to the difficulties encountered in the deployment sequence of the TDRSS [Brandel 1990]. The possibility of extending operations to incorporate a Ka band at 22.5–23.5 GHz for uplink and 22.25–27.5 GHz for downlink plus 60 GHz and laser crosslinks with ~Gbps capabilities are also being considered. Users of TDRSS have included HST, Landsat, ETS-VII and SMM. TDRSS can also track to within 100m accuracy offline for LEO satellites. The TDRS (Tracking & Data Relay Satellite) transponder for example is a small, high-gain (23 dB) phased array Ku band antenna. It has a power consumption of 5 W in the receive mode and 20.5 W in the transmit mode. It has a mass of 3.6 kg plus its diplexer of 600 g. TDRSS offers a return data rate of 2–3 Mbps.

The European Data Relay System (EDRS) is an element of the European in-orbit infrastructure programme which also includes EO (Earth Observation) satellites such as Envisat-1 and Metop-1 scientific satellites to allow direct real-time communications between scientific satellite users via EDRS [Grubilei et al 1994]. ESA's Olympus successfully conducted a data relay experiment between the ground and the Eureca spacecraft. EDRS is specifically designed to support the Envisat Polar Platform Earth Observation mission which operates in polar orbit at 800 km altitude for global Earth coverage. The first component is Artemis (Advanced Relay Technology Mission) at 16.4°E geostationary orbit which will test the intersatellite optical links through its laser data relay payload SILEX. It will be completed by the addition of a further two satellites of 2.6 tonnes each in GEO at 44°W and 59°E to give good coverage of Europe forming a polygon with vertices at Fucino, Madrid, Liverpool, Oslo, Maleno and Vienna. The EDRS satellite offers a single dual-band offset parabolic reflector antenna with a diameter of 2.85 m and f/D of 0.5 for both S and K bands and an S band deployable multiple-access phased array. Very-low-noise HEMT devices are being adopted for the phased array elements. A steerable spot antenna offers extra-Europe coverage of 5° elevation as far as America and Russia (the zone of exclusion is over the Central Pacific). The feeder link between the ground and EDRS is the Ka band of 27.5–30.0 GHz for the forward link and 17.7–20.2 GHz for the return link with 100 Mbps capability from a high-gain (44 dB) Cassegrain antenna of diameter 0.9 m. It offers a G/T of −0.8 dB/K and 8.4 dB/K respectively and an EIRP of 43.5 dBW and 49.0 dBW respectively from the west and east satellites to Liverpool. Data is delivered in BPSK or QPSK format for rf modulation with a bit error rate of 10^{-6} and an S/N ratio of 7.5 dB (signal is Viterbi/Reed–Solomon coded). The inter-orbit links cover three wavebands: S band offers 2.025–2.110 GHz for the forward link (150 kbps capability) and 2.200-2.290 GHz for the return link (3 Mbps capability); the Ka band offers 23.120–23.550 GHz for the forward link (25 Mbps capability) and 25.250–27.500 GHz for the return link (150 Mbps capability); the optical band uses a 0.8 μm semiconductor diode laser waveband in both directions differentiated by dual-pulse polarisation modulation (50 Mbps capability either way). Both the S band and Ka band use BPSK modulation but the S band can also use CDM spread spectrum techniques. The user satellite must provide an EIRP of 62.8 dBW and 45 dBW, and a G/T of 22.3 dB/K and 8 dB/K for Ka and S bands respectively. The minimum requirements for antenna size and power output for the LEO user satellite is 1 m and 30 W respectively. The EDRS pricing policy is designed to recover operating costs only and not development costs [Agonisti et al 1992]. Two possibilities have emerged:

(i)   charge for access time (regardless of data rate);
(ii) charge per bit by (data rate x access time).

The charge per bit would discourage high data rate users while charging for access spreads the cost among both high and low data rate users. The US TDRSS charges on access time at a rate of $200/minute for non-US Government users. EDRS is expected to charge in a similar manner but at a rate of $30/minute to recover the operating costs only (around 22 MAU/y). If continuous access were required the total charge would be ~11 MAU/y. ESACOM is ESA's general-purpose network system across the whole of ESA, based on TCP/IP communications protocol.

Continuous TT&C coverage can be achieved with a single ground station by using such space-based TT&C services. A robotic servicer such as ATLAS may also take advantage of dual mode operation to limit costs on both personnel and on service costs. It is conceived that during non-operational dormant phases that ATLAS would employ directly transmitted packetised TT&C bursts during the overhead pass. During operational phases, however, it would utilise the EDRS real-time link for continuous coverage without being penalised as a high data rate user.

For deep space missions, such as those involving planetary rovers and landers, rovers typically communicate using an omnidirectional whip antenna to a more powerful lander antenna which then acts as a store-and-forward relay to transmit the signals back to Earth, either via an orbiter, or directly. NASA's Deep Space Network operated by JPL for deep space missions has a limited capacity with ~2 kbps for command uplink at 2.025–2.120 GHz waveband and ~500 kbps for telemetry downlink at 2.2–2.3 GHz waveband. Its primary purpose was tracking and grew out of the JPL tracking station for Explorer 1. It was extended into a network of major tracking stations at Pasadena in California, Madrid in Spain and Canberra in Australia which are 120° apart as well as 9-m dishes in almost 30 overseas locations and five North American locations. Each of the three major complexes now comprise four types of antenna: 2×34 m diameter, one of 26 m diameter and one of 70 m diameter dishes. The 70-m antenna can detect signals as weak as 10^{-16} W characteristic of remote missions such as Voyager. The Voyagers' signals are received on the ground at a data rate of 160 bps with a power level of 10^{-12} W. The 34-m antennas are used for missions to the outer planets. The 26-m antennas are used for tracking Earth-orbiting satellites. This network was instrumental in the development of NASA's Apollo S band system which linked a worldwide network of S band tracking stations to Mission Control at Houston, Texas, through Ethernet LAN to the WAN in support of the 1969 Apollo 11 Moon landing. The largest impetus for its development was the Gemini programme which required multiple vehicle RVD manoeuvres, the tracking of two vehicles simultaneously and the uplinking of command orbital data to the spacecraft computers. The whole DSN network is connected by NASA's communications network (NASCOM) with five switching centres at Goddard Spaceflight Centre, Canberra, Honolulu, London and Madrid to interface the multiple communication trunks routed by cable or satellite circuits. It adopts distributed scheduling whereby user requests are integrated centrally far in advance and all projects are scheduled. Typically, scheduled coverage times are long, e.g. for planetary spacecraft. Telemetry and command data is modulated on the radio signals to and from the spacecraft. The tracking system provides two-way communication to make estimates of the position/velocity of the spacecraft. Very-long-baseline interferometry is used to provide data on radio sources, especially astrometric data. ESA monitors a relay at Marshal Spaceflight Centre which forms part of ESA's Interconnection Ground Subnetwork which provides the ground-based communications infrastructure for voice, video and data links. This coexists with ESA's spacecraft operations support network (OPSNET) which communicates with all ESA stations.

17.6 OPTICAL SPACE COMMUNICATION TECHNOLOGIES

There are five main candidates for the implementation of photonics [Paul et al 1994]: beam-forming networks, multicarrier multiplexing, intersatellite links and onboard C&DH. Multiple-access techniques are also amenable to optical techniques [Gagliardi & Mendez 1994]. In replacing microwave subsystems, photonics offers increased bandwidth and communications capacity through wavelength division multiple access (avoiding the need for TDMA—time division multiple access) and increased payload mass capacity due to smaller antenna sizes with lower power requirements. Optical spacecraft-to-spacecraft links imply small aperture requirements ≤25 cm diameter. They offer high bandwidth for very high data rates ≥100 Mbps with little power loss particularly for deep space communications. Data rates of 400 Mbps to 2 Gbps may be accommodated using optical techniques. For distances of 40 000 km (GEO–GEO separation distance), an optical telescope of 20 cm diameter and laser transmitter power of 200 mW are sufficient for these data rates. This is much smaller and less power-hungry than comparable microwave links at 60 GHz. All these factors impact on reducing spacecraft mass—as we noted earlier with reference to MEMS technology, microwave techniques cannot be appreciably miniaturised [Crofts et al 1994]. For space-to-ground communications, an optical ground station could be based on a 1-m telescope. Optical techniques are virtually immune to interference, do not generate sparking and are difficult to tap.

As with any communications system, optical techniques are based on transmitters and receivers, these roles being undertaken by lasers and photodetectors. Semiconductor lasers with high reliability over long lifetimes, high compactness and high efficiency of ~25% and photodetectors with heterodyne or direct detection methods are available. Although heterodyne methods are more sensitive than direct detection, they require stronger conditions on laser frequency stability and linewidth and impose complexity on the optical receiver. Hence, direct photodetection would be preferable. A new diffraction-limited semiconductor diode laser is available for space links which delivers ~1 W continuous power output. The output may be amplitude-, frequency- or phase-modulated and is suitable for low mass, efficient and reliable laser links in space between spacecraft in conjunction with Si photodiodes. For the GEO–GEO separation of 40 000 km it offers a 1-W power requirement per gigahertz of bandwidth.

SILEX (Semiconductor laser Intersatellite Link EXperiment) is an ESA programme for inter-satellite optical crosslinks for use on satellite constellations supporting mobile communications [Lutz 1997]. The SILEX device has a mass of 157 kg with a power consumption of 150 W, most of which is for the two-axis gimbal mechanism carrying the complete optical head of the telescope, transmitter, receiver, and electronics. SILEX can transmit 56 Mbps over 42 000 km using only 60 mW of power with a 25 cm diameter telescope (gain equivalent of 100 dB). The disadvantage is the requirement for very accurate pointing ~8 μrad due to the narrow width of the transmitted beam, i.e. a 400-m illuminated zone at 42 000 km. Even micro-vibrations of normal spacecraft mechanism operations disturb the pointing accuracy. To establish an optical link, search manoeuvres need to be performed further degrading performance. Optical beams of multiple laser diodes with a power consumption of 10 W can be transmitted over wider beams of 750 μrad by data relay satellite. The SILEX test device is incorporated into ESA's

Artemis (Advanced Relay & Technology Mission Satellite) to provide data transmission at 50 Mbps from LEO to GEO based on GaAlAs semiconductor laser diodes transmitting at 830 nm with a 25-cm aperture.

Optical methods offer great promise, particularly for optics-based GEO relay satellites, in reducing in-orbit servicing spacecraft mass requirements and increased data transmission bandwidth. However, they represent an immature technology as far as space-based systems are concerned.

18

Spacecraft onboard power subsystems

We shall be considering a number of issues relating to spacecraft onboard electrical power generation, storage, conversion, distribution and dissipation (the power subsystem and thermal control subsystem). Electrical power is required for all actuation mechanisms, active sensors and onboard electronics. Onboard electrical power should be optimised to provide maximum power per unit mass rather than maximum efficiency. A robotic spacecraft typically has very high power rating requirements by virtue of its use of motors. In addition, power raising and storage is particularly challenging for deep space missions such as planetary rovers. The power generation and storage subsystem are defined by the power budget as one of the critical design parameters of the spacecraft.

The spacecraft power system is a major limitation on spacecraft design due to the inherent high mass of power generation and storage devices. The electrical power system is required to generate, store, control and distribute onboard electric power. It is required to supply continuous power to the spacecraft loads throughout the mission, control and distribute the power according to average and peak loads, provide ac conversion and/or dc regulation, and suppress transient bus voltages. The average electric power required for the payload determines the size of the power generation system. The peak power which is usually ~2–3 times the average power and eclipse periods determine the size of the energy storage devices and distribution. It is the average power at EOL (end of life) which is less than BOL (beginning of life) performance that determines the size of the power system [Larson & Wertz 1992]. Long mission lifetimes >10 y demand redundant power and greater power capacities with failure protection. The spacecraft orbital parameters particularly altitude determine the major influences on the power system through the nature of incident solar radiation and degree of eclipsing.

18.1 ONBOARD POWER GENERATION

Primary power generates power while secondary power stores power for use during eclipse periods. Most unmanned spacecraft employ solar panels supplemented by rechargeable batteries for generation and storage respectively to allow continuous power despite interruptions due to eclipse of the spacecraft behind the Earth's shadow. Photovoltaic cells are light intensity measurement devices. The cell is composed of a semiconductor rectifier junction between top and bottom electrodes. When light is incident on the top surface, electrons cross the rectifier junction to produce an EMF and current is generated. Batteries are limited in energy storage capacity and lifetime and solar array power generation is limited by degradation. Military and interplanetary missions usually use RTGs (radioisotope thermoelectric generators) as they are solar illumination independent and function by converting heat from radioactive decay into electricity via the thermoelectric effect. Nuclear reactor sources generate high power/unit mass but are politically unpopular, e.g. US SNAP-10A in 1972. Power requirements vary considerably from mission to mission: high-voltage requirements are needed for telecommunication satellites with ratings of ~1–10 kW (up to 80% of which is for the transmitting antenna) and for space stations like Skylab or Mir ~10–30 kW or more. High powers for long duration missions can be met with fuel cells, RTGs, solar arrays and nuclear reactors. Currently, most solar arrays convert solar energy to electricity directly through photovoltaic processes and are suitable for <10 kW at LEO. For durations longer than 15 y, photovoltaic degradation causes large significant differences between EOL and BOL power delivery making their implementation difficult or at least wasteful of mass.

18.1.1 Solar arrays

The first photovoltaic cell was flown on Vanguard 1, the second American satellite in 1957 [Williamson 1999c]. Solar arrays require adequate solar illumination since the power output varies with the cosine of the solar incidence angle between the normal to the array surface and the sun vector up to 60°, beyond which the output power drops off more rapidly. They have the advantage that almost any desired current and voltage can be generated according to the array size but for high powers the array sizes become large causing atmospheric drag and flexure problems. Solar panels may be flexible or rigid. Rigid fold-up type panels use hinged panels of carbon-fibre-reinforced plastic. Normally, two panels are mounted symmetrically either side of the spacecraft. Panels must then be deployed from a folded position against the spacecraft structure and solar paddles offer the least mass-optimised solution. For such flat panels, the worst-case sun angle is 23.5° giving a cosine factor of 0.92. Most photovoltaic arrays are planar with the solar cells mounted with adhesive onto an insulated (Kapton with fibreglass reinforcement ~100 μm thick) hollow core Al honeycomb panel surface with carbon-fibre reinforcement. However, rapidly manoeuvring spacecraft like the ATLAS in-orbit servicer cannot use large flexible deployed arrays. Flexible arrays are based on layers of Kapton blanket laminated together with polyester adhesive and a photo-etched printed circuit to interconnect to the solar cells. Such rapidly manoeuvring spacecraft usually need body-mounted cellular arrays to eliminate the need for tracking and pointing and the consequent vibration-induced disturbances. Body-mounted arrays are typical on Mars rovers, e.g. Sojourner's upper face was mounted with solar arrays as solar energy during the day was available from overhead across a 180° arc. However, body-mounting offers a less effective solar incidence efficiency and generates a lower power/unit area than boom-mounted planar panels since not all cells are illuminated but they tend to weigh less per unit area than planar arrays. The array reduction in output per unit surface area is ~ π for cylindrical spacecraft and ~ 4 for a cubic spacecraft. Additional body-mounted panels may be deployed from folded positions against the spacecraft structure in a 'Russian-style' buttercup fashion or from a dropskirt to increase the illuminated area. If concentrators are used, stowage and deployment of the buttercup can be more complicated but this is not a problem for the dropskirt. This is a suitable approach for most spacecraft which maintain a constant attitude plane with respect to the Sun. However, for an in-orbit servicer such a ATLAS, attitude will vary so the configuration favoured here is to use re-deployable flexible solar arrays that may be furled and unfurled according to mission requirements.

Arrays are composed of rectangular solar cells of the pn junction type (Si doped with P and B impurities typically) which generates electron–hole pairs when illuminated by photons [Hubbard 1989]. They usually have dimensions 2×4 cm for efficient packing ~90% allowing for Ag-plated interconnections for minimum size and mass. Since ~50% of solar photons of 1.0–2.5 eV energies are absorbed in ~10^{-3} to 10^{-5} cm thickness of the cell, cell thicknesses are limited to ~10^{-3} cm. Tantalum peroxide Ta_2O_5 is used as an ultraviolet reflective coating to improve the cell efficiency by reducing reflective losses through reflecting the ultraviolet which is not converted to electrical energy. Series connections of cells provide the desired voltages while parallel connections of the cells provide the desired currents. A voltage of ~0.5 V/cell is generated under solar illumination. Solar cell performance is significantly degraded by radiation exposure, particularly in the Van Allen radiation belts, generating reduced output voltage and current. Degradation also occurs due to thermal cycling in and out of eclipse, micrometeoroid impingement, plume impingement from thrusters and material outgassing. For Si arrays in LEO up to 800 km, altitude power degradation averages at ~3.75%/y (of which 2.5% is due to radiation) while for GaAs cells in LEO up to 800 km power degradation is much less at ~2.75%/y (of which 1.5% is due to radiation). GaAs thus retains 78% of its nominal performance after 10 y. GaAs arrays are flight proven (e.g. UoSAT E in 1989). At GEO, degradation is more severe. Solar cells may be protected with ~100 μm thick coverslides of fused-silica glass to attenuate the radiation intensity $I \sim I_0 e^{-k\rho x}$ where k = absorption coefficient. A further advantage to coverglasses is that they can act as blue–red filters to block out ultraviolet/infrared radiation and increase emittance to reduce solar cell heating and operating temperatures. However, coverglasses cause the cells to lose 15% voltage/current output due to coverglass adhesive darkening. Photovoltaic thermal conversion efficiencies are low: for Si it is ~12% nominally and ~14% maximum; for GaAs it is ~18%; for InPh it is ~20%. The maximum cell output for Si at 1 AU ~54 W/cm^2 cell area which equates approximately to 135 W/m^2 (compare solar constant of 1358 W/m^2). If the effects of hardware such as array deployment mechanisms, support, etc. are included this reduces to ~100 W/m^2. GaAs cell BOL output is ~243 W/m^2. Operating temperatures affect array performance in that Si cell efficiency drops 0.5% per degree Celsius increase from the nominal 28°C compared with GaAs efficiency drop of 0.2%/°C. Array temperatures vary from –200°C in eclipse to +100°C in full sun and crossing the day/night terminator out of eclipse creates major voltage surges up to double the normal voltage levels due to this temperature-dependent efficiency. The array temperature average at LEO is ~67°C and at GEO ~53°C though body-mounted arrays tend to be ~5°C higher than boom-mounted arrays due to the standard deployment of optical concentrators. GaAs cells have lower temperature coefficients as well as being thermally more efficient and radiation resistant but cost seven times as much as Si cells. However, their superior performance makes them the obvious choice for long-life rugged missions like a robotic freeflyer such as ATLAS. Rather than using coverglasses to reduce solar cell degradation which are heavy, thermal annealing may be used. For Si cells, annealing occurs at an impracticable 500°C and this will eliminate damage, effectively refurbishing the cell. For GaAs cells, this annealing temperature is reduced to 125°C and this will regenerate an EOL performance close to that at BOL. This may be accomplished either with lasers for thick cells or with electron beams for thin cells, though such methods have yet to be used due to the long annealing times required. Alternatively, concentrators to maintain GaAs cells at temperatures of 125°C will not impair cell efficiency significantly but their practicality is questionable. Textured silicone concentrator coverslides (e.g. dome Fresnel lens or Cassegrainian lens) may be used to concentrate sunlight at the cells with metallised back reflectors to focus more energy into the cell, and reflect 90% of the unabsorbed incident energy from the rear side back into the cell, so increasing efficiency by ~3%.

It is possible to increase the efficiency of solar cells by using mirrors rather than lenses. Solar furnace efficiencies are higher ~20% than photovoltaic methods generating a 60% reduction in array area per unit power. Such systems require an array of mirrors to illuminate and concentrate the solar energy at each cell. Possibilities for mirror configuration include the Newtonian or Cassegrain systems, but cost and drag area of the Cassegrain favours the Newton paraboloid configuration [Simon & Nerod 1987]. For mirror reflector coatings, Ag has the highest solar reflectance at 96% but it suffers from atomic oxygen attack. Polished Al is resistant to atomic oxygen erosion and offers only slightly less solar reflectance at 92%. Alternatively, Ag could be used if protected by Al_2O_3/SiO_2 coatings to make the reflectors resistant to atomic oxygen effects.

Solar dynamic systems through the use of a working fluid to achieve greater efficiencies are most suited to space station applications rather than medium sized satellites—electric power production by thermionic or thermoelectric effects are preferred in general as they imply few moving parts.

18.1.2 Nuclear sources (RTGs)

The most common nuclear power source is the RTG (radioisotope thermoelectric generator) which uses a radioisotope surrounded by series–parallel arrays of thermoelectric couples of doped Si–Ge p-n junction between two dissimilar semiconductors to generate electricity from the subcritical spontaneous radioactive decay process. The temperature gradient between the p-n junction cells due to the slow radioactive decay outputs dc electric power. They have the advantage that they involve no moving parts. Pu-238 in the form of PuO_2 is the most common radioisotope adopted and undergoes α-decay with a half life of 87 y to generate a power output of 0.55 W/g. However, its cost is vast at $3000/W precluding its use at high powers ~kW. It is highly toxic as well as being radioactive and only Sr-90, usually in the form SrO, is an acceptable alternative which undergoes beta-decay. This has a half life of 28 y and a power output of 0.93 W/g at a cost of $250/W. RTG power output is dependent on isotope lifetimes so output decays over time as $P_t = P_0 e^{-(0.693/t_{1/2})t}$. RTGs offer favourable specific energy densities for low powers—the SNAP-27 RTG as used on Apollo ALSEP package generated 65 W in a mass of 30 kg, around 10^3 times that of batteries. However, their conversion efficiencies are low at ~5–10%. Thermal control is necessary and is achieved with a shunt radiator or fins on the outer casing to dispose of excess energy. The major problem with RTGs and nuclear sources in general, is that there is the possibility of contamination due to launch failure. They can, however, be protected from burnup during re-entry and impact on Earth. Since the fuel is in oxide form it is resistant to high temperatures and may be encased in graphite containers. In 1968 the launch of NIMBUS-81 at 300 km altitude allowed the recovery of the RTG capsules without incident, illustrating the plausibility of this concept of recovery. RTGs tend to be used for long-lived spacecraft which require power independent of the solar distance (i.e. interplanetary probes beyond Mars). They are detrimental to electronics and so are usually mounted on booms and shields—the Galileo Jupiter probe had two RTGs mounted onto opposing booms. RTG power sources are a mature technology that have powered such missions as ALSEP, Pioneers 10 and 11, the Mars Viking landers and Voyagers 1 and 2. Indeed, the Pioneer RTGs have been operating for 20 y.

The present General Purpose Heat Source (GPHS) RTG utilises Si–Ge thermoelectric elements operating at 1273 K with 6.8% efficiency with a performance of 5–6 W/g for power outputs less than 1 kW. They are being used on the Jupiter-bound Galileo to generate up to 600 W, on the solar-pole bound Ulysses, and on the Saturn-bound Cassini–Huygens probe, and will also be adopted for the Rosetta cometary rendezvous mission. The GPHS RTG module has a mass of 1.44 kg in a package $3.7 \times 3.6 \times 2.1$ in. It is capable of surviving re-entry and retaining integrity on impact at 54 m/s. Each module contains four PuO_2 fuel pellets encapsulated in iridium alloy shells and two such shells are encapsulated in graphite impact shells. A thermal insulation layer of carbon-bonded carbon fibre surrounds the graphite shells to limit thermal energy transport. Two graphite shells are enclosed in a fine-weave fabric graphite aeroshell to complete the module. A modular RTG has also been developed to reduce the mass of the non-heat source components. A thermoelectric converter multicouple offers long duration ~15 000 h and reliable performance. The multicouple comprises 20 small n-p thermoelectric elements in a close-packed, series array separated by thin glass insulation layers. There are eight thermocouple packages to each module producing 16–19 W at 28–30 V. The power output may be varied by stacking these modular RTGs. The use of ultrathin molybdenum foils separated by zirconia particles offers low mass thermal insulation. For a 300-W output, the modular RTG has a lower mass and smaller volume than the GPHS RTG, offering a specific power density of ~8 W/g. Derived systems are under development for higher efficiencies, greater power outputs and reliability for long-duration missions. Advanced thermoelectric materials are expected to yield improved performance and effort is focussed on the addition of GaP dopants to n-type Si–Ge to improve the electrical properties due to GaP ion pairing and on the reduction of thermal conductivity of p-type Si–Ge by adding 50-A sized particles which act as phonon scatterers. Such improvements may offer efficiencies ~20% and specific powers of ~10 W/g.

An alternative to using the thermocouple for energy conversion is to use thermionic energy conversion where electricity is generated through a hot electrode emitter facing a cool electrode collector in a sealed ionised gas enclosure. Electrons from the emitter flow to the collector where they condense and return to the emitter through an externally connected load. Efficiencies are higher ~10–20% but high temperatures ~1800–2400 K are required and so a nuclear reactor source is needed. They require coolant circulation at ~1000 K and are limited at present to 2-y lifetimes.

All existing RTG systems have involved direct conversion of heat energy to electricity with the electron stream acting as the working fluid moving from hot to cold reservoirs either through dissimilar materials (thermoelectric) or vacuum (thermionic). Dynamic systems use a heat source and heat exchanger to drive a turbine in a thermodynamic cycle. Dynamic RTGs (DIPS—dynamic isotope power systems) are under development for higher power levels ~1–10 kW by exploiting the higher efficiency of dynamic power conversion ~20–25% and improved specific power densities ~5–7 times that of static RTGs. This reduces mass, cost and safety risk and provides increased flexibility through greater power ratings for space exploration, particularly for outer solar system missions such as surface mobility systems on the outer planet moons. AMTEC (alkali metal thermal-to-electric conversion) is one such approach to energy conversion in dynamic RTG systems. It is a thermally regenerative electrochemical system. Liquid Na is heated by radioisotope energy sources to one side of a solid ceramic β-alumina electrolyte membrane that conducts Na^+ ions rather than electrons. The Na vapour undergoes isothermal expansion which is converted to electrical work by charge exchange in the β-alumina electrolyte. A closed recirculating Na cycle is powered by an electromagnetic pump and the Na^+ ions act as the working fluid of a Rankine cycle but without moving parts. The efficiency of conversion is ~15–30%. The solid β-alumina electrolyte separates liquid Na at 900–1300 K from a condenser at 400–700 K. A conductive metal film on the membrane collects electrons which are conducted through a load to neutralise the Na^+ ions. Na vapour condenses on the cold side and the liquid is recycled by an electromagnetic pump. A porous metal electrode of rhodium/tungsten alloy or titanium nitride produces electricity with a power density of 0.4 W/cm^2. AMTEC cells with power outputs ~10–50 W may be connected in parallel-series for high power generation over 14 000 h. For 4.5 kW, 18 GPHS stacks are required, giving a total mass of 54 kg, offering great potential, but the technology is not sufficiently developed for operational flight status.

Table 18.1. Power generation properties (adapted from Larson & Wertz 1992).

	Photovoltaic	Solar dynamic	RTG	Nuclear reactor
Power (kW)	0.2–25	1–300	0.2–10	25–100
Specific power (W/kg)	26–100	9–15	8–10	15–22
Specific cost ($/W)	2500–3000	800–1200	16k–18k	400–700
Hardness	Med.	High	V. high	V. high
Degradation	Med.	Med.	Low	Low
Storage req. (eclipse)	Yes	Yes	No	No
Sun-angle sensitivity	Med.	High	Nil	Nil
View obstruction	High	High	Low	Med.
Fuel availability	∞	∞	V. low	V. low

18.2 ONBOARD POWER STORAGE

Photovoltaics require energy storage during eclipse periods and batteries are usually selected for this purpose (other possibilities include fuel cells and flywheels) which are recharged during non-eclipse periods. Earth-orbiting satellites experience between 90 eclipses at GEO over 10 years to 5500 eclipses at LEO over 5 years during typical missions. Batteries are one of the most massive components in the spacecraft—they comprise up to ~15% of a typical comsat's dry mass or, equivalently, ~50% of the total payload mass. Power storage is also required for handling high transient loads for all spacecraft.

Inertial energy storage through flywheels are presently used for energy storage only in the form of angular momentum in momentum wheels. Such techniques have the potential for good performance with low mass and high efficiency over a wide range of power applications offering ~25 kJ/kg specific energies particularly for peak loads but they have yet to be developed to flight status [Post & Post 1973]. Flywheels are both efficient and compact with the amount of energy stored in the wheel depending on rim mass and the wheel spin velocity which may achieve ~100 000 rpm. Ultimately, the storage capacity is limited by the material's tensile strength due to centrifugal hoop stresses. High-strength, low-density carbon-fibre epoxy composites exploit the fact that energy storage capacity depends on the the square of rim speed: $K = \frac{1}{2}(\sigma_{max}/\rho)$. Fused silica materials offer 10–15 times the energy storage of steel alloys. As the typical spin rate is 5000–100 000 rpm, magnetic bearings support the wheel to eliminate friction within the evacuated housing. A 15-cm diameter flywheel offers energy densities of 88 Wh/kg. The potential is for up to 250 Wh/kg at speeds up to 600 000 rpm. Graded concentric wheels with (E/ρ) increasing outwards minimises the radial stresses. Powering terrestrial vehicles by flywheels has been suggested—a silica fibre flywheel could store enough energy, ~30 kWh, for a range of 200 miles at 60 mph. Furthermore, it could be recharged at a rapid rate of 350 kW. The flywheel has an electrical efficiency of 95% and frictional losses can be kept very low by sealing in a vacuum and using magnetic bearings—a typical rundown time for a rotor is 6–12 months. Hence, the potential for high energy density, long-term energy storage is immense, with powerful applications in spacecraft; but they have yet to be adopted in spacecraft.

Eclipses are caused by the projection of the Earth's shadow on the satellite.

The total orbit period is given by:

$$P = K_g(R_{earth} + h)^{3/2} = 1.48 \text{ h} = 89 \text{ minutes}$$

here R_{earth} = Earth radius = 6378 km, R = orbit radius = $R_{earth} + h$ = 6578 km for 200 km altitude orbit, and $K_g = 2.77 \times 10^{-6}$.

The eclipse variation depends on the Sun's declination with respect to the orbit plane and decreases with increasing declination. The total eclipse period is given by:

$$T_{eclipse} = \frac{P}{\pi} \cos^{-1} \frac{\left(1 - \dfrac{R_{earth}}{R^2}\right)^{1/2}}{\cos \delta}$$

where δ = Sun's declination. When the Sun's orbit lies in the orbit plane the eclipse period is maximised and at 200 km lasts for $T_{eclipse}$ = 37.3 minutes. The length of the shortest day T_{day} is 89–37 = 52 minutes.

Mostly, batteries used in spacecraft are rechargeable secondary batteries. Primary batteries are not rechargeable though they offer high power densities and are usable only for short duration missions <1 day. Secondary batteries are limited in lifetime due to charge/discharge cycles but are capable of sustaining ~1000s of cycles. Typically, for LEO operation they must endure ~10 000 cycles. Banks of small parallel batteries offer high voltage capabilities. AgZn batteries are strictly speaking secondary batteries but their short recharge cycle of 200–300 limits their use as such. The earliest US manned missions (Project Mercury and the Apollo Lunar Module) used AgZn batteries since they were of short duration. Although early batteries were of the AgZn variety with specific energy densities ~175 Wh/kg, Li-type primary batteries have superseded them with specific energy densities >250 Wh/kg with Li used as the negative electrode and a non-aqueous electrolyte. Li has one of the highest standard electrode potentials of any element. Secondary batteries are recharged during sunlight periods and discharge during eclipse to power the load. Rechargeable Li batteries have been under development [Dudley & Verniolle 1997]. Early Li batteries offered poor lifecycle capability due to the difficulty in re-plating the Li electrode during recharge. More recently, Li metal has been replaced by layered carbon electrodes into which Li ions can pass. Positive electrodes are typically solid solution electrodes of metal oxide/sulphide in which Li ions are free to move (the Galileo probe used $LiSO_2$ batteries, flight-qualifying them). Li ion conducting polymeric electrolytes have also been developed. Li–C cells offer great potential and can be fabricated in thin card geometries of series battery stacks. They are limited at present to 1000–2000 cycles so are suitable for GEO.

NiCd batteries with Ni positive and Cd negative electrodes in a KOH electrolyte have been the standard secondary battery for spacecraft use for many years. They have specific energy densities ~15–30 Wh/kg and a cycle limit of ~30 000, but NiH_2 batteries with specific energies of ~45–50 Wh/kg and cycle limits ~6000 have been space-qualified for GEO (first flown on USN NTS-2 satellites and also on Intelsat V and VI, HST, Olympus, Eutelsat II and others) and offer greater performance capabilities [Williamson 1999c]. Lifecycle is limited as a function of depth of discharge. NiH_2 batteries have greater lifetimes and better cycling characteristics than NiCd batteries. The NiH_2 battery is a combination between a sealed conventional rechargeable battery and a fuel cell. The cell comprises a stack of alternating Ni and Pt electrodes in a KOH electrolyte and its characteristic reaction is:

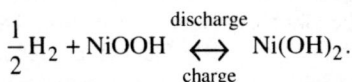

$$\frac{1}{2}H_2 + NiOOH \overset{\text{discharge}}{\underset{\text{charge}}{\longleftrightarrow}} Ni(OH)_2.$$

The cell internal pressure is a linear function of the H_2 gas content which in turn is a linear function of the state of charge. It is housed in cast Al for structural support with thermal control to maintain the temperature between 0°C and 20°C. It comes in three cell types: a 64 mm diameter cell for 5–25 A h, a 90 mm diameter cell for 30–100 A h, and a 114 mm diameter cell for 100–250 A h. The cells are connected in series in a single pressure vessel of 254 mm diameter. As a ferromagnetic material, Ni can corrupt sensitive magnetometers. If magnetometers are used, Ag–Ca batteries can provide energy densities of ~70 Wh/kg for magnetic cleanliness but they offer poor storage integrity and lifecycle performance. Ni–metal hydride batteries are in an advanced state of development with twice the energy density of NiCd and twice the volume energy density of NiH_2 batteries. NaS batteries offering ~140–210 Wh/kg densities are also under development.

The spacecraft orbital parameters, especially altitude, determine the number of charge/discharge cycles during the mission and the depth of discharge which represents the percentage of battery capacity lost during the discharge. The cell voltage output is a function of both depth of discharge and rate of discharge. The recharge energy exceeds the discharge energy by ~80%. A deeper depth of discharge shortens the life cycle. Cell capacity = power × maximum discharge time/mean discharge voltage × DoD × number of cells. At GEO, there are two solar eclipse seasons per year of 45 days around the vernal and autumn equinoxes. Each eclipse lasts 72 minutes per 24-h day during these seasons. This requires high depth of discharge of ~50%. At LEO with periods <100 minutes, there is one eclipse per orbit (~15 per day) giving a 40% eclipse time of ~35 minutes each. This requires a high cycle rating ~500/y with low depth of discharge ~10–15%. At LEO, the batteries spend ~40% of the time discharging (compare with 7% for GEO) and ~60% of the time continuously charging. For 10 000 cycles, the nickel–hydrogen battery allows 50% depth of discharge. To obtain the maximum life from secondary batteries, recondi-tioning is required. This involves very deep discharging to the point of voltage reversal followed by fast recharge to allow steady-state voltages to be maintained. This effectively erases the memory effect of repetitive cycling to a fixed depth. Battery performance can also be enhanced by applying low trickle charging prior to eclipse. Batteries are funda-mentally limited to 5–20°C operational temperatures and so are more sensitive than electronics in terms of thermal control. The use of parallel arrays of small batteries offers more graceful degradation than a few large ones.

Buffering can be achieved by using a battery discharge regulator to maintain the power bus at constant voltage during battery discharge. Hence, the peak powers can be delivered by the batteries to minimise the size of the solar array.

Hydroxyl fuel cells are a possibility for energy storage which directly converts chemical energy from a fuel into electricity and are used extensively in manned missions since they produce potable water during electricity generation without pollutant emission and offer high powers for short durations, e.g. the Shuttle orbiter carries three fuel cells, each capable of generating 2–12 kW with a specific mass of 275 W/kg under normal operation and 16 kW maximum at 27.5 V dc. They were also used on the Gemini and Apollo CSM for primary power before the Shuttle programme. Unlike batteries, they store the reactants externally. Each cell outputs 1.2 V. The fuel cell comprises a cathode and anode separated by a water electrolyte which transports ions rather than electrons. Fuel cells convert chemical energy from oxidation to electricity with efficiencies ~80%: $H_2 + O_2 \rightarrow 2H_2O$. They can be used during eclipse and can operate over extended periods for up to ~10 000 h. The fuel cell provides electrical work that is not Carnot limited and which produces only a small amount of heat. They are more efficient in energy storage than rechargeable batteries. They are applicable for high powers ~kW as well as extended operation, and are the most favourable option for electric automobiles which require typically ~10 kW [Hoogers 1998]. Proton-exchange membrane fuel cells are a recent development in which the anode, cathode and membrane electrolyte are integrated into a membrane electrode assembly. Hydrogen molecules are fed to the anode where the molecules dissociate due to the catalytic action of the Pt alloy electrodes. The solid polymer membrane conveys protons to the cathode which is supplied with oxygen to form water due to the catalytic action of the Pt alloy electrodes. The electrons generated flow through an external circuit generating electrical power. The voltage generated per fuel cell is 0.7 V and cells may be combined into a stack. KOH is a common electrolyte for operation between 50 and 150°C and have been used extensively on spacecraft. They can be made regenerable in that they can be recharged by electrolysis when integrated with an electrolyser during sunlit periods with efficiencies ~50–60%. During regeneration, water is electrolysed and hydrogen is produced at the anode and oxygen at the cathode. They have a specific energy density ~0.5 kW/kg when run at 2.6 kW but electrolytic systems for regenerative fuel cells are not yet flight-qualified though they are under development. These energy densities are more than sufficient to run electric motors (as used in robotic systems) which typically require power densities of at least 0.25 kW/kg. Since the voltage from the power source may be variable they require voltage regulation but they offer increased energy densities at higher power requirements with great operational flexibility in terms of peak power capability. They are ~2/3–1/2 times as heavy as NiH_2 batteries and ~1/3–1/5 times as heavy as NiCd batteries for the same power rating. Their major disadvantages are their requirement for active heat rejection and their large volume in comparison to NiH_2 batteries. Their fluid reactants are also difficult to handle if they need to be maintained in zero-gravity.

18.3 POWER DISTRIBUTION

The power distribution bus should be separate from the data/command distribution bus and both comprise separate wiring harnesses. A major part of harness design is in minimisation of wire length to achieve the necessary routing [Larson & Wertz 1992]. The power distribution system includes webbing, fault protection and switching gear. Wiring takes the form of coaxial cables for high-frequency ac transmission or twisted pair wiring for low frequency and dc transmission. The cabling harness interconnects the spacecraft subsystems and accounts for ~10–25% of the spacecraft power subsystem mass. Power converters must convert load voltages in the distribution system such that the loads are isolated from the bus and the power provided to the loads must be regulated. Power regulation is required to dump array overvoltages at exit from eclipse and to regulate battery charging/discharging. A shunt parallel to the array regulates the operating voltage to within a few percentage points. It offers simplified power conditioning and lower bus impedance.

Spacecraft loads may require low to high voltage dc ~5–270 V dc, high voltage single-phase ac of ~115 V_{rms} at 60 Hz, or high voltage three-phase ac of 120/440 V_{rms} at 400 Hz all of which require conversion from the standard spacecraft 28-V dc power bus. The 400 Hz frequency is commonly used on aircraft but is susceptible to acoustic noise and electromagnetic interference due to plasma coupling. Most spacecraft operate at 28-V dc but higher voltages are becoming popular for high power applications and ac especially offers much less massive high-voltage cabling—the distribution wiring harness cross section of three-phase ac is half that of a dc network for power outputs >10 kW. The use of low power high-frequency ac (300–500 kHz) would allow the use of solid-state MOSFET power switching, with little delay in switching speed, as well as allowing the use of simple transformers for voltage conversion to different load voltages. Most spacecraft require low powers <2000 W and for this the standard 28-V dc power bus is usually adopted. For higher powers ~kW, the 28-V bus is insufficient due to resistive losses in the cabling currents. For these higher powers, 50-V dc and 150-V dc are being used as well as high-frequency ac at 20 kHz and 200–300 V. However, power transistors used for shunt regulation are limited to voltages below ~100 V. Although 20 kHz ac has never been used in flight, it has a 20-y terrestrial history and offers low mass, small volume, efficient components. Such high-voltage ac would require the use of closed cycle thermal generators to produce the ac power output, although such generators are twice as efficient as solar cells. Since dc is the standard format for power transmission the 28-V dc bus is adopted for the ATLAS baseline. For redundancy, a dual-bus system is preferred to a single-bus system. A centralised distribution places power dc–dc shunt converters at each load separately for the different voltages at different loads while a decentralised distribution regulates all loads within the power system on an unregulated bus with regulation deriving from battery regulation. The unregulated system is unsuited to full eclipse operation and to missions in excess of 5 y in length. The centralised distribution is regulated and has the advantage over unregulated systems in that it is standardised and suited to larger spacecraft with high power requirements. Fault protection is provided by isolating systems with fuses so that short circuits do not drain excessive power. Power regulation and control is performed to cope with voltage variations ~20% especially between charge and discharge cycles of batteries and eclipse crossing. Batteries are charged independently rather than in parallel to reduce battery stressing which would increase their degradation. Direct energy transfer is a dissipative power control method where power not used by the loads is dissipated as heat through banks of parallel shunt regulators which act as current sinks to avoid internal power dissipation. The shunt regulators are power transistors. Shunt regulators in parallel with the power source and the load shunt away current from the loads when they do not need power. In this way, direct energy transfer can prevent batteries from overcharging. This direct energy transfer method is very efficient and dissipates little energy internally. It has high efficiency at EOL and requires fewer components than other regulation methods and lower mass.

18.4 POWER DISSIPATION (THERMAL CONTROL SUBSYSTEM)

The temperature limits for all spacecraft components must be maintained within their tolerance limits to enable them to perform effectively. The thermal control subsystem is concerned with providing the required hospitable thermal environment by providing paths for transporting and rejecting heat energy to keep components within their required temperature limits. Different systems have different requirements and thermal range limitations: electronics are limited to –10°C to 50°C; solar arrays are limited to –60°C to

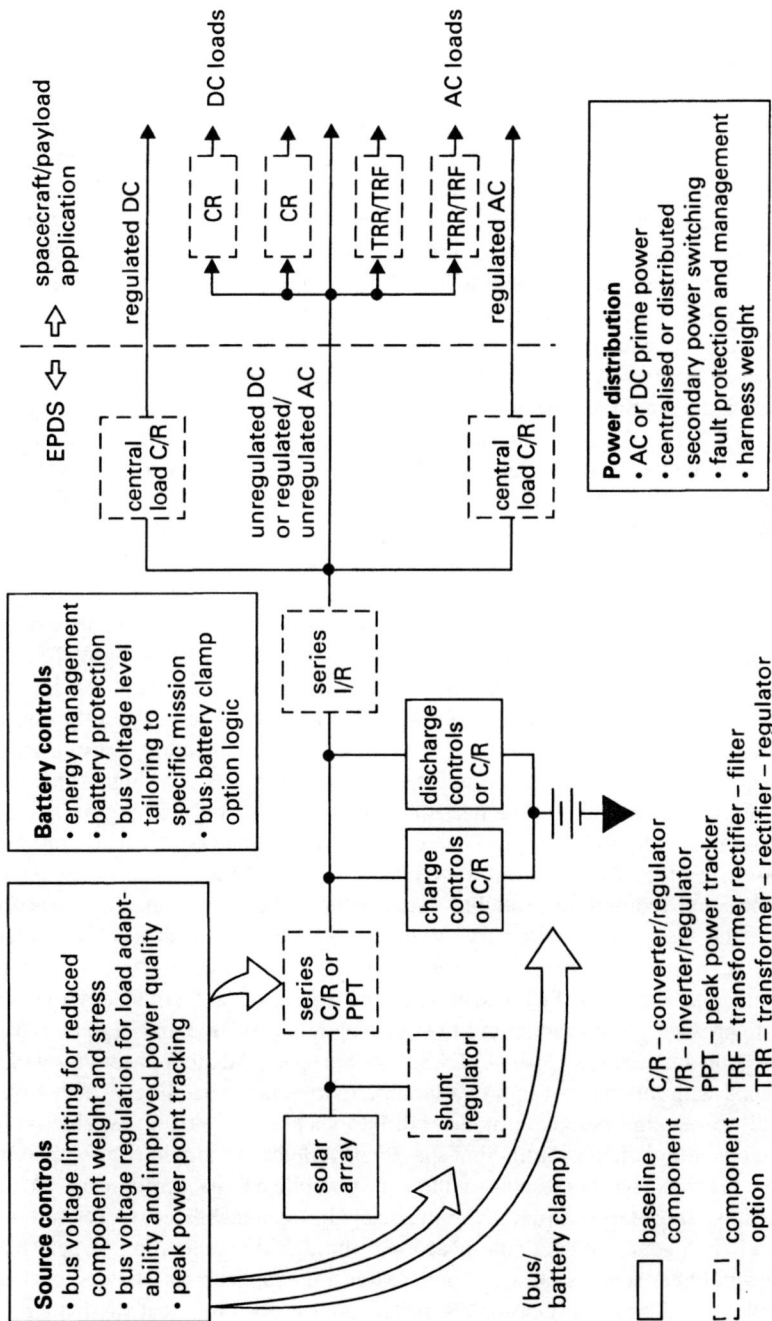

Fig. 18.1. Electrical power system configuration options (from Barter 1999). Reproduced with permission from TRW Space & Electronics Group.

60°C; liquid hydrazine is limited to 10°C to 35°C; batteries are limited to —5°C to 20°C. Indeed, batteries are usually the greatest source of energy dissipation onboard the spacecraft. Hence, batteries impose tighter thermal constraints than electronic devices. Accurate pointing requires the maintenance of temperatures to within ±5°C from nominal temperatures of antennas. Temperature control requires balancing the absorbed and radiated energy at a given temperature usually 25°C (range is typically 120–420 K).

The thermal control design is ultimately determined by the magnitude and distribution of radiation inputs from the Sun and Earth and internal heat sources (particularly if nuclear power sources are used). The number of heat sources to which the spacecraft is exposed are several. There is the direct solar flux of $Q_s = P/(4\pi^2 d^2) = 1358 \text{ W}/\text{m}^2$ from the Sun at 1 AU which radiates as a black body at a temperature of 6000 K with solar output $P = 3.8 \times 10^{25}$ W. Over 99% of the solar energy lies in the waveband 0.275 to 5.0 μm with a maximum at 0.48 μm and 40% of its output comprising visible light in the range 0.4 to 0.7 μm. The solar constant may vary by ~3.5% due to seasonal variations of Earth's ellipticity of solar orbit. Around ~30% of the direct solar flux is reflected and backscattered by the Earth. It is reflected diffusely according to Lambert's law giving an average albedo of 36 W/m^2. This albedo has a small variation ~0.02% due to variable cloud and ice cover. The Earth also acts as a thermal radiator of absorbed radiation at a black body temperature of 254 K generating an infrared emission of 237 W/m^2. Deep space provides a heat sink at a black body temperature of ~3 K. Hence, there are extreme temperature differentials from –160°C to +90°C between eclipse and sunlight. This can cause differential expansion between different regions of the spacecraft at different times. Unmanned spacecraft require special materials with very low thermal coefficients of expansion such as Invar or graphite epoxy. In LEO, atmospheric heating is also a source of thermal energy: indeed at 150 km altitude the frictional heat flux is as high as the solar flux at ~100 W/m^2/h. At 200 km, the heat flux has dropped to only 60 W/m^2/h and at 250 km it drops further to 40 W/m^2/h. The heat input will depend on the spacecraft orbit and the mission profile. Planetary infrared radiation and albedo is dependent on the satellite altitude and requires the satellite view factor to be computed. For a variable altitude orbit, the view factor will vary. View factor decreases with satellite altitude $F \approx 1/(R/R_{Earth})^2$ and this is particularly important for a spacecraft which adopts variable altitude, as the view factor will alter the spacecraft's thermal environment. Whereas Earth proximity is important at LEO, it is negligible at GEO and beyond. Hence, for a variable orbit spacecraft such as ATLAS, variable space radiators or louvres will be required to cope with the thermal input variations. Astronomical satellites which have a wide range of apogee and perigee altitudes fall into such a category of variable orbits, eclipse variations and variable Earth–Sun–spacecraft attitudes. Attitude control also has an impact on thermal control design—a three-axis stabilised spacecraft needs greater protection against short-term variations compared with spin-stabilised spacecraft. The spacecraft itself is a source of variable heat input through the generation of heat from dissipated power. Other sources of heat input are transmitters, batteries, motors, payload, electronic devices and engines. One of the major culprits for high heat dissipation are travelling wave tubes so the use of electronic amplifiers is preferred. Electric motors such as those driving robotic arms are also sources of high heat generation which must be dissipated effectively. For electronics, high thermally conductive but low electrically conductive materials are required to provide effective thermal paths to dissipate heat from these sources (e.g. BeO).

The thermal control subsystem uses passive and active methods for thermal control [De Parolis & Pinter-Vrainer 1995]. Passive control involves no moving parts and are based on controlling conductive and radiative heat paths through the selection of materials and their properties (e.g. insulation, coatings and paints). Consideration needs to be given to degradation due to the space environment. Semi-passive systems have simple temperature-activated controls to open and close conductive paths such as thermal switches. Active systems include dynamic control devices. About 95% of thermal control is passive since these techniques are less massive, require no power and are cheaper. For this reason, spacecraft are usually designed minimising the use of active elements and there is a strong reliance on passive techniques as they are usually sufficient for power dissipation ~100 W which can be spread over thermal panels for radiative rejection to deep space. Hotter spots will at least require heat pipes for thermal redistribution. Electric resistance or radioisotope heaters are sometimes used when instruments are switched off for extended periods to prevent excessive heat loss—thermostatically controlled electric heaters are used on the Shuttle RMS.

All energy absorbed from the environment and generated by the spacecraft must ultimately be dissipated to deep space which is the only heat sink available to compensate for environmental heat fluxes. Radiative heat transfer is the only mechanism by which it is possible to perform heat rejection to deep space. This involves the transmission of energy by electromagnetic waves. Thermal analysis of the spacecraft with respect to its environment is based on the heat-balance equation assuming the spacecraft to be iso-thermal. Heat loss by radiation must balance the heat sources to which the spacecraft is exposed to maintain spacecraft temperature. Radiative heat transfer of electromagnetic radiation from the spacecraft may be reduced by insulation or increased by adopting a radiator. Passive techniques depend on emissivity (relative capability of radiating heat away at a given temperature) and absorptivity (relative capability of absorbing incident energy). Different materials can provide a range of values of the ratio of absorptivity to emissivity from 10 to 0.1. Radiators require a low ratio to produce net outward flow of radiation. Absorbers require a high ratio to maintain high temperatures. Thermal insulation is used to reduce heat flow between regions to thermally isolate components. It comprises several layers of reflective material separated by vacuum so that only thermal radiation can be transmitted. The layers are made from polyimide films with surface layers of metal vacuum deposited on it. These form alternate low-emittance layers and low conductance spacer layers which may be tailored to the required properties:

$$\varepsilon = \frac{Q/A}{\sigma T_h^4 - \sigma T_c^4} \qquad \text{and} \qquad K = \frac{Q/A}{T_h - T_c}$$

both of which must be low for good insulating properties. Such multilayer thermal insula-tion is used extensively on spacecraft to control spacecraft emission and absorption of heat. They usually come in the form of multiple alternate 0.25-mm sheets of radiation reflecting aluminised Mylar or Kapton netting separated by Dacron mesh for insulation spacing. Kapton and Mylar have similar performances, but the more expensive Kapton is used (usually for the outer layers only) when flammability is a consideration.

Table. 18.2. Thermal insulation properties

	Density (g/cm^3)	Tensile modulus (N/mm^2)	Tensile strength (N/mm^2)
Mylar	1.40	4000	150
Kapton	1.42	2500	230

An outer covering of fibreglass is usually included to protect against atomic oxygen attack and micrometeoroid penetration. The effectiveness of multilayer insulation depends on the size of object being insulated. Larger objects exhibit lower effective emittance than smaller objects due to edge effects such as joints and seams which tend to have high thermal flow rates. For edges, the drugstore wrap seam offers the best perform-ance in minimising thermal leaks. The thermal conductivity of multilayer insulation with 20 layers per cm ~0.0004 W/m K (compared with glass fibre and polystryrene ~0.032 W /m K).

To maintain constant temperatures, emissivity e must balance absorptivity α—this is the Kirchhoff law $\alpha = \varepsilon$. This defines a grey body (of which the black body is a limiting case): $\alpha = \varepsilon = \text{const} \leqslant 1$. The total energy emitted per unit time per unit area of the spacecraft is given by the Stefan–Boltzmann law of radiative heat transfer, $Q = \varepsilon \sigma T^4$ where σ = Stefan–Boltzmann constant = $5.7 \times 10^{-8} \text{W}^{\prime}/\text{m K}^4$. The amount of emitted energy is equated to the amount of absorbed energy:

$$Q_s \alpha A_p = \varepsilon \sigma T^4 A$$

where A = total area = $4\pi R^2$, and A_p = projected area = πR^2.

$$\rightarrow T = \left(\frac{Q_s \alpha A_p}{\varepsilon \sigma A} \right)^{1/4} = \left(\frac{Q_s \alpha}{4 \varepsilon \sigma} \right)^{1/4} = \left(\frac{1358\alpha}{4(5.7 \times 10^{-8})\varepsilon} \right)^{1/4} \tag{18.1}$$

At room temperature $T = 290$ K, for a spherical spacecraft $\alpha/\varepsilon \sim 1.2$ (nearly a grey body). In general α and ε are functions of wavelength and depend on the spectrum of the absorbed and emitted radiation. Thermal coatings are selected according to their emittance and absorptance to achieve temperature balance. For spacecraft, α is defined with respect to solar energy at visible wavelengths and ε is defined at infrared wavelengths. Surfaces which reject radiant heat should have high reflectivity in the visible waveband to reduce the absorptance α of sunlight and have high emittance ε in the infrared waveband. It is the α/ε ratio which defines the characteristics of the surface and it can vary from near zero up to 10. Two or more coatings can be combined to provide the required properties. For polished Al, $\alpha = 0.2$ and $\varepsilon = 0.031$ hence $T = 443$ K. White paint absorbs weakly in the visible waveband but has high emissivity in the infrared waveband $\alpha = 0.2$ and $\varepsilon = 0.9$ so $T = 190$ K such that it has a cold coating. White paint is used for large surfaces such as the antenna dish but for long missions ~10 y a secondary surface Ag or Au mirror coating is required due to white paint degradation. Gold has a warm coating, since it absorbs solar flux due to its high reflectivity, and has a low emissivity $\alpha = 0.25$ and $\varepsilon = 0.045$ so $T = 430$ K. The optical solar reflector combines the Au surface with a thin transparent cover of high emissivity $\alpha \sim 0.15$ and $\varepsilon \sim 0.8$. Such second surface mirrors combine the high reflectivity/low absorptance of metallic surfaces with the high infrared emissivity of fused silica. The optical solar reflector is very expensive and a less expensive alternative is the second surface mirror comprising silver-coated Teflon with $\alpha \sim 0.08$ and $\varepsilon \sim 0.66$. Such coatings are degraded by solar ultraviolet radiation exposure beyond 2 y to $\alpha/\varepsilon \sim 0.9$. Infrared emissivity is not greatly affected as radiation damage is on a scale associated with higher wavelengths.

This analysis allows an assessment of whether a radiator to deep space is required and its size from $Q_{radiator} = A(\varepsilon \sigma T^4 - \alpha Q_s)$. Usually, the radiator is mounted on the spacecraft panels and radiate on one side only. Coatings and insulation are suitable only for fixed altitude spacecraft for which the thermal environments will be relatively fixed and reasonably well known. For variable-altitude missions, bimetallic sensor activated space radiator structural louvres can radiate heat to space. They comprise a series of polished Al blades arranged in Venetian blind configuration over a high-emittance radiator baseplate (low α/ε). Open louvre blades offer low α/ε while closed louvre blades offer high α/ε. Opening and closing of the louvre blades by bimetallic actuator springs measure theradiator baseplate temperature and modulate the angle of the blades to vary α/ε to give variable radiative heat flow to space from the baseplate. Louvres can thus provide a heat sink during solar illumination and act as heat capacitance devices during eclipse periods. For example, at 300 K open blades can reject heat ~ 50 W/m^2. They do require an adequate viewfactor to space and are generally limited to 5°C accuracy of control. The radiative viewfactor is the fraction of radiant energy directly incident on the viewing surface relative to the radiant energy leaving the surface.

Thermal management involves the redistribution of heat within the spacecraft. The spacecraft itself is used as a heat sink. Convection is the heat transport mechanism with the highest flux capacity but, since it is absent in space, only thermal conduction and radiation are available for such redistribution. In fact, internal thermal management is usually accomplished by thermal conduction. However, some Russian unmanned spacecraft sometimes use fans to force convection internally to augment conductive heat transfer but it is difficult to model. Thermal conduction between one region and another occurs if there is a thermal gradient imposed between the regions: $Q = -KA(dT/dx)$ where K = thermal conductivity. This is Fourier's law of heat conduction. Thermal conductivity of Al is 210 W/mK while Al alloy has thermal conductivity of ~ 150 W/mK (compare with Ti with $K = 20$ and Teflon with $K = 0.25$). Electronic components will require dedicated heat sinks to dispose of locally generated heat. A high thermally conducting Al alloy, such as 1050 tempered Al, offers a 70% increase in thermal conductivity over conventional Al. Such materials can be used as cold plates for mounting devices and instruments. Cold plates can act as heat pipes if they are constructed with fluid passages within the plate to transfer waste heat to an integrated radiator thereby acting as part of the structure. This type of heat sink is often used for electronic components. Heat pipes offer highly conductive paths for thermal management with no moving mechanical parts (see Fig. 18.2). They use phase change materials which store thermal energy as latent heat and are used in the range −5 to 40°C. They can provide the means for utilising fluid loop thermal bus networks for power dissipation $\sim$kW between small temperature differences such as occur in electronics units. A working liquid such as NH_3 partly fills

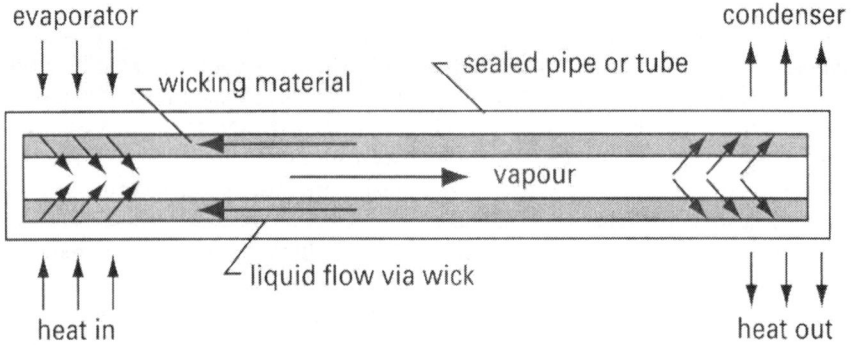

Fig. 18.2. Heat pipe (from Barter 1999). Reproduced with permission from TRW Space & Electronics Group.

an Al tube with an axially grooved cross-section ~10 mm diameter along which lies a sintered nickel wick saturated with the working fluid. The liquid absorbs latent heat and vaporises into a gas which carries the heat to a cold radiator due to the induced pressure gradient. The latent heat provides a large thermal inertia of constant magnitude. At the cold radiator the vapour condenses, surrendering the latent heat of vaporisation. The loss of liquid by evaporation results in a liquid/vapour interface and a capillary pressure is developed at the wick surface. This capillary pressure pumps the condensed liquid back to the hot region for evaporation in a continuous process. The heat pipe thus acts as a thermal conductor of high conductivity but can act only in one direction with heat flow $Q = C[(T_{evap} - T_{vap}) + (T_{vap} - T_{cond})]$, where C = pipe thermal conductance. This axial heat flow decreases as the cube root of the pipe length annd is independent of the pipe diameter. Variable conductance heat pipes may be constructed by controlling the operating temperature against the variations in thermal conditions. This may be achieved by interrupting the fluid flow by thermostatically operated valves and through a reduction in the condensation rate by introducing a non-condensible gas into the condenser. Heat pipes are one of the main methods for thermal management of localised hot spots. Heat pipes are often used for electrical devices which generate short bursts of power. They offer 200–300 times the energy transfer of a solid copper bar and have ~130–650 W/kg capabilities—a 1.27 cm diameter pipe has a capacity of 5 kW m. They are also often used with thermostat heaters to maintain battery temperatures and prevent fuel from freezing around propellant lines and valves in cold biassed regions.

Electric heaters are normally of the wire-wound resistance type, controlled thermostatically. They may be operated in a proportional control mode with temperature sensors to provide variable heat generation. They are particularly required for gimbals and joints which use motorised actuators. Coolers are required for certain sensors and subsystems. They are typically reversible heat pumps which transfer thermal energy from lower to higher temperature and so require additional energy. These are normally of the thermoelectric variety and utilise the Peltier effect which accomplishes heat transfer when an electric current flows between dissimilar metals. Thermoelectric heat pumps are usually used for tight temperature control of low power instruments.

Thermal analysis software uses either finite difference (FD) or finite element (FE) analysis. FD packages are based on network analysis similar to electrical networks and offer simplicity. FE packages are more detailed and may be coupled to structural analysis (e.g. NASA's NASTRAN). SINDA is an electrical network type thermal package which has been updated by ESA as ESATAN and is one of the commonest thermal modelling packages. The spacecraft is usually initially modelled as thin plates with solar aspect coefficient of 0.25. Thermal vacuum tests are required to test spacecraft thermal characteristics since models tend to be highly simplified as the thermal environment is difficult to predict with any accuracy. Orbital characteristics are a source of much uncertainty and variability due to eclipsing. If the orbital altitude varies there is a consequent change in the thermal environment from the Sun, Earth and deep space. Thermal tests in a vacuum chamber are standard for spacecraft. The thermal balance test simulates the operational thermal environment to test the thermal design while the thermal vacuum test at <10-5 torr verifies that the hardware meets performance requirements. A thermal balance test is expensive due to the extensive 24-h manpower required over a period of weeks (~100–200 h). It tests the thermal finite difference model of the spacecraft which defines the thermal conductance between nodes and the thermal radiation between the environment and the spacecraft. Such tests use Xenon discharge lamps for solar simulation and a cold shroud at 100 K to simulate deep space. A temperature difference of 60oC corresponds to two orders of magnitude difference in exposure rates, so a 100-h test can equate to a year's exposure in space.

18.5 POWER SYSTEM RELIABILITY

The power subsystem is a critical candidate for reliability and requires that no loss of any single component should result in more than 10% power degradation [Billerbeck & Baker 1984, O'Sullivan 1994]. Loss of power implies loss of the spacecraft so functional redundancy is required. To limit failure propagation, blocking diodes on the high voltage side of the supply lines, between the main power bus and the slip rings that feed power from the power source, prevent a single grounded slip ring from shorting out the main bus and prevent large battery currents from feeding back into grounded slip rings. For redundancy, redundant bus wiring harnesses with separate connectors for each bus should be employed. The Seasat spacecraft was lost due to a short from the single primary bus after 100 d of operation. A dual primary bus can be implemented in several ways. It is possible to have two identical power systems each supporting the complete load, e.g. Intelsat V and Intelsat VI. It is possible to employ separate TT&C systems on each bus to independently control and monitor all areas of the spacecraft. Hence, if one main power bus fails a complete TT&C capability is available for failure diagnosis and to command workarounds. This is the best approach for spacecraft housekeeping. For some payloads, however, which have a few large power loads, this is not appropriate (e.g. dual manipulators). The double insulated main bus concept involves using coaxial cable lines comprising a central conductor surrounded by an insulation layer, a metal braid and a second insulation layer. Hence, if one layer of insulation is shorted out normal operation can still continue. All loads should be fused to limit currents. However, instead of single fuses on the input power connection a parallel fuse with a series diode could be employed to provide alternative current paths. Solar arrays may be designed in modules and cross-strapped. Batteries need to employ some form of redundancy also but complete battery redundancy is costly in terms of mass. Hence, cell redundancy can be employed with little increase in battery mass. This enables repairs to be undertaken if there is a failure in the main primary system.

19

Spacecraft structural configuration

The structural subsystem mechanically supports and protects all the other spacecraft subsystems, attaches the spacecraft to the launcher and provides mountings. Structural performance is based on its provision of minimal distortion from mechanical and thermal loads. The primary requirement is that it supports these functions with minimal mass penalty [Larson & Wertz 1992]. It typically comprises a central thrust cone, usually of aluminium. The primary structure carries the spacecraft's major loads (e.g. launch loads) and provides structural stiffness while the secondary structure represents brackets and lugs to mount components such as support wires, propellant lines and components of mass less than around 5 kg to the primary structure. Rectangular components such as electronic devices and cylindrical components such as actuators use integral lugs to attach to honeycomb mounts attached to longeron frames (semi-monocoque structure). More complex shapes can be mounted using brackets. Components may also form part of the load-bearing structure where appropriate. The primary structure is determined by the launch loads while the secondary structure depends on the integration of other subsystems. The structure must function in environments which vary through the lifecycle from the manufacturing environment to EOL including hostile conditions such as welding, transport delivery by sea or air, launch and ascent, and all the conditions imposed by the space environment and operational circumstances.

The nature of the payload and mode of spacecraft attitude stabilisation most strongly influence the spacecraft configuration. Sensors require specific FOV (field of view) and pointing accuracy. Antennas must not obstruct sensors. They also require a specific FOV, such as to Earth or to GEO orbit, so sensors and antennas may require gimballing. Magnetorquers and magnetometers must be separated to avoid magnetic interference (typically boom-mounted). Avionics such as the C&DH components are highly vulnerable to the space environment so they should be mounted in the centre of the spacecraft for maximum shadow shielding but this conflicts with the in-orbit servicing requirement of easy availability of circuit boards. The propulsion system performs orbit transfer and the rocket engine modules should be placed aft of the spacecraft centre of mass and away from plume-contaminable sensors, antennas and solar cells. The thrust vector should be aligned with the spacecraft centre of mass. Batteries should be accessible for prelaunch test and replacement.

19.1 MATERIAL SELECTION

The spacecraft exploits a wide variety of materials with the strong constraint of low density to minimise mass. The spacecraft is composed of both metallic alloys and non-metallic composite materials. The most important considerations for materials selection are the specific values for fatigue strength and stiffness, thermal expansion, corrosion resistance, ease of fabrication and cost. Al alloy is the commonest structural material used for aerospace structures, together with possibly Ti, and has a stiffness/mass ratio equivalent to steel and a strength/mass ratio in excess of that of steel. Al has a low density, high strength-to-weight ratio, high electrical conductivity (~65% that of copper), good thermal conductivity resistance to corrosion and ease of fabrication. It is applicable to highly loaded structures from cryogenic to high temperature regimes up to ~200–300°C. Al alloy has a yield strength that is characterised by an standard deviation spread σ, so to minimise the probability of failure the 3σ strength should be used. All in all, Al alloy is used extensively for components and panelling and even in electronics (especially in aerospace applications) due to its diversity of use—generally, over half of the world's Al consumption derives from direct competition with traditional steel and copper usage. Al–Li alloys offer similar strengths to Al but with higher stiffness and several per cent mass reduction. Steel is used usually for parts of the propulsion system. Stainless steel has a high strength and temperature resistance and is resistant to corrosion. It is especially suited to cryogenic functions as it retains its ductility at low temperatures but it suffers from a high density. Austenitic stainless steel is commonly used for propulsion system structures and cryogenic storage structures. Ti alloys have a high specific strength and a similar stiffness to Al. They have good high-temperature capabilities up to ~450–550°C and are corrosion-resistant. They are often used for pressure vessel structures, but they are difficult to weld due to their susceptibility to crack formation. Mg alloys offer higher specific stiffness but require coating to prevent exposure to the atmosphere which corrodes such alloys. Be alloys offer the highest stiffness, high strength, high hardness and high corrosion resistance with low density. They have a high temperature tolerance with good thermal conductivity but low thermal coefficient of thermal expansion. They offer increased natural frequencies to structures, reducing potential resonance problems. They are primarily used for drive and pointing mechanisms which need accurate moving parts.

Many instruments are constructed from Be or boron carbide B_4C. The Young's modulus of both is very high. Unfortunately, Be is toxic and relatively brittle. Inconel alloys (Ni, Co, etc.) are used for high-temperature applications such as heat shields and rocket nozzles. They have a high density similar to steel but can be made into thin films with minimum mass penalty. High temperature refractory metals have high density and lack ductility (e.g. tungsten, tantalum, molybdenum). Ceramics are materials which are hard but brittle. They are poor conductors of heat and electricity, making them suitable for high temperatures. They are resistant to chemical attack, so are suited to corrosive environments. Many ceramics are metal oxides, e.g. Al_2O_3. One naturally occurring example is china clay (kaolinite, $Al_2SiO_2O_5(OH)_4$). SiC ceramic has high stiffness and low mass but is brittle and suited to only simple structures. SiC is one of the hardest substances known. SiC is a semiconductor material and offers the potential for future use in electronics for operation under adverse conditions and high-temperature conditions. SiC doped with N forms an n-type semiconductor and SiC doped with Al forms a p-type semiconductor. This allows the construction of transistors for operation at high voltages (SiC bandgap is 2.9 eV compared with Si's 1.1 eV bandgap). SiC manufacturing is under development to reduce defects generated during crystal growth. Another very strong ceramic is Si_3N_4 which can form prefabricated shapes. Niobium is a temperature-resistant material to ~1300 K with a density similar to steel, but it requires a silicide coating to prevent oxidation with the air.

Polymers have low mass, low coefficient of thermal expansion and linear stability but suffer from poor load-bearing characteristics. They are usually used for adhesives and multilayer insulation. First-generation polymers were thermoplastics and thermosettings which had low mass and low coefficients of thermal expansion but low load-bearing characteristics. Thermosetting polymers lack flexural and impact properties: during cross-linking, polymers are changed from soluble reactants into cross-linked resins, e.g. epoxy resin is one of the most common thermosettings. Thermosettings are used for load-bearing but cannot be re-melted due to their branch chain structure. Thermoplastics lack modulus and so tend to be of low rigidity, e.g. polyesters and polyimides, e.g Dacron fibre and Mylar film, so are not often used as binding resins. Thermoplastics can be re-melted and re-shaped or welded. They suffer from depolymerisation in radiation environments, but doses of ~10^6 Gy are required. The thermoplastic PVC (polyvinyl chloride) and PTFE (polytetraflouroethane) are particularly susceptible. The most commonly used plastic film is Mylar. It is a strong, transparent polymer that can be fabricated into sheets and films several micrometres thick. It is usually coated with a thick layer of Al for reflectivity and is used extensively in multilayer insulation. The polyimide Kapton has a higher strength and greater temperature tolerance and is often used for the outer layers of thermal insulation blankets. Unfortunately, it suffers severely from oxygen corrosion so Teflon coating is usually adopted. Fibreglass cloth is also used as a cover to protect against micrometeoroids.

Composite materials are heterophase systems of two or more components and their properties differ from those of the individual components [Eaton & Slachmuyders 1988]. Composites are used extensively for struts and beams but may also be used for panels. They offer superior structural properties over monolithic materials characterised by high specific stiffness E/ρ and low coefficient of thermal expansion. The matrix component offers high compression strength but low tensile strength, while the fibre offers low compression strength but high tensile strength. The matrix is usually an isotropic polymer impregnated with fibres. Composite properties are highly anisotropic in that they differ between the along-fibre direction and the across-fibre direction. The fibre direction may be unidirectional, orthogonally laid or a random multidirectional ply, this last choice offering homogeneity in properties. A chief difficulty of composites is that they tend to be brittle, but they offer superior fatigue resistance to metals. However, they do not suffer from vacuum-welding problems inherent in metals. The commonest advanced composite is graphite fibre in an epoxy resin matrix for high specific strength and high specific stiffness E/ρ, double that of metals. They have low coefficients of thermal expansion and high resistance to thermal cycling between 180°C and 200°C. Second-generation polymers are the particle-reinforced and fibre-reinforced plastics which offer high stiffness, high rigidity and low thermal expansion coefficients and low mass. They are used for basic structural elements. One of their difficulties is that their mechanical properties tend to deteriorate at increased temperatures and are generally limited to less than ~200°C (melting points typically ~260°C). They have laminate properties in direct proportion to fibre strength. Some versatility is offered through variable stacking sequence ply lay-up, fibre fraction and fibre length. Other possibilities for the matrix are graphite in polyimide and graphite in polyethersulphone matrices. Other possibilities for the fibres include boron, Kevlar, aramid and glass. Glass-reinforced plastics using epoxy matrices can form complex shapes with good structural properties and are the most widespread composite used outside the aerospace industry. The glass is typically S-glass—64% silica, 25% alumina and 11% magnesium oxide.

Graphite-epoxy offers high specific strength and stiffness. Its E/ρ ratio is around twice that of metals generally. This gives them good damping characteristics. Its low coefficient of thermal expansion minimises thermal loading stresses which arise from thermal distortion. It is a good structural material with low density and has a higher temperature capability than glass-reinforced epoxy. They are generally used for struts and mountings. The Ariane IV SPELDA fairing for dual satellite launches is composed primarily of carbon-fibre-reinforced plastic. Epoxies offer fine adhesion to the fibre in the matrix but suffer degraded mechanical properties from radiation exposure due to depolymerisation of the resin. Outgassing characteristics of composite materials are based on the matrix properties and they do not decompose appreciably below the material cure temperature (typically, ~200–300°C). Atomic oxygen attack is another problem for carbon polymeric materials in general. Around 100 μm of epoxy composite can be removed in 5 y of exposure to atomic oxygen flux (equivalent to one layer of ply). Thermal cycling instability can also be problematic as epoxy resin exhibits a 30% reduced stiffness and strength at 180°C, but the replacement of epoxy resin with cyanate esters appears to offer better resistance to microcracking from radiation exposure and thermal cycling from –180°C to 130°C. Siloxane composites which use Si rather than C may be used as matrices which are resistant to attack from atomic oxygen and to electromagnetic radiation exposure. Atomic oxygen converts Si to nonvolatile SiO_2 which acts as a passivating layer inhibiting further oxidation.

Metal matrix composites combine metallic matrices with ceramic reinforcement. One example is Al stiffened by fibres of boron or graphite for general use, or with SiC for severely high temperatures [Dermarker 1986]. They are particularly used for truss structures and antennas and offer greater stability an high thermal deformation resistance to thermal cycling over fibre-reinforced plastics in the range 300–500°C. Reinforcement volume fractions are typically 50%. The ratio of longitudinal to transverse strength varies from 2 to 15 but continuous reinforcement may be realised by combining fibres and stacking sequences for very high isotropic stiffness. They do not outgas, unlike polymer resins. Reinforced metallic composites are reinforced by short fibres, whiskers or particles, depending on the temperature of operation. Room-temperature favours particles, intermediate temperatures favour short fibres while high temperature favours whiskers. The Al matrix with boron fibres offer highly favourable mechanical characteristics but Al matrix may also be reinforced with SiC, SiO_2, Al_2O_3 or C, e.g. Nicalon comprising Al matrix and 50% SiC–SiO_2. Such materials, however, have a high cost of fabrication.

Metal–metal composites, comprising a bulk metal matrix with filaments of another metal, offer stronger and lighter composites than standard alloys and conventional composites. An Al-based metal–metal composite offers tensile strengths ~1200 MPa due to the formation of inter-metallic compounds at the interface between the matrix and the fibre. They are highly resistant to heat, and offer the flexibility of metals. For example, Cu/Nb wire of high strength in a Sn matrix offers 2000 MPa tensile strength with high conductivity. The fibres stop dislocations propagating through the matrix. Metal matrix composites are not much used, due to expense, but they would be suitable for a variety of structures including panels.

Carbon–carbon ceramic composites comprise graphite fibres in a carbon matrix and this offers high temperature stability, low thermal expansion and high resistance to oxidation at high temperatures. It is suitable for high temperature regimes but is not suitable for load-bearing structures. It is used on the US Space Shuttle for the nose cone and leading wing edges for temperatures ~2500–3000°C experienced during re-entry. SiC fibres would offer better resistance to oxidation but are costly. Indeed, ceramic and metal matrix composites are very expensive to produce and tend to be used for specialised purposes. Harnisch et al (1998) have described a new C/SiC ceramic material developed by ESA for ultra-lightweight monolithic structures. Randomly oriented SiC-coated C fibres in a phenolic matrix can be used to construct complex structures with a Youngís modulus of 270 GPa and a density of 2.7 g/cm^3.

Palsula (1993) suggested that third generation molecular composites may be constructed from flexible coil polymer macromolecules which are reinforced by molecularly dispersed rigid rod macromolecules. The molecular composite is homogeneous in that the molecules are distributed isotropically. The molecular composite retains the thermomechanical properties of the matrix while attaining the mechanical strength of the rigid polymer in all directions. It eliminates the fibre/matrix interface which can suffer from poor adhesion. Applications include vibroacoustically stable structures which utilise their passive damping characteristics of the high-strength, rigid molecule.

Spacecraft materials also have a tendency to outgas adsorbed gases when exposed to low pressures at $<10^{-3}$ torr, and the outgassing is proportional to the surface area and so radius. The emitted flux creates a self-atmosphere. The mean free path dictates the distance between collisions and specifies the distance at which 64% of the outgassed molecules are scattered, of which a large fraction ~40% will return to the original surface. Al typically results in the loss of 10 cm^{-5}/y at 550°C. Plastics suffer much loss: PTFE, Mylar and Nylon suffer 10% mass loss per year at 880°C, 200°C and 100°C respectively. Graphite-epoxy adsorbs water from the atmosphere which is released and alters the material properties, causing structural distortion. Material performance may be improved by prior baking in an inert gas followed by coating to prevent re-adsorption.

19.2 SPACECRAFT STRUCTURAL ANALYSIS

Internal packaging is important for compactness and ease of repair and several approaches are possible [Larson & Wertz 1992]:

(i)   dual shear plate—flat honeycomb plates are bolted to the carrier and outer shear
      plates are inserted into the bus frame from the outside. Both shear plates are bolted
      to the bus. This packaging approach is rigid and sturdy with highly efficient
      packing density. Thermal transfer capability is good but there is a large piece count.
(ii) Inner shelf—inner flat shelves or bulkheads are mounted orthogonally to the axis of
      a cylindrical or polygonal spacecraft as supports for components particularly elec-
      tronics. It it less volumetrically efficient but it is the most adaptable to standard
      electronics. Heat rejection is a problem since components are mounted near the
      centreline away from the walls.
(iii) Skin panel/frame—the basic structural frame of the bus is closed with panels that
      form part of the load-bearing structure. Panels are hinged along one side to swing
      out for easy access. The panels provide mounting structures for electronics on the
      plates directly for thermal control. It is less rigid and less efficient than the dual
      shear plate approach but offers ease of access for repair.

The spacecraft size is limited by the dimensions of the payload shroud fairing of the
launch vehicle that protects the spacecraft during launch. The payload fairing is typically
cylindrical. However, if the standard payload fairing is unsuitable many expendable
launchers may develop special fairings at cost. Ariane IV fairings come in modular sizes
with a standard fairing diameter of 2.9 m: 8.6 m, 9.6 m and 11.1 m in length for a single
launch. The equivalent SPELDA fairing dimensions for lower section dual launch are
3.3 m, 2.8 m and 1.8 m. The cargo bay of the US Space Shuttle is also cylindrical with a
length of 18.3 m and diameter of 4.6 m. The Shuttle pricing policy makes vertical
mounting cheaper than horizontal mounting so that short fat designs are preferable. There
does exist the possibility of an aft cargo container being mounted behind the external tank
for larger payloads but this requires the tank to be carried into orbit rather than being
jettisoned and this imposes a performance penalty. Such limitations on the spacecraft
geometry and size introduce the necessity for deployable structures. Deployable elements
must be stowed or folded during launch for later deployment in orbit. The operating
profile rate determines the deployment rate and the maximum angular acceleration:
operating torque, $\tau = \alpha I$ plus 20% friction. Solar panels should ideally be body-mounted
but high power requirements necessitate the use of deployable arrays. Solar arrays may
be flat rigid panels that fold or flexible arrays that roll up. Deployable booms come in
several types. The simplest and heaviest is the long rigid boom of tubular construction
with one or more hinged joints. The boom is folded for launch. In space, springs deploy
the boom. Typically the stowed length is one-third of the deployed length. The use of
composite materials keep the natural frequency as high as possible. The astromast boom
is 15 m in length deployed and 1 m stowed. Three fibreglass longerons are stiffened at
intervals by fibreglass interstruts. The boom is deployed by strain energy in the coiled
longerons. Be–Cu cables are attached to a motor to control the deployment rate. The
cycle rate is limited. Stem-type booms comprise of metal/composite strips welded along
the edges. The strips are stored by rolling up on a wheel and stow like tapes. Strips return
to their original cross-section on deployment. They become rigid at high L/A (load per
unit area). They are capable of many cycles and high repeatability precision, e.g. manipulator arm on the Viking lander on Mars.

It is necessary to reinforce areas which have high strain energy to stiffen structures. Tensile loads only occur in the primary structures via joints or attachments. Compressive loadings that act directly on the primary structure are the most catastrophic failures in that they can cause buckling. The spacecraft structure includes skin panels, trusses, brackets, boxes and pressure vessels. Monocoque Al honeycomb skin panels provide a shell strucure and do not have attached stiffening and are suitable only if loads are uniformly distributed. Semi-monocoque panels are Al honeycomb skin panels with stiffness imparted by longitudinal frame members (often of Ti) distributed at intervals. Sandwich structures are lightweight and provide a shear resistant core bonded to two outer face sheets of Al. Sandwich panels act as I-beams for structural stiffness. The core is usually either corrugated or honeycomb cells built from ribbons of Al with longitudinal stiffness greater than transverse stiffness. Thermal insulation requirements tend to favour the corrugated core over the honeycomb core. Antenna support structures are often constructed from carbon-fibre-reinforced plastic tubes. A truss is an assembly of axial members which are stable under applied loads—they are particularly suited to manipulator link construction (e.g. the ROTEX arm was constructed from composite trusses).

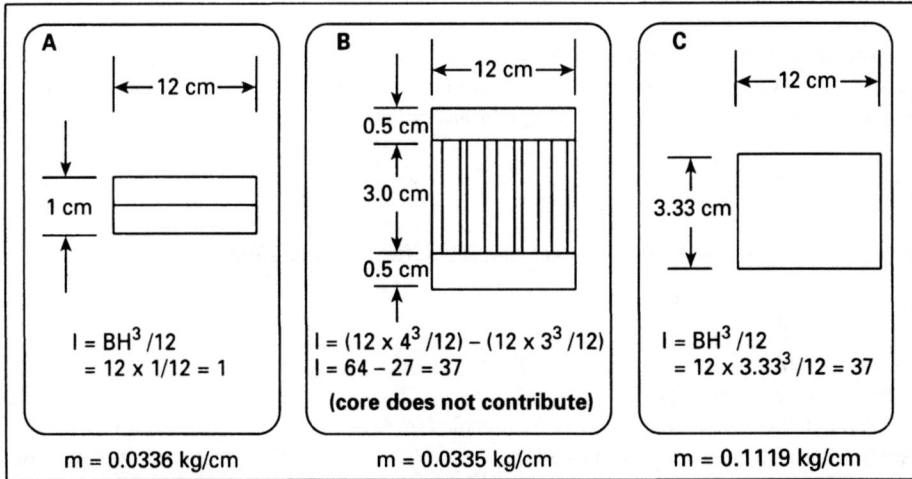

Fig. 19.1. Sandwich and monocoque construction example (fromWertz & Larson 1999).

The spacecraft is subjected to major mechanical loads during the launch for short periods ~10 minutes whereas the mechanical loads in orbit during the $\geqslant 10$ y operational life are low. These launch loads necessitate the incorporation of a thrust cone as the kernel of the structural design of the spacecraft. Hence, the spacecraft comprises the central tube or cone as the main load-bearing part of the structure within a surrounding box-type structure with the propulsion system typically being located within the thrust cone. Typically, the central structure has a semi-monocoque Al construction while the panels of the box are constructed from Al sandwich panels. The launch loads comprise a steady-state load and a simple harmonic component at the natural frequency of the launcher structure. Axial loads are caused by the launch vehicle ascent and lateral loads through steering and wind gusts. The engine thrust exerts a steady-state linearacceleration along its thrust axis which increases as the booster is depleted of fuel and this thrust is transmitted through the launch vehicle interface to the payload. Thermal loads also occur due to atmospheric heating of the nose fairing during ascent. Shuttle longitudinal accelerations are limited to less than 3 g while that for Ariane IV may be as high as 5 g. Typically, the maximum axial acceleration ~6 g and the maximum lateral acceleration ~3 g. The primary structure is generally modelled as a cylindrical tubular beam which is mass loaded by the total launch mass. The lateral load produces a moment causing compression at the end of the beam. The addition of the axial load gives the size of the primary structure. The launch vehicle's natural damping frequency responds to engine oscillations and aerodynamic effects such as buffeting during ascent and is typically 5–100 Hz. The spacecraft structure must avoid the launch vehicle's natural frequency of oscillation and other possible sources of resonance, e.g. interaction of the spacecraft and launch vehicle control system. The resonant frequency occurs when the largest amplitude of displacement occurs for a given amplitude of input. It occurs when any additional perturbing energy is retained by the system so that the vibration amplitude increases theoretically to infinite values. Damping provides the means to dissipate the energy as heat. The magnitude of damping increases the dissipation rate by increasing the range of resonant frequencies. The dynamic response is determined by the natural frequency and the mode shapes of the natural frequency:

$$f_n = \frac{1}{2\pi} \sqrt{\frac{k}{M}} \tag{19.1}$$

where $k = Mg/\delta$ = stiffness constant, and δ = deflection. For a monocoque shell

$$f_n = \frac{1}{2\pi} \sqrt{\frac{EI}{\mu L^3}} \tag{19.2}$$

where μ = mass/unit length.

The axial and transverse natural frequencies of Ariane IV and Space Shuttle launchers are similar at 30–35 Hz and 10–15 Hz respectively. Hence, the spacecraft payload must have natural frequencies exceeding 35 Hz in the thrust axis and 15 Hz in the lateral axis (give the different mounting configurations for the two launchers). Transients are generated at various phases during the ascent to orbit particularly at multistage separation and fairing jettison which are transmitted as random acoustic noise via the shroud: at engine burnout, at staging due to pyrotechnic shocks from explosive bolts at stage separation, and at next stage ignition.

The first staging transients are usually the highest load due to the greater thrust/weight ratio. Lateral transients are also generated by aerodynamic effects. Random vibrations are also generated from the engines. All liquid propulsion vehicles impose high acoustic vibrations exceeding the linear accelerations. The Space Shuttle exhibits the worst structural vibration effects since the payload is mounted near the engines. At lift off the acoustic noise is highest due to noise reflection from the ground. Acoustic noise is also generated in the transonic regime. Structures with high surface area and low mass tend to respond unfavourable to acoustic noise. The Ariane IV power spectrum yields a maximum of 140 dB at 500 Hz during launch, particularly during transonic flight, and this may be approximated by a spectrum with an rms acceleration of 7.3 g. Shock loads due to pyrotechnic bolt firing can be as high as 20g in the kHz range. The steady-state and transient accelerations comprise the initial loads. All materials have a fatigue life that depends on the stress levels and the number of load cycles but typically reliability is ~99% for structures. Failure may occur through yielding or buckling. Designs which accommodate Ariane IV environments will generally be acceptable for the Shuttle environment. The interface to the launcher is through mounting points which form load paths. Such mounting tends to decouple the payload from the launcher. However, the decoupling is not complete. The system may be regarded as a free (booster) – constrained (payload) or as a constrained (booster) – constrained (payload) dynamic system which may be analysed through component mode synthesis [Engels et al 1984]. The Shuttle cargo bay has a series of points along two longerons at the sill and along the keel to allow the 50-kg spacecraft adapter to be mounted into the Shuttle. The Shuttle attachment system minimises loads from the Shuttle by allowing one degree of freedom at each attachment point. The payload itself dampens vibrations by virtue of its own mass and low mass payloads tend to exhibit high peak accelerations.

The primary structure of the spacecraft is essentially a thin-walled structure. The dynamic properties of the structure are governed by three parameters:

Young's modulus of elasticity

$$E = \frac{\sigma}{\varepsilon} \tag{19.3}$$

where $\sigma = F/A =$ stress, and $\varepsilon = \Delta l/l =$ strain.

Poisson's ratio

$$v = \frac{\varepsilon_l}{\varepsilon_a} \tag{19.4}$$

where $\varepsilon_l =$ lateral strain, and $\varepsilon_a =$ axial strain.

Modulus of rigidity

$$G = \frac{\tau}{\gamma} \tag{19.5}$$

where $\tau =$ shear stress, and $\gamma =$ shear strain.

Young's modulus, Poisson's ratio and modulus of rigidity are all related by $E = 2G(1 + v)$ for isotropic materials such as metal.

Materials which yield much without failing are ductile and can survive local concentrations of strain without failing until it reaches the ultimate tensile strength F_u. If a beam is cantilevered a lateral force exerts both shear forces and bending moments such that the maximum bending moment $M = Pl$ where $P =$ shear force. The ability to resist bending loads is dependent on the second moment of inertia of the cross-sectional area

$$I_{axis} = \int_A y^2 dA$$

where y = distance from centroid to dA [Larson & Wertz 1992].

For a rectangular $b \times h$ panel:

$$I = \frac{bH^3}{12} \tag{19.6}$$

where b = breadth, and h = height.

For a thin-walled hollow tube:

$$I = \pi r^3 t \tag{19.7}$$

where r = internal radius, and t = thickness.

For an I-bar:

$$I = \frac{BH^3}{12} - \frac{bh^3}{12} \tag{19.8}$$

where B = outer breadth of I, H = outer height of I, $b/2$ = inner breadth of I, and h = inner height of I.

Bending stresses at a distance x from the origin $\sigma = Mx/T$.

Column buckling is the tendency of an axially loaded beam to deflect laterally due to compression. A material will fail at the critical Euler buckling load

$$F_{cr} = \frac{\pi^2 EI}{L^2}$$

where L = effective length dependent on the end conditions of the beam (i.e. fixed or free). The maximum stress induced by a suddenly applied impact load is typically twice that produced by the same load applied gradually, i.e.

$$F_{cr} = 2 \frac{\pi^2 EI}{L^2} \tag{19.9}$$

Buckling stress for a skin panel in compression:

$$\sigma_{cr} = \left(\frac{k\pi^2}{12}\right)\frac{E}{1-r^2}\left(\frac{t}{b}\right)^2 \tag{19.10}$$

where k = geometric coefficient dependent on r/t, r = radius of curvature, t = panel thickness, and b = panel width.

Structural efficiency

$$\frac{F_{cr}}{W} = const\left(\frac{F_{cr}}{L^2}\right)^{2/3}\left(\frac{E^{1/3}}{\rho}\right) \tag{19.11}$$

where F_{cr} = critical buckling load, and W = mass = $btl\rho$ = $const(\rho/E^{1/3})L^{5/3}F_{cr}^{1/3}$.

For a monocoque cylindrical shell to buckle under compression, cylindrical critical buckling stress

$$\sigma_{cr} = \frac{0.68Et}{R} \qquad (19.12)$$

where t = shell thickness, and R = shell radius.

$$\frac{F_{cr}}{W} = const\left(\frac{E^{1/2}}{\rho}\right)\left(\frac{F_{cr}^{1/2}}{L^2}\right) \qquad \text{where} \qquad W = bL\rho = const\, L^2\rho\left(\frac{F_{cr}}{E}\right)^{1/2}$$

$$(19.13)$$

Critical buckling load,

$$F_{cr} = \sigma_{cr}A = \sigma_{cr}2\pi Rt = 1.28\pi Et^2 \text{ for a cylinder.} \qquad (1914)$$

A finite element (FE) model of the spacecraft (NASTRAN being an example) is used to derive the equations of motion of the spacecraft of the form:

Launch loads,

$$F = M\ddot{x} + B\dot{x} + Kx \qquad (19.15)$$

where x = degree of freedom, M = mass matrix, B = damping matrix, and K = stiffness matrix.

The solution to such an equation gives the spacecraft response to the launch loads. The natural frequency and mode shapes of the spacecraft may be found by solving the eigenvalue problem: $(-w_n^2 M + K)x = 0$. Modal synthesis may be used to reduce the number of degrees of freedom for the FE model. As well as modelling, physical tests are required since models are seldom sophisticated enough for complete confidence. Such mechanical testing tests the structural model of the spacecraft. Such tests would include acceleration tests, random acoustic vibration tests and shock tests. The spacecraft is tested by subjecting it to sinusoidal, acoustic and shock tests to qualify it for flight. Static load tests are applied at various points on the structure in a centrifuge. Sinusoidal vibration tests involve applying a swept sine across a frequency range to the spacecraft. Random vibration testing involves applying acoustic excitation via microphones in a testing chamber. Shock tests usually involve sudden movements to a platform on which the spacecraft is mounted. For transportation, spacecraft are usually air cushioned in trucks with jet aircraft being preferable for low vibration and acoustics.

19.3 SMART MATERIALS AND STRUCTURES

Smart or adaptive materials offer the potential for adaptive structures for manipulator actuation, antenna reflector profiles, active noise and vibration control of structures and in-orbit spacecraft health monitoring, e.g. piezoceramic transducers for active vibration suppression or piezoelectric transducers for surface shape control. Vibration control can be accomplished by stiffening or damping the structure through the generation of internal forces to force the system towards stability. Smart materials are modelled on biological systems with sensors acting as a perceptual processing system, actuators acting like muscles and real-time microprocessor-based control system acting like a nervous system [Vincent 1992]. Smart materials are materials that perform functions dependent on environmental changes. The sensors and actuators are embedded in the material during fabrication as an integral part of the structure. A definition of a smart or adaptive structure is: 'a structural system whose geometric and inherent characteristics can be changed beneficially to meet mission requirements either through remote commands or automatically in response to external stimulation' [Wada 1990]. They come in three main types: passive smart structures, reactive smart structures and intelligent structures [Davidson 1992]. Passive smart structures employ structurally integrated optical microsensors to determine the state of the structure. Reactive smart structures use embedded distributed optical microsensors and microactuators to effect a change in stiffness, shape or other aspect of the structure. Intelligent structures will be capable of adaptive learning. The key is that smart materials offer rapid response capability to a dynamic environment. Smart materials are in fact an extension of composite materials whereby the sensors and actuators are embedded in the material. Autonomous self-adaptation enables optimal performance over a range of operational conditions in an unstructured environment. Future developments will concern the implementation of control system for flexible behaviour.

Sensors that may be embedded into the composite material during manufacture may be optical waveguides or microelectronic devices of which fibre-optic systems appear to be the most promising due to their capability to withstand the strains of the materials manufacturing process. They offer the possibility of distributed continuous monitoring of the interior of the composite at all stages of the structure's life as they are stretched or compressed, i.e. structures with optical fibres. They can also serve to reinforce the composite. Coatings must be temperature-resistant, e.g. polyimide is stable up to 200°C. Fibre-optic sensors can detect a range of physical parameters through distortion of the propagation path of the optical signal: temperature, strain, pressure, electric and magnetic fields by the interaction of the parameter to produce a modulation of amplitude, phase or polarisation of a transmitted signal. Interferometric phase sensors detect the influence of physical parameters on the phase of coherent laser light in a single-mode fibre with respect to reference arm fibres. The sensed information is integrated strain and temperature information. Phase shifts are caused by a change in the strain and temperature state in the fibre and may be detected as fringe shifts in an interferometric pattern. If a fibre is operated at a wavelength less than the cut-off wavelength the fibre may be used as a modal interferometer to detect intensity changes due to strain or temperature variations. Polarimetric sensors in a fibre can also be used to detect strain/temperature changes. By incorporating elliptical cores orthogonal polarisation modes will propagate at different velocities, generating a phase difference as a function of strain and/or temperature. Dual-core optical fibres operating at two different wavelengths offer the possibility of separating strain and temperature information. To make measurements as a function of position along the fibre length the fibre is sensitised at discrete points along the length to create larger signals from each sensing element by for example Bragg reflection or partially reflective splices. One molecule thick Langmuir–Blodgett films such as organic dyes may be doped with rare metal ions to create light emitting layers. By changing the film thickness or molecular orientations the optical properties such as wavelength of light transmission can be altered.

Actuation mechanisms vary but they generally involve changes in stiffness in response to environmental conditions without the use of moving parts: shape memory alloys, electrorheological fluids and piezoelectric solids. The commonest shape memory alloy is the NiTi alloy known as NITINOL (others include Cu–Zn–Al and Cu–Al–Ni alloys). When plastically deformed in the low-temperature thermoelastic martensitic phase, the alloy will return to its original shape reversibly by heating above a characteristic temperature to the austenitic phase (which is tailorable between 0–100°C) and if restrained from regaining its memory shape will induce stresses of ~700 MPa. Shape change can occur over very small temperature ranges through the transition temperature. Thermocouples may be embedded in composites to alter the temperature of the material but as they are metal conductors, temperature change can be induced by resistance heating. Shape memory alloys may be embedded as 200–400 μm plastically elongated wires or springs constrained from recovering their normal memorised length during fabrication. If the fibres are heated for instance by a current, they will generate a uniformly distributed shear load along the fibre to alter the material's modal response and elastic stiffness or if offset from the structure's central axis the structure will deform. Such systems could be used to control vibrations in large composite structures such as space platforms. The fibres may be arranged such that they induce a state of strain and the structure thereby altering the structure's stiffness, natural frequency and modal responses. Other possibilities include latch release mechanisms, fibre-optic switching, microvalves of thin film NiTi on etched silicon. Shape memory alloys require cooling to undergo reverse transformation and have low efficiency.

Electrorheological (ER) fluids may be used as actuators due to their reversible phase change in material characteristics especially bulk viscosity and flow rate when subjected to electrostatic potentials. Many ER fluids are based on micrometre-sized hydrophilic particles suspended in a hydrophobic nonconducting liquid in random orientations, e.g. particles of crystalline Al silicate zeolites suspended in a nonconducting dry fluid such as silicone oil which can function up to 250°C. On the application of an electric field ~2–4 kV/mm across the ER layer, its molecular structure changes to a solid as the particles orient themselves into a regular chain of columns in milliseconds imparting solidity. The energy dissipation characteristics are thereby altered and the natural frequencies and resonance responses can be changed to damp out vibrations in composite structures by controlling the voltage imposed on the material. ER fluid voids in advanced composites offer this capability. Other possibilities include gear transmission without moving parts, shock absorbers and hydraulic actuators. The chief difficulties are fluid/particle separation due to sedimentation (not a problem in space), electrophoresis and evaporation. Particle abrasion can cause wear. They tend to have high sensitivity to temperature.

Piezoelectric ceramics and polymers change dimensions on the application of an electric field, e.g. PZT (lead zirconium titanate), PVDF (polyvinyledene fluoride) and lead niobate. A voltage applied across the material generates longitudinal and transverse strains such that the material can act as an actuator. Since the Curie temperature must be higher than the processing temperature of the composite to retain its piezoelectric properties ceramics such as PZT with $T_{curie} = 360$ K offer a wider operating temperature range than polymeric materials for use as embedded fibres in a composite. They can give precise motion with repeatable rapid oscillations $\sim 10^3$ Hz when wed with ac electricity. Piezoelectric materials can also act as sensors since an applied load will generate a measurable current. They are used in STM (scanning tunnelling microscope) tips and in ultrasonic transducers, and extensively in MEMS devices. Electrostrictive and magnetostrictive materials appear to give enhanced performance for active damping over piezoelectric materials. Electrostrictive materials include lead magnesium niobate. Magnetostrictive materials exhibit shape changes when subjected to magnetic fields generating large forces over small distances to create moving mechanisms activated by an external magnetic field from a coil wound core, e.g. rare-earth elements combined with fluorine such as $TbCl_2$ (terfenol). Many are temperature-limited and are brittle. They offer fast operation at low voltages. Examples include miniature linear inchworm motors and low-power sonar transducers. Piezoceramics have also been demonstrated on micro-robots with piezoceramic legs—a femur provides vertical motion while the tibia provides horizontal motion.

Electrohydrodynamic (EHD) motion occurs when particles of a polar fluid are subjected to an electric field to generate fluid flow pressure. EHD provides direct conversion of electricity into hydraulic fluid flow without moving parts. They require high operating voltages with low currents and can generate high volume flows. A microscale solid-state ethanol pump constructed from charged grids of etched Si ~ 3 mm $\times$ 3 mm $\times$ 30 μm in size has been demonstrated to generate pressures up to 2480 Pa at 700 V and a flow rate of up to 14 ml/min. Electroactive polymers have been suggested for use as muscle-like actuators which contract under electrical excitation [Bar-Cohen et al 1998]. Fluorinated ion-exchange membranes with platinum electrodes chemically deposited on both sides of the membrane are generally adopted. These 0.2-mm film polymers bend when exposed to an applied electric field. Back-to-back pairs will contract longitudinally while thickening in diameter. An applied voltage of 5 V will contract the electroactive polymer by 10%. A 0.1-g film can carry 6 g with low power consumption ~ 20–30 mW. Such devices have been the basis of serpentine 'snakebots' that adopt undulation for locomotion. Although their force delivery is limited, they have mass, power and response speed advantages. Simple two- and four-fingered grippers have been demonstrated. Phase change, thermomechanical and shape memory alloys require heat as the primary driving force and are inefficient with low cycle rates due to difficulties in heat transport.

Intelligent structures will require microprocessor-based control of large numbers of instruments to generate flexible behaviour on the basis of the ambient conditions such as for counteracting destabilising effects. Optical neural networks of simple nonlinear processing elements capable of adaptive learning offer great promise for direct integration into composite materials [Coghlan 1992]. This effectively offers the integration of avionics into the structures of aerospace systems. Critical issues for the integration of sensors, actuators and circuits include the maintenance of the structural integrity of the composite with the impregnation of the sensor–actuator package. They often require different dimensions to reinforcement fibres particularly in terms of diameter.

Diamond fibre yields a 50% increased stiffness over SiC fibres with a lightweight material. Polyimide coatings may minimise the degradation effects of integration on mechanical integrity. It is unlikely that such integration could extend from polymeric composite to metal matrix composites, due to their high temperature processing requirements, unless sapphire fibres can be developed. A small number of active elements may be distributed throughout a structure to provide damping. Changes in material properties may be effected according to [Thompson et al 1992]:

$$M\ddot{x} + C\dot{x} + Kx = Q$$

where x = displacement, Q = load, $M(t)$ = mass distribution, $K(t)$ = stiffness, and $C(t)$ = energy dissipation. These parameters may be varied to alter the structural properties for active damping by controlling vibration amplitudes, natural resonance frequencies, mode shares, etc.

Rather than discrete variability in actuation mechanisms, continuous variability in actuation response offers greater potential and may be accomplished using embedded multiple actuators in conjunction with fibre-optic sensors [Ghandi et al 1991]. Electrorheological fluids and piezoelectric materials exhibit more continuous behaviour than shape memory alloys which tend to exhibit discrete behaviour. Non-space applications include the control of aerodynamic, hydrodynamic and optical surfaces and vibration suppression in structures such as robot manipulators. Space-based applications are similar but vibration control of structures by active damping and spacecraft health monitoring are the major foreseen applications. Their major impact is in their adoption in MEMS devices of low mass.

Adaptive materials are likely to contribute substantially to these approaches to reliability specifically with regard to structure for *in situ* self-diagnosis and repair. Failures may be prevented by the ability to monitor strain, fatigue, damage and delaminations. Optical fibres can continuously monitor composite structures at all stages of their life from manufacture, testing and operational lifetime (health monitoring and biometrics). Sensors must be able to withstand the manufacturing environment without degradation. Self-healing fibres have been developed for composites consisting of hollow porous glass fibres filled with adhesive. Shearing of such fibres by crack propagation releases the resin to fill the cracks. Corrosion may be offset by wrapping plastic based-fibres around metal reinforcement fibres. Changes in acidity dissolve the fibre coatings releasing chemicals to halt the decay.

20

Commercial and legal constraints on robotic space missions

In this chapter, we examine relevant financial and legal issues that impact on space missions including planetary missions, and in-orbit servicing missions in particular. All space missions including robotic ones occur in a commercial and legal environment imposing constraints and limitations as real as those imposed by technology. Unlike most areas of industrial activity, space activities have traditionally been dominated by government involvement which introduces a political element, both national and international, to space missions. More recently, however, commercial considerations have become important as certain types of space mission have become subject to the requirement to be financially profitable, often under the jurisdiction of the private sector [MacArthur 1984].

The *raison d'être* for a dedicated robotic spacecraft such as ATLAS was outlined earlier on the basis that spacecraft maintenance is more cost-effective than replacing spacecraft [Robertson et al 1988]. Commercial application and economic return is the only standard against which space programmes can survive in the modern market-driven world. Space applications have gone some way to indicating commercial viability by providing practical services of economic value to humans on Earth. The use of Earth-oriented satellites for telecommunications, navigation, weather prediction and environmental monitoring have value to the economy of the world far exceeding the cost of their development, yielding some $40b/y with demand doubling every 5 years. Private investment is becoming increasingly important for financing application-oriented, commercial space programmes. Market-oriented space applications comprise some 30% of space activities in Europe. All commercial applications of space technology to date and for the foreseeable future will be in global services. In 1996, KPMG Peat Marwick declared that commercial revenues from space activities exceeded for the first time that from government investment, in total exceeding $77b. Of 150 payloads launched in 1997, 75 were commercial.

20.1 ATLAS COST–BENEFIT ANALYSIS

Recently, many space projects have been required to be commercially beneficial without extensive support from governmental space agencies. Indeed, public funding for space programs has decreased as a fraction of total world space industry turnover ~$100b/y. All commercial applications of space technology to date and for the foreseeable future will be in global services. Ordinarily technologies take ~ 20–30 y to develop before they become commercially viable and only after ~30–40 y do they become mature technologies. This implies massive initial state support and this has certainly been the case with satellite telecommunications and more recently with Earth Observation. ATLAS as a commercial venture is deemed to be an exception to this rule such that commercial return is expected to be available after ~5–10 y of operational service. This section presents a cost–benefit analysis in support of this statement.

Cost–benefit analysis is a primary tool in assessing investment projects to ascertain the degree of benefits over cost achievable from the project. It is clearly desirable for benefits to exceed costs. Since with any R&D project there are uncertainties in cost estimates, uncertainties due to the complexity of secondary benefits (e.g. economic cost reduction through increased employment, stimulation to further investment and productivity, etc.) present market prices are used to value costs and primary benefits to a first approximation. Further, it is not possible to transform all costs and benefits to monetary value, e.g. social benefits and costs. For instance, increased quality of life or hazardous pollution effects as social impacts are impossible to quantify. It is questionable whether it is possible to assign monetary value to human life such as by a Pareto optimum. Such social effects are monetary incommensurables and cannot be equated with monetary value but do have valid non-monetary units of measure. Other things of value even defy any unit of measure; an example is national prestige, which is intangible. Cost–benefit analysis used in assessing the desirability or undesirability of expenditure of public funds is a branch of welfare economics. It cannot give any evaluation of the goals of a nation, such as the determination to compete or dominate in high-technology products. Furthermore, secondary effects may be difficult to ascertain, such as indirect benefits on the rest of the economy, and such benefits may be enormous, larger even the direct benefits. Technology transfer to other fields is one example. Secondary costs are even more difficult because it is often not possible to ascertain the deleterious effects of a project, particularly in the long term, e.g. environmental degradation. Such effects cannot be included in the cost–benefit analysis unless they are well-defined and quantifiable.

Cost–benefit analysis, then, is merely an economic aid to be used with other methods which are concerned with policy and priorities. The main outcome to be ascertained using cost–benefit analysis is the survivability of the enterprise as evidenced by its capability for generating revenue which exceeds its costs over long periods of time. As with any investment project, however, benefits will show a time profile and an initial outlay of capital will be required with possible later outlays. Harmon (1982) considered space robotics for in-orbit manipulation as a highly favourable applications area of space technology for investment as it is characterised by the lowest difficulty–demand product with the potential for the highest payoffs whilst offering a marketable service as well as advancing high technology. Capital cost is defined as that required to build and manufacture a product. For space projects in general the capital cost investment tends to be large. Opportunity cost is not a real cost as such but represents what is forgone as a result of not taking advantage of a good investment quickly, i.e. the loss of return due to delay. An example of this might be that other financiers may take up the investment.

Space lifecycle costs are divided into three phases: RDT&E (research, development, test and evaluation), production, including the production of flight units and launch costs, and finally operations and support. Typically, ongoing development consumes 30% of total cost, flight model 15% and ground segment costs ~15%. The AOCS is the most expensive hardware subsystem, followed by the power subsystem, structure, TT&C/ OBDH subsystems. Payload costs vary widely but a robotics payload is likely to be significant. The RDT&E phase is a non-recurrent cost. The prototype approach is one where the qualification test unit is refurbished for flight as the flight unit but this increases the non-recurrent RDT&E cost by ~30% though it reduces production costs substantially as there is no dead-end hardware. Furthermore even the refurbishment requirement may be eliminated by eliminating the need for a hardware model prototype altogether. Major cost reductions may be effected in reducing hardware testing through the use of computer simulations and CAD/CAM for geometric and dynamic analysis, structural analysis, control analysis, visual animated simulations. The Boeing 777 was developed with no model prototype. Further cost cuts can be made. Since management comprises 30% of RDT&E costs, such should be ruthlessly minimised through a flat management structure.

Product assurance and testing accounts for 20% each of RDT&E costs and the elimination of a non-flight prototype reduces those factors to half their values. All in all these factors should offset the 30% increase in RDT&E costs for the flight unit. Finally as this is a commercial spacecraft it is likely to be less expensive ~80% than a government-built spacecraft. A heuristic method of capital cost estimation was made based on average production cost/unit mass of spacecraft. Cost estimating relationships (CER) are log-linear laws of the form: Cost, $C = AxP^x$, where P = parameter and x is 0.5 for RDT&E costs and 0.75 for production costs. Such were not used since their validity is suspect regarding innovative designs like ATLAS. The production cost per unit mass of a spacecraft varies from \$100 000/kg for commercial spacecraft to \$150 000/kg for interplanetary probes. An average gives \$125 000/kg which for a 1.5-tonne spacecraft gives a total production cost of \$175m in round figures. Now RDT&E costs are usually around two to three times the production costs. This would generate a total RDT&E cost of \$350m. This cost however is based on a 10-y technology development schedule (similar to that for a warship) which for ESA is divided into six phases:

— Phase 0: this is the exploration study which defines the mission opportunity and initial technological requirements and typically lasts 3 y.
— Phase A: this is the pre-development phase which demonstrates the mission feasibility and identifies key tradeoffs and typically lasts 1 y.
— Phase B: this is the development phase which shows flight suitability of a project through in-orbit demonstration (or in-orbit hardware simulation) and defines the detailed project design. This lasts 2 y.
—Phase C: this is the qualification phase whereby flight hardware and technology is

constructed and flight qualified. This lasts 3 y.
— Phase D: this is the launch development phase which prepares the project for launch and typically lasts around 1 y.
— Phase E: this is the launch and post-launch operations phase.

The coverage here equates to Phase 0 and much of Phase A of the ATLAS proposal, or equivalently to a business plan. A business plan comprises the following elements:

(i)   technical description of the product/service in terms of user benefits and services (essentially, the systems design example, ATLAS);
(ii)   market analysis based on estimating the size of the need for the product/service (in this case, in-orbit failures);
(iii) competitive analysis based on market share for similar competitive products or services (in this case, EVA astronauts);
(iv)   strength, weaknesses, opportunities and threats analysis;
(v)   marketing strategy in capturing the available market segment;
(vi)   financial planning and risk analysis such as net present value (NPV) cost–benefit analysis.

There is a new philosophy to space projects of minimising the development period exemplified by the NASA 'faster, cheaper, better' Discovery programme for space exploration. NASA's Discovery programme is focussed on small spacecraft missions for planetary exploration with a development schedule of 36 months or less with a budget ceiling of $150m or less (not including the cost of the launch vehicle). It emphasises the use of new, advanced technologies and off-the-shelf hardware [Editors, 1998]. The first Discovery mission was the Near Earth Asteroid Rendezvous (NEAR) mission launched in 1996 to the asteroid Eros. However, on approach to the Eros, it tumbled after an aborted engine firing. However, communication was re-established and another encounter will occur in 2000. The second Discovery mission was the Mars Pathfinder mission launched in 1996 at a cost of $265m. The Lunar Prospector was the third Discovery mission launched in 1998 with a cost of only $63m and a development schedule of only 22 months. Hence, it is quite conceivable that a 5-y timeline may be projected typical of that from initial specification to beginning of production for a new automobile. The short lead time from conception to launch is based on the premise that most of the technologies are essentially in place. This is necessary to ensure that costs do not soar, e.g. Envisat which is one of a series of polar-orbiting environmental platforms for Earth Observation following on from ERS-1 and 2 had a lead time of 15 y imposing a huge $1.5b cost. Similarly, the Cassini spacecraft launched to Saturn involved a staff of 5000 over 15 y at a cost of $3.4b. An aggressive business methodology can reduce the cost of spacecraft development by up to 50%—Lockheed developed the all Ti SR-71 aircraft in only 22 months [Mandell 1992]. Another earlier example of rapid, low-cost development is the Clementine mission launched in 1994 as a BMDO (Ballistic Missile Defence Organisation, formerly Strategic Defence Initiative, SDI) project by the US Naval Research Lab to test sensors and spacecraft components by producing the first lunar polar maps. The design philosophy was a departure from the established spacecraft design procedures in that it utilised miniaturised sensors and electronics and lightweight structural materials.

Its seven new lightweight sensor technologies included three ultraviolet/visible CCD cameras, a near and longwave infrared CCD cameras cooled by Stirling cycle cryocoolers, a lidar Nd:YAG altimeter for mapping, and ring laser and fibre-optic gyroscopes within a graphite epoxy structure. Furthermore, the design departed from the usual practice of using military standard components by using commercial standard lightweight components. It employed a 16-bit 10-krad (Si) radiation hardened computer with a 1.7 Mips processing power for housekeeping operations and a 20-krad (Si) radiation hardened 32-bit RISC processor with 18 Mips processing power for image processing and onboard autonomy. A GaAs on Ge solar array provided 360 W power in conjunction with high-power ultra-lightweight nickel–hydrogen batteries. Its orbit mass was 1690 kg (dry mass of 235 kg, 223 kg liquid fuel, the rest being the solid rocket booster to give it the 550 m/s Δv for lunar orbit insertion from LEO). Despite the advanced technologies used and its relatively high mass as a macro-spacecraft, it was developed, built and launched in 2 y by a team of 55 engineers for less than $80m. It successfully completed its lunar survey of the Moon's poles in testing its sensors but was unable to perform the planned flyby of the near-Earth asteroid Geographos 1620 due to a thruster sticking open on leaving lunar orbit which induced a spin and necessitated burning up most of the spacecraft's manoeuvring fuel. It was then returned to a near-Earth orbit. Project funding is highly dependent on the schedule since most of the cost of R&D comprises salaries. The cost of hardware is small compared to time since hardware already has skilled labour time added to its purchase price. Long projects cost more because salaries are paid for longer, interest on borrowed money is lost, and storage and maintenance costs for additional storage is incurred. By reducing the R&D phase from 6 y to 2 y, the RDT&E would normally last 6 y. RMS, ROTEX and ETS-VI may be regarded as potential demonstrators of the feasibility of robotic systems in space but a specific Phase B technology demonstrator would be required. Particularly relevant for hardware validation of ATLAS on the ground is the existence of ESA's Docking Dynamics Test Facility mock-up and their European Proximity Operations Simulator mock-up as well as their extensive robotic hardware facilities. ESA has also constructed a neutral buoyancy tank to allow neutral buoyancy simulation. The tank forms part of the IVA (intravehicular activity) Underwater Testing Programme. The tank is 6 m deep and 10 m in diameter. Alternatively this could be accomplished using small satellites for in-orbit technology testing. UoSat is in the microsat range (~10–50 kg) and is a LEO rated spacecraft costing ~£0.5m while the STRV is a minisat (~50–500 kg) and is a GEO-rated spacecraft costing ~£5m. Lessons can be learned from smallsat programmes in cost-cutting. The programme durations are short ~1.5–2 y designed by a small spacecraft team of 15–30. The 150-kg STRV is in fact designed to investigate the in-orbit performance of new advanced space technologies.

An additional factor enabling the realisation of a short development period is that no fundamental advances in technology are required. Such developments are one of the greatest hindrances to meeting schedule deadlines: the Space Shuttle required considerable advances in materials technology to accommodate the aerothermal environments of re-entry ~1 MW/m^3 without the use of ablative materials. Major problems were encountered with the adhesive for bonding the various tile layers together until a silicone-based adhesive was developed generating huge cost overruns. Costs may be minimised by using off-the-shelf technology, a philosophy underlying the BMDO's DC-X SSTO prototype.

DC-X (Delta Clipper eXperimental) was a successful one-third-scale, sub-orbital prototype testbed SSTO (single stage to orbit) cone-shaped launcher developed under the US
Ballistic Missile Defence Organisation, BMDO (formerly Strategic Defence Initiative
SDI). It had a budget of only $65m but used advanced lightweight composite structures to
minimise its 600-tonne mass of which only 50 tonnes was structure. The full-scale version should have a payload capability of 10 tonnes to LEO and 5 tonnes to polar orbit.
DC-X successfully flew to 1200 ft altitude, translated 350 ft and descended to its launch
pad in 72 s. It was built in under 2 y and used F-15 inertial navigation systems, commercial GPS receiver, sensors from the F-18 and a flight control system developed by
Honeywell for airliners. It was controlled from a portable launch pad set up comprising a
trailer-based Flight Operations Control Centre, the launch pad, fuel-handling facilities
and communications equipment. This enabled it to have a turnaround time of less than 1
week. Couple these design philosophies with the slimming of administrative costs and the
elimination of non-flight hardware, a conservative estimate of RDT&E costs for ATLAS
is taken as $225m. The total cost is then projected as $400m as the baseline.

The next largest cost are launch fees and launch insurance. Launch costs are estimated
by the cost per unit mass to LEO:

	Max payload to LEO (kg)	Unit cost ($m)	Cost ($K/kg)	
Ariane IV	17 800	125	6.5	
Shuttle	23 000	210	8.2	

Ariane IV offers the lowest cost per unit mass with ~$7000/kg (1990 prices—cost per
unit mass to GEO increases to ~$50K/kg). A more reliable estimate comes from the
Shuttle pricing algorithm:

$190m $\times$ (spacecraft mass/(0.75 $\times$ shuttle capacity)) =

$190m $\times$ (1500/0.75 $\times$ 23000) = $16.5m (1992 figures)

The Shuttle is no longer used for commercial launches, and more recent figures for
representative launch costs must be used. A dedicated launch costs $50–100m, but a 1.5-
tonne spacecraft will likely share the launch shroud with other spacecraft so we assume at
a cost of $20 000/kg for launch, $30m. Insurance costs are around 15–20% of the
spacecraft cost so with insurance at 15% of $400m the insurance bill is $60m. Hence, the
total cost is just under $500m ($400m + launch + insurance) which is taken as the total
capital investment cost over a 5-y investment period from conception to launch given
current estimates in launch reliability. Funding would be given year-by-year and costs are
usually heaviest in the first two years such that by the midpoint of the project ~60% of
the costs have been consumed. Hence, the investment funding profile is perceived to be:

Year 1: $125m—RDT&E
Year 2: $100m—RDT&E
Year 3: $50m—Production
Year 4: $50m—Production
Year 5: $175m—Launch

The next stage concerns the operational costs of running ground station staff and maintenance which can usually be equated to software costs. This covers the cost of maintaining
the product and its service flow over the length of its life. Ground software costs are
~$175/line of code [Larson & Wertz 1992]. Assuming 100 000 lines of ground software
code, this gives the ground software costs at ~$17.2m. Maintenance of this code is
quantified by:

$$C = 0.1 \times (SW + EQ + F) \times \$17.2m \text{ per year}$$

where SW = software cost fraction = 1, EQ = equipment cost = 0.81, and F = facilities
cost fraction = 0.18.

Hence, $C = \$3.4m$ per year. Additional costs are for labour priced at $130 000/y for
contract staff and $90 000/y for government labour. Assuming ten staff of each (cf. Cray
Systems CLEO mobile ground station system requires six personnel):

Labour costs = $1.3m + $0.9m = $2.2m/y.

The total yearly operating cost is therefore $5.6m/y. In addition, in-orbit insurance at 2%
of the satellite cost gives an additional $8m/y overhead (except the first year of
operations which is covered by launch insurance). This totals almost $15m/y to operating
costs (except during the year of launch). These costs are recurrent and run through the
length of life which may be up to 20 y with maintainability. Additional costs include
EDRSS rental during operational periods only—dormant phases will operate through a
direct packetised command/telemetry mode. Finally, there is the necessity to refuel
ATLAS periodically. Docking alone is estimated to use around 40 kg of hydrazine per
dock although this would be relaxed through the use of manipulators. The major consumption will, however, be in-orbit manoeuvres. Such refuelling may take advantage of
various launch opportunities offered by launch companies for launching small payloads.
NASA has its GAS (Get Away Special) programme to replace Space Shuttle balancing
ballast with small <68 kg payloads fixed to the payload bay wall for $50 000 (these
payloads are not usually deployed). NASA also has its Complex Self-Contained Payload
programme for slightly larger payloads of <114 kg for $150 000. Arianespace offers its
ASAP (Auxiliary Payloads) facility which can launch up to six microsats on Ariane VI
underneath a single main satellite. There are six mounting points in a circle of radius
2.9 m for minisatellites of less than 50 kg up to a total of 200 kg. For larger minisatellites,
adapter cones must be used to sandwich between the larger spacecraft. Launch costs are
very low ~15% normal launch costs ~$700/kg. Hence, each 200 kg of fuel launched as
fuel canisters would cost $140 000. Four such launches would completely refuel ATLAS
at a cost of $560 000, i.e. around $0.5m. Such refuelling would depend on the number of
operations undertaken and across what time scales so represents a negligible cost as far as
the minimalist scenario adopted here is concerned. The payloads are flown on standby
with no guaranteed launch date. Even with 'no deals', the cost to launch 200 kg is ~$1m.
ATLAS must also use the EDRS system during operational phases (it will use a store-
and-forward strategy when dormant) which costs $30/minute (TDRSS is too expensive
at $200/minute). This amounts to $0.43m for a continuous 10 day operation. The implementation of a space-based support platform as an orbital 'repair shop' for highly
complex operations, repairs and overhauls would be expensive both in its production and
its operation. Such would provide a thermally controlled environment for the storage of

ORUs, consumables, and equipment for ATLAS as well as providing a stable work area. This implies the need for dedicated subsystems including electrical power, thermal control, communications, data processing and attitude control.

Once operational, revenue will be generated and this must be maximised. Selling price is a function of perceived value of the product or service and this is increased mostly by R&D which generates product innovation. Productivity increases through R&D are usually much larger than the savings in manufacturing costs like labour [Suh 1984]. Indeed the economy is driven by new innovations. In essence cost relates to time in that all human activities take time and value is placed on that time particularly that invested in a project. ATLAS can offer in-orbit servicing capabilities at lower cost than the only other supplier of in-orbit servicing capabilities: NASA through EVA from the Space Shuttle. There are other systems being proposed such as Ranger and Robonaut but these systems are generally viewed as support capabilities for the ISS. ATLAS should be marketed as an operational service for the repair and maintenance of space assets and it could retrieve ailing spacecraft much more cheaply and regularly than the Space Shuttle. Indeed, ATLAS would be a considerable asset to the ISS as it could provide more flexible support deployment than Robonaut (which has no propulsion capability). The space market is expected to undergo rapid expansion particularly at equatorial and polar LEO over the next few decades and EVA will be limited to equatorial LEO. An important part of the success of any commercial venture depends on how well it is marketed. Marketing is the process of identifying and servicing customer needs and is one of the most important functions in any business. It involves identifying the customer, communicating the product or service to the customer, assessing the competition and assessing the product price. The space industry differs from traditional industries in that there is usually extensive government involvement, political influences which can often overshadow financial ones, emphasis on high technology, reliance on direct distribution and on public relations and trade shows rather than broadcast advertising. It is important that ATLAS is marketed in a multitude of ways and not just as a commercial enterprise. Initially, marketing methods used are of the 'push' type with promotion through advertising, but the addition of 'pull' approaches by targetting potential clients provides a robust marketing philosophy. As a virtual monopoly, ATLAS offers the possibility of simultaneous penetration and skimming pricing. This monopoly arises primarily due to the control of the service but may be maintained by virtue of the large capital investment requirement and through patent ownership. Penetration pricing is offered since the price is low in comparison to the competition to enable securing of the market, and the skimming pricing is offered since the price is high enough to recover R&D costs fairly rapidly and generate high profits. It is perceived that an average revenue of $150m per mission is appropriate. This was suggested by the Intelsat VI repair mission which cost the insurers $150m and by launch insurance costs. Other lesser satellite malfunctions requiring interdiction which have been documented support this price. BS-3A in August 1990 suffered a partial power loss and required electrical unit replacement; Superbird A in December 1990 suffered a thruster malfunction and loss of control and required unit replacement, re-orientation and refuelling; Anik E2 in April 1991 suffered antenna deployment failure and recovered 81 days later with EVA grapple assistance; these three 'minor' failures involved a $200m insurance claim suggesting ~$70m costs for even minor repairs. In the space of 12 months from mid-1990 to mid-1991, four repair operations were required with a market value ~$300m for that year. Numerous other cases of transponder failure, power loss, control loss and thruster malfunction have been documented. This suggests that the $150m price tag per average operation is a reasonable price and ~$75m per simple operation.

The next question concerns the number of operations that may be required per year. Launch rates have been constant until recently since the mid-1960s at ~40 launches per year of which 60% are presently military launches. More recently, commercial launch rates alone have been increasing to 25–30/y. The majority of such commercial launches have been for the communications sector. Geographically, the spread has been: USSR ~80%, USA ~12%, others ~8%. Even if Russian launches ceased, there will still be ~8 launches/y. Each launch places on average ~4 satellites in orbit placing ~30 satellites per year into orbit. This is the baseline taken. In fact, the launch rate is expected to increase with the advent of mobile comsats in LEO and Earth Observation satellites to polar orbit to around ~100 launches per year (each with four satellite payloads) so the baseline represents a highly conservative figure. Launch failure rates of ~5% as given by reliability figure for launchers suggest that 10% of the 15% satellite insurance rate are in-orbit failures such as failure during checkout, etc. Hence, if 10% of satellites are expected to fail or give suboptimal performance around three satellites will fail in-orbit per year. This correlates well with the 1990/1991 actual failure rates in LEO alone. Of those expected three failures, it is assumed somewhat arbitrarily that at least one will be an average complexity operation repairable remotely in-orbit by ATLAS or at least two will be simple operations remotely repairable by ATLAS. Hence, the total annual revenue is conservatively taken to be $150m/y. Note that this does not specifically include satellite refurbishment through refuelling and module replacement. This is specifically for repair purposes. The requirement for refuelling and updating existing space assets is likely to be a primary utilisation of ATLAS particularly for scientific satellites as they tend to be expensive yet have short lifetimes due to fuel and other limitations. Refuelling a scientific satellite represents a cost-effective way of extending scientific return and is likely to be a lucrative but as yet unquantifiable source of additional revenue.

The net present discounted value for a stream of net benefits $N_0 \dots N_n$ is defined from the concept of compound interest:

$$\text{Compound interest value } N = P(1+r)^n \rightarrow P = \frac{N}{(1+r^n)}$$

Hence [Mishan 1971, Prest & Turvey 1965, Sallaberger 1992, et al 1976],

$$N = N_0 + \frac{N_1}{(1+r)} + \frac{N_2}{(1+r)^2} + \dots + \frac{N_n}{(1+r)^n} = \sum_{t=0}^{n} \frac{N_t}{(1+r)^t}$$

where P = net present value (NPV), N = net benefit = $B - C$ where B = benefit > 0 and C = cost < 0, and r = rate of discount or compound interest rate ~10%.

The project is attractive if $N > 0$ but for high risk ventures $N > 40\%$ is usually required for venture capitalists. The NPV is based on the time discounted value of money in that money is worth more now than the same amount in the future since it may be invested to earn a return. The interest rate depends on the source of payment defining the time cost of capital. The US Office of Management & Budget recommends a rate of 10% for government projects and 15% for private projects though Greenberg (1992) suggests that in reality 3–5% is more appropriate as this represents the real rate of return defined as the difference between yield from long-term, high-value securities and the inflation rate. This tends to assume that markets are secure and that little technological risk is involved with investment over a 5–10-y period. Non-assured markets with high technical risk projects tend to push up the cost of capital to 30% or so. ATLAS offers an assured market which has been outlined conservatively. Furthermore, although the technical risk may be regarded as high in that no pre-existing dedicated robotic spacecraft system exists, it may also be regarded as low in that it is an Earth-orbiting mission involving standard spacecraft bus design and that robotic manipulator systems are reasonably well established and well proven in space (e.g. Shuttle RMS). More powerfully, as will be alluded to later, it is essential that government investment for technology demonstration is available to make the ATLAS project feasible in terms of commerciality. By assuming a partnership between government and the private sector with 51% funding from the government with $r = 10\%$ while 49% funding from private sources requires $r = 15\%$. Hence, the interest rate is given as $r = 0.5(0.10) + 0.49(0.15) = 0.125$, i.e. 12.5%. We assume here that the cost–benefit analysis is that perceived by the private investor (see Fig. 20.1) [Ellery 1996]. The breakeven point is 12 y after the initial investment or equivalently after 7 y of operational service. After 20 y from the start of the project, or equivalently after 15 y of operational service, the ATLAS programme has generated a 45% discounted return on investment (ROI), above the 40% requirement for high-risk ventures. Furthermore, the high risk is offset by the insurance cover from launch throughout its operational life, so it is debatable whether the 40% ROI is a necessary qualification. A major contribution to the attractiveness of an investment like ATLAS is that the investment risk is effectively insured through the launch and in-orbit insurance and can include loss of revenue. Private financing is generally obtained through banks and venture capitalists. Bankers do not fund speculative ventures but venture capitalists have a limited time horizon—they are normally limited to a 5-y investment period after which they expect their 40% return on investment (ROI). Furthermore, their interest tends to be close to the stock market. Only 4% of venture capital is invested in early stage small companies in the UK (~£80m). Later stages of the project tend to be the attractive periods for private investors once advanced sales contracts have been obtained and sufficient assets have been created. Project financing is a technique whereby bankers lend money which is reimbursed on the basis of cash flow generated once cash flow has been established. Bank facilities such as overdraft are of dubious value due to the high interest rates in accessing such banking assets. Banks tend to invest in 'lifestyle business' such as customer-oriented ventures like shops rather than technology-based small firms. However, the largest source of private investment comes from the sale of shares. Capital is a scarce resource and many projects must compete for capital investment and it is allocated on the basis of risk/reward tradeoffs. Lenders expect to be repaid while investors expect to make a return. However, there is no shortage of commercial funding schemes for terrestrial infrastructure projects such as dams, road and rail networks, power plant installations, bridges and so on. The

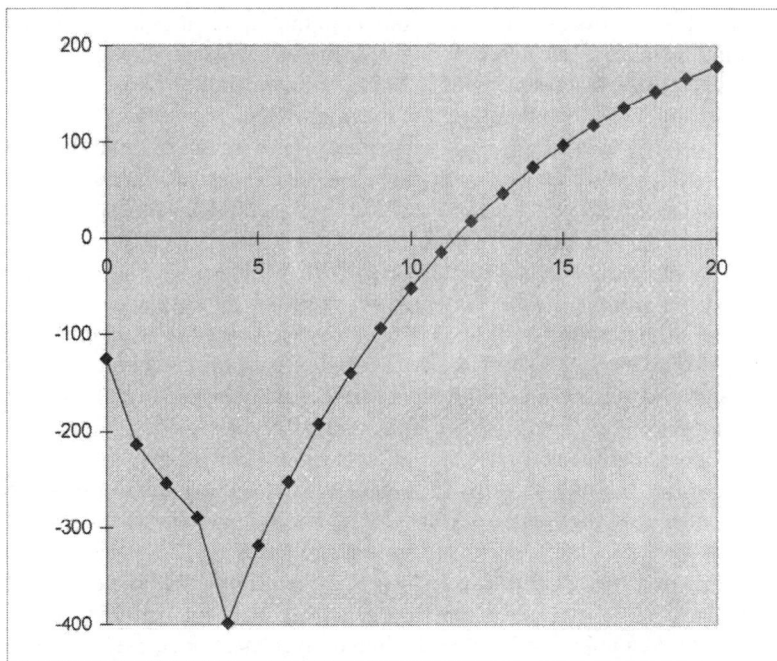

Fig. 20.1. Funding profile for purely private investment (from Ellery 1996).

French rail transport infrastructure programme required an investment of $230b over 20 y (cf. the Apollo programme which cost $80B over 10 y).

Although ATLAS may be regarded as a space infrastructure programme, unlike terrestrial infrastructure programmes with large investment requirements ~$10b/y and very long-term but low rates of return, ATLAS offers rapid high returns from high investments ~$100m/y. The government can provide public funds at the initial stages as investment bankers to reduce the perceived risk of such space ventures and encourage private investors for stock perhaps later augmented by funds from the private lending banks. Only short-term financing should be in the form of loans while longer-term financing should be in the form of share sales priced on the Stock Exchange. Government finance is particularly important where the R&D component tends to be the largest fraction of the total cost, as with most space projects. Space business ventures involve large capital investment, long payback periods and high risk. Government has a responsibility for the development of infrastructure to enable commercial exploitation and ATLAS represents a space infrastructure effort. Uncertainty (i.e. perceived risk) may be reduced by technology demonstration programmes which otherwise effectively reduce the expected return on investment. In economic theory the reward for uncertainty bearing is profit but in actuality private investors are highly risk averse. With government funding for demonstration programmes, this reduces the initial capital burden as well as reducing the perceived risk. Furthermore, government investment is the only way to ensure commercial survivability—over 80% of technology-based business startups (especially those concerned with electronic components) fail in the first year of trading in the UK. However, there are firms that provide most of their revenue from exported products. For ATLAS, the government as a 51% shareholder should invest the first $255m over the initial three years to open up a new market and new opportunities for private investors. In the third year, stocks may be issued in the market to invite private investment. The government should have controlling interest to ensure that a significant proportion of the profits are re-invested into research rather than just churned out as dividend payments. UK companies maximise dividend payments to shareholders—all firm profits are taxed at ~35% while dividends to shareholders (comprising 33% of profit) are taxed at only 25%. Shareholderships are usually diffuse and are generally invested for no more than two or three years. Furthermore, their high expectations tend towards susceptibility to lucrative takeover offers. In 1992, UK companies overall spent twice as much on dividend payments to shareholders than on their internal R&D, compared with the top 200 international companies which spent three times as much on internal R&D than on dividend payments (some as much as 15 times). The top 300 companies returned 4.5% of their sales revenue into R&D compared with 2.5% characteristic of British companies. The problem is to balance the dividends to shareholders to provide incentive to invest and the need for the business to reinvest in R&D, marketing and new assets. Contractual investment should imply responsibility and obligations in their property.

Assuming government investment in the initial stages of the ATLAS project, the cost–benefit analysis shown in Fig. 20.2 [Ellery 1996] indicates the importance of this initial government investment. The payback period is within 5 y (or just over 2 y of operational service) just within the time horizon of venture capitalists. Furthermore, the discounted ROI to the private investor after 17 y (15 y of operational service) is 252%. With the appropriate corporation tax rate of 35–40%, the tax returns alone on profits back to the Treasury would effectively repay the government's original investment. If the government required repayment of the investment through a government loan this would have a direct effect in reducing the likelihood of private sector investment. However, it is apparent that the tax returns to the Treasury would exceed the original government outlay, even when discounted eliminating any argument favouring a government loan. Albus (1990) also suggested that government financing could be achieved through selling 5-y industrial development bonds and through the introduction of a savings tax. Capital raised through savings tax could be placed in industrial development funds which could be repaid with interest well above the inflation rate to the savers when the bonds mature. A savings tax would furthermore place the heaviest burden on the wealthy as a progressive surcharge on income tax. All the capital gained could be invested in state R&D programmes such as the initial ATLAS investment which would in turn create new productive capacity through the development of new technologies and industries.

Foxell (1994) suggested that the emphasis on NPV (net present value) in cost–benefit analysis is one of the major reasons that corporate management is bedevilled by short-termism. It can certainly kill promising long-term projects since it tends to ignore cash flow after the payment period. Furthermore, future profits are assigned less value than immediate costs which has the effect of diminishing long-term benefits at the expense of short-term costs. It is the ultimate threat to long term investment. It is important to

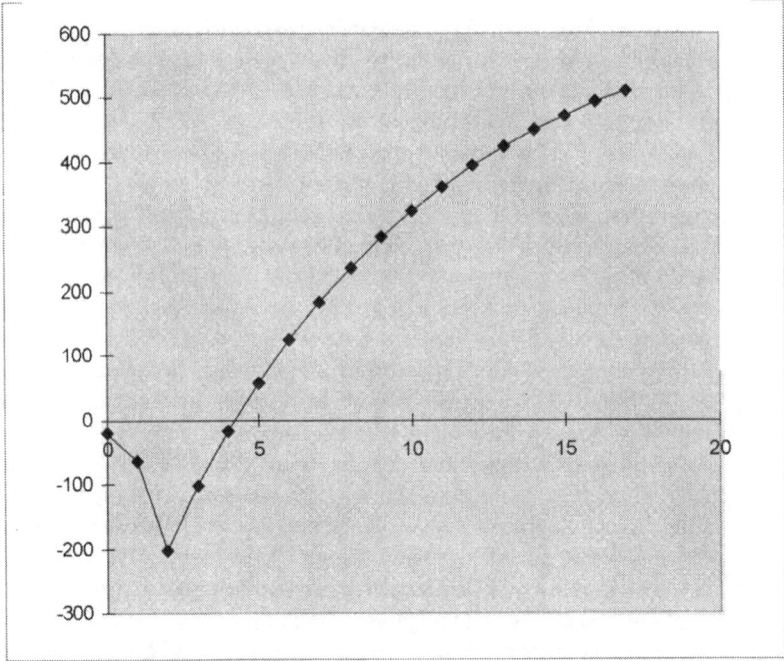

Fig. 20.2. Funding profile from public/private partnership investment (from Ellery 1996).

remember that NPV cost–benefit analysis is an accountant's tool and, as a tool, has well-defined applicability but does not represent the sole criterion for decision-making.

For more than one product unit of ATLAS, production costs drop per unit such that the production cost for n units $= TFU \times L$, where TFU = theoretical first unit cost and L = learning curve factor $= n^b$ where $b = 1 - \ln(100/S)/\ln 2$ and S = learning curve slope. Economies of scale are essential for cheaper prices. S represents the percentage reduction in cumulative average cost when the number of production units is doubled. For less than 10 units $S = 95\%$ and $b = 0.926$:

Number of units	Production cost ($TFU \times L$)	Average cost	Unit cost
1	1	1	1
2	1.9	0.95	0.9
3	2.77	0.92	0.87
4	3.61	0.9	0.84

Spacecraft production runs are unusual but ESA's Cluster mission is an example where four spacecraft were required for simultaneous *in situ* investigation of the Earth's magnetosphere in almost identical, highly eccentric, polar orbits (apogee of 20 Earth radii and perigee of four Earth radii). They required four identical sets of eleven electrical and magnetic instruments on four identical spacecraft. This constituted a low production run of a custom design to be produced in the time normally allowed for one spacecraft. Hence, Cluster occupies a situation between the one-off product typical of scientific satellites and the 'mass production' run of some telecommunications satellites. At the component level, however, the mass production character becomes more evident. In this case, standard off-the-shelf components are characteristic and this reduces development time and risk. Modular design allows parallel workflow and parallel integration reducing the production time. A variant of this is a Eureca-type mode of deployment which involves re-utilisation of flight hardware and ground infrastructure which minimises procurement costs and gives short lead times to launch. Costs for reflight are limited to satellite refurbishment and payload integration, launch and recovery, and, mission operations. Turnaround between flights would be around 2 years. Reflights are estimated at 20 times cheaper than starting from scratch and building new satellites for each mission. The longest period during the spacecraft lifecycle is taken for systems-level testing. These issues are relevant for any production run of ATLAS.

In-orbit or ground-based spares would not generate additional revenue, so the cost would not be justified. However, a small number of units dedicated to different orbits would generate a requirement with the potential for increased earnings. Another source of potential revenue is through the sale of individual hardware units to reduce the individual cost: the sale of units to the US or NATO military offers productions runs ~1–5 items (typical aerospace production run) and so reduce the individual unit cost. Each unit could be sold at $175m minimum to recover production costs (the savings due to production of multiple units would comprise the profit) or at a higher price to regain some of the R&D costs (e.g. $300m representing $125m profit per unit). The sale of such units offer large cash injections above that generated by operating the ATLAS in-orbit service. The military would benefit greatly from the sale of such units as an additional level of reliability to their present and future space assets and networks including the BMDO (formerly SDI) programme to give them flexibility of operations they could not otherwise achieve.

20.2 ATLAS AND TECHNOLOGY TRANSFER

Government-funded space agencies are strongly geared towards technology transfer to commercial applications of space technology [ESA 1993, Heirronus-Leuba et al 1993, Hertzfeld 1992]. ATLAS is an example of a Schumpeterian model of innovation whereby new products and innovations open up new markets and better production methods. Technological progress increases productivity in two ways: product augmentation in generating new capital goods, and labour augmentation by reducing labour inputs and increasing outputs through economies of scale. Such technological innovation generates new products and methods which are applicable to industrial sectors beyond those for which they were developed. Some 20% of JPL's rover and telerobotics programme is devoted to non-space applications. The corporate philosophy should be based on maximising the market/product breadth to create spin-off (technology transfer) opportunities to other non-space industries. The spin-off potential of ATLAS is larger than the in-orbit servicing market. Gibson (1989) specified information technology (including robotics) and advanced materials as the two most suitable space research activities that can be applied to other sectors of the economy—the USA is aware of the importance of this industry as it has invested proportionally more in IT than any other country. The European Community are also determined to develop high technology products in competition with the USA, whose lead has been eroding. The EC launched its commitment to information technology in 1984 with ESPRIT (European Strategic Program for Research into Information Technology), a 10-y programme to maintain European industrial competitiveness in the world market. The total budget was ~3b ECU. R&D into information technology is generally viewed as the key to the future of economies of the future.

Industrial R&D investment is particularly critical in the post-Cold War era which has been characterised by massive cuts in defence spending, including military R&D which has always been the primary driver in technological developments and innovations. The space industry in general provides an important source of technological advance with the general retraction of the defence industry. For instance virtual reality (VR) is a spin-off from the defence sector to the entertainment's industry. It is an example of a development for military application which has been adapted for another commercial application. As a result it has spawned an almost cult status as 'cyberspace'. It is doubtful if VR would ever have been developed for the entertainment industry in response to market forces. This illustrates the importance and the power of technology transfer. In-house software developed to support robotic operations may have significant commercial value when adapted for the leisure market. There is a new trend towards 'service-on-demand' business products with practical applications. Turnkey services are based on profitable but secure markets, and such products are becoming common.

As a key component in high technology, automation and robotics (A&R) may be regarded as a means to maintain technological leadership in the world. A&R technologies were defined to encompass the whole of computer science technologies with specific reference to: robotics, remote teleoperation, sensors, advanced control systems, knowledge-based expert systems, man–machine interfaces, planning systems, voice recognition and natural language understanding, computer vision, distributed processing, software engineering and validation, and autonomous systems design. Technology transfer domains for robotics and automation are numerous. Automated industrial assembly and inspection is an obvious candidate, particularly in the automotive industry which dominates industrial applications. Robots can take the human worker from hazardous processes such as materials handling tasks involving die-cast machines, furnaces and heavy presses whilst increasing productivity due to shorter time cycles and the reduction of work in progress. Manufacturing accounts for around one-third of GNP of industrial activity in a typical Western nation while services account for around one-half of GNP. The rest of manufacturing comprises the extractive industries (agriculture, mining, oil) and construction, to which A&R is also directly applicable, offering increased productivity, safety and reduced costs [Merchant 1985]. Productivity improvement and growth generally involves innovation for new products with high growth markets and so it is the manufacturing industry (mostly defence rather than civil) which contributes to R&D expenditure rather than the services industry. Services typically grow from manufacturing products. This is particularly true of high-quality manufacture of high-cost items. The importance of manufacturing industry is critical to the health of industrialised economies.

Robots have been used in the manufacturing environment to reduce the cost of manu-facturing of products as well as improving their quality. Around 75% of manufacturing is in batch volume mode and so suited to automation. The repercussions on employment have been to increase the demand for skills and a shift towards services. Manufacturing productivity increases as 16% labour, 27% capital and 59% technology. The robot offers higher reliability in that it remains in the service of the employer, has higher efficiency in that the robot performs at higher speeds ~120%, offers higher quality in that the robot performs consistently to a standard matchable by humans only after 20-y training and experience. Assembly, inspection and parts fabrication are the most labour-intensive tasks in manufacturing (especially assembly) and the mechanisation of these tasks would have a major impact on industrial efficiency [Nevins & Whitney 1980]. CIM (computer-integrated manufacturing) systems have the greatest impact on industrial efficiency: they reduce design costs by ~15–30%, reduce lead time by ~30–60%, increase product quality by ~200–500%, increase productivity by ~200–3500%, reduce work in progress by ~30–60% and reduce personnel costs by ~5–20%. All in all there can be a several hundred per cent improvement from a single installation [Albus 1990].

The importance of information technology is that it has effects not just in the future of service industries but in the future of the manufacturing industry as well particularly through computer-integrated manufacture (CIM) [Spur 1984, 1988; Merchant 1985, 1988]. The computer comprises the central component in the modern automated factory which will provide the basis of future manufacturing industries. CIM offers variable-program, computer-controlled manufacturing of variable products, on-line process optimisation of production control, dynamic scheduling operations control of jigs, tools, etc. and dynamic coordination of scheduling and resources. Essentially, CIM provides the integration of energy, material and information flows for continuous production with shorter lead times, shorter set-up times, reduction in manualism and the minimisation of errors. The primary objective of CIM is the maximisation of output with respect to the input within the integrated system. At present, IT has contributed geometric processing and design through CAD and production planning through CAM and quality control through CAQ. CAM provides operational efficiency by integrating production control of CNC machines and general-purpose robots. The introduction of advanced manufacturing techniques such as laser and plasma cutting and high-speed grinding and milling machines offer adaptable techniques amenable to IT planning and control. CAM allows optimisation for high processing rates for short product lifecycles. Automated machines produce high-quality goods at high productivity rates with low lead times. CAD/CAM shortens the development cycle, increases the diversity of alternative designs, reduces development and coordination costs, reduces manufacturing costs, and produces higher-quality products. Limited volume output of high-value, complex products which can be modified rapidly differ from the mass-production of low-quality goods. These products are high-cost, customised capital-intensive goods and systems (rather than components). They are tailor-made for specific customers or made in small batches. They form a large part of the industrial base and blur the distinction between the manufacturing and service sectors of industry. This adaptability is necessary to increase the rate of change of product design and the ability to react quickly to market changes. Production will move away from the mass-production principle in favour of adaptation capability to industrial standards. The automotive industry is the largest single manufacturing industry in the world—individual products are made to custom order but at mass-production costs. Consumers demand greater variety and diversity of product choice.

Potential medical applications of robotics include leg/arm prosthetics and tele-surgery [Kavanagh 1994, Kosugi et al 1988]. The nuclear industry requires remote and automated manipulation capabilities such as inspection, repair, maintenance and decommissioning for nuclear reactors, irradiated fuel reprocessing plants, particle accelerators, nuclear waste burial and nuclear fusion research facilities. Similar arguments may apply other industries such as chemical/pharmaceutical handling and manufacturing, telechir mining [Kassler 1985] and civil construction, and automated agriculture. Emergency services also have much to gain from automation and robotics technology: explosives ordnance (bomb disposal), fire-fighting and chemical spillage reaction may all be enhanced by these technologies. The military may be expected to be one of the most intensive users of robotics and automation technology [Robinson 1986]: the cruise missile is essentially a robot with real time adaptive capabilities. Even the service sector could substantially benefit from A&R technology advances: transportation safety, underwater pollution monitoring and cleanup, toxic and radioactive waste handling, security systems, hospital support, education and entertainment, and the leisure industry.

20.2.1 Oceanographic applications

Any space robot is likely to have an underwater neutral buoyancy vehicle simulator to shadow operations in space. Such technologies have direct application in oceanographic applications. Furthermore, such applications provide a testbed for technologies relevant to the development of planetary hydrobots for Titan and Europa for instance. Around 85% of the Earth's ocean bed lies below 2.5 km depth. Oceanographic exploration is one of the fastest-growing applications of robotics. Unmanned ROV activity accounts for 90% of oceanic exploration compared with the 10% of manned coverage primarily because of the vast expense of manned systems. Underwater prospecting and exploration is undertaken by submersible unmanned ROVs (remotely operated vehicles).

NASA Ames has developed TROV (Telepresence Remotely Operated Vehicle) based on the Phantom II ROV for operation under sea ice off the US McMurdo Science Station on Antarctica to conduct benthic surveys up 300-m depth using virtual telepresence techniques. It is equipped with two pan/tilt stereo camera pairs, acoustic navigation, a science sensor suite and a 3-degree-of-freedom manipulator. ROVs generally comprise a frame, usually of Al tubing, to support equipment, thrusters, sensors, robotic manipulators and the electronics to control the vehicle. Unmanned undersea robots are generally controlled from surface ships through high-bandwidth, fibre-optic cables, but more modern ROVs are employing remote communications. The high power requirements for operating thrusters, manipulators, etc. are generally supplied through umbilicals, but this limits their operational flexibility, so lead-acid batteries are used for medium power supply and represent a major limitation on their deployment—fuel cell technology is essential for greater flexibility of operation. Their submerged speeds are ~10 knots (compared with 50 knots for a hydrofoil). One example is Jason, a 1.2-tonne tethered ROV which explored the RMS Titanic close to its maximum operating depth of 6 km.

Trojan is a UK-built general-purpose offshore subsea support vehicle designed by Slingsby Engineering Ltd [Liddle 1986]. It is a subsea manoeuvring system mounted with dual manipulators designed for depths of 1 km extendable to 3 km. Its role covers the whole range of subsea operations from drilling support and inspection to construction. Tools include cable cutters, cutting wheels, grinders and wrenches. Specific capabilities include visual observation, NDT measurements, latch release and attach, component replacement, recovery of objects and debris, make and break hydraulic connectors, placement and removal of nuts and bolts, and specialised tool deployment, i.e. very similar in scope to those in in-orbit servicing. Its mass of 1.8 tonnes includes 90 kg of payload, of which 68 kg is for tools in a $2.2 \times 1.6 \times 1.6$ m volume. Constructed from twin pressure vessels of Al alloy, it provides 2.5 knots maximum speed from four vectored horizontal thrusters and three vertical thrusters, all identical with 305 mm propellers offering pitch, roll, yaw and depth control capability. It includes a 7-degree-of-freedom master/slave position control feedback manipulator with a 45-kg payload capability and reach of 1.3 m, and a 5-degree-of-freedom grab-bar. Electronics include fluxgate compass, gyrocompass, bathymetric pressure transducer, echo sounder, cameras with 250 W lights and 27 kHz acoustic pinger (sonar). Servocontrol loops allow automatic control of depth, attitude and heading to provide a stable platform. The ship-based surface control cabin of mass 1850 kg in a $3.7 \times 2.5 \times 2.5$ m volume includes graphic displays, single joystick control and a three-phase ac power supply. An umbilical provides telemetry via optical fibres from the ROV and power distribution control. A separate A-frame winch provides physical lift control of depth. Electric power is fed down the umbilical and Trojan has an onboard power pack to convert to hydraulic power using axial piston pumps driven by electric motors. The pumps can give variable flow and a high-pressure manifold distributes output power to all services onboard. Data communications between subsystems is distributed among dedicated subsystem microprocessors with the surface computer acting as the master unit. The Trojan programme took 10 months from conception to operational service. Trojan overcomes some of the difficulties inherent in autonomous submarines. Autonomous subs usually require battery power limiting their range to 80 km or so. Autonomous subs also require navigation systems based on dead reckoning using gyroscopes and accelerometers which need to be recalibrated periodically by resurfacing.

There are certain fundamental differences between a robotic spacecraft and a robotic submersible. The marine environment is generally one of the most corrosive environments for engineering structures [Campbell 1985]. The ocean is a strong electrolyte with many materials in solution and in suspension. Carbon steels are affected uniformly by corrosion ~0.07 mm/y, the rate being only weakly dependent on salinity and temperature. Initial corrosion is high but it decreases to a constant rate after a year or so due to marine growth which reduces access of dissolved oxygen to the structure's surface. Materials that maintain a passive oxide film, such as Al, reduce corrosion, but deep localised pitting can occur depending on the chloride content of the water. Pitting is particularly severe for propeller shafts, pumps and valves. If the velocity of motion of the structure is higher than ~3 m/s, a corrosion rate is maximised caused by shear forces of high fluid flow rate removing protective film. This is particularly true in regions of turbulence such as valves and pumps. Cavitation damage due to the formation and collapse of gas bubbles in the water from pump and propeller action generating acoustic pulses is also an important example of this phenomenon. Usually, the structure is protected by organic coatings and paints. The difference of electrical potential that exists between dissimilar metals immersed in electrolytic seawater results in a corrosion current between them. This is galvanic corrosion. Cathodic protection involves depressing the corrosion potential of iron/steel of –0.9 V by connecting a sacrificial anode of Zn alloy or Al alloy with a corrosion potential of –1.0 V to make steel the cathode of the corrosion cell. Both cathodic protection and paints are commonly employed to protect structures.

Sonar is a fundamental sensor modality used in underwater operations and is the commonest imaging and detection sensor used in subsea operations—it is the only practical sensor for turgid water imaging. Sonar comprises an array of pressure transducers. One of the chief difficulties is sound reverberation and clutter due to boundaries formed by the sea surface, seafloor, bubbles and objects. The inverse solution problem is that, given the properties of the echoes and of the sensor, it is necessary to derive the properties of the scatterers. This information extraction involves solution of the wave equation in the scattering process around complex shapes. The mapping of large regions of the ocean floor from hull-mounted, narrow, multiple beam sonar is restricted to 1/10 000 scale maps. For greater resolution, sonar must be used closer to the ocean bottom, so side-scan sonars which transmit acoustic pulses and receive reflected echoes are used for smaller-scale mapping, typically of length 0.5–3 km. Sound travels at constant speed 1.5 km/s through water, dependent on temperature and salinity. Above 20 kHz, range is limited to 1 km as higher sound frequencies are absorbed by water. Phased arrays of hydrophones can provide imaging of limited resolution.

Ambient sonar noise may be passively detected to yield imaging from scattering effects. Synthetic aperture sonar can image the ocean floor to resolutions 20 times finer than conventional sonar. The Sea Marc I developed by International Submarine Technology Ltd is a system of sonar transmitters encased in a fish to be towed 100–400 m above the ocean floor. It transmits two horizontal beams with a frequency of 27–30 kHz to map the sea floor in sections 5 km wide.

Long-range communications is a critical problem for autonomous submersibles. The usual means for communication with submarines is through VLF radio but VLF radiation is rapidly attenuated by seawater above 10 Hz, so submarine antennae must be near the surface, placing operational constraints on speed, depth and course. Acoustic links are also possible but it offers limited bandwidth ~1 Mbps [Baggerver 1984]. Transmitter and receiver are usually hydrophones. Long-range acoustic transmission is possible via the deep sound channel. The acoustic vertical sound-speed profile initially decreases with depth due to the decreasing temperature to a minimum at 1 km depth. Below this depth, temperature is near-constant but increasing pressure causes sound speed to increase. Sound entering this channel is trapped and low frequencies propagate over very long ranges. Logarithmic transmission losses dominate near the source due to spreading and linear transmission loss is dominant at long distances due to absorption. For ranges up to 10 km and a data rate of 2×10^3 bps across 500 Hz bandwidth, the operating carrier frequency must be less than 10 kHz. This 10 kHz frequency is a common parameter selection. Present tethered arrangements offer a high risk of entanglement and limitation to ice-free ocean surfaces, and acoustic frequency links do not allow high data rate transfer. However, space-based blue-green lasers have been suggested to solve the communication problem [Wiener & Karp 1980].

Generally, ATLAS provides a focus for information technology research and technology development but there are specific application areas for technology transfer such as oceanographic ROV design which can be directly spun-off from the need for neutral buoyancy simulator vehicles for the robotic spacecraft.

20.3 LEGAL ENVIRONMENT OF SPACE ACTIVITIES

All spacecraft operate within a legal environment and, as such, space-related law provides constraints on the design and operation of all spacecraft.

20.3.1 UN treaty space law
International space law regulates the affairs and relations between states and determines the consequences of actions concerning the exploration and utilisation of outer space. However, international law differs from national law in that there is no legislature which can bind all states to accept treaties and there is no executive that can force international law. The Vienna Convention on the Law of Treaties (1969) defines a treaty as '... international agreement concluded between States in written form and governed by international law, whether embodied in a single instrument or in two or more related instruments and whatever its particular designation'. Boycotts and sanctions can provide a fragmentary enforcement process and the Gulf War was an example of enforcement action under the UN charter. This was an exceptional case rather than the norm. International law is founded on two modes of introduction—treaties which are formed by negotiation and bind only signatory states, and customary law which has the status of international 'common law' whereby behaviours are generally accepted through widespread practise. Customary law binds all states without exception. This is a cultural heritage from the English tradition of common law. The effectiveness of international treaty is critically dependent on the commitment of nations to adhere to treaties [Lyall 1992a,b].

Curiously there is no strict delimitation between air and space though 100–110 km altitude is generally and informally accepted. International air law is based on the absolute sovereignty of airspace above any state (Article 1 of the Chicago Convention on International Civil Aviation of 1944) whereas international space law is based on the negation of any sovereignty in space. In this way, space law resembles the UN Convention on the Law of the Sea which conform to the principle of freedom of the high seas but which establishes the extension of national sovereignty to territorial limits of 12 nautical miles from land, and with economic sovereignty within exclusive economic zones within 200 nautical miles of land. Foreign states, however, have right of passage through the 200-mile exclusive economic zone. The lessons of the law of the sea is instructive. During the seventeenth century, the European nations in their pursuit of maritime exploration and colonisation had recognised freedom of the high seas beyond national sovereignty limits to a three nautical mile territorial zone offshore. This situation lasted until 1945 when President Truman (the Truman Proclamation) illegally declared exclusive US jurisdiction over the US continental shelf. In 1947 and subsequently, a number of Latin American countries declared their national rights to resources of the ocean and seabed out to 200 nautical miles of their coastlines. Then followed Africa. In 1958, the UN convened its first conference on the law of the sea to establish the present convention after the third conference in 1982 after much political negotiation (particularly the USA which wished to maintain freedom of passage of its ballistic nuclear submarine fleet). Similarities of international space law also exists with the Antarctic Treaty 1959 which prohibits territorial sovereignty on that continent and the basis of cooperation for scientific research. The 1991 Protocol on Environmental Protection of Antarctica is a supplement to the Antarctic Treaty which constrains all human activity in Antarctica. Article 7 of the Protocol prohibits any activity relating to mineral resource exploitation through a moratorium in force for 50 y. It subsumes most of the Antarctic Treaty. The Treaty guaranteed freedom of scientific research and exchange of data, forbids nuclear explosions and the disposal of radioactive waste, prohibits military activities and opens all installations to international inspection. In addition, it has been subsequently supplanted by a number of Conventions which are separate from the Treaty itself. The basic principle of non-sovereignty in space law was questioned by the Bogota Delegation of 1976 which attempted without success to claim sovereignty of equatorial orbital segments (specifically segments of equatorial GEO) as the natural resources of developing equatorial states—this objection was overruled on the basis of customary law that until then it had been accepted that the orbit is free for utilisation as observed since 1957.

Most space law is treaty-based in five treaties [van Bogart 1986, Lyall 1992b]: the Outer Space Treaty of 1967, the Rescue Agreement of 1968, the Liability Convention of 1972, the Registration Convention of 1976, and the Moon Agreement of 1984. All treaties are drafted through consensus by the UN General Assembly COPUOS (Committee on the Peaceful Uses of Outer Space) set up in 1959. It comprises two subcommittees: the legal subcommittee and the scientific and technical subcommittee. Both operate through consensus rather than through votes.

The Outer Space Treaty of 1967 provides the basis for all subsequent treaties of space law, so much so that it has become a part of customary law. Hence, although it has 97 ratifications and 27 signatories, it binds all states without exception. It states that outer space should be used only for the benefit of Mankind and to the benefit of all states. Exploration is free to all states without national appropriation (Articles I and II). All activities in space should comply with the UN charter in the interests of international peace (defined as non-aggression thereby allowing defence) and security. This provides the central premise of the Outer Space Treaty.

The Agreement governing the Activities of States on the Moon and Other Celestial Bodies of 1979, which incidentally defines celestial bodies as those within the solar system only, expands on the themes in the Outer Space Treaty in such a way that of the few states (eight) that have ratified it, none are spacefaring states. It appears that 'for the benefit of Mankind' does not mean that an investor spacefaring state is obliged to relinquish the commercial and economic advantage that it may have as a consequence of those space activities, which is what the Moon Treaty appears to imply. By stating that the moon and natural resources are the common heritage of Mankind and that there should be equitable sharing of the benefits of those resources seems to bar commercial application.

However, this particular Treaty is very much open to interpretation and if the Outer Space Treaty were interpreted in a similar manner, telecommunications would not be commercially possible. A similar situation occurred with the drafting and collapse of the 1988 Convention on the Regulation of Mining Activities on Antarctica to formalise a voluntary moratorium proposing limited exploitation that had been observed since 1972. Similarly, the law of the Sea Convention of 1982 set up a Seabed Authority to control and exploit seabed resources beyond national limits. Most major national states have not signed it. It appears that when commerce is involved, principles undergo major convolutions. Articles V to VIII of the Outer Space Treaty are elaborated by later treaties. Article IX states that all spacefaring states must avoid harmful contamination of both outer space and the Earth as 'global commons' beyond the jurisdiction of any nation. This has major implications for robotic sample return missions to planetary bodies. Article XI retains the UN authority by placing an obligation to report all space activities to the UN Secretary-General.

Articles V to VIII of the Outer Space Treaty have been expanded upon in subsequent treaties and these treaties have a direct impact on the nature of any robot servicing spacecraft. Article V stated that there was an obligation to assist astronauts as envoys of Mankind in the event of accident and inform the UN Secretary General of possible dangers to astronauts. Article VIII states that there is an obligation to return space objects if found to the owner state. The Treaty on the Rescue and Return of Astronauts and the Return of Objects Launched into Outer Space of 1968 (ratified by 67 nations) was more detailed. It states that astronauts and space objects found beyond the limits of the launch states must be returned to the launch state promptly. At all times the launch state retain jurisdiction over astronauts and their space objects. All astronauts as envoys of Mankind have diplomatic protection. In the event of an accident relevant states should afford all possible assistance. If a space object falls on foreign territory, that foreign state should notify the launch state and the UN. Recovery must be instigated by the launch state but the recovery process falls under the jurisdiction of the foreign state. Assistance offered by the launch state may or may not be accepted. Indeed the offer of the USSR to assist the Canadian clean-up of Cosmos 954 in 1978 was refused on the grounds that the USA had all ready offered to assist, an offer which had been accepted. However, the compensation to the cost must be shouldered by the launch state. The re-entry of Cosmos 954 over Canada in 1978 spread reactor debris over 124 000 km² of the north-west region. The total cost of the cleanup cost Canada $14m of which $8m was for personnel and equipment. Canada claimed $6m but the USSR compensated $3m arguing that they could have done the clean-up much more cheaply. The 50% settlement has set a precedent for future situations. It is suspected that the 7-tonne Russian Mars-96 spacecraft also re-entered Earth's atmosphere and crashed in the Andes in 1997. Its second fourth-stage firing failed to eject the probe from its 160 km parking orbit into its Mars trajectory, entering an Earth elliptical orbit with a maximum altitude of 1500 km. It carried some 200 g of plutonium which has yet to be recovered. Other examples of re-entry include the US Skylab re-entry over the west coast of Australia in 1979 and the re-entry of the Soviet Salyut 7 which fell over Argentina in 1991. This treaty has considerable implications for the operation of salvage operations in space. Since launch states retain jurisdiction of space objects outside their state confines there is no question of gaining ownership through salvage. Evidently if salvage became commercially viable, new treaties would have to come to terms with the nature of ownership and its transition from state control and ownership to another legal entity. This has considerable implications for robotic debris removal of inoperative space assets as there is no commercial imperative to encourage such a capability. As it stands, claim of ownership of spacecraft as salvage would constitute an act of international piracy.

Article VI of the Outer Space Treaty declares that launch states bear complete responsibility of their space activities and that private corporations must obtain from their respective states special licenses to indulge in space activities. Similarly Article VIII states that the launch state retains jurisdiction over its space objects. Again the Convention on the Registration of Objects Launched into Outer Space of 1974 makes this explicit (ratified by 37 nations). All space objects must be registered. The state of registry which may be either the launching state or the launch-procuring state (referred to as the launch state) has jurisdiction over its space objects. The registration provides a form of identification and liability for damage. The registry implies title of the satellite to include ownership and that control of the satellite by TT&C by contractors are answerable to the owners. The register must include the spacecraft's launch date and site, orbital parameters (nodal period, inclination, apogee and perigee altitudes) and its function (article II). Military satellites circumvent this by placing their satellites in an initial orbit (reported to the UN) and then manoeuvre them to their operational orbits (not reported to the UN). This is not so unusual as military satellites have a tendency to be retasked to different orbits during their operational lifetimes. The UN Secretary-General is also obliged to maintain a register to whom every state is obliged to supply the necessary information for that express purpose (Articles IV and V). This has similarities to the UN Convention on the Law of the Sea 1958 whereby registration of a sea-going vessel entails the right of the vessel to use the registration state's flag as a form of identification and the flag state has complete jurisdiction onboard its ship. Similarly the Chicago Convention on International Civil Aviation of 1944 states that aircraft have the nationality of the state where it is registered and that the registration state is responsible for all activities of the aircraft.

Article VII of the Outer Space Treaty states that the launch state is liable for damage to any state caused by a space object. This may be regarded as a space equivalent to the legal liabilities of any supplier of goods and services with regard to safety. In the UK, contraventions of the Health & Safety at Work Act (1974) and the Consumer Protection Act (1987) are minimal. The HSAW places a duty on employers to ensure the protection of the health and safety of their employees and any other persons who may be affected by their activities such as people who use those products or services at work. The CPA enacts the European Directive on Product Liability of 1985 which imposes a general safety requirement. Personal loss of health or possessions due to failure of a product to perform as specified incurs automatic rights to compensation. The Convention on International Liability for Damage caused by Space Objects of 1972 elucidates this victim-oriented liability imposed by Article VII of the OST (ratified by 35 nations). The launch state is absolutely liable to compensate for damage caused by its space objects on Earth or in sovereign airspace to foreign states (Article II). Damage to the launch state by its own space objects is governed by its own national civil legislation (Articles VII and XI). There is usually national liability legislation to public services which cause danger to private concerns regardless of fault. When damage is caused by a space object to another space object in space liability is relative according to fault established by causal link (Article III). If, however, damage resulted from negligence, insufficient information or wilful act on the part of the claimant, liability is exonerated (Article VI). Under certain conditions the national law of tort may be applied which requires a proof of negligence. Such liability must be negotiated through diplomatic channels. If diplomatic channels do not settle the issue then the matter may be referred to a claims commission comprising the claimant state, launch state and a jointly chosen chairman as arbiter who casts the decider vote (Articles X, XIV–XX). Liability may be joint and several between two or more registered states according to fault (Articles IV and V). Launch states may thus negotiate special agreements regarding the apportionment of liability. In-orbit servicers such as ATLAS should include a waiver of liability clause in the contract to the customer (owner of the space asset to be serviced) to exonerate from liability for damage to target spacecraft in the same way that launch contracts pass liability to the procurer so that the procurer bears the risk.

The carriage of payload by launchers obliges the sponsor to pay a share of costs to the launching state. Hence inter-party liability will apply. Procurement of launch establishes liability in this way. Once separation from the launcher is achieved and the payload is operational autonomously the launch-procurer becomes the flag, and so liable, state. Procurement according to section 308 of the NASA Act of 1979 requires the procurer to obtain liability insurance against damage to their property or to a third party during launch, operations or recovery. No mention is made of post-operational phases after control has been relinquished. At present there is no clear directive concerning space debris—it is possible that liability for damage may ensue by national civil legislation but there is no legal obligation to boost into post-operational graveyard orbits and de-orbit for re-entry. This is an unclear issue but it certainly appears that if a reasonable means of clearing away debris were readily available (such as ATLAS operating in an in-orbit removal capacity) states may become liable for debris damage in space on the basis of negligence—such an interpretation could provide a commercial incentive to take responsibility for debris mitigation.

The high cost of many space activities has motivated many nations to prefer cooperative ventures [Aldridge 1993, Logsden et al 1993]. There are pros and cons to cooperation and it typically involves bargaining between cooperation and competition between economic regions of the world. The advantages of cooperation are: cost sharing, access to technology and scientific knowledge and the strengthening of relationships between nations. The disadvantages of cooperation are: loss of autonomy, an increase in the overall total cost due to increased managerial complexity, the enhancement of future competition and political difficulties in broken commitments. US legislation acts as a considerable barrier to other nations in indulging in cooperative ventures however. The Buy America Act of 1933 prohibits the buying of foreign goods rather than services and imposes high tariffs on foreign goods to favour domestic goods. NASA, however, operates an additional policy of not using foreign services which are based on foreign goods. Foreign is defined such that if more than 50% of the product originates from domestic sources then it is not foreign. The DoD is an exception to this as it may buy from NATO countries, e.g. the USMC operates British Harrier jump jets as their main aviation element. There are several ways to interpret this as regards ATLAS. First,

ATLAS offers a service which is not available within the USA except through EVA in-orbit servicing at great expense as a comparable capability. EVA in-orbit servicing has been a major linchpin in NASA's justification for manned spaceflight (prior to ISS). It would be important for ATLAS marketeers to allow NASA to treat utilisation of ATLAS as a cooperative venture effectively shouldering the role of the now defunct FTS in support of ISS activities. The second point is that the DoD have fewer restrictions as regards imports. Any NATO country can export military products and/or services to the USA. Regarding the export of components to foreign countries from the USA, an extensive set of legislation controls exports. The Export Administration Act of 1979 requires licensing of all dual-use technical data and commodities and licensing is subject to national security, foreign policy and short supply considerations on a case-by-case basis. Three lists are used: the Industrial List of core technologies critical to Western security including dual-use technologies; the Munitions List of inherently military technologies; and the Atomic Energy List of nuclear technologies. Military technology is heavily subject to control under the Arms Export Control Act of 1976 through the issue of licenses—military technology includes dual-use items such as satellites, computer systems, rockets and technical data and services pertaining to such articles. However, once again, the rules concerning the export of military technology such as spacecraft components do not apply to NATO countries [Smith 1990]. The COCOM (Co-ordinating Committee for Multilateral Export Controls) controls the export of strategic technologies which contribute to military capabilities. It is a multilateral organisation of NATO countries and is currently releasing control of computer systems and inviting republics of the former Soviet Union to participate. Missile technology is subject to control to maintain missile non-proliferation through the multilateral Missile Technology Control Regime (comprising 18 members including the USA). Missiles capable of carrying 500 kg of payload to a range of 300 km are defined as missiles and so include space launcher technology as well as Cruise missiles. To be fair to the USA, it does have the highest space expenditure and space expertise in the world (with the possible exception of Russia) and this favours cooperative ventures in space projects for other nations with an inevitable transfer of technological intellectual property from the US to other nations in the cooperative. Several forms of cooperation are possible: exchange of data from separate missions (e.g. weather data in the World Weather Watch); coordination of separate missions (e.g. the Comet Halley armada encounters); increasing the mission capabilities by instrumentation from different nations (e.g. Cassini–Huygens mission); joint missions (e.g. Apollo/Soyuz mission); joint facility operations (e.g. ISS); new institutions (e.g. Intelsat, ESA). Scientific missions such as the robotic exploration of the solar system and astronomical observation satellites have been the dominant areas of international cooperation and have been heralded as the 'success story' of international co-operation. However, although large space projects are amenable to cooperation, a small space project such as ATLAS is best pursued commercially and autonomously to maximise the profit return as coordination between different states inevitably involves financial overheads.

20.3.2 Space insurance
Insurance is a necessary and expensive element of a space project. CT Bowring Space

Projects Ltd (1993) divides insurance covers into three segments: transit and pre-launch, launch/early orbit, and in-orbit insurance segments. Transit and pre-launch covers damage to the satellite and delays from the commencement of loading operations for transit until lift-off (usually ~2–3 months) including shipment. The insurance premium rates are ~0.5% of the spacecraft value. The launch/early orbit phase covers from lift-off through to commissioning to between 180 days and 360 days after launch. Commissioning involves placing the satellite in its operational configuration and orbit and subsystem function confirmation through in-orbit checkout tests, orbit manoeuvres and deployment functions. The extended period from 180–360 days included in this phase of insurance allows the satellite's function to be evaluated during eclipse. This covers infant mortality failures such as launch failure, failure of kick motors, and deployment failures being typical. This launch phase insurance premium rate is ~16–20% of the satellite cost with the cost being split with ~9–13% for launch and ~8–11% for post-separation phase. Usually the cover is for 12 months after launch (since this is only marginally more expensive than 180-day cover) to cover the total loss of satellite capability (defined as the failure of 50% of spacecraft subsystems) and/or partial loss. The purchase of full coverage policy including both the combined partial and total loss cover in the launch phase is the most cost-effective approach to space insurance. Limits are usually imposed for launch insurance cover. The US Commercial Space Launch Act of 1988 limits launch insurance to $100m for commercial launches on expendable vehicles with a maximum return of $500m. In 1996, the insurance industry paid losses of $600m while insurance premium incomes amounted to $850m. Large satellites are insured for $150–$200m, often higher than launch costs ~$50–$100m. Estimated gross premium incomes for insurers from 1980 to 1996 was ~$5b. Scientific spacecraft are not usually insured as they are generally government-funded. Consequently, when in 1996, the first Ariane V maiden launch failed, the cost of the four ESA Cluster spacecraft was not recovered [Clubb 1999]. The launch phase from lift-off to in-orbit checkout is the most expensive portion of space insurance. This launch phase insurance premium is 10–20 times greater than most terrestrial ventures and 10 times higher than airline insurance. Third-party launch liability covers the satellite owner against damage to the launching organisation arising out of damage caused by the satellite. The insurance premium rate is ~0.1–0.2% of the satellite value. Usually launch manufacturers include in the launch contract a guarantee to refly the mission in the event of launch failure.

Usually insurance coverage is also required for the in-orbit operational phase of the mission, particularly for communications satellites due to their commercial earning value. This will thus apply to ATLAS. In-orbit insurance covers the period from the expiration of the launch cover to 12 months and insures against subsequent loss or damage to the satellite and provides for the replacement and relaunch of the spacecraft, loss of revenue and the fulfilment of contractual obligations. Total and partial loss coverage for a healthy satellite is ~1.75–2.25% of the satellite cost, but for a satellite with health anomalies this would increase to 3–4%. In-orbit insurance is a complex problem since partial in-orbit failure involves complex problems in accurate loss assessments. Different kinds of insurance payments can be given for the loss of propellant, loss of electrical power and other failures according to their severity on payload functioning and assessment of payload value. Communications satellites have the highest in-orbit insurance demand and require insurance against loss of operating profits. In addition the operating company may wish to negotiate indemnities against any manufacturing defects by subcontractors (e.g. the Hubble Space Telescope syndrome).

The mid-1980s involved considerable loss to underwriters due to launch vehicle failures and reduced capacity, down to $100m/launch, and high insurance premiums up to ~25–30% of the satellite cost (in 1986). Prior to 1984, insurance premiums were ~5–6% for the Shuttle and ~8–10% for Ariane launches. In 1984, claims exceeded $250m when both Westar VI and Palapa B2 rocket engines failed on release from the Shuttle. In 1986, the Shuttle *Challenger* disaster was followed by the loss of two Titan 34Ds, a Delta, an Atlas, an Ariane and a Proton. By 1985, launch insurance rates has climbed to 20%. The market has since stabilised with an overall capacity of $500m/y. The UK comprises 15% of the space insurance market capacity with only the USA and France having higher capacities at 28% and 20% respectively. Indeed fewer than ten insurance companies control over 80% of the capacity. It is evident that ATLAS may contribute significantly to offsetting insurance company losses that have tended to characterise space insurance by providing a means for inexpensive repair. This may even reduce in-orbit insurance costs and the checkout portion of launch insurance—the potential effect may be as high as 50%.

20.3.3 Law of contract
All business activities are based on the notion of contract as a form of negotiated exchange during business activities. This section in no way represents legal authority.

A contract represents a legally binding and enforceable agreement (written or spoken) between two parties for the purpose of insurance, purchasing and selling. For a contract to exist, there must an offer, acceptance and consideration. Initially an offer, written or spoken, is made which indicates a commitment to comply to an exchange on acceptance of the offer. An advertisement is regarded as an invitation to an offer through subsequent negotiation. The offer defines the subject-matter, the parties involved, the offer price and sometimes the time for acceptance and execution of the offer—these constitute the terms of the contract which may be expressed or implied (generally expressed for a written contract). The conditions of the offer must be properly communicated to the offeree before the contract is completed. Prior to acceptance of the offer, an offer may be revoked at any time. Acceptance by the offeree becomes binding once it is received by the offerer—agreement of the contract forms the basis the contract. If the conditions of the offer are modified by the receiver, the original offer is rejected and the altered conditions become the counter-offer. Acceptance of the offer involves either bilateral contract, by promising to perform an action, e.g. payment, or unilateral contract, by performance of the action, e.g. contract for work to be done. Consideration describes the concept of the exchange of something of value to distinguish contracts from gifts, e.g. the promise to pay for a product or service. Such consideration is a necessary property of the contract and the consideration must have been agreed upon by the parties involved. The consideration does not have to be equal in value but it must be real. The UK Sale of Goods Act 1979 imposes certain obligations on the seller of goods undergoing change of ownership with regard to the description of the good, quality of the good and suitability of the good for its purpose.

There are several potential defences to contract that either define the contract as void or provide the option to make the contract voidable and so non-enforceable. There are certain specific requirements for the contract to be valid:

1. The parties must have legal capacity by virtue of competence to enter into the contract, i.e. not mentally diminished, intoxicated, or of minority age.
2. One party must have made an offer while the other party must have accepted that offer.
3. The agreement of the contract must be genuine without the incidence of mutual mistake of fact (namely, subject-matter of the contract, identity of a party named in the contract, or nature of the contract), misrepresentation of a statement of fact (either innocent or fraudulent), duress through coercion (such as the threat of violence, criminal proceedings or economic pressure) or undue influence (typically due to inequality of bargaining power). The insurance contract represents a contract of the utmost good faith (contract *uberrimae fidei*) wherein the insured has a duty to volunteer all relevant facts to the insurer—this is the only case where such a duty is placed on the purchaser as the relevant facts are known only by the purchaser (and can therefore be subject to misrepresentation).
4. The contract must be supported by consideration. If a party contracts to perform an action which is impossible according to our present scientific knowledge, the contract is void by virtue of lack of real consideration for the other party. If the party contracts to perform an action which can be performed, but the party cannot fulfil the obligation, the contract is valid and subject to liability for its breach. In the case where performance of the contract subsequently becomes impossible, the obligated party should have included terms in the contract to safeguard against the contingency that disallows completion of the contract—the only situations in which a contract can be discharged is if it subsequently becomes illegal, the subject-matter no longer exists, or an expected subject-matter did not subsequently materialise.
5. The subject-matter of the contract of exchange must be legal—if the subject-matter of the exchange is illegal (e.g. contract that injures public service such as bribery, contract trading with an enemy state in war-time, contract breaking the laws of a friendly country, contract invoking unreasonable restraint of trade) the contract is non-enforceable in a court of law.

Specialty contracts are those made by deed, i.e. a written instrument which has been signed, sealed and delivered, or a simple contract (contracts which are not deeds). The deed is complete only on delivery. A specialty contract is exempt from the consideration requirement, and is used formally to illustrate the importance of the transaction (e.g. lease or title to land). Some types of contract must necessarily be written to be valid: marine insurance, bills of exchange, credit agreements, transfer of shares and asssignment of copyright. There other contracts which do not have to be in writing but enforcement of the contract by courts will require written memorandum including the names of the parties, the subject-matter of the contract, the consideration, and the signature of defendant—these are the contract of guarantee and contract for the disposition of land (such as mortgage).

A corporation is a group of individuals focussed towards a common goal, and represents a legal entity in its own right independent of the individual comprising the corporation. As legal entities, they can assume legal responsibility and possess legal rights. Most commonly, they owe their legal existence to following the procedure dictated by the UK Companies Acts (1948–1981) which allow individuals to register their intention with a public official to form a company. Such a company as a legal entity can make contracts and be held legally accountable. The Fourteenth Amendment of the US Constitution guarantees due process to corporations as 'people'. Chartered corporations have unlimited contractual capacity. An agent of a company may act on behalf of that company with regard to contract through the agent's relationship with that company. Yet the corporation is morally a virtual entity as, in reality, corporations act only through human agents. Although theoretically a debatable issue, in practice this distinction is rarely questioned.

A contract ends by mutual agreement, once the timespan specified is spent, the project is completed, or on breach of the contract when one party does not fulfil their obligation specified in the contract—a major breach which affects the subject-matter and outcome of the contract may result in liability to damages. The type and degree of damages due to the plaintiff on breach of contract by the defendant depends on a number of issues. Some contracts include provision for breach of contract and specifies liquidated damages awardable in the event of a breach—unliquidated damages are those awarded by the court. Liquidated damages must represent a fair estimate of losses incurred on the breach of contract. However, the ability to include contractual clauses to exclude or restrict liability for breach of contract is strongly limited by the UK Unfair Contract Terms Act of 1977—this represents a significant stumbling block for contractual waivers of liability in the event of incomplete or accidental damage during in-orbit servicing operations. A breach of contract may be due to negligence due to a lack of reasonable care, or lack of reasonable skill (be it accidental or intentional). Liability for death or personal injury from negligence cannot be excluded or restricted within the contract. Only terms which are fair and reasonable can be included to restrict liability on the basis that all relevant circumstances are known to both parties. Regard must also be taken of the resources available to the party seeking to restrict liability, and whether insurance cover was available. Hence, restriction of liability in contract terms will be invalid unless they are reasonable. A breach of contract resulting in injury (legal injury rather than physical injury) usually invokes monetary damages to the injured party. Typically, the injured party is entitled to a money difference between the cost promised and the cost required to complete the promise. The injured party must, however, take reasonable action to mitigate the extent of injury towards completion. Protection against liability for loss or damage will only be valid if fair or reasonable caution were exercised with regard to the execution of the contract. The contract may specify terms for arbitration of disputes resultant from the contract in advance with each party specifying an arbitrator who in term mutually specify a third arbitrator.

21

Conclusions

In conclusion, we take a look to the future of space exploration, and the central role that robotics and automation will play. We have seen that robotics is an enabling technology at the present state of development of space technology, both for scientific exploration of the solar system and for commercial development as space infrastructure in the near-Earth environment. We will see that automation and robotics will be critical to the future of spaceflight, however it is played out, be it Earth-oriented as in the ISS, lunar-based or in pursuit of the manned occupation of Mars. As we shall see, the Space Exploration Initiative (SEI), although consigned to the wastepaper basket as an explicit programme, it is very much alive as an implicit programme as it represents a logical programme of space exploration development. The success of such future missions will hinge critically on advanced robotics and automation technology—the need for automated and/or telerobotic *in situ* resource utilisation will decide whether manned Moon or Mars missions occur in the near future, and the ISS in terms of developing such manufacturing technologies to maturity will play a critical role in acting as medium for technology demonstration of such. Furthermore, advanced space applications of robotics will drive aggressive strides in the technological development of robotics generally, imposed by the extreme remoteness of the environments involved, the diversity of hostile environments, the utility of *in situ* resources and manufacturing in reducing space mission costs, and the need for extensive autonomy and reliability.

21.1 INTERNATIONAL SPACE STATION (ISS) AND MICROGRAVITY MANUFACTURING

The Space Station Programme was initiated on the twentieth anniversary of the 1969 Apollo 11 manned lunar landing by the US President George Bush: 'I am proposing a long-range, continuing commitment. First, for the coming decade—for the 1990s—Space Station Freedom—our critical next step in all our space endeavours. And next—for the new century—back to the Moon. Back to the future. And this time, back to stay. And then—a journey into tomorrow—a journey to another planet—a manned mission to Mars'. If this was meant to echo Kennedy's 1962 speech following Yuri Gagarin's historic first manned orbital flight in space in Vostok 1 in 1961 which advocated a manned lunar landing before the end of that decade, he should have noted the $80b cost of the Apollo mission. Furthermore, the Apollo programme was a component of foreign policy as a result of the Cold War at its height. Bush's rhetoric became the basis of NASA's Space Exploration Initiative as an integrated set of activities culminating in the establishment of a permanently manned lunar base and eventually the manned exploration of Mars. Central to these developments was a series of unmanned robotic missions to generate the required data to support the later manned missions. Rather than being projected as opening up new commercial markets and ventures, the SEI was marketed as a search for new technologies and as a great adventure. The SEI came complete with a $30b/y price tag over 15 years to a total expenditure requirement of $500b. Bush's unequivocal statement of support for the SEI, however, was short-lived in the face of economic difficulties in the USA particularly in the 1980s climate of financial frugality, illustrating the fragility and susceptibility of space programmes to dissolution once political motives in support of them recede. The chief difficulty is that any long-term objectives such as the SEI cannot be guaranteed given the short-term horizon and volatility of political will. The failure of the SEI was due to its huge cost and the implication for long-term commitment. Justification of highly costly manned spaceflight is difficult in economic terms. In 1993 President Clinton closed NASA's Office of Exploration. The only surviving element of the SEI is the International Space Station (ISS) programme. The Apollo mission to the moon was a success in its aim only because of the unstinting political support from the US Government motivated by the Cold War with the USSR. However, even the Apollo project has not emerged unscathed in recent historical analysis. Despite the fact that the Apollo programme was a spectacular success and that the event is ingrained in the human psyche, its high cost has subsequently attracted criticism as being self-indulgent. However, taken into the perspective of the Vietnam War with its huge cost in human life and its $800b expense, this criticism is hollow. Even so, the Cold War is now over, since the collapse of the USSR in 1991, and there is little need today for political posturing of militarily derived prowess on the part of the USA. Even today, NASA's annual budget of $14b is only 5% that of the US military budget. But, as we emerge from the end of the twentieth century and into the twenty-first century, one enduring image that we bring with us as a symbol of the incredible achievements of the twentieth century is that of men standing on the Moon.

Political priorities have changed—governments today are more concerned with earth-oriented issues such as regional conflict resolution, economic growth and combating environmental degradation (this last only half-heartedly), and indeed, space activities have contributed to all these activities. Traditionally, space exploration has been justified as a biological urge to expand our domain of existence made possible by our techno-logical inventiveness. This process has enabled Man to settle regions previously inimical to human life with ever greater population densities through the manipulation of our environment. Space exploration is seen as an extension of this process of exploitation. No longer is the appeal to pioneering exploration as being in our nature and as a manifesta-tion of human vigour sufficient justification in a society driven by market forces and where the economy takes precedence over all other social issues (except war). The only way that space programmes can attain robustness to political fancies is to justify themselves economically or to maintain a strong and consistent level of public support. Yet despite the fact that the doctrinal 'accountant-speak' philosophy of short-term rationalisation so dominant in the 1980s yielded no great surge in corporate profitability or the human condition, the balance sheet is still the bottom line. Space activities have gone some way to commercial application by providing practical services of economic value to human beings on Earth. Satellite communications have made significant contributions to the Earth as a global village. Earth observation satellites provide a unique tool to study and monitor the Earth globally. In reality, the history of exploration has not been so much a process of adventure (except by subsequent historical re-interpretation), but one of essential economic growth spurred by the potential for commercial reward. Columbus did not set out to discover the New World in 1492 for the sake of exploration—he wished to open up a westward seaborne trade route to Cathay (China) which would be more rapid than the existing land 'silk road' trade route across Europe and Asia and which would circumvent the Muslim-dominated lands. Indeed, by 1504, Breton fishermen were trawling cod off the Newfoundland coast. A similar motivation lay at the heart of the American pioneering trek westwards during the nineteenth century—gold. However, the historical reinterpretation process is instructive—humanity is quite ready to reinterpret historical activities in terms of more noble values, yet not as a justification for present-day activities—it appears that noble values recede when it comes to immediate financial commitments. The nobility of the human spirit it seems is only a virtue in the past, and that achievements of the present and those of the future are viewed as mere cost entries in a national or international nominal ledger.

The wisdom of developing the International Space Station has been regarded in various quarters as debatable—there is little doubt that it was backed by NASA on the basis as a logical extension to NASA's manned STS programme and its successes as an in-orbit servicing facility through astronaut EVA during the 1980s and 1990s. NASA's defence of its manned capability is motivated by the fact that the human spaceflight was the central and most dramatic component of the 'Space Race', a component it naturally wished to retain. NASA is quite justified in that goal as recent history has shown that loss of technological capability is rarely recovered—the UK's edge in Europe in launch capability in the early 1970s appears to have been lost permanently.

Over 15 years, five major space station redesigns have cost over $12.2b up to 1995 without orbiting any hardware. It had been estimated that ISS would cost a further $17.4b to complete in 2002. US Congress had tried on a number of occasions to kill the Space Station programme without success. The Space Station's present budget is $2.1b/y culminating in a total cost of some $40b. It has now been recast as the International Space Station (ISS, originally named Freedom and subsequently Alpha) to symbolise post-Cold War cooperation between Russia and America in particular. The ISS offers a unique platform for forging international relations for involved nations [Massevitch 1989]. ISS is a fully multinational effort with the USA leading with Russia as its main partner. Providing Russia with the opportunity to contribute their space capabilities to the ISS eased US fears of uncontrolled proliferation of sensitive dual-use technology to unfriendly third parties. Furthermore, their expertise in long-duration manned spaceflight is extensive gained from the Mir space station launched in 1986. Furthermore, Russian involvement was originally perceived to save money. NASA paid $728m to Russia between 1994 and

1998 but the financial burden is expected to increase by a further $660m due to Russia's debilitating economic problems. Furthermore, the delay of the Russian service module with its re-boost capability to prevent ISS orbit decay over 500 days and European ERA manipulator arm has necessitated the USA in developing its own contingency, the Interim Control Module.

The ISS will be a permanent manned facility orbiting at an altitude of 400 km and inclination of 51.6° for 85% coverage of the Earth but within reach of LEO launchers. It comprises seven pressurised laboratory modules, two pressurised crew habitation modules for life support, and two logistics supply modules with robotic and power functions—all in all, over 30 separate modules. The main modules and interconnecting nodes will be mounted onto an integrated truss structure. It is 110 × 75 m in size with a mass of 418 tonnes. For a seven-man crew, it will provide a total enclosed pressurised volume of 11 311 m^3 and solar array power of 189 kW. The US elements include three laboratory modules, four of the eight solar arrays giving 110 kW, a habitation module, an unpressurised logistics module and a centrifuge module. It is also developing its 10-tonne X-38 re-entry vehicle with a seven-passenger capacity as a backup crew return vehicle for operational service in 2003. Russia is providing two laboratory modules, a habitation service module with propulsion capability for ISS re-boost, an unpressurised Progress logistics transport vehicle and a Soyuz emergency crew transfer vehicle with a passenger capacity of three for service until 2003. Canada is providing the two external robotic arms. Japan is providing its Japanese Experimental Module as one of the laboratory module with an exposed exterior platform to be launched in 2002. ESA's pressurised Columbus Orbital Facility is the main European contribution to the International Space Station as one of the laboratory modules to be launched in 2003. This is a multipurpose cylindrical pressurised laboratory (6.7 m in length by 4.46 m diameter) permanently attached as part of the ISS. It has a capacity of ten international standard payload racks for materials science, life science and other experiments. The other main element is the Automated Transfer Vehicle. The ISS will use momentum wheels, thrusters, gyroscopes and GPS receivers for attitude control and communications will be via TDRSS through a 70-kbps S-band uplink and 43-Mbps Ku-band downlink. The ISS's primary purpose is for scientific microgravity experiments in life sciences, materials sciences and remote observation of the Earth and astronomical objects. Around 60% of scientific experiments are devoted to materials science. Both laboratory racks and externally mounted sites are available—30% of ISS payloads are open to commercial users. The first phase of the ISS was the Shuttle–Mir programme which began in 1995 involving US astronauts serving onboard Mir and Shuttle–Mir docking manoeuvres. ISS will require 45 US and Russian flights with 1700 h of EVA to assemble the most complex engineering construction ever attempted plus 50 more launches every three months for supplies, crew rotation and fuel delivery. The Shuttle launches alone will cost $20b in total. The first permanent crew will board the ISS in 2000.

As an Earth-orbiting facility, ISS has utility for Earth-oriented applications as a multi-purpose platform: military observation (such as nuclear non-proliferation), large-scale Earth observation (especially unforeseen situations such as natural disasters for insurance assessments), mobile satellite communications technology testbed, detection of resources for exploitation such as mineral and oil deposits, and land utilisation policy. All these functions are essentially commercial but would probably be accomplished more efficiently through unmanned dedicated platforms.

The ISS is being sold as an experimental platform for microgravity research. The development of microgravity processing has been slow due to the speculative nature of the endeavour and the high costs. Commitment has been fragmentary due to the lack of an existing infrastructure of freeflying space platforms and guaranteed access to space. Several microgravity experiments have been performed on the Space Shuttle and Mir space station, and will subsequently be performed on the ISS. IML-2 (International Microgravity Lab) onboard ESA's Spacelab in 1994 was a highly successful microgravity experiment mission lasting 15 days. It involved 82 biotechnology, protein crystal growth, alloy solidification and near-critical-point experiments. The shuttle maintained an altitude of 300 km and the overall noise level from 0 to 50 Hz was $10^{-3} g$. This mission also demonstrated remote telescience operations to allow ground based users to participate in the payload operations in ESOC. Such missions provide the model for the commercial US Spacehab based on ESA's Spacelab which has been developed as a pressurised module for the shuttle cargo bay with experimental lockers. It has a pressurised volume of 30 m^3 capable of accommodating 1.36 tonnes of payload to fit into the shuttle payload bay. It provides 50 mid-deck lockers for microgravity experiments. A shirtsleeve environment is provided for the two payload specialists. Each locker can accommodate 27 kg of payload in a volume of 0.057 m^3. The lockers are available for rent but they rely on shuttle flights being extended to a month.

The major commercial application pleaded for the ISS is microgravity materials processing. The principal effect of microgravity is to reduce the importance of density differences in manufacturing such that buoyancy-driven convection is minimised. Drop towers, parabolic flights and sounding rockets are limited to microgravity durations of the order of seconds to minutes. Materials processing for the manufacture of products in the microgravity environment of space $\sim 10^{-4}$–$10^{-5} g$ is a potential application of space technology that will be investigated on the ISS. It offers the potential for new products and production methods with higher quality and efficiency. It offers reduced sedimentation of particles in fluids, reduced buoyancy of gases in liquids and reduced mixing by convection in fluids. Thermal convection is buoyancy-driven and generated through surface tension and gravity characterised through the Bond number of the fluid defined as the ratio of gravitational force to surface tension forces. If the Bond number is much less than unity, gravitational effects are negligible. The main contribution to heat transfer in melts under microgravity is due to Marangoni convection which occurs when Marangoni convection heat flow exceeds the thermal conductivity capacity of the melt [Chun & Wuest 1978]. In this case, the effects of temperature differences affect surface tension— Marangoni convection is a thermocapillary effect caused by surface tension at the liquid/ gas interface in the presence of a thermal gradient. Surface tension minimises the surface area-to-volume ratio of the liquid by adopting a spherical shape. It moves from hot to cold regions as the interface drives the motion of the fluid such that bubbles grow rapidly with a few stable large bubbles which do not disperse through the liquid. Hotter regions of the liquid tend to spread at the expense of the cooler surface areas, thereby driving mass cellular convection. The dominance of surface tension allows the fluid to assume perfectly spherical shapes in the absence of gravity. The potential for containerless processing eliminates wall temperature gradients in space. The major potential is for high structural homogeneity due to low levels of convection providing pure diffusion growth for crystallisation with low microsegregation and sedimentation. The reduction of sedimentation in alloy processing is highly desirable especially where the alloy materials are highly disparate, e.g. Bi–Mn, Au–Al, Al–In. The reduction in crystal growth fluctuations in microgravity imply increased material purity due to the lack of constraint on the floating liquid zone length in zone refining [Chun & Wuest 1978]. Defects and dopant inhomogeneity in the crystal lattice of semiconductor crystals adversely affect their performance. Hence, it is desirable to grow defect-free semiconductor crystals without gravity-driven turbulent convection. Microgravity minimises lattice defects and impurity segregation. The almost pure vacuum of space $\sim 10^{-10}$ torr also favours molecular beam epitaxy of vaporisation and deposition of layers of semiconductor materials.

Microgravity also has certain physical effects on biological materials which affect transport and growth processes of living organisms. ESA's Biorack has been used extensively to study the effects of microgravity on biological life forms. Certain results are evident in unicellular organisms such as bacteria, algae and protozoa:

(i)   enhanced cell proliferation and cell division;
(ii)   circadian periodicity maintained;
(iii)   reduction in glucose consumption;
(iv)   increase in antibiotic resistance in bacteria;
(v)   immunity system lymphocyte activation reduced by 90%;
(vi)   biosynthesis of interferon is five times greater;

For higher plants, other results are evident:

(i)   normal seed germination;
(ii)   loss of shoot/root relative orientation;
(iii)   no major morphological disturbance in plant development for 1+ weeks;
(iv)   after 2 weeks or so, functional integrity begins to become affected by changes in reproduction, viability and functional integrity generally.

The potential for biotechnology is vast. The biotechnology industry is founded on electrophoretic separation such that an electric field applied to two electrodes imparts a force on charged biological materials. The Earth's gravity field however causes partial remixing since heat from the applied current generates thermal differentials and so gravity-driven convection. Microgravity conditions increase the electrophoretic efficiency. Protein crystallisation enables three dimensional structure to be determined using X-ray and neutron diffraction techniques. The growth of such crystals requires stabilisation of individual molecules to allow aggregation through weak interactions. Unfortunately, gravitation can overpower such weak interactions through sedimentation and convection, a problem overcome by microgravity conditions providing larger and better structures. Such research is perceived to lead to large-scale space manufacturing of high-value products. Protein crystal growth studies should advance knowledge for the development of designed drugs (with an estimated market potential of $100m–$1b). The first commercial products to derive from microgravity research are likely to be high value products in small production batches such as designer drugs, proteins and semiconductor crystals, and advanced structural composite materials. For example, the protein α-interferon used to treat hepatitis remains in the bloodstream for several hours longer if manufactured in space than on the ground, offering a potential for commercial utility [Beardsley 1996, Williamson 1999b].

Microgravity experiments are strongly dependent on the manned space programme with high costs. As far as microgravity studies are concerned, it is generally considered that unmanned robotic platforms such as Eureca are superior as crew movements onboard the ISS jeopardise the microgravity condition of $10^{-6}g$ to the extent that experimental results may be invalidated and Eureca-type platforms are a good deal cheaper. Furthermore, gas venting from the ISS will degrade the near-vacuum environment. Indeed, the Space Shuttle which was originally conceived to reduce launch costs to LEO to \$300/kg by virtue of its re-usability (unsuccessfully—it still costs \$20 000/kg), has itself become a miniature re-usable space station with Spacelab and Spacehab. At least 20 scientific organisations around the world have suggested that the ISS as an experimental platform is a waste of money. The long flight design and integration timescales for space experiments in conjunction with improved terrestrial techniques for crystal growth limits the prospects for such technological advances on the ISS. The Eureca platform may be resupplied or refurbished orbitally. Eureca-1's primary payload of 16 experiments is for microgravity experiments in materials science and life science and 80% of the payload is reserved for that purpose with 20% for space science and space technology purposes [Nelleson 1992]. The experiments on Eureca 1 were: the Automated Mirror Furnace Facility to study the growth of semiconductor crystals; the Multifurnace Assembly to study the growth characteristics of alloy materials; the Solution Growth Facility to study the growth of organic crystals of low solubility; the Protein Crystallisation Facility to study the growth of protein crystals; the Exobiological Radiator Assembly to study the effects of exposure of biological materials such as bacterial spores to the space environment to determine their survivability. These experiments comprised the core facilities and two other microgravity instruments were included: the High Precision Thermostat provided uniform temperatures for calorimetry with ~100 μm accuracies; the Surface Force Adhesion Facility to examine liquid behaviour in microgravity. Space science experiments are particularly suited to the long observation periods and to the opportunities for Sun-pointing solar observation. The Solar Spectrum instrument and Solar Constant instrument were included to examine the characteristics and possible effects of the sun on the Earth's climate. Other astronomical instruments included: the Occultation Radiometer to measure the vertical variation of aerosol gases in the Earth's atmosphere; the Wide Angle Telescope to detect and locate X-ray events; the Timeband Capture Cell to study microparticle populations in space similar to LDEF. Space technology experiments were also included: an inter-orbit communications package provided the communications link via the geostationary European Olympus communications satellite as the relay; an rf ion thruster assembly with a thrust of 10 mN was tested; an advanced GaAs solar array was also included for in-orbit demonstration. Hence, Eureca was a multiple payload spacecraft. It provides a multi-role capability and its capabilities for in-orbit demonstration suggests its use to demonstrate in-orbit refuelling and in-orbit manipulation. Eureca 2 and Eureca 3 are cooperative flights with NASA such that Shuttle launches and retrievals were provided free of charge in return for a share of the payload resources.

There is little doubt that in-orbit servicing of spacecraft will be a major role for the ISS, building on EVA operations of the Space Shuttle and its own construction process. The only leverage provided by the ISS is in the assembly, checkout and provisioning of spacecraft that are too large for launch into LEO by a single booster, but in-orbit assembly robots could provide that facility at lower cost. An in-orbit servicing robotic system such as ATLAS will become a baseline for further robotic and automated activities in space. General-purpose manipulation represents the most varied and diverse sets of tasks that can be performed in space. A subset of this includes the servicing of scientific payloads on dedicated platforms to maintain experiments in controlled environments such as through reagent replenishment, product harvesting and product sampling. Indeed, Sheshkin (1985) proposed that an unmanned space platform controlled through telepresence from the ground with automated capabilities would be a lower-cost alternative to a permanent manned space station. The development of a robotic infrastructure such as ATLAS would reduce the technological risk of investment in microgravity research. Maintenance of individual microgravity platforms by robotic resupply and refurbishment increases the number and variety of experiments per platform, and platform mission lifetimes. It has been suggested that a space platform capable of mounting three Spacelab-type pallets ~1800 kg for experimental payloads (especially for microgravity materials processing) would alleviate the problems of the corruption of the microgravity environment in space caused by human activity in manned missions. Pallet handling and the addition of new pallets would require manipulators. Resupply of the platform and experiment changeout may be accomplished autonomously. More immediately, the NASA Robotic Operated Materials Processing System Project is a GAS experiment to determine the feasibility of producing semiconductor materials autonomously.

Microgravity manufacturing of alloys, pharmaceuticals and semiconductor crystals may become one area of future commercial benefit which will rely critically on the development of robust robotics and automation technology. The future of space-based manufacturing will ultimately depend on lessons learned in general-purpose manipulation tasks in zero-gravity. This commercial application would mark the start of space industrialisation and the advent of large space platforms dedicated to manufacturing controlled robotically. All space activities require transport of materials from Earth which is ultimately dependent on launch costs. For microgravity product applications such as pharmaceuticals, this cost must be recouped from the commercial market. Microgravity platforms offer potential for future commercial application, but it requires an in-orbit infrastructure which is presently limited to communication platforms in GEO and short-duration manned flights. It has been calculated that space manufacturing could generate ~$42b/y in sales, of which pharmaceutical production alone would account for $29b/y [Sepehri 1987]. Generally, highly product-specific cost items are amenable for space-based manufacturing—the breakeven product price is $25/g for launch costs of $10/g—pharmaceuticals and semiconductor crystals presently command ~$100/g. The raw material cost of high-grade silicon is $25/g compared with the processed wafer price of $250/g. Microgravity may be seen as the catalyst to orbital manufacturing facilities. NASA's Wake Shield Facility demonstrated the potential for automated production of GaAs crystals by molecular beam epitaxy at a rate of 10 million wafers per year with a potential market of $20b. The space shuttle external tank offers 2000 m^3 of pressurised volume and provides a potential manufacturing facility, repair shop or storage facility if it is transported all the way into orbit rather than being jettisoned prior to orbit insertion. As real estate it could be fitted out in orbit with customer pallets, experiments, manufacturing facilities and externally attached pallets. It may be that a space station facility could provide the basis for a semi-automated Earth-orbiting space manufacturing facility as envisioned by O'Neill (1976). A space manufacturing facility allows the production of large and fragile products for use in microgravity.

As far as ISS being a stepping-stone to other celestial bodies such as the Moon or Mars is concerned, it is worth noting that this concept is also debatable as the Apollo missions show that direct access is perfectly feasible, and similar approaches have been suggested for Mars. However, such sprints could remain just that—sprints with no consolidation to establish permanent presence, re-enacting the same desertion of the Moon after 1972. ISS will play a major role in determining the long-term exposure of humans to microgravity. This will be critical in determining the plausibility of any future manned mission to Mars. The primary justification of the ISS programme is to learn how human beings can live in the microgravity conditions of space on a regular long-term basis to support long-duration manned space activities, particularly those required for an extended manned mission to Mars—it is an essential process of defining a consolidated manned infrastructure in space.

In attempting to appeal to multiple ideals with the ISS, the Shuttle programme provides a lesson in loss of focus due to lack of clear objectives. As an attempt to satisfy all possible space development scenarios with a flexible short-term goal, the Space Station programme was retrieved from the SEI as a compromise. This multipurpose approach to the utility of ISS leaves open NASA's choice of longer-term goals. Therein lie the dangers of attempting to promote a jack of all trades to appeal to as many visions as possible. The only solution is a series of small, well-defined intermediate goals, each of independent value and significance but providing an incremental approach to long-term objectives. It is true, however, that the ISS offers a general-purpose platform which can accomplish multiple functions rapidly and in parallel. It possesses the advantage of rapid re-tasking and flexibility on a short timescale. There is little doubt that if NASA had relinquished its expertise in manned spaceflight by concentrating on more politically and economically supportable activities such as unmanned Earth-oriented applications, 'man in space' as a symbol of human achievement would eventually have become a forgotten and minor chapter of squandered opportunity by a weak and ineffectual people in the history of the human race. For, as we have seen, it is in future history that our actions of today will be judged—as a people of vision and tenacity or as a people enslaved by navel-watching and triviality. The thrifty of vision are condemned to the annals of the forgotten. It is the adventurous of spirit who breathe life into the annals of the legendary.

21.2 FUTURE LUNAR MISSIONS AND LUNAR RESOURCE UTILISATION

Beyond ISS, there is little doubt that the focus of attention will be beyond Earth orbit. Robotic planetary missions to both the Moon and Mars and beyond have recently established renewed interest in the search for life. A permanent return to the Moon represents an attainable goal through the initial establishment of robotic and automated lunar stations. This would follow naturally from the earlier US space programme with the eventual pursuit of manned missions at a later date once the supporting lunar infrastructure had been established. Such a lunar return would involve no requirement for major new developments in technology beyond that available today. From 1969 to 1972, beginning with the historic Apollo 11 mission landing on the Sea of Tranquillity in July 1969, six Apollo missions (Apollo 13 aborted the landing *en route* to the Moon after an oxygen tank explosion) placed 12 men on the surface of the Moon at a cost of $80b (or equivalently, 5% of the annual US federal budget). This was preceded by two sets of US robotic spacecraft, the hard landing Ranger and the soft landing Surveyor series. The USSR sent robotic spacecraft to the Moon during the same period. NASA at present has no plans to return to the Moon due to its over-commitments to other programmes in the light of its tightening resources and its commitment to the International Space Station programme. NASA's present budget is $15b/y, most of which is spent on its STS operations and the Space Station programme. Resources for new missions or initiatives such as lunar colonisation or industrialisation are not presently available. However, there is little doubt that before any comprehensive mission to Mars could be undertaken, attention must be focussed on the Moon to develop the basic technologies needed to pursue such a mission. The basic lunar mission involves launch into GTO followed by injection into lunar transfer orbit (Δv of 650 m/s), capture into high elliptical lunar orbit (Δv of 150 m/s), 90° inclination change for polar orbit (Δv of 105 m/s), circularisation into low lunar orbit at 100 km altitude (Δv of 650 m/s), lunar landing on the surface for the lander (Δv of 2.25 m/s). A return mission also requires relaunch back to Earth (Δv of 2.9 km/s). For the main orbital manoeuvres, a restartable bipropellant engine of high thrust is required. The ultimate goals could be the establishment of resource exploitation or use as a testbed for a future manned mission to Mars, or both. Such a return to the Moon would be costly in capital, risk and time—the return on investment is likely to be considerably deferred beyond the 10–15-year return on investment period capacity of private enterprise. Furthermore, the capacity for economic utilisation of lunar resources is as yet undeveloped. Only when capital barriers are reduced will private participation be possible—public investment should seek to create greater opportunities for private sector exploitation, e.g. public-funded lunar facilities could be leased to the private sector [Sterner & Benavoya 1994]. Most lunar colonisation scenarios are perceived to be conducted across three broad fronts. Firstly, robotic exploration for lunar investigation and surveying for (secondly) the establishment of an automated teleoperated/automated infrastructure on the lunar surface would be essential; but as activities on the Moon become more complex, it is certain that (thirdly) humans would follow to construct and man permanent bases. A lunar polar orbiter for mapping and imaging of the lunar surface (e.g. Lunar Surveyor) has already been achieved. The next step would be an unmanned lunar rover of 6–8 tonnes equipped with manipulators, 1 m depth drilling devices, analysis equipment and a series of surface stations. Lander instrumentation could include seismometers, multispectral cameras, magnetometers, heat flow probes and a number of analytic spectrometers. A lunar rover explorer with manipulators plus tools for drilling, TV cameras, sample analysis spectrometers and soil-analysis packages similar to ALSEP

(Apollo Lunar Surface Experiment Packages) could be deployed by the rover to provide more detailed and geographically diverse data. Japan's SELENE (SELenological and ENgineering Explorer) to be landed in 2003 is the only likely foreseeable such mission. A lunar sample return mission (similar to the original Rosetta mission) to return 50 kg of soil samples in a small return capsule using a drill, surface sampler and robotic arm would provide augmentation of those recovered by the Apollo lunar landings. The first manned lunar stations are likely to be similar to polar stations and/or offshore oil rigs. This would require habitats and experimental modules employing environmental control. Kozlov & Shevchenko (1995) suggested a mobile lunar base comprising of three vertical cylindrical with a diameter of 5 m and a height of 7 m set into a triangular girder construction as a kind of manned version of Lunokhod 1 and 2 with three leg-mounted tracks—it would not be significantly larger than the Apollo lunar excursion module and would be deployed automatically prior to the manned flight. As the primary drivers are launch vehicle payload restrictions and the interface with the lunar lander, such a system would require assembly in Earth orbit by teleoperated robots or ISS-based EVA.

A permanent re-usable lunar transfer space tug of similar size to the Apollo CSM/LEM spacecraft of 30 tonnes and 15 tonnes respectively could provide the transportation infrastructure between lunar orbit and Earth orbit. It may be chemically propelled initially for robotic lunar activities evolving to a nuclear-electric propulsion system with a CSM segment of 125 tonnes and LEM segment of 100 tonnes for manned activities—this would require heavy lift launcher capability. There is a lunar injection window from Earth orbit every 6–11 days (average 9 days) that lasts less than one day. The translunar trip takes 3–5 days across the 384 000 km distance. The translunar injection should be targeted to lunar inclination to avoid severe performance limits. The lunar orbit inclination is driven by the latitude of the landing site. For near-equatorial sites an equatorial orbit provides the best access with near-continuous access for latitudes greater than 85°. The advantage of the L_1 and L_2 libration points is that they provide continuous access to the entire lunar surface. Mission transit time is longer with 4–9 days for Earth–lunar transfer and 2–3 days to the lunar surface from the libration points. Earth-based tracking and communication limit landing sites to 165° E/W longitude. Autonomous operation for routine productions, with remote control interrupt facility for non-routine operations offers the optimal control strategy.

The major operational problem of the lunar environment is the pervasiveness of lunar regolith dust which adheres electrostatically to equipment [Johnson et al 1995]. Around 25% by mass of regolith is less than 20 μm in size and around 10% is less than 10 μm. It appears that a levitation process occurs due to the day/night terminator causing ultraviolet photoconductive charge differences. This causes dust ~5–10 μm to be lifted around 3–30 cm above the local horizon generating a column density of 5 grains/cm^2. Furthermore, most operational activities will scatter dust up to 1 km from the location of that activity. Dust mitigation will be an important part of lunar operations, particularly due to its degradation effects on solar arrays and radiator surfaces. Optical surfaces are also particularly vulnerable. Electrostatic fencing, shields and baffles may be appropriate techniques for such mitigation. The ultra-high vacuum of 10^{-9} to 10^{-12} torr may cause outgassing of equipment, particularly synthetic materials such as seals or bearings, making them brittle. Another consideration is the 14 lunar day/night cycle which imposes power generation problems during the night portion. Industrial operations limited to lunar day-time (equivalently, 50% utilisation of equipment) may be economically unacceptable. To overcome this, it may be necessary to use nuclear power sources ~kW or move the lunar processing plant into lunar orbit. Alternatively, a photovoltaic array may be used in conjunction with regenerative fuel cell storage, though it is not clear if such a system is operable over such long active/dormant cycles for power ratings greater than 50 kW. This day/night cycle causes temperature variations from 110°C during the day to −170°C at night. Even during the day, temperature differences of up to 280°C may occur between sunlit and shaded regions of equipment surfaces. These factors impose considerable thermal control problems. A manned lunar base would require regolith to be excavated to cover the structure against meteoritic impact and radiation exposure from solar, cosmic and secondary radiations. The permissible doses for radiation workers is 5 rem/y and 200 rem for life accumulated exposure (this last is the dose that increases the risk of dying from cancer by 3%). Such doses require a regolith depth of 400 g/cm^3 (3.5 m thick). Shielding is likely to be the most massive component of a manned lunar base, so the use of regolith represents a considerable mass saving. As radiation levels increase to 1000 rem at solar maximum, temporary storm shelters would require 700 g/cm^2 shielding—6 m equivalent depth of regolith. Generally it is expected that any manned lunar base would comprise eight astronauts. The Shuttle provides 2–5 m^2 living quarters area per person but the Space Station will provide 10 m^2. For a manned lunar base, 15 m^2 represents a reasonable living area per person with 6-month stay times. Such a base could be constructed from two Space Station-like modules, each of 60–70 m^2 internal area and 25–30 tonnes mass capable of a crew of eight plus a power generation facility. Extension and expansion may be attained by providing additional modules. Power is required throughout the lunar day/night cycle—each habitation module requires 15 kW and each laboratory module requires 30 kW; a FBR (fast breeder reactor) nuclear reactor with liquid lithium cooling and a Brayton cycle gas turbine system using an He–Xe working fluid would be suitable. Current space-rated nuclear power plant designs have average power outputs of 300 kW with an operational lifetimes of 10 years in a mass of 7.5 tonnes, i.e. 0.04 kW/kg.

Jones (1992) compared the possible terrestrial analogues to space economic development. The Antarctic, although laden with mineral wealth, has been subject to little prospecting due to the cost of coping with its permanent thick ice cover—after Roald Amundsen and Robert Scott reached the South Pole in 1911, permanent occupation of the Antarctic did not begin until 1942. In North Alaska and Canada, however, resource development has been driven by short-term economics such that only highly concentrated deposits of high-value minerals (usually oil) have been exploited. The Siberian Arctic was developed heavily under Stalin as a consequence of the political desire of the USSR to attain resource self-sufficiency. The British settlement of the Sydney region of Australia as a penal colony created a local market for goods only due to the high transport costs to Europe. Local development was followed by the development of export of wool due to its high value. Furthermore, as pointed out by French (1995), historically ocean exploration was possible only because the ships were replenished along the way with consumables whereas today's space missions carry all their supplies from the point of departure imposing huge mass penalties to space vehicles. The lesson is that the only practical way to accomplish two-way exploration missions is to reprovision consumables from local resources [Bank & Kassing 1993].

These scenarios have relevance to lunar, planetary and asteroidal utilisation and industrialisation, though Martian colonisation is difficult to justify economically as yet. Space can provide in the long term vast quantities of raw materials, abundant clean solar energy, safe waste disposal and the industrial environment to relieve the Earth's biosphere of heavy industrialisation and pollution that has ravaged it since the Industrial Revolution [Maitra 1989]. The Club of Rome, based on Forrester's mathematical model of the world economy, advocated the notion of limits to economic growth which can only be relieved by expansion into space to acquire new resources and it is projected that this process must be achieved by 2040–2070. The Forrester world dynamics model simulated the world's ecology through five major variables: population, natural resources, pollution, capital investment in agriculture and extraction, and capital investment in technology. To maintain quality of life, population and resource depletion must be stabilised such that they do not exceed the technological substitution rate. Whereas the Earth's resources are finite and not evenly distributed throughout the world, imposing problems in the rational distribution of wealth, the space environment has abundant resources where the rational distribution of resources becomes a non-zero-sum game beyond Malthusian limits of exponential growth of population in conjunction with exploitable resources growing linearly with technological advances [Parkinson 1991]. At present, the high cost of space transportation requires research and development funding to come from governments until an adequate infrastructure of reliable space transport and product handling facilities has been developed for private investors to utilise to provide a quantitative benefit to taxpayers. Furthermore, this high transportation cost will be the major factor in determining the cost of any extraterrestrial products. Establishing a space infrastructure requires a re-useable launcher technology with a minimum capability of 100 tonnes and cost of $400/kg (assuming launch of a 350-tonne minimal 'base' supplied with 2 kW of power per person). Water is probably the most important commodity that exists in space—the Moon, Mars and asteroids all have a complement of water of indeterminate size. Both lunar and asteroidal material invoke low transport costs due to the low gravity of these bodies.

Although most extraterrestrial utilisation will be for local space markets, lunar industrialisation is a long-term goal which may be approached in incremental steps and which ultimately may have potential for commercial returns. This process of lunar commercialisation has all ready begun—Lunacorp Inc from Virginia is undertaking the first privately funded lunar mission by launching and landing on the lunar surface, a pair of unmanned teleoperated robotic rovers. Telepresence based on virtual reality technology will allow customers to drive the rover on the lunar surface remotely from Earth. It will also conduct scientific exploration over two years, visiting the Apollo 11, Ranger 8, Surveyor 5, Apollo 17 and Lunokhod 2 landing sites, and the lunar north pole, in a 1000-km traverse, collecting data, including ice-deposit characterisation, on the lunar environment. Such commercial investment by private sources is essential for the further development and exploitation of any lunar infrastructure. The cost drivers to such exploitation are the cost of lunar transportation and lunar operations. All lunar equipment must have low mass/unit output, be highly automated, and be repairable from the Moon. The cost of transporting goods from the lunar surface to GEO ~2.44 km/s^2 is significantly less than that from the Earth's surface to GEO, due to the much lower potential energy differential—the local space market takes less than 5% of the energy to lift payloads from the Moon as it does from Earth. The linear asynchronous electric motor (electromagnetic mass driver) has been advocated for launch of bulk cargos from the lunar surface [O'Neill 1978, O'Neill & Kolm 1980]. A mass driver with 40-cm drive coils is capable of launch accelerations up to 1800g (though structural considerations of drive coil stresses restrict accelerations to 100g). The cargo is accelerated on the lunar surface to high velocity catapulting them into space. Such a mass driver of 160-m length could impart heavy cargos to a lunar escape velocity of 2.44 km/s to the Earth–Moon L$_2$ libration point some 63 000 km beyond the lunar farside and L$_1$ libration point beyond the lunar nearside with high accuracy [O'Neill 1978, O'Neill et al 1978, O'Neill 1980]. Mass catchers at these libration points, typically large, high-strength Kevlar nets some 100 m in diameter, capture and decelerate the launched payloads from which there is ease of access from the moon and the Earth. The mass driver comprises a pipe formed by a stack of superconducting coils within which a series of buckets are supported and guided by sequentially pulsed magnetic fields through the coils which accelerate each bucket through the pipe by currents discharged sequentially in the outer drive coils. It can accelerate up to 10 km/s^2 to lunar orbit along a track of 2.7 km. It can transport up to 300 times its own mass in 5 years. Around 80% of the mass of the mass driver is in power generation and distribution. Lunar scientific activities such as astronomical telescopes to study the solar system and the universe in general could piggyback lunar industrial activities at lower cost so that instrument cost rather than lunar transport cost becomes the dominant financial investment. Any return to the Moon must minimise the mass of material transported there, and nearly all the material needed for the construction of a lunar base are available on the lunar surface and could be exploited by bootstrapping. Such utilisation of local materials would reduce the cost of lunar operations considerably [Happel 1993].

The lunar surface comprises two regional types, the highlands and the marias. Highlands are rugged and heavily cratered, comprising 85% of the surface. Marias are basaltic plains. Lunar regolith can provide radiation protection with 400 g/cm^2 of soil. The Surveyor spacecraft soft landing in 1966 and subsequent manned lunar landings characterised the soil as relatively dense, fine-grained and weakly cohesive soil with a bulk density of 1.8 g/cm^3. Lunar mineral approximately comprises an average of 21% silicon (for solar cell and electronics manufacture) and 42% oxygen (for oxidiser), 13% iron, 7% aluminium, 6% magnesium and 3% titanium (for metallic structures). On Earth, only three of the most important metals (Al, Fe and Mg) are present in amounts over 2%. All other useful metals occur in amounts below 0.1%. Metal is generally present in their most basic forms, i.e. oxides, sulphides and silicates.

Fabrication techniques would be potentially easier on the Moon due to the low gravity and vacuum conditions. Self-contained automated materials processing plants could extract over 99% pure elements from feedstock. The extraction of metal on Earth comprises some 35% of its energy cost, of which the major component is the mining. Open-pit mining is the cheapest form of mining, generating rock fragments of ~1 m in size. The US Bureau of Mines has researched into teleoperations applied for continuous mining in deep thin coal seams. Drilling, boring and coring are the basic mining tasks and may be accomplished by standard terrestrial methods of quarrying through extraction, lifting and hauling. Standard drilling techniques such as percussion drilling is directly applicable. Astronaut-held tools for these activities were used during the manned lunar landings. Automated and teleoperated lunar equipment must require virtually no upkeep, maintenance and repairs. Bucket-wheel excavators, movable conveyors and spreaders may provide the basic capabilities for mining and earth moving. All basic tasks may be accomplished by three vehicles: a rover, a truck haulage vehicle, and an excavator loader vehicle. Typical power requirements for such vehicles are around 30 kW per vehicle. Tracked equipment is not suitable for lunar operations due to high uncertainty of many moving parts and extensive maintenance requirements. Wheels offer greater mobility and versatility. The rover could operate for 16 hours, with 8-hour battery recharging from solar arrays every 24 hours during the lunar day, and remain dormant during the lunar night. Maximum speed should be limited to 0.3 km/h teleoperated and it should be capable of negotiating 0.4-m obstacles. The Boeing rover had a mass of 2.8 tonnes and a design life of 15 years. A self-contained mobile lunar miner which excavates lunar regolith to a depth of 3 m by loading regolith at the front of the vehicle with a slide-bucket system has been proposed to separate out particles of less than 50 μm, heat them to 700°C and collect the gases in high-pressure cylinders with subsequent cryogenic fractionation while ejecting the processed regolith back to refill trenches in the lunar surface.

Mineral processing involves milling, crushing and grinding to small particle sizes (1 mm–100 μm sizes) in several stages to physically separate out the valuable mineral concentrate from the waste rocky gangue. This process is critical as it determines the recovery fraction but it also consumes up to 50% of the energy input to material processing. Washing of the crushed rock to remove sticky clays is not necessary on the Moon. The equipment to mine lunar resources may be based on terrestrial techniques but without the emphasis on water/air for cooling and the reliance on gravity implied in beneficiation. Separation of the concentrate from the tailings may involve a variety of techniques, most of which involve the use of large amounts of water. Separation by differential movement of minerals due to specific gravity differences in a fluid suspension (gravity concentration) and surface affinity of minerals for rising air bubbles in a fluid reagent (froth flotation), typically toxic such as tetrabromoethane diluted with carbon tetrachloride, are the most important techniques with versatile capabilities but are totally unsuitable for lunar operations as they require fluid media [Podnicks & Sickmeier 1994]. Ferromagnetic (such magnetite Fe_3O_4) and non-ferrous paramagnetic materials (such as ilmenite $FeTiO_3$, rutile TiO_2, pyrrholite FeS and haematite Fe_2O_3) can be separated magnetically, and any mineral can be separated by electrostatic concentration as all minerals show differences in electrical conductivity; both methods should be suitable for lunar beneficiation. Material production may be automated and mining may be remote-controlled from Earth.

As well as regolith, lunar construction cement represents a further development in materials processing. Blocks could be made from a lunar brick factory based on current ceramics processing technology capable of producing 200 000 bricks per year [Sullivan & McKay 1991]. The feedstock is mined and coarse rock removed by physical filtering. Much lunar surface rock is all ready pulverised so it does not require crushing. The resultant soil would be loaded into a mould and pressed, followed by heating to 1100°C to produce the sintered bricks. Such a factory would have a mass of 25–40 tonnes with a power requirement of 375 kW. The Japanese Shimitzu corporations have proposed the mixing of lunar regolith for plagioclase feldspar, $CaAl_2Si_2O_8$, as concrete to form concrete hexagonal module blocks. The lunar cement of high Ca content would require water from Earth (hydrogen is virtually absent on the lunar surface). Furthermore, concrete requires a pressurised environment and is brittle, requiring reinforcement. It is, however, versatile but energy-intensive to produce. Cost pricing of lunar products will vary with production volume which is presumed to increase with time, as is production efficiency. This will cause a shift from raw materials such as lunar regolith to higher-value products such as lunar cement. Imports would need to be minimised as these will contribute to the cost of products. Costs are perceived to drop from $1100/kg to $40/kg for raw materials, $700/kg to $350/kg for simple fabricated products, $120/kg to $55/kg for lunar oxygen over time [Koelle & Johenning 1994]. Hence, space resources will initially be almost exclusively used in space, and this requires support and funding from governments to develop the basic infrastructure.

All chemical processing techniques are based on electrolysis or thermochemical reduction [Koelle 1982]. Smelting of metallic ore is the commonest terrestrial technique but is energy-intensive and implies high mass and volume. Carbothermic reduction is a general closed-cycle chemical processing technique. Raw material in the form of ferrous oxide and/or metallic silicates (the minerals olivine, pyroxene and plagioclase are metal silicates that are the commonest minerals throughout the solar system) is subject to hot methane (methane is a much stronger reducing agent than hydrogen but carbon can also be used) which reduces the material to metal, producing CO and H_2 gas which may be fed into a Sabatier reactor to produce methane and water [Rosenberg 1997]. The methane may be recycled while the water may be electrolysed into oxygen and hydrogen:

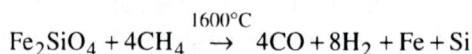

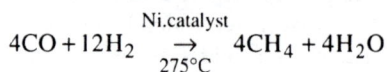

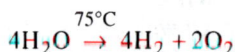

$$Fe_2SiO_4 + 4CH_4 \xrightarrow{1600°C} 4CO + 8H_2 + Fe + Si$$

$$4CO + 12H_2 \xrightarrow[275°C]{Ni.catalyst} 4CH_4 + 4H_2O$$

$$4H_2O \xrightarrow{75°C} 4H_2 + 2O_2$$

This process does not require prior beneficiation of the feedstock, simplifying raw material handling, mining and transportation. However, the process is energy-intensive requiring 0.8 kW/annual tonne of oxygen. The estimated plant mass is 20 tonnes plus 6 tonnes of power generation equipment per 100 tonnes of oxygen produced per year. Around 30% of the mass of raw feedstock is converted to oxygen. Byproducts of this process are metals including iron, silicon and ceramic. A plant of 60 tonnes can produce 200 tonnes/year with a power requirement of 1.2 MW—a typical rover can yield 5 tonnes of regolith feedstock per day, sufficient for such a throughput. Ilmenite which comprises 20% of the mare basalt, such as Mare Tranquillitas, is an even richer source of iron and titanium for aerospace structural materials. This is little different from the situation on Earth in which iron production involves the extraction of oxygen from iron oxide ore. Silicon has been extracted from simulated lunar anorthosite $CaAl_2Si_2O_8$ (which may be separated out from lunar soil by magnetic separation) with purities sufficient for solar cell production in the laboratory. The recent discovery of $\sim 10^9$ tonnes of ice in the polar regions of the Moon primarily in permanently shadowed regions of craters by Clementine in 1994 and Lunar Prospector in 1998 suggests a potential for commercial utilisation. The most prominent of local space market products is lunar oxygen for propellant, fuel cells and life support—oxidiser comprises five-sixth of the total LOX/LH propellant mass. The mass of a lunar oxygen plant with a capacity of 50 tonnes of oxygen per lunar day would be around 100 tonnes. A 5 tonne per year plant has a mass of 780 kg comprising of 560 kg of oxygen liquefaction and storage facilities and 220 kg chemical processing plant with a total power requirement of 40 kW (of which 37 kW is for chemical processing [Rosenberg 1997]. A first pilot plant on the Moon to produce lunar oxygen will most likely use hydrogen reduction of ilmenite. It is a closed process but requires transport of an initial mass of hydrogen. If lunar soil with the mineral ilmenite is placed in a particle bed reactor and hydrogen gas is passed through the soil to reduce the ilmenite and yield water vapour which may be electrolytically treated to yield oxygen. This involves the reduction of Fe^{2+} oxide in ilmenite minerals by hydrogen at 1100°C to form metallic Fe and titania (TiO_2). Water is produced which is then converted to hydrogen and oxygen by electrolysis at 750°C with the hydrogen recycled. Basically, the process involves reduction of FeO part of the ilmenite $FeTiO_3$ leaving titania and Fe with O_2 volatile release comprising 10% of the reaction products:

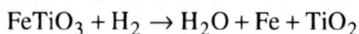

$$FeTiO_3 + H_2 \rightarrow H_2O + Fe + TiO_2$$

$$H_2O \rightarrow H_2 + 0.5O_2$$

Around 300 kg of soil is required to produce 1 kg of oxygen. A pilot plant with production rate of 2 tonnes/month has a mass of 25 tonnes and a power requirement of 180 kW. The production rate increases linearly with mass and power requirement [Koelle & Lo 1997]. A 20-tonne/month plant has a mass of 100 tonnes with a power requirement of 1080 kW. A 2500-tonne per year plant has a mass of 50 tonnes with a power requirement of 4000 kW. The power subsystem comprises 33% of the total mass of the system. The liquefaction subsystem is to liquefy and store the oxygen extracted and comprises 5% of the total mass and power of the plant as a whole. Lunar-derived oxygen depots could be established on Earth-orbiting platforms in LEO/GEO to permit refuelling access points to Earth-orbiting spacecraft such as OTVs (orbital transfer vehicles).

Nakamura et al (1994) proposed an optical waveguide solar energy system for processing lunar material. Solar radiation is collected by a concentrator arrays of parabolic reflectors which transmit the solar energy to optical fibre transmission line bundles which transfer the energy directly into thermal reactor through holes in the spherical reactor cavity wall at 1400–2500 K with efficiencies of 75–85%. The furnace is evacuated to reduce convective heat losses. Regenerative cooling of the cavity wall and a multilayer radiation shield at the cavity wall substantially reduces heat losses. This system avoided uneven heating and vaporisation losses that characterises direct solar heating methods. For aluminium extraction, electrochemical processing is the only feasible route for materials processing. Aluminium is a useful structural material and is suitable as an electric conductor. It can be shaped easily by casting, rolling, drawing and welding. It can be sprayed, vapour-deposited and atomised and its low melting point of 660°C gives it advantages over iron. Lunar soil may be dissolved in a molten fluoride electrolyte or may be melted at 1430°C as a molten silicate electrolyte. This provides a lunar analogy to the electrolytic Hall–Herault process for aluminium extraction on Earth from bauxite which produces 1 tonne of Al metal from 4 tonnes of bauxite. Around 40 MW can produce 20 kilotonnes of Al in two years. There are alternatives to electrolytic methods. Over 90% of all chemical manufacturing involves catalysis but it is desirable to reduce the use of Earth-based materials that would require costly launch and transportation [Waldron 1988]. Carbo-chlorination reduction, for instance, reduces peak temperature requirements by 1000°C over carbothermic reduction of mineral compounds. Alternatively, soil may be dissolved in HF acid. The dissolved metals may be removed from the glassy silicates with HF acid leaching reagent which is recycled.

Lunar glass may be constructed for use as high strength fibres ~700 MPa for reinforcement in matrix materials. Such a matrix material could be bulk basaltic glass of low strength also from the lunar surface. Glasses may be formed by melting the regolith and rapidly cooling it. Indeed, in some areas at the maria/highland interface regions, up to 40% of the regolith is pyroclastic glass. The lack of water would increase the strength of the lunar glass. Cast basalt is manufactured by melting regolith in a furnace and cooling it slowly so that it crystallises. This offers high compression strength/moderate tensile strength, ten times that of concrete, suitable for structural elements. Its chief disadvantage is that it is brittle and reinforcement by metal rods overcomes this. However, it does require high temperatures due to the regolith's high melting point. A fleet of tractors may construct electronic power collection circuits constructed from lunar iron extract, thin film solar cells, glass coversheets and the main structural elements. Mobile glass manufacturing processors to make lunar basalt regolith into fused fibreglass structures to maximise local resources *in situ*. Thin film solar conversion cells of moderate efficiency ~5–10% may be readily produced from lunar resources based on amorphous silicon. Sunlight collected by solar concentrators could be converted to electric power for solid-state microwave integrated circuit converters (MICCs), each of which forms an individual beam to contribute to the composite beam. Large phased-array microwave technology is well-developed.

All in all, the most likely materials to be used are metals and cast basalt for construction with sintered regolith for shielding and glass for fibre-reinforced ceramics. The greatest uncertainty lies in the machine shop operation, in the conversion of processed materials into finished products. Robots may be used for fabrication, inspection, assembly and general servicing and maintenance. However, with a roundtrip communications delay of 2.7 s, supervisory teleoperation from Earth is feasible in conjunction with automated chemical plant technology and expert systems.

Nuclear fusion could be through D/T or D/He-3 burning. The latter has yet to be developed but, while the requirements for D/He-3 fusion are more demanding in terms of reaction temperature, it offers three orders of magnitude reduced neutron flux compared to the D/T reaction. Indeed, it generates less than 0.01% of the radioactive waste of fission reactors [Schmitt 1994]. There is no danger of meltdown due to the absence of high-level radioactive heat sources. In addition, the energy of the fusion reaction can be converted directly to electricity at twice the efficiency of fission plants. The D/He-3 reaction produces protons and alpha-particles rather than neutrons, so direct conversion with efficiencies of 70% are possible. The low neutron flux allows a D/He-3 plant to operate for 30–40 years without reactor wall replacement and residual radioactivity disposal would be similar to that required for radioactive hospital waste. The chief difficulty with D/He-3 fusion is that He-3 is virtually absent from Earth. However, it has been estimated that there are around 1×10^6 Mt of He-3 loosely embedded in the lunar regolith of the basaltic maria implanted by solar wind ions at a density of 30 ppb or equivalently 10.8 mg/m^3 of the surface layer (dependent on the maturity of the lunar regolith)—compared with some 7×10^{22} kg of He-3 in Jupiterís atmosphere [Sved et al 1995]. Hence, 30×10^6 kg of regolith must be processed to yield 1 kg of He-3. The world's present annual electricity supply could be supplied by 200 Mt of He-3 plus 70 Mt of D—this would require mining of 0.05% of the lunar surface per year. If regolith is heated to 700°C or above, volatiles are released, including He-3, H$_2$ and reacted compounds such as H$_2$O, CH$_4$, CO$_2$, etc. The Boeing rover could collect 33 kg of He-3 per year by mining 1 km^2 with a 50% recovery rate. The extraction of He-3 into the lunar regolith by lunar mining offers a potential exception to local market limitations—it has a value of $3m/kg with 1 kg of He-3 capable of producing 10 MW of electricity for a year. If sales reached the global oil revenue of $400b/y then the terrestrial economy would become dependent on a lunar economy. He-3 offers the future possibility of a practically limitless source of energy from nuclear fusion with almost no nuclear waste.

Wireless power transmission from space offers a means to provide abundantly available energy without environmental degradation by fossil fuel combustion—solar energy in space is 10–20 times as intense as it is on the Earth's surface. Earth orbiting SPS (solar power system) satellites in GEO are based on the concept of collecting solar power which is converted using solar cells into electricity which is fed to a microwave generator which forms part of a planar phased-array antenna. This transmitting antenna directs a microwave beam of lower power density to receiver antennas (rectennas) on Earth which reconvert the microwave energy back to electricity for the grid [Glaser 1968]. SPS generates no pollution, harmful radiations, or contamination of the Earth's surface. Heat is rejected directly to space with only useful energy being delivered to the surface site. As 95% of the mass of an Earth-based power station is for structural support in the gravity field, this overhead is eliminated by the SPS concept which can provide a mass/power density of ~1 tonne/MW compared with 10^3 tonnes/MW for ground-based power stations [Grey 1976]. The solar-electric conversion process may be solid-state photovoltaic solar arrays, but their efficiencies are limited to less than 20%. At 1AU, the solar collector area must be large—a 5 km diameter parabolic dish can supply 2.5×10^7 kW. Electric power output to the concentrator area increases with receiver temperature and a high-temperature receiver can operate a heat engine from a waste heat radiator to enhance efficiency. Photovoltaic organic polymer semiconductors such as oxozole offer the promise of high photovoltaic con-version efficiencies. Thermodynamic conversion using the Brayton/Stirling cycle working fluids offer efficiencies as high as 30% but involve moving parts, heavy turbomachinery and radiators for heat rejection. Amplifiers operating at more than 90% efficiencies can operate the microwave radiation for transmission. Power transmission at S band or below, such as at 2.45 GHz, avoids atmospheric absorption and rectenna efficiency is high ~90%. The power density at the rectenna centre is 23 mW/cm^2 which drops to 1 mW/cm^2 at the rectenna edge (equivalent to microwave oven emission at a distance of 5 cm) within accepted standards for radiation exposure. Aircraft overflying the rectenna should be protected by the metallic skin of the fuselage and the short transit time across the rectenna. A typical SPS system could supply 10 000 GW to the grid. SPS designs are very large and employ microwave power transmission to ground rectennas to generate electrical power outputs. A single 6-GW GaAs planar solar array system would have a mass of 35 000 tonnes and surface reflector area of 53 km^2 and surface cell area of 27 km^2. The unit cost has been estimated at $12b. The cost of a 5000-GW SPS of mass 10 kt has been estimated at $1300/kW. Some 27% of that cost is for launch transportation to GEO. To compete with fossil fuel and nuclear power systems, the cost of launch should be reduced to $100/kg. SPS 2000 is a Japanese project to demonstrate the SPS concept in a 1000 km orbit generating a 3-km diameter footprint with a power output of 300 kW, initially projected to grow to 10 MW. A European project is also being proposed (global solar energy concept) to demonstrate in-orbit assembly of large flexible structures and explore biological aspects of power beaming at 1 MW. It will be assembled in LEO by astronauts (or in-orbit servicers) and boosted to GEO by electric thrusters. At present there appear to be no major technological difficulties to the development of such an orbital infrastructure. SPS is difficult to justify on the basis of environmental restrictions on launch operations needed to lift large numbers of 35-tonne 5-GW SPS satellites.

Lunar power offers a potential terrestrial application—the Space Studies Institute indicated that the cost of SPS could be reduced by 98% by using lunar resources to construct 95% of the space segment of SPSS due to the high cost of launching materials into space from Earth—this reduces the cost of SPSS from $1300/kW to $250/kW. Solar power arrays could be implemented on the lunar surface, beaming the power back to Earth as microwaves by large antennae [Criswell & Waldron 1993]. Virtually any conceivable structure could be placed on the Moon as the lunar soil has an ultimate bearing capacity of 3000–11 000 kPa [Johnson & Chua 1993]. The adoption of lunar surface SPS allows the adoption of very large transmitter diameters ~100 km of phased arrays to reduce Earth-based rectenna diameters of 10 km required of GEO SPSS to less than 1 km diameters. The baseline LPS architecture comprises a pair of solar power stations that beam power directly to Earth-based rectennas when in Earth–Moon line of sight. Power storage during non-viewing (16 hours/day) and lunar eclipse periods (3 hours) allow continuous power. Over-sizing the lunar transmitters would allow reduction of 10–20 km rectenna sizes required for SPS satellites to ~100 m. As power storage is expensive, the addition of Earth-orbiting microwave mirrors overcomes the non-viewing periods except for new Moon phases. Such reflectors would allow a three-fold reduction in rectenna size. Solar collectors could be stationed at GEO as relay stations to direct the microwaves to Earth-based rectennas. Furthermore, performance enhancements may be obtained by placing sunlight reflecting mirrors in lunar polar orbit to illuminate the lunar transmitter stations during the two-week lunar night and its new Moon periods. A coincident lunar base has been suggested as a means to reduce the costs of demonstration of LPS by providing much of the infrastructure. Lunar Power Systems (LPS) could provide power outputs of 20 TW by 2050 at low cost with less environmental impact [Criswell &

Waldron 1993]. Such a system could be initiated over 5–10 y by a small scale 1–100 TW demonstrator. If implemented soon after 2000 with growth capacity, LPS could supply the required 20 000 GW of terrestrial power by 2050 at a selling price of $0.1/kWh to generate profits ~$15 000b/y. Two 10 000-GW LPS plant units ould be produced per year, based on an estimated manufacturing mass throughput of 30 kg/s. This could sustain a population of 10 billion at 2 kW/person. Its growth capacity offers stable growth and stable energy costs. The total outlay prior to full profitability was estimated at $100b, but the initial R&D investment costs would amount to only $3b. The major operating cost once fully operational would be Earth-based rectennas ~76.4% or $6m/GW y but these are most suited to private operation.

21.3 FUTURE MARS MISSIONS AND MARTIAN RESOURCE UTILISATION

There is little doubt that a manned mission to Mars is a principal long-term objective of NASA and space exploration as a whole, implicitly or explicitly. In fact, in reality, it represents a beginning, as once a stable foothold has been established on Mars, the rest of the solar system opens up as feasible real estate. Mars is the most Earth-like planet in the solar system with a day–night cycle similar to Earth's and the principal reason for its investigation is the search for water and possibly fossil biota. The exploration of Mars depends on three fundamental criteria—technical feasibility, political feasibility and financial feasibility, of which the last two are the most difficult to comply with. There are two main classes of Mars mission. Opposition class missions involve high Δv and so large amounts of propellant usage for rapid Mars missions in 1.5 years, of which only 30 days are spent on the Martian surface—these missions are costly with low payoff. They subject the crew to substantial cosmic ray flux in interplanetary space. Most Mars missions are baselined around the minimum-energy Hohmann transfer conjunction class mission of 260 days between Earth and Mars. At arrival at Mars, Mars is 75° behind the Earth, but to leave Mars it must be 75° ahead of Earth. Therefore, a Martian stay-time of at least one Earth-year is necessary. With a little more propellant for greater Δv, the Earth–Mars transfer can be accomplished in 190 days each way allowing a 500-day surface stay. Around 50% of the mission resides on the Martian surface, maximising the payoff. C3 is defined as the square of the velocity of departure from Earth. A C3 of 15 represents the minimum Δv of 4.15 km/s for a Mars mission. However, a C3 of 25 reduces the interplanetary flight time by 70 days by increasing the Δv by 0.42 km/s to 4.6 km/s. A further advantage of the C3 of 25 is that a free return trajectory is available to the mission if the landing is aborted.

First of all, any comprehensive Mars mission will involve extensive automated robotic precursor missions [Rea et al 1990]. Such a mission will involve a communications/ remote sensing orbiter with stereo imager, UV/IR spectrometer and microwave radar instrument for surface/subsurface mapping, a lander for *in situ* analysis of rock and soil geochemistry and a rover segment for traversability. Aerocapture at Mars into an elliptical Martian parking orbit would be essential as it reduces launch mass by 40%. The aerobrake and thermal protection mass fraction is around 15% [Tauber 1993]. A 5-km/s hyperbolic velocity produces a Mars atmospheric entry velocity of 7 km/s compared with 8 km/s entry experienced by the Space Shuttle. The Mariner Mark II spacecraft bus is designed to carry and eject landers enclosed in aeroshells which enter the Martian atmosphere followed by parachute deployment. Volatile and chemical composition analysis can be performed *in situ* using instruments such as alpha–proton–X-ray spectrometer and magnetometer for geochemical analysis, but rovers are required to obtain good surface range. Such rovers could deploy multiple surface stations to construct a distributed network of seismic and other instruments to probe the Martian internal structure, and meteorological instruments to probe the Martian atmospheric processes. Penetrators on the surface stations could perform subsurface analysis.

Sample return missions may employ an ascent vehicle to return a capsule of samples back to Earth for detailed laboratory analysis. Part of the return propellant for the sample return ascent module could be manufactured on Mars [Shafirovich & Goldschleger 1995]. Liquid oxygen could be extracted from its 95.3% carbon dioxide atmosphere (the rest being 2.7% nitrogen, 1.6% argon, 0.13% oxygen, 0.07% carbon monoxide and 0.03% water vapour) but this would require a chemical plant to be transported to the surface. However, Martian carbon dioxide could be used as an oxidiser with a metal such as Mg or Be transported from Earth. Although Be offers the highest specific impulse, it is highly toxic, favouring Mg, but offers a specific impulse of only 190 s. In this case, a carbon dioxide liquefaction compressor needs to be transported to the surface. Zubrin et al (1995) has suggested that rocket propellant may be manufactured on the surface of Mars to support a roundtrip mission. The Martian atmosphere could provide feedstock for the manufacture of oxygen to reduce the mass and complexity of any manned or unmanned return Mars mission. It has been demonstrated with a chemical synthesis unit utilising the Sabatier/electrolysis cycle that CH_4/O_2 bipropellant can be produced from the Martian CO_2 atmosphere with hydrogen transported from Earth (or possibly from local water resources). Indeed, the Sabatier reaction has been used in manned spacecraft as CO_2 scrubbers. Hydrogen from Earth and carbon dioxide from the Martian atmosphere mixed in a stoichimetric ratio of 4.0 in a Sabatier reactor will produce CH_4/H_2O with 94% efficiency in a 1:2 molar ratio with the required purity: $CO_2 + 4H_2 \rightarrow CH_4 + 2H_2O$ at 400°C. This is exothermic and occurs spontaneously with an Ni catalyst. The methane is liquefied and water is electrolysed: $2H_2O \rightarrow 2H_2 + O_2$. The oxygen is liquefied and hydrogen recycled back into the Sabatier reactor. Modern electrolysers are composed of layers of solid electrolyte impregnated plastic separated by metal meshes. They are used in nuclear submarines. Around 50% of the hydrogen required is recycled through the methanator—for every 1 kg of hydrogen consumed, 18 kg of propellant/oxidiser are produced. Ruthenium-on-alumina catalyst offers superior performance below 300°C and does not produce the toxic nickel carbonyl products involved with nickel catalysts. The oxygen/methane ratio can be used as rocket fuel with a specific impulse of 340–375 s. A practical output of 0.72 kg of propellant per day produces 360 kg of propellant after a 500-day conjunction period, sufficient to return 4 kg of samples from the surface of Mars direct from Mars without the need for a Mars orbiter. The processing device had a mass of 20 kg consuming 160 W. The basis of the laboratory demonstrator was the Sabatier unit and a solid polymer electrolyser. The addition of a pump for carbon dioxide acquisition and compressor for propellant refrigeration would provide a complete system. The multi-stage axial flow pump is suited to high volume flow rates in compressing the 8 mbar ambient pressure to 1 bar. The compressor must implement a filtration system to sift out fine super-oxidising dust particles of the Martian atmosphere. Bag filters can be used which are periodically pneumatically cleared of accumulated dust by high-pressure nozzles—which is simple and uses minimal power. Thermal cycling between the average temperature of —60°C at the equator (which varies from 15°C in the day to –100°C at night) and operating temperatures could be a problem. A sorption pump containing pellets of the clay zeolite could absorb up to 40% of its own mass of CO_2 when cold (during the night) and release it when heated by a 50-kW power source to the reaction chamber, i.e. a sorbent mass of 2.5 kg to yield 1 kg of carbon dioxide. Zeolite is an alumino-silicate ceramic exhibiting a regular molecular array of apertures with selective adsoptivity, e.g. zeolite A is used for desiccation. The carbon dioxide would subsequently be condensed and liquefied in another chamber with a bank of tubes with arrays of cooling fins—nitrogen and argon constituents would remain gaseous and be vented. There are certain difficulties in long-term storage of hydrogen for transport to Mars, but this may be offset by the high specific impulse of methane/oxygen propellant compared with carbon monoxide/oxygen propellant. Heavy insulation with multilayer insulation can reduce boil-off of liquid hydrogen in space to less than 1% per month during inter-planetary transit, and gelling with a small amount of methane to prevent leakage reduces boil-off by a further 40% by suppressing convection within the tank.

The Sabatier-electrolysis process produces O_2/CH_4 in a ratio of 2:1, but the oxidiser/propellant ratio needs to be 3:1 for higher specific impulses. The Martian soil contains trace amounts of oxygen but to supply 1 kg of oxygen, 40 m^3 (or equivalently 65 tonnes) of soil would have to be humidified. One option is direct reduction of carbon dioxide. Heating carbon dioxide to 1100°C causes the gas to dissociate:

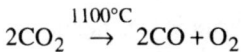

$$2CO_2 \xrightarrow{1100°C} 2CO + O_2$$

The ceramic, zirconia (zirconium oxide) doped with yttrium is a solid electrolyte that can be used in an electrolytic cell to convert CO_2 input at the cathode to be released as O_2 at the anode to provide the additional oxygen. This process, however, yields small outputs of oxygen. Alternatively, some of the methane could be pyrolised into graphite and hydrogen: $CH_4 \rightarrow C + 2H_2$. The hydrogen could be recycled back into the Sabatier reaction. Water is certainly present in the Martian polar caps, the soil and the atmosphere which could be used to generate both hydrogen and oxygen by electrolysis. It is unlikely, however, that water would be available in sufficient quantities that are readily accessible on Mars—the amount of water in the atmosphere is small, polar near-pure water ice caps (especially the north pole) are inaccessible to lower latitudes, and extraction of subsurface permafrost may be difficult to mine. The largest amount of water vapour in the atmos-phere lies in northern high latitudes above 60° at low elevations during summer. It has been suggested that water vapour at concentrations ~0.03% could be extracted from the Martian atmosphere using a Water Vapour Adsorption Reactor which uses microporous zeolite as an exothermic molecular sieve [Williams et al 1995]. Zeolite is a hydrophilic ceramic that can separate water vapour from the Martian atmosphere as the atmospheric gases are pumped through it without the use of moving parts inherent in compression cooling methods and with lower energy requirements. The zeolite of choice is UOP Molecular Sieve 3A which has an aperture size of 3A slightly larger than a water molecule, thereby excluding carbon dioxide, nitrogen and argon. Regeneration of the adsorption bed involves heating the bed uniformly to 400 K with microwave to release a utilisable fraction of the adsorbed gas ~5% the mass of the absorbent bed leaving a residual fraction ~20% of the mass of the absorbent bed. The low concentration of water vapour implies that $100\,000\ m^3$ or equivalently $1.1\ m^3/s$ of Martian atmosphere is required to be processed to yield just 1 kg of water per day. To achieve a mass flow rate of $0.0094\ kg/s$ and volume flow rate of $0.562\ m^3/s$ by a complete system of mass 30 kg to generate 0.5 kg of water per day requires 140 W of power.

A human mission to Mars is generally regarded as essential for effective exploration, particularly in the search for life, be it extant or ancient. Such evidence is likely to be sparse and microscopic hidden depth, possibly several hundred metres or even a couple of kilometres below the surface. Robotic rovers are unlikely to be successful in hunting for fossils given their limitations. Humans in the field are the only option in providing high mobility, wide area coverage and, most importantly, high adaptability and flexible decision-making. Landis [1995] has suggested that the Martian moons Deimos and Phobos represent a reasonable stepping stone towards a manned mission to Mars. Both moons are carbonaceous chondrites so they are generally believed to be captured asteroids rich in carbon compounds suitable for mining [O'Leary 1988]. Deimos is the outermost moon of Mars in near-synchronous orbit which provides less than 0.2 s delay for teleoperative activities on the Martian surface. A mission to the Martian moons employs half the propellant and hardware that a Mars mission requires, so half the cost and risk. It represents a technology demonstrator for a manned mission to Mars. Initial exploration of Mars by manned missions could be accomplished with small crews with an estimated cost of $20b.

The biggest problem for any manned mission to Mars is the fragility of the crew. In the 200-day Mars transit, the crew are susceptible to solar flare emissions requiring the adoption of a real-time warning system and a radiation storm shelter in the centre of the spacecraft for maximum shielding for a stay of up to several days by the crew. Solar flares are unpredictable, occurring on average once per year during solar maximum and 0.2/year during solar minimum, and warning can be as short as 30 minutes—it is possible that an onboard solar X-ray telescope may be necessary to compensate for communication delays from Earth or loss of communications altogether. Furthermore, on the surface of Mars, its thin atmosphere of $20\ g/cm^2$ and negligible magnetic field provide little protection from solar radiation which is about 25% that of interplanetary space. Bagged Martian regolith would be required to provide adequate protection of $\sim 35\ g/cm^2$. Most Mars mission scenarios involve an accumulated radiation exposure of 75 rem which is significant (acceptable limits are given by 5 rem/year and 200 rem accumulated dose). Every 60 rem adds a 1% additional risk of fatal cancer. Mir astronauts were exposed to 0.15 Sv compared with radiation safety limits of 0.05 Sv/y for civilians. A 0.1 Sv exposure will result in an additional 0.8% chance of contracting a fatal cancer. Other factors in long-duration spaceflight include muscle atrophy and bone mineral leaching during the flight and poor orthostatic tolerance on the Martian surface. Although most physiological changes can be compensated for by extensive physical exercise regimens and tend to stabilise after 3 months or so, bone loss continues at 1–2%/month [Nicogossian et al 1988, Leach 1990]. Psychological problems are likely to be significant due to isolation, confinement and constant danger leading to depression, lethargy or paranoia. There is immune system suppression which can increase susceptibility to sickness. And finally, accidents due to unforeseen events are inevitable. There is little doubt that medical aspects of long duration spaceflight typical of a manned Mars mission will dominate, to the extent that astronauts may be limited in their capability to perform adequately on arrival at Mars.

The baseline manned Mars mission is as follows. The initial launches at the first mission opportunity in 2011 deliver three cargo vessels to Mars (some in-orbit assembly would be required depending on the payload capability of the launchers—the present heavy lift launcher is the Titan IIIE-Centaur). The propulsion system adopted for trans-Mars injection is generally a solid or gas core nuclear thermal rocket (NTR) engine, though solar/nuclear-electric propulsion spiral trajectory is a popular variant for the unmanned cargo phases which are not restricted to minimise passes through the Van Allen radiation belts. A NERVA-derivative would be perfectly feasible. Aerocapture followed by aerobraking, parachute deployment and retro-rockets would be required for a soft Martian landing at 500 m/s—a triconic aerobrake shape has sufficient L/D ratio of 0.6 for aerocapture. The Earth Return Vehicle is aerocaptured into Mars orbit where it remains. Mars Ascent Vehicle and Mars Habitat-1 are also aerocaptured but enter the Martian atmosphere to land on the Martian surface. At the second launch opportunity in 2014, the manned Mars Habitat-2 with a crew of 6–8 is launched to Mars, followed by aerocapture, atmospheric entry and surface landing near the previously landed resources. Each person requires 400 kg of food, water and oxygen for a 200-day Mars trajectory. The crew remain on the Martian surface for typically 500 days, after which they ascend in the Mars Ascent Vehicle which provides a Δv of 4.2 km/s into Mars orbit due to the Martian surface gravity of $0.38g$, rendezvous and dock with the Earth Return Vehicle which provides a Δv of 2.2 km/s for the return of both docked vehicles to Earth. The low average solar flux at the surface of Mars of 120–230 W/m^2 (50% that from Earth), due to its distance of 1.52 AU from the Sun, implies the need for nuclear power generation for surface activities, e.g. a closed Brayton cycle nuclear power source rather than solar power generation which would be susceptible to the annual Martian dust-storms which can absorb 85% of the incident radiation. A long-range rover capability would be essential—batteries are heavy, RTGs are power-limited, so either solar power regenerative fuel cells or combustion engines are possible options. Combustion engines have high power/mass ratios but they are fuel intensive. A crew rotation and resupply mission could be launched at every launch opportunity every two years to maintain a permanent presence. The vast majority of mass launched is propellant. Bruckner et al (1995) have suggested that *in situ* resource utilisation could be used to produce fuel and oxidiser for a Mars landing vehicle/hopper capable of 35 km ballistic hops at a cost of $200m. The Valles Marineris represents an ideal landing site as it represents a 'Grand Canyon' cross-section of Martian geology.

It has been suggested that direct missions to Mars are also possible in the near term using lunar-sized transportation systems, without the need for in-orbit assembly of Mars-bound spacecraft, through the judicious use of *in situ* propellant production on the surface of Mars [Zubrin & Weaver 1993, 1995, Zubrin et al 1991, Bruckner et al 1995]. Furthermore, such a Mars mission has no requirement for an LEO in-orbit infrastructure. The primary motivation for this concept is to eliminate the need for multiple heavy launch vehicle launches—it does, however, require a single heavy-lift launch vehicle of Saturn V class with a payload capability of 200 tonnes for trans-Mars injection and subsequent Mars landing. The Russian Energya with a payload capacity of 100 tonnes could suffice. The first launch is for a 6-tonne unfuelled two-stage Mars Ascent Vehicle (MAV), 15-tonne provisioned Earth Return Vehicle plus 15 tonnes of propellant, 4 tonnes of liquid hydrogen feedstock payload, a 1.5-tonne truck-mounted 50-kW nuclear reactor, several 1-tonne rovers and a 0.5-tonne automated methane/oxygen chemical processing plant, all of which are landed on the Martian surface via a C3 of 15 mission (Δv of 4.15 km/s)—in total 45 tonnes. On landing on Mars, the chemical processing plant begins to manufacture fuel. The ERV contains life support and consumables for the 8-month return back to Earth for four astronauts. Some 24 tonnes of the 36 tonnes of manufactured methane/oxygen bipropellant is for ERV fuel, the other 12 tonnes being fuel for the rovers. A 1-tonne rover requires 0.5 kg of methane/oxygen to travel 1 km. In addition, 18 tonnes of water is produced for consumption by the subsequent crew. Once propellant manufacture is completed, two more heavy-lift launches inject another provisioned ERV as backup and chemical plant/MAV into trans-Mars trajectory with a 30-tonne, four-man habitat module with 200 m^3 of living space provisioned with 9 tonnes of consumables for 800 days (4.8 tonnes of food, 4.2 tonnes of water and 320 kg of oxygen) plus a 2-tonne methane–oxygen powered pressurised rover via a C3 of 25 mission (Δv of 4.6 km/s). The ERV structure is similar to the habitat module but streamlined in mass. Artificial gravity could be supplied by tethering the empty heavy lift vehicle upper stage and the habitat module with a 1500 m tether and spinning them up at 1 rpm during the outboard journey only to provide 0.38g (Mars surface gravity). The second propellant plant and ERV are emplaced for the follow-up mission. Landing is automated and guided by a guiding radio beacon on the ERV on the surface. The crew remains on the surface for 500 days. The habitat and MAV modules are powered by two independent 12-kW DIPS (dynamic isotope power systems) power systems. With 12 tonnes of propellant for surface opera-tions, 24 000 km of ground range can be covered in the rovers. The four-man crew comprises two flight engineers and two field scientists (nominally, a geologist and a biogeochemist) with cross-training in pilot and medical skills aided with expert systems. The crew returns to Earth directly from the Martian surface in the fully fuelled ERV on a trans-Earth trajectory, leaving the habitat module as part of a Martian infrastructure for subsequent missions launched every two years, thereby building up a growing Martian infrastructure. A variant of this mission is a semi-direct implementation in which a fully fuelled ERV is delivered to elliptical Mars orbit and an unfuelled Mars Ascent Vehicle is delivered to the surface. On return to Earth, the crewed MAV is launched from the Martian surface to rendezvous with the ERV in Mars orbit, both of which return to Earth on a trans-Earth trajectory with a Δv of 1.73 km/s. The chief advantage of the semi-direct version is that a larger ERV can be launched. Another variant is the hydrid direct imple-mentation in which the ERV delivered to Mars orbit is unfuelled with propellant supplied by the surface chemical plant and subsequently launched within the MAV. The total estimated cost of the Mars Direct mission is $20b compared with the $400b for the NASA baseline mission.

Asteroidal utilisation is unlikely to be economic in the near future, though it does offer potential for space-based metal mining without Earth degradation [O'Leary 1977, 1979, Lewis 1995]. That said, SpaceDev is planning to send the Near Earth Asteroid Prospector with an ejectable payload to the asteroid 4660 Nereus in 2002 at a cost of $50m. Space Dev will claim ownership and then sell the data recovered. Asteroids represent a vast resource of useful material. The colonisation of Mars will be essential to develop asteroid utilisation. Mars has the potential to become self-sufficient based on asteroid mining and trade of high-grade metal ore from both the asteroid belt between 1.8 and 3.5 AU and its moons Phobos and Deimos [O'Leary 1988]. Metal could be extracted by carbonyl reduction at 200°C yielding 40% utilisation of the feedstock. The residual 60% of the material comprises of 30% Si which may be used for electronics fabrication, 30% metal alloy, 20% refractory material and 20% wastage. The production of metal by melting Fe–Ni metal from asteriodal material has been estimated to require 15–25% less energy than that required to produce metal from lunar oxides [O'Leary 1977]. At present, human consumption amounts to 10 tonnes of Fe and 70 tonnes of building material per person per year, totalling a usage of 10^{14} kg/y increasing at a rate of 3%/y— indeed, it has been estimated that 10^{15} kg of material (equivalent to 500 km^3 volume, i.e. a 5-km radius asteroid) has been consumed since the Industrial Revolution. A 1-km diameter Ni/Fe asteroid is worth $5000b, containing 400 000 tonnes of high-grade metal Ni, Co and Pt group ores at much higher concentrations than terrestrial ores [Lewis 1995]. The advantage of asteroids is that a 1-km sized asteroid has a small escape velocity ~1 m/s. The cost of extracting those metals is estimated at 20% of the asteroid's value. Several problems emerge from this scenario. Firstly, it ignores the impact on the market value of such large quantities of metal which would depress the market value of the asteroid to $320b. Furthermore, most Ni and Pt group metals are exported by the USSR and South Africa, totalling $7b/y sales globally. This is too small for a large investment. The main problem is in despinning the asteroid so that docking and mining can be conducted. A network of fibres and pylons attached to the asteroid would enable some form of rocket thrust to despin the asteroid. Solar-electric propulsion would be an option for this or an electromagnetic mass driver which uses asteroid material as reaction mass to provide a Δv of 3 km/s to high Earth orbit. Carbonaceous chondrates could yield hydrocarbons which may be processed into hydrocarbon plastics or reduced for hydrogen. Some 15% of near-Earth asteroids are easier to reach than the Moon. Furthermore, soft landing on the Moon, limited solar energy availability on the Moon, and transport of lunar material for processing increase the cost of lunar resource exploitation. Asteroids are richer in metallic Fe and Ni and poorer in Al and Ti than the Moon. Carbonaceous asteroids are rich in CHN which are almost absent on the Moon. The near-Earth Apollo asteroid 1982DB has been suggested as a candidate, due to its suitable orbital characteristics of 0.95 AU perihelion, 0.375 eccentricity and 1.4° inclination, which implies a one way Δv of 4.64 km/s [Parkinson 1991]. Such near-Earth objects have short life expectancies ~30–100 My and are believed to be extinct comet nuclei. Comet nuclei may have some icy volatiles that have not been outgassed (water, carbon dioxide, methane and ammonia predominantly). Methane and other hydrocarbons may be extracted—oil represents one of the most valuable mineral commodities on Earth. Some 50% of near-Earth asteroids are C-type asteroids with significant ~5–20% water in the form of hydrated salts and water-rich clays, possibly as high as 60% in the form of permafrost. Water in the form of ice or hydrated minerals could be evaporated from the asteroid using a solar furnace based on a mirror to direct solar illumination onto the asteroid. Solar array generated electricity from 28 W/m^2 solar illumination could electrolyse the water into hydrogen and oxygen to fuel a tanker spacecraft. Alternatively, water could be used as the propellant through nuclear-thermal or solar-thermal propulsion to heat the water in a thrust chamber.

We have outlined the potential future of manned space exploration—indeed, details are likely to vary but the general progression is likely to follow the route outlined above. A central enabling capability will be automation and robotics technologies which will play a fundamental role in manned space activities. The goal of Mars will be a beginning rather than an end for space exploration it will establish a foothold infrastructure that will enable the colonisation and with it exploitation and control of our whole solar system, and then ultimately to the stars [Jaffe & Norton 1980].

References

Abdullah C et al (1991) 'Survey of robust control for rigid robots' *IEEE Control Systems* (Feb), 24–29

Abidi M & Gonzalez R (1990) 'Use of multisensor data for robotic application' *IEEE Trans Rob & Autom* **6**(2), 159–177

Abidi M et al (1991) 'Autonomous robotic inspection and manipulation using multisensor feedback' *IEEE Comp* (Apr), 17–30

Adams R et al (1987) 'Remote repair demonstration of Solar Maximum main electronics box' *Proc 1st European In-Orbit Operations Technology Symp* (ESA SP-272), 227–323

Agonisti A et al (1992) 'Use of EDRS by astronomy satellites and other scientific missions' *ESA J* 16, 217–231

Agrawal B (1980) *Design of Geosynchronous Spacecraft*, Prentice Hall

Agre P & Chapman D (1990) 'What are plans for?' *Rob & Auton Syst* 6, 17–34

AIAA (1991) 'AIAA guidelines for serviceable spacecraft grasping/berthing/docking interfaces' *ANSI/AIAA R-XXX-1991*

Aiello L et al (1986) 'Representation and the use of metaknowledge' *Proc IEEE* **74** (10), 1304–1321

Alberts T & Soloway D (1988) 'Force control of a multi-arm robot system' *Proc IEEE Int Conf Rob & Autom*, 1490–1496

Albus J (1975) 'A new approach to manipulator control: the cerebellar model articulation controller (CMAC)' *Trans ASME Jour Dynamic Syst, Meas & Cont* 97 (Sept), 220–227

Albus J (1979a) 'Model of the brain for robot control I' *Byte* (Jun), 10–34

Albus J (1979b) 'Model of the brain for robot control II' *Byte* (Jul), 54–95

Albus J (1984) 'Robotics' in *NATO ASI F11 Robotics & Artificial Intelligence*, 65–93

Albus J (1990) 'Robotics: where has it been and where is it going?' *Rob & Auton Syst* 6, 198–219

Albus J & Lumia R (1988) 'Teleoperation & autonomy for space robots' *Robotics & Auton Syst* 4, 27–33

Albus J, McCain H & Lumia R (1987) 'NASA/NBS Standard Reference Model for Telerobotic Control Systems Architecture' *NASA TN-1235*

Albus J, Lumia R & McCain (1988) 'Hierarchical control of intelligent machines applied to space station telerobots' *IEEE Trans Aero & Elect Syst* **24**(5), 535–541

Aldridge E (1993) 'International space cooperation: learning from the past; planning for the future' *Rep AIAA Wshop preprint* (Mar)

Aldridge H (1999) *private communication*

Aleksander I & Morton H (1987) 'Artificial intelligence: an engineering perspective' *Proc IEE* **134D**(4), 218–223

Aloimonos Y & Rosenfeld (1991) 'Computer vision' *Sci* **253**, 1249–1254

Allen J & Litman D (1986) 'Plans, goals and language' *Proc IEEE* **74** (7), 939–948

Amati N et al (1999) 'Twin rigid frame walking microrovers: a perspective for miniaturisation' *Jour Brit Inter Soc* **52**, 10–30

Anderson D & Rockoff L (1990) 'Telerobotics as an EVA tool' *Space Station Technologies, NASA SP-830*, 43–49

Anderson T & Donath M (1990) 'Animal behaviour as a paradigm for developing robot autonomy' *Rob & Auton Syst* **6**, 145–168

Andre G, Berger G & Elfving A (1990) 'BIAS: a bi-arm servicer' *ESA Rep N7406/87/NL/MAC*

Angell J et al (1983) 'Silicon micromechanical devices' *Sci Am* **248** (4), 26–47

Angle C & Brooks R (1990) 'Small planetary rovers' *preprint*

Angold N et al (1993) 'Ulysses operations—planning for the unknown' *ESA Bull* **72**, 44–51

Antsaklis P et al (1991) 'Introduction to autonomous control systems' *IEEE Cont Syst Mag* (June), 5–13

Arbib M (1976) 'AI: cooperative computation and man–machine symbiosis' *IEEE Trans Comp* **25** (12), 1346–1352

Arbib M & Caplan D (1979) 'Neurolinguistics must be computational' *Beh & Brain Sci* **2**, 449–483

Arduyersky V et al (1982) 'Some physical aspects of industrial crystallisation in microgravity' *Acta Astron* **9** (9), 583–588

Arkin R (1987) 'Motor schema based mobile robot navigation' *Int J Rob Res*, 92–112

Asade H, Kanade T & Takeyama I (1983) 'Control of a direct drive arm' *Trans ASME J DSMC* **105** (Sept), 136–142

Asare H & Wilson D (1987) 'Evaluation of three model reference adaptive control algorithms for robotic manipulators' *Proc IEEE Int Conf Rob & Autom*, 595–601

Ashford D (1991) 'Commercial demand for space stations' *J Brit Interplan Soc* **44**, 269–274

Astrom K (1987) 'Adaptive feedback control' *Proc IEEE* **75** (2), 185–217

Athans M (1971) 'Role and use of stochastic LQG problem in control systems design' *IEEE Trans Autom Cont* **16** (6), 529–551

Aviziennis A (1976) 'Fault tolerant computers' *IEEE Trans Comp* **25** (12), 1304–1311

Backes P (1993) 'Supervised autonomy for space telerobotics' *preprint*

Backes P & Tso K (1990) 'Autonomous single arm ORU changeout—strategies, control issues and implementation' *Robot & Auton Syst* **6**, 221–241

Backes P et al (1994) 'A prototype ground-remote telerobot control system' *Robotica* **12**, 481–490

Backes P et al (1998) 'Prototype ground-remote telerobot control system' *preprint*

Baggerver A (1984) 'Acoustic telemetry—an overview' *IEEE Journ Ocean Eng* **9** (4), 229–234

Baillieul (1980) 'Chaotic motion in nonlinear feedback systems' *IEEE Trans Circ & Syst* **27** (11), 996–997

Baillieul J (1985) 'Kinematic programming alternatives for redundant manipulators' *Proc IEEE Int Conf Rob & Autom*, 722–728

Bailly M et al (1985) 'CCD imaging sensor' *IFAC Autom Cont in Space*, 231–238

Baker V et al (1991) 'Ancient oceans, ice sheets and the hydrological cycle on Mars' *Nat* **352**, 589–594

Baldassini F (1968) 'Strapdown inertial guidance systems study' *AGARD CP-43*, 2.1–2.20

Ballard D (1991) 'Animate vision' *Artif Intell* **48**, 57–86

Ballard D et al (1983) 'Parallel visual computation' *Nature* **306** (3), 21–26

Bank V & Kassing D (1993) 'Technologies for automatic lunar exploration missions' *ESA Bull* **74**, 29–35

Bar-Cohen Y et al (1998) 'Flexible, low mass robotic arm actuated by electro-active polymers and operated equivalently to the human arm and hand' *Robotics* **98**, 3rd Conf Robotics for Challenging Environments, New Mexico

Bar-Cohen Y et al (1999) 'Ultrasonic/sonic drilling/coring (USDC) for in-situ planetary applications' *preprint*

Barer & Filipenko (1994) *preprint*

Bargellini P (1978) 'Principles and evolution of satellite communications' *Acta Astron* **5**, 135–149

Barrow H & Tenenbaum J (1981) 'Computational vision' *Proc IEEE* **69** (5), 572–595

Barter N (1999) 'TRW Space Data' TRW Space & Electronics Group, USA (5th Ed)

Bate R, Mueller D & White J (1986) 'Fundamentals of Astrodynamics' Dover Publications, New York

Bayard D & Wen J (1987) 'Simple robust control laws for robot manipulators: Pt II, adaptive case' *Proc Wshop on Space Telerobotics* **III**, 231–243

Beardsley T (1996) 'Science in the sky' *Sci Am* (Jan)

Beech M et al (1995) 'Potential danger from meteor storm activity' *QJ Roy Astron Soc* **36**, 127–152

Beer R et al (1990) 'Biological perspective on autonomous agent design' *Rob & Auton Syst* **6**, 169–186

Bejczy A (1979) 'Advanced teleoperators' *Astro & Aero* (May), 20–26

Bejczy A (1980) 'Sensors, controls, and man-machine interface for advanced teleoperation' *Sci* **208**, 1327–1335

Bejczy A & Tarn T (1986) 'Robot control as a systems control problem' *IFAC Theory of Robots*, Venice, 1–3

Berliner H (1978) 'Computer chess' *Nature* **274**, 745–749

Bhargava V (1983) 'Forward error correction schemes for digital communications' *IEEE Comm Mag* (Jan), 11–19

Biederman I (1987) 'Recognition-by-components: a theory of human image understanding' *Psych Rev* **94** (2), 18–64

Billerbeck W & Baker W (1984) 'Design of reliable power systems for communications satellites' *AIAA 84-1134*

Binford T (1982) 'Survey of model-based image analysis systems' *Int Journ Rob Res* **1** (1), 18–64

Bonner S & Shin K (1982) 'Comparative study of robot languages' *IEEE Comp* (Dec), 82–96

Boston P et al (1992) 'On the possibility of chemosynthetic ecosystems in sub-surface habitats on Mars' *Icarus* **95**, 300–308

Boudreault R (1988) 'Design and economics of freeflying platforms for space manufacturing' *Acta Astron* **17** (4), 415–420

Bouquet F & Koprowski K (1982) 'Radiation effects on spacecraft materials for Jupiter and near-Earth orbits' *IEEE Trans Nucl Sci* **29** (6), 1629–1631

Bowring Space Projects Ltd, C T (1993) 'Guide to satellite insurance market and its current status' *preprint*

Boyd R & Clark B (1992) 'Exploration strategies and the astronaut's toolset' *Journ Brit Inter Soc* **45**, 195–202

Brady M (1985) 'Artificial intelligence and robotics' *Artif Intell* **26**, 79–121

Brandel D (1990) 'Advanced tracking and data relay satellite system: next generation' *AIAA 90-0894-CP*

Brodie P & Giardina C (1975) 'New techniques for low cost strapdown inertial systems' *AGARD CP-116*, 2.1–2.13

Bronez M, Clarke M & Quinon A (1986) 'Requirements for development for a freeflying robot— ROBIN' *IEEE Int Conf Rob & Autom*, 667–672

Brooks R (1983) 'Solving Findpath problems by good representation of freespace' *IEEE Trans Syst, Man & Cyber* **13** (3), 190–197

Brooks R (1986) 'Robust layered control system for a mobile robot' *IEEE J Rob & Autom* **2** (1), 14–23

Brooks R (1989) 'Robot that walks: emergent behaviours from a carefully evolved network' *Neural Comp* **1**, 253–262

Brooks R (1990) 'Intelligence without representation' *Artif Intell* **47**, 189–159

Brooks R (1990) 'Elephants don't play chess' *Rob & Auton Syst* **6**, 3–15

Brooks T (1992) 'Operator aids for telerobotic assembly and sensing in space' *Proc IEEE Int Conf Rob & Autom*, 886–891

Brown E (1994) 'Space policy roundup' *Earth–Space Rev* **3** (4), 5–7

Bruckner A et al (1995) 'Mars in-situ propellant technology demonstrator mission' *Journ Brit Inter Soc* **48**, 37–346

Bruhm H (1987) 'Remote manipulation in orbital construction, servicing and repair missions: is one arm enough?' *Proc 1st European In-Orbit Operations Technology Symp* (ESA SP-272), 217–225

Burkhalter B & Sharpe H (1995) 'Lunar roving vehicle: historical origins, development and deployment' *Journ Brit Inter Soc* **48**, 199–212

Burrow J (1961) 'Momentum wheel dumping using magnetic torquers' *ARS J* (Dec), 1776–1778

Caldwell D et al (1994) 'Telepresence: visual, audio and tactile feedback and control of a twin armed mobile robot' *Proc IEEE Int Conf Rob & Autom*, 244–249

Calzolari G et al (1998) 'Improving Rosetta ́s return link margin' *ESA Bull* **95** (Aug), 82–86

Camarinha-Matos L (1987) 'Plan generation in robotics: state of the art and prospects' *Rob & Auton Syst* **3**, 297–328

Campbell H (1985) 'The corrosion factor in marine environments' *Metals & Materials* (Aug), 479–483

Camponella (1986) 'Existing satellite systems and networks' *AIAA 86-1204*

Carignan C & Akin D (1988) 'Cooperative control of two arms in transport of material load in zero gravity' *IEEE J Rob & Autom* **4** (4), 414–419

Carr M (1986) 'Mars: a water-rich planet?' *Icarus* **68**, 187–216

Carr M (1987) 'Water on Mars' *Nat* **326**, 30–35

Chambers A & Nagel D (1985) 'Pilots of the future: human or computer?' *Comm Assoc Comp Mach* **28** (11), 1187–1199

Chand S & Doty K (1985) 'Online polynomial trajectories for robot manipulators' *Int J Rob Res* **4** (2), 38–48

Chande E & Newcomb R (1985) 'Decision tree for inflight data processing for robot spacecraft trajectory guidance' *Proc IEEE Int Conf Rob & Autom*, 414–419

Chapman D (1987) 'Planning for conjunctive goals' *Artif Intell* **32**, 333–377

Charniak E & McDermott D (1985) *Introduction to Artificial Intelligence*, Addison-Wesley

Clarke J (1989) 'Principles of applications of SQUIDS' *Proc IEEE* **77** (8), 1208–1223

Chicarro A et al (1993) 'Marsnet—a network of stations on the surface of Mars' *ESA J* **17**, 225–237

Chicarro A et al (1994) 'INTERMARSNET—an international network of stations on Mars for global Martian characterisation' *ESA J* **18**, 207–218

Christ U et al (1998) 'Multimedia services for interactive space mission telescience' *ESA Bull* **96**, 63–69

Chubb W et al (1975) 'Flight performance of Skylab attitude and pointing control systems' *NASA TM D-8003* (June)

Chun C & Wuest H (1978) 'A microgravity simulation of the Marangoni convection' *Acta Astron* **5** (9), 681–686

Clarke J (1989) 'Principles and applications of SQUIDS' *Proc IEEE* **77** (8), 1208–1223

Clarke K (1975) 'Survey of electric propulsion capability' *J Space* **12** (11), 641–654

Clarke M (1985) 'Recent advances in teleoperation: implications for the Space Station' *Proc Space Tech Conf*, Anaheim, 4.1–4.10

Claros V (1992) 'TS1—transportable telemetry, tracking and command' *ESA J* **16**, 57–64

Claudinon B et al (1985) 'Control techniques for rendezvous' *IFAC Autom Cont in Space*, Toulouse, 287–294

Clearwater S et al (1991) 'Cooperative selection of constraint satisfaction problems' *Science* **254** (Nov), 1181–1183

Clemint V et al (1993) 'Interpretation of remotely sensed images in a context of multisensor fusion using a multispecialist architecture' *IEEE Trans Geoscience & Remote Sensing* **31** (4), 779–790

Clubb I (1999) 'Overview of the space insurance industry' *Jour Brit Inter Soc* **52**, 227–231

Cockell C (1995) 'Polar exploration of Mars' *Jour Brit Interplan Soc* **48**, 355–364

Coghlan A (1992) 'Smart ways to treat materials' *N Sci* (Jul), 27–29

Cohen A & Erickson J (1985) 'Future uses of machine intelligence and robotics for the Space Station and its implications for the US economy' *IEEE Trans Robot & Autom* **1** (3), 117–123

Cohendet P (1992) 'Economic effects of space programs' ISU Business & Management Lecture Notes

Collette R & Herdan B (1977) 'Design problems of spacecraft for communication missions' *Proc IEEE* **65** (3), 342–356

Colucci F (1990) 'Freedom under review' *Space* **6** (6), 28–33

Cook A & Marsh H (1988) 'Introduction to spread spectrum' *IEEE Comm Mag* (Mar), 8–16

Cooper A & Chow W (1976) 'Development of onboard space computer systems' *IBM J* (Jun), 5–19

Cornelisse J et al (1979) *Rocket propulsion and spacecraft dynamics*, Pitman, London

Cotavelo M et al (1990) 'Radiation environment of the Inmos transputer family' *Proc ESA Elect Comp Conf*, Noordwijk (ESA SP-313), 435–440

Courtois P (1985) 'On time and space decomposition of complex structures' *Comm Assoc Comp Mach* **28** (6), 590–603

Craig J, Hsu & Sastry S (1986) 'Adaptive control of mechanical manipulators' *Proc IEEE Int Conf Rob & Autom*, 190–195

Craig J, Hsu P & Sastry S (1987) 'Adaptive control of mechanical manipulators' *Int J Rob Res* **6** (2), 16–28

Crawford I (1998) 'The scientific case for human spaceflight' *Astron & Geophys* (Dec), 6.14–6.18

Criswell D & Waldron R (1993) 'International lunar base and lunar-based power system to supply Earth with electric power' *Acta Astron* **29** (6), 469–480

Crofts D et al (1994) 'Current capabilities and photonics trends in optical intersatellite links' *AIAA 94-1161-CP*

Crowther R (1994) 'Trackable debris population' *J Brit Interplan Soc* **47**, 128–133

Dahl V (1983) 'Logic programming as a representation of knowledge' *IEEE Comp* (Oct), 106–111

Daley S & Gill K (1986) 'Design study for self-organising fuzzy logic controllers' *Proc. IMechE* **200**C (1), 59–69

Das H et al (1999) 'Robot manipulator technologies for planetary exploration' Proc 6th Ann. Int. Simp. Smart structures & materials (No. 3668-17), Calif, USA

Davidson R (1992) 'Smart composites: where are they going?' *Mater & Des* **13** (2), 87–91

Davis R (1985) 'Logic programming and Prolog: a tutorial' *IEEE Software* (Sept), 53–62

Davis R (1987) 'In-orbit and laboratory exchange of ORU's designed/not designed for servicing' *Proc 1st European In-Orbit Operations Technology Symp* (ESA SP-272), 123–126

Davis R & Comacho M (1984) 'Application of logic programming to the generation of plans for robots' *Robotica* **2**, 137–146

Day D (1993) 'Review of recent American military space operations' *J Brit Interplan Soc* **46**, 459–476

De Aragon M (1994) 'Future applications of micro/nano technologies in space systems' *ESA Bull* **85**, 65–72

De Aragon A et al (1998) 'Future satellite services, concepts and technologies' *ESA Bull* **95**, 99–107

De Garis H (1991) 'Genetic programming: building artificial nervous systems with genetically programmed neural network modules' in *Neural and Intelligent Systems Integration* (ed. Sonak B), John Wiley 207–234

De Parolis M & Pinter-Vrainer W (1995) 'Current and future techniques for spacecraft thermal control' *ESA Bull* **87**, 73–83

de Peuter W (1994) 'Magnetic gearing for robotics' *Preparing for the Future (ESA)* **4** (3), 1–3

de Peuter W et al (1993) 'Satellite servicing in GEO by robotic service vehicle' *ESA Bull* **78**, 33–39

de Silva C (1991) 'Trajectory design for robotic manipulation in space' *J Guid & Cont* **14** (3), 670–674

Decker K (1987) 'Distributed problem solving techniques: a survey' *IEEE Trans Syst, Man & Cyber* **17** (5), 729–740

Delpech M & Marrette M (1985) 'Feasibility of time delay computation for a space teleoperative task' *IFAC Autom Cont in Space*, Toulouse, 279–286

Denavit J & Hartenburg R (1955) 'Kinematics notation for lower pair mechanisms based on matrices' *Trans ASME J Appl Mech* **77**, 215–221

Denning P & Tichy W (1990) 'Highly parallel computation' *Sci* **250**, 1217–1222

Deptovich T & Stoughton C (1989) 'General approach for manipulator system specification, design and validation' *Proc IEEE Int Conf Rob & Autom*, 1402–1407

Dermarker S (1986) 'Metal matrix composites' *Met & Mat* (Mar), 144–146

Dodds D (1988) 'Fuzziness in knowledge-based robotic systems' *Fuzzy Sets & Syst* **20**, 179–193

Doengi F (1998) 'Lander shock alleviation techniques' *ESA Bull* **93**, 51–60

Donnart J & Meyer J (1996) 'Learning reactive and planning rules in a motivationally autonomous animat' *IEEE Trans Sys Man & Cyber* **26** (3), 381–393

Dorigo M (1993) 'Genetics-based machine learning and behaviour-based robotics: a new synthesis' *IEEE Trans Syst, Man & Cyber* **23** (1), 141–154

Dorigo M (1995) 'Alecsys and the autonoMouse: learning to control a real robot by distributed classifier systems' *Mach Learning Journ* **19** (3)

Dougherty H, Lebstock & Rodden (1971) 'Attitude stabilisation of synchronous communi-cations satellites employing narrow beam antennas' *J Space & Rock* **8**, 834–842

Drabble B (1991) 'Spacecraft command and control using AI techniques' *J Brit Interplan Soc* **44**, 251–254

Dreike P & McCoy H (1998) 'Co-simulating software and hardware in embedded systems' *Embed Syst Prog* (June)

Dreyfus H (1967) 'Alchemy & AI' *RAND Corp p-3244*

Dropport P (1990) 'Effects of microgravity on the skeletal system—a review' *Journ Brit Interplan Soc* **43**, 19–24

Du Moutier L & Collett C (1991) 'Using lunar resources—the next step' *ESA Bull* **77**, 59–68

Dubowsky S & Desforges D (1979) 'Application of model reference adaptive control to robotic manipulators' *Trans ASME J DSMC* **101** (Sept), 193–200

Dubowsky S, Vance E & Torres M (1989) 'Control of space manipulators subject to spacecraft attitude control saturation limits' *Proc NASA Wshop on Space Telerobotics IV*, 409–418

Duda R, Hart P & Nillson N (1976) 'Subjective Bayesian network for rule-based inference systems' *Proc AFIPS Comp Conf* **45**, 1072–1082

Dudley G & Verniolle J (1997) 'Secondary lithium batteries for spacecraft' *ESA Bull* **90**, 50–54

Eaton D & Slachmuyders E (1988) 'Use of advanced materials in space structure applications' *Acta Astron* **17** (8), 863–874

Eaton L et al (1994) 'Impact of superconductive electronics on space commmunications' *AIAA 94-0989-CP*

Editors (1998) 'The Future of Space Exploration' *Scientific American* presentation magazine

Elfving A (1990a) 'Bi-arm servicer: final report (executive summary)' *ESA Rep MATRA/DAS/VEL/285/89*

Elfving A (1990b) 'Bi-arm servicer: systems descrpition handbook' *ESA Rep MATRA/DAS/VEL/279/89*

Ellery A (1993) 'Lunar surface characterisation' in *International Lunar Farside Observatory ISU Design Project Report*

Ellery A (1994) 'Resolved motion control of space manipulators' *Proc 45th IAF Congress*, Tel Aviv, Israel ST 94-W2-574

Ellery A (1996) 'Systems design and control of a freeflying space robotic manipulator system (ATLAS) for in-orbit satellite servicing operations' *PhD thesis*, Cranfield Institute of Technology (now Cranfield University)

Ellery A & MacGuire P (1993) 'Robotic rover delivery and emplacement of radio dipole antennas to the lunar surface' in *International Lunar Farside Observatory ISU Design Project Report*

Elliott C (1992) 'Space agencies and their programmes' *Smith System Eng Rep PM-92/1041/1.0*

Engels R et al (1984) 'Survey of payload integration methods' *J Space* **21** (5), 417–424

Erickson J (1987) 'Manned spacecraft automation and robotics' *Proc IEEE* **75** (3), 417–426

ESA (1986) 'Teleoperation control study: a final report' *ESA CR(P) 2413*

ESA (1993) 'Technology from space: successful technology transfers from European space research' *ESA Cat No 1*

Everith H (1976) 'Telecommunications—the resource not depleted by use: a historical and philosophical review' *Proc IEEE* **64** (9), 1292–1299

Fahlman S (1974) 'Planning system for robot construction tasks' *Artif Intell* **5**, 1–49

Fallin E (1975) 'Optimal intersatellite transfers for in-orbit servicing missions' *J Space* **12** (9), 565–568

Fearn D (1982) 'Review of future orbit transfer technology' *J Brit Interplan Soc* **35**, 304–325

Feather N (1967) 'Expectancy-value model of information seeking behaviour' *Psych Rev* **74** (5), 342–360

Featherstone R (1983) 'Position and velocity transformations between robot end effector coordinates and joints' *Int J Rob Res* **2** (2), 35–45

Fikes R & Kehlar T (1985) 'Role of frame based representation in reasoning' *Comm Assoc Comp Mach* **28** (9), 904–920

Fikes R, Hart P & Nilsson N (1972) 'Learning and executing generalised robot plans' *Artif Intell* **3**, 251–288

Flynn P & Jain A (1991) 'CAD based computer vision: from CAD models to relational graphs' *IEEE Trans Patt Anal & Mach Intell* **13** (2), 114–132

Forrest S (1987) 'Optoelectronic integrated circuits' *Proc IEEE* **75** (11), 1488–1496

Foxell C (1994) 'Pendulum of industry' *Physics World* (June), 32–36

Freitas A & Gilbreath W (1980) 'Advanced automation for space missions' *NASA CP-2255*

French J (1995) 'Concepts for in-situ resource utilisation on Mars: a personal historical perspective' *Journ Brit Inter Soc* **48**, 311–313

French R & Boyce B (1985) 'Satellite servicing by teleoperators' *Trans ASME J Bas Eng Ind* **107** (Feb), 49–54

Fritz W et al (1989) 'The autonomous intelligent system' *Rob & Auton Syst* **5**, 109–125

Fu K (1982) 'Pattern recognition for automatic visual inspection' *IEEE Comp* (Dec), 34–40

Fu K & Rosenfeld A (1976) 'Pattern recognition & image processing' *IEEE Trans Comp* **25** (12), 1336–1343

Fu K, Gonzalez R & Lee C (1987) *Robotics: Control, Sensing, Vision and Intelligence* McGraw-Hill, Singapore

Gagliardi R & Mendez A (1994) 'Photonic data routing in satellite payloads' *AIAA 94-1109-CP*

Game G (1993) 'Application of neural networks to vibration suppression in spacecraft structures' *Proc 2nd Int Conf Dyn & Cont Struct in Space*, Cranfield, 529–543

Garg D (1989) 'Multiarm coordination and control' *ASME DSC* **15**, 27–35

Garner T & Ross A (1991) 'Satellite control throughout complete lifecycle' *ESA Bull* **72**, 107–109

Garrett I (1983) 'Towards fundamental limits in optical fibre communication' *IEEE J Lightwave Tech* **1** (1), 131–138

Gary (1987) 'Spacecraft attitude real time determination system based on multiprocessor distributed architecture' *IFAC 10th Triennial World Cong* (Mar), 37–47

Gat E et al (1994) 'Behaviour control for robotic exploration of planetary surfaces' *IEEE Trans Rob & Autom* **10** (4), 490–503

Genesareth M & Ginsberg M (1985) 'Logic programming' *Comm Assoc Comp Mach* **28** (9), 933–941

Georgeff M & Lansky L (1986) 'Procedural knowledge' *Proc IEEE* **74** (10), 1388–1397

Gersho A (1977) 'Quantisation' *IEEE Comm Syst Mag* (Sept), 16–29

Geshke C (1983) 'System for programming and controlling sensor-based robot manipulators' *IEEE Trans Patt Anal & Mach Intell* **5** (1), 1–7

Gewal M et al (1991) 'Application of Kalman filter to calibration and alignment of inertial navigation systems' *IEEE Trans Autom Cont* **36** (1), 4–12

Ghandi H et al (1991) 'Smart materials and structures incorporating hybrid actuator and sensing systems' *ASME Smart Structures & Materials* **24**, 151–156

Gibson R (1989) 'European community: crossroads in space' *Commission of European Communities EUR 14010*

Gilbertson R & Busch J (1996) 'Survey of micro-actuator technologies for future spacecraft missions' *Jour Brit Inter Soc* **49**, 129–138

Glaser P (1968) 'Power from the Sun: its future' *Sci* **162**, 857–861

Gonzalez R & Safabakhsh R (1986) 'Computer vision techniques for industrial applications and robot control' *IEEE Comp* **15** (12), 17–32

Goodwin G et al (1984) 'Perspective on the convergence of adaptive control algorithms' *Automatica* **20** (5), 519–531

Gottlieb A & Schwartz B (1982) 'Networks and algorithms for very large scale parallel computation' *IEEE Comp* (Jan), 26–36

Gough M (1993) 'Space instrument neural networks for real-time data analysis' *IEEE Trans Geoscience & Remote Sensing* **31** (6), 1264–1268

Gough M & Wooliscroft L (1989) 'Microprocessors in space instrumentation' *Space Technol* **9** (3), 305–313

Graham J (1989) 'Special computer architectures for robotics: tutorial and survey' *IEEE Trans Rob & Autom* **5** (5), 543–554

Graves J (1995) 'Ranger telerobotic flight experiment progress update' SSI Princeton Space Manufacturing Conf. (MAY)

Greenberg J (1992) 'Financial investment analysis' in *Space Economics* (ed. Greenberg J & Hurtzfeld H), *Progress in Astronomy* vol. 144 (AIAA)

Grew J & Jones S (1999) 'Advanced technology for mission planning and scheduling' *Jour Brit Inter Soc* **52**, 49–52

Grey J (1976) 'Outlook for space power' *Astron & Aeron* (Oct), 29–36

Grossberg S (1980) 'How does the brain build a cognitive code?' *Psych Rev* **87** (1), 1–51

Grossman D et al (1985) 'Value of independent robot arms' *Rob & CIM* **2** (2), 135–141

Grubilei R et al (1994) 'European data relay system: space segment' *AIAA 94-0906-CP*

Gruver et al (1984) 'Industrial robot programming languages: comparative evaluation' *IEEE Trans Syst, Man & Cyber* **14** (4), 560–565

Guerin M & Lane H (1991) 'Commonality—key word for ground segment infrastructure' *ESA Bull* **72**, 88–93

Gupta A & Roth B (1982) 'Design considerations for manipulator workspace' *Trans ASME J Mech Des* **104** (Oct), 704–711

Gupta A & Toong H (1983) 'Microprocessors—the first 12 years' *Proc IEEE* **71** (1), 1236–1255

Habinc S et al (1998) 'TEAMSAT's data handling systems' *ESA Bull* **95** (Aug), 148–151

Hackett J & Shah M (1990) 'Multisensor fusion: a perspective' *Proc IEEE Int Conf Rob & Autom* 1324–1329

Hamman R (1985) 'Design techniques for robots (space application)' *Rob & Auton Syst* **1**, 223–250

Handelman D (1990) 'Integrating neural networks and knowledge-based systems for intelligent robotic control' *IEEE Control Syst Mag* (Apr), 77–86

Handley S (1993) 'Genetic planner: the automatic generation of plans for a mobile robot via genetic programming' *Proc Int Symp Intell Cont*, 190–195

Hanson J et al (1983) 'Generation and evaluation of workspaces of manipulators' *Int J Rob Res* **2** (3), 22–31

Happel J (1993) 'Indigenous materials for lunar construction' *ASME App Mech Rev* **46** (6), 313–324

Hardacre S (1999) *private communication*

Harmon L (1982) 'Automated tactile sensing' *Int J Robot Res* **1** (2), 3–33

Harmon S et al (1986) 'Sensor data fusion through a distributed blackboard' *Proc IEEE Int J Rob & Autom*, 1449–1454

Harnisch B et al (1998) 'Ultralightweight C/SiC mirrors and structures' *ESA Bull* **95**, 108–112

Harrison D et al (1993) 'Micromachining a miniaturised capillary electrophoresis-based chemical analysis system on a chip' *Sci* **261**, 895–897

Hassan H & Jones J (1997) 'Huygens probe' *ESA Bull* **92**, 33–44

Hassan H et al (1994) 'Huygens—a technical and programmatic overview' *ESA Bull* **77**, 21–30

Hauck S (1997) 'Roles of FPGA's in reprogrammable systems' *preprint*

Hayati S (1986) 'Hybrid position/force control of multiarm cooperating robots' *Proc IEEE Int Conf Rob & Autom*, 82–89

Hayati S (1988) 'Position and force control of coordinated multiple arms' *IEEE Trans Aero & Elect Syst* **24** (5), 584–590

Hayes P (1973) 'Frame problem and related problems in AI' *Int J Cong Artif Intell*, 223–230

Hayes-Roth F (1984) 'Knowledge based expert systems: a tutorial' *IEEE Comp* (Sept), 11–28

Hayes-Roth B (1985a) 'Blackboard architecture for control' *Artif Intell* **26**, 251–321

Hayes-Roth F (1985b) 'Rule-based systems' *Comm Assoc Comp Mach* **28** (9), 921–932

Hayes-Roth F & Jacobstein N (1994) 'State of knowledge-based systems' *Comm Assoc Comp Mach* **37**, 27–39

Haynes L et al (1982) 'A survey of highly parallel computing' *IEEE Comp* (Jan), 9–24

Hecht M (1973) 'Figure of merit for fault tolerant space computers' *IEEE Trans Comp* 22 (3), 246–251

Hedley D (1986) 'Design characteristics of the Shuttle Remote Manipulator Arm' *Trans Soc Automotive Engines* **7**, 1249–1254

Heer E (1978) 'New lustre for space robots and automation' *Astro & Aero* (Sept), 48–50

Heimel H & Schulz H (1985) 'Large wheel actuator study: final report' *ESA CR(P)-2265*

Hein G, Stevenson S & Siro J (1976) 'Cost benefit analysis of space technology' *NASA TM-X-3453*

Heironnus-Leuba A et al (1993) 'Return from space—ESA's Technology Transfer programme' *ESA Bull* **74**, 46–51

Hemami A (1985) 'Control and programming of two armed robots' *Proc Robots 9 Conf* **2**, 16.38–16.58

Hemami A (1986a) 'Collision free generation of two armed robots' *Proc Robots 10 Conf*, 3.41–3.50

Hemami A (1986b) 'Kinematics of two armed robots' *IEEEE J Rob & Autom* **2** (4), 225–228

Henderson T & Gruyen R (1990) 'Logical behaviours' *J Robot Syst* **7** (3), 309–336

Hendler J et al (1990) 'AI planning: systems and techniques' *AI Mag* (Summer), 61–77

Hendrix G & Sacerdoti E (1982) 'Natural language processing—the field in perspective' *Byte* (Sept), 304–352

Hertzfeld H (1992) 'Measuring returns to space research & development' in *Space Economics, Prog In Astron* **144**

Hinton S (1988) 'Architectural considerations for photonic switching networks' *IEEE J Sel Areas Comm* **6** (7), 1209–1226

Hirzinger G (1987) 'Sensory feedback in robotics—state of the art in research and industry' *IFAC 10th Triennial World Cong*, (Mar), 193–206

Hirzinger G (1993) 'Multisensory shared autonomy and telesensor programming—key issues in space robotics' *Rob & Auton Syst* **11**, 141–162

Ho J (1977) 'Direct path method for flexible multibody spacecraft dynamics' *J Space* **14** (2), 102–110

Hoag S et al (1999) 'Use of WWW technology for mission control systems' *ESA Bull* **97**, 49–55

Hobbs J et al (1996) 'Achieving improved mission robustness' *preprint*

Hogan J et al (1985) 'Impedance control: an approach to manipulation—Part I–III' *ASME J Dyn Syst Meas & Cont* **107** (Mar), 1–24

Hollerbach J & Sahar G (1983) 'Wrist partitioned inverse kinmematics accelerations and manipulator dynamics' *Int J Rob Res* **2** (4), 61–76

Hollerbach J (1980) 'Recursive Lagrangian formulation of manipulator dynamics and comaparative study of dynamics formulation complexity' *IEEE Trans Syst, Man & Cyber* **10** (11), 730–736

Hollerbach J & Suh K (1987) 'Redundancy resolution of manipulators through torque optimisation' *Int J Rob & Autom* **3** (4), 308–316

Holliday M (1993) 'Human or robot? Technical and political considerations' *ISU Lecture Notes*, Huntsville

Hoogers G (1998) 'Fuel cells: power for the future' *Physics World* (Aug), 31–36

Hooker W (1970) 'Set of r dynamical attitude equations for arbitrary n-body satellite with r rotational degrees of freedom' *AIAA J* **8** (7), 1205–1207

Hooker W & Margulies G (1965) 'Dynamical attitude equations for n-body satellite' *J Astron Sci* **12** (4), 123–128

Horak D (1984) 'Simplified modelling and computational scheme for manipulator dynamics' *Trans ASME J DSMC* **106**, 350–352

Howard P (1995) 'SCOS II development' *J Brit Interplan Soc* **48**, 163–168

Hsia T (1986) 'Adaptive control of robot manipulators—review' *Proc IEEE Int Conf Rob & Autom*, 183–189

Hubbard H (1989) 'Photovoltaics today and tomorrow' *Sci* **244**, 297–304

Huddleston M (1991) *private communication*

Hughes P (1979) 'Dynamics of a chain of flexible bodies' *J Astron Sci* **27** (4), 359–380

Husbands P et al (1995a) 'Use of genetic algorithms for the development of sensorimotor control systems' *preprint*

Husbands P, Harvey I & Cliff D (1995b) 'Circle in the round: state space attractors for evolved sight robots' *Rob & Auton Syst* **15** 83–106

Hutchings B & Wirthlin M (1996) 'Implementation approaches for reconfigurable logic applications' *preprint*

Huttenbach R (1992) 'Life support & habitability manual' ESA PSS-03-406 (921338)

Hwang K (1987) 'Advanced parallel processing with supercomputer architectures' *Proc IEEE* **75** (10), 1348–1379

Irwin P et al (1999) 'Atmosphere of Mars' *Jour Brit Interplan Soc* **52**, 209–216

Iwata T et al (1991) 'Dynamic control of freeflying robot for capturing manoeuvres' *AIAA 91-2824-CP*

Jaffe L & Norton H (1980) 'Prelude to interstellar flight' *Aeron & Astron* (Jan), 38–44

Jain A (1991) 'Unified formulation of dynamics of serial link rigid multibody systems' *J Guid & Cont* **14** (3), 531–542

Jakosky B (1998) *Search for Life on Other Planets* Cambridge University Press, Cambridge, UK

Jakosky B & Jones J (1994) 'Evolution of water on Mars' *Nat* **370**, 328–329

Jarvis R (1983) 'Perspective on range finding techniques for computer vision' *IEEE Trans Patt Anal & Mach Intell* **5** (2), 122–139

Jerkovsky W (1978) 'Structure of multibody dynamics equations' *J Guid & Cont* **1** (3), 173–182

Jewell C et al (1992) 'Very low temperature cryogenics and related space applications' *ESA Bull* **75**, 55–61

Johnson D & Hill J (1985) 'Kalman filter approach to sensor based robot control' *IEEE J Rob & Autom* **1** (3), 159–162

Johnson S & Chua K (1993) 'Properties and mechanics of lunar regolith' *ASME App Mech Rev* **46** (6), 285–293

Johnson S et al (1995) 'Lunar dust, lunar observation and other operations on the Moon' *Journ Brit Inter Soc* **48**, 87–92

Jones C & Watson J (1985) 'Application of intelligent robotic welding system for fabrication of aerospace hardware' *Proc Space Tech Conf*, Anaheim, 4.25–4.34

Jones E (1992) 'Putting space resources to work' *Acta Astron* **26** (1), 15–18

Jones, Holton & Stratton (1982) 'Semiconductors: the key to computational plenty' *Proc IEEE* **70** (12), 1380–1408

Junkins J & Turner J (1986) *Optimal Spacecraft Rotational Manoeuvres* Elsevier, Amsterdam

Kaidy J (1986) 'Linear quadratic tracker for control moment gyro based attitude control of the Space Station' *AIAA 86-1194*

Kalaycroglu S & Jaifu S (1992) 'Ground based control of space station Freedom robots' *Proc IEEE Int Conf Rob & Autom*, 2796–2798

Kamm L (1961) 'Magnetorquer—a satellite orientation device' *ARS J* **31** (6), 813–815

Kane T & Levinson D (1980) 'Formulation of equations of motion for complex spacecraft' *J Guid & Cont* **6** (2), 99–122

Kane T & Levinson D (1983) 'Use of Kane's dynamical equations in robotics' *Int J Rob Res* **2** (3), 3–21

Kapitsa A et al (1996) 'A large deep freshwater laake beneath the ice of East Antarctica' *Nat* **381**, 684–686

Kaplan M (1976) *Modern Spacecraft Dynamics & Control*, John Wiley

Kargel J & Strom R (1996) 'Global climatic change on Mars' *Sci Am* (Nov)

Kassler (1985) 'Robots and mining' *Robotica* **3**, 13–19

Katbib A (1988) 'Nonlinear adaptive control of robotic manipulators—hyperstability approach' *Rob & Auton Syst* **4**, 265–273

Kavanagh T (1994) 'Application of image-directed robotics in otolaryngological surgery' *Larygoscope* **104** (Mar), 283–293

Kawato M et al (1988) 'Hierarchical neural network model for voluntary movement with applications to robotics' *IEEE Cont Syst Mag* (Apr), 8–15

Kayton M (1989) 'Avionics for manned spacecraft' *IEEE Trans Aero & Elect Syst* **25** (6), 786–827

Kelly R & Bonner S (1985) 'Understanding, uncertainty and robot task execution' *IFAC Robot Control*, Syroco, 331–335

Kennel H (1982) 'Steering law for parallel mounted double gimballed CMG's' *NASA TM-82390*

Kerns S & Shafer B (1988) 'Design of radiation hardened IC's for space: compendium of approaches' *Proc IEEE* **76** (11), 1470–1508

Khatib O (1985) 'Real time obstacle avoidancefor manipulators and mobile robots' *Proc IEEE Int Conf Rob & Autom*, 500–505

Khatib O (1987) 'Unified approach for motion and force control of robotic manipulators: operational space formulation' *Trans ASME J Bas Eng* (Mar), 35–45

Kim S (1989) 'Systematic approach to intelligent system design' *Rob & CIM* **6** (2), 143–155

Kirkpatrick S, Gellatt C & Vecchi M (1983) 'Optimisation by simulated annealing' *Sci* **320**, 671–680

Klaes, L (1999) *private communication*

Klein C & Huang C (1983) 'Review of pseudoinverse control for kinematically redundant manipulators' *IEEE Syst, Man & Cyber* **13** (3), 245–250

Klinkrad H & Jehn R (1992) 'Space debris environment of the Earth' *ESA J* **16**, 1–11

Klisker S (1986) 'Expert systems: an overview' *IEEE J Oceanic Eng* **11** (4), 442–448

Klumpp A (1976) 'Singularity free extraction of quaternions from direction cosine matrix' *J Space & Rock* **13** (12), 754–755

Koelle H (1982) 'Preliminary analysis of a baseline system model for lunar manufacturing' *Acta Astron* **9** (6/7), 401–413

Koelle H & Johenning B (1994) 'Cost estimations for lunar products and their respective commercial prices' *Acta Astron* **32** (3), 227–237

Koelle H & Lo R (1997) 'Production of lunar propellant' *Jour Brit Inter Soc* **50**, 353–360

Kohlhase C & Peterson C (1997) 'Cassini mission to Saturn and Titan' *ESA Bull* **92**, 55–68

Koivo A & Guo T (1983) 'Adaptive linear controller for robotic manipulators' *IEEE Trans Autom Cont* **28** (2), 288–297

Kondo K et al (1993) 'A study of microsat communications systems' *Acta Astron* **30**, 247–252

Konigstein R et al (1989) 'Computed torque control of freeflying cooperating arm robot' *Proc NASA Workshop on Space Telerobotics* **V**, 235–243

Kopf C (1989) 'Dynamic two arm hybrid position/force control' *Rob & Auton Syst* **5**, 369–376

Korf R (1982) *Space Robotics* (preprint)

Kosha J & Kanade T (1988) 'Experimental evalution of nonlinear feedback and feedforward control schemes for manipulators' *Int J Robot Res* **7** (1),18–26

Kosugi Y et al (1988) 'Articulated neurosurgical navigation system using MRI and CT images' *IEEE Trans Biomed Eng* **35** (2), 147–152

Koza J (1993) 'Evolution of subsumption using genetic programming' *preprint*

Kozlov I & Shevchenko V (1995) 'Mobile lunar base project' *Journ Brit Inter Soc* **48**, 49–54

Kraft G & Campagna D (1990) 'Comparison of CMAC neural network control and two traditional adaptive control systems' *IEEE Control Syst Mag* (Apr), 26–43

Krishen K (1987) 'Advanced communication, tracking, robotic vision technology for space application' *Proc EASCON*, Washington DC, 143–153

Krishen K (1989) 'Robotic vision technology and algorithms for space application' *Acta Astron* **19** (10), 813–826

Kroghe B & Thorpe C (1986) 'Integrated path planning and dynamic steering control for autonomous vehicles' *IEEE Int Conf Rob & Autom*, 1664–1667

Kumar A & Waldron K (1981) 'Workspace of a mechanical manipulator' *Trans ASME J Mech Des* **103** (3), 665–672

Kung H (1982) 'Why systolic architectures?' *IEEE Comp* (Jan), 37–46

Kung S & Hwang J (1989) 'Neural network architectures for robotic application' *IEEE Trans Rob & Autom* **5** (5), 641–656

Kuperstein M (1987) 'Adaptive visual-motor coordination in multi-jointed robots using parallel architectures' *Proc IEEE Int Conf Rob & Autom*, 1595–1602

Kuritsky M & Goldstein M (1983) 'Inertial navigation' *Proc IEEE* **71** (10), 1156–1176

Kweon I et al (1992) 'Behaviour based mobile robot using active sensor fusion' *Proc IEEE Int Conf Rob & Autom*, 1675–1682

Lambert M (1993) 'Shielding against natural and man-made space debris: a growing challenge' *ESA J* **17**, 31–40

Lambreton J-P et al (1995) 'Huygens—the science, payload and mission profile' *ESA Bull* **77**, 31–41

Landau I (1981) 'Model reference adaptive control and stochastic self-tuning regulator—a unified approach' *Trans ASME J DSMC* **103** (Dec), 404–416

Landau I (1985) 'Adaptive control techniques for robotic manipulators—state of the art' *IFAC Robot Control*, Syroco, 17–25

Landis G (1995) 'Footsteps to Mars: incremental approach to Mars exploration' *Journ Brit Inter Soc* **48**, 367–32

Landzettal K et al (1995) 'Telerobotic concepts for FSS' *preprint*

Larson W & Wertz J (1992) *Space Mission Analysis and Design*, Kluwer, London

Larson E et al (1986) 'Beyond the era of materials' *Sci Am* **254** (6), 34–41

Latombe J (1984) 'Automatic synthesis of robot programs from CAD specifications' *NATO ASI F11 Rob & Artif Intell*, 199–218

Lavery D (1994) 'Perspectives on future space robots' *Aero Amer* (May), 32–37

Lawes R (1998) 'Microsystems and how to access the technology' *Journ Brit Inter Soc* **51**, 127–134

Leach C (1990) 'Medical considerations for extending the human presence in space' *Acta Astron* **21** (9), 659–66

Leahy M (1990) 'Model based control of industrial manipulators: experimental analysis' *J Robot Syst* **7** (5), 741–758

Leahy M et al (1987) 'Experimental evaluation of PUMA manipulator model reference adaptive controller' *Proc IEEE 26th Conf Dec & Cont*, 2196–2201

Lee B & Lee C (1987) 'Collision-free motion planning of two robots' *IEEE Trans Syst, Man & Cyber* **17** (1), 21–32

Lee C (1982) 'Robot arm kinematics, dynamics and control' *IEEE Comp* **15** (12), 62–80

Lee C & Lee B (1984) 'Resolved motion adaptive control for mechanical manipulators' *Trans ASME J DSMC* **106** (Jun), 134–142

Lee C & Ziegler M (1984) 'Geometric approach in solving kinematics of the PUMA robot' *IEEE Trans Aero & Elect Syst* **20** (6), 695–706

Lee T & Yang D (1983) 'On the evaluation of manipulator workspaces' *Trans ASME J Mech Des* **105** (Mar), 70–77

Lefferts E et al (1982) 'Kalman filtering for spacecraft attitude estimation' *AIAA J Guidance* **5** (5), 417–427

Lenat D & Feigenbaum E (1991) 'On the threshold of knowledge' *Artif Intell* **47**, 185–250

Levi P (1987) 'Principles of planning and control concepts for autonomous mobile robots' *Proc IEEE Int Conf Rob & Autom*, 874–881

Lewis J (1995) 'Planetary resources for extraterrestrial technology' *Quart Journ Roy Astron Soc* **36**, 445–448

Liddle D (1986) 'Trojan: a remotely operated vehicle' *IEEE Journ Ocean Eng* **11** (3), 364–372

Liebermann L & Wesley H (1977) 'AUTOPASS: an automated programming system for computer controlled mechanical assembly' *IBM J Res & Dev* **2** (4), 324–333

Liegeois A (1977) 'Automatic supervisory control of the configuration and behaviour of multibody mechanisms' *IEEE Trans Syst Man & Cyber* **7** (12), 868–871

Lim D & Seraji H (1996) 'Configuration control of a mobile dextrous robot: real-time implementation and experimentation' *preprint*

Lim J & Chyung D (1987) 'Resolved position control for two cooperating robot arms' *Robotica* **5**, 9–15

Lin C & Chang P (1983) 'Joint trajectories of mechanical manipulators' *IEEE Trans Syst Man & Cyber* **13** (6), 1094–1102

Lin F et al (1983) 'Formulation and optimisation of cubic polynomial joint trajectories for industrial robots' *IEEE Trans Autom Cont* **28** (12), 189–197

Lin Z et al (1989) 'Online robot trajectory planning for catching a moving object' *Proc IEEE Int Conf Rob & Autom* 1726–1731

Lindberg R, Longman R & Zedd M (1986) 'Kinematics and reaction moment compensation for spaceborne elbow manipulator' *AIAA 86-0250*

Liska D (1986) 'Two degree of freedom CMG for high accuracy attitude control' *J Space & Rock* **5** (Jan), 74–83

Logsden J et al (1993) 'Partners in space: international cooperation in space—strategies for the new century' *CREST Project Report*

Longman R (1988) 'Kinetics and workspace of robot mounted on satellite that is free to rotate and translate' *AIAA 88-4097-CP*

Longman R, Lindberg R & Zedd M (1987) 'Satellite-mounted robot manipulators—new kinematics and reaction compensation' *Int J Robot Res* **6** (3), 87–103

Lozano-Perez T (1981) 'Automatic planning of manipulator transfer movements' *IEEE Trans Syst, Man & Cyber* **11** (10), 681–698

Lozano-Perez T (1983a) 'Robot programming' *Proc IEEE* **71** (7), 821–8421

Lozano-Perez T (1983b) 'Spatial planning: configuration space approach' *IEEE Com* **32** (2), 108–120

Lozano-Perez T et al (1987) 'HANDEY: a robot system that recognises, plans and manipulates' *IEEE Int Conf Rob & Autom*, 843–849

Luh J (1983a) 'Anatomy of industrial robots and their controls' *IEEE Trans Autom Cont* **28** (2), 133–153

Luh J (1983b) 'Conventional controller design for industrial robots—a tutorial' *IEEE Trans Syst, Man & Cyber* **13** (3), 298–316

Luh J & Lin C (1981) 'Optimum path planning for mechanical manipulators' *Trans ASME J DSMC* **102** (Jun), 142–151

Luh J & Lin C (1984) 'Approximate joint trajectories for control of industrial robots along cartesian paths' *IEEE Trans Syst, Man & Cyber* **14** (3), 444–450

Luh J & Zheng Y (1986) 'Computation of input generalisation forces for robots with closed kinematic chain mechanisms' *IEEE J Rob & Autom* **1** (2), 95–103

Luh J, Walker M & Paul R (1980) 'On-line computational scheme for mechanical manipulators' *Trans ASME J Dyn Syst Meas & Cont* **102** (Jun), 69–76

Luh J, Walker M & Paul R(1980) 'Resolved acceleration control of mechanical manipulators' *IEEE Trans Autom Cont* **25** (3), 236–241

Lumia R & Wavering A (1989) 'Trajectory generation for space telerobots' *Proc NASA Wshop on Space Telerobots II*, 123–131

Luo R & Kay M (1989) 'Multisensor integration and fusion in intelligent systems' *IEEE Syst, Man & Cyber* **19** (5), 901–927

Luo G & Saridis G (1985) 'LQ design of PID controllers for robot arms' *IEEE J Rob & Autom* **1** (3), 152–158

Lutz H (1997) 'Optical communications in space—20 years of ESA effort' *ESA Bull* **91** (Aug), 25–36

Lutz H & Armandillo E (1992) 'Laser based remote sensing from space' *ESA Bull* **66**, 73–79

Lyall F (1992a) *private communication*

Lyall F(1992b) *Space Law* preprint

Maes P (1990) 'Situated agents can have goals' *Rob & Auton Syst* **6**, 49–70

MacArthur J (1984) 'Space: the finance sector' *Phil Trans R Soc Lon* **A312**, 75–81

MacInnes B & Lin C (1986) 'Kinematics and dynamics in robotics: tutorial based upon classical concepts of vectorial mechanics' *IEEE J Rob & Autom* **2** (4), 181–187

MacInnes C (1992) 'Optimum orbit selection for two-vehicle rendezvous' *ESA J* **16**, 447–454

MacInnes C (1993) 'Analytical model for the catastrophic production of orbital debris' *ESA J* **17**, 293–305

Maiers J & Sherif Y (1985) 'Applications of fuzzy set theory' *IEEE Trans Syst Man & Cyber* **15** (1), 175–189

Maimon O & Nof S (1985) 'Coordination of robot sharing assembly tasks' *Trans ASME J DSMC* **197**, 299–307

Maimone M (1999) *private communication*

Maitra A (1989) 'Benefits from out of this world' *Acta Astron* **19** (9), 757–759

Malcolm C & Smithers T (1990) 'Symbol grounding via a hybrid architecture in an autonomous assembly system' *Rob & Auton Syst* **6**, 123–144

Mandell H (1992) 'CER for space programs' in *Space Economics* (ed. Greenberg J & Hertzfeld H), *Progress in Astronomy*, vol. 144 (AIAA)

Marcyk J & Bellazzi A (1989) 'Dynamics and control of freeflying inspection and maintenance vehicle with manipulators' *Proc 2nd European In-Orbit Operations Technology Symp* (ESA SP-297), 413–427

Marr D & Poggio T (1976) 'Cooperative computation of stereo disparity' *Sci* **194**, 283–287

Martin A (1994) 'Review of spacecraft/plasma interaction and effects on space systems' *J Brit Interplan Soc* **47**, 134–142

Mason M (1981) 'Compliance and force control for computer-controlled manipulators' *IEEE Trans Syst Man & Cyber* **11** (6), 418–432

Massevitch A (1989) 'Influence of space exploration on the development of mankind' *Acta Astron* **19** (9), 755–757

Masutani Y, Mujazaki F & Arimoto S (1989) 'Sensory feedback control for space manipulators' *Proc IEEE Int Conf Robot & Autom*, 1346–1351

Mataric M (1992) 'Integration of representation into goal-driven behaviour-based robots' *IEEE Trans Rob & Autom* **8** (3), 304–312

Mataric M (1995) 'Issues and approaches in the design of collective autonomous agents' *Rob & Auton Syst* **16**, 321–331

Mataric M & Cliff D (1997) 'Artificial evolution of control systems' *preprint*

Matijevic J (1998) 'Mars Pathfinder microrover—implementing a low cost planetary mission experiment' *IAA-L-0510*

Matthews J & Bates D (1978) 'Future applications of low cost strapdown laser inertial navigation system' *AGARD CP-220*, 1.1–1.13

McBarron J (1994) 'Past present and future of the US EVA program' *Acta Astron* **32** (1), 5–14

McBarron J et al (1994) 'US pre-breathe protocol' *Acta Astron* **32** (1), 75–78

McCarthy J (1961) 'Programs with common sense' *Mechanisation of Thought Processes* **1**, 77–84 (HMSO)

McCarthy J & Hayes P (1969) 'Some philosophical problems from the stanpoint of AI' *Mach Intell* **4**, 463–502

McInnes B & Lin C (1986) 'Kinematics and dynamics in robotics: tutorial based upon classical concepts of vectorial mechanics' *IEEE J. Rob & Autom* **2** (4), 181–187

Mehmud B (1989) 'Relevance of space activities to society' *Acta Astron* **19** (9), 759–761

Meier W & Graf J (1991) 'Two arm robot system based on trajectory optimisation and hybrid control including experimental evaluation' *IEEE Int Conf Rob & Autom*, 2618–2623

Merchant M (1985) 'Computer integrated manufacturing as the basis for the factory of the future' *Rob & CIM* **3** (2), 89–99

Merchant M (1988) 'Precepts and sciences in manufacturing' *Rob & CIM* **4** (1/2), 1–6

Meyer T & McKay C (1989) 'Resources of Mars for human settlement' *Jour Brit Inter Soc* **42**, 147–160

Meystel A (1988) 'Intelligent control in robotics' *J Rob Syst* **5** (4), 269–308

Michie D (1973) 'Machines and the theory of intelligence' *Nat* **241**, 507–512

Mickelson W (1967) 'Auxiliary primary electric propulsion, present and future' *J Space & Rock* **4** (11), 1409–1422

Mickle M (1993) 'EVA Operations' *NASA Mission Operations Directorate, NASA-JSC/DF42*

Miller D (1990a) 'Minirovers for Mars exploration' *preprint*

Miller D (1990b) 'Multiple behaviour-controlled microrobots for planetary surface missions' *IEEE preprint*

Miller D (1992) 'Reducing software mass through behaviour control' *Proc Cooperative Intelligent Robotics in Space III*, SPIE Cambridge Symp, Cambridge, MA

Miller D & Varsi G (1993) 'Microtechnology for planetary exploration' *Acta Astron* **29** (7), 561–567

Miller D et al (1989) 'Autonomous navigation and control of a Mars rover' *IFAC Autom Control*, Tsukuba, 111–114

Miller D et al (1992) 'Reactive navigation through rough terrain: experimental results' *Proc 10th National Conf on AI*

Minsky M (1961) 'Steps towards AI' *Proc IRE* (Jun), 8–30

Mirchandani M & Figueiredo M (1998) 'Technology evolution in spacecraft communications' *IAF-98-M.2.03*

Mishan E (1971) *Cost Benefit Analysis* George Allen & Unwin, London

Moore R et al (1987) 'Expert systems methodology for real-time process control' *IFAC 10th Triennial World Cong*, Munich, 279–286

Morgan W (1994) 'Potential uses of superconductivity in communications satellites' *AIAA 94-0990-CP*

Morrison J & Nguyen T (1998) 'Onboard software for the Mars Pathfinder microrover' *IAA-L-0504P*

Mowforth P & Bratko I (1987) 'Artificial intelligence and robotics: flexibility and integration' *Robotica* **5**, 2618–2623

Moya M & Seraji H (1987) 'Robot control systems: survey' *Rob & Auton Syst* **3**, 329–351

Mufti I (185) 'Model reference adaptive control for manipulators—a review' *IFAC Robot Control*, Syroco, 111–115

Murphy S Wen & Saridis (1991) 'Simulation of cooperating robot manipulators on a mobile platform' *IEEE Trans Rob & Autom* **7** (4), 468–477

Nagashima F & Nakaruma Y (1992) 'Efficient computation scheme for the kinematics and inverse dynamics of a satellite-based manipulator' *Proc IEEE Int Conf Robot & Autom*, 905–912

Nakaruma T et al (1994) 'Optical waveguide over solar energy systems for lunar material processing' *SSI Update* **20** (4)

Nakaruma Y & Mukherjee R (1989) 'Nonholonomic path planning of space robots via bidirectional approach' *IEEE Trans Robot & Autom* **7** (4), 500–514

Narendra K & Parthasavathy K (1990) 'Identification and control of dynamic systems using neural networks' *IEEE Trans Neural Net* **1** (1), 4–26

NASA Telerobotics Unit (1988) 'Telerobotics: problems and research needs' *IEEE Trans Aero & Elect Syst* **24** (5), 542–551

Nau D (1983) 'Expert computer systems' *IEEE Comp* (Feb), 63–73

Nelleson W (1992) 'Eureca project—from concept to launch' *ESA Bull* **70** (May), 17–25

Nenchev D Umetani & Yoshida (1992) 'Analysis of a redundant freeflying spacecraft manipulator system' *IEEE Trans Rob & Autom* **8** (1), 1–6

Nevins J (1986) 'Information control aspects of sensor systems for intelligent robotics' *IFAC Robot Control*, Syroco, 11–16

Nevins J & Whitney D (1980) 'Assembly research' *Automatica* **16** (6), 595–613

Nevins J et al (1987) 'Integrated approach to spacecraft design for robotic servicing' *AIAA 87-1672*

Newell A & Simon H (1976) 'Computer science as empirical enquiry: symbols and search' *Comm Assoc Comp Mach* **19**, 113–126

Newell A, Shaw J & Simon H (1958) 'Elements of a theory of human problem solving' *Psych Rev* **65**, 151–166

Nguyen D & Widrow B (1990) 'Neural network for self-learning control systems' *IEEE Control Syst Mag* (Apr), 18–23

Nicogossian A et al (1988) 'Assessment of the efficiency of medical countermeasures in spaceflight' *Acta Aston* **17** (2), 195–198

Nicosia S & Tomei P (1984) 'Model reference adaptive control algorithms for industrial robots' *Automatica* **20** (5), 635–644

Nitta K et al (1991) 'Various problems in lunar infrastructure habitat construction scenarios' *Acta Astron* **25** (10), 647–657

Nitzan D (1985) 'Development of intelligent robots: achievements and issues' *IEEE J Rob & Autom* **1** (1), 3–13

Norci A & Stanley J (1989) 'Adaptive human-computer inteface: literature survey and perspective' *IEEE Syst Man & Cyber* **19** (2), 399–408

Noyce R (1977) 'Microelectronics' *Sci Am* **237** (3), 63–69

O'Connor B & Morine L (1969) 'Description of CMG and its application to space vehicle control' *J Space & Rock* (3), 225–231

O'Leary B (1977) 'Mining the Apollo and Amor asteroids' *Sci* **197**, 363–366

O'Leary B (1979) 'Asteroidal resources for space manufacturing' *Acta Astron* **6**, 1467–1480

O'Leary B (1988) 'Asteroid mining and the moons of Mars' *Acta Astron* **17** (4), 457–462

Oliviera E et al (1991) 'Multiagent environment in robotics' *Robotica* **9**, 431–440

O'Neill G (1976) 'Engineering a space manufacturing facility' *Aero & Astro* **14** (Oct), 20–28

O'Neill G (1978) 'Low (profile) road to space manufacturing' *Astron & Aeron* **16** (3), 18–32

O'Neill G (1980) 'New routes to manufacturing in space' *Astron & Aeron* **18** (10), 11–14

O'Neill G & Kolm H (1980) 'High acceleration mass drivers' *Acta Atron* **7**, 1229–1238

O'Neill G, Billingham J & Gilbreath W (1978) 'Space resources and space settlements' *NASA SP-428*

Orin D & Oh S (1981) 'Control of force distribution in robotic mechanisms containing closed kinematic chains' *Trans ASME J DSMC* **102**, 134–141

Orin D & Schrader W (1984) 'Efficient computation of Jacobian for robotic manipulators' *Int J Robot Res* **3** (4), 66–75

Orin D et al (1979) 'Kinematics and kinetics analysis of open chain linkages utilising Newton-Euler methods' *Mathem Biosci* **43**, 107–130

Ortega R & Spong M (1988) 'Adaptive motion control of rigid robots: tutorial' *Proc 27th Conf Dec & Cont*, 1575–1585

O'Sullivan R (1994) 'Space power electronics—design drivers' *ESA J* **18**, 1–23

Palsula S (1993) 'Molecular composites: third generation polymers for aerospace applications' *ESA J* **17**, 133–145

Pan D & Sharp R (1991) 'Fast motion control robot manipulators with inclusion of actuator dynamics' *Proc IMechE* **204C** (5), 341–348

Panin F (1992) 'New latching mechanism' *ESA J* **16**, 363–372

Papadopoulos E & Dubowsky S (1989) 'On dynamic singularities in the control of free-floating manipulators' *Trans ASME J Dyn Syst & Cont* **15**, 45–52 (Winter Annual Meeting)

Papadopoulos E & Dubowsky S (1990) On the nature of control algorithms for space manipulators' *Proc IEEE Int Conf Rob & Autom*, 1102–1108

Papadopoulos E & Dubowsky S (1991a) 'Coordinated manipulator /spacecraft control for space robotic systems' *Proc IEEE Int Conf Rob & Autom*, 1696–1701

Papadopoulos E & Dubowsky S (1991b) 'On the nature of control algorithms for freefloating manipulators' *IEEE Trans Robot & Autom* **7** (6), 750–758

Pappalardo R et al (1999) 'The hidden ocean of Europa' *Sci Am* (Oct), 34–43

Paradiso J (1991) 'Application of a directed search to global steering of single gimballed CMG's' *AIAA 91-2718-CP*

Parkinson R (1991) 'Space economy of 2050 AD' *Journ Brit Inter Soc* **44**, 111–120

Parrish J & Akin D (1996) 'Ranger telerobotic flight experiment: missions, technologies and pragmatics' *Robotics for Challenging Environments: Proc RCEII Conf*, 136–142

Parrish J (1998) 'Ranger Telerobotic Shuttle Experiment (RTSX): status report' *preprint*

Parrish J (1999) private communication

Partridge D (1981) 'Computational theorising as the tool for resolving wicked problems' *IEEE Syst Man & Cyber* **11** (4), 318–322

Pasupathy S (1979) 'Minimum shift keying: spectrally efficient modulation' *IEEE Comm Mag* (Jul), 14–22

Patterson D (1985) 'Advanced reduced instruction set computers' *Comm Assoc Comp Mach* **28** (1), 8–21

Pau L (1987) 'Knowledge representation approaches in sensor fusion' *IFAC 10th Triennial World Cong*, Munich, 323–327

Paul D et al (1994) 'Feasibility of a photonic satellite payload' *AIAA 94-1108-CP*

Paul R (1979) 'Manipulator cartesian path control' *IEEE Trans Syst, Man & Cyber* **9** (11), 702–711

Paul R (1981) *Robot Manipulators: Mathematics, Programming and Control* MIT Press, Cambridge, MA

Paul R (1987) 'Problems and research issues associated with hybrid control of force and displacement' *Proc IEEE Int Conf Robot & Autom*, 1966–1971

Paul R, Shimano B & Meyer C (1981a) 'Differential kinematic control equations for simple manipulators' *IEEE Trans Syst, Man & Cyber* **11** (6), 456–460

Paul R, Shimano B & Meyer C (1981b) 'Kinematic control equations for simple manipulators' *IEEE Trans Syst Man & Cyber* **11** (6), 449–455

Payton D et al (1990) 'Plan guided reaction' *IEEE Trans Syst, Man & Cyber* **20** (6), 1370–1382

Petersen K (1982) 'Silicon as a mechanical material' *Proc IEEE* **70** (5), 420–457

Philips Electronics Ltd (1989) 'Infrared detectors' *Mil Tech* **2**, 56–64

Pin F et al (1992) 'On the design and development of a human–robot synergistic system' *Rob & Auton Syst* **10**, 161–184

Pittelkau M (1988) 'Adaptive load-sharing force control for two manipulators' *Proc IEEE Int Conf Rob & Autom*, 498–503

Pivirotto D (1993) 'MESUR Pathfinder microrover flight experiment' *Mars V Conf*, Boulder, Colorado, *preprint*

Podnicks E & Sickmeier J (1994) 'Role of mining in lunar base development' *Journ Brit Inter Soc* **47**, 543–548

Poggio T et al (1985) 'Computational vision and regularisation theory' *Nature* **317**, 314–319

Porill J (1988) 'Optimal combination and constraints for geometric sensor data' *Int J Rob Res* **7** (6), 66–77

Post R & Post S (1973) 'Flywheels' *Sci Am* **229** (6), 17–23

Post S & Sage (1990) 'Overview of automated reasoning' *IEEE Syst, Man & Cyber* **20** (1), 202–224

Potschke J & Hohenstein K (1982) 'Preparation of dispersed alloys under microgravity conditions' *Acta Astron* **9** (4), 261–264

Powell B et al (1971) 'Synthesis of double gimbal CMG systems for spacecraft attitude control' *AIAA 71-937*

Prest A & Turvey R (1965) 'Cost–benefit analysis: a survey' *Econ J* **75** (Dec), 683–735

Price K & Jorasch R (1990) 'Role of communications satellites in the fibre optic era' *AIAA 90-0792-CP*

Pringle P (1966) 'On the stability of a body with connected moving parts' *AIAA J* **4** (8), 1394–1404

Pritchard W (1977) 'Satellite communications—an overview of the problems and programs' *Proc IEEE* **65** (3), 294–307

Procyk T & Mamdani E (1979) 'Linguistic self-organising process controller' *Automatica* **15**, 15–30

Raibert M & Craig J (1981) 'Hybrid position/force control of manipulators' *Trans ASME J Energy Res Technol* **102** (Jun), 126–133

Rando N et al (1999) 'S-Cam: a technology demonstrator for the astronomy of the future' *ESA Bull* **98** (Jun), 67–74

Raol J & Sinha N (1985) 'On the orbit determination problem' *IEEE Trans Aero & Elect Syst* **21** (3), 274–290

Raulin F et al (1990) 'Titan and exobiological aspects of the Cassini–Huygens mission' *Jour Brit Int Soc*, 257–271

Rea D et al (1990) 'International Mars Exploration programme' *Acta Astron* **22**, 255–260

Reddy M et al (1992) 'Effect of LEO atomic oxygen environment on solar array materials' *ESA J* **16**, 193–208

Rembold V & Dillman R (1985) 'Artificial intelligence in robotics' *IFAC Robot Control*, Syroco, 1–10

Rennel D (1980) 'Distributed fault tolerant computer system' *IEEE Comp* (Mar), 55–64

Rex D et al (1989) 'Space debris—origin, evolution and collision mechanics' *Acta Astron* **20**, 209–216

Reynolds R et al (1983) 'Man-made debris in LEO—threat to future space operations' *J Space* **20** (3), 279–285

Richards D (1985) 'Biological strategies for communication' *IEEE Comm Syst Mag* **23** (6), 10–18

Risbeth H (1991) 'Ionospheric science and geomagnetism' *Quart J Roy Astron Soc* **32**, 409–421

Robertson W et al (1988) 'Cost effectiveness of on-orbit servicing for large constellations' *AIAA 88-3519*

Robinson R (1986) 'National defence applications of autonomous underwater vehicles' *IEEE Journ Ocean Eng* **11** (4), 462–467

Rockoff L & Anderson D (1990) 'Freeflyers for Space Station EVA operations' *Space Station Advanced Technologies*, NASA SP-830, 59–64

Rodgers E & Stroll P (1986) 'EEPROM for spacecraft operations' *IEEE Aero App Conf Digest*, Colorado, 1–9

Rodriguez G (1987) 'Kalman filtering, smoothing and recursive robot arm forward and inverse dynamics' *IEEE J Rob & Autom* **3** (6), 624–639

Rosenberg S (1997) 'Lunar resource utilisation' *Jour Brit Inter Soc* **50**, 337–352

Rosenfeld A (1981) 'Image pattern recognition' *Proc IEEE* **69** (5), 596–605

Rosenfeld A (1986) 'Robot vision' *NATO ASI F33 Machine Intelligience & Knowledge Engineering*, 1–19

Rosenfeld A & Weszka J (1976) 'Picture recognition and scene analysis' *IEEE Comp* (May), 28–38

Roth B (1985) 'Overview of advanced robotics: manipulation' *Int Conf Advanced Robotics*, 569–580

Rothery D (1994) 'Large icy planetary bodies' *Geoscientist* **5** (2), 19–20

Rouse W & Cody W (1987) 'On the design of man–machine systems: principles, practices and prospects' *Proc IFAC 10th Tiriennial World Cong*, Munich, 281–288

Rudnicky A et al (1994) 'Survey of current speech technology' *Comm Assoc Comp Mach* **37** (3), 52–57

Russel P & Price K (1990) 'Servicing communications satellites in geostationary orbit' *AIAA 90-0830-CP*

Rycroft M (1989) 'Solar terrestrial physics: a review' *Phil Trans R Soc* **A328**, 39–42

Sabroff A (1968) 'Advanced stabilisation and attitude control techniques' *J Space & Rock* **5** (12), 1377–1392

Sackheim A & Rosenthal R (1994) 'Electric propulsion: major advancement in space transportation for comsats' *AIAA 94-1012-CP*

Salisbury J (1980) 'Active stiffness control of a manipulator in cartesian coordinates' *Proc IEEE Conf Dec & Cont*, 95–100

Salisbury J (1988) 'Issues in human/computer control of dextrous remote hands' *IEEE Trans Aero & Elect Syst* **24** (5), 591–596

Salisbury J & Craig J (1982) 'Articulated hands: force control and kinematic issues' *Int J Robot Res* **1** (1), 4–17

Sallaberger C (1992) 'Profitability analysis' *ISU preprint*

Salutin A & Bainum P (1983) 'Analysis of a double gimballed reaction wheel spaceraft attitude stabilisation system' *Acta Astron* **40** (2), 55–66

Sanderson A, Peshkin M & Homem-De-Mollo L (1988) 'Task planning for robotic manipulation in space applications' *IEEE Trans Aero & Elect Syst* **24** (5), 619–628

Sargent P & Derby B (1982) 'Advanced alloy and metal/ceramic composites from lunar materials' *Acta Astron* **9** (9), 593–595

Saridis G (1979) 'Towards the realisation of intelligent controls' *Proc IEEE* **67** (8), 1115–1132

Saridis G (1983) 'Intelligent robotic control' *IEEE Trans Autom Cont* **28** (5), 547–557

Saridis G (1988) 'Analytical formulation of the principle of increasing precision with decreasing intelligence for intelligent machines' *IFAC Robot Control*, Karlsruhe, 529–534

Sato T & Hirai S (1987) 'Language aided robotic teleoperation system (LARTS) for advanced teleoperation' *IEEE J Rob & Autom* **3** (5), 476–481

Sato Y et al (1993) 'Resolving attitude disturbance while teleoperating a space manipulator' *Proc IEEE Int Conf Rob & Autom*, 516–523

Scarrott G (1980) 'From computing slave to knowledgable servant: the evolution of the computer' *Proc Roy Soc* **A366**, 1–30

Schenker P (1988) 'NASA R&D for space telerobotics' *IEEE Trans Aero & Elect Syst* **24** (5), 523–534

Schmidt C et al (1978) 'Plan recognition problem: an intersection of psychology and AI' *Artif Intell* **11**, 45–83

Schmidt R et al (1999) 'ESA's Mars Express mission—Europe on its way to Mars' *ESA Bull* **98** (Jun), 56–68

Schmitt H (1994) 'Lunar industrialisation: how to begin?' *Journ Brit Inter Soc* **47**, 527–530

Schroer B (1988) 'Telerobotic issues in space applications' *Rob & Auton Syst* **4**, 233–244

Schuyer P et al (1992) 'Probing the Earth from space—Aristoteles mission' *ESA Bull* **72**, 67–75

Schwarz J & Sharir M (1988) 'Survey of motion planning and related geometrical algorithms' *Artif Intell* **37**, 157–169

Schwenn G & Hechler M (1993) 'Rosetta—ESA's planetary cornerstone mission' *ESA Bull* **77**, 7–18

Segre A (1991) 'Learning how to plan' *Rob & Auton Syst* **8**, 93–111

Sepehri M ((1987) 'Resupply models for space logistics and influence on design' *AIAA 87-0657*

Seraji H (1987a) 'Adaptive force and position control of manipulators' *J Rob Syst* **4** (4), 551–578

Seraji H (1987b) 'Design of force/position controllers for manipulators' *AIAA 87-2267*

Seraji H (1987c) 'Adaptive control strategies for cooperative dual arm manipulators' *J Robot Syst* **4** (5), 653–684

Seraji H & Long M (1993) 'Motion control of 7 degree of freedom arms: the configuration approach' *IEEE Trans Rob & Autom* **9** (2), 125–19

Shafer S et al (1986) 'Architecture for sensor fusion in mobile robots' *Proc IEEE Int Conf Rob & Autom*, 2002–2010

Shafirovitch E & Goldschleger U (1995) 'Mars multi-sample return sample' *Journ Brit Inter Soc* **48**, 315–319

Shepard R (1984) 'Ecological constraints on internal representation: resonant kinematics of perceiving, imagining, thinking and dreaming' *Psych Rev* **91** (4), 417–447

Sheshkin T (1985) 'Unmanned platform as an initial capability in space' *Proc Space Tech Conf*, Anaheim, 4.35–4.42

Shimano B & Roth B (1975) 'On force sensing information and its use in controlling manipulators' *Proc 8th Int Symp Ind Robotics*, 119–126

Siedman L (1992) 'Towards a policy for space robotics' *AIAA 92-1718*

Siewiorek D (1984) 'Architecture of fault tolerant computers' *IEEE Comp* (Aug), 9–17

Sieworek D (1991) 'Architecture of fault tolerant computers: a historical perspective' *Proc IEEE* **79** (12), 1710–1731

Sijmati J & Schneck P (1984) 'Supercomputing' *IEEE Comp* (Oct), 97–112

Silbey A (1986) 'Survey of advanced microprocessors and HLL computer architectures' *IEEE Comp* (Aug), 72–77

Silver W (1982) 'On equivalence of Lagrangian and Newton–Euler dynamics for manipulators' *Int J Robot Res* **1** (2), 60–69

Silverman B et al (1989) 'Blackboard system generator: alternatives distributed problem solving paradigm' *IEEE Syst Man & Cyber* **19** (2), 334–354

Simmons R et al (1990) 'Six dimensional trajectory solver for autonomous proximity operations' *AIAA 90-3459-CP*

Simon H (1978) 'Rationality as a process and as the product of thought' *Am Econ Rev* **68** (2), 1–16

Simon H (1991) 'AI: where has it been and where is it going?' *IEEE Trans Know & Data Eng* **3** (2), 128–136

Simon W & Nerod D (1987) 'Manned spacecraft electrical power systems' *Proc IEEE* **75** (3), 277–307

Simonett D (1982) 'Development and principles of remote sensing' in *Manual of Remote Sensing* (ed. Colwell R), 1–35

Sklar B (1983a) 'A structured overview of digital communications—tutorial review I' *IEEE Comm Mag* (Aug), 4–17

Sklar B (1983b) 'A structured overview of digital communications—tutorial review II' *IEEE Comm Mag* (Oct), 6–21

Smith G (1990) 'Doing business with Europe: space and defence issues in the 1990's' *AIAA 90-3584*

Smith I & Cutts J (1999) 'Floating in space' *Sci Am* (Nov)

Smith R (1980) 'Contract net protocol: high level communication and control in a distributed problem solver' *IEEE Trans Comp* **29** (2), 1104–11113

Smith T (1991) 'Semiconductor science: the fourth generation' *Proc IEEE* **79** (8), 406–418

Snyder L (1982) 'Introduction to the configurable highly parallel computer' *IEEE Comp* (Jan), 47–56

Soloway D & Alberts T (1989) 'Comparison of joint space versus task force level distribution optimisation for multi-arm manipulator systems' *Proc NASA Wshop on Space Telerobotics IV*, 413–443

Somlo J & Cat P (1988) 'Robust adaptive control of robot manipulators' *IFAC Robot Control*, Karlsruhe, 151–156

Spiering V et al (1998) 'Technologies and microstructures for separation techniques in chemical analysis' *Journ Brit Inter Soc* **51**, 133–136

Spofford J & Akin D (1988) 'Redundancy control of freeflying telerobots' *AIAA 88-4094-CP*

Spudis P (1999) 'Robots v. humans in space' *Sci Am Sp Iss 'The Future of Space Exploration'*

Spur G (1984) 'Growth, crisis and the future of the factory' *Rob & CIM* **1** (1), 21–37

Spur G (1988) 'Advanced manufacturing systems' *Rob & Cim* **4** (1/2), 7–12

Squyres S & Kasting J (1994) 'Early Mars: how warm and how wet?' *Sci* **265**, 744–749

Srour T & McGarrity J (1988) 'Radiation effects on microelectronics in space' *Proc IEEE* **76** (11), 1443–1469

Stark L et al (1987) 'Telerobotics: display, control and communications problems' *IEEE J Rob & Autom* **3** (1), 67–75

Stassinopoulos E & Raymond J (1988) 'Space radiation environment for electronics' *Proc IEEE* **74** (1), 1443–1464

Stengel R (1991) 'Intelligent fault tolerant control' *IEEE Control Syst Mag* (Jun), 14–22

Sterner E (1994) 'Dual use technology for near term lunar exploration: Clementine program' *J Brit Interplan Soc* **47**, 521–526

Sterner E & Benavoya H (1994) 'Lunar industrialisation and colonisation: towards a policy framework' *Journ Brit Inter Soc* **47**, 516–520

Stone H (1998) 'Mars Pathfinder microrover—a small, low cost, low power spacecraft' *pre-print*

Stone R (1992) 'VR & telepresence' *Robotica* **10**, 461–467

Suh N (1984) 'Factory of the future' *Rob & CIM* **1** (1), 39–49

Sullivan T & McKay D (1991) 'Using space resources' *NASA Johnson Space Centre Report*

Surkov Y (1997) *Exploration of Terrestrial Planets from Spaceccraft* Praxis–Wiley, Chichester, UK

Sutherland J (1986) 'Assessing AI's contribution to decision technology' *IEEE Trans Syst Man & Cyber* **16** (1), 3–20

Sved J et al (1995) 'Commercial lunar helium-3 fusion power infrastructure' *Journ Brit Inter Soc* **48**, 55–61

Syromiatnikov V (1992) 'Manipulator system for module redocking on Mir orbital complex' *Proc IEEE Int Conf Rob & Autom*, 913–918

Tabert M & Menus G (1986) 'Aerothermodynamics of transatmospheric vehicles' *ASAA* **86** 1257

Tao J, Luh H & Zheng Y (1987) 'Compliant coordination control of two moving industrial robots' *Proc 26th Conf Dec & Cont*, 186–191

Tarn T, Bejczy A & Yun X (1987) 'Design of dynamic control of two cooperating robot arms: closed chain configuration' *Proc IEEE Int Conf Rob & Autom*, 7–13

Tarn T, Bejczy A & Yun X (1988) 'New nonlinear control algorithms for multiple robot arms' *IEEE Trans Aero & Elect Syst* **24** (5), 571–582

Tauber M (1993) 'Aerobrake design studies for manned Mars missions' *J Space & Rock* **30** (6), 656–664

Taylor J et al (1987) 'Rule based real time control systems' *Proc 26th Conf Dec & Cont*, 1923–1928

Taylor R (1979) 'Planning and execution of straight line manipulator trajectories' *IBM J Res & Dev* **23** (4), 253–264

Taylor S et al (1998) 'Standard payload computer for the ISS' *ESA Bull* **93** (Feb), 10–18

Theis D (1983) 'Spacecraft computers: state of the art survey' *IEEE Comp* (Apr), 85–97

Thomlinson J et al (1987) 'SEU and total dose response of the Inmos transputer' *IEEE Trans Nucl Sci* **34** (6), 1803–1807

Thomopoulos S (1990) 'Sensor integration and data fusion' *J Robot Syst* **7** (3), 337–372

Thompson B et al (1992) 'Introduction to smart materials and structures' *Mat & Des* **13** (1), 3–9

Thomson W (1986) *Introduction to Space Dynamics* Dover Publications, New York

Tong R (1977) 'Control engineering review of fuzzy systems' *Automatica* **13**, 559–569

Thornton C (1997) 'Adaptation of obstacle avoidance in the form of emerging environmental dynamics' *preprint*

Tosunoglu S & Temar D (1987) 'Survey of adaptive control technology in robotics' *Proc NASA Wshop on Space Telerobotics III*, 205–214

Tourassis V & Neumann C (1985) 'Robust nonlinear feedback control with multiple sensors' *IEE Proc* **132D** (4), 134–143

Trembley P (1994) 'EVA safety design guidelines' *Acta Astron* **32** (1), 59–68

Trivedi M et al (1990) 'Developing robotic systems with multiple sensors' *IEEE Trans Syst, Man & Cyber* **20** (6), 1285–1301

Tsin Y & Sani A (1983) 'Algorithms for workspace of a general n-R robot' *Trans ASME J Mech Des* **105** (Mar), 52–57

Uchiyama M & Dauchez P (1988) 'Symmetric hybrid position/force control scheme for coordination of two robots' *Proc IEEE Int Conf Robot & Autom*, 351–356

Uchiyama M et al (1987) 'Hybrid position/force control for coordination of two arm robots' *Proc IEEE Conf Rob & Autom*, 1242–1247

Umetani Y & Yoshida K (1989) 'Resolved motion rate control of space manipulators using a generalised Jacobian matrix' *IEEE Trans Robot & Autom* **5** (3), 303–314

Underwood C (2000) 'COTS-based spacecraft in the LEO environment—17 years of experience with UoSAT microsatellites' *Journ Brit Inter Soc* **53**, 89–96

Utreja L (1993) 'The lunar environment' *Trans ASME J App Mech Rev* **46** (6), 278–284

Uttal W (1989) 'Teleoperators' *Sci Am* (Dec), 74–79

Vadali S, Oh H & Walker S (1989) 'Preferred gimbal angles for single gimbal CMG's' *AIAA 89-3477-CP*

Vafa Z (1990) 'Space manipulator motions with no satellite attitude disturbances' *Proc IEEE Int Conf Rob & Autom*, 1770–1775

Vafa Z & Dubowsky S (1987) 'On dynamics of manipulators in space using the virtual manipulator approach' *Proc IEEE Int Conf Rob & Autom*, 579–585

Vafa Z & Dubowsky S (1990) 'Kinematics and dynamics of space manipulators: the virtual manipulator approach' *Int J Robot Res* **9** (4), 3–21

Valavannis K & Saridis G (1985) 'Analytic design of intelligent machines' *IFAC Robot Control*, Syroco, 139–144

Valavannis K & Saridis G (1988) 'Information theoretic modelling of intelligent robotic systems' *IEEE Trans Syst Man & Cyber* **18** (6), 852–872

Valavannis K & Yuan P (1988) 'Hardware and software for intelligent robotic control systems' *J Intell & Rob Syst* **1**, 343–373

van Bogart E (1986) *Aspects of space law* Kluwer, Amsterdam

Vandenbusshe F (1999) 'SOHO's recovery—an unprecedented success story' *ESA Bull* **97** (Mar), 39–47

Varsi G (1990) 'Telerobotics for the efficient utilisation of space' *J Brit Interplan Soc* **43**, 273–280

Varsi G (1991) 'Advances in space robotics' *Acta Astronaut.* **25** (4), 199–207

Verdant M & Schwehm G (1998) 'The International Rosetta Mission' *ESA Bull* **93** (Feb), 39–50

Vere S (1983) 'Planning in time: windows and durations for activities and goals' *IEEE Trans Patt Anal Mach Intell* **5** (3), 246–260

Vijaykumar R & Arbib M (1987) 'Problem decomposition for assembly planning' *Proc IEEE Int Conf Rob & Autom*, 1361–1366

Vincent J (1992) 'The future: towards intelligent materials and structures' *Metas & Mat* (June), 13–15

Visentori G & Didot F (1999) 'Testing space robotics on the Japanese ETS-VII satellite' *ESA Bull* **99** (Sept), 61–65

Volpe R & Ivler R (1994) 'Survey and experimental evaluation of proximity sensors for space robotics' *IEEE Int Conf Rob & Autom*

Volz R et al (1984) 'CAD, robot programming and ADA' *NATO ASI F11 Robotics & Artificial Intelligence*, 229–246

Wada B (1990) 'Adaptive structures: an overview' *J Space & Rock* **27** (3), 330–336

Wah B et al (1989) 'Computers for symbolic processing' *Proc IEEE* **74** (4), 509–539

Wakeling J (1999) 'New satellite services for the next millenium' *Jour Brit Inter Soc* **52**, 217–222

Walberg G (1985) 'Survey of aeroassisted orbit transfer' *J Space* **22** (1), 3–18

Waldron R (1988) 'Lunar manufacturing: survey of products and processes' *Acta Astron* **17** (7), 691–708

Walker M & Orin D (1982) 'Efficient dynamic computer simulation of robotic mechanisms' *Trans ASME J DSMC* **104** (Sept), 205–211

Walker M & Wee L (1991a) 'Adaptive control of space based robot manipulator' *IEEE Trans Rob & Autom* **7** (6), 828–835

Walker M & Wee L (1991) 'Adaptive control strategy for space based robot manipulators' *Proc IEEE Int Conf Rob & Autom*, 1673–1680

Walker P (1985) 'Transputer' *Byte* (May), 219–235

Wandell H (1989) 'Principles of modern fighter radars' *Mil Tech* **2**, 27–36

Wang K & Lien T (1988) 'Structure, design & kinematics of robot manipulators' *Robotica* **6**, 299–309

Wang P (1987) 'Control strategy for a dual arm maneouvrable space robot' *Proc NASA Wshop Space Telerobotics*, 256–266

Weaver L & Wickman L (1993) 'Principles of EVA design' *ISU preprint*

Weisbin C & Montemerlo M (1992) 'NASA's telerobotic research programme' *Proc IEEE Int Conf Rob & Autom*, 2653–2663

Weisel W (1989) *Spaceflight Dynamics*, McGraw-Hill, New York

Wen J & Murphy S (1991) 'Stability analysis of position and force control for robot arms' *IEEE Trans Autom Cont* **36** (3), 365–371

Wertz J (1978) *Spacecraft Attitude Determination and Control* Reidel, Dordrecht

Wertz J & Larson W (1999) *Space Mission Analysis and Design*, 3rd edn, Kluwer, London

Wertz J et al (1988) 'Reducing the cost and risk of orbit transfer' *J Space* **25** (1), 75–80

West H & Asade H (1985) 'Method for the design of hybrid/position force controllers for manipulators constrained by contact with the environment' *Proc IEEE Int Conf Rob & Autom*, 251–259

Weybrauch R (1980) 'Promulga to a theory of mechanised formal reasoning' *Artif Intell* **13** 133–170

White G (1976) 'Speech recognition: tutorial overview' *IEEE Comp* (May), 40–53

White G (1990) 'Natural language understanding and speech recognition' *Comm Assoc Comp Mach* **33** (8), 72–82

Whitney D (1969) 'Resolved motion rate control of manipulators and human prostheses' *IEEE Trans Man-Mach Syst* **10** (2), 47–53

Whitney D (1972) 'Mathematics of coordinated control of prosthetic arms and manipulators' *Trans ASME J Dyn Syst Meas & Cont* **122** (Dec), 303–309

Whitney D (1977) 'Force feedback control of manipulators fine motions' *ASME Trans J Dyn Syst Meas & Cont* (June), 91–97

Whitney D (1987) 'Historical perspective and state of the art in robot force control' *Int J Rob Res* **6** (1), 3–14

Whittaker W et al (1997) 'Atacama Desert Trek: a planetary analog field experiment' *Proc Int Symp AI, Robotics & Autom for Space*, Tokyo, Japan

Wiener T & Karp S (1980) 'Role of blue-green laser system in strategic submarine communications' *IEEE Trans Commun* **28** (9), 1602–1607

Wilcox B et al (1992) 'Robotic vehicles for planetary exploration' *Proc IEEE In Conf Rob & Autom*, 175–180

Wilkins D (1989) 'From HF radio to unified S band: historical review of the development of space communications' *Acta Astron* **19** (2), 171–190

Wilkinson J et al (2000) 'Preliminary assessment of new orbital debris shielding for unmanned satellites' *Journ Brit Inter Sci* **53** 111–116

Williams J et al (1995) 'Design of a water vapour adsorption reactor for Martian in situ resource utilisation' *Jour Brit Inter Soc* **48**, 347–354

Williamson M (1998) 'Satellite constellations in the ascendant' *IEE Rev* (Spet), 209–213

Williamson M (1999a) 'Can satellites unblock the internet?' *IEE Rev* (May), 107–111

Williamson M (1999b) 'New star in orbit' *IEE Rev* (Sept), 201–205

Williamson M (1999c) 'Watts in space' *IEE Rev* (Jan), 19–23

Williamson R (1992) *private communication (ECSL Workshop, INMARSAT, London)*

Wilson S (1991) 'Animat path to AI' *preprint*

Winchell D (1987) 'Selecting affordable levels of support from spare unit suppliers for HST orbit maintenance' *AIAA 87-0696*

Winograd T (1980) 'What does it mean to understand language?' *Cog Sci* **4** 209–241

Winton A et al (1995) 'The transponder—a key element in ESA's spacecraft TTC systems' *ESA Bull* **86**, 72–79

Wojtalik (1987) 'HST systems engineering' *IFAC 10th Triennial World Cong*, Munich, 63–68

Wolber W & Wise K (1979) 'Sensor development in the miccrocomputer age' *IEEE Trans Elect Dev* **26** (12), 1864–1874

Wu C & Paul R (1982) 'Resolved motion force control for robot manipulators' *IEEE Trans Syst Man & Cyber* **12** (3), 266–275

Xu Y (1993) 'Measure of dynamic coupling of space robot system' *Proc IEEE Int Conf Rob & Autom*, 615–620

Xu Y et al (1992) 'Control system of self mobile space robot manipulator' *Proc IEEE Int Conf Rob & Autom*, 866–871

Yabuta T & Yamada T (1990) 'Possibility of neural network controllers for robotic manipulation' *Proc IEEE Int Conf Rob & Autom*, 1686–1691

Yang D & Lee T (1983) 'On the workspace of mechanical manipulators' *ASME J Mech Des* **105** (Mar), 62–69

Yarber G et al (1966) 'Control moment gyro optimisation study' *NASA CR-400* (March)

Yoerger D et al (1990) 'Influence of thruster dynamics on underwater vehicle behaviour' *IEEE Trans Oceanic Eng* **15** (3), 167–177

Yoshida K et al (1991) 'Dual arm coordination of space freeflying robot' *Proc IEEE Int Conf Rob & Autom*, 2516–2521

Yoshida K et al (1992) 'Modelling of collision dynamics for space free floating links with extended generalisation inertia tensor' *Proc IEEE Conf Rob & Autom*, 899–904

Yuh S (1990) 'Modelling and control of underwater robotic vehicles' *IEEE Trans Syst Man & Cyber* **20** (6), 1475–1483

Zalzak A & Morris A (1991) 'Distributed robot control on transputer network' *IEE Proc* **E138** (4), 169-176

Zeigler B (1989) 'DEVS representation of dynamical systems: event based intelligent control' *Proc IEEE* **77** (1), 72–80

Zheng Y (1989) 'Kinematics and dynamics of two industrial robots in assembly' *Proc IEEE Int Conf Rob & Autom*, 1360–1365

Zheng Y & Luh J (1986) 'Joint torques for the control of two coordinated moving robots' *Proc IEE Int Conf Rob & Autom*, 1375–1380

Zheng Y & Luh J (1988) 'Optimal load distribution for two industrial robots handling a single object' *Proc IEEE Int Conf Rob & Autom*, 344–349

Zheng Y & Paul R (1985) 'Hybrid control of robot manipulators' *Proc IEEE Int Conf Rob & Autom*, 602–606

Zubrin R (1992) 'Long range mobility on Mars' *Journ Brit Inter Soc* **45**, 203–210

Zubrin R (1995) 'Economic viability of Mars colonisation' *Journ Brit Inter Soc* **48**, 407–414

Zubrin R & Weaver D (1993) 'Practical methods for near-term piloted Mars missions' *AIAA 93-2089*

Zubrin R & Weaver D (1995) 'Practical methods for near-term human exploration of Mars' *Jour Brit Inter Soc* **48**, 287–300

Zubrin R, Baker D & Gwynne O (1991) 'Mars Direct: a simple, robust and cost effective architecture for the space exploration initiative' *AIAA 91-0328*

Zubrin R et al (1995) 'Report on the construction and operation of a Mars in-situ propellant production unit' *Journ Brit Inter Soc* **48**, 327–336

Index

Lightning Source UK Ltd.
Milton Keynes UK
27 September 2010

160425UK00006B/74/P

9 781852 331641